MONOGRAPHS ON THE
PHYSICS AND CHEMISTRY OF
MATERIALS

MONOGRAPHS ON THE PHYSICS AND CHEMISTRY OF MATERIALS

Neutron diffraction (*Third edition*) G. E. Bacon

Strong solids (*Third edition*) A. Kelly and N. H. Macmillan

Optical spectroscopy of inorganic solids B. Henderson and G. F. Imbusch

Chemistry of the metal–gas interface M. W. Roberts and C. S. McKee

Quantum theory of collective phenomena G. L. Sewell

Experimental high-resolution electron microscopy Second edition J. C. H. Spence

Theory of defects in solids A. M. Stoneham

Experimental techniques in low-temperature physics Guy K. White

Principles of dielectrics B. K. P. Scaife

Surface analytical techniques J. C. Rivière

Surface Analytical Techniques

J. C. RIVIÈRE

Institut für Physikalische Chemie,
Universität Innsbruck, Austria

(previously Materials Development Division,
Harwell Laboratory, Oxfordshire, U.K.)

CLARENDON PRESS · OXFORD

1990

Oxford University Press, Walton Street, Oxford OX2 6DP
Oxford New York Toronto
Delhi Bombay Calcutta Madras Karachi
Petaling Jaya Singapore Hong Kong Tokyo
Nairobi Dar es Salaam Cape Town
Melbourne Auckland
and associated companies in
Berlin Ibadan

Oxford is a trade mark of Oxford University Press

Published in the United States
by Oxford University Press, New York

British Library Cataloguing in Publication Data
Rivière, J. C.
Surface analytical techniques.
1. Materials. Interfaces & surfaces. Analysis
I. Title II. Series
530.4
ISBN 0-19-851370-4

Library of Congress Cataloging in Publication Data
Rivière, J. C.
Surface analytical techniques / J. C. Rivière.
p. cm.—(Monographs on the physics and chemistry of materials)
Includes bibliographical references.
1. Surfaces (Technology)—Analysis. 2. Surface chemistry.
I. Title. II. Series.
QD506.R58 1990 530.4′17—dc20 89-26609
ISBN 0-19-851370-4

Typeset by
Cotswold Typesetting Ltd, Gloucester
Printed in Great Britain by
Courier International Ltd
Tiptree, Essex

To my wife Sheila

PREFACE

The original suggestion for this book came from Professor Sir Peter Hirsch, who felt that there was a need for a description of surface analytical techniques in terms that would be familiar to any scientist with a physical or chemical background, but yet would also be useful to surface scientists themselves. As envisaged initially, the book was to concentrate only on the half dozen or so major surface analytical techniques. During the writing, however, it became obvious that, for several reasons, the less commonly used techniques could not be ignored. There was no certainty that with continued development a 'minor' technique might not become a 'major' one; there was an increasing tendency to bring several techniques to bear on a problem, so that 'minor' and 'major' ones were often to be found together on the same instrument; and in any case the physical principles underlying some of the 'minor' techniques were usually just as, if not more, interesting than those involved in the 'major' ones and deserved to be described. In this book, therefore, I have attempted to be as comprehensive as possible within the space limitations, by bringing together descriptions of all the surface analytical (compositional) techniques known at the time of writing. The original terms of reference have been adhered to, which means that a good prior level of physical and chemical knowledge is assumed, although lengthy mathematical expositions have been eschewed. It should be useful both to those from other fields who wish to inform themselves about surface analysis, and to those already in the field who would like to know more about other surface analytical techniques.

Innsbruck J.C.R.
1990

ACKNOWLEDGEMENT

I would like to set on record my enormous debt to Anita Harvey, who typed this book from beginning to end. Her ability to convert page after page of my untidy scrawl into neat, virtually error-free, text, in the minimum of time, was nothing short of miraculous. What is more, her cheerfulness and helpfulness remained undiminished throughout. The preparation of the book was, thereby made so much easier and she must take great credit from the final result.

CONTENTS

1	**Introduction**	1
2	**Resumé of physical principles**	5
2.1	Electron excitation	5
2.2	Photon excitation	11
2.3	Ion excitation	14
2.4	Other techniques	19
2.5	Surface specificity	22
3	**Instrumentation**	27
3.1	Vacuum conditions	27
3.2	Sources	28
3.3	Analysers	53
3.4	Sample insertion and handling	66
3.5	Depth profiling	70
4	**Electron excitation: Auger electron spectroscopy (AES) and scanning Auger microscopy (SAM)**	81
4.1	Operation	81
4.2	Spectral features and interpretation	85
4.3	Quantification	97
4.4	Applications	102
5	**Electron excitation: electron energy loss spectroscopy (ELS), core electron energy loss spectroscopy (CEELS), and high resolution electron energy loss spectroscopy (HREELS)**	125
5.1	Operation	125
5.2	Spectral features and interpretation	126
5.3	Quantification	141
5.4	Applications	144
6	**Electron excitation: soft X-ray appearance potential spectroscopy (SXAPS), Auger electron appearance potential spectroscopy (AEAPS), and dis-appearance potential spectroscopy (DAPS)**	167
6.1	Operation	167
6.2	Theory and spectral identification	172
6.3	Applications	181

7 Electron excitation: inverse photoemission spectroscopy (IPES) 193

7.1 Operation 193
7.2 Theory and spectral identification 197
7.3 Applications 202

8 Electron excitation: cathodoluminescence spectroscopy (CLS) 210

8.1 Operation 210
8.2 Theory and spectral identification 210
8.3 Applications 212

9 Electron excitation: electron stimulated desorption (ESD) and electron simulated desorption ion angular distribution (ESDIAD) 219

9.1 Operation 219
9.2 Theory and spectral interpretation 223
9.3 Applications 229

10 Photon excitation: X-ray photoelectron spectroscopy (XPS) and X-ray excited Auger electron spectroscopy (XAES) 238

10.1 Operation 238
10.2 Spectral interpretation 258
10.3 Theory 265
10.4 Quantification 274
10.5 The Auger Parameter 278
10.6 Fine structure in Auger spectra studied by XAES 280
10.7 Applications 283

11 Photon excitation: ultra-violet photoelectron spectroscopy (UPS) and synchrotron radiation photoelectron spectroscopy (SRPS) 314

11.1 Operation 314
11.2 Theory and spectral interpretation 318
11.3 Quantification 338
11.4 Applications 340

12 Photon excitation: reflection–absorption infra-red spectroscopy (RAIRS) and surface-enhanced Raman spectroscopy (SERS) 357

12.1 Reflection–absorption infra-red spectroscopy 357
12.1.1 Operation 357
12.1.2 Theory and spectral interpretation 362
12.1.3 Quantification 367

12.1.4 Applications 369

12.2 Surface-enhanced Raman spectroscopy 373

12.2.1 Operation 374

12.2.2 Theory and spectral interpretation 375

12.2.3 Quantification 378

12.2.4 Applications 378

13 Ion excitation: ion-excited Auger electron spectroscopy (IAES) and proton-excited Auger electron spectroscopy (PAES) 386

13.1 Operation 386

13.2 Theory and spectral interpretation 391

13.3 Quantification and applications 401

14 Ion excitation: ion neutralization spectroscopy (INS) and metastable quenching spectroscopy (MQS) 403

14.1 Ion neutralization spectroscopy 403

14.1.1 Operation 403

14.1.2 Theory and data processing 406

14.1.3 Spectral interpretation 410

14.2 Metastable quenching spectroscopy 417

15 Ion excitation: ion beam spectrochemical analysis (IBSCA) and glow discharge optical spectroscopy (GDOS) 423

15.1 Ion beam spectro-chemical analysis 423

15.1.1 Operation 423

15.1.2 Theory and quantification 425

15.1.3 Applications 429

15.2 Glow discharge optical spectroscopy 431

15.2.1 Operation 432

15.2.2 Quantification 435

15.2.3 Applications 439

16 Ion excitation: ion scattering spectroscopy (ISS) 445

16.1 Operation 445

16.2 Theory and spectral interpretation 450

16.3 Quantification 454

16.4 Applications 459

16.5 Surface structural analysis 467

17 Ion excitation: static secondary ion mass spectrometry (SSIMS) 470

17.1 Operation 470

17.2 Spectral identification 480

17.3 Theory 482
17.4 Quantification 504
17.5 Applications 514

18 Ion excitation: secondary neutral mass spectrometry (SNMS) and glow discharge mass spectrometry (GDMS) 528

18.1 Secondary neutral mass spectrometry 528
18.1.1 Operation 528
18.1.2 Theory 545
18.1.3 Spectral interpretation 552
18.1.4 Quantification 554
18.1.5 Applications 559
18.2 Glow discharge mass spectrometry 563
18.2.1 Operation 563
18.2.2 Theory and spectral identification 567
18.2.3 Quantification 567
18.2.4 Applications 572

19 Neutral excitation: fast atom bombardment mass spectrometry (FABMS) 578

19.1 Operation 578
19.2 Theory 580
19.3 Spectral interpretation and quantification 585
19.4 Applications 585

20 High field excitation: inelastic electron tunnelling spectroscopy (IETS) 593

20.1 Operation 593
20.2 Theory 598
20.3 Spectral identification 604
20.4 Quantification 607
20.5 Applications 612

21 High field excitation: atom probe field ion microscopy (APFIM) 620

21.1 Operation 620
21.2 Theory 629
21.3 Spectral interpretation 631
21.4 Quantification 632
21.5 Applications 633

22 High field excitation: scanning tunnelling microscopy and spectroscopy (STM and STS) 648

22.1 Operation 648
22.2 Theory and image interpretation 655
22.3 Quantification 660
22.4 Applications 660

23 Thermal excitation: thermal desorption spectroscopy (TDS) 671

23.1 Operation 671
23.2 Theory and interpretation 673
23.3 Quantification 678
23.4 Applications 681

Index 691

1
Introduction

SURFACE analysis in its most general sense continues to be a growth area of remarkable vitality, and this is true particularly of that part of it that is concerned with attempting to establish the composition of a surface. Twenty or so years ago knowledge of the nature and number of elements at or near a surface could be obtained only by indirect means, and the development of surface science can now be seen in hindsight to have been hindered for that reason. The present situation could hardly be in greater contrast. If a surface scientist requires compositional information he now has a very considerable battery of techniques at his disposal, from which he may pick and choose according to the type of information he wants, and to the area and depth from which he wants it. Such is the range of choice that he may even exercise personal preference if he wishes, though few can afford that luxury, since generally the technique(s) must be matched to the problem and if there is one thing that most of them have in common it is high capital cost.

It is hardly necessary today to justify the existence of, and need for, surface compositional techniques as did some of the early reviews. In whatever field, either technological or basic, that information about a surface or an interface has been needed, there one will find these techniques applied. If they had been successful in no fields other than those of corrosion, catalysis, microelectronics, and brittle fracture, then their contributions to those alone would constitute more than ample justification, but reference to the now vast literature shows that their application has indeed extended very much further. What *is* necessary, however, in view of the remarkable proliferation of these techniques in the last few years, is clarification. Even to those working in surface analysis the galaxy of acronyms is often bewildering, so how much more bewildering must it appear to the outsider.

This volume, therefore, aims to achieve such clarification by providing a description of the physical principle, the method of operation, the nature of the information, the range of application and the advantages and disadvantages of each of the current techniques of surface compositional analysis. The term 'compositional' is chosen rather than 'elemental' as being less restrictive, since although most of the techniques will give specific elemental information, there are some that do not but yet give chemical information that is complementary to the purely elemental. Techniques for studying the structure of surfaces are thus excluded, although of course occasional reference will have to be made to results obtained by them when discussing some of the more basic applications. A further restriction in the subject matter is to techniques that are surface-*specific* in the nature of their information. There are certain other techniques with enough sensitivity to be capable of analysing material if present only at a surface, but they could not state that it was restricted to the surface. They would be regarded as surface-sensitive but not surface-specific. In this context it is necessary to be clear as to what is meant by a 'surface'. There are various physical and chemical definitions of a surface, but here the techniques themselves define the term, since without exception they are concerned with only the first few atom layers, extending to at most about 5 nm from the 'true' surface, that is, the point at which the material actually terminates. This may seem like a rather circular definition, but in fact the specificity of these techniques covers just that region over which the effect of termination of the material decays; the transition from surface to bulk properties occurs within that distance.

As has been found by most other authors and reviewers in this field, it is convenient to group the techniques according to the nature of the incident radiation. They fall, therefore, mostly under the headings of electron excitation, photon excitation, ion excitation, and neutral excitation, and they will be treated here in that order of classification, apart from those few that cannot be so classified. It is also conventional, because convenient, to introduce the techniques to be discussed not as a long list, but in the form of a matrix, in which the columns refer to the nature of the excitation and the rows to the nature of the emission. Such a matrix is given in Table 1.1, which thus acts as a guide to the techniques to be discussed here. Under the table is a list of the meanings of the acronyms, and some of the alternative acronyms that have appeared in the literature. It is possible that other alternatives can be found in remote places, but it is believed that those in the table are the most common.

Table 1.1
Surface-specific analytical techniques

Emission	Excitation			
	Electrons	Photons	Ions	Neutrals
Electrons	AES, SAM ELS, HREELS CEELS AEAPS	XPS UPS SRPS XAES	IAES, PAES INS MQS	
Photons	SXAPS, DAPS CLS IPES	RAIRS, SERS	IBSCA GDOS	
Ions	ESD, ESDIAD	PSD	SSIMS ISS GDMS	FABMS
Neutrals	ESD	PSD	SNMS	

Guide to acronyms and their alternatives:

AES	Auger Electron Spectroscopy
SAM	Scanning Auger Microscopy
ELS	Electron (Energy) Loss Spectroscopy
HREELS	High Resolution Electron Energy Loss Spectroscopy
(or LEELS	Low Energy Electron (Energy) Loss Spectroscopy)
(or VLS	Vibrational Loss Spectroscopy)
CEELS	Core-Electron Energy Loss Spectroscopy
(or ILS	Ionization Loss Spectroscopy)
AEAPS	Auger Electron Appearance Potential Spectroscopy
SXAPS	Soft X-ray Appearance Potential Spectroscopy
(or APS	Appearance Potential Spectroscopy)
DAPS	Dis-Appearance Potential Spectroscopy
CLS	Cathodoluminescence Spectroscopy
(or EIL	Electron-Induced Luminescence)
IPES	Inverse Photo-Emission Spectroscopy
ESD	Electron Stimulated Desorption
ESDIAD	Electron Stimulated Desorption of Ions Angular Distribution
XPS	X-ray Photoelectron Spectroscopy
(or ESCA	Electron Spectroscopy for Chemical Analysis)
UPS	Ultra-violet Photoelectron Spectroscopy
(or PES	Photo-Emission Spectroscopy)
SRPS	Synchrotron Radiation Photoelectron Spectroscopy
XAES	X-ray (Excited) Auger Electron Spectroscopy
RAIRS	Reflection–Absorption Infra-Red Spectroscopy
(or IRAS	Infra-Red Reflection–Absorption Spectroscopy)
SERS	Surface-Enhanced Raman Scattering
PSD	Photon Stimulated Desorption
IAES	Ion (Excited) Auger Electron Spectroscopy
PAES	Proton (or Particle) (Excited) Auger Electron Spectroscopy
INS	Ion Neutralization Spectroscopy
MQS	Metastable Quenching Spectroscopy
IBSCA	Ion Beam Spectro-Chemical Analysis
(or SCANIIR	Surface Composition by Analysis of Neutral and Ion Impact Radiation)
GDOS	Glow Discharge Optical Spectroscopy
SSIMS	Static Secondary Ion Mass Spectrometry
ISS	Ion Scattering Spectroscopy
(or LEIS	Low Energy Ion Scattering)
GDMS	Glow Discharge Mass Spectrometry

SNMS	Secondary Neutral Mass Spectrometry
FABMS	Fast Atom Bombardment Mass Spectrometry

Other techniques not classifiable in Table 1.1:

APFIM	Atom Probe Field Ion Microscopy
IETS	Inelastic Electron Tunnelling Spectroscopy
STM	Scanning Tunnelling Microscopy
STS	Scanning Tunnelling Spectroscopy
TDS	Thermal Desorption Spectroscopy

Some of these techniques have angle-resolved (AR) versions, which, apart from ESDIAD, have not been listed as separate techniques. Thus one will find in the literature ARUPS (or ARPES), ARAES, etc.

2
Resumé of physical principles

2.1 Electron excitation

2.1.1 *Auger electron spectroscopy (AES), scanning Auger microscopy (SAM)*

The surface is irradiated with an electron beam of sufficient energy, 1 to 30 keV, to ionize one or more shallow core levels in surface atoms. Following ionization the atom relaxes either by the ejection of a characteristic X-ray photon (fluorescence) or by the ejection of an Auger electron, but for shallow levels the latter process is by far the more probable. The Auger process for an atom in a solid is shown schematically in Fig. 2.1. The final state of the atom is doubly ionized and in the example shown the energy of the Auger electron is

$$E_{\mathrm{KL_1L_{2,3}}} = E_{\mathrm{K}} - E_{\mathrm{L_1}} - E^*_{\mathrm{L_{2,3}}} \tag{2.1}$$

where E_i are the binding energies of the ith atomic energy levels and $E^*_{L_{2,3}}$ is starred because it is the binding energy of an electron in the $\mathrm{L_{2,3}}$ level in the presence of a hole in the $\mathrm{L_1}$ level, that is, not the same as $E_{\mathrm{L_{2,3}}}$.

Equation (2.1) shows that the Auger energy is a function only of atomic energy levels, so that measurement of the Auger energies from a surface gives direct elemental identification. Even if the electrons involved originate in the valence band of the material, analysis is still possible since in eqn (2.1) the binding energy of the initially ionized core level is always the dominant term.

2.1.2 *Core-electron energy loss spectroscopy (CEELS)*

When a primary electron interacts with an electron in a core level resulting in ionization, it can give up all or only part of its energy in the process. Thus there appears in the spectrum a step at a distance from the primary energy equal to the binding energy of the core level, and the step is followed by a long tail stretching to zero energy, although in fact the step feature is so weak that the tail is not generally observable. However, the step can be amplified or differentiated, and then gives directly accessible information about the binding energy of that core level. Such features should be observable in every Auger spectrum, but the region within a few hundred eV of the elastic peak (i.e. that of primaries scattered with no loss in energy) is rarely used since the Auger features themselves are always much more intense. In experiments where core loss features have been studied, the range of primary energy has been between 0.5 and 3 keV, i.e. at the lower end of the range for AES.

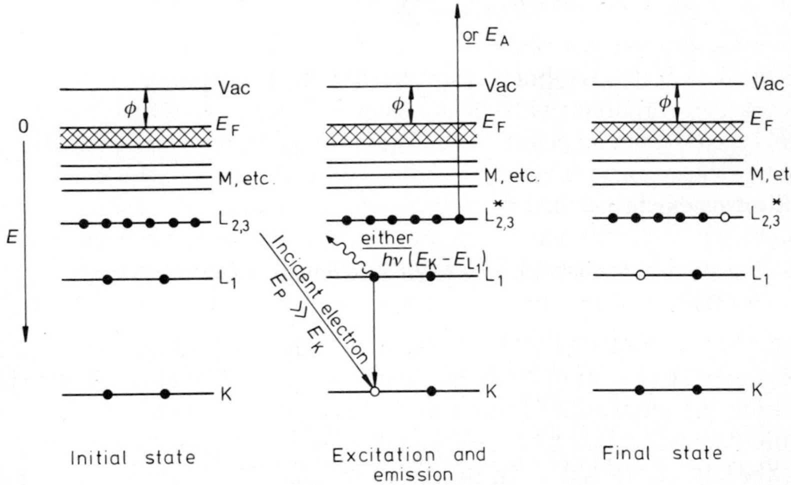

FIG. 2.1. Schematic energy diagram of an Auger process taking place in a solid. On the left, the initial state of an atom in the solid. The valence band is shown filled up to the Fermi level, the work function is ϕ, and the core levels, K, L, M, etc., are indicated, fully occupied. In the middle, the consequences of creating a hole in a core level, in this case the K shell, by an incident electron of energy greater than the binding energy E_K. The hole is filled by an electron from a shallower level, shown here as the L_1; the energy $(E_K - E_{L_1})$ so released can then be dissipated either in a radiative process with the emission of an X-ray photon, or be transferred in a non-radiative process to a second electron, shown here as in the $L_{2,3}$ shell, which is then ejected if $(E_K - E_{L_1} - E^*_{L_{2,3}})$ is positive. The latter electron is the Auger electron and its energy E_A is characteristic of the atom. The binding energy of $L_{2,3}$ is starred because it is the energy of an electron in that level when there is a core hole in an inner level, and is not equal exactly to $E_{L_{2,3}}$. On the right, the final state of the atom after the Auger process, with two holes. The transition depicted would be termed conventionally the $KL_1L_{2,3}$. Electrons can also be transferred from the valence band, and transitions involving the valence band are often the most intense.

2.1.3 *Electron (energy) loss spectroscopy (ELS)*

The energy losses studied in this technique are those suffered by an electron at the primary energy in (a) stimulating electronic transitions within the valence band of the solid, i.e. of only a few eV, and (b) stimulating collective oscillations of the electron sea of the solid, the oscillations having characteristic frequencies or energies called plasmon frequencies. The latter usually extend to several tens of eV, and it is possible to distinguish between plasmon frequencies characteristic of the bulk structure and those characteristic of the surface layer with a quasi-two-dimensional structure. Since all the loss features are by definition associated with the primary energy, by varying the primary energy from high to low, i.e. by varying the average escape depth from deep to shallow, it is possible to accentuate losses associated with the surface compared to those with the bulk. Thus the range of primary energies used is typically 50 eV to 500 eV, and the features are those within about 30 eV of the primary energy, the valence band losses of course being very close to the primary energy.

2.1.4 *High resolution electron energy loss spectroscopy (HREELS)*

A molecule in the gas phase has certain degrees of freedom of motion associated with it, i.e. vibrational, rotational and translational. When that molecule meets a surface with which it interacts chemically so that adsorption occurs, there is then no possibility of translational degrees of freedom, and all that are left are various vibrational degrees of freedom either similar to, or modified from, the original molecular vibrational and rotational degrees of freedom. These vibrational modes are of course of low energy, generally less than 3 eV, and in the gas phase can be studied by infra-red spectroscopy. If now an electron of low primary energy, typically 3–5 eV, approaches the surface with the attached molecule, it can interact with the molecular field when it arrives within a certain distance of the surface, and be reflected (scattered) from the surface having lost a discrete amount of energy corresponding to one of the characteristic vibrational frequencies. Analysis of the energy distribution of the scattered electrons thus provides values of the characteristic frequencies and hence information about the way the molecule is attached to the surface.

2.1.5 *Auger electron appearance potential spectroscopy (AEAPS)*

Primary electrons of energy sufficient to ionize a shallow core level are allowed to irradiate a surface, just as in AES. However, the primary energy is not in this case kept constant, but ramped from zero up to whatever energy is appropriate to the sample atoms, while the primary *current* to the sample is held constant. As the incident energy is increased, it passes through various thresholds for ionization of core levels, and as each threshold is crossed, the secondary electron yield increases as a result of Auger recombination of the core hole. The change in the yield is a measure of the probability of excitation of an electron in a core level to an empty state above the Fermi level, by the incident electron. The yield changes are small, but can be detected by appropriate differentiation techniques.

2.1.6 *Soft X-ray appearance potential spectroscopy (SXAPS)*

In operation very similar to AEAPS, in that the primary electron energy is ramped upwards from zero, passing through ionization thresholds as it increases. However, in this technique it is the soft X-rays produced as a result of recombination of the core hole that are detected. At each threshold the production of soft X-rays increases sharply. The energy level diagram describing the physical processes in both AEAPS and SXAPS is shown in Fig. 2.2, from Park.[1] Above the threshold for emission of a soft X-ray photon characteristic of the particular core level ionized, the excitation probability is a function of the density of unoccupied states, but a complicated function since two electrons are involved, the incident electron and the excited core electron.

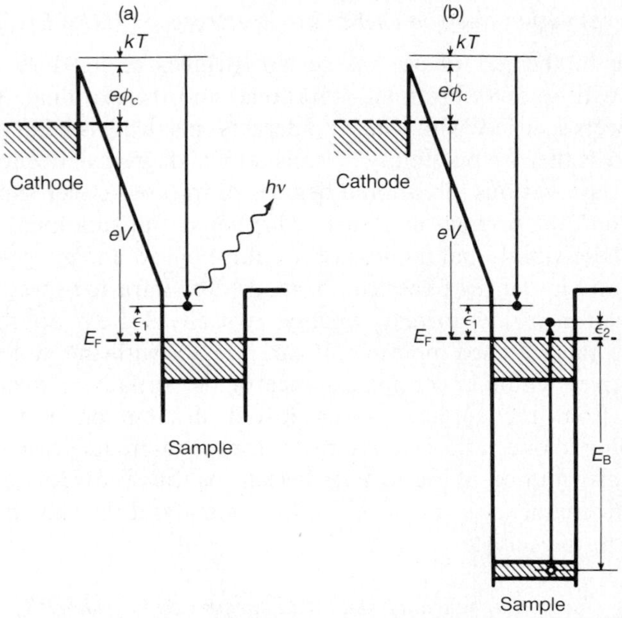

FIG. 2.2. Schematic energy diagrams of the physical processes leading to the emission of continuum, in (a), and of characteristic X-ray, in (b), radiation. In (a) the incident electron is captured in an empty but allowed state distant ϵ_1 above the Fermi level, followed by radiative decay with the emission of bremsstrahlung radiation of energy $h\nu = -eV + e\phi_c + kT - \epsilon_1$, where ϕ_c is the work function of the cathode. Variation of the incident energy eV then maps the densities of unoccupied states, in the technique of IPES. In (b) a core level has been ionized by the excitation of the electron to the state ϵ_2, where the energy ϵ_2 is given by $\epsilon_2 = eV + e\phi_c + kT - \epsilon_1 - E_B$. Here E_B is the binding energy of the electron in the core level. The core hole may decay by emission of an X-ray photon (detected in SXAPS) or by an Auger process (detected in AEAPS), the process in each case having a threshold at an incident energy $eV_{crit} = E_B - e\phi_c - kT$. Above threshold the excitation probabilities are functions of the densities of unoccupied allowed states. From Park.[1]

Both electrons must find available states above the Fermi level, and the shape of the spectral structure above the threshold is thus proportional to the self-convolution of the density of unfilled states. AEAPS and SXAPS are therefore probes of the unoccupied densities of states above the Fermi level, but extraction of information about the one-electron DOS is not mathematically easy.

2.1.7 *Disappearance potential spectroscopy* (*DAPS*)

As the name of this technique suggests, it is in a sense the inverse of AEAPS and SXAPS. It is based on the fact that those primary electrons that have lost energy by ionization of a core hole will of course be lost from the current of reflected electrons at the primary energy. The reflected current of electrons at the primary energy, or, as it is more usually termed, the elastically reflected electron current, should therefore show sharp *decreases* as successive ionization thresholds are crossed. Again, as in AEAPS, it is necessary to maintain the primary current from the electron source constant as a function of the ramped voltage. The percentage changes in the elastically reflected current are small, but measurable. Since it is the reflected electrons that are studied, there is in this case no influence from the processes of decay of the core hole, and it is a direct measurement of the density of unoccupied states.

2.1.8 *Inverse photoemission spectroscopy (IPES)*

The physical principles are best illustrated with the help of Fig. 2.3, taken from Woodruff *et al.*[2] At the left-hand side of the diagram is shown schematically a hypothetical density of states, occupied up to the Fermi level, unoccupied above, of a solid. Electrons of energy E irradiate the solid and decay radiatively. If the detecting equipment is set to detect only those photons at a particular energy, $\hbar\omega$ in the figure, then as E is ramped from zero upwards the variations in the photon flux should duplicate the variation in the density of unoccupied states. As will be seen later under UPS, the photoemission process employs an incoming photon to excite electrons from the *occupied* part of the density of states. The photons produced by the radiative decay following capture of the incident electron are of course known collectively as bremsstrahlung radiation. Where the selected photon energy is low, typically ~ 10 eV, the technique is more accurately known as ultra-violet bremsstrahlung spectroscopy, whereas at higher photon energies in the soft X-ray range it is known as characteristic isochromat spectroscopy.

2.1.9 *Cathodoluminescence spectroscopy (CLS)*

When some solids are irradiated with electrons, the recombination of the ionized core hole by a radiative channel, although of much lower probability for shallow core levels than Auger decay, can give rise to photons in the visible as well as in the soft X-ray region. The lower energy photons are produced by a transfer of energy from one core level to another by a cascade process, and the probability of visible light emission will depend on the precise arrangement of atomic energy levels in relation to the primary energy. The process is known also as cathodoluminescence. The technique requires simply the same sort of variable energy electron source as is used in AES, and a photon detector such as a sensitive photometer. It is not clear what the exact nature of the physical information is that is provided by the light emission, but the emission has been found to be very dependent both in intensity and spectral characteristics on the nature and concentration of adsorbed species on a surface. Thus the emission from atomically clean metals is low, but from the same metals after oxidation is greatly increased.

2.1.10 *Electron stimulated desorption (ESD), electron stimulated desorption of ions angular distribution (ESDIAD)*

If a surface covered with one or more adsorbed species is irradiated with electrons of a suitable energy, interaction of the electrons with the adsorbate

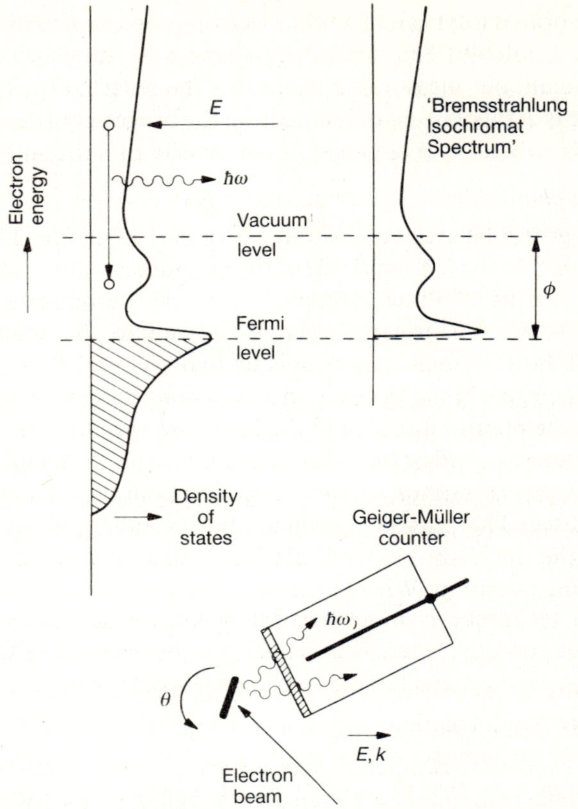

FIG. 2.3. Schematic energy diagram of the physical mechanism of inverse photoemission spectroscopy, from Woodruff *et al.*[2] On the left are shown the densities of states in a notional metal, with a valence band partially filled up to the Fermi level, and unoccupied but allowed states extending upwards. An electron of energy E is decelerated by stopping in an unoccupied state, and decays radiatively by emission of a bremsstrahlung photon. If a particular photon energy $\hbar\omega$ is selectively detected and the incident energy E is swept, then the variations in the emitted flux should map the density of empty states, as shown on the right. Below is indicated the experimental approach using a Geiger–Müller counter.

can lead to the ejection of both ions and neutrals. The interaction takes the form of adsorption of the electron by an adsorbed molecule or a substrate–adsorbate complex leading to excitation of the adsorbing species into an excited state by a Franck–Condon process. If the excited state is an anti-bonding state, then the species may be ejected from the surface as an ion, an excited neutral, or a ground state neutral. If the ejected species are ions they may be detected directly in a mass spectrometer, if neutrals they must be ionized before entering the mass spectrometer. Separate identification and separation of excited and ground state species must be carried out with a device such as a time-of-flight mass spectrometer. The information obtainable from the technique is clearly chemical, but the interpretation is difficult.

The variation of ESD in which the angular distributions of emitted ions are measured is called ESDIAD. With a sufficiently narrow angle of acceptance into the mass spectrometer, patterns of ion intensity as a function of direction in the irradiated surface are obtained which give information about the arrangement of adsorbed species with respect to the underlying substrate.

2.2 Photon excitation

2.2.1 *X-ray photoelectron spectroscopy (XPS), ultra-violet photoelectron spectroscopy (UPS), synchrotron radiation photoelectron spectroscopy (SRPS)*

In this group of techniques the surface is irradiated with photons, soft X-ray in the case of XPS, hard ultra-violet in the case of UPS, and from synchrotron radiation in the case of SRPS. Interaction of a photon of energy $h\nu$ with an electron in a level of binding energy E_B leads to the transfer of the entire photon energy to the electron, with the resultant ejection of a photoelectron of kinetic energy

$$E_{kin} = h\nu - E_B. \quad (2.2)$$

The process is shown schematically in Fig. 2.4.

The initially bound electron may be in a core level or in the occupied portion of the valence band. Ejection from core levels is the basis of XPS; since no two elements have the same set of electronic binding energies, measurement of the kinetic energies of the photoelectrons provides an elemental analysis. Furthermore, any changes in the binding energy of an electron in a level as a consequence of changes in the chemical environment of the atom are reflected in changes in the photoelectron kinetic energies.

Although ejection from the valence band can be observed in XPS, the photoelectrons from valence levels are of course of the highest kinetic energy, and are not therefore fully characteristic of the surface. In addition the inherent line width of X-ray sources is not quite narrow enough to reveal all the fine structure observable in valence band spectra. The principal valence band spectroscopy is therefore UPS, in which the valence photoelectrons have low kinetic energies, and the exciting radiation lines are sharp. Since synchrotron radiation energy is tunable from a few tens to a few hundreds of *eV*, SRPS is able to cover the range from UPS to XPS continuously, and according to the photon energy chosen either valence band or core-level photoelectrons may be observed.

2.2.2 *X-ray (excited) Auger electron spectroscopy (XAES)*

The creation of a core hole by absorption of an X-ray photon, as in XPS, leads to the possibility of recombination of the hole by an Auger process, just as

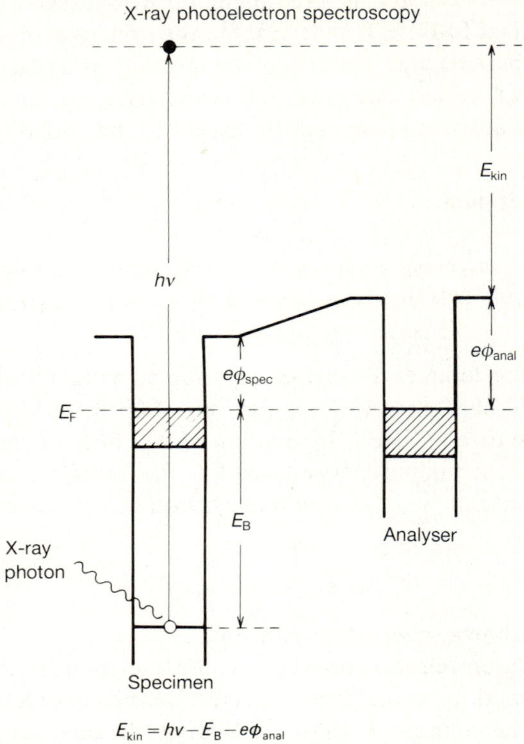

FIG. 2.4. Schematic energy diagram of the physical mechanism of X-ray photoemission and of photoelectron collection. An X-ray photon of energy $h\nu > E_B$ interacts with an electron in a core level, of binding energy E_B, causing photoionization and ejection of a photoelectron. By the Einstein principle the photon must give up all its energy in the interaction, thus the kinetic energy of the photoelectron as measured by the analyser is $E_{kin} = h\nu - E_B - e\phi_{anal}$, where ϕ_{anal} is the effective work function of the analyser.

described in Fig. 2.1, under AES. Thus in an XPS spectrum, as will be seen later, there are peaks due to the ejection of electrons by Auger emission as well as to photoemission, so that in a sense XAES is really only a part of XPS. However, in an analysis by XPS the Auger peaks are still largely ignored, and XAES has been regarded as a separate technique because of the way in which it has been used. This is as an extension of AES, either where the fine structure associated with certain Auger transitions needs the higher energy resolution used in XPS, or where the possibly damaging effect of electron irradiation in a surface reaction must be eliminated. Another, expanding, use of XAES is in conjunction with XPS to provide a measurement of the Auger Parameter, to be discussed later, which is a combination of photoelectron and Auger peak positions to give for many elements a more sensitive indicator of the chemical state than the photoelectron peak energy alone.

2.2.3 *Reflection–absorption infra-red spectroscopy (RAIRS)*

Infra-red radiation is reflected from a plane surface through an adsorbed layer, in which the reflected light loses intensity as a result of absorption in the layer. The intensity losses are discrete and occur at those light frequencies that match vibrational modes at the surface. The vibrations are those occurring either within the adsorbed species or consequent on interaction of the adsorbed atoms or molecules with the substrate. The application to surface studies is restricted to substrates that are highly reflecting in the infra-red, and to adsorbates that have high absorption cross-sections for the characteristic vibrations. For sufficiently high absorption the angle of incidence must be close to grazing, and plane polarized light must be used.

2.2.4 *Surface enhanced Raman spectroscopy (SERS)*

When an intense source of monochromatic light, e.g. a laser, irradiates a surface with adsorbed molecules, the incident light undergoes Raman scattering. Collision with a molecule in a vibrational ground state raises the molecule to a higher vibrational level, and the scattered photon is then of lower energy than that of the incident light photon. This is also known as Stokes radiation. If the molecule were already in an excited vibrational state, the collision with the incident photon may give rise to scattered light of energy higher than that of the incident radiation, and this is known as anti-Stokes radiation; since at normal temperatures few molecules are in excited states, the anti-Stokes is generally much weaker than the Stokes radiation. The Raman spectra are recorded by measuring the scattered intensity of the Stokes radiation as a function of the energy difference between the incident and scattered energies. At energies corresponding to vibrational transitions within the adsorbed molecule discrete lines, or Raman shifts, are observed.

If conventional Raman spectroscopy is attempted to be applied to monolayer adsorption on a smooth metal surface it is found that the scattered intensity is much too low to be of any use. It has been found, however, that if the metal surface is rough then it is possible for the signal intensity to be increased by as much as a factor of 10^6; the precise scale of the roughness required for this enhancement is a matter for debate, and may indeed vary from one surface or material to another, but it is generally atomic or superatomic. This variation of Raman spectroscopy is called SERS. Some Raman frequencies in the enhanced spectrum can correspond to those allowed in infra-red spectroscopy but normally forbidden Raman modes. The information from SERS is thus complementary to both infra-red spectroscopy and to LEELS.

2.2.5 *Photon stimulated desorption (PSD)*

The electronic excitation process leading to photo-desorption is similar to that already described for electron-stimulated desorption (ESD). Capture of a photon causes a surface molecule or complex to go to an excited state, which, if of anti-bonding character, leads to desorption. Both ions and neutrals can be desorbed, and can be detected in a mass spectrometer as described before. Experimentally the technique is difficult because the heating effect of the photon flux usually dwarfs that of the electronic excitation by some orders of magnitude. Low photon flux densities must therefore be used, which means that the desorbed ion and neutral current is also very low. The detection circuitry has thus to be extremely stable, since changes in partial pressure lower than 10^{-11} torr have to be measured.

2.3. Ion excitation

2.3.1 *Ion (excited) Auger electron spectroscopy (IAES), proton (excited) Auger electron spectroscopy (PAES)*

Auger emission following creation of a core hole by either incident electrons or incident photons has already been described. Since incident ions, be they noble gas ions, protons, or α-particles, can also create core holes it follows that Auger emission as a result of ion irradiation also occurs. As well as the normal Auger features found also in AES and XAES, additional peaks are found in the IAES spectrum, due to Auger transitions taking place in atoms or small clusters that have been sputtered from the surface in excited states. Further, in PAES, satellite features appear associated with the normal Auger peaks, owing to the fact that the very high energy proton and α-particle projectiles can cause simultaneous ionization of several core levels.

2.3.2 *Ion neutralization spectroscopy (INS), metastable quenching spectroscopy (MQS)*

If an ionized atom is allowed to approach a metal surface very slowly, i.e. if it has a kinetic energy of only a few eV, there is time as it passes within the interaction distance (a few Å) from the surface for it to form a temporary ion–atom pair with the nearest surface atom. One of the paths of de-excitation of the excited system represented by this pair is by an Auger type recombination. The possibility of Auger de-excitation or neutralization depends on the correspondence of unfilled electron levels in the ion and of filled electronic levels in the surface atom of the solid, as illustrated schematically in Fig. 2.5 from Hagstrum.[3] On the left of the figure is the energy level diagram of the solid, with a filled valence band stretching from B to F, and in the middle and on the right are two atomic wells, one of the He^+ core and one of the He^{2+} core. Where the atomic levels lie within the range of the filled levels of the solid, as is the case for He^{2+}, de-excitation can occur simply by tunnelling of electrons from the filled band into the empty atomic levels, but where the atomic levels lie outside that range, then de-excitation can take place

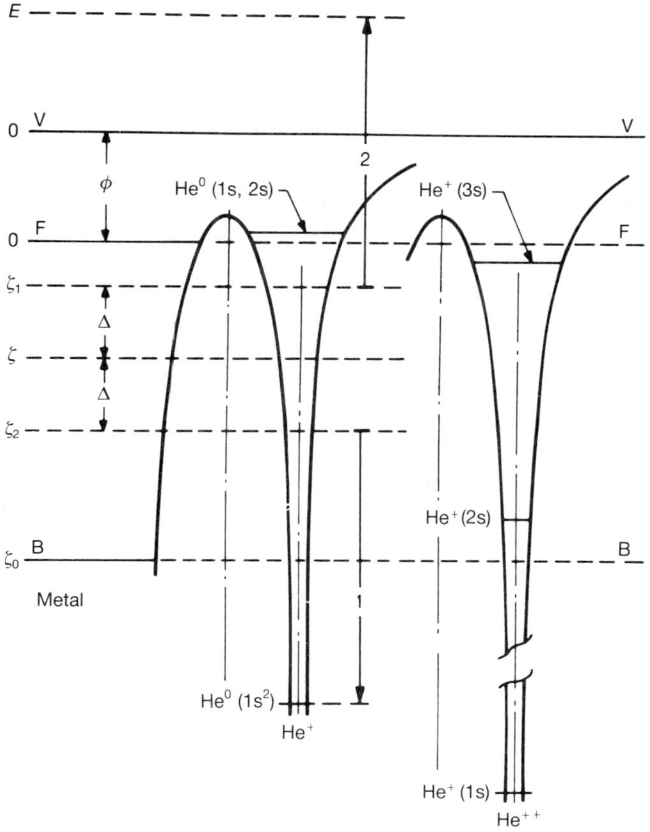

FIG. 2.5. Schematic energy diagram of the physical processes taking place in INS, depicted by Hagstrum.[3] On the left, the notional band structure of a metal, in which the filled part of the valence band extends from F to B. Two valence levels ζ_1 and ζ_2 are each separated by Δ from their mean position ζ. ϕ is the work function.

In the middle and on the right are the schematic atomic wells of ions He^+ and He^{++}, respectively, approaching the metal surface. For He^{++} the atomic levels in the ion fall within the range of allowed levels in the solid, and electron tunnelling can thus take place resonantly, leading to decay by resonance ionization. If, as in He^+, the atomic levels fall outside the allowed range in the solid, then resonance ionization cannot take place, but decay occurs via an Auger process in which an electron from the metal (in level ζ_2 in the illustration) falls into the core hole, and the excess energy given to a second electron (ζ_1), which is then ejected if the excess energy is greater than twice the work function. The latter, Auger, process forms the basis of INS.

by an Auger process, shown as transitions 1 followed by 2 to produce an emitted electron of kinetic energy E. The illustration shows the levels ζ_1 and ζ_2 as being the origins of the two electrons, and since these two levels could be anywhere between B and F, the observed spectrum of E is a self-convolution of the distribution of levels in the valence band, together with the transition probabilities across the band. Because the de-excitation involves an atom right at the surface, a density of states extracted from the technique is characteristic of the surface itself.

If an atom approaching a surface very slowly is not ionized but in a metastable state, that is, if it is excited but electrically neutral, and cannot return to the ground state by photon emission, then interaction with the surface gives rise to electron emission just as in INS. According to the relative positions of the excited atomic level and of empty states above the Fermi level of the surface, different emission mechanisms operate. When the excited state can resonate with the empty states, then resonance ionization occurs, which is followed by Auger neutralization, and the resultant electron emission spectrum is virtually the same as in INS. On the other hand, if the excited level corresponds to filled levels in the surface, resonance cannot occur and Auger de-excitation takes place by the filling of the hole in the inner atomic level by an electron from the occupied densities of states in the surface, followed by ejection by an Auger process. Since only one electron from the filled surface states is involved, the spectrum reflects directly the local one-electron density of states, and no deconvolution is necessary as in INS.

2.3.3 *Ion beam spectro-chemical analysis* (*IBSCA*)

When an energetic ion beam interacts with a surface a great variety of different types of secondary particle can be produced, including secondary electrons, Auger electrons, positive secondary ions, negative secondary ions, photons, neutrals, and excited neutrals. One of the channels for de-excitation or decay of ions and excited neutrals is by the emission of photons of characteristic energies. The emission spectrum consists of a series of sharp lines, reminiscent of that from an arc discharge, from which it can be deduced that de-excitation takes place in the gas phase, probably just above the surface. Identification of the elemental composition of the ion bombarded surface can therefore be made by reference to standard wavelength tables. The principal difference between IBSCA and GDOS, to be described below, is that in the former a mono-energetic collimated beam of primary ions is used, as in SSIMS or in depth profiling.

2.3.4 *Glow discharge optical spectroscopy* (*GDOS*)

The surface to be analysed is made the cathode in a diode-type discharge cell through which argon is flowed. Atoms (*inter alia*) are sputtered from the surface, and as they enter the glow discharge are either ionized or optically excited by collisions with the ions and electrons in the discharge plasma. Subsequent de-excitation occurs principally by photon emission, as in IBSCA, the radiative spectrum having lines characteristic of the sputtered atom and hence of the surface. Although the sputtering rate in the discharge is high, the inherent sensitivity of the technique is also high, and thin surface layers can be detected.

2.3.5 *Static secondary ion mass spectrometry* (*SSIMS*)

As already mentioned under IBSCA, bombardment of a solid with energetic ions leads to a large variety of secondary emitted particles. The techniques of SIMS use the positive and negative secondary ion emission for analysis. It is important to note that according to the way in which SIMS is used, it may or may not be regarded as surface-specific. In the variation called *dynamic* SIMS (often simply called SIMS) the primary current density is high, and the surface is eroded at such a rate that information about the original surface is lost almost at once. The surface-specific variation is called *static* SIMS simply because the current density is reduced to such a low level that the erosion rate is only a fraction of a monolayer a minute. Since modern quadrupole mass spectrometers are very sensitive, there is plenty of time to analyse the original surface as it is removed—hence the name static. Quasi-static would be more accurate.

2.3.6 *Ion scattering spectroscopy* (*ISS*)

Inert gas ions have a very strong electron affinity; when such ions collide with a surface there is therefore a high probability that the majority will be neutralized after one collision. After two or three collisions the number of original ions left will be negligible, although this number will vary according to the ion, the material and the angle of incidence. If therefore a beam of inert gas ions is directed at a surface, and the energies of those ions of the same mass and charge, that are emitted from the surface, are analysed, then the analysis will refer almost entirely to those ions that have been scattered after only one collision, that is, from the outermost layer of atoms. When the primary ions are of low mass and low energy, e.g. He^+ or Ne^+ at 0.5 to 4 keV, the interaction time is much shorter than the vibration time of a surface atom, so that the collision can be approximated to that between the ion and a free atom, i.e. a billiard-ball type collision. Under these circumstances there is a relationship between the energy and mass of the primary ion, and those of the surface atom, which for a scattering angle of 90° reduces to

$$E_S/E_P = (M_{AT} - M_{ION})/(M_{AT} + M_{ION}) \qquad (2.3)$$

where E_S is the energy of scattered ion, E_P the energy of primary ion, M_{AT} the mass of surface atom, and M_{ION} the mass of primary ion. If the ion energy ratio E_S/E_P is plotted as a function of scatterd ion current, there will be a series of peaks at particular values of the ratio, which can be identified in terms of masses of surface atoms, thus providing an elemental analysis.

2.3.7 *Glow discharge mass spectrometry* (*GDMS*)

This technique is closely related to that of GDOS already mentioned. Instead of analysing the spectrum of the light emitted by the de-excitation of ions or excited neutrals formed in the glow discharge, GDMS extracts the ions themselves into a region of lower pressure than that in the discharge and analyses their masses in a quadrupole mass spectrometer. It is important to note the difference between GDMS and SSIMS. In the latter the secondary ions ejected from the surface by the primary beam are mass-analysed directly; in GDMS the positive ions formed near the surface by sputtering cannot leave the surface because of the direction of the electric field, whereas the neutrals can. The positive ions analysed in GDMS are therefore formed from the neutrals in the discharge, and because most of the material sputtered from a surface is in the form of neutrals, GDMS has a very high sensitivity. However, the parallels between IBSCA and GDOS on the one hand, and SSIMS and GDMS on the other, are obvious.

2.3.8 *Secondary neutral mass spectrometry* (*SNMS*)

Again, there are close similarities in principle if not always in practice between SNMS and GDMS. As in GDMS it is the emitted secondary neutral contribution to the sputtered flux that is of interest. In this case the neutrals are postionized after leaving the surface by a resonant low pressure high frequency plasma, that is, they are passed through a hot Maxwellian electron gas before being analysed in a mass spectrometer. Primary ions for sputtering are produced either in the conventional way for SSIMS by an external ion gun, in which case the primary current density can be varied over several orders of magnitude without needing to change the postionization conditions in the plasma, or by exposure to the ion component of the plasma itself, in which case the current density is high. The latter configuration is clearly very similar to that in GDMS. The advantage of the latter mode is that the bombarding ion energy can be varied over a much greater range than with a conventional ion gun.

2.3.9 *Fast atom bombardment mass spectrometry* (*FABMS*)

Very similar to SSIMS in practice except that energetic primary neutral atoms are used for sputtering. As before, the sputtered ions are mass analysed in a mass spectrometer. The neutrals are produced in a special source whose design and operation are aimed at a high proportion of energetic neutrals in the beam; any ions produced at the same time are deflected away from the sample. The technique has turned out to be advantageous for the analysis of non-volatile organics, e.g. polymers, and the problems of specimen charging when using charged particles as primary probes are avoided, so that analysis of insulating surfaces is easier.

2.4 Other techniques

2.4.1 *Atom probe field ion microscopy (APFIM)*

If a material can be fashioned into a very fine point, of radius not more than a few hundred Å, the application of a potential of a few keV between it and a counter-electrode imposes a very high field strength at the tip itself. When the tip is made the positive electrode, and the field strength approaches a few V/Å, it is possible to exceed the binding energy of an adsorbed atom on the tip or even the cohesive energy of the material itself. This is known as field desorption. If the volume around the tip is filled with helium (usually) to a certain pressure, field ionization of helium atoms occurs and the positive ions are then repelled from the region of the tip surface under the influence of the applied field. They travel in straight lines from the point on the tip at which they were formed, and if a fluorescent screen of suitable shape is placed at the correct distance an image is formed, basically an atomic image, of the arrangement of atoms at the tip, either adsorbed or of the tip material. The technique is the well known one of Field Ion Microscopy (FIM).

The APFIM variation uses both field desorption and FIM. In it the fluorescent screen is provided with a small hole, accurately on the axes of the tip and the centre of curvature of the screen. An FIM image is formed and the tip is tilted by means of a manipulator until the image of a particular spot on it whose analysis is required is just contained within the hole. A voltage pulse is then applied to the tip to take the field strength momentarily into the field desorption region, whereupon an atom or cluster of atoms is desorbed, and passes through the hole into the drift tube of a time-of-flight mass spectrometer. By using very fast counting circuitry, well shaped pulses with extremely short (~ 1 ns) rise times, and a drift tube of sufficient length (~ 1 m), the mass-to-charge ratio of the atom or cluster can be measured, and hence an analysis obtained.

2.4.2 *Inelastic electron tunnelling spectroscopy (IETS)*

Unlike all the other techniques in Table 1.1, this is not a technique that requires a vacuum environment, although there is no obvious reason why it should not be used in vacuum if required. If two metals are separated by a thin insulating layer, and a voltage is applied across them, electrons can tunnel through the insulator. The arrangement is shown schematically in Fig. 2.6(a) from Khanna and Lambe.[4] Electrons from metal I on the left can make transitions into empty states in metal II on the right without losing energy, and this is known as elastic tunnelling. However, in the gap between the metals, perhaps at one or other of the interfaces, there may be impurity states, or molecules trapped that have characteristic vibrational energies $h\nu$. The tunnelling electron can give up some of its energy to the state or the vibrational mode and still find empty states in metal II to which to go. This is called inelastic tunnelling. It will occur only when the applied field is greater than $h\nu$, and if the current through the sandwich is recorded as a function of the applied voltage, there will be increases in the current at each value of $h\nu$ corresponding to a new state or another vibrational mode, as shown in Fig. 2.6(b). For increased visibility of the threshold, the current–voltage plot is double-differentiated, as seen in Figs 2.6(c) and (d). The technique thus provides a set of values of $h\nu$, from which deductions can be made both of the nature of the states or molecules present in the sandwich, and of the way in which molecules are adsorbed there, and in that respect is similar and complementary to HREELS, SERS and infra-red spectroscopy.

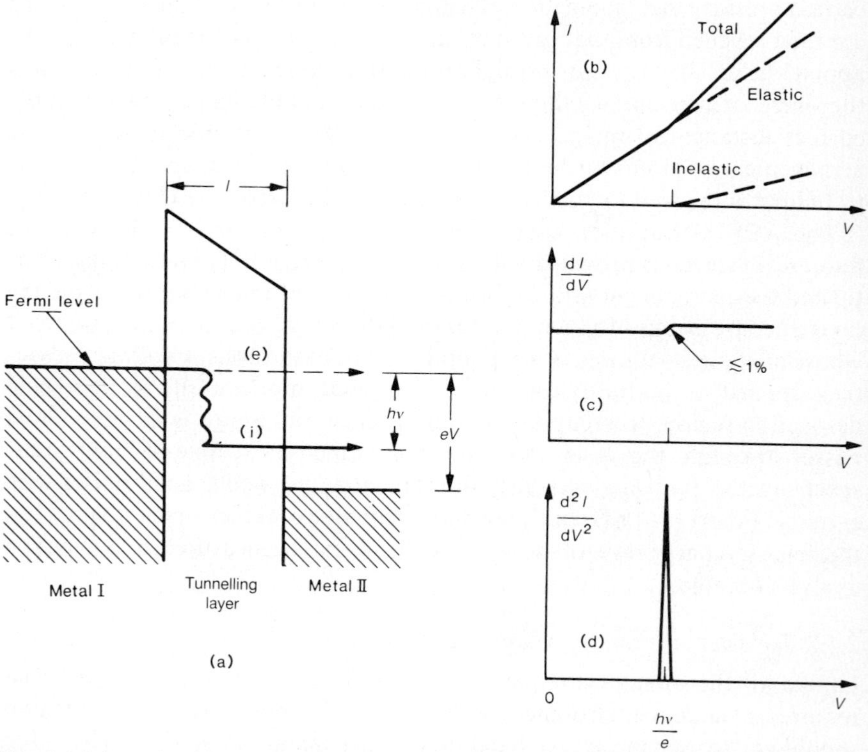

FIG. 2.6. Schematic of the physical principles and mode of recording of IETS, from Khanna and Lambe.[4] In (a) two metals I and II are separated by a very thin dielectric layer of thickness l, typically 2.5–3.0 nm. Application of an applied voltage across such a layer between the metals can then stimulate electron tunnelling from I to II, in which electrons can make transitions into empty states in II without losing energy. Such tunnelling is called elastic. If, however, there are adsorbed materials, or impurity states, somewhere in the insulating gap between the metals, possessing some characteristic vibrational frequency ν, a tunnelling electron can give up some energy to the vibrational mode and still find empty states in II. This is called inelastic tunnelling. Obviously it will occur only when eV is greater than $h\nu$. A plot of tunnelling current vs. applied voltage will then show increases at each vibrational threshold, as in (b). To increase the visual resolution, the plot is double-differentiated, as in (d). The double differential then corresponds to a spectrum of vibrational state energies.

2.4.3 *Scanning tunnelling microscopy (STM), scanning tunnelling spectroscopy (STS)*

STM is, ostensibly, concerned with the imaging of surface topography and would not strictly speaking be a technique that fell within the definition of this volume, but in actual fact the interpretation of STM images is so closely bound up with the nature and distribution of surface states that STM and STS should really be considered merely as different aspects of the same technique (STMS?).

In the two techniques a metal tip is brought very close (0.5–1.0 nm) to a surface, sufficiently close that at applied voltages of 2 mV–2 V between tip and surface an electron tunnelling current becomes measurable. In STM the tip is then scanned across the surface, while either the height of the tip is changed to keep the current constant, or the current is varied to maintain the height constant. Since, to a first approximation, the tunnelling current I and applied voltage V_T are related by

$$I \propto V_T \exp(-A\phi^{1/2}s) \tag{2.4}$$

where A is a constant, ϕ the average work function, and s the tip/surface separation, the variations in either I or s at constant V_T will provide a three-dimensional map related to the atomic contours of the surface. The lateral resolution is of the order of 0.5 nm. The approximate relationship between recorded variation and topography will hold only if the work function remains constant; if there are patches of different work function then the observed variations will not be mainly topographical in origin.

However, the one-dimensional approximation on which expression (2.4) is based, is valid only for very low applied voltages. The tunnelling current is in fact a function of the three-dimensional structure of the surface, in other words the local band structure. This can be exploited by recording the tunnelling current at some chosen fixed position on the surface as a function of the applied voltage, V_T, while maintaining the separation s constant. The I–V_T dependence will then reveal local surface electronic properties; this is the spectroscopic part of STS. If a prominent spectroscopic feature at a partiular value of V_T is identified, then a scanning image of the distribution of that feature can be produced by fixing V_T at that value, fixing s as well, and recording variations of I as the tip is scanned. Since such a procedure is exactly what is carried out in STM, it will be appreciated therefore that STM and STS are inseparable, and that STM images that genuinely reflect surface topography are merely those recorded at values of V_T at which there are no major spectroscopic contributions.

2.4.4 *Thermal desorption spectroscopy (TDS)*

Simple in principle, this technique is not so easy to achieve in practice. It consists of allowing a gas or mixture of gases to adsorb on a metal foil for a chosen length of time, and then heating the foil to a high temperature in a controlled way, while monitoring the current of a particular ion or group of ions desorbed from the surface as a function of temperature. The ions are of course selected in a quadrupole mass spectrometer. According to the various binding energies of the adsorbed species on the metal surface, peaks in the desorbed ion current will appear at characteristic temperatures. The difficulties arise in ensuring (a) that the metal foil heats uniformly, (b) that the rate of temperature rise is strictly linear, and (c) that different rates of heating can be chosen while still maintaining linearity. For unambiguous interpretation all of these are essential.

2.5 Surface specificity

Both the title and the Introduction to this volume insist that the above-mentioned techniques are surface-*specific*, therefore it is necessary to examine why they are considered to be so, i.e. the physical reasons.

2.5.1 *Techniques involving secondary electron emission (AES, SAM, XPS, UPS, SRPS, XAES, IAES, PAES, INS, MQS)*

Any electron in a solid has a probability of travelling a certain distance through the solid before suffering an inelastic collision and losing some or all of its energy. If this probability is set equal to e^{-1}, and if the kinetic energy of the electron at any moment is E, then the average distance travelled in a matrix M before a collision leading to loss of energy is known as the *inelastic mean free path* (imfp) $\lambda_M(E)$. The imfp is a function only of the matrix and of the electron energy. Many measurements of λ have been made, and the compilation by Seah and Dench[5] is given in Fig. 2.7, in terms of atomic monolayers as a function of electron kinetic energy. The very important consequence of the dependence of λ upon energy is that in the regimes of secondary energies used in the above techniques, the imfp is of the order of only a very few monolayers. For example, in AES a typical regime would be from 20 eV to 1000 eV, over which λ varies from about two to about six monolayers; similarly, in XPS a typical regime might be from 500 eV to 1500 eV, giving a range of λ from about four to about 8 monolayers. In other words, in all these techniques, the secondary electron *must originate at or very close to the surface if it is to escape into the vacuum and be detected.* It is this phenomenon that makes these techniques surface-specific.

If the secondary electrons are detected along a particular direction making an angle θ to the surface, then the characteristic average depth of origin is $\lambda \sin \theta$, which is generally called the *escape depth.* As will be seen in a later chapter, variation of θ in order to vary the escape depth is a method of obtaining compositional information as a function of depth for very thin films.

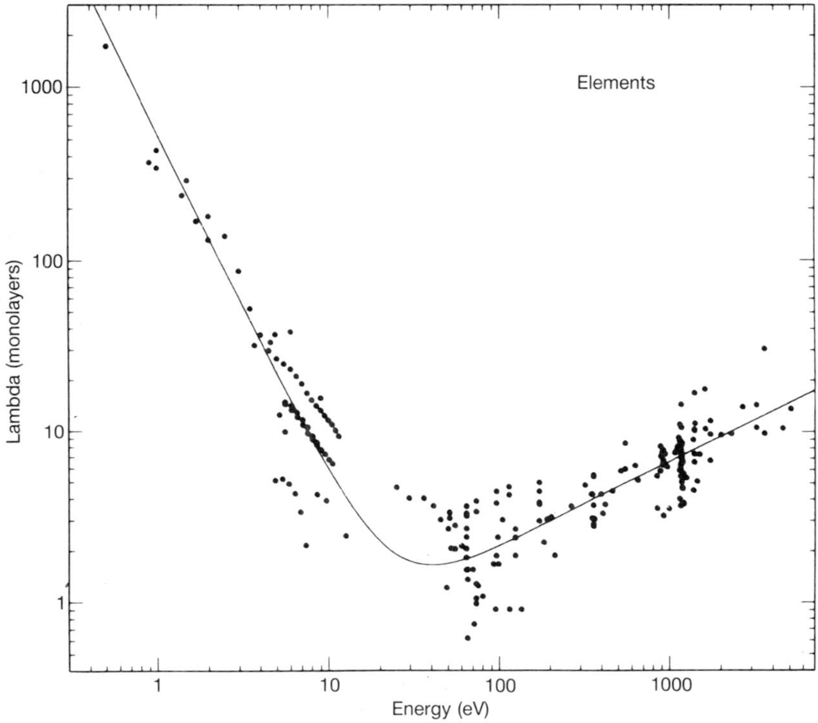

FIG. 2.7. The compilation by Seah and Dench[5] of measurements of the inelastic mean free path, λ, in terms of monolayers, as a function of electron kinetic energy. The solid line is the least squares fit to the general relation

$$\lambda = AE^{-2} + BE^{-1/2}$$

where $A = 538$ and $B = 0.210$ for elements.

2.5.2 *Techniques involving energy losses and ionization thresholds (ELS, CEELS, AEAPS, SXAPS, DAPS, IPES)*

In these techniques it is not the kinetic energies of the secondary electrons that decide the surface specificity, but those of the primary electrons. Considerations of the imfp still hold, and Fig. 2.7 still applies, since there is of course no physical differentiation possible between the scattering processes affecting primary electrons and secondary electrons of the same energy, in a solid. Thus in the loss spectroscopies ELS and CEELS, in which the loss peaks are of course 'tied' to the primary energy, the loss information can be made more or less surface-specific simply by moving the primary energy to the left or to the right, respectively, in Fig. 2.7. In the ionization threshold spectroscopies AEAPS, SXAPS, and DAPS, the primary energy is ramped from near zero up to a few hundred eV to cross various thresholds for ionization, so that again the information depth is increasing according to Fig. 2.7, but even so restricted to a few monolayers at most. Similarly in inverse photoemission (BIS, CIS), the primary energies used in both the ultra-violet and characteristic isochromat recordings are low, and their swept range is narrow. Thus these techniques are also surface-specific.

2.5.3 *HREELS*

The physical principles involved in HREELS make it somewhat of a special case in the context of surface specificity. The primary electrons are of such low energy, only a few eV, that according to Fig. 2.7 their imfp's inside a solid would be considerable, perhaps as much as 100 monolayers. The loss mechanism, however, does not involve a primary electron after it has entered the solid, but when it interacts with the dipole field of a molecule adsorbed on the surface, and is scattered with an energy loss corresponding to a molecular vibrational mode. Some primary electrons, probably most, will indeed enter the solid, but they are not the ones with which the analysis is concerned. By definition, therefore, the surface specificity of HREELS must be the ultimate, i.e. the first monolayer, since the peaks observed in the spectrum can have arisen only by scattering off the adsorbed molecules on the surface.

2.5.4 *CLS*

Although the precise mechanisms operating in this technique, and the exact nature of the information provided, are not too clear, it is possible to decide on two counts that it must be surface-specific. Firstly, the luminescence can be stimulated by the same magnitude of primary electron energy, ~3 keV, as is often used in AES, so that the information depth cannot be greater than about 10 monolayers. Secondly, the luminescence is strongly enhanced by the adsorption of oxygen (and to a lesser extent other molecules) on a surface, to a coverage of a monolayer as observed by AES; it has been established (see later chapter) that oxygen molecules physically adsorbed cause much greater enhancement than chemisorbed oxygen. Under these circumstances the technique must be highly surface-specific.

2.5.5 *Desorption techniques (ESD, ESDIAD, PSD, TDS)*

By their very nature these techniques must be surface-specific, since the desorbed entities can come only from the surface. Without adsorbed or physisorbed material present at the surface there is no desorption signal, and since in general such material adsorbs to a thickness of only one or two monolayers, the surface specificity is high.

2.5.6 *Optical techniques (RAIRS, SERS)*

Since the absorption features observed in RAIRS (or FT-RAIRS) are due entirely to the presence of a monolayer or less of adsorbed material on the reflecting surface, it follows that the technique must be included amongst those what are the most highly surface specific. It ranks in this respect with its complementary technique HREELS in being monolayer-specific.

The degree of surface-specificity of SERS is rather more problematical since there have not as yet appeared a large number of applications of SERS in UHV, and the estimated average magnitude of the surface roughness needed for the enhancement varies according to the physical model used for the process. Some experimenters estimate the average roughness in their experiments to be between 100–1000 Å, but there are many others where the figure is below 100 Å, often 20–80 Å. For the time being, then, the technique should be regarded as quasi-surface-specific until more information is available.

2.5.7 *Techniques involving secondary ions or neutrals (IBSCA, GDOS, SSIMS, GDMS, SNMS, FABMS)*

Although the penetration depth of primary ions, and the consequent trail of destruction that they leave, varies with their energy and can be up to many tens of atom layers, the process by which secondary ions and neutrals are ejected from the surface implies that nearly all such species originate in the outer one or two layers. This is because part of the energy deposited in the solid by the primary ion finds its way back to the surface by a cascade process, and, if sufficient when it reaches the surface, ejects an ion or neutral, rather like a row of ninepins knocking each other over in a successive manner. The ninepin at the end of the row is the ejected ion (or neutral), the intermediate ones being unable to leave the row because of their neighbours. These techniques are therefore inherently surface-specific. Some of the above techniques are not necessarily normally used in a surface-specific way, if the rate of erosion by incident ions is too fast for the detection system to capture the analysis of the first few layers. GDOS and GDMS are, like dynamic SIMS, often used in the fast erosion mode, but they are also perfectly capable of being slowed down so that the initial surface composition can be analysed. Normal operation of the other techniques, however, is in a slow erosion mode, and they are regarded as highly surface-specific.

2.5.8 *ISS*

The description of the principles of ISS given previously under 2.3.6 shows unambiguously that the technique must be highly surface-specific. Since virtually all incoming inert gas ions will be neutralized after only a couple of collisions with surface atoms, it follows that the scattering is almost entirely from the top layer of atoms of the solid. It is in fact one of the most surface-specific techniques in Table 1.1.

2.5.9 *APFIM*

It hardly needs saying, because it is obvious from the description in 2.4.1, that APFIM also represents the ultimate in surface-specificity, since the field desorbed atoms can come only from the first atom layer either of the tip material or of adsorbed material.

2.5.10 *IETS*

The information depth in IETS depends entirely on the individual experiment, that is, to what thickness an adsorbed layer has been applied to the insulator surface before the sandwich, or junction as it is more usually called, is completed by the superposition of the second metal (see Fig. 2.6(a)). In most cases the thickness is only a monolayer, but in others it may be two or three monolayers. The technique is thus basically as surface-specific as most of the others discussed here.

2.5.11 *STM and STS*

As described in section 2.4.3, the information obtainable from STM/STS is a convolution of topographic, and surface-density-of-state spectroscopic, contributions and which of these factors dominates in any one system will be a function of the applied voltage. For the spectroscopic contribution, which is the one of primary interest here, although the applied voltages are low the applied field is in fact high, and can approach field emission conditions. Under such conditions the tunnelling electrons carry information about electronic properties of the surface, in the immediate vicinity of the potential barrier; just as field emission microscopy provides a map of local work function variations, so STS maps relate to very local electronic states. The ability to image dangling bond states in silicon is an example of this. The information depth is likely therefore to be no greater than 1–2 monolayers.

References

1. Park, R. L. (1975). *Surf. Sci.* **48**, 80.
2. Woodruff, D. P., Johnson, P. D., and Smith, N. V. (1983). *J. Vac. Sci. Technol.* **A1**, 1104.
3. Hagstrum, H. D. (1970). *J. Res. NBS* **74A**, 433.
4. Khanna, S. K. and Lambe, J. (1983). *Science* **220**, 1345.
5. Seah, M. P. and Dench, W. A. (1979). *Surf. Interface Anal.* **1**, 2.

3
Instrumentation

ANALYSIS of the properties of emitted particles is central to all the techniques in Table 1.1, and it is therefore important to give details not only of the characteristics of the analysers themselves but also of the other instrumental aspects that are involved in the analysis. These must include, where appropriate, the necessary vacuum conditions, the sources of incident or exciting particles, the various ancillary devices usually found, and the methods of sample insertion, positioning and treatment. Some of these are common to the majority of the listed techniques, but where they are not the techniques to which they apply will be indicated.

3.1 Vacuum conditions

Although a surface from which particles are emitted needs a surrounding vacuum if the particles are to reach an analyser without being scattered by gas molecules, the vacuum requirement for that reason alone is not stringent. Vacua in the region of 10^{-5}–10^{-6} torr would be adequate. A much more stringent requirement is imposed by the surface specificity of the techniques themselves. With typical sampling depths of a few atom layers, and element sensitivities ranging from 10^{-1}–10^{-5} atom layers, it is clear that the techniques are very sensitive to surface contamination. In many cases very small amounts of contamination can affect the course of an experiment, or the interpretation of analytical results, very drastically, and the principal source of contamination is from the residual gas in the vacuum system. What this means in practice is that in order to have enough time to conduct an experiment, or make a measurement, before accumulation from the gas phase interferes, the base pressure should be about 10^{-10} torr. Such a pressure is said to be in the region of ultra-high vacuum (UHV).

The achievement of vacua in the low UHV region was once an art but is now routine. To do so it is necessary to accelerate the rate at which gas molecules leave internal surfaces, called the outgassing rate, by raising the temperature of the whole system, i.e. by baking. A standard bake-out procedure would be 150°C–200°C for several hours; overnight bake-out is often the most convenient. If the system is not baked after pumping from atmospheric pressure, then typical outgassing rates would prevent UHV from being reached for a long time, if ever.

Since bake-out is necessary, the nature of the materials that can be used for constructional purposes is restricted. Such materials should not disintegrate, or outgas excessively, or oxidize readily, or lose their strength, at the elevated temperature, and must retain their integrity in those respects even after repeated bake-out. For these reasons the principal constructional material used universally is stainless steel, although occasionally a manufacturer will use μ-metal for the fabrication of a vessel if magnetic screening of internal components is important. Similarly it is not possible to use the types of elastomer material used so commonly in unbaked systems, for seals between components of the system, since such materials not only deteriorate on baking but also become porous. Instead the seals are of metal, normally of copper but sometimes of gold. The seal is made by compressing the soft copper or gold between flanges, either with knife-edges in the case of copper or flat in the case of the gold. Copper of course oxidizes on bake-out but only on the outer exposed surface; the sealing face is not affected.

Other materials that can be used between UHV and atmosphere include borosilicate glass or quartz for windows, and alumina for electrical isolation of current or voltage connections to be made to the interior of the system. Inside the system any material can be used that does not contain constituents volatile either at room or baking temperatures; brass, for example, that contains volatile zinc, is not permitted, but there are obviously many metals that are.

For pumping to UHV there are four types of pump in common use, either singly or in combination—diffusion, sputter ion, turbomolecular and sublimation. Each type has its advantages and disadvantages, which can be found in any treatise on vacuum technology, but in general the most trouble-free and least temperamental is the diffusion pump. An efficient liquid nitrogen cooled trap (to prevent backstreaming of the oil and to trap water vapour) is needed with a diffusion pump, and the system must be so constructed that the trap can be baked along with the rest of the system. Sublimation pumps, in which titanium is evaporated on to a cooled surface, are used in combination with one of the others to achieve and maintain vacua better than 10^{-10} torr.

The above description applies to the achievement of UHV as a base pressure in the vacuum systems in which all the techniques of Table 1.1 would be operated. It will be clear, however, from the preceding chapter that a few of the techniques (e.g. GDOS, GDMS) operate at a working pressure considerably greater than UHV. Nevertheless, since the increase in pressure to the working level is caused by the admission of an inert, non-contaminating, gas such as argon, it is still necessary to take as the starting-point a clean, uncontaminated, system.

3.2 Sources

3.2.1 *Electron sources*

The techniques listed in Table 1.1 under electron excitation can be divided into those (ELS, HREELS, CEELS, IPES, SXAPS) in which the energy spread in the incident beam governs the energy resolution attainable in the technique, and those (all the others) in which it does not. The latter can be divided again into those techniques whose application often requires high spatial resolution (AES, SAM) and those in which the nature of the information available means that spatial resolution is of secondary importance. In the first group, however, it is only in HREELS that the energy resolution requirement is so stringent that special sources have had to be developed for the technique, and it can thus be said that electron source development for surface analysis has gone in the directions either of energy resolution or of spatial resolution, but not both together. These two directions will be discussed first.

Monochromatic sources As described in Chapter 2, HREELS measures the discrete energy losses due to the excitation of various vibrational modes of molecules adsorbed on surfaces. Since each of these modes has an associated characteristic energy, the energies being typically in the range 20–2000 meV, and with widths <0.1 meV, it follows both that the primary exciting electron energy must be low, <5 eV, and that the energy width must be reduced to as near the inherent vibrational width as possible. A tungsten filament operated at a typical temperature of 2200 K emits electrons with a thermal spread of about 0.5 eV, and such a source used unmodified would be useless for HREELS. What is done in practice, therefore, is to pass the electrons from such a filament through the same type of energy selector as is used to analyse electron energies, the selector in this case acting as an electron monochromator.

A typical arrangement for HREELS is shown in Fig. 3.1, from Nishijima *et al.*[1] It is symmetrical, as all HREELS systems are, in that the same energy selector is used for both monochromator and analyser, in this particular example the 127° cylindrical electrostatic deflector of Hughes and Rojansky[2]. The electron gun, contained in the enclosure 6, consists of a tungsten hairpin filament (1), a Wehnelt electrode (2), and a series of anodes (3–5), and produces an electron beam that is focused at the entrance slit (7) of the monochromator. Anodes 4 and 5 are split in directions normal to each other and allow beam deflection in the X and Y directions for exact alignment of the beam in the slit. Typical operating conditions of the gun are ~ 1 eV beam energy and ~ 1 μA beam current, with an energy spread of ~ 0.6 eV.

The monochromator (8, 9, 10) and analyser (20, 21, 22) were both chosen in this example to have a mean radius of 22.5 mm, and electrode (8, 9 and 20, 21) separation of 7 mm. The performance of the monochromator is then such that, if a slit (7, 11) size of 0.25 mm is used and the potential difference between the inner (8) and outer (9) electrodes is 0.40 V, the effective pass energy is 0.63 eV and the energy resolution at full-width-half-maximum (FWHM) of the elastic peak is 7 meV. On leaving the exit slit (11), the

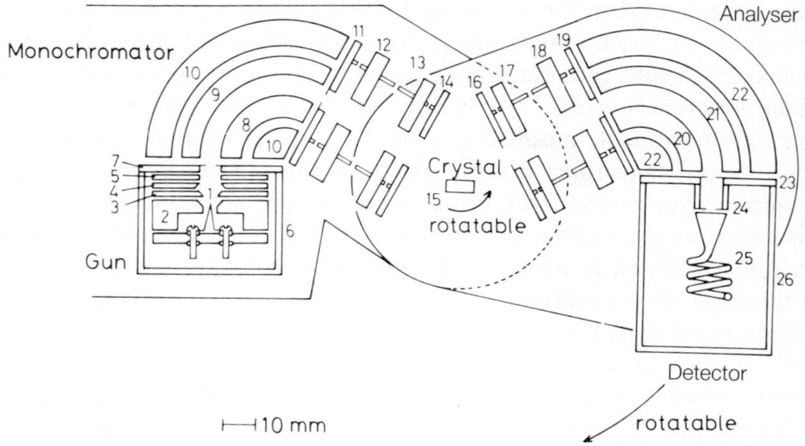

FIG. 3.1 Experimental system constructed for HREELS by Nishijima *et al.*[1] (1) Filament. (2) Wehnelt electrode. (3)–(5) Anodes. (6) Electron gun enclosure. (7) Entrance slit to monochromator. (8) and (9) Inner and outer 127° cylindrical electrodes of monochromator. (10) Guard electrode. (11) Exit slit from monochromator. (12)–(14) Accelerating and focusing lens system. (15) Sample. (16)–(18) Decelerating and focusing lens system. (19) Entrance slit to analyser. (20) and (21) Inner and outer 127° cylindrical electrodes of analyser. (22) Guard electrode. (23) Exit slit of analyser. (24) Electron suppressor. (25) Electron multiplier detector. (26) Detector enclosure.
Analyser and detector can be rotated through 80° about the centre of the sample, and the sample can itself be rotated.

electrons are accelerated in a lens system (11–14) to the chosen primary energy; the lens also has a focus at the sample (15) position. Electrodes 12 and 13 are also split in opposite directions to provide *X* and *Y* deflection if required.

On leaving the sample surface, secondary electrons are decelerated in another lens system (16–19), and are focused at the entrance slit (19) of the analyser. Once more *X* and *Y* deflection for exact alignment is provided by the splitting of 17 and 18 into opposing sectors. Construction and operation of the analyser is similar to that of the monochromator, and the performance is therefore similar as well. The sample and electrodes 14 and 16 are held at ground potential. After passage through the analyser the energy-selected electrons are collected and multiplied in a high gain electron multiplier (25). Electrode 24 is inserted to suppress electrons produced by internal reflection within the analyser. The whole analyser plus detector system can be rotated through an angle of 80° centred on the point of impingement of the primary beam on the specimen, to allow measurement of angular dependence of the emission. Similarly the sample itself can be rotated in order to vary the angle of incidence.

Because of the low primary energies used in HREELS, and the requirement for high energy resolution, it is vital to ensure that the whole volume of the system is virtually free of magnetic and electrostatic fields. All metallic components are made of non-magnetic materials and any ceramic insulators that have to be used unavoidably are positioned out of line of sight of the electron trajectories. The entire system is encased in several layers of μ-metal, which reduces the earth's field by virtue of its high permeability. By these means the field at the sample can be reduced to less than 2 mG.

A further necessary precaution is the suppression of secondary electrons. To this end all electrode surfaces are coated with fine soot to reduce secondary emission, and the electrode systems are surrounded by metal boxes (6, 10, 22, 26) to act as shields against the ingress of stray electrons.

The HREELS system shown in Fig. 3.1 would have a typical performance of primary energy variable from 2.5 to 5.0 eV, sample current about 2×10^{-10} A, and energy resolution at FWHM in the elastically reflected peak of about 8 meV.

Unmonochromatized sources Where the only purpose of an incident electron beam is to produce ionization in core levels in order to trigger a relaxation mechanism in an atom, then the energy spread in the beam is unimportant. The same is true if the only function of the beam is to cause ejection of species adsorbed on a surface into the gas phase, where one or other of the properties of the species can be measured. Thus those techniques that depend on ejection of Auger electrons following core ionization, namely AES and SAM, or on electron desorption, viz. CLS, EID, ESDIAD, and ESD, do not inherently require monochromatic electron sources.

In the remaining techniques listed in Table 1.1 under electron excitation (ELS, CEELS, AEAPS, SXAPS, DAPS, and IPES) the information is extracted through observation either of loss features or of threshold features, and, as already remarked above, the resolution in the spectrum is thus governed by, *inter alia*, the incident beam width. However, in all these techniques the actual energy resolution requirement is not as stringent as in HREELS simply because the experimentally observed features are themselves broad, and what this means in practice is that the thermal spread of 0.3–0.5 eV is acceptable. Thus the electron sources used in them are also unmonochromatized.

Since electron beams can be focused into a fine probe, a secondary requirement for electron sources in the techniques of AES and SAM is the production of as small a spot size on the sample as possible, to achieve maximum spatial resolution. This is secondary in the sense that it does not derive directly from the physical process used in the technique. On the other hand, as much if not more effort has been devoted technologically to the development of electron sources of high spatial resolution as to those of high energy resolution. In the description that follows, therefore, the electron sources used in AES and SAM will be described first, followed more briefly by those used in the other techniques.

Electron emitters are usually one of two types, the thermionic and the field emission. In the former a material is heated to a high enough temperature to give some electrons sufficient additional energy to surmount the work function barrier at the surface and escape into the surrounding vacuum. The current of thermionically emitted electrons is governed by the Richardson equation

$$J = A(1-\bar{r})T^2 \exp(-e\phi/kT) \qquad \text{A cm}^{-2} \tag{3.1}$$

where J = current density
A = Richardson constant, ideally equal to 120 A cm^{-2} deg^{-2}
$\bar{r}$ = zero-field reflection coefficient for incident electrons
T = absolute temperature
ϕ = work function.

Emitting materials are normally refractory metals such as tungsten or tantalum, and are shaped into thin filaments or ribbons.

Field emission does not impart any additional energy to the electrons, but extracts electron current by reducing the work function barrrier. Application of a high electric field, of many tens of kilovolts, will reduce the barrier sufficiently to allow electrons to 'tunnel' through it into the vacuum. To achieve the necessary high fields the emitting material must be fashioned into a very fine point, of radius typically 50 nm, and again tungsten is used universally as the material since it can be so fashioned and yet withstand the high applied fields. In this case the emitted current follows the Fowler–Nordheim expression

$$J = (1.55 \times 10^{-6} E^2/\phi)\exp\{-6.86 \times 10^7 \phi^{3/2}\theta(x)/E\} \qquad \text{A cm}^{-2} \tag{3.2}$$

where J = current density
E = field strength
$\theta(x)$ = Nordheim elliptic function
$x = 3.62 \times 10^{-4} E^{1/2}/\phi$.

Expression 3.2 shows that the current density is strongly dependent on the work function, and therefore the needle-like emitter is fabricated as a single crystal with a plane of low work function parallel to the tip. Field emission sources are known as 'high brightness' sources since the emitting area is so small and the emission concentrated into such a small solid angle.

The principal drawbacks in the use of field emitters are that their surfaces must be kept clean, which means that they must always operate in very good UHV, and that they tend to be unreliable. For these reasons the great majority of electron sources are still thermionic emitters, generally of tungsten.

Electron sources used in conventional AES, i.e. in the primary energy range 5–10 keV, have always been of electrostatic design. An electron gun using a thermionic filament, typical of those provided by most manufacturers, is shown diagrammatically in Fig. 3.2 (courtesy of VG Scientific). When correctly aligned and shielded from stray magnetic fields, it is capable of providing a spot size on the specimen of $\sim 0.5\ \mu m$ at 10 keV and a beam current of $\sim 10^{-8}$ A. The beam can be rastered over the specimen surface to provide topographical or scanning Auger images of the surface at moderate to poor spatial resolution, but the gun would not be regarded as adequate for modern SAM imaging. It can, and has, been used for ELS and CEELS by operation at low primary energies, but it is not designed to provide the necessary high currents at such energies, and those techniques are usually performed with rather simpler guns (see below).

For SAM the thrust in design is towards ever smaller spot sizes, commensurate with maintaining enough beam current to provide adequate sensitivity for the production of a usable Auger signal. To reduce the spot size beyond that obtainable with an optimized electrostatic gun such as that shown in Fig. 3.2 involves going to an electromagnetic focusing system, and to higher primary beam energies. Thus in SAM primary energies are typically in the range 25–30 keV, and an electron gun for a SAM system might look like the commercial design of Fig. 3.3 (courtesy of VG Scientific). At 25 keV and 10^{-10} A beam current the optimum spot size on the specimen would be ~ 35 nm. The lens windings are completely enclosed, i.e. not exposed to the vacuum, and of course additional insulation has to be provided because of the high voltage. The ultimate performance depends to a large extent on the accurate machining and alignment of the pole pieces, on the general overall mechanical stability, and on the ability to eliminate entirely any alternating magnetic fields from the gun and specimen region. As improvements in detection methods increase the sensitivity of detection of Auger electrons, so the guns for SAM will be able to reduce the current in the beam and hence the spot size in the specimen. It is difficult at the moment to see when the limit of the technology will be reached.

The techniques of ELS and CEELS do not require spatial resolution, but they do require an electron source that can be operated at primary energies low compared to those used in conventional AES, that provides high stability of the chosen primary energy, and a reasonably high current, of the order of 50 μA. The stability is largely a function of the power supplies, but the other requirements can be met by the type of electron gun used either in Low Energy Electron Diffraction (LEED) or in the earlier stages of development of AES. A type[3] that was popular for AES in the early 1970s, and that is suitable for the loss spectroscopies (except HREELS, obviously) is shown with its associated circuitry in Fig. 3.4. It can be operated between 50 and 2500 eV at currents of up to several tens of μA, if needed, and the area irradiated is about 500 μm. Although it has deflecting plates at the end of the lens system, they would not be used in ELS or CEELS. The energy spread in the beam is of course thermal, typically 0.5 eV, but this can be reduced if an indirectly heated low

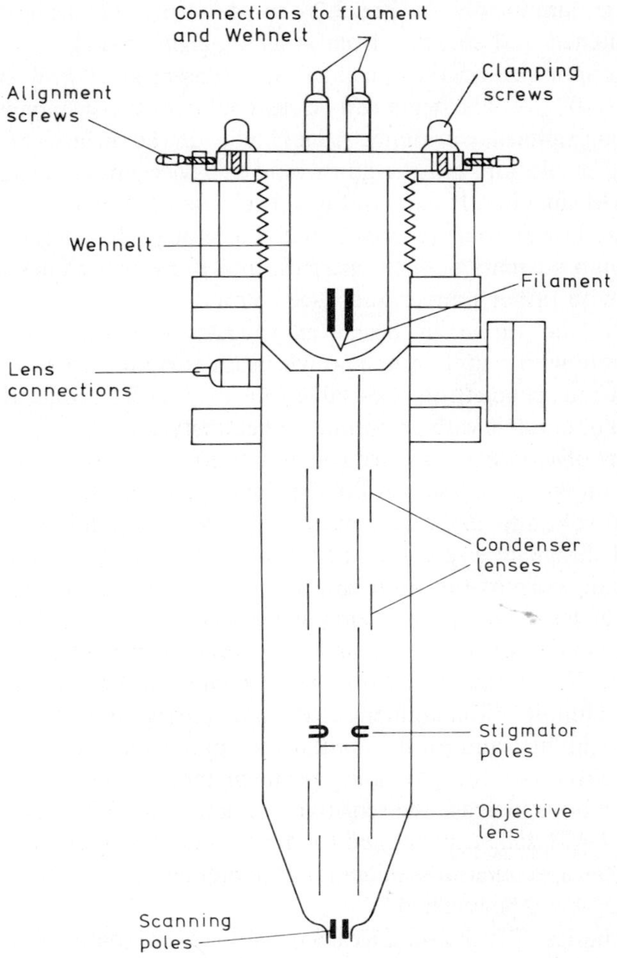

FIG. 3.2 Thermionic electron source with electrostatic focusing for use in techniques requiring incident energies in the 2–10 keV range. At 10 keV and a beam current of 10 nA it can be focused to a spot about 0.5 μm in diameter on the specimen, and rastered if required to produce a SAM image, although the spatial resolution is in general inadequate for SAM. (Courtesy of VG Scientific.)

temperature cathode is used as shown in the figure, and such a reduction is desirable for those techniques since it is the energy spread that largely governs the energy resolution obtainable.

The above source is also suitable for IPES if directional information is required. For angle-integrated information it is adequate simply to have a hot filament near the specimen surface, but for angular resolution the incident electron beam must be focused, the spot size of 500 μm quoted above being quite sufficient.

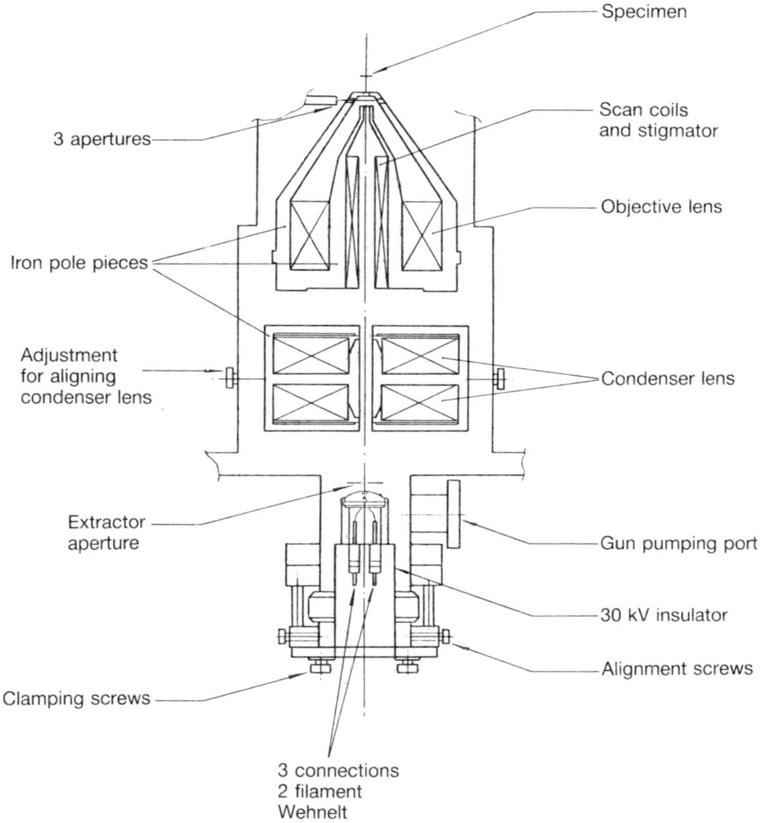

FIG. 3.3 Thermionic electron source with electromagnetic focusing for use in the techniques of AES and SAM. Primary energies are in the range 25–30 keV and currents typically 0.1 nA. The final aperture can be selected by remote operation from outside the system, and the aperture to specimen optimum distance is about 10 mm. At 25 keV and 0.1 nA the optimum spot size is about 35 nm.

For the techniques of AEAPS, SXAPS, and DAPS, directional information has almost never been required, and the electron source consists of a hot filament in front of the sample, as indicated schematically in Fig. 3.5 for SXAPS, from Park and Houston.[4] The energy resolution in the techniques is a function not just of the thermal spread of the electron energies from the hot filament, but also of the potential gradient from one end of the filament to the other, and this effect must be minimized. Ways of minimizing it have included choice of filament cross-section to give a resistivity such that the desired temperature can be reached with only a very low applied voltage, and heating by alternating current of relatively high frequency and voltage, but very short duty cycle.[5] The latter method, for example, reduces the energy spread resulting from the heating voltage to less than 0.2 eV. The total energy spread

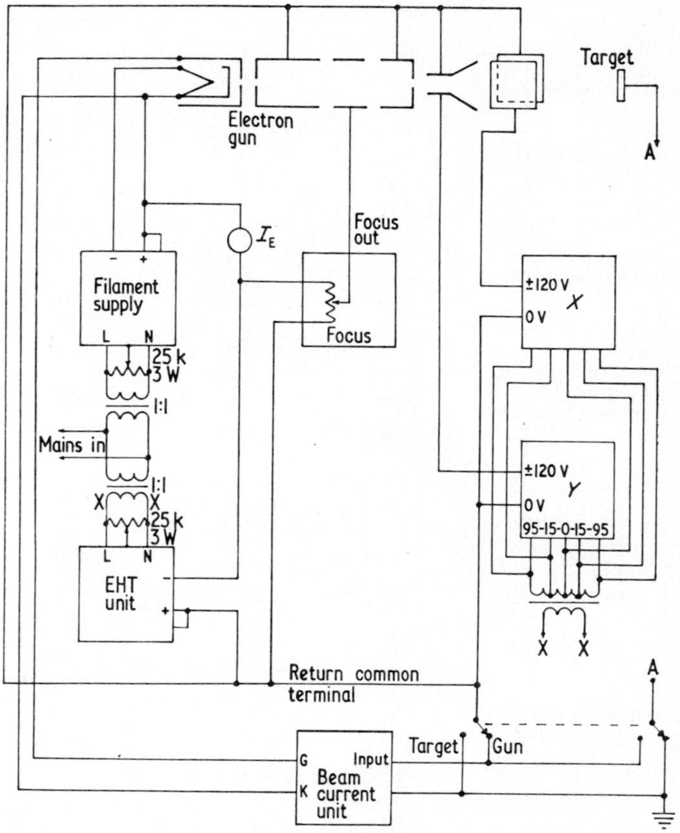

FIG. 3.4 Electron gun and associated control circuitry intended principally for LEED, but used also for AES in the earlier days of the latter technique, as described by Baines *et al.*[3] The electron emitter is an indirectly heated cathode, and the beam is extracted through a Wehnelt cylinder into a simple focusing system, followed by *X–Y* deflection plates. The control circuitry in this case is designed to maintain the beam current constant while the accelerating potential is varied, when recording LEED intensity vs. voltage plots. At typical operating conditions of 1.5 keV beam energy and 7.5 μA beam current, and with 10 V on the Wehnelt cylinder, the spot size on the target is 0.5 mm.

can also be reduced by using a low temperature cathode material such as LaB_6.[6]

3.2.2 *Photon sources*

Ultraviolet sources Sources of UV radiation for surface analysis depend on electronic transitions taking place in a rare gas discharge. Rare gases are chosen because of general convenience in handling and of their compatibility with UHV requirements. The most commonly used rare gases are He and Ne, with the former being by far the more predominant. The line radiation, for

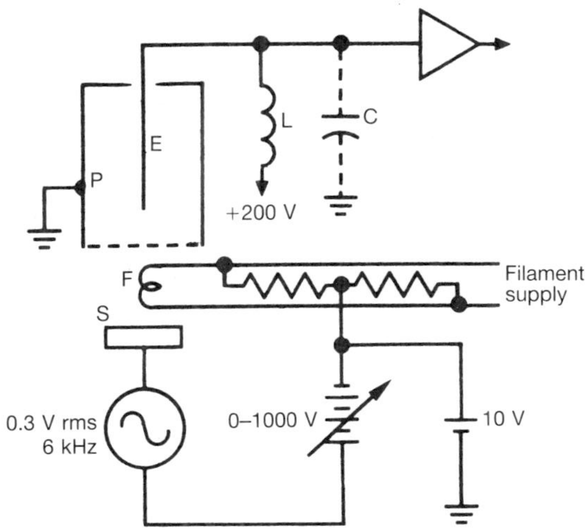

FIG. 3.5 Soft X-ray appearance potential spectrometer used by Park and Houston,[4] showing the very simple type of electron source that is adequate for SXAPS. The potential gradient from one end of the emitting region to the other must be minimized in order to optimize the energy resolution in the technique. In this example the filament F is a thick ribbon of tungsten with the narrow edge facing the sample S; the ribbon is thinned slightly over a short length to localize the emission. The potential drop across the emitting region is then a few tenths of a volt.

example, from He arises from the $2p \rightarrow 1s$ transition occurring either in the neutral atom, in which case it is called HeI, or in the singly ionized state, when it is called HeII; the respective energies are 21.21 eV and 40.8 eV. The corresponding NeI and NeII lines are at 16.8 eV and 26.9 eV. In both He and Ne the line widths are of the order 20 meV, thus there is negligible contribution to photoelectron line broadening from the source in UPS. Satellite lines do occur in UV light sources but are only a very small percentage of the intensity of the principal lines.

The most universally used type of UV source is the cold cathode discharge, of which a commercial design is shown in Fig. 3.6 (courtesy of Leybold Ltd). Since the pressure in the capillary where the discharge takes place is of the order of 1 torr, and the light is being beamed without a window directly into UHV, there must clearly be much differential pumping, and the source in Fig. 3.6 has two stages of such pumping. The proportions of HeI and HeII, for example, in the UV light depend on the precise pressure in the discharge region, with a lower pressure required for a usable flux of HeII. Typical pressures in the UHV system during the use of HeI and HeII would be $\sim 7 \times 10^{-9}$ torr and $\sim 5 \times 10^{-10}$ torr, respectively. In order to cause the discharge to 'strike', a voltage of > 1 kV is applied across the capillary while the pressure inside is slowly increased; normally an over-pressure is required. Once it has struck, the pressure is reduced slowly to the level appropriate to the line to be used. The capillary bore is of the order of 1–2 mm, which is also

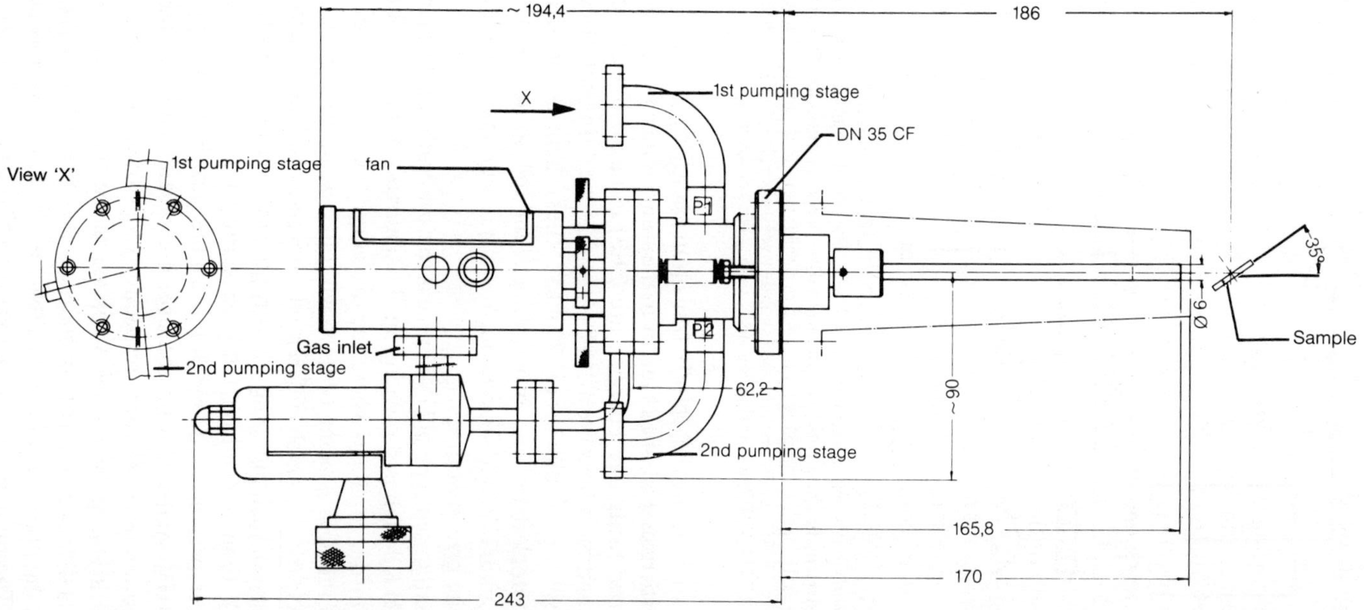

FIG. 3.6 Diagram of a commercial ultraviolet source of the cold cathode discharge type used universally in UPS, by courtesy of Leybold Ltd. Helium is admitted via a leak valve into the capillary tube where the discharge takes place, at a presssure of about 1 torr. The internal surface of the capillary, whose bore is about 2 mm, must be very smooth and the bore accurately linear, otherwise the discharge will be quenched too easily. The discharge is struck with an applied voltage of 1.0–1.2 kV, following which the gas pressure can be reduced slowly to the required region, since the proportions of HeI and HeII radiation depend on the pressure in the discharge. Since the UV beam enters the chamber without passing through a window, differential pumping is required. Stage 1 is normally via a fast rotary pump, stage 2 via a diffusion pump or a turbomolecular pump. The source must be cooled when in operation, via a small fan.

therefore the approximate diameter of the UV pencil beam striking the specimen.

X-ray sources In choosing an anode material for an X-ray source suitable for XPS, there are certain conditions that have to be satisfied. The most important is energy resolution. Equation 2.2 giving the relationship between the kinetic energy of an ejected photoelectron, the characteristic line energy of the X-ray photon, and the binding energy of the core electron, indicates that the line width of E_{kin} will depend, *inter alia*, on the line width of $h\nu$. As described in section 2.2.1 the experimenter is continually seeking chemical information in XPS by detailed study of the photoelectron spectra of individual elements, a study that needs an energy resolution of the order of 1.0 eV. In order to avoid any limitation on the available resolution originating in the line source, it is therefore necessary to use sources of line widths significantly less than 1.0 eV. Another condition is of course that the anode material must be a good conductor, otherwise heat transfer from the point of impingement of the electrons that stimulate the X-rays will be insufficient and the anode might melt. A further condition is that the material must be compatible with UHV.

In Table 3.1 are listed a few energies and widths of some characteristic X-ray lines from some possible candidate materials.

Table 3.1
Energies and widths of some characteristic X-ray lines

Line	Energy (eV)	Width* (eV)
Y Mζ	132.3	0.47
Zr Mζ	151.4	0.77
Nb Mζ	171.4	1.21
Mg K_α	1253.6	0.70
Al K_α	1486.6	0.85
Si K_α	1739.5	1.00
Y L_α	1922.6	1.50
Zr L_α	2042.4	1.70

*Defined as the full width at half maximum height of the line.

There are very few available sources with lines of sufficiently small width. The yttrium and zirconium *Mζ* are narrow enough, but their low line energies do not make them useful for general application since they would not be capable of exciting a sufficient number of photoelectron lines, although they have been used in some special applications. Silicon is not easy to apply to an anode surface and in any case is a poor conductor, so one is left with the magnesium and aluminium K_α lines. It is in fact these two that are used so universally in XPS.

For efficient production of Mg and Al K_α radiation it is necessary to use exciting electron energies at least an order of magnitude higher, and in most modern spectrometers an accelerating potential of 15 kV is used. For optimum sensitivity in the technique it is desirable both to operate the source at as high an electron bombarding current as the anode will stand, and to bring the anode as close as possible to the sample surface. At the same time the anode must be cooled adequately and the source must be compact. These requirements have led to the designs of source shown in Fig. 3.7. To allow rapid heat transfer to the forced flow of cooling water, the anode block and water tubes are made of copper. The anode material is deposited on the end of the block to a thickness of $\sim 10\ \mu m$, which excludes L_α radiation from the underlying copper but allows adequate heat transfer. All modern sources are in fact multi-anode sources, with two or more anode materials deposited on different faces of an angled block, as in Fig. 3.7(b), from Yates *et al.*[7] Such an arrangement requires a separate filament for each anode material, and it is possible therefore to change rapidly from one line to another merely by flicking a switch. A dual anode would normally consist of Mg and Al, but special combinations such as Al/Zr are sometimes useful.

It is necessary to interpose a window between the anode and the sample to screen the latter from stray electrons, from heating effects, and from any possible contamination from the anode region. The material must obviously be transparent to Mg and Al K_α radiation, and as thin as possible so that the soft X-ray flux is not unduly attenuated. The material normally used is Al foil of thickness about $2\ \mu m$. Care has to be taken on pumping-down the spectrometer not to put too great a differential pressure across the Al window, otherwise it will rupture.

The characteristic line radiation is not the only useful radiation from Mg and Al sources. As Fig. 3.8 shows, there is a not insignificant flux above the characteristic line, called bremsstrahlung radiation, which consists of a continuous background extending up to the bombarding electron energy. Although of no use in XPS, this radiation is useful for exciting Auger transitions from core levels lying too deep to be ionized by the characteristic radiation, and forms part of the technique of XAES.

Up to now the discussion of X-ray line sources has been concerned with the 'natural', or unmonochromatized, radiation. It should be realized that the X-ray radiation from an Mg or Al source, although normally labelled K_α for the sake of brevity, is in fact more complex than that label suggests. The principal line is itself an unresolved doublet, $K_{\alpha_{1,2}}$, and there are satellite lines $K_{\alpha_{3,4}}$ and K_β. The $K_{\alpha_{3,4}}$ can sometimes be obstructive in interpretation of spectra since it is situated some ~ 10 eV on the high energy side of the $K_{\alpha_{1,2}}$ and is roughly 10% of its intensity. Again, the K_β, lying about ~ 70 eV to the high

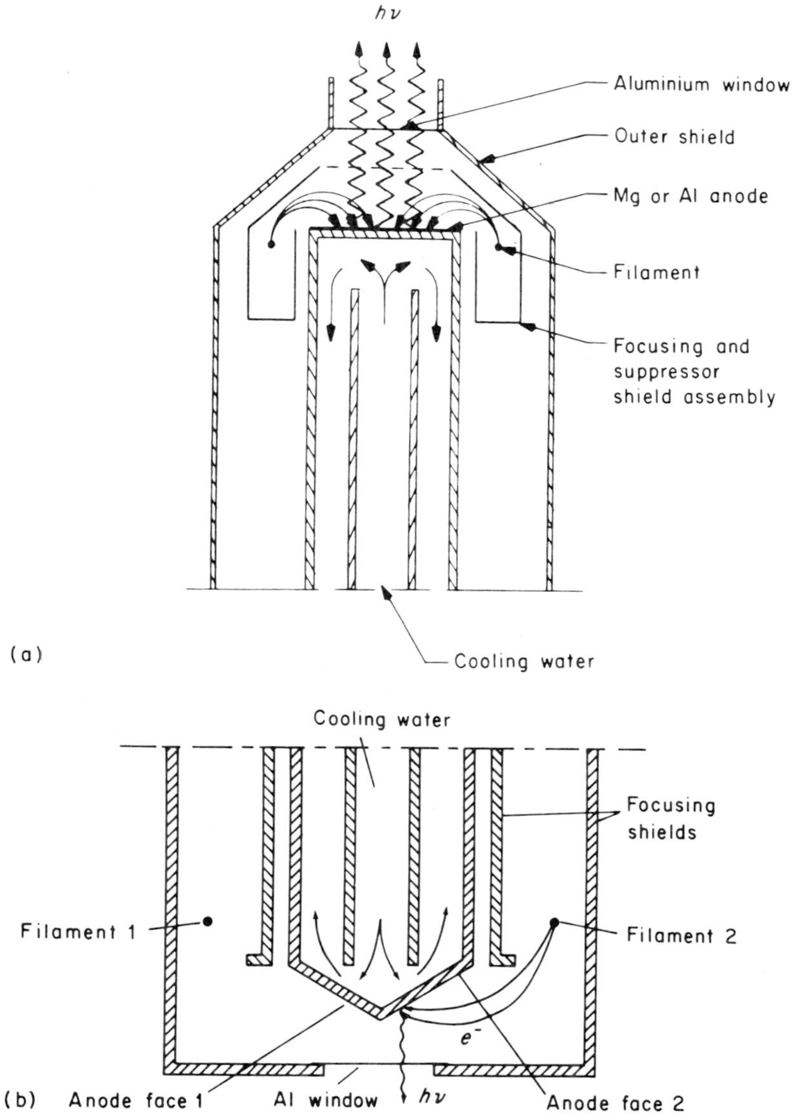

FIG. 3.7 Two designs of soft X-ray sources for use in XPS. In both cases electrons from a filament are accelerated onto an anode at 10–15 kV positive potential, thereby exciting soft X-radiation containing both characteristic lines and bremsstrahlung background (see Fig. 3.8). Cooling water is forced through copper pipes to the end of the anode, at several bar pressure and with flow rates of the order of $3\ l\ min^{-1}$. (a) Single-anode type (courtesy of Perkin–Elmer Physical Electronics Division) in which a thick ($\approx 10\ \mu m$) film of either aluminium or magnesium is deposited on a flat anode surface. The filament is surrounded by a cylindrical focusing shield at the same potential. (b) Dual-anode type of Yates *et al.*[7] in which the tapered end of the anode has aluminium on one face and magnesium on the other, with separate filaments for each. Focusing is similar to that in the single anode.

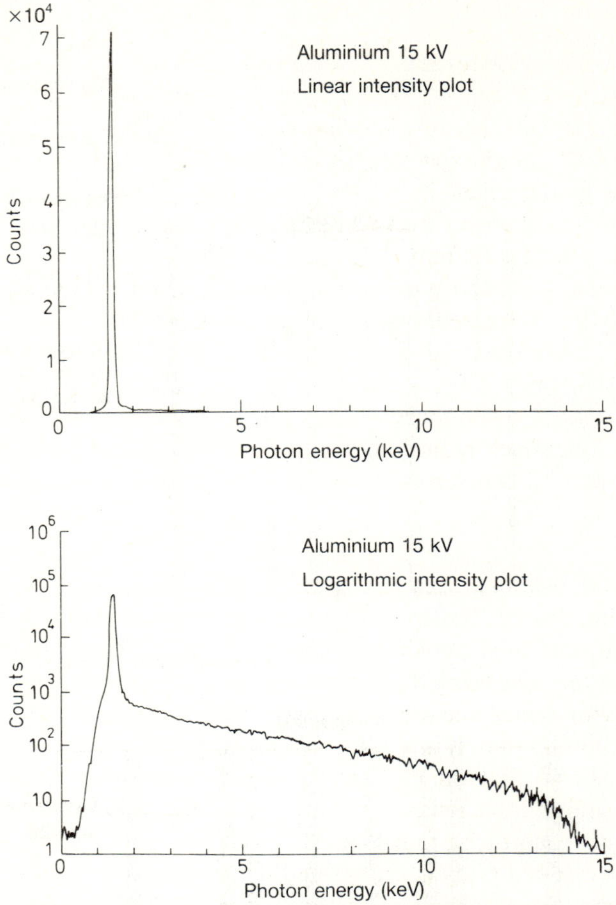

FIG. 3.8 Radiation emitted from an aluminium target when bombarded by electrons accelerated to 15 kV, on linear (above) and logarithmic (below) plots of intensity at each photon energy. Although the characteristic sharp $K_{\alpha_{1,2}}$ line is the most intense feature in the spectrum, nevertheless there is bremsstrahlung radiation extending all the way up to the electron energy, that is useful for excitation of Auger transitions.

energy side, can lead to misinterpretation. Removal of the satellites, improvement of signal-to-background by elimination of the bremsstrahlung continuum, and selection of an individual line from the principal doublet can be achieved by monochromatization.

The principle is shown in Fig. 3.9 (courtesy of Kratos Ltd). The source is placed at a particular point on a Rowland sphere, on the other side of which is placed a disperser, normally an accurately ground quartz crystal. The X-rays emitted from the source are diffracted by the crystal, and if the sample is at the appropriate point on the Rowland sphere, only the K_{α_1} component is focused on it. Diffraction of course occurs by the Bragg relation

$$n\lambda = 2d \sin \theta \tag{3.3}$$

where n = diffraction order, λ = wavelength of X-rays, d = crystal spacing, θ = Bragg angle.

Quartz is chosen because for first-order diffraction of Al K_{α_1} X-rays, for which $\lambda = 0.83$ nm, the spacing of the $10\bar{1}0$ planes is 0.425 nm, and the Bragg angle thus a convenient 78.5°. A popular size for a monochromator would involve a 0.5 m diameter Rowland sphere, for which the dispersion along the edge of the sphere is 1.6 mm eV^{-1}. Since the line width of the Al K_{α_1} radiation is less than 0.4 eV, the bombarding electron beam on the anode must be sufficiently finely focused, and the sample sufficiently accurately positioned, that the width of the irradiated area on the sample in the dispersion plane is less than 0.6 mm.

The flux from a monochromator, for the same power dissipated in the source, will be much reduced compared to the unmonochromatized source, but the detectability in terms of signal-to-background, and lack of satellites, is improved.

Synchrotron radiation All the sources discussed to this point have been discrete line sources (except, of course, the bremsstrahlung tail). Although convenient, and used universally, they have certain drawbacks in XPS, the most important one being that the photoionization cross-sections for different core levels for a fixed line will differ markedly, some being near maximum and others near minimum. It is always possible to go to another line source, but it has been shown that the choices are very limited. A continuously tunable source of sufficient intensity would be ideal, and in fact such a source exists in the form of synchrotron radiation.

When a charged body is accelerated it emits radiation, and the process is particularly efficient for an electron because of its small mass. In a synchrotron electrons are accelerated to near-relativistic speeds around an approximately circular ring or torus in an UHV by the action of pulsed magnetic fields. As they are accelerated they emit light in a continuous spectrum whose shape depends on the radius of curvature and the electron energy. The emitted radiation is concentrated in a narrow cone tangential to the electron orbit, and is therefore ideal for passing into a monochromator for energy selection. The general shape of the spectrum is shown in Fig. 3.10, after Farge and Duke.[8] A synchrotron torus would typically be of diameter many tens of metres, and its energy and current several GeV and 1 A, respectively, so that it is a large device with a large amount of energy stored in it. A typical energy near the maximum intensity would be 1–2 keV, so it is ideally suited for XPS. There is also a significant intensity down to much lower energies, i.e. down to the UPS operating regime. Synchrotrons are so large that they have to be nationally owned and would-be users have to arrange to take their experiments to them.

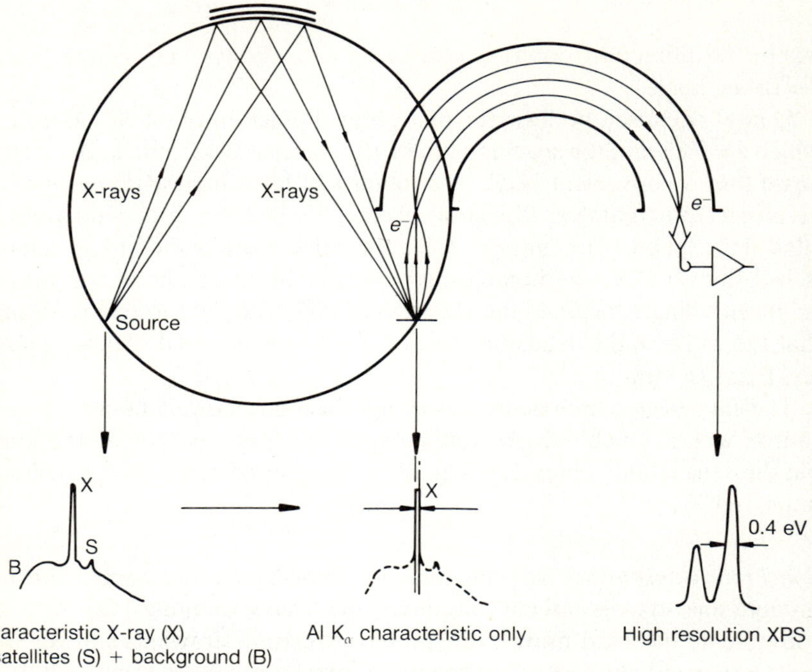

FIG. 3.9 Principle of monochromatization of Al K_α radiation as applied in XPS. The X-ray source is positioned on one point of a Rowland sphere, typically of diameter 0.5 m, and on the other side of the sphere is a bent and ground quartz crystal, or sometimes a group of crystals, matching the sphericity. X-rays are diffracted by the crystal, and if the sample is placed at the appropriate point on the Rowland sphere, then only the K_{α_1} is focused onto it. The K_{α_2} component will be dispersed further along the surface of the sphere, but since for a 0.5 m Rowland sphere the dispersion is 1.6 mm eV^{-1}, both source and sample must be positioned very accurately. Higher energy satellites such as the $K_{\alpha_{3,4}}$ are dispersed much further away, and the background is removed, as indicated at the bottom of the figure (Courtesy of Kratos Ltd.)

Laser radiation A description of the mode of operation of a laser is outside the scope of this volume, and the reader is invited to consult one of the many reference books on the subject.

3.2.3 *Ion sources*

Low energy sources 1–10 *keV* (*SSIMS*, *ISS*, *SNMS*, *IAES*, *IBSCA*) Many experiments, particularly the earlier ones, in these techniques were carried out using the same rather simple type of ion source as is still frequently used for cleaning sample surfaces by ion bombardment, and for depth profiling. This source produces ionization of the noble gas (usually argon) by a discharge initiated by the combined action of magnetic and electrostatic fields, i.e. a Penning discharge. A typical example of this type is shown in Fig. 3.11

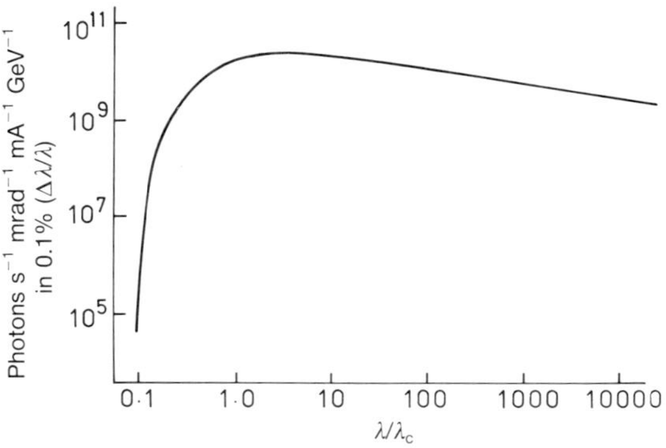

FIG. 3.10 Normalized spectrum of photon energies emitted from a synchrotron, adapted from Farge and Duke.[8] The intensity (vertical) scale is a function only of electron energy and current within the synchrotron ring, while the wavelength (horizontal) scale is a function of the critical wavelength λ_c. If photon energies of ≈ 1000 eV, as required for XPS, are to be obtained near the maximum, then λ_c should be of the order of 1 Å. To achieve that, the electrons orbiting the ring should have energies of several GeV, and the radius of their orbits should be several metres.

(courtesy of VG Scientific). The noble gas is piped directly into the discharge chamber through the back of the gun, and the positive ions are extracted via the accelerating potential past a rudimentary focusing electrode. Guns of this type operate between 1 and 10 keV, with an optimum spot size of greater than 1 mm diameter, and with no provision for rastering the beam. Although probably suitable for low energy IAES and IBSCA, they are not really suitable for SSIMS since it is difficult to produce a sufficiently low ion current from them, nor for ISS since the energy spread in the beam is too great.

A more sophisticated source is illustrated in Fig. 3.12 (courtesy of Leybold Ltd.). Ion generation takes place in a collision chamber, and in this case it is not a cold cathode discharge that is employed, but a thermal emitter. Electrons from a circular thoriated iridium filament are accelerated to a grid through about 100 V, and ionize gas molecules present in the filament/grid volume. The anode and a surrounding electron repeller electrode are both shaped to allow electrons to pass through the volume many times, thus increasing the ionization probability. Positive ions are extracted from the volume by a voltage of about 150 V and then accelerated to the chosen final energy between 0.5 and 5 keV. The ion beam is first focused down to a diameter of about 0.1 mm with a single element lens, onto an aperture of diameter 1 mm, and then from the aperture via two tubular lenses onto the specimen situated some 200 mm from the source volume. The aperture has such a small conductivity for the source gas that the pressure drop across it is 3–4 orders of magnitude. At the end of the gun two pairs of mutually orthogonal plates allow the ion beam to be rastered on the specimen surface over a typical area of 5×5 mm. The optimum spot size at the working distance is about 10 μm. The gun is useful for SSIMS because the current in the ion

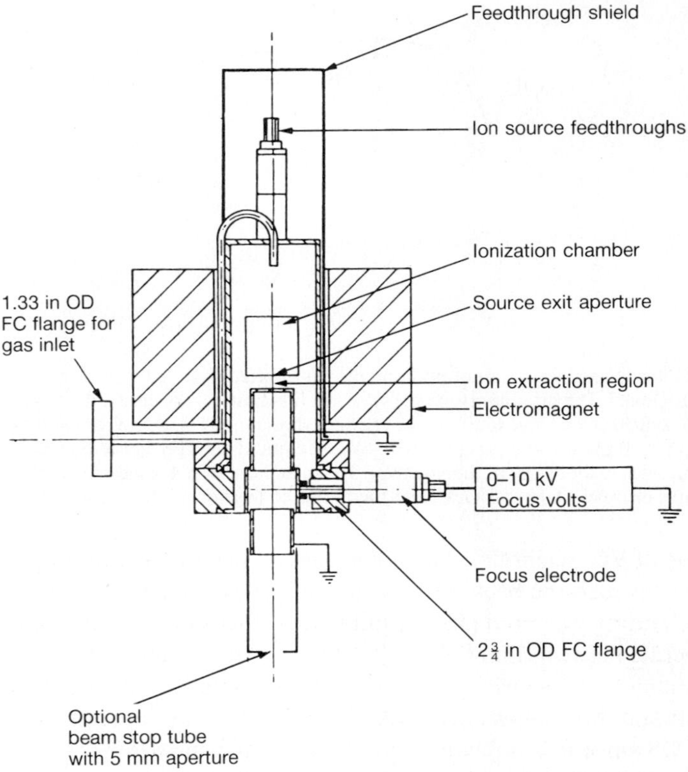

FIG. 3.11 Discharge type ion source used mainly for cleaning sample surfaces and for depth profiling. The noble gas, usually argon, is fed directly into the ionization chamber, where a discharge is initiated by the combined action of magnetic and electrostatic fields. Positive ions are extracted via the accelerating voltage through a small aperture, and focused crudely on their way to the sample. The area bombarded is usually large, i.e. several mm^2, and rastering is not possible. Ion energies can be varied in the range 1–10 keV, with beam currents 10–100 μA. The source also produces many excited neutral atoms in the beam. (Courtesy of VG Scientific.)

beam can be reduced to a few nA by reducing the emission from the filament, and for ISS because the energy spread in the beam is only a few eV.

A third type of low-energy ion source that is becoming popular in surface analysis, even though it has a long history in other applications, e.g. accelerators, is the duoplasmatron. A commercial instrument is shown schematically in Fig. 3.13 (courtesy of VG Scientific). As the name of the source suggests, the discharge is separated into two regions. The gas is fed into a hollow cathode, inside which the first discharge is struck by the application of a high voltage. An intermediate anode is hollow and cone-shaped and also forms a magnetic pole piece, and it acts to compress the discharge in the second region, i.e. acts as a magnetic pinch. The result is very efficient ionization of the gas to produce a dense plasma, which is allowed to expand

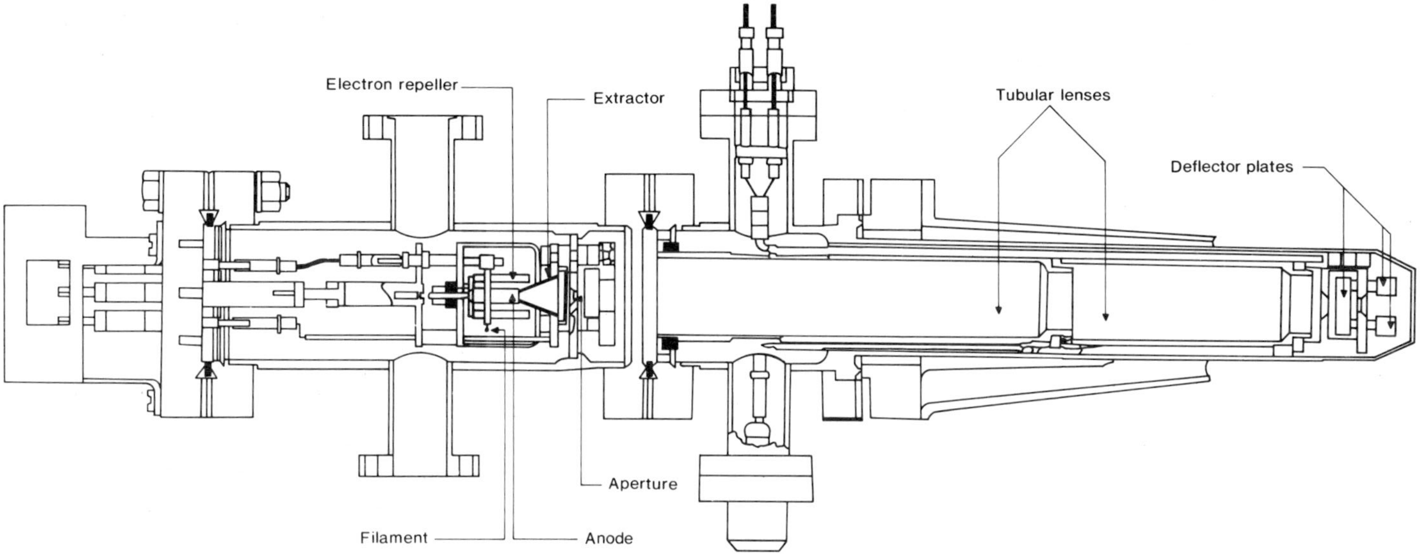

FIG. 3.12 Ion source using a thermionic emitter in a collision chamber for ionization. The emitter is a circular thoriated iridium filament, capable of operating in oxidizing or otherwise harmful ambients, from which electrons are accelerated to a grid anode. The anode is surrounded by an electron repeller, both shaped to allow the electrons to follow long trajectories, thus increasing ionization probability. Positive ions are extracted from the chamber through 150 V and then accelerated through an aperture to the chosen energy between 0.5 and 5.0 keV. Two tubular lenses focus the beam onto the specimen, the overall distance from ionization chamber to specimen being about 200 mm. Two pairs of mutually perpendicular deflection plates at the end of the gun allow rastering of the beam over areas that can be selected from 1×1 mm^2 to 10×10 mm^2. At 5 keV the optimum spot size is ≈ 10 μm. Beam current can be selected from 1 nA to 10 μA by varying the filament emission. (Courtesy of Leybold Ltd)

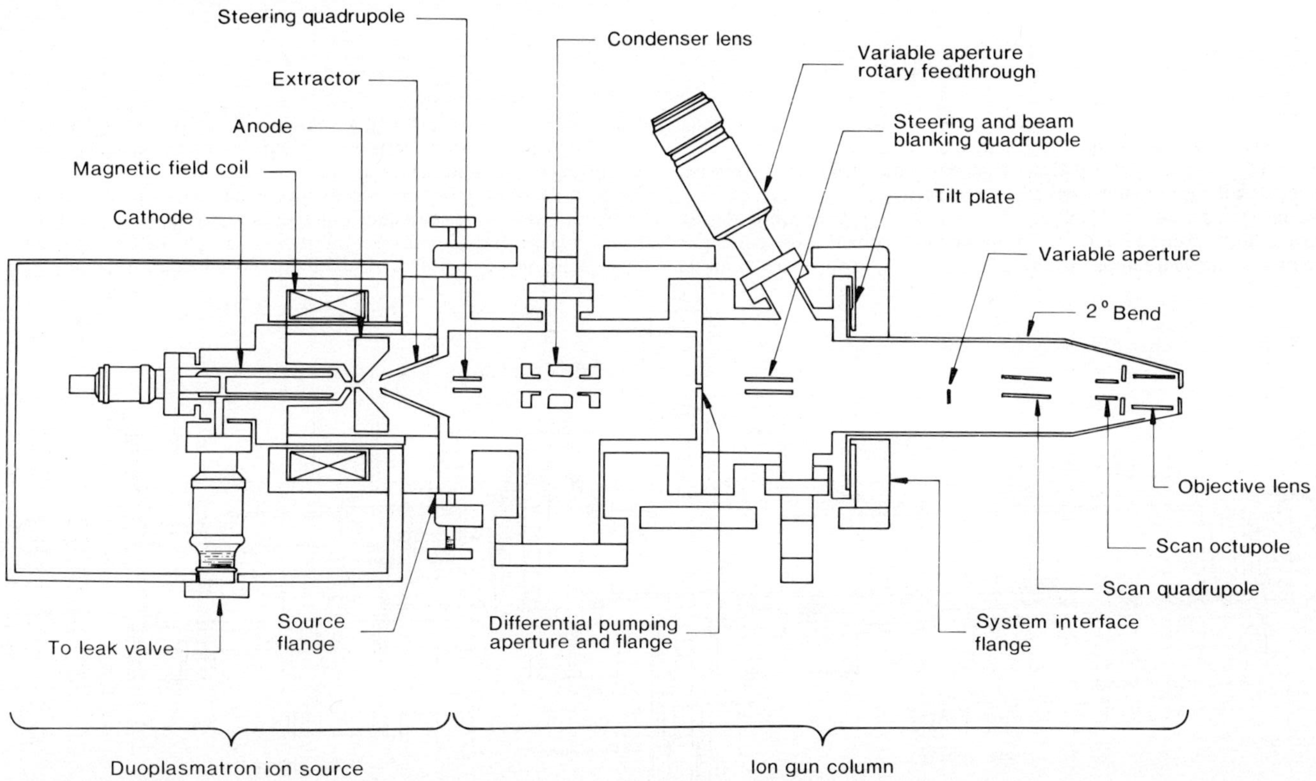

FIG. 3.13 Example of a duoplasmatron type ion source, by courtesy of VG Scientific. Noble gas is fed into the back of the source into a hollow cathode, inside which a cold cathode discharge is initiated by application of high voltage in the range 5–10 keV. Just beyond the cathode a hollow, cone-shaped anode is also a magnetic pole piece, and it compresses the discharge into the second discharge region. The ionized gas forms a dense plasma, which expands through a small orifice in the anode into an extraction region. According to the degree of focusing required, the beam then passes through either a single or a double lens system, providing spot sizes of ≈ 0.5 mm or ≈ 20 μm, respectively. Scan quadrupoles allow rastering of the beam, which, for the fine spot, is over an area of ≈ 0.5 mm^2. Currents available on the specimen are ≈ 20 μA or ≈ 1 μA according to the mode of operation.

through a small orifice in the anode itself, the latter also completing the magnetic circuit. Positive ions are then extracted from the plasma by application of a voltage between the anode and an extractor electrode. According to the degree of focusing required on the specimen, the extractor is then followed either by a single lens system with scan plates, providing a 0.5 mm spot, or a double lens system that provides a spot size of ~20 μm, with scan quadrupoles. This source can also be operated at a low beam current, particularly the model fitted with a double lens, and is therefore suitable for SSIMS; however, the energy spread may be too great for useful ISS.

The latter two sources would be entirely suitable for low energy IAES and IBSCA.

Medium energy sources, 10–100 keV (IAES, IBSCA) The techniques of IAES and IBSCA have also been operated at primary ion energies greater than 10 keV. For high energies special sources have been designed and built by individual experimenters,[9] such as the one shown in Fig. 3.14. The ions are

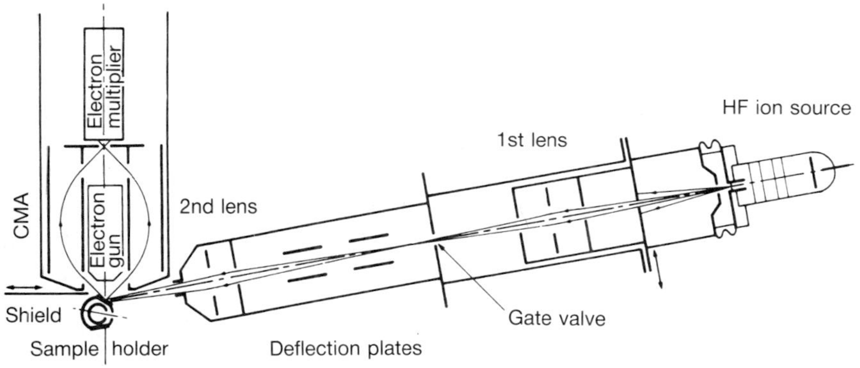

FIG. 3.14 Schematic diagram of the ion source and detector used for IAES by Viaris de Lesegno and Hennequin.[9] Ions are created by high frequency discharge in a source region at high pressure, and extracted through a succession of small apertures, which limit the conductance of gas to the analysis chamber. A lens system focuses the ion beam onto an aperture of diameter 0.6 mm, following which the beam can be deflected by a set of plates, and then a second lens focuses the beam onto the specimen. The gun can operate at ion energies up to 50 keV, and the beam currents can be varied between 1 μA and 14 μA. The irradiated area is about 0.3 × 0.5 mm.

created in a high frequency source, which must of necessity operate at a relatively high gas pressure; half-way along the device is a small aperture of diameter 0.6 mm onto which the first lens system focuses the beam extracted from the high frequency region. The aperture acts as a differential pumping barrier, since its conductivity for gas flow will be very low. After the aperture the beam can be deflected by a set of X–Y plates, and is then finally focused on to the specimen.The ion energy can be varied from a few keV to about 50 keV, so in fact it can span both low and medium energy regions.

A totally different and quite recently developed type of source, that also spans the low and high energy regions, is the liquid metal ion source (LMIS). The principle[10] is shown in Fig. 3.15. A metal needle, usually of a refractory metal such as tungsten or molybdenum, with a tip radius between 1 and 10 μm dips into a reservoir of liquid metal. If the liquid metal wets the needle satisfactorily, it will be drawn up over the needle and its tip by capillary action. An extractor electrode with an aperture centred over the needle is placed a short distance away, and a voltage between 5 and 12 keV is applied between needle and extractor. Under the combined forces of surface tension and electrostatic stress, the liquid is drawn up into a protrusion at the end of the needle, and ions, as well as other species, are formed in the region just beyond the protrusion. The positive ions are extracted through the aperture in the extractor. The precise mechanisms for the formation of the ions are still a matter for debate. The most common metal used in the source is gallium, which melts at 35°C, but indium, tin, bismuth, gold, copper, and caesium have also been used.

The LMIS is useful mostly as a 'high brightness' source, that is, for providing a large current into a small area. Spot sizes of less than 0.5 μm are easily achievable. This particular source has not yet had much application in surface-specific analysis, because of the high currents involved, but with proper design and development it should be capable of being operated at much lower currents.

Proton and α-particle sources, 0.3–4.0 *MeV* (*PAES*) Irradiation with very high energy (MeV) particles cannot be regarded as a common excitation method for surface analysis, and indeed those interested in using such high energies do so not because they are intending to use them for analytical purposes but for studies in the basic physics of ion–atom collisions. Nevertheless, the Auger transitions involved are still those of low energy, and if the technique were to be applied to surfaces it would be regarded as surface-specific.

The experimental arrangement that has been used by Kobayashi *et al.*[11] is shown in Fig. 3.16. A hydrogen or helium ion beam is extracted from a 4 MeV Van de Graaff accelerator, and passed through a gaseous charge exchanger, where the ions are converted to protons or α-particles, respectively. After magnetic deflection and focusing, the beam is collimated to 2 mm diameter and enters a vacuum chamber containing the gas to be studied. The gas is at about 2×10^{-3} torr pressure. The beam energy width is about 5 keV and the currents used 0.1–1 μA.

GDOS and GDMS In neither of these techniques is there any separately identifiable ion source as such. The ions are produced in a glow discharge in the volume immediately above the specimen surface, and the only electrode

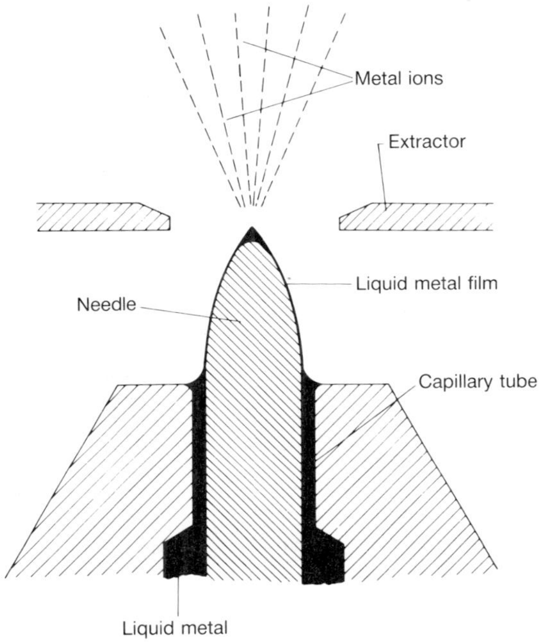

FIG. 3.15 Principle of operation of a liquid metal ion source, as depicted by Prewett and Jefferies.[10] A needle of tungsten or molybdenum dips into a reservoir of liquid metal through a capillary tube. Provided the liquid metal wets the needle, it is drawn up along the needle and over the tip. A voltage of 4–12 kV applied between the needle and an extractor electrode held a short distance away then draws the liquid up into a cusp under the combined forces of surface tension and electrostatic stress. Ions, as well as many other species, are formed in the region a few nm beyond the cusp, and are accelerated through the aperture in the extractor. High currents, as much as 100 μA, can be drawn from the source. Because of the very small volume of emission, the ion beam can be focused to a fine spot.

that could be called beam defining is a hollow anode placed very close to the specimen. The geometry of the Grimm[12] cell used by Berneron and Charbonnier[13] for GDOS, is shown in Fig. 3.17. The distance between the anode and cathode (the specimen) is about 0.5 mm, with the inner diameter of the anode about 8 mm. After evacuation the cell is backfilled with argon gas to a pressure of a few torr, when application of a DC voltage of about 1 keV between anode and cathode establishes a discharge. Positive ions are accelerated to the specimen by virtue of the potential gradient. The current densities are high, and surface analysis can be obtained only by operation of the discharge lamp for a few seconds.

3.2.4 *Neutral sources*

The simplest type of neutral, or more correctly, fast atom, source, is based on the design shown schematically in Fig. 3.18, called the saddle field source. The body of the source is the cathode and is at earth potential, and an anode wire

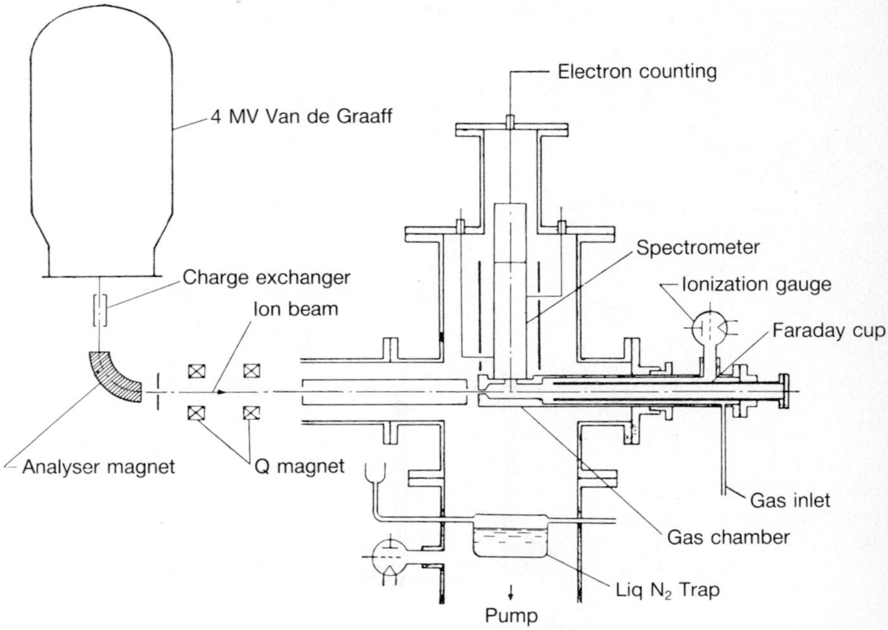

FIG. 3.16 Experimental arrangement used by Kobayashi *et al.*[11] for the production of either protons or α-particles, at energies of several MeV. A 4 MeV Van de Graaff accelerator produces beams of either hydrogen or helium ions, which are passed through a gaseous charge exchanger to convert them to protons and α-particles, respectively. After magnetic deflection and focusing the collimated beam of 2 mm diameter enters the target chamber, where the target is in fact a pure gas fed into a special mini-chamber at a pressure of 2×10^{-3} torr. Beam energies used varied from 0.5–2.6 MeV, with a spread of 5 keV, and a current of 0.1–1 μA.

or wires (usually a pair) pass through the centre of the cathode, i.e. diametrically if it is spherical, axially if it is cylindrical. Argon gas is fed into the back of the cathode body directly from a supply, to a pressure of about 10^{-4} torr. Application of a DC voltage between 2 and 10 kV to the anode sets up a discharge. Because of the geometry of the source electrons formed in the discharge are constricted to travel in a figure-of-eight path, as shown, describing long trajectories before final capture at the anode; it is that shape that gives the source its name. The probability of ionization of the argon is therefore high. Positive ions formed in the discharge travel along straight lines to the cathode, and if an aperture is placed in the cathode wall, emerge as a collimated beam. There are also neutralization processes taking place in the source, and according to the voltage applied to the anode, the emergent beam will contain a greater or lesser proportion of neutrals to ions, lower voltages favouring neutrals. Neutralization is thought to be most favourable at the two electron trajectory reversal points, where the electrons have slowed down to velocities to match those of the ions. By selecting operating conditions to

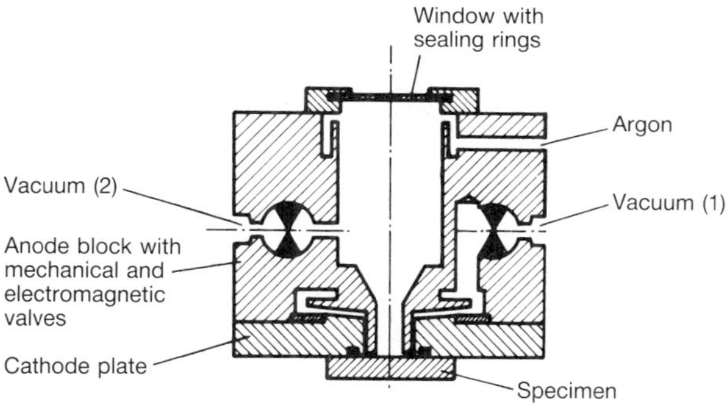

FIG. 3.17 Geometry of the Grimm[12]-type discharge cell used for GDOS by Berneron and Charbonnier.[13] The specimen is clamped to the cathode block, and in fact forms the cathode during the discharge. A specially shaped anode comes to within 0.5 mm of the specimen, allowing an area of ≈8 mm diameter on the specimen surface to be eroded. The cell is evacuated and then back-filled with argon to a pressure of a few torr. Application of about 1 kV between anode and cathode (specimen) initiates a glow discharge. Positive ions are accelerated to the specimen and erode it, the sputtered neutral material then passing into the discharge volume, where it becomes excited and decays with emission of light of wavelengths characteristic of the material. The light is then analysed in a monochromator.

produce the highest proportions of neutrals, and then by placing a grid electrode in front of the aperture at the same potential as the ions, thus deflecting them out of the beam, an intense beam of neutrals is produced. It has even been found in the larger sources that by careful design and shielding of the cathode, electron capture by ions can be complete inside the source, so that the deflecting electrode is not then necessary.

Details of the recently developed charge exchange neutral sources can be found in chapter 19.

3.3 Analysers

3.3.1 *Electron energy analysers (AES, SAM, ELS, HREELS, CEELS, XPS, UPS, SRPS, XAES, IAES, PAES, INS)*

In all the above techniques the device which measures the energies of electrons ejected from a surface is the principal component of the technique and is called an analyser. It is also often called a spectrometer but here we shall reserve the term spectrometer for the whole instrument, i.e. vacuum system, source, analyser together. As pointed out in the descriptions of the basic principles involved in the techniques, the electron kinetic energies to be analysed are invariably low, some very low, and analysers based on deflection in a magnetic field are therefore not suitable. Not only is it difficult to design devices that need magnetic fields in UHV, but at low kinetic energies magnetic analysers are not easy to operate. Thus all the analysers described here are electrostatic,

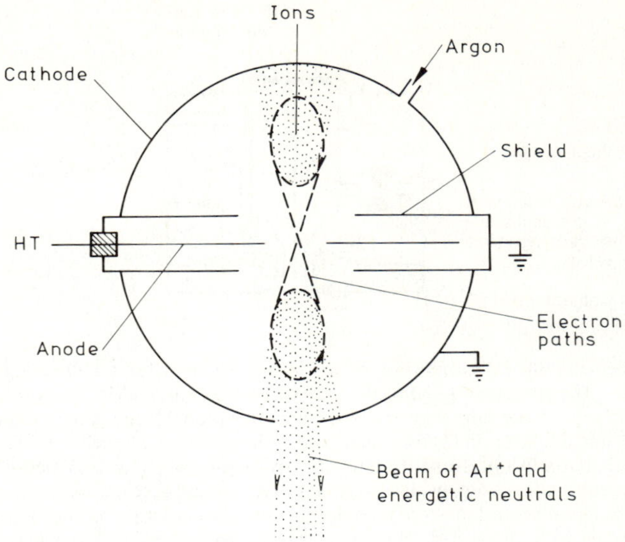

FIG. 3.18 Principle of operation of the saddle field ion source that also produces a high (≈50%) proportion of excited neutral, or fast, atoms. The source can take different shapes according to the application. The body acts as the cathode and is at earth potential. An anode wire or wires (often a parallel pair) pass through the centre of the cathode, i.e. either diametrically as in the figure, or axially if the body is cylindrical. Argon is fed into the back of the source to a pressure of about 10^{-4} torr, and a voltage of 2–10 kV applied between anode and cathode. A discharge is initiated, and because of the geometry electrons from the discharge travel in approximate figure-of-eight trajectories, with most of the excitation taking place at the ends of the loops. Positive ions and excited neutrals travel in straight lines to the cathode, and can emerge from an aperture as a collimated beam. If a fast atom beam is required, the ions can subsequently be deflected.

mostly of the deflection or dispersive type, but also retarding field. Before their description, it is necessary to discuss the definition of energy resolution and the requirements of the various techniques.

3.3.2 *Energy resolution requirements*

The two definitions of energy resolution in current use are those of the absolute resolution and of the relative resolution. The absolute resolution ΔE is defined and measured as the full-width-at-half-maximum (FWHM) of a chosen observed peak in the secondary electron distribution. The peak is normally chosen as the one that appears to be the most free from broadening effects and artefacts, and would probably be that of least width in the spectrum. In techniques using primary electron irradiation it is nearly always the elastically reflected peak. Sometimes the absolute resolution is measured in terms of the width ΔE_B of a peak at its base, called the base width, and in an ideal situation $\Delta E_B = 2\Delta E$. In electron spectroscopy, however, the latter is hardly ever used.

The relative resolution is defined as

$$R = \Delta E / E_0 \tag{3.4}$$

where E_0 is the kinetic energy of the peak position. R is generally expressed as a percentage, i.e. $(\Delta E/E_0) \times 100$, but it is not uncommon to hear the relative resolution expressed also as the resolving power, which is simply the reciprocal of R, i.e.

$$\rho = 1/R = E_0/\Delta E. \tag{3.5}$$

Thus the absolute resolution is independent of peak position, but the relative resolution can be specified only by reference to a particular kinetic energy. It will be seen now that each of the different electron spectroscopy techniques has its own energy resolution requirements.

In the photoelectron spectroscopic techniques (XPS, UPS, SRPS) the analysis is usually concerned just as much with identification of different chemical states as it is with elemental composition, so that not only must the energy resolution be good enough to enable the peaks corresponding to the various states of an element to be separable, but the resolution should be the same for all peaks in the spectrum. That is to say, the absolute resolution should be constant. Now the natural (i.e. unmonochromatized) line widths of the standard X-ray sources are 0.70 eV for Mg K_α and 0.85 eV for Al K_α. If these line widths were to be matched by the absolute resolution at the maximum available photoelectron energies (1253.6 eV for Mg K_α, 1486.6 eV for Al K_α), a relative resolution of $\sim 6 \times 10^{-4}$ would be required. This is possible instrumentally, but would be accompanied by too much loss in sensitivity. Furthermore if the X-ray sources were to be monochromatized the relative resolution would have to improve to $\sim 1 \times 10^{-4}$, which is not easy to achieve in a simple electrostatic deflection analyser.

It is therefore standard procedure to retard the kinetic energies of the photoelectrons, either to a chosen analyser energy or by a chosen ratio, which are kept fixed during the acquisition of any one spectrum. Thus if an analyser energy, or pass energy, of 50 eV were to be selected, an absolute resolution of 0.7 eV would require a relative resolution of only $\sim 10^{-2}$. Retardation thus allows the same absolute resolution to be maintained for a lower relative resolution, but the penalty is that the area analysed on the sample surface is then changing with kinetic energy. The chosen pass energy mode is called the CAT, or constant analyser transmission mode, and the other the CRR, or constant retard ratio mode.

To the list of photoelectron techniques that need good absolute resolution should also be added XAES, since in that technique the experimenter is normally interested not so much in chemical differences as in detailed fine structure in the Auger peaks. Since that technique is in some ways an integral part of XPS it does not need separate consideration here.

In UPS and SRPS the line width of the exciting radiation, either ultra-violet or from a synchrotron, respectively, is very narrow, only a few tens of meV, and is not normally the limiting factor. The inherent line widths of the photoelectron peaks themselves, convoluted with whatever instrumental broadening functions might be present, are the limitations on resolution, and for most purposes an absolute resolution of about 50 meV is adequate. Thus the analyser pass energy for these techniques would be set to much lower values than for XPS.

The Auger spectroscopies (AES, SAM, IAES, PAES) depend for their emission only on the creation of a core hole, by whatever means, and the natural line width of any exciting radiation, or the energy spread of incident particles, is irrelevant. The energy resolution requirements are set by the inherent widths of the Auger peaks themselves. Since most Auger peaks are several eV wide, there is no need for the same absolute resolution as in XPS, UPS, etc., an average absolute resolution of 2–3 eV being adequate. (Where there is fine structure of narrower width to be studied, it is normal to use XAES). AES is indeed normally operated under resolution conditions that allow high sensitivity at high Auger energies (e.g. >1000 eV) and high absolute resolution at low Auger energies (e.g. <150 eV), and this is achieved by using constant relative resolution rather than constant absolute resolution. A typical figure for relative resolution used would be 0.3%, i.e. a resolving power of ~ 330.

ELS and CEELS are usually performed in an AES system. Since they are loss spectroscopies the line widths, or rather, sharpnesses of steps, are linked to the line width of the incident electron beam. It is found in practice that the energy spread (~ 0.5 eV) in the electron guns used in conventional AES is perfectly adequate for these techniques, and the energy resolution requirements are thus only marginally more exacting than in AES.

For HREELS the situation is very different. In order to resolve the often closely spaced peaks resulting from vibrational losses from the very low primary energy (3–5 eV), the line width of the incident electron beam must be reduced to the absolute minimum technically possible. Since typical vibrational line widths are of the order 0.5–1 meV, ideally the incident line width should be less than that. The best that has been achieved so far is ~ 3 meV. Thus there is a need for very high absolute resolution, and analyser pass energies of less than 1 eV are used.

In AEAPS no energy resolution requirements are involved since electron energies are not measured, only secondary electron current. The Auger electron energies in INS are very low, and the same resolution considerations apply as, for instance, in UPS, which is complementary to it.

3.3.3 *Retarding field analyser* (*RFA*)

As the name suggests, an RFA is a high-pass energy filter that places a variable potential barrier in the paths of electrons, and only those electrons with energies high enough to surmount the barrier pass through to the detector.

If the barrier takes the form of a fine grid or mesh, as is usual, with the variable potential applied to it, there will be unacceptable energy aberrations in the spectrum unless the great majority of the electrons approach it at normal incidence. Where the electron source is virtually a point, as in incident electron methods, the only geometry that will minimize such aberrations is spherical, centred on the point of origin of the electrons. Just

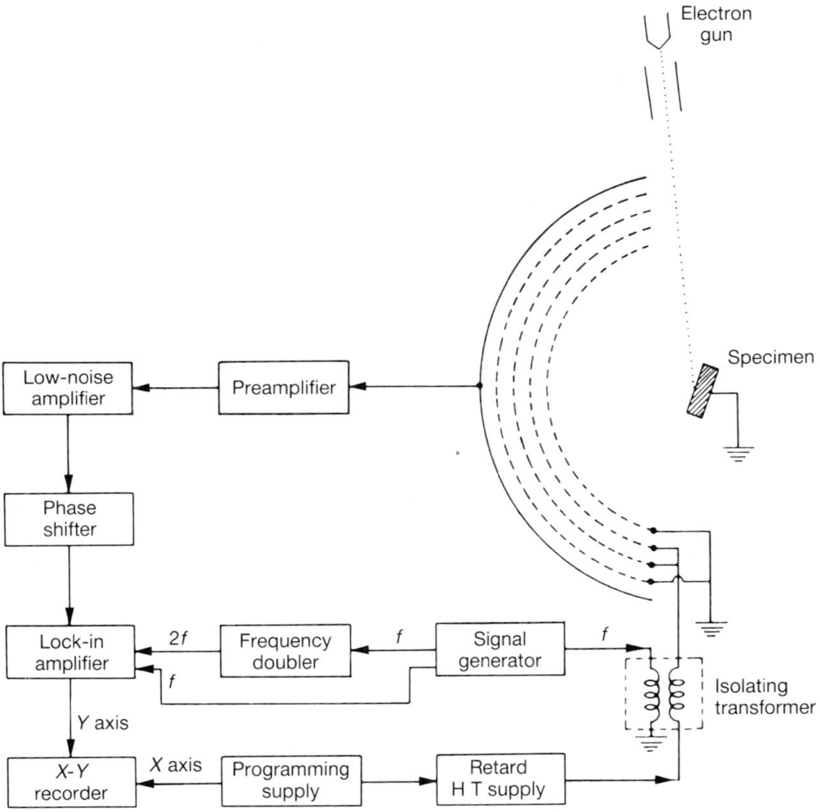

FIG. 3.19 Schematic of a Retarding-Field Analyser (RFA) still used occasionally for AES, when combined with LEED studies. For LEED a special electron gun would protrude through the centre of the spherical screen/grid structure, but is not shown here for clarity. The point of emission of secondary electrons from the specimen surface is the centre of the sections of spheres formed by the four grids and the screen. The first and fourth grids are grounded, in order to maintain field-free conditions near the specimen, and to minimize field penetration between grids and screen, respectively. The second and third grids are connected together and have the retarding potential applied to them from an HT supply. As the potential is swept it has applied to it a small modulation between 0.5 and 2.0 V p–p at some kHz frequency f, chosen to avoid resonances. The modulated component of the collected current is then amplified at frequency $2f$ to provide the differential energy distribution $dN(E)/dE$.

such an analyser has been in existence for many years, and its mode of operation is shown in Fig. 3.19.

The spherical retarding-field analyser of the type shown there was used, and still is, for Low Energy Electron Diffraction (LEED). For that technique a low energy electron gun protrudes through the centre of the collector and grids, but for energy analysis alone, without LEED, it is better removed as shown, to avoid aberrations. Electrons from the source on the sample surface drift in straight lines across the space between the sample and first grid, which is grounded; the space is thus field-free. The second and third grids are connected together and have the retarding potential applied to them. The reason for using two grids there is to reduce the lens effect suffered by electrons in passing through a finite mesh. Those electrons with energies high enough to pass through these grids then pass also through the fourth grid, which is grounded, and are collected on a solid spherical collector. A small negative potential may be applied to the collector to ensure collection of all the electrons. The grounded fourth grid is inserted to shield the collector from capacitative coupling with the modulation applied to the retard grids, to be described below.

Superimposed on the retarding voltage via an isolating transformer is a small modulation voltage, chosen between 0.5 and 2.0 V p–p normally, at a few kHz frequency. The frequency is also variable and is chosen to avoid resonances with other possible frequencies in the system, such as that of the mains. The collector current is of course then modulated at the same frequency. This current is compared in a phase-sensitive detector (also called a lock-in amplifier or a synchronous detector) with a reference signal from the signal source. Amplification of the component of collector current at the fundamental frequency f then gives the energy distribution $N(E)$ of electrons, while amplification of the component at frequency $2f$ gives the differential distribution $\mathrm{d}N(E)/\mathrm{d}E$. These relationships can be derived quite easily with some simple algebra.

If sufficient care in construction is taken, an RFA of diameter about 130 mm can have a relative resolution of 0.3%, but for most commercial systems the resolutions normally obtainable are nearer 0.5–1.0%. Despite its simplicity, the RFA has one overriding disadvantage, and that is, that at any one retarding potential, all electrons of energies greater than that potential will reach the collector. Shot noise is thus generated by many electrons at energies other than those being analysed, resulting in poor signal-to-noise characteristics. There is also, of course, no possibility of carrying out angle-resolved measurements because of the very large solid angle subtended by the collector. Nevertheless, for those using LEED for structural studies, the RFA provides a convenient way, via AES, of checking the cleanliness of a surface.

3.3.4 *Deflection analysers*

Cylindrical mirror analyser (CMA) This analyser and the Concentric Hemispherical Analyser (CHA) to be described next are by far the two most commonly used analysers for electron spectroscopy. The CMA was adopted for use in this field at the time in the early history of AES that workers were looking for a substitute for the RFA, whose drawbacks were obvious. It was found then, and experience has confirmed that it still is, ideal for AES and SAM. Over the years many commercial variations have appeared, but the principle of operation is exactly the same as it was originally.

The CMA is shown diagrammatically in Fig. 3.20. It consists of two co-

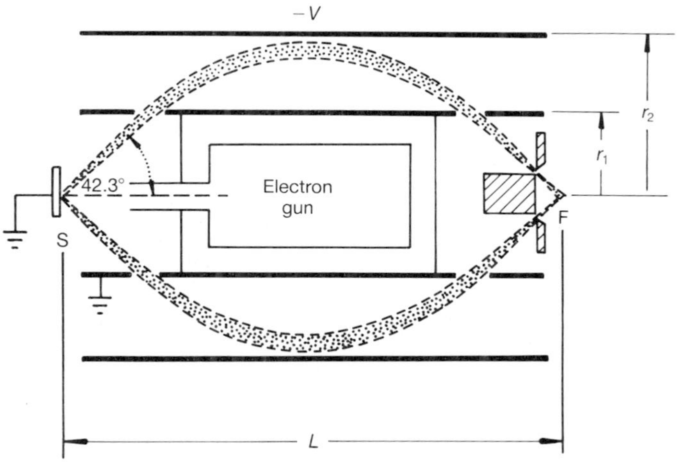

FIG. 3.20 Schematic diagram of a single-pass Cylindrical Mirror Analyser (CMA). Two cylinders of radii r_1 and r_2 are positioned co-axially. Annular entrance and exit apertures are cut in the inner cylinder in such places that electrons leaving the surface of the specimen S, on axis, pass into the intra-cylinder space at an angle of $42.3 \pm 6°$, and pass out again to the focus F. The relationship between the deflection potential $-V$ placed on the outer cylinder, with the inner earthed, and the electron energy E_0 is given by expression (3.6). The position of minimum electron trace width is in front of the focus and off-axis, and a limiting aperture (shown on an exaggerated scale) is placed at that point to improve the resolution. L is the distance, equal to 6.1 r_1, between source S and focus F. The exciting electron gun is accommodated internally. Typical dimensions would be $r_1 = 3$–4 cm, $r_2 = 7$–8 cm.

axial cylinders of radii r_1 (inner) and r_2 (outer) positioned accurately co-axial. Annular entrance and exit apertures are cut in the inner cylinder, in the appropriate places and of the appropriate width (see below). The potential of the inner cylinder is normally earthed for AES, although for the version that has been used for XPS the analyser is floating above earth, and a deflecting potential of $-V$ is placed on the outer cylinder. Electrons emitted from a source S on axis, at an angle α, pass through the entrance aperture and those of a particular energy E_0 are deflected by the outer cylinder back through the exit aperture to a focus F also on the axis. For the special case where $\alpha = 42.3°$, the first-order spherical aberrations vanish, and the CMA becomes a second-order focusing instrument, in which case the relationship between the electronic energy E_0 and the deflecting potential becomes

$$E_0/eV = 1.31/\ln(r_2/r_1). \qquad (3.6)$$

The apertures must of course have a finite width, and the latter is chosen as a compromise between adequate energy resolution and optimum sensitivity. A typical aperture width would be $\Delta\alpha = 6°$, i.e. the acceptance angle into the analyser is $42.3 \pm 6°$. Under these conditions the electron paths are not infinitely narrow but consist of annuli, and it can be shown that the minimum trace width (near the focus) is

$$\omega_m = 7.76\, r_1 (\Delta\alpha)^3. \qquad (3.7)$$

The position of the minimum trace is slightly in front of the focus and off-axis. The distance off-axis is

$$r_c = 5.28 r_1 (\Delta\alpha)^2. \qquad (3.8)$$

It is found that if an additional limiting annular aperture is placed at the position of minimum trace width, and made equal to the width, the energy resolution is improved without loss in sensitivity. For the special angle $\alpha = 42.3°$, the distance L between source and focus is 6.1 r_1, and the maximum radial distance off-axis reached by the electrons traversing the analyser is

$$r_m = r_1 \exp\{1.31 \sin^2(\alpha + \Delta\alpha\}. \qquad (3.9)$$

Thus all the parameters scale with r_1, the radius of the inner cylinder, and once that is chosen the other dimensions follow automatically.

It is essential that electrons passing through the CMA 'see' only uniform radial fields, otherwise severe aberrations will result. The two apertures are therefore covered with a fine high-transparency mesh to shield the inside of the analyser from the sample region, and in addition the fields at the open ends of the analyser, which must be of finite length, have to be terminated correctly. This involves the insertion of fringe plates at each end, whose design and applied potentials ensure a smooth transition from the deflecting potential V to earth.

Merely scanning the potential $-V$, and recording the output as a function of electron energy, provides the energy distribution of electrons passing through the CMA. It should be noted, however, that the transmission of the CMA varies as E_0, so that the distribution recorded is not $N(E)$ (as in the RFA), but $EN(E)$. Modulation of V with a small sinusoidal voltage can be used to extract a differential distribution, but in this case the phase-sensitive detector is tuned to the fundamental frequency and not the second harmonic as in the RFA. Again the recorded differential distribution will be $EN'(E)$ and not $N'(E)$.

The relative resolution of a CMA is given by the approximate expression

$$\Delta E/E_0 = 0.18w/r_1 + 1.39(\Delta\alpha)^3 \tag{3.10}$$

where w is the notional slit width. When $\Delta\alpha$ is relatively wide (6° is usual) the slit width in eqn 3.10 is replaced by the source size. For example, for a CMA with an inner radius r_1 of 3 cm, a semi-angular aperture of 6°, and an effective source size of 5 μm, the theoretical relative resolution would be 0.15%. This will be somewhat degraded in practice by inaccuracies in fabrication and by stray magnetic fields.

In AES the source size is small (and becoming smaller) and the simple (i.e. single-pass) CMA with its very high transmission has been very effective. In XPS, however, the sample is normally flooded with soft X-rays, and the photoelectron source area is very large. Equation 3.10 shows that in that case the relative resolution would be too poor for XPS; in any case XPS needs better absolute resolution than AES. What is important in XPS is not the transmission of an analyser, but the luminosity, which is the product of transmission, entrance solid angle and entrance area. Thus the single-pass CMA is not suitable for XPS.

High luminosity combined with adequate absolute resolution has been achieved by Palmberg[14] by putting two CMA's in series, so that the exit aperture for the first is the entrance aperture to the second, and placing spherical pre-retardation grids in front of the entrance aperture. The result is called the double-pass CMA. It is shown schematically in Fig. 3.21, adapted from Palmberg.[14] For XPS the analyser itself is operated at a chosen band pass energy, to which the kinetic energies of the photoelectrons are retarded by the grids. The effect of pre-retardation to the pass energy E_p from E_0 is to improve the energy resolution by a factor (E_p/E_0). At the same time, however, the source volume is collapsed by a factor $(E_p/E_0)^{1/2}$; to maintain sufficient overall luminosity for XPS, the transmission in the second stage of the double-pass CMA is therefore increased by opening out the aperture of that stage.

For AES analysis, the pre-retardation grids are grounded, an external control (see the figure) allows the second stage apertures to be decreased, and the inner cylinder is also grounded as in the single-pass instrument.

All the critical parameters for a CMA scale with r_1, the radius of the inner cylinder, so that for any given pass energy, simply increasing the size of a double-pass CMA will lead to improved energy resolution. Because of its large acceptance solid angle, the CMA cannot accept an input lens, so that its development for electron spectroscopy would eventually be limited.

Concentric hemispherical analyser (*CHA*) If the CMA could be said to have grown up with AES, then the CHA has grown up with XPS and UPS, although of course the principles of operation of both were known long before

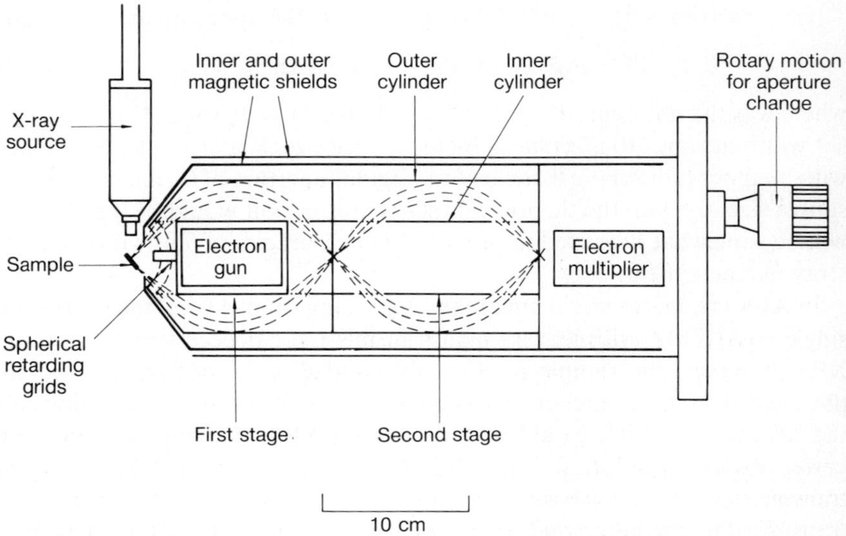

FIG. 3.21 Double-pass CMA designed by Palmberg[14] for both AES and XPS. The exit aperture for the first stage is the entrance aperture for the second. In front of the first-stage entrance are two spherical grids, centred on the specimen, to retard electrons entering the analyser to a constant pass energy, when used in the XPS mode. For AES the grids are grounded. The entrance angle is again $42.3 \pm 6°$. The X-ray source must be external since the space inside the first stage is occupied by the electron source. The transmission in the second stage can be varied via a variable aperture between the stages, operated by an external rotary drive.

surface analysis was born. Unlike the CMA, which was the second type of analyser to be used in AES (i.e. after the RFA), the CHA has been used from the beginning of XPS. It is even tending to replace the CMA for AES analysis in some instruments.

The CHA consists of two concentrically positioned hemispheres, as illustrated[15] in Fig. 3.22, of radii R_1 (inner) and R_2 (outer). A deflecting potential $-V$ is applied between the spheres so that the outer is negative and the inner positive. There will be a median equipotential surface between the hemispheres of radius R_0, and ideally $R_0 = (R_1 + R_2)/2$. The entrance and exit slits are centred on R_0. For the ideal situation where an electron is travelling on the circular orbit of radius R_0 and with kinetic energy E_0, the relationship between E_0 and the deflecting potential is

$$eV = E_0(R_2/R_1 - R_1/R_2) \tag{3.11}$$

if the angle between the entrance and exit aperture is 180°. As with the CMA, it is desirable to work with as large an entrance angle α (equivalent to $\Delta\alpha$ in Fig. 3.20) as possible commensurate with adequate energy resolution, and the compromise often reached is to choose α so that

$$\alpha^2 \approx w/2R_0 \tag{3.12}$$

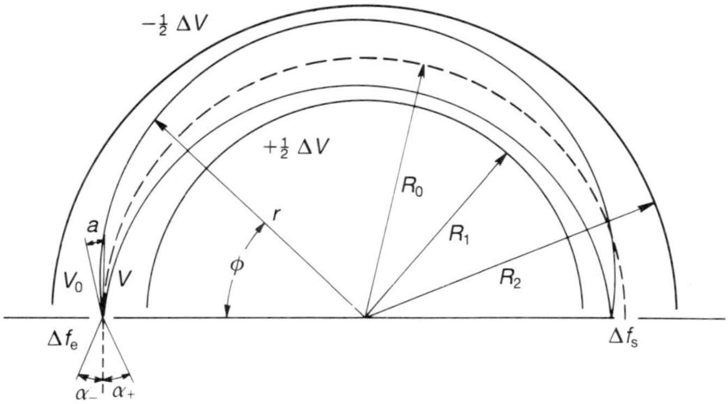

FIG. 3.22 Schematic of the Concentric Hemispherical Analyser (CHA). It consists of two concentric hemispheres of radii R_1 and R_2, between which a deflecting potential $-V$ is applied so that the outer is negative and the inner positive. The median equipotential surface has radius R_0, and the entrance and exit slits are centred on R_0. ϕ and r are the angular and radial coordinates, respectively, of an electron of energy E_0 entering the analyser at an angle α. Δf_e and Δf_s are the widths of the entrance and exit slits. After Roy and Carette.[15]

where w is the slit width.

As mentioned earlier, for reasons connected with the resolution requirements of the electron spectroscopies, it is customary to pre-retard electrons as they enter the analyser either to a chosen fixed analyser energy called the pass energy, or by a chosen ratio. Retardation when used with a CHA is usually accomplished by placing planar grids across the entrance slit, but may also be achieved in a lens system. Lens systems are also used in other ways to improve the overall efficiency of detection and analysis. A simple transfer lens has the advantage that the specimen does not have to be positioned close to the analyser entrance slit, since the analysed area on the specimen is instead projected on to the slit, making available a much greater working volume around the specimen. According to whether the fixed analyser energy (called CAT, constant analyser transmission) or the fixed retard ratio mode (called CRR, constant retard ratio) is required, the lens has to be made more or less complex. In the CRR mode the transmission between the specimen and the analyser slit is constant, whereas in the CAT mode it is varying. The latter needs a three-element lens, but the former can be performed with a single element.

The relative resolution of a CHA is given by the approximate expression

$$\Delta E/E_0 = w/2R_0 + \alpha^2/4 \qquad (3.13)$$

where w is the slit width. With the compromise given by expression (3.12), the relative resolution can be written

$$\Delta E/E_0 = 0.63\ w/R_0. \qquad (3.14)$$

In other words, if the dimensions (i.e. R_0) of the analyser are fixed, then there is an inverse relationship between E_0, which in the normal CAT mode is the pass energy, and the slit width w. As an example, if $R_0 = 10$ cm is chosen, which is a typical value, and an absolute resolution of 1.0 eV is required, then a pass energy of 50 eV would go with a slit width of 3.2 mm.

Commercially produced CHA's are used very widely indeed in XPS, UPS, SRPS and XAES. Special versions are also employed as both incident electron monochromators (see paragraph under 3.2.1) and very low energy electron energy analysers in HREELS, and will be described in the chapter dealing with that technique.

3.3.2 *Ion mass analysers (ESD, ESDIAD, PDS, SSIMS, GDMS, FABMS, SNMS)*

Because it is compact, does not require bending or focusing magnets, and is entirely UHV compatible, the device used universally for ion mass analysis in the surface characterization field is the quadrupole mass spectrometer. As its name suggests, it consists of four circular rods, or poles, held exactly co-axial and equally spaced in a rectangular array. From the strictly theoretical point of view, the rods should not be circular but of hyperbolic section, but the error involved in making them circular is small. Typical dimensions for a QMS of moderate mass resolution would be of rods about 6 mm diameter and about 120 mm long, with adjacent rods separated about 5 mm. One of the ways of maintaining the rods very accurately parallel, after machining and grinding each so that it is straight within a minute tolerance, is to separate them with sapphire balls, which can also be ground to extreme accuracy.

A cross-section through a quadrupole is shown schematically in Fig. 3.23, in which the hyperbolic potential distribution is also indicated.

Two voltages are applies to the rods, a DC voltage V_1 and an RF voltage V_0. The instantaneous potential in the space between hyperbolic rods would then be

$$\Psi = (V_1 + V_0 \cos \omega t)(x^2 - y^2)/r_0^2 \qquad (3.15)$$

where ω is the RF frequency, x and y are the lateral cartesian co-ordinates of an ion between the rods, and r_0 is the distance from the axis to the nearest point on each rod. When a singly charged ion of mass m enters the space between the rods, it is accelerated by the electrostatic field, and its equations of motion are

$$m\,\mathrm{d}^2x/\mathrm{d}t^2 + (2e/r_0^2)(V_1 + V_0 \cos \omega t)x = 0 \qquad (3.16)$$

$$m\,\mathrm{d}^2y/\mathrm{d}t^2 - (2e/r_0^2)(V_1 + V_0 \cos \omega t)y = 0 \qquad (3.17)$$

$$m\,\mathrm{d}^2z/\mathrm{d}t^2 = 0 \qquad (3.18)$$

z being the axial co-ordinate. Integration of the last of these equations gives

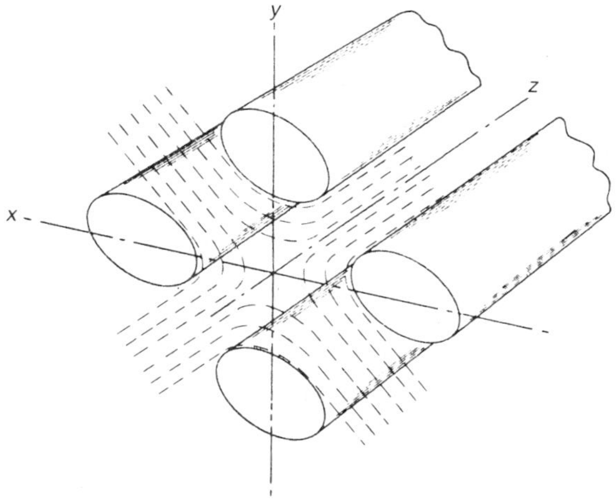

FIG. 3.23 Cross-section through the rods of a quadrupole mass spectrometer. The x, y, and z coordinates appearing in eqns (3.15) to (3.18) are shown, and the hyperbolic potential distribution when equal and opposite potentials are applied to opposed rod pairs, is also indicated.

the result that the axial velocity of the ion is constant, and not affected by the applied voltages. This velocity is that at the entrance to the analyser.

The mass filtering capability arises from the lateral trajectory characteristics determined by eqns 3.16 and 3.17. These have the form of the well known Mathieu equation. In mathematical terms, the solutions of the Mathieu equation fall into bands or regions of real solutions separated by regions of imaginary solutions. What this means in practical terms in the QMS is that for certain combinations of V_1 and V_0 the amplitude of the ion trajectory diverges rapidly, whereupon it hits one of the poles (or passes through to other parts) and is lost. For other combinations of V_1 and V_0 for an ion of that particular mass the trajectory is stable and the ion is focused at the end of the rod assembly into a multiplier. By careful choice of dimensions, RF frequency, and the ratio of V_1 to V_0, it can be arranged that for a fixed set of conditions a very small range of ion charge-to-mass ratios can be transmitted by the quadrupole, that is, the mass resolution can be varied to suit the requirements of the experiment. It should be noted that in this device there is roughly inverse proportionality between mass resolution and sensitivity; the better the resolution selected, the lower the sensitivity at that charge-to-mass ratio, and vice versa.

3.3.3 *Photon detectors*

Since in those techniques in Table 1.1 in which emitted photons must be detected, the detector or photon wavelength analyser is usually outside the vacuum system and of conventional type, the description of such detectors is regarded as outside the scope of this volume. The method and type used in each case will be mentioned in the appropriate chapter on the individual technique.

3.4 Sample insertion and handling

In this section, 'sample insertion' refers to those techniques operated in a UHV system, which is indeed all of them except IETS, and, normally, GDOS and GDMS. The method of sample insertion in APFIM is too specialized to be discussed here, and will be mentioned in the detailed discussion of the technique later.

The method to be used for inserting a sample to be analysed depends to a certain extent on how it is to be handled or treated subsequently. Generally speaking a sample that needs carefully controlled heating or cooling is intended to be the subject of an extended study rather than of a routine analysis, and must therefore be mounted individually. This means direct attachment to the specimen stage of a manipulator such as that shown in Fig. 3.24, which enables the specimen to be positioned accurately either in the analysis position or in a position for treatment such as ion bombardment. Attachments to such a manipulator for heating or cooling the specimen are available from most manufacturers, but it is not particularly difficult to design and build them in the laboratory.

Insertion of a sample by mounting directly on a manipulator is naturally time-consuming since the analysis chamber has to be brought to atmospheric pressure to remove the manipulator, and thus subsequently re-pumped and re-baked to achieve UHV again. Modern surface analysis systems all have insertion facilities that do not disturb the UHV in the analysis chamber during transfer of samples. In some cases these take the form of a shaft or probe that slides in close-fitting sealing rings in and out of the analysis chamber via a fore-chamber, with the sample mounted on a special carrier at the end of the probe. The carrier is designed to locate in, and be gripped by, a sample carousel that can be rotated to bring successive samples into the analysis position. In another design the sample is transferred successively from fore-chamber to intermediate chamber to analysis chamber via trolleys running through gate valves from one chamber to the next. It is possible with these facilities to introduce a sample from atmosphere to UHV in a very few minutes without disturbing the UHV, always assuming that the sample itself is not so gassy that UHV cannot be maintained.

Once the sample is inside the analysis chamber, in its final position on the manipulator, it may be necessary to treat it in various ways, in particular it may need to be cleaned before any experiments can be started. Cleaning procedures will be different for different materials. In a few cases heating alone will suffice, in others ion bombardment alone, but in general it is found that a

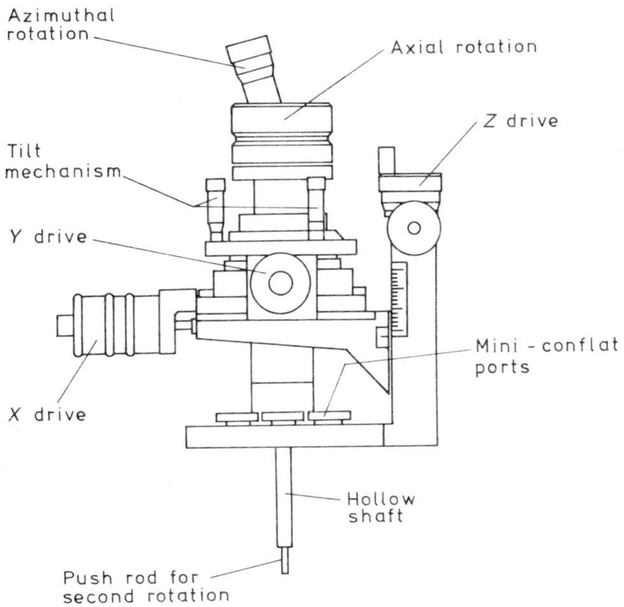

FIG. 3.24 Drawing of a universal manipulator of the type used for accurate positioning of samples in UHV. The sample would be placed on an attachment that fastens to the hollow shaft protruding below the main flange. The shaft itself can be rotated about its axis, and a second rotation of the sample is possible via the push rod down the centre of the shaft actuating some suitable mechanism. The manipulator head can be driven via micrometer screws in the *X* and *Y* directions, and a further drive pushes the head in and out in the *Z* direction. On some manipulators additional small screw drives allow a limited degree of tilt of the head. On the main flange a series of mini-flanges allow passage to the sample of heating and thermocouple leads, or of cooling pipes. (Courtesy of VG Scientific.)

combination of these two is needed. The reasons for having to use both heating and ion erosion are that although heating will remove some impurities it often causes others to appear at the surface by segregation from the bulk, while ion erosion is normally capable of removing segregated impurities but tends to leave behind traces of carbon and/or oxygen when present as carbides or oxides. Where the sample is a single crystal, it is usually necessary to anneal out the damage caused by ion bombardment in any case. It should be realized that the use of ion bombardment frequently introduces artefacts into an analysis by changing the relative concentrations of species in alloys or compounds, by changing the chemical state of some species due to reduction, or by introducing topographical changes. These will be discussed further in the section on depth profiling.

In special cases, i.e. where the material is fairly soft, a very effective method of cleaning *in vacuo* has been found to be that of scraping the surface with a mechanical tool. The tool can take the form either of a razor blade edge or of a sintered fine-grained grinding tool, in both cases fastened to a flexible bellows so that the tool can be moved across the sample surface. Care has to be taken, of course, that transfer of material does not take place from one sample to the next via the tool, and it is standard practice to change the scraping area frequently, either by replacement of the razor blade or, in the case of the grinding tool, by rotating it to a fresh area.

An even more specialized method of preparing a clean surface, in the context of general surface analysis, is by evaporation onto a substrate from a heated filament or crucible. A full description of the various methods used for the evaporation of metals, alloys and compounds is not regarded as within the scope of this volume.

Yet another specialized method, but one that is used very frequently in metallurgical, and occasionally in ceramic, applications, is that of fracture in vacuum. Some materials will cleave easily along certain crystal planes, but in general polycrystalline metals and alloys need both substantial force to produce fracture, and a reduction in temperature, often to $\sim -100°C$, to induce brittleness. Fracture may then occur either through individual grains (transgranular) or along grain boundaries (intergranular), producing in both cases a surface free from contamination apart from possible segregated impurities from the bulk. The latter are in fact often the object of the study, the concentrations and distribution of such impurities as a function of previous heat treatment being of immense importance in the field of metallurgical (and ceramic) brittle fracture. The breaking of the specimen obviously cannot be carried out when it is on a standard stub or carrier, or in the analysis position, and special fracture stages such as the one shown in Fig. 3.25 must be used. Broken pieces from the sample are captured and transferred to the analysis position as quickly as possible after fracture; although the fracture surface is clean when first prepared, it is highly reactive and any analysis must be performed before contamination has a chance to accumulate from the UHV residual gases.

Once a clean surface has been prepared, by whatever method, the experimenter would normally wish to observe the changes in the properties (compositional, chemical, electronic, etc.) of the surface as a result of reaction. The reaction may be with gases or vapours, with liquids, or with other solids. Liquids are difficult to handle in UHV and there are very few published instances of the study of liquid–solid reactions using surface analytical techniques. Many gases and vapours, on the other hand, are easy to handle, and on most surface analysis systems there is provision for the admission in a controlled way of gases either singly, sequentially, or as a mixture. A gas admission facility would consist of a series of containers (either glass or metal) of spectroscopically pure gases, each attached via a demountable joint and a control valve to a manifold. An example of such an arrangement using pressurized metal containers is shown in Fig. 3.26 (courtesy Messer–Greisheim Ltd). The manifold is then attached to a suitable entry point on the

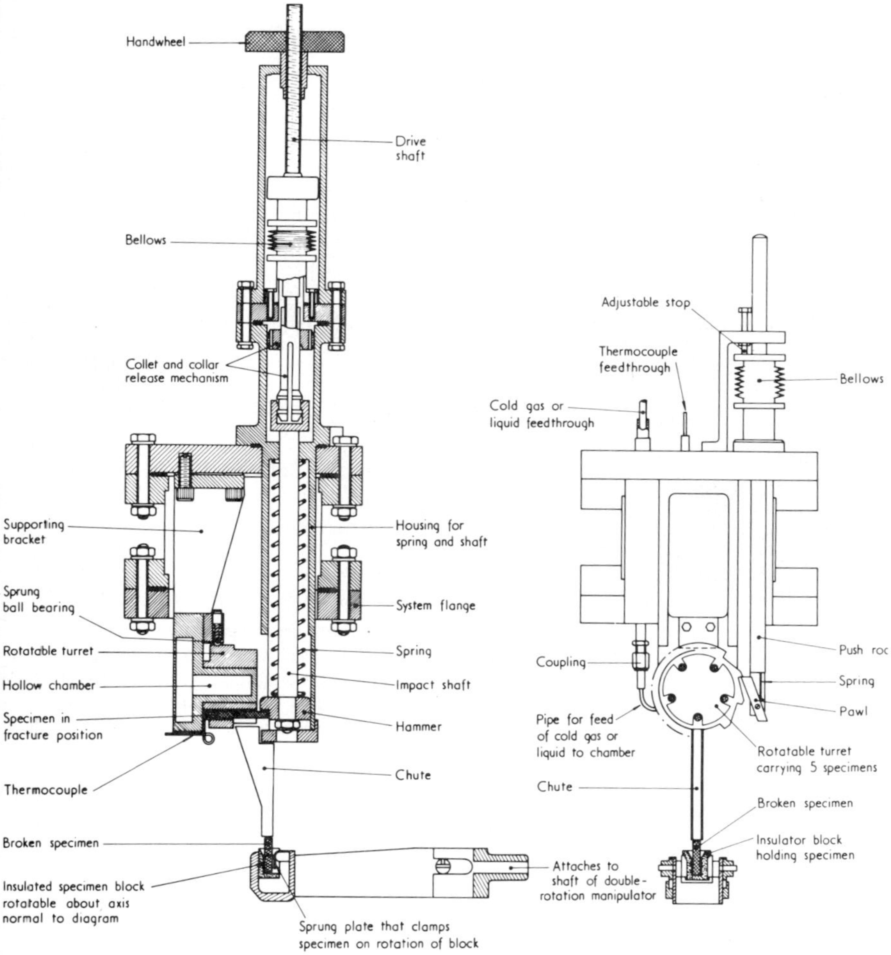

FIG. 3.25 Stage for performing fracture of brittle samples in UHV. In this design a rotatable turret carries five specimens, as seen in the right-hand drawing. The specimens project slightly from the turret as seen on the left. When in the fracture position a specimen is broken by impact on its projecting end of a hammer driven by a powerful spring; the spring is wound upwards until a release mechanism allows it to be driven downwards. The broken piece falls down a chute into a waiting catcher, which then grips it and transfers it to the position for AES analysis. The specimens can be cooled to liquid nitrogen temperature before breaking.

surface analysis system via a high quality leak valve. If the pressure in the manifold is likely to approach atmospheric or above during gas mixing, an additional control valve is advisable before the leak valve, for safety reasons. For accurate mixing of gases there should be a pressure gauge in the manifold, preferably an absolute one such as the differential manometer type, whose

FIG. 3.26 Arrangement of pressurized pure gas containers for admission of a gas or a mixture of gases to a system. Each container has its own valve (shown more clearly on the one standing up at the back) and is connected to a manifold consisting of small bore pipes. An additional central valve would normally be placed in the line to the system leading off to the right, and a pressure manometer is also desirable. (Courtesy of Messer–Griesheim Ltd.)

method of measurement is independent of the nature of the gas. It is also very advisable to install on the manifold a small quadrupole mass spectrometer for checking the purity of the gases.

3.5 Depth profiling

As pointed out at the end of the previous chapter, the techniques being described here are all surface-specific, the various types of information coming from average depths between 0.5 and 5.0 nm. Very frequently indeed, however, the experimenter or the technologist or whoever else has brought a problem for surface analytical study will need information from depths greater, often considerably greater, than the above range of average escape depths. Most often he will need the information as a quasi-continuous function of depth, that is, he will want to know how a particular property, usually the composition, is changing as one proceeds from the surface to the near-surface layers and then into the bulk. According to the thickness of the layer or film at the surface with properties different from the bulk there are various methods of attempting to measure the variation from the surface inwards, or, as it is commonly called, the depth profile.

3.5.1 *Variation of take-off angle*

In section 2.4.1, concerned with the reasons for surface-specificity of these techniques, the connection between the inelastic mean free path λ and the average escape depth of secondary electrons was given as

$$d = \lambda \sin \alpha \tag{3.19}$$

where α is the angle to the surface along which the electrons are detected. The relationship is illustrated simply in Fig. 3.27. If therefore the take-off angle α is

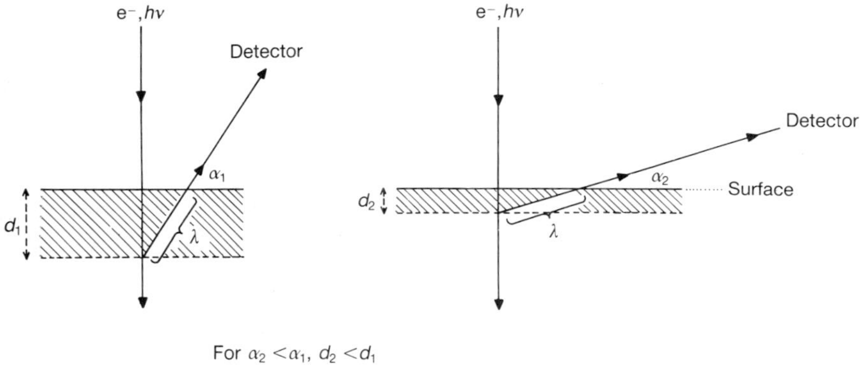

FIG. 3.27 Illustration of the principle of obtaining information about compositional variation with depth, by variation of electron take-off angle. λ is the inelastic mean free path. By reducing α from α_1 to α_2, the average escape depth is reduced from d_1 to d_2, thereby making the analysis more surface-specific.

varied, the average escape depth, or sampled depth, varies too; in particular, if α is made small the depth sampled is also small and the information is more characteristic of the outer surface layers than if α were close to 90°. In the situation where there is a very thin film of one material on another, or where a material is in a different chemical state right at the surface compared with the bulk, a succession of analyses at progressively decreasing take-off angles should enable the surface and the bulk components to be separately identified and analysed. This is known as depth profiling by take-off angle variation.

Clearly the method will work only for surface layers or films whose thickness is of the same magnitude as the inelastic mean free path, and only where the surface is flat. Nevertheless it is finding considerable application in, for instance, study of the very thin native oxides on some semiconductors.

3.5.2 *Ion bombardment*

In the vast majority of situations requiring knowledge of a depth profile the thickness of the surface layer or film is too great for the take-off variation method, and recourse has to be made to methods that actually remove the surface physically. The universally employed method is that of sputtering by noble gas ion bombardment, the gas usually used being argon. In practice it is very simple. Either the surface is eroded at a constant rate while a fast technique such as AES or SSIMS is used to record the quasi-instantaneous composition simultaneously and continuously, or, if a slower technique such as XPS is used, then the depth profile information is recorded in a step-wise fashion, i.e. by alternate cycles of bombardment and analysis.

The ion guns used in depth profiling are of the low energy types already described under ion sources (3.2.3), since the range of energies is from 0.5 to 10 keV, the region 4–5 keV being the most popular. Current densities employed range from 1 to 100 $\mu A/cm^2$, again the most popular region being 10–20 $\mu A/cm^2$. As described in section 3.2.3, some ion gun beams can be focused to a fine spot, in which case they are used in a rastered mode in depth profiling in order to produce an evenly eroded crater in the surface. The latter is particularly important when the analysis technique being used is, like XPS, lacking in spatial resolution and in which the uniformly eroded area must be larger than the analysed area if a sensible depth profile is to be achieved.

Although, as remarked, the practice of ion bombardment depth profiling is simple, the interpretation in an attempt to arrive at a true composition as a function of depth is complicated. The raw data obtained during profiling are in the form of signal intensities of the elements detectable as a function of ion bombardment dose, i.e. the product of ion current and time. The problems are, firstly to relate the dose to the depth, and secondly, to relate the signal intensities back to the undistorted concentrations. These are not independent since they depend on the same physical phenomena occurring during sputtering.

The sputtering rate in a direction along the surface normal, taken to be the z direction, is given by

$$dz/dt = I_p MS/\rho Ne \text{ m s}^{-1} \tag{3.20}$$

where I_p = primary ion current density in A m^{-2}
M = average atomic number
S = average sputtering yield in atom/ion
ρ = average density in kg m^{-3}
N = Avogadro's number
e = electron charge.

In expression 3.20, the only constant factors in most situations are I_p, N, and e, although of course for a pure element the other factors are also constant. In general M, ρ, and S are functions of composition, and will not usually remain constant during the depth profile of a multi-element surface. The main reason for this can be deduced from the elemental sputtering yields predicted by Seah[16] and shown in Fig. 3.28. Across the periodic table the sputtering yields vary dramatically, by more than an order of magnitude in fact, and it follows therefore that in an alloy or a compound containing two or more elements

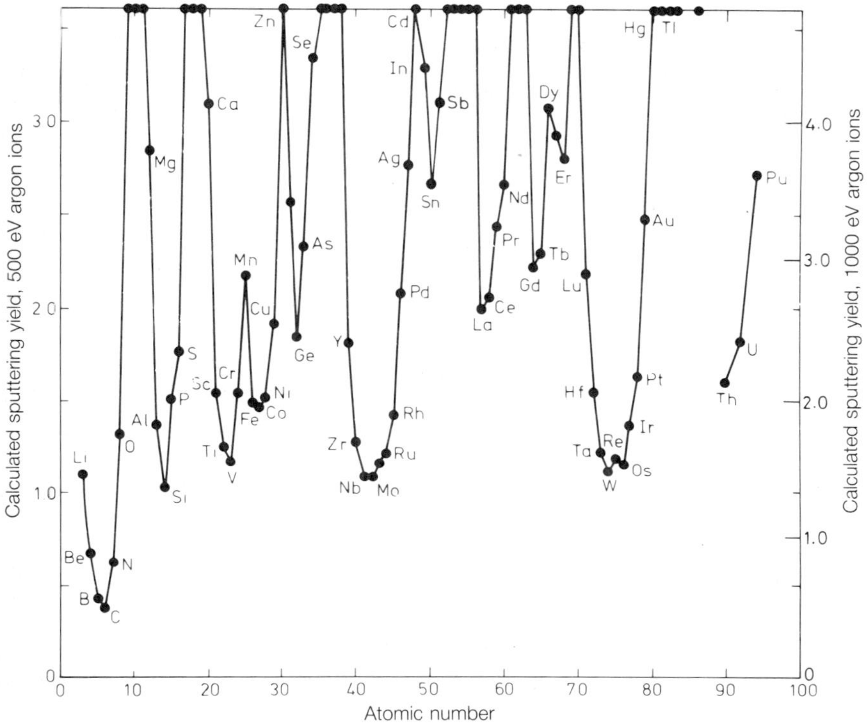

FIG. 3.28 Values of the sputtering yields for all elements calculated for two energies of the incident argon ions by Seah.[16] The points on the upper axis correspond to those elements for whom the sputtering yields were too high to fit on the same scale as the others. These were for 500 eV ions: Na, 3.70; K, 6.17; Zn, 5.07; Rb, 9.20; Cd, 7.20; Te, 4.34; Cs, 11.47; Ba, 4.73; Sm, 4.35; Eu, 5.02; Yb, 6.09; Tm, 5.12; Pb, 4.81; Bi, 4.35. These values should be multiplied by 1.33 for 1000 eV ions.

those with the highest sputtering yield will tend to be removed preferentially. It is of course risky extrapolating from the elemental yields in the figure to what might be expected for the same element in quite a different matrix, but qualitatively the preferential removal has been confirmed very often.

As the relative numbers of atoms of different elements (i.e. the concentrations) in the surface change as a result of differential sputtering so too must M, ρ, and S, and therefore the sputtering rate from eqn 3.20 must vary with time. Since knowledge of the variation of S for the same element in a variety of different matrices is still only fragmentary, eqn 3.20 cannot be used very often to estimate the sputtering rate, and other empirical methods must be used. What is needed, of course, is a way of measuring the instantaneous sputtering rate at all points in the depth profile, and that has been attempted by measuring the residual film thickness by optical means, but it would be both inconvenient and time-consuming to have to do that in a conventional surface analytical system every time a depth profile is recorded. The compromise reached by most workers is to measure instead an *overall* sputtering rate by observing the time at fixed current density necessary to sputter through a film or layer whose thickness is known reasonably accurately from other types of measurement or from its method of preparation. The drastic assumptions made there are that the instantaneous sputtering rate does not differ much from the overall rate and that the film or layer used for the calibration is sufficiently similar in structure and composition to the sample to be analysed. It is also difficult if not impossible to produce calibratory films that will match every analytical situation, and no one has tried to do so. As a result the situation with regard to the relationship between ion bombardment dose and depth removed is highly unsatisfactory, and depth scales on profiles must in general be taken to be very approximate; the errors in some cases could easily amount to factors of 3–5. Indeed there is a tendency among some workers not to attempt to give a depth scale at all, but to present a profile in terms of the bombardment dose.

At any stage in a depth profile the instantaneous actual surface composition can be measured, e.g. by AES, within the errors of quantification of the method of analysis used. However, for the same reasons as above, the actual observed composition in the presence of sputtering cannot be the real one that originally existed at that particular depth, and the problem is how to work back from what is observed to the true composition. In simple cases this is possible. If in a binary alloy containing metals A and B it is assumed that the sputtering yields S_A, S_B are independent of the bulk concentrations C_{bA} and C_{bB}, then the surface concentrations C_{sA} and C_{sB} are inversely proportional to the sputtering yields, i.e.

$$C_{sA}/C_{sB} = S_B C_{bA}/S_A C_{bB}. \tag{3.21}$$

Thus the element with lower sputtering yield is enriched in the surface, as already stated above. The enrichment continues until the respective sputtering yields are balanced by the respective numbers of atoms in the surface, when a steady state is reached. If τ is the characteristic time taken to establish this steady state, τ of course being a function of S_A and S_B, then the instantaneous surface concentration of A before the steady state is reached is

$$C_{sA}(t) = (C_{bA} - C_{sA})\exp(-t/\tau) + C_{sA}. \tag{3.22}$$

It is thus possible to work back at any one point in the depth profile to the true original composition. The assumption that the sputtering yields are independent of the bulk concentrations holds for some metallic solid solutions, but is far from generally true. Matrix effects can be severe and unpredictable. With multi-component systems families of equations similar to 3.22 can be set up, but the basic assumption about independence of sputtering yields becomes even more dubious. In the present state of knowledge it has been found more profitable to arrive at approximately correct depth profiles simply by building on experience, that is, by studying the profiles recorded from a particular material or from a narrow range of compositions after being given many different treatments, e.g. oxidation, and aiming at internal self-consistency using other physical and chemical knowledge of that material. That is a relative approach; the absolute one seems unusable at the moment.

Ion bombardment introduces other artefacts as well, particularly into compounds. In many cases more or less severe chemical reduction of elements to lower oxidation states has been observed by XPS. Some examples are given in Figs 3.29 and 3.30. In the former, the O 1s photoelectron peak is recorded[17] by XPS for three different calcium compounds, before and after argon ion bombardment. In each case some reduction to calcium oxide has taken place, the degree of reduction differing with the compound, the sulphate being the least and the nitrate the most, susceptible. The authors compared the percentage reduction of each compound with various thermodynamic quantities and found a correlation only with the free energy of formation, which was difficult to understand. The other example, in Fig. 3.30, shows[18]

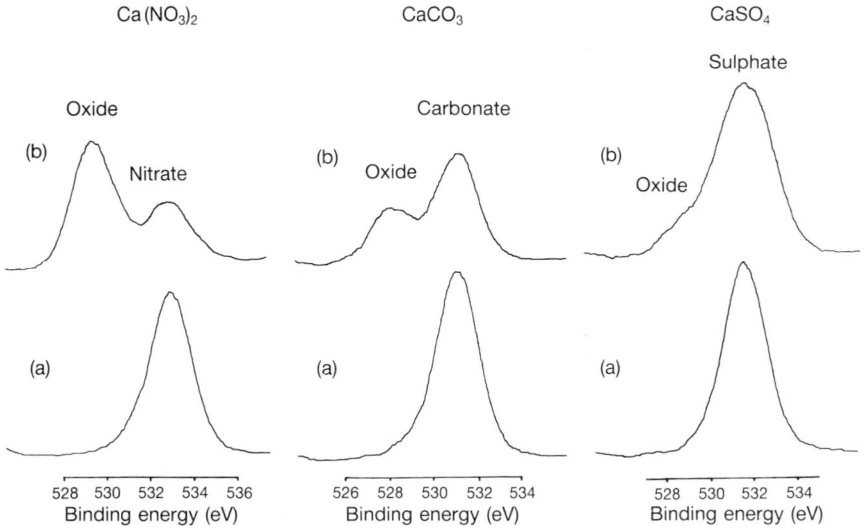

FIG. 3.29 Example of sample degradation by ion bombardment, from Christie *et al.*[17] Spectra (a) are the O 1s peaks from the three calcium compounds given at the top of the figure, before bombardment. Spectra (b) are the same after bombardment with argon ions at 3 keV to a total dose of 7.5×10^{16} ions cm^{-2}. The order of susceptibility to degradation goes nitrate > carbonate > sulphate.

the effect on Ta_2O_5 of sputtering with 3 keV argon ions for many minutes. Structure appears on the low binding energy side of the Ta 4f peaks after three minutes, which quickly develops into a new set of peaks. The intermediate spectra could be shown to be made up of contributions from sub-oxides containing Ta^{2+} and Ta^{4+} as well as from the expected Ta^{5+}.

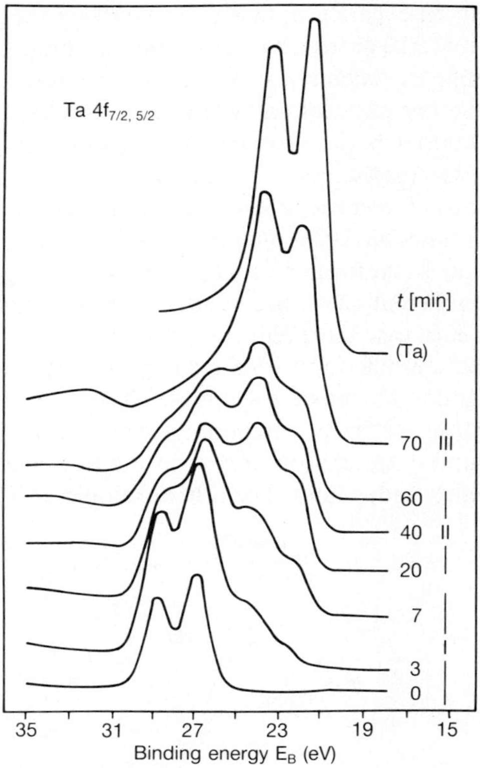

FIG. 3.30 Effect of bombardment of a Ta_2O_5 film of thickness 30 nm on Ta foil, by argon ions at 3 keV, recorded by Hofmann and Sanz.[18] In the figure the Ta 4f photoelectron spectra are shown as a function of time of bombardment. Additional structure appears after only three minutes, and from 20 to 60 minutes stabilizes into a multicomponent peak. Analysis of this structure reveals that during bombardment of the bulk of the film the three species Ta^{5+}, Ta^{4+}, and Ta^{2+} were all present owing to reduction of the Ta_2O_5 by the ion beam.

There are many other published instances of ion beam reduction in the literature, but as yet no unified theory that could be applied to take account of reduction during depth profiling. All that can be done at the moment is to be aware that such reduction might occur and if it is suspected, then to make tests on standard materials relevant to the analytical problem in order to check the extent of the changes.

The other artefact introduced by ion bombardment is that of topographical change, that is, the surface becomes rough or rougher. A surface that is already rough, e.g. a fracture surface or that of an irregular coating, will tend to become rougher owing to the fact that sputtering efficiency is greater at higher incident angles, and any asperities will thus be accentuated. Another cause of induced roughness is the presence of inclusions in the material, of sputtering yield much less than that of the surrounding matrix, which leads to

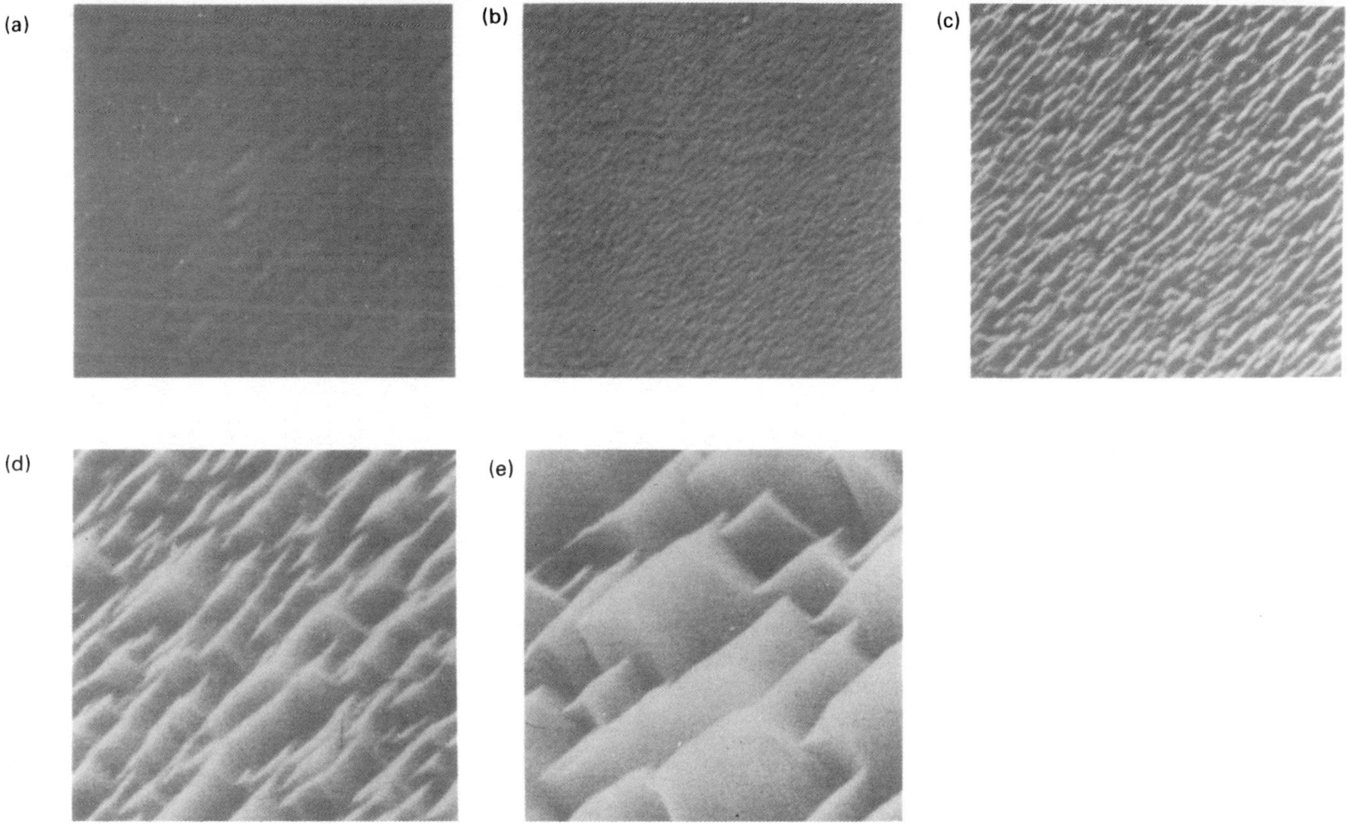

FIG. 3.31 Sequence of SEM micrographs of a flat silicon surface during increasing doses of ion bombardment, from the work of Lewis *et al.*[19] At high total doses, ridges composed of prismatic structures appear. (a) 2×10^{18} ions cm^{-2}; mag. 20,000. (b) 4×10^{18} ions cm^{-2}; mag. 5,000. (c) 1×10^{19} ions cm^{-2}; mag. 5,000. (d) 5×10^{19} cm^{-2}; mag. 5,000. (e) 8×10^{19} ions cm^{-2}; mag. 5,000.

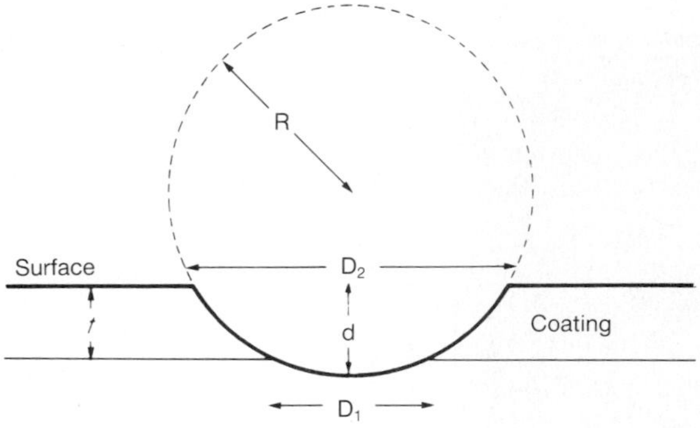

FIG. 3.32 Schematic of the method of profiling a surface film by ball cratering, originated by Walls *et al.*[20] A grinding ball of radius R, made of highly polished stainless steel, is coated with 0.25 μm diamond paste, and allowed to rotate against the sample surface. The speed of rotation and the pressure of the ball on the surface can be varied. A circular crater of diameter D_2 is ground, exposing an area of the substrate of diameter D_1. The depth d of the crater is then $D_2^2/8R$, and the thickness t of the film or coating is $(D_2^2 - D_1^2)/8R$. The sample is rinsed in an appropriate solvent to get rid of the diamond paste, and given a light ion bombardment once it is inside the spectrometer. The depth profile is then derived by traversing the walls of the crater with an electron beam of high spatial resolution, recording AES spectra at the same time. The relationship between distance traversed and the depth of the film is given by expression (3.23).

'shadowing' of material behind the inclusion in the direction of the incident beam. It is possible to minimize these effects either by using two ion guns simultaneously from different directions, or better still by rotating the sample continuously during bombardment, but not many workers seem to have adopted this approach.

Topographical changes can also be induced on perfectly flat substrates purely statistically; irregularities that appear on an atomic scale during the first arrival of ions at a surface due to the statistical distribution of intensity in the ion beam, become progressively accentuated. This effect is a function of the ion current density, or fluence, as might be expected. Figure 3.31 shows the effect on a flat silicon substrate of bombardment at different fluences.[19] At high fluences ridges composed of prismatic structures appear.

Despite the possibility, indeed likelihood, of one or more of the above artefacts being present during a depth profile by ion bombardment, there is really no alternative method at present of attempting to derive compositional information from depths greater than the average escape depths. There is a practical useful limit to the depth to which it is worthwhile profiling by ion bombardment, governed both by the induced artefacts and by the statistical nature of the removal of material. Atom layers are not removed uniformly by an ion beam; parts of the second and third and even deeper layers will be exposed before the first layer has completely disappeared, and the effect becomes worse with increasing depth. The result is that the depth resolution degrades progressively with the continuation of profiling, to the point where it is not worth going any further. For most materials this point will be met at depths between 1 and 2 μm.

If a film or a layer is thicker than ~2 μm and profile information is required, one must resort to mechanical methods of profiling. These consist simply of grinding a section across the layer at an angle sufficiently low that the profile through the layer is spread out over a comparatively large distance. For many years, taper sectioning at a low angle, 1–2°, has been a standard technique in the electron probe micro-analyser, but a more controllable variation of the technique has been found[20] to be that of ball cratering. A highly polished steel ball of diameter typically 30 mm is coated with fine diamond paste, and allowed to rotate against the specimen surface. The pressure on the surface can be varied as can the speed of rotation. A circular crater is ground through the layer, and by observation of the crater through a microscope it can be decided when to stop the profiling. The geometry is shown diagrammatically in Fig. 3.32. If D_2 is the crater diameter the depth of the crater is $D_2{}^2/8R$, R being the radius of the grinding ball. Once a desired depth has been reached, the specimen is removed, rinsed in an appropriate solvent to get rid of excess diamond paste, and inserted in an AES spectrometer. After a light ion bombardment to minimize contamination, the depth profile is measured by traversing the walls of the crater with the incident electron beam of high spatial resolution used in AES. The distance x traversed by the beam is related to the depth z at that point by the relationship

$$z = d - R + \{R^2 - (D_2 - x)^2\}^{1/2}. \tag{3.23}$$

Naturally, the depth resolution achievable by this method is limited by the roughness induced by the grinding, but for film and layer thicknesses greater than 2 μm it is to be preferred to ion bombardment. There are no problems in this case over establishing a depth scale; eqn 3.23 gives an exact and absolute measure of the depth.

References

1. Nishijima, M., Masuda, S., Kobayashi, H., and Onchi, M. (1982). *Rev. Sci. Instrum.* **53**, 790.
2. Hughes, A. L. and Rojansky, V. (1929). *Phys. Rev.* **34**, 284.
3. Baines, M., Dean, E. M., and Wilson, J. M. (1975). *J. Phys. E: Sci. Instrum.* **8**, 305.
4. Park, R. L. and Houston, J. E. (1972). *Phys. Rev. B.* **6**, 1073.
5. Chopra, D., Martin, G., Naraghi, H., and Martinez, L. (1981). *J. Vac. Sci. Technol.*, **18**, 44.
6. Nyberg, C. (1975). *Surf. Sci.*, **52**, 1.
7. Yates, K., Barrie, A., and Street, F. J. (1973). *J. Phys. E: Sci. Instrum.* **6**, 130.
8. Farge, Y. and Duke, P. J. (eds.), *European Synchrotron Radiation Facility, Suppl. 1*, European Science Foundation, 1979.
9. Viaris de Lesegno, P. and Hennequin, J-F. (1978). *Surf. Sci.* **80**, 656.

10. Prewett, P. D. and Jefferies, D. K. (1980). *J. Phys. D: Appl. Phys.*, **13**, 1747.
11. Kobayashi, N., Maeda, N., Hori, H., and Sakisaka, M. (1976). *J. Phys. Soc. Japan* **40**, 1421.
12. Grimm, W. (1968). *Spectrochim. Acta B* **23**, 443.
13. Berneron, R. and Charbonnier, J. C. (1981). *Surf. Interface Anal.* **3**, 134.
14. Palmberg, P. W. (1975). *J. Vac. Sci. Technol.* **12**, 379.
15. Roy, D. and Carette, J-D. (1971). *Can. J. Phys.* **49**, 2138.
16. Seah, M. P. (1981). *Thin Solid Films* **81**, 279.
17. Christie, A. B., Lee, J., Sutherland, I., and Walls, J. M. (1983). *Appl. Surf. Sci.* **15**, 224.
18. Hofmann, S. and Sanz, J. M. (1983). *Fresenius Z. Anal. Chem.* **314**, 215.
19. Lewis, G. W., Kiriakides, G., Carter, G., and Nobes, M. J. (1982). *Surf. Interface Anal.* **4**, 141.
20. Walls, J. M., Hall, D. D., and Sykes, D. E. (1979). *Surf. Interface Anal.* **1**, 204.

4
Electron excitation: Auger electron spectroscopy (AES) and scanning Auger microscopy (SAM)

4.1 Operation

In conventional AES an electrostatically focused electron gun of the type shown in Fig. 3.2 irradiates the sample with electrons in the energy range 5–10 keV and at currents between 10^{-8} and 10^{-5} A. Although such a gun can produce a spot size of ~ 0.5 μm and can be rastered to provide low resolution SEM and SAM images, SAM needs better spatial resolution to be useful and therefore employs an electromagnetically focused gun of the type of Fig. 3.3. It would operate typically in the range 10–30 keV, with the optimum spot size of ~ 35 nm available at the higher energies, viz. 25–30 keV.

The secondary electrons ejected from the sample surface will have all kinetic energies from the primary energy to zero, as shown in the lower curve of Fig. 4.1. Superimposed on the continuous inelastic distribution will be peaks due to Auger emission at energies characteristic of the sample material, according to the process described in Fig. 2.1. The technique of AES still to an overwhelming extent records the Auger peaks as the differential distribution of which a portion is shown in the upper part of Fig. 2.1. Although the primary energy may be several keV, the kinetic energy range normally covered in AES is typically from 20–1000 eV. There are two reasons for this. Firstly, nearly all elements will have usable Auger features in that region, and there is little point when carrying out a routine analysis in using peaks at higher energies, particularly since the electrons of higher energy will have greater average escape depths and therefore be less typical of the surface. Secondly, the variation of ionization cross-section of a core level with primary energy shows a maximum at a primary energy about 3–5 times the core-level binding energy, as seen in Fig. 4.2 from Casnati *et al.*[1] Above the maximum the cross-section is fairly constant. Thus for primary energies in the conventional range 5–10 keV, the cross-sections for ionization of core levels of binding energies much greater than 1000 eV would start to decrease, but for all core levels of binding energies lower than 1000 eV the cross-section is much the same.

Three types of analyser have been used to measure the energy distribution of the secondary electrons. In the early years of the development of AES there were a large number of LEED systems in use around the world, employing

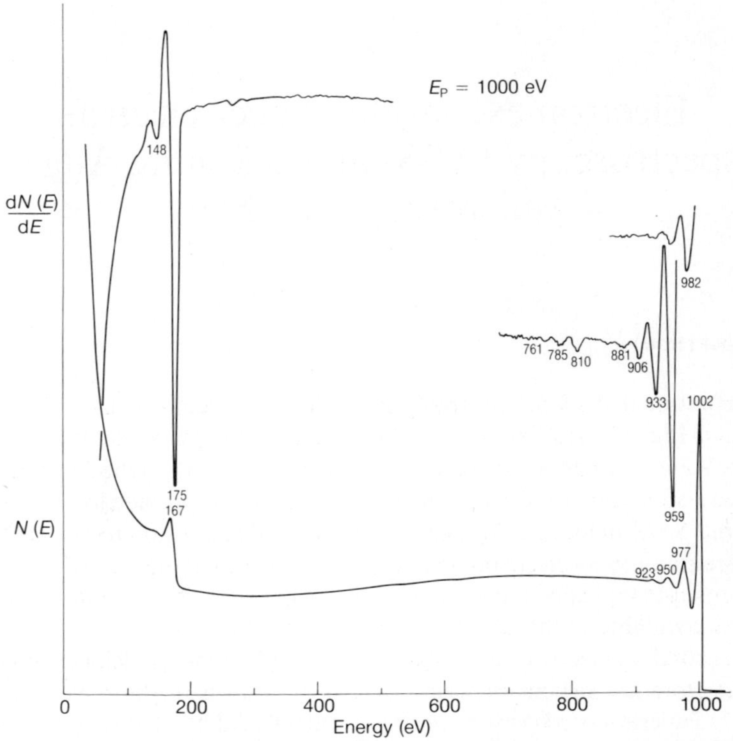

FIG. 4.1 Secondary electron distribution $N(E)$ from a boron sample at a primary energy of 1000 eV, and, above, the differential distribution $dN(E)/dE$ from selected parts of the spectrum. In the $N(E)$ distribution can be seen, from right to left, the elastic peak at 1002 eV, successive loss peaks for a plasmon energy of ≈ 27 eV, the boron KLL Auger peak at 167 eV, and the steeply rising 'true' secondary electron peak. In the $dN(E)/dE$ distribution, the position of the Auger peak is taken by convention to be that of the maximum negative excursion, i.e. at 175 eV for boron. Differentiating also accentuates the minor peaks, e.g. the plasmon losses can then be seen extending to 881 eV, while the ionization loss peak associated with ionization of the B K shell becomes visible at 810 eV, with attendant plasmon losses.

spherical grid and collector optics. It was realised by Palmberg and Rhodin[2] that such an arrangement could be used as a retarding field energy analyser (RFA) and its mode of operation has been described in Section 3.3.3. There are still many LEED systems in existence, in which AES is also performed, but the inherently poor signal-to-noise characteristics of the LEED RFA meant that its development as an AES analyser was severely limited.

The search for an analyser with better characteristics for AES soon led Palmberg *et al.*[3] to the Cylindrical Mirror Analyser (CMA), and the CMA has remained the principal analyser used for AES ever since. Its configuration and mode of operation have been discussed under Section 3.3.4. In the single-pass form it is almost ideal for AES because of its very high transmission; invariably in AES the highly focused primary electron beam irradiates an area much

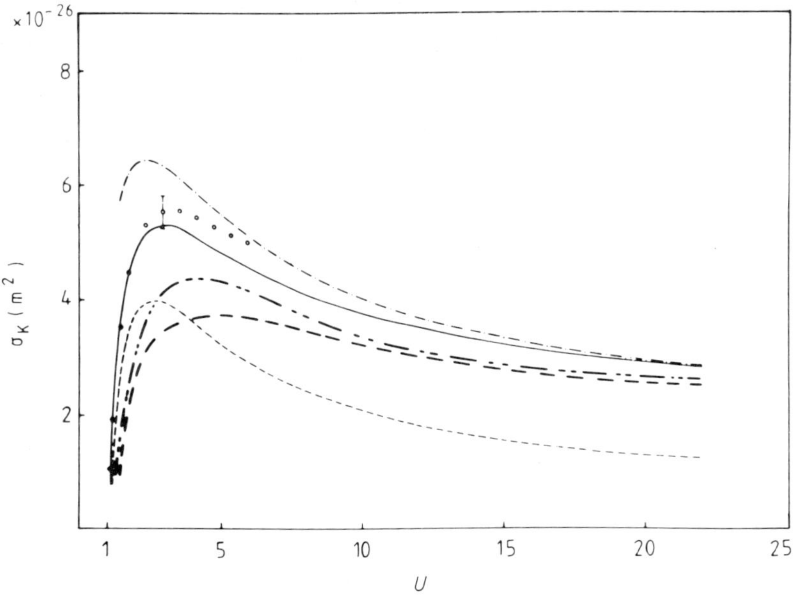

FIG. 4.2 Cross-section for ionization by electron impact of the Ni K shell, as a function of the reduced energy U; $U = E_p/E_K$, E_p being the primary electron energy, and E_K the binding energy of an electron in the Ni K shell. The open circles are experimental points, and the other curves are all either semi-empirical or theoretical expressions for the cross-section. From the work of Casnati *et al.*,[1] whose own theoretical expression gave the solid line in the figure. Note that the cross-section goes through a maximum at a primary energy some 3–4 times the core-level binding energy.

smaller than the acceptance area of the analyser, so that it is transmission and not luminosity of an analyser that is important for the technique. A typical arrangement for recording of Auger spectra in either the $E\,N(E)$ or the $E\,\mathrm{d}N(E)/\mathrm{d}E$ modes is shown in Fig. 4.3, from Landolt and Mathieu.[4] Sweeping the ramp voltage on the outer cylinder gives the $E\,N(E)$ distribution, while modulation of the ramp voltage by a small sinusoidal voltage, followed by phase-sensitive detection at the frequency of modulation, gives the $E\,\mathrm{d}N(E)/\mathrm{d}E$ distribution. Primary electrons are normally produced from a gun situated inside the analyser on axis, but external guns have also been used, although use of the latter involves angles of incidence far from the normal.

With the rapidly increasing use of surface analytical systems that combine two or more techniques, the Concentric Hemispherical Analyser (CHA) that was developed originally for XPS has also come to be used extensively for AES. A description of the mode of operation of a CHA is given under Section 3.3.4. When used for AES, it is essential to operate at as high a transmission as possible, which means opening up the entrance and exit slits to their widest, to the point where there are really no effective slits at all. In addition AES is recorded normally at a constant relative resolution (see

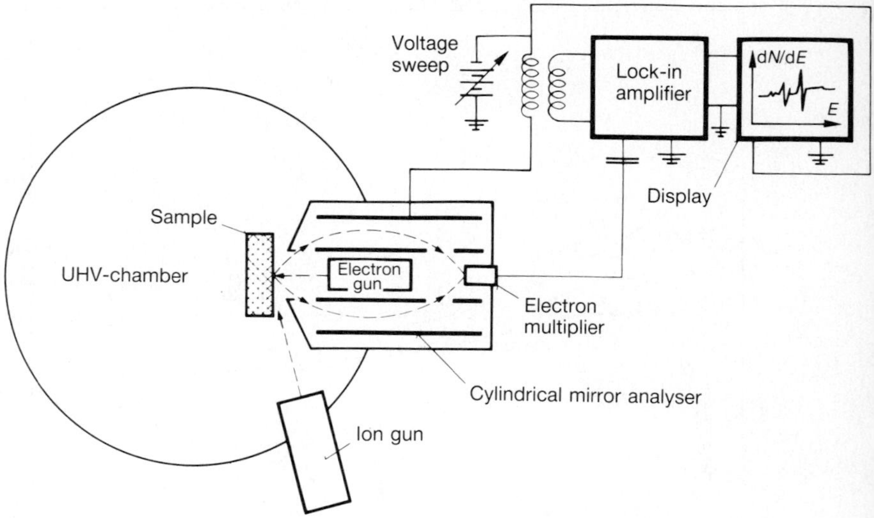

FIG. 4.3 Typical experimental arrangement for AES using a single-pass CMA with an integral electron gun. A small modulation is imposed on the voltage sweep applied to the outer cylinder of the CMA, and the second harmonic of the detected current is amplified in the lock-in amplifier (phase-sensitive detector) to provide the differential distribution $dN(E)/dE$. From Landolt and Mathieu.[4]

Section 3.3.2) so that a CHA must be operated in the Constant Retard Ratio, or CRR mode. These are in fact compromises in the optimum way in which a CHA is designed to work, and it must be accepted that inherently the CMA is better suited to AES than the CHA. Nevertheless, it is highly convenient to be able to use the same energy analyser for both techniques, just as will be seen to be the case for XPS also, where it is the CMA that has been adapted for use in that technique. One big advantage that a CHA with input lens system has over a CMA is the much greater working distance available between the sample and the nearest point of the analyser; a typical focal distance from the entrance to a CMA would be ~ 10 mm, whereas from the end of a lens more like ~ 50 mm. As a result, it is possible to install a greater number of additional probes and other devices in the vicinity of the sample when working with a lens system.

In the variation of AES known as Scanning Auger Microscopy (SAM), the incident electron beam is focused electromagnetically to a fine spot by a gun of the type shown in Fig. 3.3. This operates at a higher energy, 25–30 keV, and a lower current, ≤ 1 nA, than in conventional AES, in order to produce the fine spot; the minimum spot size achieved so far is about 35 nm. If the beam is rastered over the surface, and the deflection potential on the CMA is set to cover a very narrow range of energies (an energy 'window') that encompasses the range of an Auger peak from an element on the surface, then if the intensity of a visual display is modulated by the intensity of the Auger peak, a two-dimensional distribution or 'map' of that element over the surface will be obtained. A secondary electron image can also be obtained in the same way as in an SEM by collecting secondary electrons in a scintillator detector. Since the intensity in an Auger peak will be a function not only of the Auger peak intensity itself but of the background on which it is sitting as well, and since that background can be affected significantly by the topography (roughness) of the surface, it is usual in modern SAM systems to attempt to eliminate the effects of topography by dividing the magnitude of the peak intensity by the magnitude of the background just to the high energy side of it. This can be done automatically by a computer as the map is acquired.

4.2 Spectral features and interpretation

4.2.1 *Nomenclature*

The Auger transition shown in Fig. 2.1 was said there to be called $KL_1L_{2,3}$. For the atom whose levels were depicted in that figure there would clearly be other transitions possible in the KLL series, namely KL_1L_1 and $KL_{2,3}L_{2,3}$. If either or both of the electrons taking part had come from the valence band shown just below the Fermi level, then the transitions KL_1V, $KL_{2,3}V$, and KVV are possible, where of course V is used conventionally to denote a valence band electron. Similarly further up the periodic table one has many other possibilities, based on primary ionization of L, M, N, etc., core levels, such as LMM, LMN, MNN, and so on.

The above notation is used for Auger transitions for historical reasons, dating back to the discovery of the effect by Auger,[5] and is the notation used by X-ray spectroscopists. It is based on the description of orbital motion of an electron around an atomic nucleus given by the so-called *j–j* coupling scheme. In that scheme the total electronic angular momentum is obtained by summing vectorially the individual electronic spin and angular momenta. Thus for any one electron the total angular momentum is characterized by a quantum number j, where $j=l+s$, l being the electronic orbital angular momentum quantum number and s the electronic spin momentum quantum number. Since l can take the values 0, 1, 2, 3, 4, . . . and s either of the values $\pm\frac{1}{2}$, clearly $j=\frac{1}{2}, \frac{3}{2}, \frac{5}{2}$, etc. In the *j–j* notation, states with principal quantum number $n=1, 2, 3, 4, \ldots$ are designated K, L, M, N, . . . , respectively, and states with various combinations of $l=0, 1, 2, 3, \ldots$ and $j=\frac{1}{2}, \frac{3}{2}, \frac{5}{2}, \ldots$ are given the suffixes 1, 2, 3, 4, ... The combinations are listed in Table 4.1. On this basis there would, for example, be six possible Auger transitions in the KLL series.

The alternative way of labelling the states in *j–j* coupling is by the spectroscopic notation, which is the one used in XPS, and shown in the last column of Table 4.1. States with $l=0, 1, 2, 3, \ldots$, are designated, s, p, d,

Table 4.1
X-ray and spectroscopic notation in *j–j* coupling

Quantum numbers					
n	l	j	X-ray suffix	X-ray level	Spectroscopic level
1	0	$\frac{1}{2}$	1	K	$1s_{1/2}$
2	0	$\frac{1}{2}$	1	L_1	$2s_{1/2}$
2	1	$\frac{1}{2}$	2	L_2	$2p_{1/2}$
2	1	$\frac{3}{2}$	3	L_3	$2p_{3/2}$
3	0	$\frac{1}{2}$	1	M_1	$3s_{1/2}$
3	1	$\frac{1}{2}$	2	M_2	$3p_{1/2}$
3	1	$\frac{3}{2}$	3	M_3	$3p_{3/2}$
3	2	$\frac{3}{2}$	4	M_4	$3d_{3/2}$
3	2	$\frac{5}{2}$	5	M_5	$3d_{5/2}$
	etc.		etc.	etc.	etc.

f, . . . , respectively, and the letter is preceded by the principal quantum number n. The j values are thus appended as suffixes. Thus the state written M_5 in the X-ray notation, where $n=3$, $l=2$, $j=\frac{5}{2}$ is written $3d_{5/2}$ in the spectroscopic.

Strictly speaking, *j–j* coupling is applicable only to elements of high atomic number ($Z > \sim 75$). Its use for all elements in AES is a convenient simplification, and also takes account of the fact that in most experimental systems and problems the fine structure associated with many Auger peaks is either washed out by the use of poor energy resolution, or not of interest. For elements of low atomic number ($Z < \sim 20$) the coupling scheme called *L–S* (sometimes also Russell–Saunders) is a better description of orbital electronic motion. Instead of summing the individual orbital angular and spin momenta before carrying out the total summation for all electrons, all the electronic orbital angular momenta are summed first, then all the electronic spin momenta, and the two totals are summed vectorially to give the total angular momentum. The two total electronic summations are given the quantum numbers $L(=\sum l)$ and $S(=\sum s)$, and the vector sum of those then given the quantum number J, equal to $|L \pm S|$. Since L and S can take the integral values 0, 1, 2, 3, . . . , J can take any integral value between $|L-S|$ and $|L+S|$. Hence *L–S* coupling. The notation is in the form $^{(2S+1)}L$. States with $L=0$, 1, 2, 3, . . . , are designated in capital letters S, P, D, F, . . . , respectively, while the total spin quantum number S enters as the prefix $(2S+1)$. (Do not confuse the state S (for $L=0$) with the quantum number S.) The combinations are shown for the KLL series in Table 4.2, where also the configuration in spectroscopic terms is given. *L–S* coupling again predicts six possible transitions in the KLL series, although one is forbidden.

Table 4.2
Notation in *L-S* coupling

Transition	Configuration	*L*	*S*	Term
KL_1L_1	$2s^02p^6$	0	0	^1S
$KL_1L_{2,3}$	$2s^12p^5$	1	0	^1P
		1	1	^3P
$KL_{2,3}L_{2,3}$	$2s^22p^4$	0	0	^1S
		[0	0	^3P]*
		2	0	^1D

*Forbidden on grounds of conservation of parity.

In the middle of the periodic table neither of the above coupling schemes is adequate, and an intermediate scheme is appropriate. Each (*L–S*) term is split into multiplets of different *J* values, the term symbols being written as $^{(2S+1)}L_J$. In this scheme there are ten possible Auger transitions predicted in the KLL series, as seen in Table 4.3, but one of these is forbidden as before on grounds of conservation of parity. This notation can be found in the publications of physicists recording high resolution Auger data for comparison with theoretical prediction, and is not normally encountered.

Table 4.3
Notation in intermediate coupling

Transition	Configuration	*L–S* term	*L*	*S*	*J*	IC term
KL_1L_1	$2s^02p^6$	^1S	0	0	0	$^1S_0$
		^1P	1	0	1	$^1P_1$
			1	1	0	$^3P_0$
$KL_1L_{2,3}$	$2s^12p^5$	^3P	1	1	1	$^3P_1$
			1	1	2	$^3P_2$
		^1S	0	0	0	$^1S_0$
			1	1	0	$^3P_0$
$KL_{2,3}L_{2,3}$	$2s^22p^4$	^3P	[1	1	1	$^3P_1$]*
			1	1	2	$^3P_3$
		^1D	2	0	2	$^1D_2$

*Forbidden on grounds of conservation of parity.

4.2.2 *Characteristic Auger series*

Since the maximum primary energy used in conventional AES is about 10 keV, and since the ionization cross-section is maximum at a primary energy some 3–5 times the primary energy, it follows that there are limitations on the depths of core levels that can be ionized efficiently. For this reason Auger transitions based on K shell ionization are limited in usefulness up to about silicon, on L_3 shell ionization up to about strontium, on M_5 ionization up to about osmium, and so on. What this means in practice is that in each region of the periodic table there are characteristic Auger transitions that are the most prominent under the conventional operating conditions. For the light elements the characteristic transitions are the KLL series, examples of which are seen in Fig. 4.4 taken from Davis *et al.*[6] Some of the structure on the low

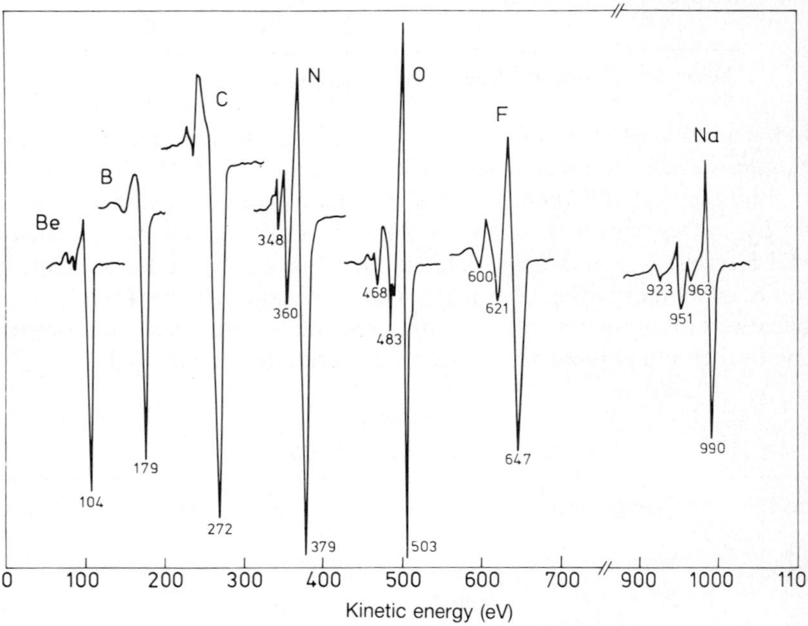

FIG. 4.4 Auger spectra in the differentiated mode showing the KLL Auger series characteristic of the very light elements. The major peak arises from the $KL_{2,3}L_{2,3}$ transition, and peaks due to minor Auger transitions as well can be seen on their low energy sides, although there will be associated plasmon loss peaks as well. According to the chemical state of the element, the shapes of the major peaks can undergo considerable changes. Compiled from Davis *et al.*[6]

energy side is due to minor Auger peaks and some to plasmon losses (see a later section). Slightly further up the periodic table, amongst the first series of transition metals, a highly characteristic Auger spectrum appears in the form of a prominent triplet of peaks, the LMM series, consisting of $L_3M_{2,3}M_{2,3}$, $L_3M_{2,3}V$, and L_3VV, in increasing order of kinetic energy. These spectra are shown for chromium, manganese, and iron in Fig. 4.5, also from Davis *et al.*[6] Notice also the progressively increasing $M_{2,3}VV$ at low energies. Again, associated with the triplet are various satellite features on the low energy side. Just to higher energy of the L_3VV can be seen the L_2VV peak. Still further up the periodic table, into the second transition series, another characteristic

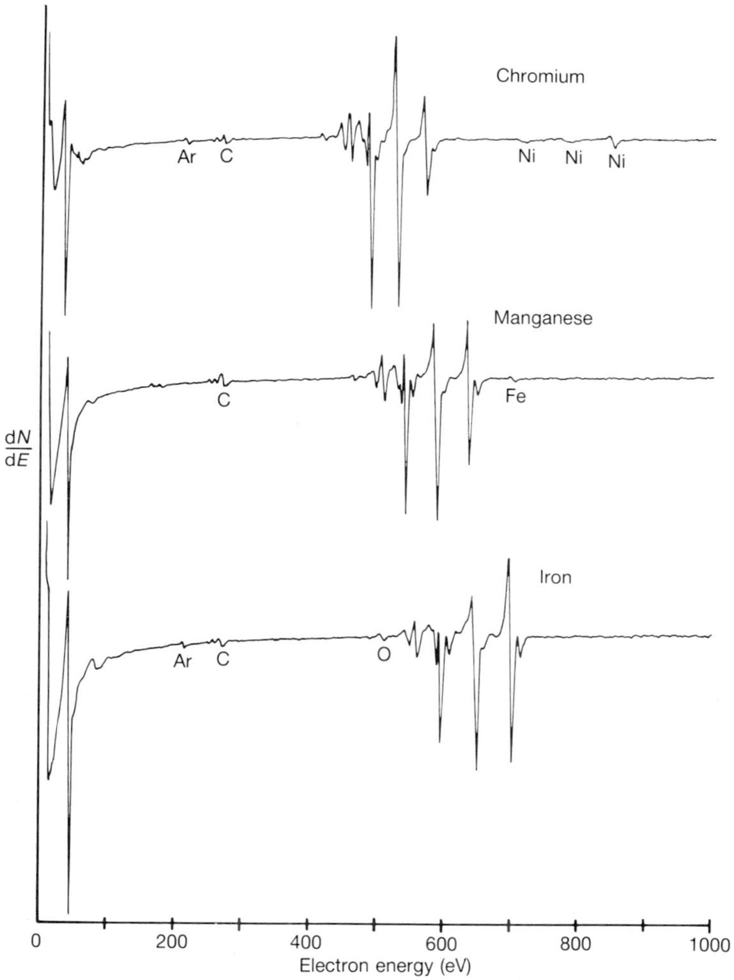

FIG. 4.5 Auger spectra in the differentiated mode from three metals in the first transition series, showing the very characteristic triplet of Auger peaks based on the LMM transitions. In increasing order of energy, these are labelled conventionally $L_3M_{2,3}M_{2,3}$, $L_3M_{2,3}V$, and L_3VV. The much less intense transitions based on L_2 ionization can be seen on the high energy sides in some cases, while at energies just below the triplet are visible other minor transitions. At very low energy there is another major peak arising from the $M_{2,3}VV$ Auger transition. Peaks due to impurities are marked. Compiled from Davis *et al.*[6]

'fingerprint' spectrum is found, based on the MNN Auger transitions. Examples are given in Fig. 4.6, again from Davis *et al.*[6] Here the 'fingerprint' is the closely spaced doublet, not quite resolved for rhodium and palladium, but striking from silver to tin, due to the $M_{4,5}N_{4,5}N_{4,5}$ transitions, the doublet arising from the separation of M_4 and M_5.

In other parts of the periodic table there are Auger spectra characteristic of

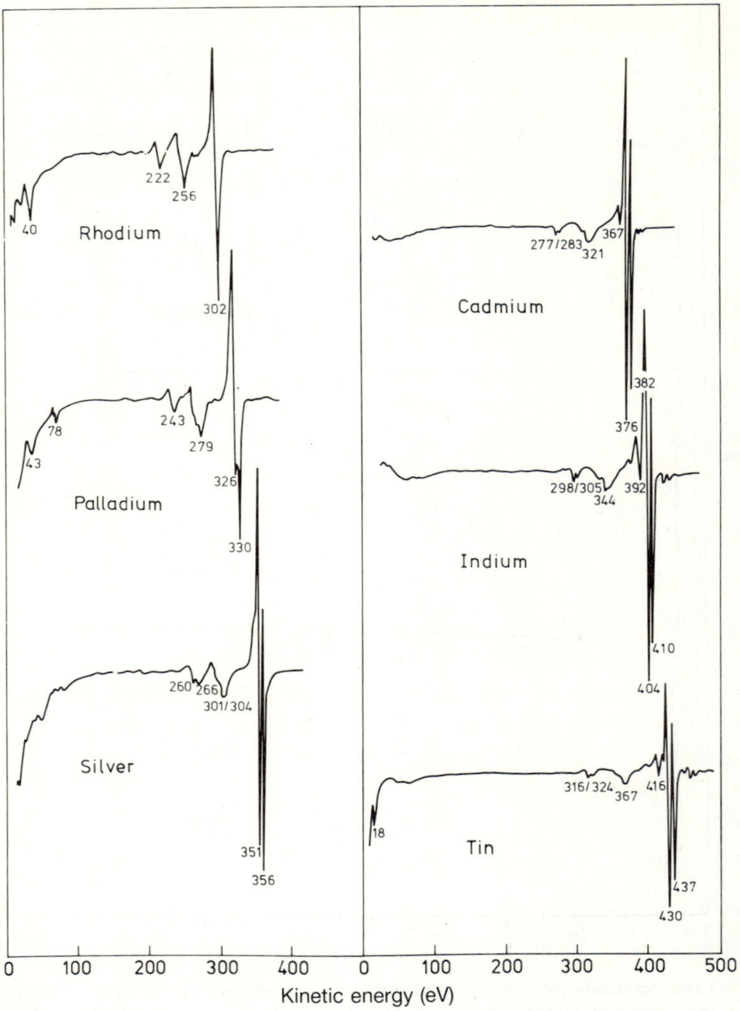

FIG. 4.6 Auger spectra in the differential mode from six metals in the second transition series. The characteristic of this series is the closely spaced doublet, not quite resolved for rhodium and palladium, based on the $M_{4,5}N_{4,5}N_{4,5}$ Auger transitions. Peaks due to other, minor, Auger transitions, are also present, and in a few cases a very low energy peak probably based on $N_{4,5}$ ionization is visible. Compiled from Davis *et al.*[6]

the elements in those regions, but none quite as immediately recognizable as the above. One aspect of AES that will be obvious, and it applies to most other spectroscopic techniques, is that when several elements are present at a surface, there is the likelihood of overlap of spectral features. Figure 4.5 makes that quite clear. For the medium to heavy elements this is not a great problem, since if there are several prominent peaks it is usually possible to find one that is clear of interference, but for the light elements, e.g. those of Fig. 4.4, the only major Auger peak available may coincide with one from a heavier element. Thus it is difficult to analyse nitrogen in the presence of titanium, chlorine, and argon with molybdenum, sulphur with zirconium, carbon with ruthenium, etc. In such cases recourse must be made to spectrum subtraction techniques, in which a standard elemental spectrum is subtracted from the composite one.

4.2.3 *Auger energies*

Nowadays identification of an unknown Auger peak in a spectrum can usually be made quickly by reference to one or other of the atlases of Auger spectra available commercially. Before such compilations appeared, however, there was no alternative but to calculate the energies of expected Auger transitions as best one could. Expression 2.1 for the kinetic energy of an Auger transition shows that the difficulty in calculation arises through the last term on the right, the binding energy of an electron in an outer level in the presence of a hole in a deeper core level. Various approximations to the energy have appeared, the most successful at the time being that of Chung and Jenkins,[7] which is

$$E_{ABC}(Z) = E_A(Z) - \tfrac{1}{2}\{E_B(Z) + E_B(Z+1)\} - \tfrac{1}{2}\{E_C(Z) + E_C(Z+1)\} \quad (4.1)$$

where $E_i(Z)$ are the binding energies of electrons in the ith levels in an element of atomic number Z, and $E_i(Z+1)$ of the same levels in the next higher element up the periodic table. Use of expression 4.1 produces values of Auger energies accurate enough for many purposes, but it has little theoretical justification. It was used as the basis for the compilation of calculated energies made by Coghlan and Clausing.[8]

The correct physical expression for the Auger energy is

$$E_{ABC}(Z) = E_A(Z) - E_B(Z) - E_C(Z) - \mathscr{F}(BC{:}x) + R_x^{in} + R_x^{ex} \quad (4.2)$$

where $\mathscr{F}(BC{:}x)$ is the energy of interaction between the holes in B and C in the final atomic state x and R_x are relaxation energies. The latter terms arise from the presence of the core hole, which necessitates additional screening of the nucleus. To achieve this the outer electronic orbitals contract inwards towards the nucleus, or 'relax'. As the relaxation occurs, there is increased interaction between the outer and inner orbitals, through Coulomb and exchange integrals. The term R_x^{in}, or the intra-atomic relaxation energy, arises *within* the atom and will be present even in an isolated atom. When the atom is part of a solid, additional screening electrons are available from the local environment *outside* the atom, in which case R_x^{ex}, or the extra-atomic relaxation energy, appears. In the calculation of E_{ABC} from 4.2 it is normal to use experimentally measured binding energies for the E_i and calculated energies for the other terms. The calculated values have now become very accurate, and various tabulations have been published, in particular the extensive one by Larkins.[9]

It is worth pointing out that in a differential spectrum, i.e. $E\,dN(E)/dE$, the convention is that the energy of an Auger peak is taken as that of the minimum in the high energy negative-going excursion, as labelled in Fig. 4.4 to 4.6. Such a conventional energy does not correspond to the actual position of the peak in the undifferentiated spectrum, but is convenient. With improved sensitivity and better methods of background subtraction there is an increasing tendency to use the undifferentiated, or $E\,N(E)$, spectrum.

4.2.4 *Auger intensities*

If all Auger transitions had the same probability, that is, if all the Auger features energetically possible in the spectrum from an element appeared at the same intensity, the technique of AES would not be workable. It is precisely because only a few of the possible transitions are actually measurable that it can provide an elemental analysis. The differences in intensity between different Auger transitions in the same element arise partly through the different electronic populations of the levels involved, but largely through the variations in transition probability due to greater or lesser overlap of electronic orbitals. Unlike the situation for Auger energies, there are no useful approximations to Auger intensities, or even semi-empirical expressions for their calculation. In general the theory for the prediction of intensities is not in as good a state as it is for energies.

4.2.5 *Fine structure in Auger features*

Intrinsic fine structure The Auger spectra from some metals show line shapes whose widths are far too narrow to be typical of Auger emission from an atom in a solid. At the same time the spectrum in the energy region around the expected energy of the transition based on simple *j–j* coupling notation is complex, with a great deal of fine structure. An example, that of cadmium, is shown in Fig. 4.7 from the work of Aksela and Aksela.[10] Auger spectra of this type are called 'quasi-atomic', because both the transition width and the fine structure are characteristic of the same transitions occurring in the free atom. The explanation for the effect is that the Coulomb interaction energy between the two holes in the final doubly ionized state of the atom (see Fig. 2.1) is so large that the holes cannot decay rapidly as they can do in other metals. As a result they become localized on the ionized atom for long enough for there to be significant electronic correlation between the holes, as in an atom. This correlation gives rise to a variety of final-state spectroscopic terms, whose notation is given by the term symbols in intermediate coupling similar to those in Table 4.3 which has the KLL series as an example. In Fig. 4.7 these final states, or multiplets, have been calculated with respect to position and relative intensity, and given a notional experimental broadening. When they are added together the resultant theoretical spectral envelope agrees excellently with the observed one. If an element with quasi-atomic Auger features is

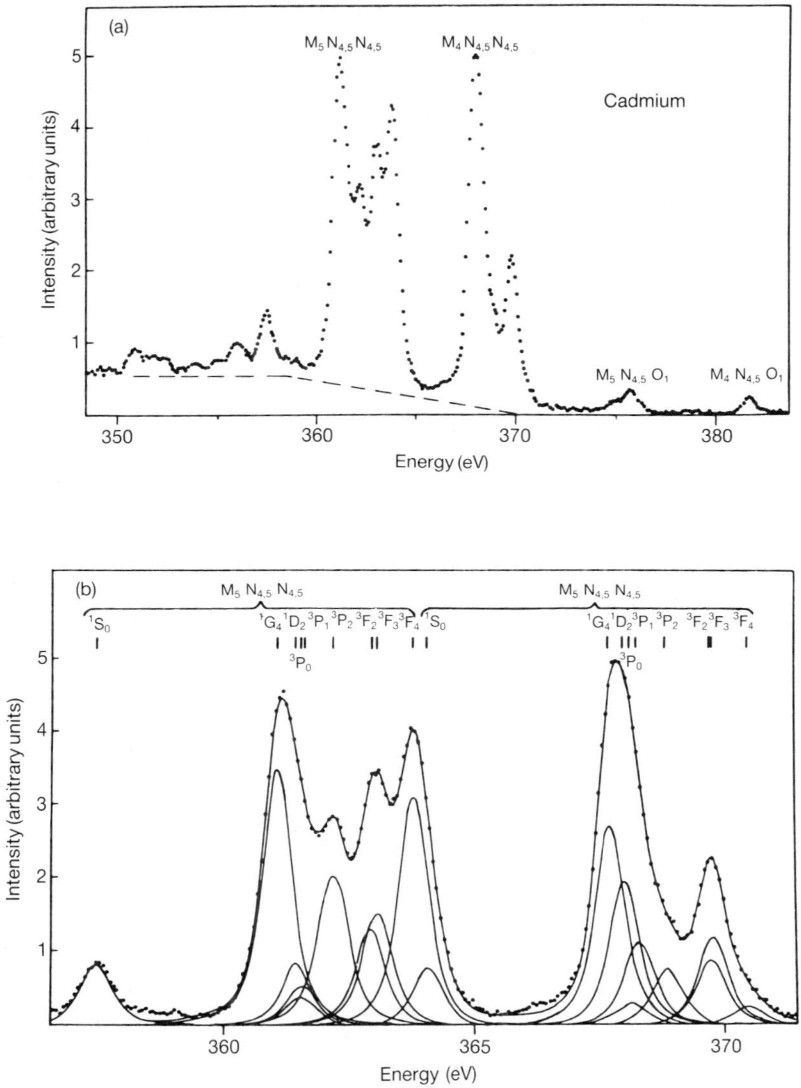

FIG. 4.7 (a) Auger spectra in the undifferentiated mode from cadmium metal, recorded by Aksela and Aksela.[10] Much higher energy resolution was used than in Fig. 4.6, and the $M_{4,5}N_{4,5}N_{4,5}$ Auger series can be seen to possess a large amount of fine structure. Spectra of this type are called 'quasi-atomic', because the fine structure is characteristic of transitions occurring in a free atom rather than one embedded in a solid. The dashed line shows the background that was subtracted to allow comparison of theory and experiment. (b) Comparison of the experimental Auger spectrum (points) with a theoretically calculated spectrum based on quasi-atomicity. The atomic multiplet structure has been calculated to give line positions and relative intensities, and the lines have then been broadened with Gaussian functions, and added together. The resultant solid line follows the experimental points almost exactly.

combined in a compound or alloyed with another element not showing the same effect, the quasi-atomicity may be reduced or lost.

Chemical effects Fine structure can also originate in chemical effects. Sometimes this has the same origin as the 'chemical shift' (see Chapter 10) found in XPS, i.e. the change in the binding energy of core-level electrons due to a change in chemical environment. If two or more chemical states of the same element are present on a surface, e.g. if there is a very thin oxide layer on an underlying metal substrate, then the Auger spectrum of that element will consist of the spectra of the different states shifted from each other and superimposed. However, composite Auger spectra of this type have not been reported very often in the literature.

More commonly observed is fine structure in the Auger spectra of non-metallic elements present at surfaces due to differences in chemical environments. These differences are not due to straight core-level chemical shifts, as above, but to the variations in the local density of electronic states in the vicinity of the non-metallic element. The fine structure is always associated with transitions of the type XVV, where the initial ionization takes place in the core level (X = K, L, M, etc.) of the element, but the two electrons involved in the transition come from the local valence band. Since the local density of states will vary from one substrate or compound to another, so will the fine structure in the Auger peak. An example of the nature of the fine structure to be found is given for carbon in Fig. 4.8, the spectra being taken from several authors.[11] Similar effects have been seen in the Auger spectra of sulphur, nitrogen, phosphorus and oxygen.

4.2.6 *Inelastic loss features*

Plasmon losses The 'sea' of conduction electrons in a solid can be excited into waves, or collective oscillations, which have frequencies characteristic of the nature (structure, electron density, etc.) of the solid, and thus need characteristic energies for their excitation. Any electron with enough energy travelling through a solid can excite these oscillations, and an electron that has given up a certain amount of energy in so doing is said to have suffered a plasmon loss. Such a loss occurring under the influence of the three-dimensional nature of the solid is called a 'bulk' plasmon loss, while near the surface the loss associated with the quasi-two-dimensional nature of the surface is called a 'surface' plasmon loss. Theoretically, and frequently in practice too, the energy of a surface plasmon loss is $2^{-1/2}$ times that of a bulk plasmon loss.

Thus there will nearly always be associated with the energies of electrons being ejected from a solid a series of plasmon losses, manifesting themselves as more or less prominent peaks on the low kinetic energy side of the parent peak. Since electrons that have suffered a plasmon loss can themselves excite

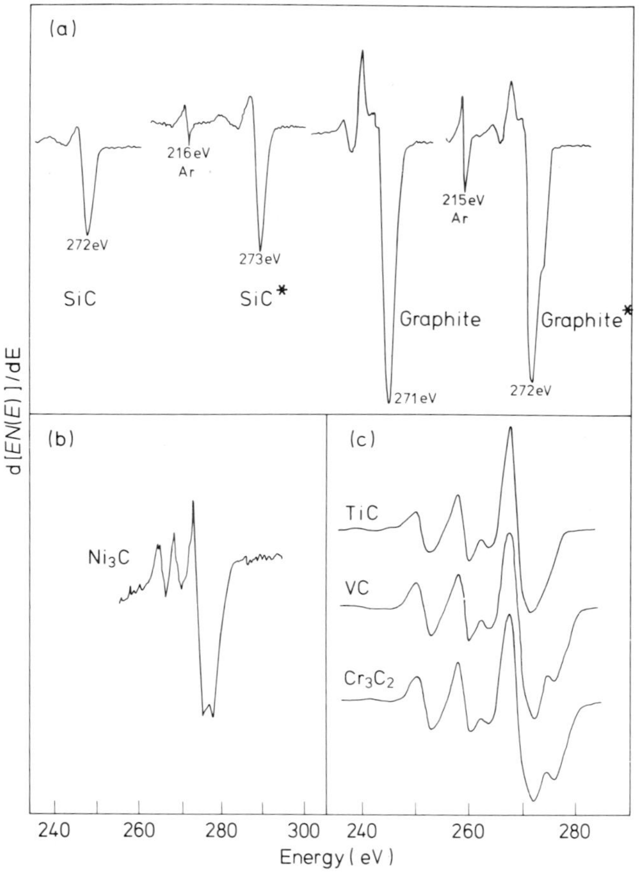

FIG. 4.8 Examples of the effects of different chemical states on the KLL Auger spectrum of carbon. (a) Shapes of the carbon KLL Auger peak in SiC and in graphite; the asterisks refer to ion bombarded surfaces, in which the argon LVV peak can also be seen. From Kny.[11] (b) Shape of the carbon KLL Auger peak in Ni_3C after argon ion bombardment. From Kleefeld and Levinson.[11] (c) Shapes of the carbon KLL Auger peak in TiC, VC, and Cr_3C_2. The principal minima have been aligned for better comparison. From Smith and Levinson.[11]

plasmons and thus suffer further losses, it is possible to find peaks at multiples of the plasmon loss energy, and also at the combined bulk and surface plasmon loss energies, although of course the multiple loss peaks become rapidly weaker as the number of successive events increases. By reducing the energy of the primary electrons, and thus their average penetration depth, the surface plasmon loss peaks can be enhanced compared to the bulk plasmon loss peaks; this is part of the technique of ELS (see next chapter).

A good example of a spectrum showing bulk and surface plasmon losses, combined losses, and multiples, is that from aluminium recorded by Dufour *et al.*,[12] in Fig. 4.9, using both electron and X-ray excitation. It can be seen that,

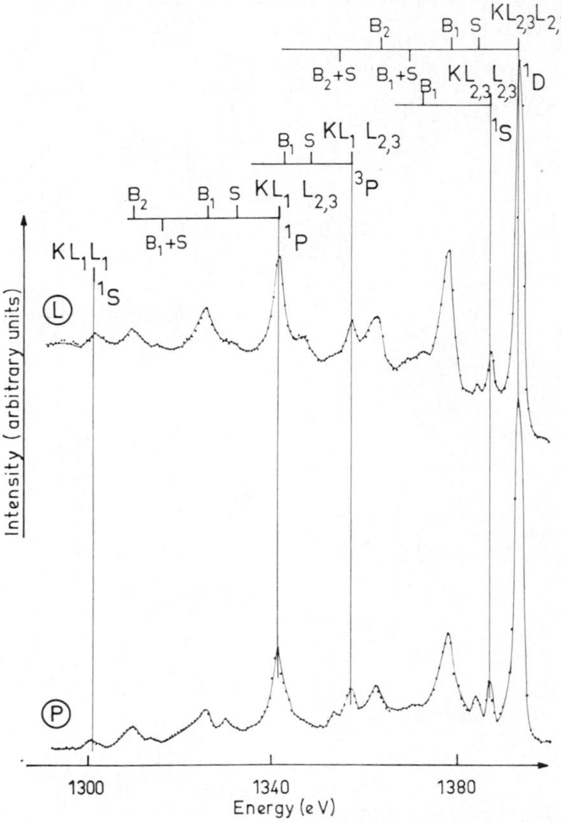

FIG. 4.9 Example of the surface and bulk plasmon losses associated with Auger peaks, from Dufour *et al.*[12] The principal peak to the right is the aluminium $KL_{2,3}L_{2,3}\,^1D$ Auger peak, induced by photons in the upper spectrum (L) and by electrons in the lower spectrum (P), both in the undifferentiated mode. The minor Auger transitions $KL_{2,3}L_{2,3}\,^1S$ are indicated. Associated with each peak are bulk (B) and surface (S) plasmon loss peaks. Multiple losses are shown as B_1, B_2, . . . , and combined losses as (B + S). The same losses appear irrespective of the method of excitation.

as expected, the same plasmon losses appear irrespective of the method of excitation, although the relative magnitudes of the loss peaks are not exactly the same in all cases.

Core losses As described in Section 2.1.3, there should appear in every Auger spectrum weak features, not so much peaks as steps in the undifferentiated spectrum, due to ionization of core levels. Since these features are the basis of the technique of CEELS, they will be discussed in more detail in the next chapter. It is sufficient to note here that they are in practice rarely seen in published Auger spectra simply because the energy region chosen for analysis (i.e. typically 20–1000 eV, as remarked above) is too far below the typical primary energies of 5–10 keV.

4.2.7 *Diffraction (or temperature-dependent) features*

In the event that the sample being analysed is a single crystal, or perhaps a polycrystalline material with large grain size, and the area being electron irradiated is very well ordered, it is possible to find features in the secondary electron spectrum arising from diffraction effects. According to McDonnell *et al.*[13] some secondary electrons originating close to or just below such a surface will be at the correct energies to satisfy the Bragg condition for the particular crystal direction along which they are emerging. This has the effect of making emergence at those particular energies easier, i.e. effectively increasing the inelastic mean free path, with the result that many more electrons at those energies are ejected than those at adjacent energies, and peaks at those energies appear in the spectrum. They could be mistaken for Auger peaks, but have characteristics quite different in that their intensities decrease much more rapidly with increasing temperature than do those of Auger peaks, and they tend also to peak sharply in intensity as functions of angles of incidence and take-off.

An example of such a peak at low energy, for a well-ordered CoTi (110) surface is shown in Fig. 4.10, from Rivière.[14] That particular peak might have been mistaken for an aluminium $L_{2,3}VV$ Auger peak except for its temperature and angular dependences, and for the lack of any sign of the high energy aluminium Auger peaks.

4.3 Quantification

If the core level X in an atom of type A in a solid is ionized by a primary electron of energy E_p, giving rise to an Auger electron following the transition XYZ, and if the electron is to be detected externally in an electron spectrometer, then the current of Auger electrons of energy E_{XYZ} from A can be written down as

$$I_{XYZ} = I_0 \rho \gamma_{XYZ} \sigma(E_p : E_X) \{1 + r_M(E_X : \alpha)\} T(E_{XYZ}) D(E_{XYZ}) \int_0^\infty N_A(z) \exp \{-z/\lambda_M(E_{XYZ}) \cos\theta\} \, dz. \tag{4.3}$$

Here ρ is a surface roughness factor, γ_{XYZ} is the probability of the Auger transition XYZ, σ is the cross-section for ionization of a core level with binding E_X by an electron of energy E_p, r_M is the back-scattering factor that takes account of the additional ionization of E_X in the matrix M by inelastic electrons of energies between E_X and E_p, α is the angle that the incident beam makes to the surface normal, T is the transmission of the spectrometer for

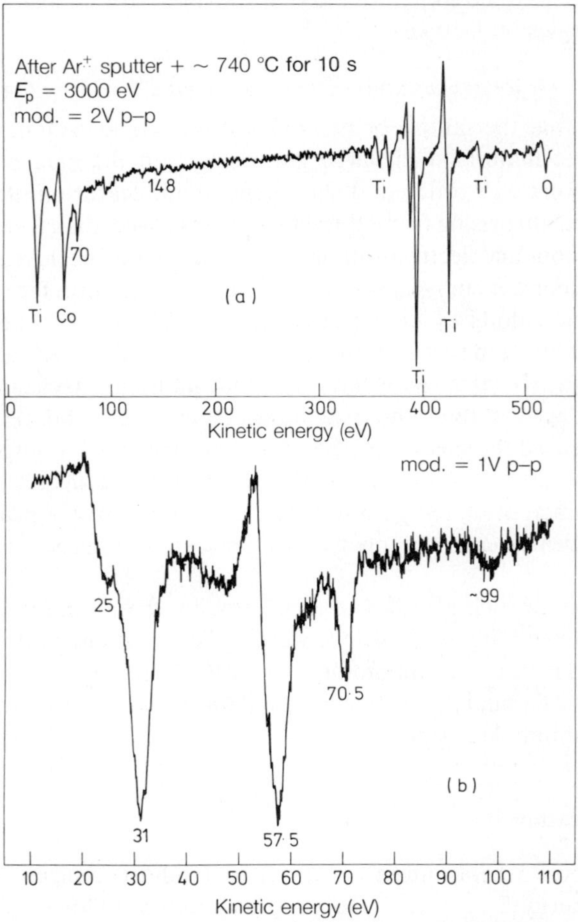

FIG. 4.10 Example of a feature in an Auger spectrum arising from diffraction effects, from Rivière.[14] The annealed CoTi(110) surface is very well ordered, and a peak at 70.5 eV appears that is not part of the Auger spectrum of either Co or Ti. The peak is very temperature and surface-condition sensitive, and is due to diffraction channelling of emergent secondary electrons.[13]

electrons of energy E_{XYZ}, D is the efficiency of the detector (i.e. the electron multiplier) at the energy E_{XYZ}, $N_A(z)$ is the distribution of atoms of type A with depth z, λ_M is the inelastic mean free path of electrons of energy E_{XYZ} in the matrix M, and θ is the angle of emission to the surface normal. The various terms appearing in 4.3 need some discussion.

If the ionization cross-section σ of a core level at a particular primary energy is actually known, then of course it can be used directly in 4.3, but generally it is not, and recourse must be had to mathematical approximations to σ. Several of these have been suggested from time to time, but at the moment the best agreement with experimental measurements is given by Gryzinski's[15] cross-section, which has the form

$$\sigma(E_p:E_X)=\frac{\pi e^4 K_X}{E_X{}^2 U}\left(\frac{U-1}{U+1}\right)^{3/2}\left\{1+\frac{2}{3}\left(1-\frac{1}{2U}\right)\ln(2.7+\sqrt{U-1})\right\} \quad (4.4)$$

where K_X is the number of electrons in the X-shell, and U is the so-called overvoltage, i.e. $U=E_p/E_X$. Although the various cross-sections that have been employed have different mathematical forms it should be noted that they all include the factor $E_X{}^{-2}$. The Gryzinski cross-section is shown plotted as the double-dashed line in Fig. 4.2.

The most important matrix effect in expression (4.3) is probably the back-scattering factor r_M; often one finds it defined instead as $R=(1+r_M)$. In earlier AES quantification attempts an expression derived by Reuter[16] from electron probe microanalysis measurements at 20 keV was used, viz.

$$R=1+2.8(1-0.9E_X/E_p)\eta \quad (4.5)$$

where $\eta=-0.0254+0.016Z-1.86\times10^{-4}Z^2+8.3\times10^{-7}Z^3$, Z being the atomic number. However, extension of the application of this expression to energies more appropriate to AES, i.e. below 10 keV, and at angles of incidence other than normal led to inaccuracies. As a result Ichimura *et al.*[17] carried out very extensive Monte Carlo calculations of the back-scattering coefficient at a variety of angles of incidence of the primary beam, and the best fit to their results gives the following relationships

$$R=1+(2.34-2.1Z^{0.14})U^{-0.35}+(2.58Z^{0.14}-2.98) \quad \text{for } \psi=0^\circ$$

$$R=1+(0.462-0.777Z^{0.20})U^{-0.32}+(1.15Z^{0.20}-1.05) \quad \text{for } \psi=30^\circ$$

$$R=1+(1.21-1.39Z^{0.13})U^{-0.33}+(1.94Z^{0.13}-1.88) \quad \text{for } \psi=45^\circ \quad (4.6)$$

Their calculated values and the best fits to them using expressions (4.6) are shown in Fig. 4.11, for a primary energy of 10 keV, at normal incidence.

The other matrix-dependent term in expression (4.3) is the inelastic mean free path λ_M. On the basis of a very extensive compilation of measurements of the imfp and its related quantity the average escape depth, Seah and Dench[18] have derived best fit relationships between λ_M and electron kinetic energy E for elements, and for inorganic and organic compounds. These relationships are:

For elements

$$\lambda_M=538E^{-2}+0.41(aE)^{1/2} \text{ monolayers}$$

For inorganic compounds

$$\lambda_M=2170E^{-2}+0.72(aE)^{1/2} \text{ monolayers}$$

For organic compounds

$$\lambda_M=49E^{-2}+0.11E^{1/2} \text{ mg m}^{-2} \quad (4.7)$$

where a = monolayer thickness in nm.

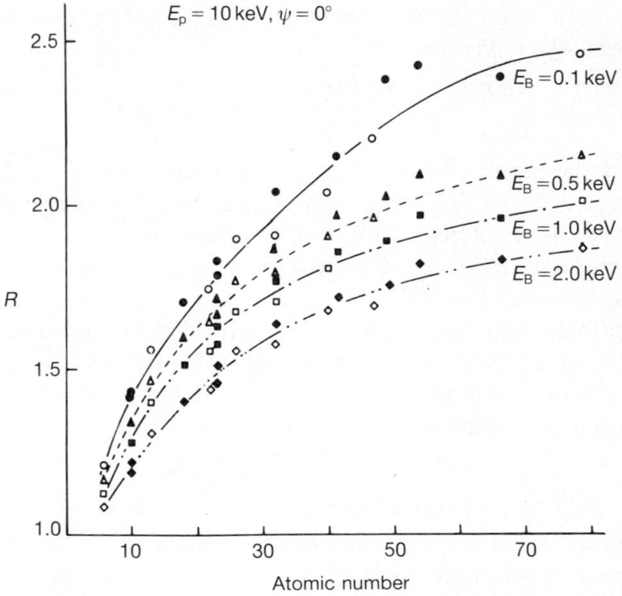

FIG. 4.11 Variation of the backscattering factor R with atomic number for chosen indicated values of core-level binding energy E_B, calculated by Ichimura *et al.*[17] for a variety of mono-atomic (open symbols) and poly-atomic (solid symbols) materials. The lines drawn are best fits to the theoretical points.

For a fixed set of experimental operating conditions and for a particular Auger transition many of the terms in expression (4.3) are constant, and it may be abbreviated to

$$I_{XYZ} = K\sigma R \int_0^\infty N_A(z)\exp(z/\lambda\cos\theta)\,dz \tag{4.8}$$

If the usual assumption is made that element A is distributed homogeneously within the analysed depth, with an average atomic density $\bar{N}_A$, then (4.8) can be simplified to

$$I_{XYZ} = K\sigma R\bar{N}_A\lambda\cos\theta \tag{4.9}$$

where K includes all the terms that remain constant during any one experimental determination. The $\cos\theta$ term will of course normally be constant also, since its value will depend on the angular acceptance of the particular analyser being used.

Normal practice in AES is not to attempt to evaluate the terms in (4.9) absolutely, but to arrive at a relative quantification by using standards. If the average atomic density of atoms A in the standard is $\bar{N}_A{}^s$, then an expression similar to (4.9) can be written down for the standard, whereupon the ratio of the atomic densities in the sample and in the standard can be written as

$$\bar{N}_{\mathrm{A}}/\bar{N}_{\mathrm{A}}^{\mathrm{s}} = \frac{I_{XYZ}R^{\mathrm{s}}\lambda^{\mathrm{s}}}{I_{XYZ}^{\mathrm{s}}R\lambda} \tag{4.10}$$

since the cross-section σ will be the same for both. Expression (4.10) can be abbreviated again to

$$\bar{N}_{\mathrm{A}} = S_{\mathrm{A}} I_{XYZ} \tag{4.11}$$

where

$$S_{\mathrm{A}} = \frac{\bar{N}_{\mathrm{s}}^{\mathrm{A}} R^{\mathrm{s}}\lambda^{\mathrm{s}}}{I_{XYZ}^{\mathrm{s}}R\lambda},$$

and can be regarded as a relative sensitivity factor. These factors can either be evaluated using expressions (4.6) and (4.7) as appropriate, or derived empirically, as has been done by several workers. The average surface concentration of the ith element in a multi-elemental surface composition is thus

$$\bar{C}_i = \bar{N}_i \Big/ \sum_i \bar{N}_i. \tag{4.12}$$

For the quantification of metallic alloys containing between two and four constituents, the above procedure leads to quite good accuracy (e.g. see Minni[19]) of the order of ± 5 at.%. When one comes to the quantification of non-metals, however, the situation is much worse, for a variety of reasons. The non-metallic elements are not so likely to be uniformly distributed in the analysed depth, provision of standards for them is difficult since it usually involves using compounds rather than pure elements, and the accurate measurement of the Auger current I_{XYZ} is much more problematic. Probably the most difficult element to quantify accurately is carbon since it can exist in so many different forms, each with a characteristic Auger peak shape.

One of the popular methods of AES quantification, since it involves least effort and understanding, is the use of the 'standard' spectra published in various handbooks,[6,20,21] mostly for pure elements. The operating conditions are stated, although not always with sufficient clarity, and each part of each spectrum has a stated amplification factor, so that in principle all the spectra are inter-comparable from the point of view of relative sensitivities. Problems always arise, however, for various reasons; not all the surfaces of the pure elements are clean, some being quite heavily oxidized, which has a profound effect on peak intensities; it is not always possible to match the exact energy resolution used for the standard; and strictly speaking the spectra should be used only in comparison with those obtained on similar analysers, because of differing transmission functions. Nevertheless the method continues to be used widely, and the quantification based on it is probably sufficiently accurate in measuring *changes* in element concentration, even if in absolute terms it is often far from accurate. A comparison of this approach with the 'semi-first-principles' approach mentioned above has been made by Mroczkowski and Lichtman[22] for several binary alloys.

4.4 Applications

The range of applications of AES and SAM is now so diverse, and the number of publications so large, that any attempt to describe them in a reasonably short space must involve drastic selectivity on the part of the author. The following examples are therefore just a few taken from amongst the many thousands, selected because each brings out one or another aspect of the use of these techniques to solve problems and to contribute to surface science.

4.4.1 *Interfaces*

In the semiconductor device industry, knowledge of the characteristics of thin films on semiconductors and of the interfacial regions between these films and the underlying material, has become increasingly important as the required thicknesses of the films continue to reduce. These films may consist of the residual oxide on a semiconductor left behind after some standard surface treatment in production, or of the product of a surface reaction designed to create a Schottky barrier as part of the construction of a complicated integrated circuit. The information needed about the film will be not only elemental and chemical, but also structural in the sense that the abruptness of the interface is one of the factors that governs the eventual properties of the device.

One of the many studies made on the oxide films on GaAs, grown either anodically or thermally, was that of Wang *et al.*[23] Their oxide films were grown thermally in oxygen at 500°C on (100) oriented GaAs wafers, and were about 180 nm thick. The variation in composition through the film to the substrate crystal, and the nature of the interface, were studied by simultaneous AES and ion profiling with argon ions. Such profiles had been published many times before, and what Wang *et al.* were particularly concerned to investigate was the apparent dip in the Ga concentration just as the interface was reached, that appeared in all the previous profiles. In their raw data this dip was also seen, as evident in the upper part of Fig. 4.12, where the atomic concentrations of As, Ga and O are plotted as a function of bombarding time. (It is interesting here that they found that the relative sensitivity factors for these elements taken from one of the handbooks gave unreasonable compositions, so they devised their own set of factors from bulk GaAs and Ga_2O_3.)

At various stages in the bombardment, marked along the Ga profile, the Ga $L_3M_{4,5}M_{4,5}$ line shape was carefully recorded, and the shapes at these stages are given in the lower half of the figure. Not only was the shape changing as the interface was traversed, but the position of the peak was shifting more

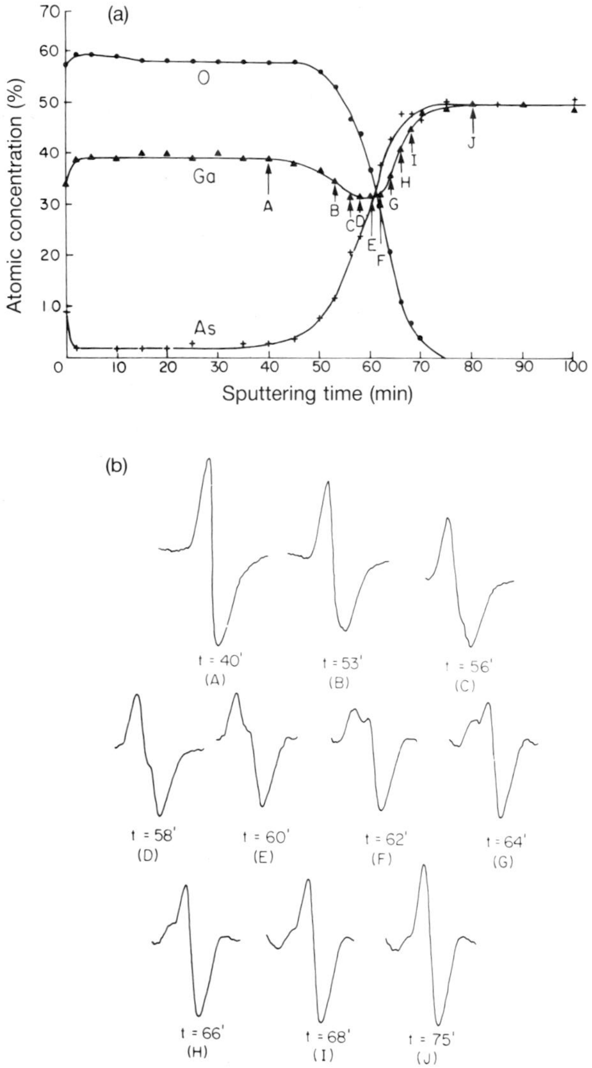

FIG. 4.12 (a) Uncorrected atomic concentration profile using the peak-to-peak heights of the principal O, Ga, and As Auger peaks during profiling of an oxide film on GaAs(100), recorded by Wang *et al.*[23] The letters A–J at various points correspond to the Ga $L_3M_{4,5}M_{4,5}$ line shapes shown in the lower half of the figure. (b) Shapes of the Ga $L_3M_{4,5}M_{4,5}$ Auger peak recorded at the various points in the depth profile indicated in (a). The sputtering times for each peak are given.

than 6 eV as well. In fact in the interface there are *two* Ga peaks, one from the oxide and one from the GaAs, with the result that the peak-to-peak height on which the profile is based is severely reduced, leading to the apparent dip in Ga concentration. When the authors manipulated the data by integrating the peaks, subtracting an inelastic electron background, and then resolving the interface peaks into their two components, a much more realistic picture of the interface emerged, as shown in Fig. 4.13. The Ga in the oxidized state decreases as that in GaAs increases, and there is no longer an apparent excess of As over Ga as can be seen in Fig. 4.12(a). The As peak was treated in the same way.

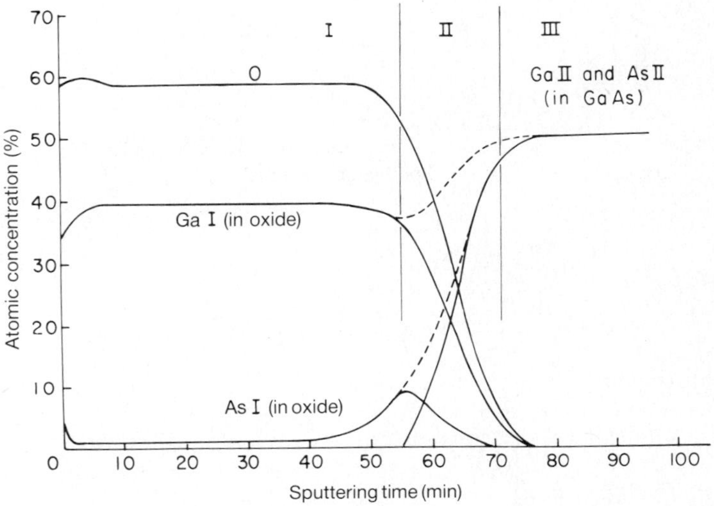

FIG. 4.13 Atomic concentration profile during profiling of a thermally grown oxide film on GaAs(100), corrected for the contributions to the Ga and As Auger peaks from oxide and semiconductor components. The sharp dip in Ga concentration in the interface region seen in Fig. 4.12 can now be seen to be only apparent, since the oxide concentration drops sharply at the interface as expected, while the semiconductor contribution appears and grows. At the start of the interface there is a layer of arsenic oxide. From Wang *et al.*[23]

Another type of interface of technological interest is that between the metallic matrix and the fibre surfaces in fibre-reinforced materials. In such materials there must be good adhesion between the matrix and the fibres, and the properties of the contact region, or interface, are therefore important. One of the substances being favoured for use in fibre form is SiC, and it would be advantageous to be able to use it as reinforcement in aluminium. Unfortunately there is no interaction leading to adhesion between SiC and Al, and an intermediate 'sandwich' must be found that will interact with both. For interaction with SiC it is likely that a silicide-forming metal would be most suitable, and to that end the reactivity of SiC with Pd was studied by Bermudez.[24]

The study consisted first of depositing small amounts of Pd on a clean SiC surface at room temperature, while monitoring the Auger spectra to detect any reaction that might be occurring, then annealing at 600°C to simulate the temperature reached in the production of fibre-reinforced material, and finally ion profiling the resultant surface layer to establish the width of the interface region. Figure 4.14 shows the changes that were found in the Si $L_{2,3}VV$ Auger peak and its surroundings during the deposition of Pd up to a coverage equivalent to $3\frac{1}{2}$ monolayers (i.e. $\theta = 3.5$). Clean SiC has a single Si $L_{2,3}VV$ peak, whereas the various Pd silicides all have a 5-peak structure in the same energy region, as seen in Fig. 4.15. The characteristic 5-peak spectrum was observed on SiC after deposition of only $3\frac{1}{2}$ monolayers of Pd, and indeed it can be seen in Fig. 4.14 that even after only half an equivalent monolayer the Si $L_{2,3}VV$ peak was already showing incipient extra fine structure. In the unannealed state, the c/d ratio (see Fig. 4.14) was between those of Pd_3Si and Pd_4Si but after annealing to 600°C it fell to a value between those of Pd_2Si and Pd_3Si. Thus reaction between Pd and SiC to form one or more Pd silicides takes place spontaneously at room temperature. The 5-peak structure in silicides arises from changes in the Si valence band due to bonding with Pd, and the allocations are indicated briefly in Fig. 4.15, where S refers to the Si 3s component, and P_b and P_a to the bonding and antibonding combinations of Si 3p and Pd 4d orbitals. The X sign of course represents convolution of the components, since the $L_{2,3}VV$ transition involves two electrons from the valence band.

The result of ion profiling through the reaction layer is shown in Fig. 4.16, in which the quantity plotted for the Si $L_{2,3}VV$ intensity is the sum 'c + d' of the fine structure respective intensities marked in Fig. 4.14. For the Pd film of equivalent thickness 1.8 nm a silicide layer of thickness about 3–4 nm has been formed, with an interfacial region spread over almost the same distance. For thicker films the c/d ratio remained constant until the substrate $L_{2,3}VV$ spectrum appeared, indicating a silicide formed of uniform composition. The results of the study confirm that a Pd coating on SiC fibres should promote good adhesion between the fibres and the metallic matrix during fibre reinforcement.

4.4.2 *Metallurgical*

Many AES and SAM applications involve trouble-shooting in connection with production and fabrication problems, and this is particularly true of the metallurgical industry. Production is usually of high volume and rapid, and any faults that occur must be solved as quickly as possible since they are generally expensive. Often the faults are due either to the absence of a desirable thin film (e.g. protective or lubricating) or the presence of an undesirable thin film (e.g. discolouration or foreign material), and can thus be solved only by surface analysis.

A typical example of a problem in fabrication has been described by Stout,[25] in this case from the wire industry. The SEM pictures in Fig. 4.17 show, above, a cross-section of a 0.5 mm wire with a steel core coated first with copper and

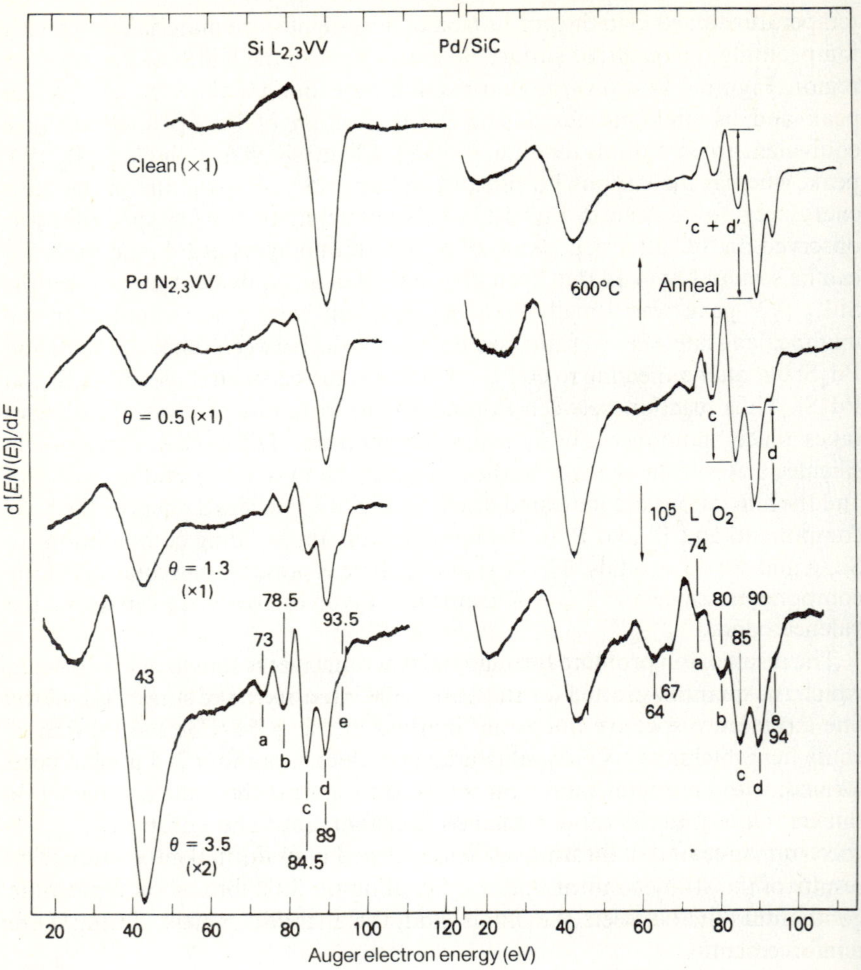

FIG. 4.14 The Pd $N_{2,3}VV$ and Si $L_{2,3}VV$ Auger spectra from a SiC surface as a function of the deposition of Pd, recorded by Bermudez.[24] At the top is the Si $L_{2,3}VV$ spectrum from clean SiC. The Pd coverages are indicated, up to $\theta = 3.5$. As the Pd coverage grows, the Si $L_{2,3}VV$ Auger spectrum changes progressively to that characteristic of a silicide, with a 5-peak structure. On the right are shown the effects of annealing at 600°C (upper) and of exposure to oxygen (lower). The method used of measuring the peak-to-peak heights of the c and d peaks, or of their combined height, is shown.

then with an outer layer of solder; and below, a cross-section of such a wire welded to a stainless steel cap on an electrical component. In the latter, the copper layer is on the outside and the steel core of the wire has penetrated slightly into the stainless steel, forming a ring structure. It was found that in production a small but unacceptable proportion of components were failing because of lack of strength at the weld, failure occurring at the weld interface.

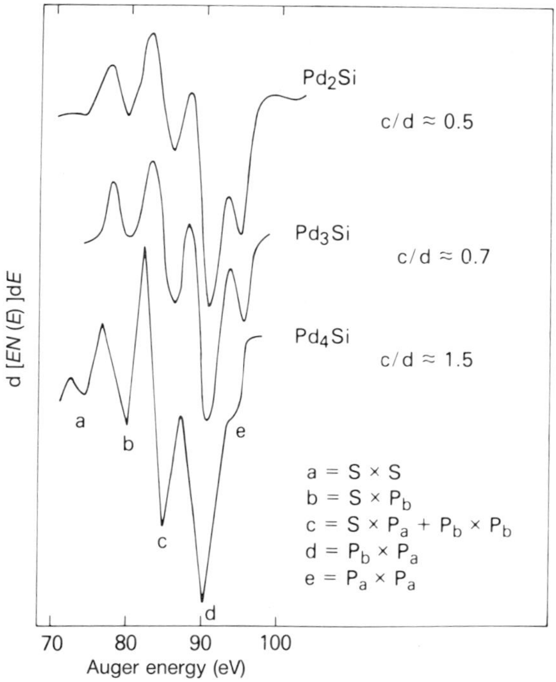

FIG. 4.15 The 5-peak structures characteristic of the Auger spectra of the various palladium silicides, in the differential mode, from Bermudez.[24] Correlation between the various peaks and structure in the Si valence band partial densities of states is shown in the right hand corner. S refers to the 3s component, P_b and P_a to bonding and antibonding combinations of Si 3p and Pd 4d orbitals.

The reason for the failures was established conclusively by SAM, as seen in Fig. 4.18. The upper row of micrographs a, b, and c refer to the wire side of the weld, and the lower row, d, e, and f to the cap side, and in each row there is an SEM micrograph of the failed interface with Auger maps of Cu and Fe. Two faults are evident. There is a strong Cu signal on both sides in the weld region, where there should be no Cu at all, and the penetration ring is poorly formed in several places. The reason for the presence of Cu was quickly traced to the way in which the wire was cut to prepare an end for welding; the cutting operation tended to smear the solder and Cu across the cut end of the wire. Change in the way that the cutting was performed, and use of sharper tools, eliminated the failure problem.

The study of impurities at grain boundaries in metals was one of the earliest applications of AES and SAM, and this general field still represents a high proportion of the total number of applications of the techniques that are published each year. In most cases metallurgists are interested in analysing the types and amounts of elements segregating to grain boundaries that are present as original impurities in the bulk material, but another interesting part

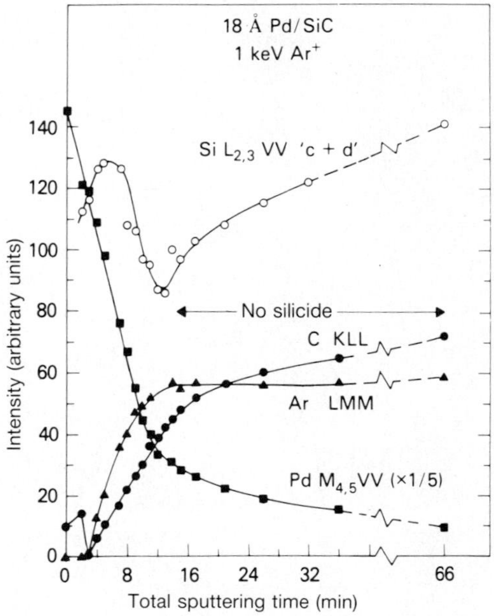

FIG. 4.16 Ion profile through a 1.8 nm thick film of Pd on SiC, measured by Bermudez.[24] The 'c+d' intensity for the Si $L_{2,3}VV$ can be understood by reference to Fig. 4.14. The argon profile arises from argon implanted during the bombardment. In the region labelled 'No silicide', the Si $L_{2,3}VV$ line shape was that of clean SiC.

of the same field concerns the penetration of elements into the grain boundaries by attack from the outside. Obviously if such a penetration were sufficiently extensive, disintegration of the penetrated metal would be a possibility. It was this possibility that led Ruedl *et al.*[26] to study the compatibility of Cr–Mn austenitic stainless steels with Li; such steels could be alternative structural materials in fusion reactors to the normal Cr–Ni stainless steels, for reasons of reduced radioactivity and helium production rate, but clearly they must be compatible with Li.

Ruedl *et al.* therefore subjected specimens of several Cr–Mn commercial steels to Li metal at 600°C for many thousands of hours, and analysed them afterwards using a variety of techniques including AES and SAM. Their work is interesting on two counts, one from the metallurgical point of view, and the other from the AES point of view, since not only is Li the lightest element that can in principle be analysed by the technique, but its usable Auger peaks are at such low kinetic energies that they can easily overlap with peaks from other elements. Indeed on the Li-exposed surface of the steel, Li was identified only by comparison of the Auger spectrum from a Li-rich area (a in Fig. 4.19) with that from a Li-deficient area (b in the same figure). By tuning the analyser very carefully to the Li signal, the SAM image (lower part of Fig. 4.19) revealed that

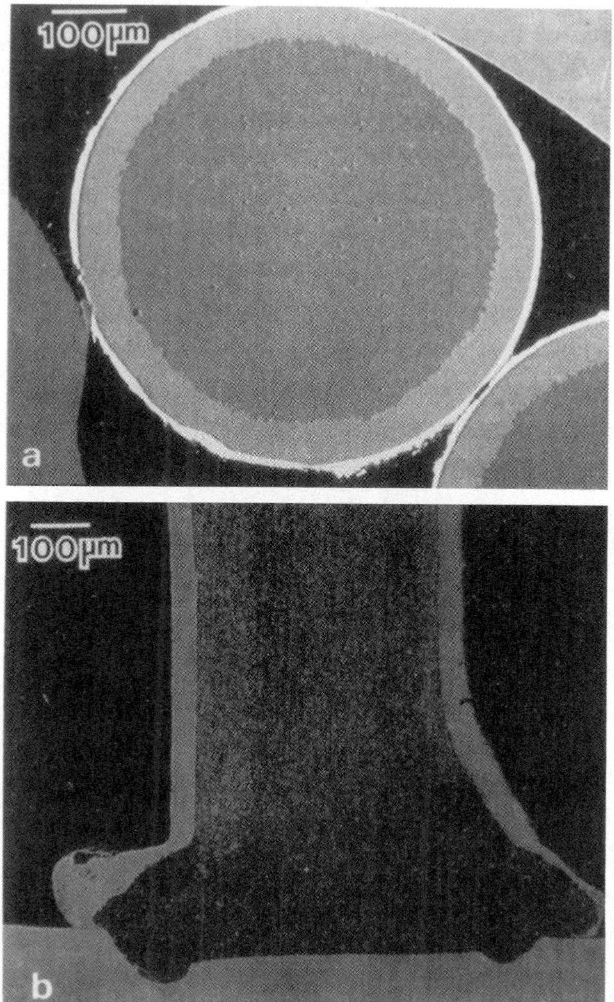

FIG. 4.17 SEM micrographs of (a) a cross-section of a steel wire coated with successive layers of copper and solder, and of (b) a cross-section of the wire butt-welded to a stainless steel cap. The steel core in (b) has penetrated slightly into the cap. From Stout.[25]

the Li-rich regions were indeed the grain boundaries of the steel, showing that the steel had been attacked by penetration of Li down the boundaries. The Auger spectra shown in Fig. 4.19 also indicate that a Li compound is formed in the grain boundaries, since there is a substantial amount of oxygen associated with it, and that Mn has been removed preferentially from the grain surfaces, since the Mn Auger peaks are much reduced compared to those from the unattacked surface. The study thus shows that the Cr–Mn stainless steels are

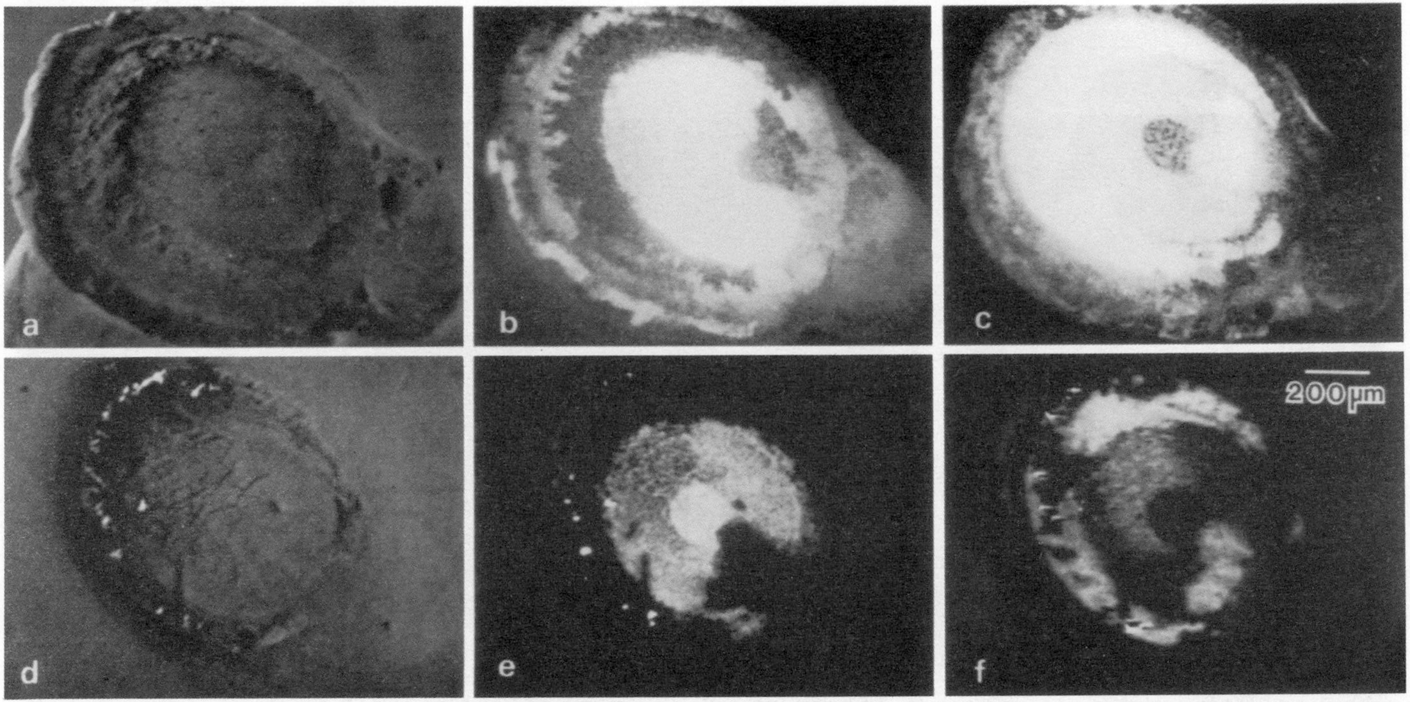

FIG. 4.18 (a) and (d), SEM micrographs of a failed wire-to-cap weld, from the wire and cap sides, respectively. (b) and (c), SAM maps in Cu and Fe, respectively, from the area in (a). (e) and (f), the same from the area in (d). From Stout.[25]

There are strong Cu signals from both sides of the weld, where there should not be any Cu, and the penetration ring of Fe is poorly formed. The solution was found to lie in the use of sharper cutting tools that did not smear the Cu across the end of the wire.

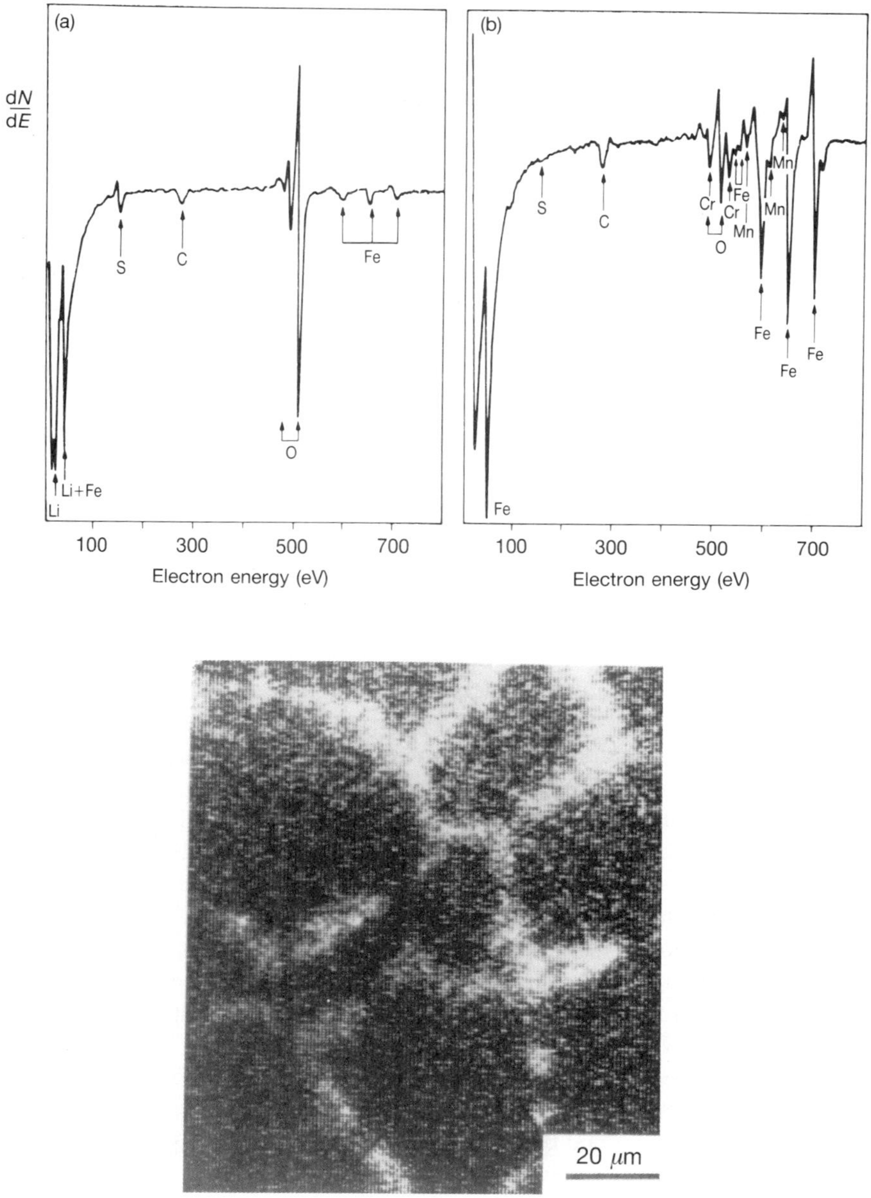

FIG. 4.19 AES and SAM study by Ruedl *et al.*[26] of grain boundary attack of an Fe–17.5Mn–10Cr stainless steel by lithium. Above. Auger spectra in the differentiated mode recorded from (a) the grain boundary region, and (b) the surrounding matrix of the steel after exposure to Li at 873 K for 4000 h. The presence of lithium can be identified only by careful comparison of the two spectra, and the presence of a doublet at very low energies. Below. The SAM image from the same surface, obtained by tuning the analyser very carefully to the energy of the Li peak in spectrum (a). The Li is concentrated heavily in the grain boundaries.

not compatible with Li unless some special protective surface treatment can be devised.

4.4.3 *Oxidation*

Although XPS has been brought to bear more often than AES in the fields of oxidation and corrosion, because of the additional chemical information available, nevertheless there have been situations in which the spatial resolution achievable in AES and SAM has given valuable information. Just such a situation arose in connection with the mechanism of spalling of the oxide scale from the oxidized surface of the 20Cr–25Ni steel alloy used as the containment for the fuel in the advanced gas cooled nuclear reactor. Oxidation occurs at high temperatures ($\sim 850°C$) in CO_2 and the surface eventually starts flaking away, which is clearly highly undesirable. Various theories had been put forward as to exactly where in the oxide scale the fracture leading to spalling occurs, and the problem was studied by Allen *et al.*[27]

The principal problem was to duplicate in the laboratory the spalling conditions in the reactor so that the interface corresponding to the weak point in the scale could be analysed. Conventional metallography and ion profiling from the external surface had been tried, but without any certainty that any of the various interfaces observed was the one involved in spalling. However, a new technique was discovered[28] that caused the laboratory-produced scale to separate along the line of weakness, by coating the oxidized specimen with sputter ion-plated nickel at elevated temperatures and allowing the mechanical stresses due to differential expansion on cooling to pull the scale away from the underlying metal. With the correct interface thus exposed, surface analytical techniques could then be applied to each side of it to gain information about the reasons for spalling. In fact, the separation appeared to take place at the interface between the metallic alloy and the scale, since the alloy side had a distinctly metallic appearance. However, the surfaces of both sides were very rough, with such a variety of different features that it was clear that surface analysis with high spatial resolution was certainly required, so Allen *et al.* used first AES in the point analysis mode and then SAM.

The point analyses by AES on each side are shown in Figs. 4.20 and 4.21. Note that on the metal side the SEM picture shows the alloy grains clearly delineated by grooves, and many pits and hollows within the grains themselves. Note also that the topography of the scale side matches that of the metal side almost exactly, with ridges where there are grooves, and particles where there are pits. The point analyses reveal considerable differences in composition across the surfaces of each side, but the most interesting observation is the substantial amounts of Si found at several points, more than 30 at.% compared with the bulk concentration in the alloy of about 1 at.%. The concentrations of the alloying elements were also varying considerably from point to point.

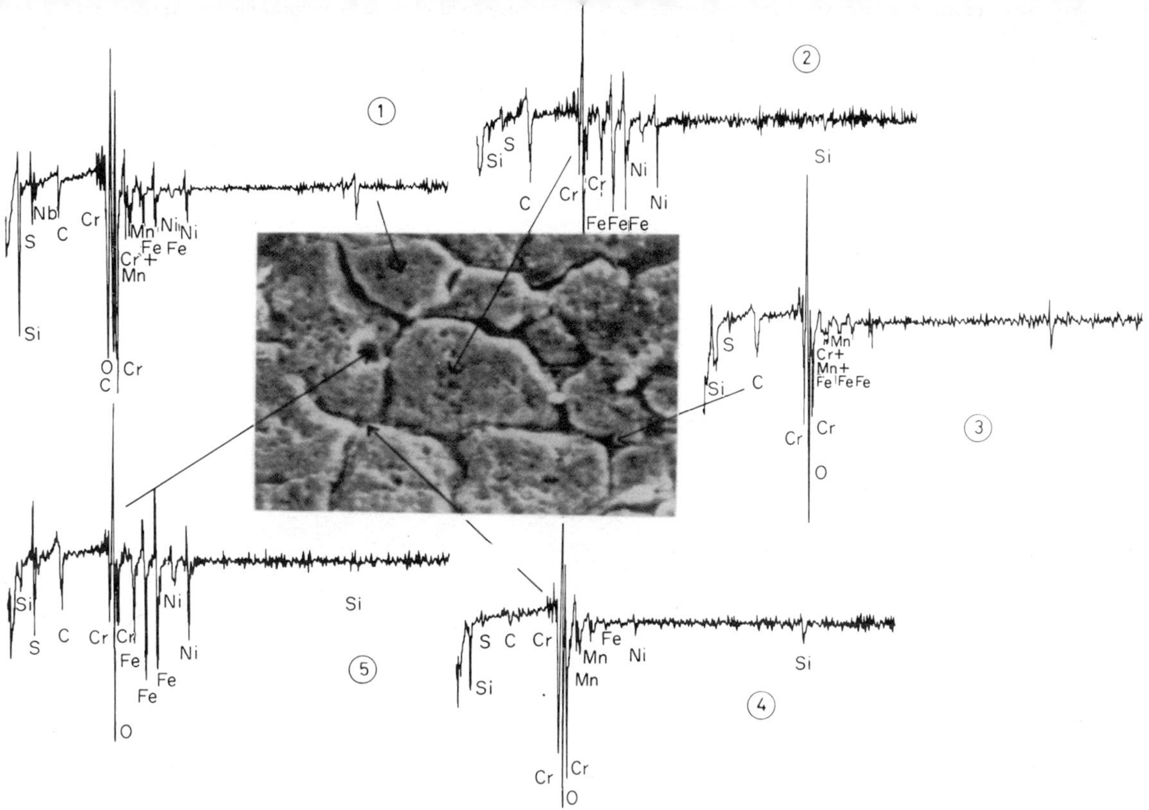

FIG. 4.20 Point analyses by AES at the various positions marked in the SEM picture of the metal side of a mechanically separated[28] metal/oxide interface. The oxide had been formed on the surface of an Fe–20Cr–25Ni–1Nb stainless steel by oxidation in $CO_2 + 1\%$ CO at 850°C for 72 h. The alloy grains are delineated clearly by grooves, and there are many pits and hollows within the grains. Note the occasional presence of large amounts of Si, e.g. at position ①. From Allen *et al.*[27]

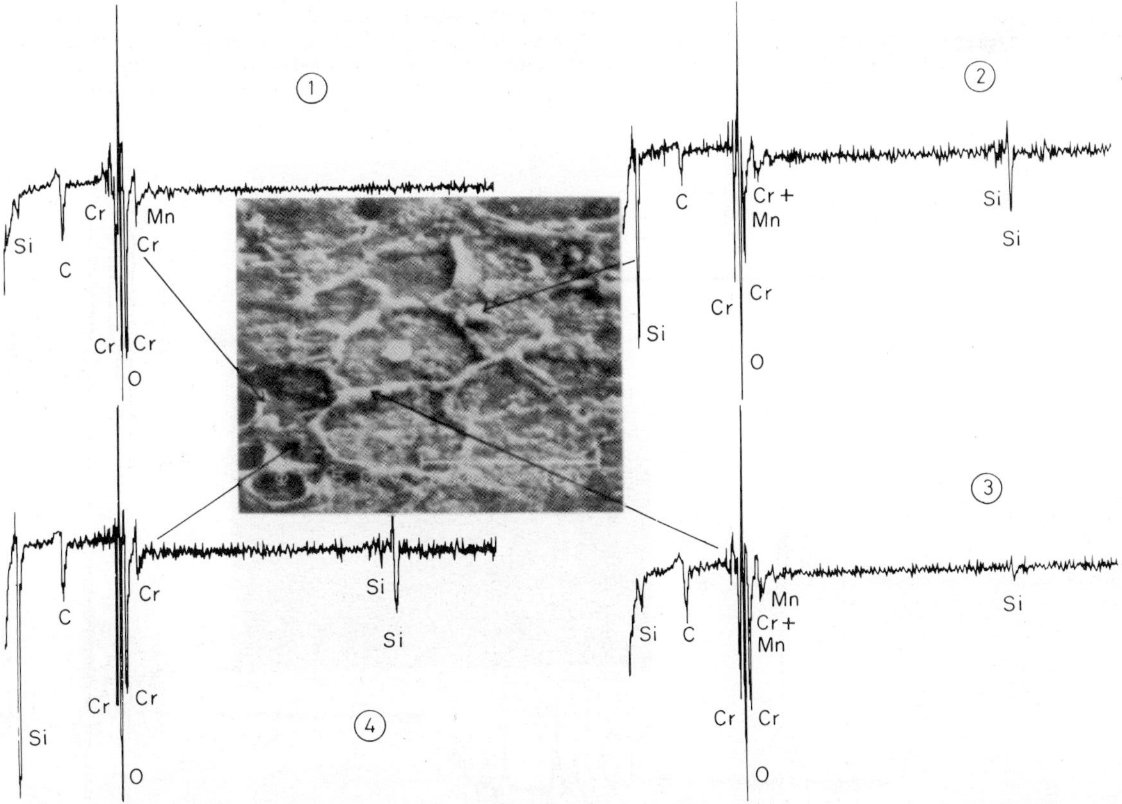

FIG. 4.21 Same type of analysis as in Fig. 4.20, but this time from the oxide scale side of the metal/oxide interface. Here the grains are delineated by ridges, which can be shown by comparison of corresponding SEM micrograph pairs to match exactly the grooves on the exposed metal side. There is an even higher concentration of Si on the oxide side, more than 30 at.% in places, particularly on the ridges and in the particles. From Allen *et al.*[27]

In order to identify more precisely any correlations between elements that might exist, a series of SAM images of the two sides was recorded for all the important elements, and the positive and negative correlations so found are displayed in Figs 4.22 and 4.23. On the metal side there is a close correlation

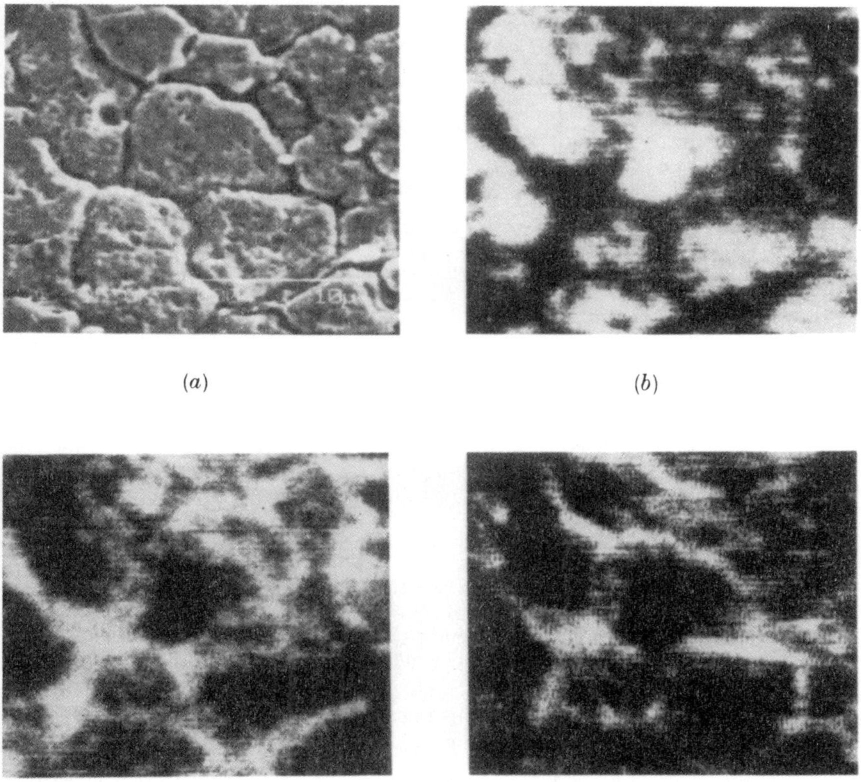

(*a*) (*b*) (*c*) (*d*)

FIG. 4.22 (a) SEM micrograph of the metal side of the metal/oxide interface (same area as in Fig. 4.20). (b), (c), and (d), SAM maps of the same area in Fe, Cr, and Si, respectively. The Cr and Si are clearly concentrated in the grain boundary grooves, in which there is almost no Fe. From Allen *et al.*[27]

between Cr and Si, which are both found mostly in the grain boundaries, whereas Fe is found almost entirely in the grains. On the scale side there is no such correlation; Cr again tended to be concentrated at the grain boundaries, this time on the ridges rather than in the grooves, while Si was concentrated in the centres of the grains.

From these observations, the authors were able to build up a model for the chemical structure just near the interface and into the scale. The alloy surface is

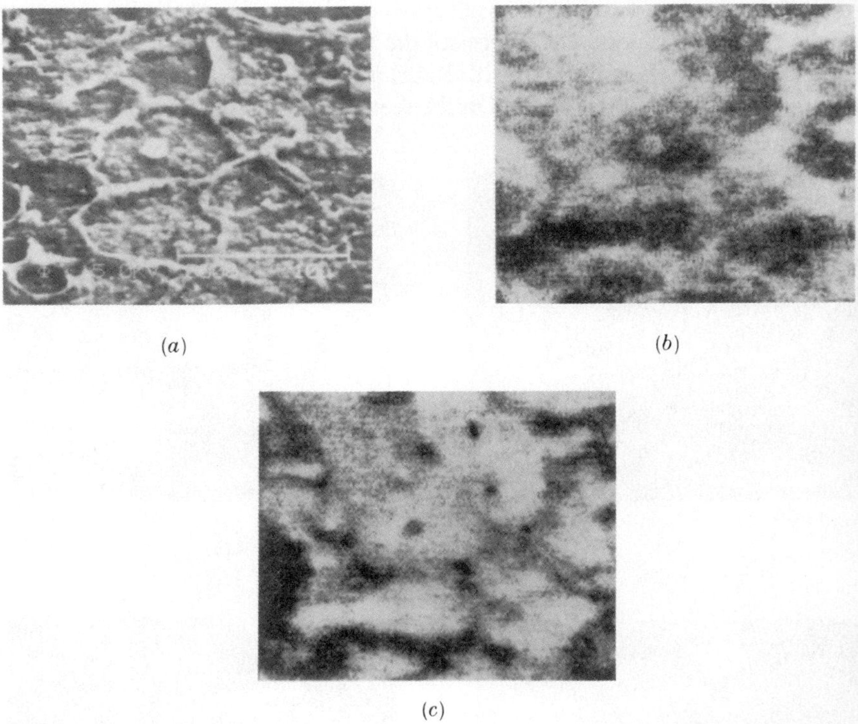

FIG. 4.23 (a) SEM micrograph of the oxide side of the metal/oxide interface (same area as in Fig. 4.21). (b) and (c), SAM maps in Cr and Si, respectively. Si is much more widely distributed than on the metal side and is concentrated more in the centres of the grains. Cr, on the other hand, is still associated more or less with the grain boundaries. From Allen *et al.*[27]

covered with a thin layer of silicaceous material, probably SiO_2, only 5–10 nm thick, and containing some Cr_2O_3. Between it and the alloy there may be an even thinner layer of a Fe–Cr–Si oxide. Above the silicaceous layer is a thick layer of Cr_2O_3 containing a little Fe and Si, then the outermost layer which is of $MnCr_2O_4$, separated from the Cr_2O_3 by a carbon-rich layer. The spalling can then be envisaged as occurring by fracture at the thin silicaceous layer, which is brittle and provides few keying points for the scale. At grain boundaries, however, the steep angle into the boundary provides a strong key for the oxide, and the fracture just above grain boundaries occurs in the Cr_2O_3 layer itself, leaving a plug behind in the boundary. The fact that the exposed alloy surface appears metallic and stays so is then explained by the presence of the thin protective silicaceous layer.

4.4.4 *Materials characterization*

AES and SAM have been used very frequently both on their own and to supplement other techniques in providing information about the nature of bonding, distribution of impurities between phases, origin of porosity, effects of radiation, etc., in solid materials. They have also to a limited extent been applied to liquid metals, where the vapour pressure at the melting point is not a limiting factor, to study surface segregation and changes in bonding in going from solid to liquid and vice versa. Recent studies of various aspects of ceramic and glassy materials have been particularly interesting, and a few examples are described here.

Metallic glasses, or metglasses for short, are a fairly recent development. They are generally formed by very rapid cooling of alloys of first-row transition elements such as Fe and Ni with non-metals such as P, Be, B, Si, etc., and the nature of the chemical bonding in them has attracted much attention. For instance, the metglass system $Fe_{82}B_{18-x}Be_x$ was investigated with AES by Bevolo *et al.*[29] The system is ferromagnetic but shows an anomalous peak in magnetic moment near $x=4$ in the composition range.

Compositions with $x=4$ and 5 respectively were prepared, and their B and Be Auger spectra recorded at high energy resolution, after the surfaces were cleaned by ion bombardment. The results are shown in Fig. 4.24, with a comparative spectrum for elemental B as well. The B KVV Auger spectra from the metglasses show little or no change between $x=4$ and 5, but do show additional fine structure compared to the elemental spectrum. The change in the Be KVV Auger spectrum between $x=4$ and $x=5$ is in contrast quite dramatic. There is an enormous apparent shift of about 14 eV to lower energy, and the peak width also reduces substantially. Clearly some drastic changes in the local bonding configuration of the Be atom occur on going from 4 to 5 at.% Be, which is just the composition at which the magnetic moment anomaly is observed.

The first interpretation of the change in the Be spectrum was that of a change in Auger transition character from quasi-atomic (see under Section 4.2.5) to band-like (the opposite of quasi-atomic) on going from $x=5$ to $x=4$. The glass-forming non-metals are all s–p elements and their CVV-type Auger spectra are expected to be band-like. However, the objections to this interpretation are that a similar line shape change from Be to BeO does not require such an explanation, and that a narrow valence band width is required for quasi-atomic character to appear, which these metglasses do not have. The more probable interpretation, one that can account for both the Be and B spectra, is based on the known presence of a weak peak in the valence band spectra of the FeB metglasses about 10 eV below the Fermi level, ascribed to an Fe–B bonding state. In a CVV transition involving two electrons from the valence band, the self-fold of a state at 10 eV would give rise to a feature 20 eV below the onset of the principal transition, and just such a weak feature is seen at $\sim$165 eV in the B spectrum, i.e. about 20 eV below the onset at $\sim$185 eV. Now the Be 2s–p states peak not at the Fermi level but about 3 eV below, so that the weak peak associated with Fe–B bonding is in fact 7 eV below the

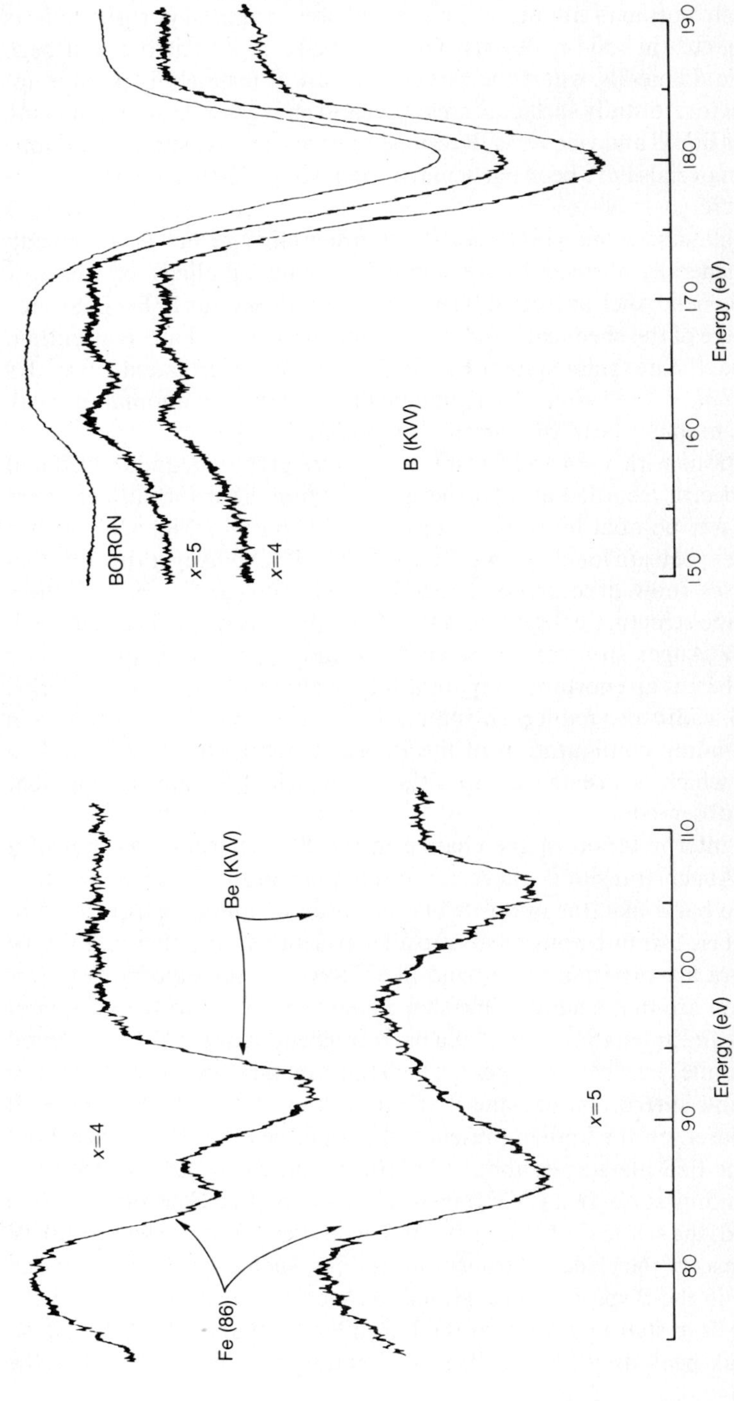

FIG. 4.24 Details of the energetic regions of the Be (left) and B (right) KVV Auger spectra from clean surfaces of the metglasses $Fe_{82}B_{14}Be_4$ and $Fe_{82}B_{13}Be_5$, with a spectrum from elemental B as well. Although the B spectra show little or no change on going from Be = 4 to Be = 5, the changes in the Be spectra are dramatic, in particular a shift of some 14 eV. The magnetic moment anomaly appears at the composition for which Be = 5. The changes are discussed in the text. From Bevolo *et al.*[29]

maximum in the Be density of states, and the self-fold of the peak would then give rise to a feature 14 eV below the principal Be Auger peak. This energy difference is exactly the shift observed when going from $x=5$ to $x=4$. Furthermore, since the principal Auger feature at 106 eV disappears entirely, the only explanation can be that the Be 2s–p states have been entirely depopulated, and the only Auger transition that can be sustained is based on the weak Fe–B bonding state! The implication is also that there must be considerable overlap between the Fe–B bonding state and the Be 1s core state. Thus AES is able in this case to provide a convincing picture of the changes in bonding taking place in the metglass, providing information about the reasons for the coincident changes in magnetic moment.

If a ceramic is not too insulating, it is normally possible to perform SAM without undue charging problems, and the technique has helped in situations where the distribution of elements must be known. Many ceramics for industrial purposes are made by sintering fine powders of the compound, but almost invariably the sintering process must be helped by the addition of small quantities of additives that seem to act as a sort of glue to hold the powder grains together. It is important to know not only how uniformly distributed the additive is around the grains but also whether any unwanted impurities have crept into the sintered compact during the sintering process. These were the questions to which Hamminger *et al.*[30] wanted answers when they examined the microstructure of sintered SiC with AES and SAM amongst several other microanalytical techniques. In the sintering of SiC the additives used as aids are either B and C or Al and C, and the sintering can then be carried out at high temperature but without applied pressure. The sintered compacts were fractured just before insertion in the spectrometer, to expose the surfaces of the original powder grains in the bulk of the compact, and lightly ion beam cleaned after insertion. Elemental distributions of some of the principal elements found on one of the surfaces are shown in Fig. 4.25. Several conclusions can be drawn at once. Clearly the boron is not distributed very uniformly in this particular compact, since the B map in (a) shows high concentration of the element in some areas, and none in others. Additionally there is evidence of K and Na, particularly the latter, which is also associated with the B. Point analyses at the individual features marked (A) and (B) showed also the presence of Si and S in localized areas. There seems little doubt on this evidence that, as the authors state, 'the manufacturing process has not been fully optimized'.

Another industrial use of ceramic materials in which it is important to ensure that elements are distributed correctly amongst the various phases present is that of incorporation of radioactive waste in glass-ceramics. Various glass-ceramic materials have been suggested for the purpose, and there is much surface analytical activity involved not only in checking the above distributions but also in monitoring surface reactions and leaching during

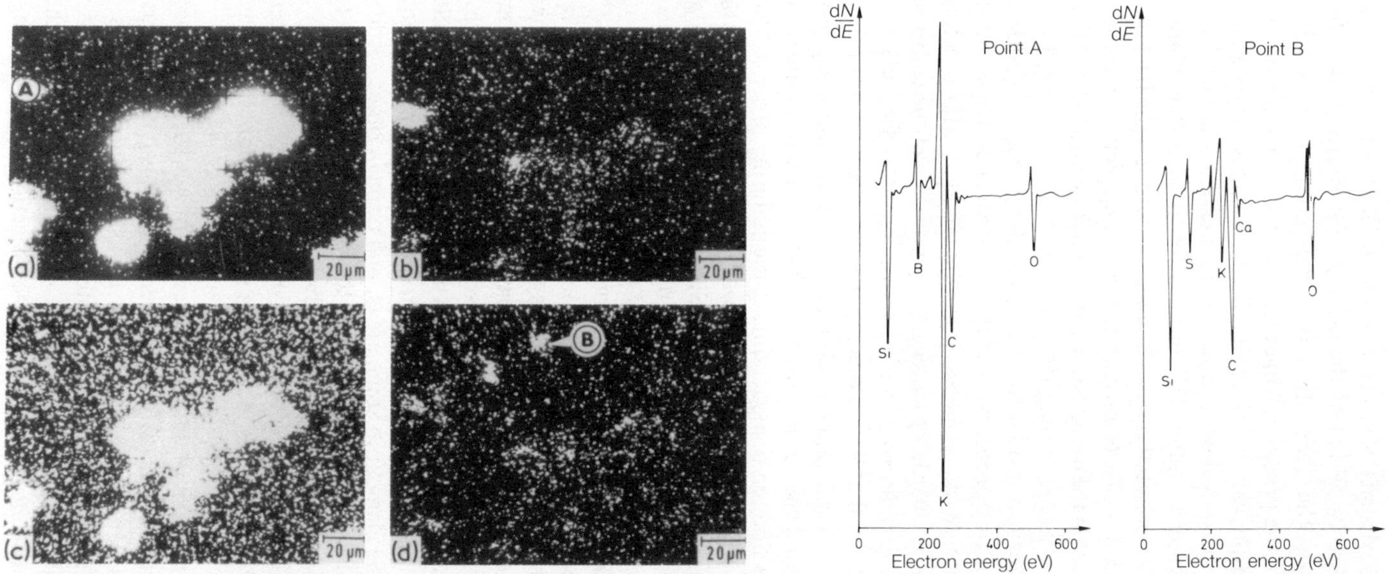

FIG. 4.25 Left-hand side. SAM maps from the fracture surface of SiC sintered with addition of B. (a) B, (b) K, (c) Na, and (d) O. For some areas B and Na concentrations correlate, for others B and K concentrations. Right-hand side. Analysis by AES at the points marked A and B in the boron and oxygen maps on the left. As well as the elements mapped in the SAM images, there are also significant levels of Si and S at different places on the surface. From Hamminger *et al.*[30]

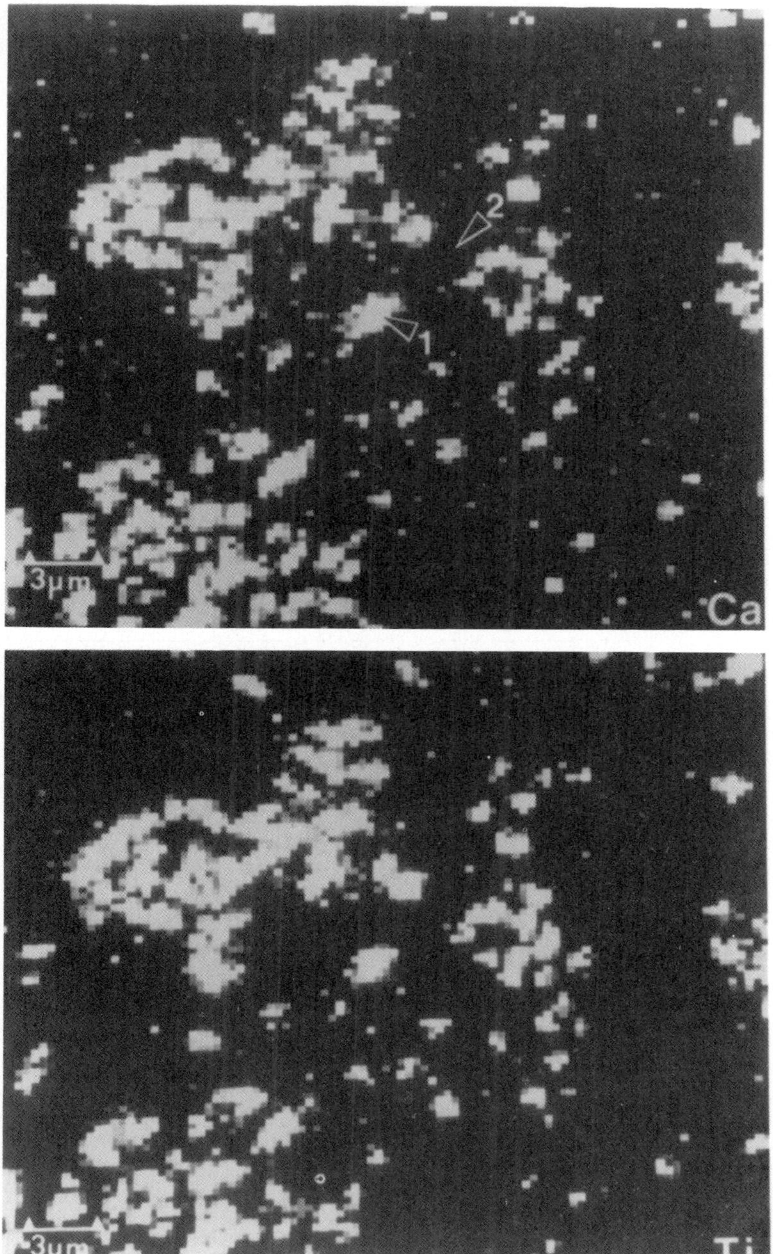

FIG. 4.26 SAM images in Ca and Ti recorded by Hocking *et al.*[31] from the fracture surface of a glass–ceramic made from 57% SiO_2–15% TiO_2–14% CaO–5% Al_2O_3–7% Na_2O–2% La_2O_3. There is a strong correlation between the two distributions, suggesting a sphene crystalline phase.

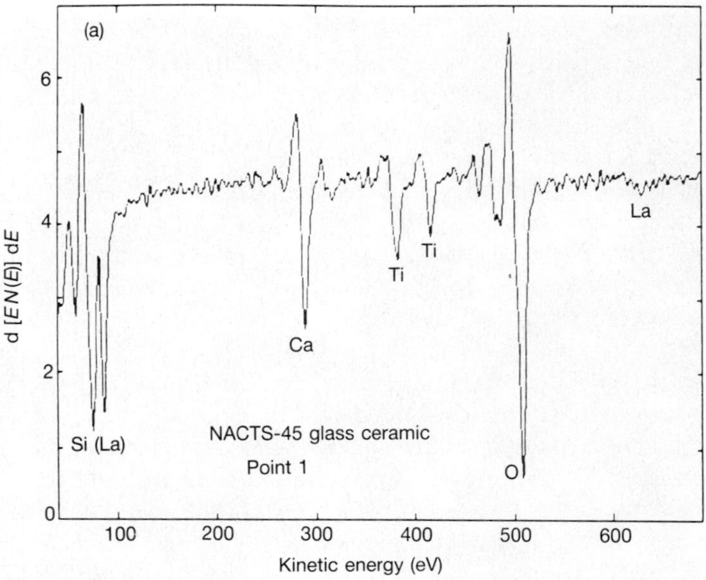

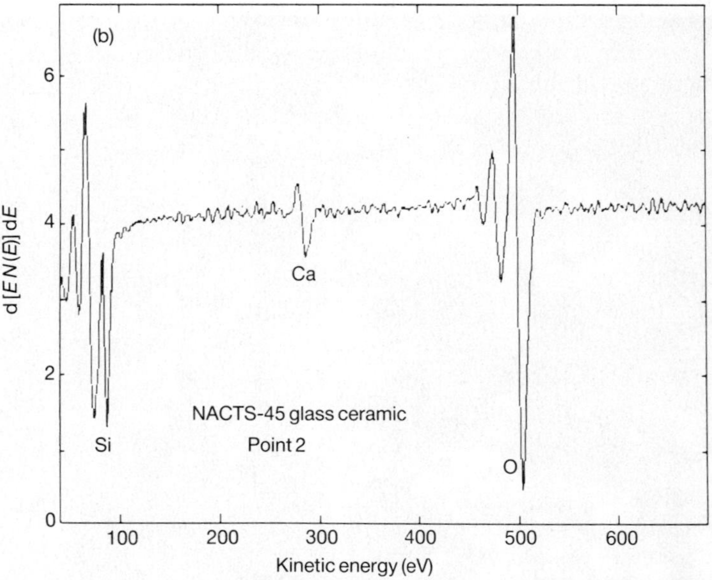

FIG. 4.27 Analyses by AES at the points marked 1 and 2 in Fig. 4.26, corresponding to a crystallite and to the matrix, respectively, made by Hocking *et al.*[31] It can be seen clearly that the matrix surface contains no Ti and only a little Ca, whereas the crystallite has significant amounts of Ti and Ca. Notice also the presence of small amounts of La in the crystallite, indicating preferential incorporation of La into the sphene phase.

simulated exposure to a natural environment such as would be encountered during storage. One of the candidate glass-ceramics is based on the mineral sphene, which is a crystalline calcium titanosilicate $CaTiSiO_5$; the proposal is to grow sphene crystallites in a sodium alumino-silicate glass matrix, the final product being two-phase. The theory of waste retention is based on the possibility of extensive substitution of calcium and titanium by many other types of cation, e.g. the radioactive decay products. The material was studied by Hocking *et al.*[31] to establish the way in which lanthanum, a decay product, was partitioned between the phases. For the simulation non-radioactive lanthanum was of course used. With such materials there are almost certain to be charging problems with incident electron beams as used in AES and SAM, and the authors had to take a variety of preventive measures to enable the analyses to be made at all. These included embedding a small chip of the glass-ceramic in a gold film sputtered onto adhesive conducting tape, and then using the incident electron beam at low angle of incidence to increase the secondary electron yield. Even so, Auger peaks were shifted by several eV.

The distribution of Ca and Ti in the surface of the material is shown in the Auger maps of Fig. 4.26, which provide good evidence for restriction of the distribution of those elements entirely to the sphene crystallites. The partitioning of La between the sphene and the matrix was investigated by performing point analyses in the two regions as shown by the arrows marked 1 and 2. The Auger spectra at these points are displayed in Fig. 4.27. The upper spectrum, from a sphene crystallite shows only the constituent elements making up the calcium titanosilicate, plus an indication of the presence of La. From the matrix there are again in the spectrum only those elements characteristic of the matrix, but this time no La. Line scans across a sphene particle, and across a region of matrix, with the analyser tuned to the La MNN transition energy, confirmed the presence of La in the sphene but not in the matrix. A similar result was found for another ceramic of the same composition but doped with P as well as La; both P and La were found to be restricted to the sphene phase. Thus AES and SAM were able to provide some preliminary confirmation of the mode of incorporation of potential radioactive waste products in this proposed storage medium.

References

1. Casnati, E., Tartari, A., and Baraldi, C. (1982). *J. Phys. B: At. Mol. Phys.* **15**, 155.
2. Palmberg, P. W. and Rhodin, T. N. (1968). *J. Appl. Phys.* **39**, 2425.
3. Palmberg, P. W., Bohn, G. K., and Tracy, J. C. (1969). *Appl. Phys. Lett.* **15**, 254.
4. Landolt, D. and Mathieu, H. J. (1980). *Oberfläche-Surface* **21**, 8.
5. Auger, P. (1925). *J. Phys. Radium* **6**, 205.
6. Davis, L. E., MacDonald, N. C., Palmberg, P. W., Riach, G. E., and Weber, R. E. (1976). *Handbook of Auger Electron Spectroscopy*. Physical Electronic Industries, Inc., Minnesota.

7. Chung, M. F. and Jenkins, L. H. (1970). *Surf. Sci.* **21**, 253.
8. Coghlan, W. A. and Clausing, R. E. (1973). *Atomic Data*, **5**, 317.
9. Larkins, F. P. (1977). *Atomic Data and Nuclear Tables* **20**, 311.
10. Aksela, H. and Aksela, S. (1974). *J. Phys. B: Atom. Mol. Phys.*, **7**, 1262.
11. Kny, E. (1980). *J. Vac. Sci. Technol.* **17**, 658; Kleefeld, J. and Levinson, L. L. (1979). *Thin Solid Films* **64**, 389; Smith, M. A. and Levinson, L. L. (1977). *Phys. Rev.* **B16**, 1365.
12. Dufour, G., Mariot, J. M., Nilsson-Jatko, P. E., and Karnatak, R. C. (1976). *Phys. Scripta* **13**, 370.
13. McDonnell, L., Powell, B. D., and Woodruff, D. P. (1973). *Surf. Sci.* **40**, 669.
14. Rivière, J. C. (1985). *Surf. Sci.* **161**, 521.
15. Gryzinski, M. (1965). *Phys. Rev.* **A138**, 305.
16. Reuter, W. (1972). *Proc. 6th Int. Conf. X-ray Optics and Microanalysis*, University of Tokyo, (eds. G. Shinoda, K. Kohra, and T. Ichinokawa), p. 21.
17. Ichimura, S., Shimizu, R., and Langeron, J. P. (1983). *Surf. Sci.* **124**, L49.
18. Seah, M. P. and Dench, W. A. (1979). *Surf. Interface Anal.* **1**, 2.
19. Minni, E. (1983). *Applic. Surf. Sci.* **15**, 270.
20. McGuire, G. E. (1979). *Auger Electron Spectroscopy Reference Manual.* Plenum Press, New York.
21. Sekine, T., Nagasawa, Y., Kudoh, M., Sakai, Y., Parkes, A. S., Geller, J. D., Mogami, A., and Hirota, H. (1982). *Handbook of Auger Electron Spectroscopy.* JEOL Inc., Tokyo.
22. Mroczkowski, S. and Lichtman, D. (1983). *Surf. Sci.* **127**, 119.
23. Wang, X., Reyes-Mena, A., and Lichtman, D. (1982). *J. Electrochem. Soc.*, **129**, 851.
24. Bermudez, V. (1983). *Applic. Surf. Sci.* **17**, 12.
25. Stout, D. A. (1983). *Applic. Surf. Sci.* **15**, 166.
26. Ruedl, E., Coen, V., Sasaki, T., and Kolbe, H. (1982). *J. Nucl. Mater.* **110**, 28.
27. Allen, G. C., Wild, R. K., and Weiss, M. (1983). *Phil. Mag.* **48**, 373.
28. Coad, J. P., Tappin, G., and Rivière, J. C. (1982). *Surf. Sci.* **117**, 629.
29. Bevolo, A. J., Severin, C. S., and Chen, C. W. (1982). *J. Vac. Sci. Technol.* **20**, 852.
30. Hamminger, R., Grathwohl, G., and Thümmler, F. (1983). *J. Mater. Sci.* **18**, 353.
31. Hocking, W. H., Hayward, P. J., Watson, D. G., and Allen, G. C. (1984). *Phil. Mag.* **49**, 637.

5

Electron excitation: electron energy loss spectroscopy (ELS), core electron energy loss spectroscopy (CEELS), and high resolution electron energy loss spectroscopy (HREELS)

5.1 Operation

In none of these loss spectroscopies is spatial resolution required, but they do require electron sources capable of variable primary energy down to very low energies, with good stability. For ELS and CEELS, energy spreads of ~0.5 eV, i.e. thermal, are acceptable, but as described already under Section 3.2.1, HREELS requires specifically designed sources in order to achieve as narrow a spread as possible, the best current value being ~3 meV. Thus for ELS and CEELS, electron sources of the type shown in Fig. 3.4 would be adequate, but for HREELS a very accurately monochromatized source of the type already described in detail under Section 3.2.1 is necessary. The need in HREELS for careful screening of the whole analytical system of source and analyser from strong magnetic fields has already been emphasized, as has the practice of minimizing currents of unwanted secondary electrons by coating internal surfaces with a low secondary emitter such as carbon.

Secondary electron energy analysis in ELS and CEELS is performed in either of the analysers CMA or CHA (under Section 3.3.4) used routinely for AES and XPS. Spectra are recorded in either the direct, i.e. $E\,N(E)$, or differentiated, $E\,\mathrm{d}N(E)/\mathrm{d}E$, modes, rather according to the whim of the experimenter, although in general it is more usual to find ELS spectra in the direct mode and CEELS spectra in the differentiated mode. The kinetic energy regions covered in spectral recording in the two techniques are somewhat different because of the differing ways in which the techniques are applied. In ELS the information sought is very often of a comparative nature, loss structure within 50 eV of the elastic peak energy being compared between surface and bulk by varying the primary energy from 1500 eV down to ~100 eV, so that the kinetic energy range of interest would be 1450–1510 eV at the higher primary energy and 50–100 eV at the lower. In CEELS the primary energy must be high enough to be able to ionize the core levels of interest with enough efficiency, and in addition the ionization thresholds will often be several hundred eV from the elastic peak. Thus primary energies of 2–4 keV are typical but the kinetic energy range recorded would not usually be continuous up to the elastic peak since it would be only the structure in the region of the threshold itself that would be of interest.

The type of analyser used for HREELS has also been described in detail under Section 3.2.1. The very low kinetic energies of the scattered electrons, and the need for good resolution with low background requires a design of analyser dedicated to this type of loss spectroscopy. Spectra are always recorded in the $E\,N(E)$ mode, with primary electron energies in the range 2–5 eV.

5.2 Spectral features and interpretation

5.2.1 *ELS*

When ELS is performed with fast (> 10 keV) electrons by passage through a thin foil, both the elastically and inelastically scattered electrons are observed in transmission, and electronic excitations stimulated within the solid can be studied by analysis of momentum and energy of the transmitted electrons. When using much lower primary energies, however, the secondary electrons are observed in a reflection mode, which implies at once there must be two scattering events, an inelastic and an elastic, in either sequence, because the inelastic event is always strongly peaked in the forward direction. If there were not a second event, the secondary electrons could not reach the detector.

According to the first Born approximation, the intensities of losses due to bulk excitation can be described by the dielectric loss function

$$\mathrm{Im}\,\epsilon^{-1} = \epsilon_2/(\epsilon_1^2 + \epsilon_2^2) \tag{5.1}$$

where ϵ is the complex dielectric constant

$$\epsilon(\omega, q) = \epsilon_1(\omega, q) + \mathrm{i}\epsilon_2(\omega, q). \tag{5.2}$$

At the surface the effect of the sudden termination of the solid results in a corresponding loss function $\mathrm{Im}\,(\epsilon + 1)^{-1}$.

Now in specular reflection there is no momentum transfer perpendicular to the direction of the elastic beam (the so-called optical limit with $q_\perp = 0$) and the loss function can be made up from $\epsilon_1(\omega, 0)$ and $\epsilon_2(\omega, 0)$ measured optically. Features in ϵ_1 and ϵ_2 can be identified with losses due to collective plasmon excitations, while peaks in ϵ_2 can also be associated with losses due to vertical interband transitions. Thus ϵ_2 basically maps the joint density of states for such transitions according to

$$\epsilon_2(\omega, 0) = K\omega^{-2}|M_{nn'}|^2 J_{nn'}(\omega) \tag{5.3}$$

where $M_{nn'}$ is the dipole matrix element of the transition between n and n' states, and $J_{nn'}(\omega)$ is the JDOS of the initial n and final n' states at the excitation frequency ω.

In non-specular reflection ($q_\perp \neq 0$), electron excitation allows non-vertical transitions and some optically forbidden transitions to occur. By carrying out angle-resolved measurements at varying energies, i.e. by varying q and ω, it is therefore possible to map the dispersion of plasmon excitations, and to obtain information about band shapes for comparison with theory. Most experiments, however, have been performed in the specular reflection angle-integrated mode.

The correlation between structure in the dielectric loss function calculated from optical data, and the measured electron energy loss spectrum in specular reflection, for CaO, is shown in Fig. 5.1, from Protheroe *et al.*[1] Agreement

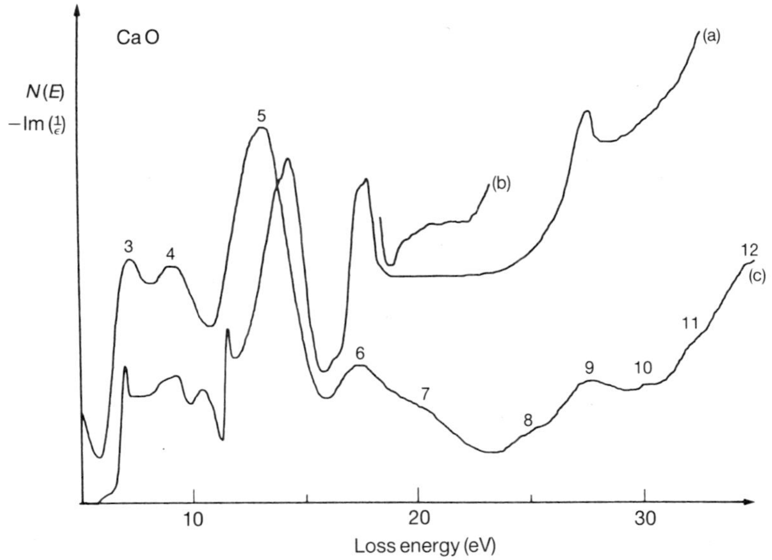

FIG. 5.1. Correlation of the structure in the dielectric loss function $-\text{Im}(I/\epsilon)$ calculated from optical data, curve (a), and the measured electron energy loss spectrum at a primary energy of 200 eV, curve (c), for CaO, from Protheroe *et al.*[1] Curve (b) is a portion of the calculated spectrum using higher digitizer resolution. See text for a discussion of the similarities and differences between calculation and experiment.

between the positions of the principal features is quite good, except that peak 5 is not particularly well reproduced. The band gap in CaO is at ~8 eV, i.e. at the onset of peak 3 in the figure. With a primary energy for the energy loss measurement of 200 eV it is to be expected that the surface plasmon loss would be prominent, and peak 5 is in fact identified as this surface loss, with peak 6 as the bulk plasmon loss. The ratio of their energies is approximately $2^{1/2}$, as expected if $\epsilon = 1$ for a surface interfacing with a vacuum. Strictly speaking the $2^{1/2}$ relation should apply only to free electron metals, but it seems to apply in other situations as well. Peak 9 is then the second surface plasmon loss and peak 12 (not shown in full in the figure) is the second bulk plasmon loss. Peaks 3 and 4 were found not to be affected by electron bombardment or by oxygen exposure, but to be enhanced at glancing angle detection, and are thus intrinsic and surface specific. They are believed to be due to surface excitons characteristic of the oxide structure.

A similar correlation between optical data and the energy loss spectrum is shown for gadolinium[2] in Fig. 5.2; the primary excitation energy was 250 eV.

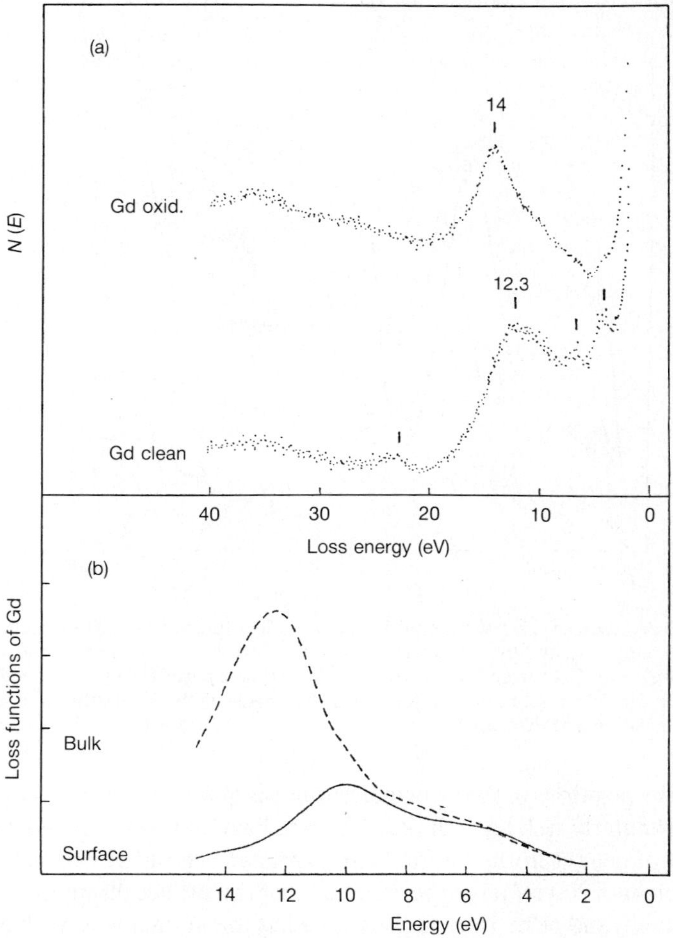

FIG. 5.2. Comparison for gadolinium of the electron loss spectrum (a), measured by Netzer *et al.*,[2] with the bulk and surface loss functions (b) from optical data. The surface plasmon loss is present in the measured spectrum but only as a tail on the bulk loss. Note the shift in the bulk plasmon loss from 12.3 eV to 14.0 eV due to increased electron density in the valence band of the oxide compared with that in the metal.

In this case it is the bulk plasmon loss in clean Gd that is the prominent feature, and its observed energy in the loss spectrum is in exact agreement with the calculated peak from the optical data. The surface plasmon loss is present but as an enhanced tail to the bulk loss. The minor features marked in the energy loss spectrum have their counterparts also in the ϵ_2 spectrum, probably arising from interband transitions involving 5d and 6s conduction electrons. They are quenched rapidly on oxidation, which is diagnostic. Oxidation also causes the bulk plasmon to shift from 12.3 eV to 14.0 eV due to increased electron density in the valence band of Gd_2O_3 compared to that in the metal.

The ELS spectrum will thus contain peaks due to bulk plasmon losses, to surface plasmon losses if the primary energy is low enough (usually <200 eV), and to intraband and interband transitions. The intensity of the latter may well also depend on the primary energy, as seen in Fig. 5.3 for Al,[3] where at 140 eV primary energy both bulk (or volume) and surface plasmon losses are strongly stimulated on the {100} surface, but at 30 eV the strongest loss feature is an interband transition at about 6 eV. Note also from this figure that exposure of Al to enormous doses of CO at room temperature has virtually no effect on the loss spectrum, indicating no interaction of that gas with the metal at room temperature. It is possible from free electron theory to predict[4] the energies of bulk plasmons from the expression

$$E_{bp} = 2eh(\pi n/m)^{1/2} \tag{5.4}$$

where n is the electronic density in the conductance band and m is the electron mass. The surface plasmon energy is then

$$E_{sp} = E_{bp}/(\epsilon + 1)^{1/2} \tag{5.5}$$

that is $E_{sp} = E_{bp}/\sqrt{2}$ for a free electron metal bounded by vacuum.

It is also possible to find in an ELS spectrum features due to ionization of core levels if they are very shallow (i.e. of binding energy $< \sim 50$ eV) although their study is of course the realm of the technique of CEELS.

5.2.2 *CEELS*

When a primary electron of sufficient energy interacts with an electron in a core level, the resultant excitation can be either to an unoccupied continuum state, in which case complete ionization occurs, or to localized or partly localized final states if available in the atom. Since the primary electron need not give up all its energy in the interaction (unlike a photon) and since after interaction in the solid the erstwhile primary and the secondary become indistinguishable, the loss feature observed in the spectrum should appear as a step at the ionization threshold, with a tail of secondaries at all energies from the threshold down to zero. However, because of the much greater cross-section near threshold than further away, loss features appear more usually as peaks, followed by a tail.

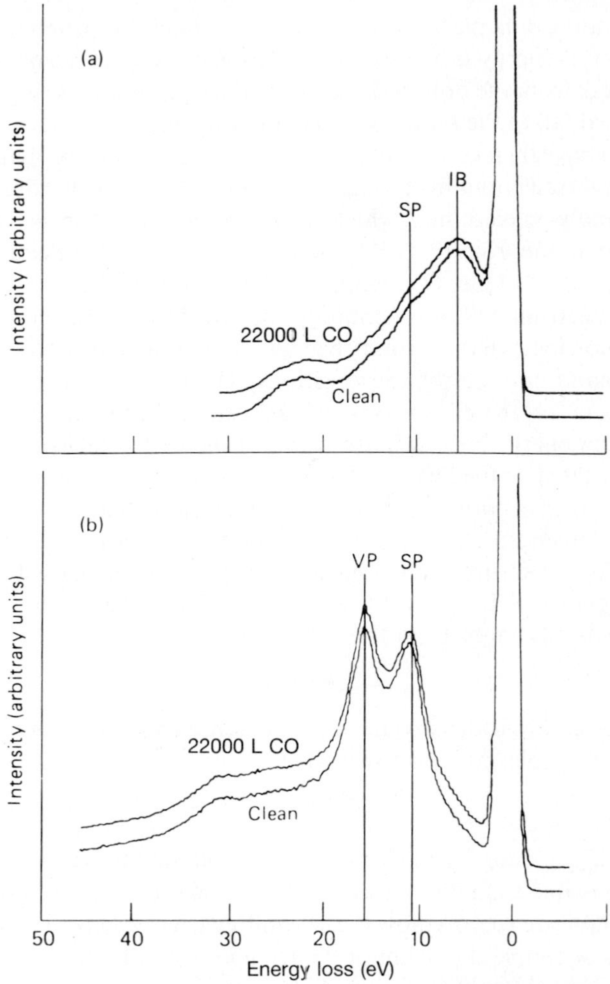

FIG. 5.3. The electron loss spectra from Al(100) at primary energies in (a) of 30 eV, and in (b) of 140 eV, measured by Bargeron and Nall.[3] Lower curves for clean Al, upper curves after exposure to 22,000 L of CO. At 140 eV primary energy both bulk and surface plasmon losses are very prominent, but at 30 eV only the surface loss remains barely visible, the strongest feature at that energy being an interband transition at about 6 eV.

An example of a core-level ionization loss feature is shown[5] in Fig. 5.4, for the Ca 2p levels in $Ca_2V_2O_7$ at a primary energy of 1000 eV. The feature in spectrum I looks very much like a broad peak in this case, but close examination of the background curvature reveals that it is in fact a step. Differentiation or double-differentiation, i.e. spectra I' and I'', respectively, shows up more clearly the doublet nature of the loss feature, corresponding to the $2p_{3/2}$ and $2p_{1/2}$ ionizations. In this case there is no fine structure associated

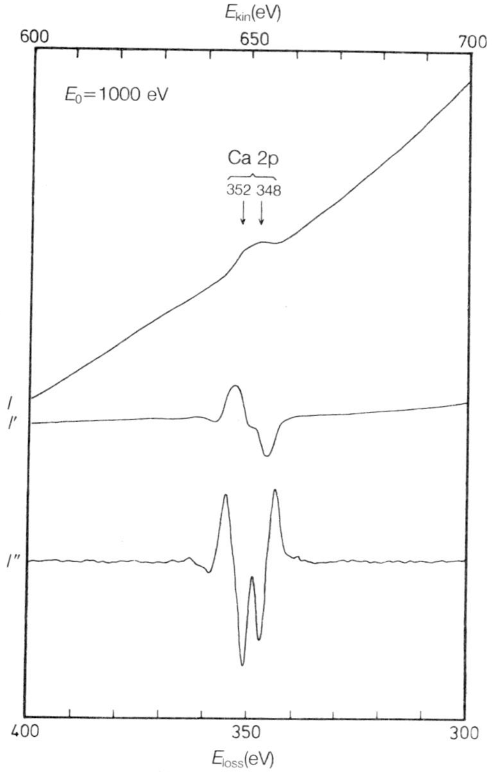

FIG. 5.4. Example of features in the energy spectrum due to losses as a result of ionization of core levels, from Curelaru *et al.*[5] The spectra are shown in the undifferentiated (upper), singly differentiated (middle), and doubly differentiated (lower) forms, and show the losses due to ionization of the Ca $2p_{3/2}$ and $2p_{1/2}$ core levels in $Ca_2V_2O_7$, at a primary energy of 1000 eV. Double differentiation is normally used in CEELS to enhance the resolution of the losses.

with the threshold, there being no localized states available for the excited electron.

Where localized final states do exist, the range of fine structure associated with the ionization threshold can be extensive. The prime examples of materials in which such fine structure is readily observable are the lanthanide rare earths. Their loss spectra have been studied exhaustively by Netzer and co-workers, and an example from one of their papers[6] is given in Fig. 5.5. There the 4d core-level excitation loss spectra are recorded at high and at low primary energies for the four lightest lanthanides. The rich additional structure arises from the coupling of 4d holes left after excitation with 4f electrons or holes, the 4f electrons in the lanthanides being strongly localized even though in some cases very close in energy to the valence bands. The vertical lines are theoretical calculations of the final state energies. Study of the

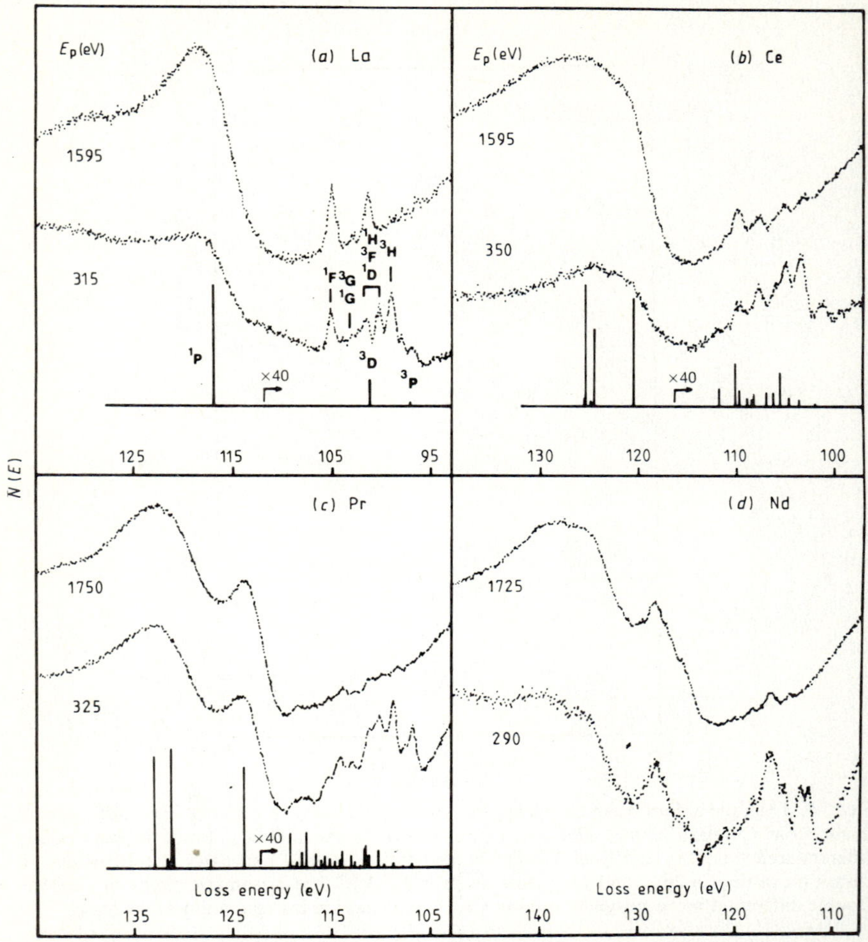

FIG. 5.5. The core-level ionization losses for the lanthanides La, Ce, Pr, and Nd measured at high and low primary energies, by Strasser *et al.*[6] Above the main excitation threshold there is a great deal of additional fine structure due to the coupling of the 4d holes left after excitation, with 4f electrons or holes, since the latter are strongly localized in the lanthanides.

variation of fine structure with primary energy and with surface reactions has been most fruitful in elucidating electronic interactions in these and other metals.

5.2.4 *HREELS*

When an electron of low energy (2–5 eV) approaches a surface, it can be scattered inelastically by two basic mechanisms. Within a certain distance of the surface it can interact with the dipole field associated with a particular surface vibration, e.g. either the vibrations of the surface atoms of the substrate itself, or one or other of the characteristic vibrations of an adsorbed molecule. Alternatively the electron can be scattered inelastically from an atomic core at the surface, again either of a substrate or of an adsorbate atom. The former is called *dipole* scattering and the latter *impact* scattering.

Virtually all HREELS observations have been concerned with the first mechanism, that of dipole scattering. The reasons for this are concerned with the dipole selection rules. The scattering process is regarded as taking place by either of two paths; an inelastic reflection followed by interaction with the vibrating dipoles through the long-range Coulombic field, or vice versa. Since in conducting materials an electric field is generated normal to the surface only those vibrations that can induce a dipole moment change normal to the surface are excited. The consequence is that the current of inelastically dipole scattered electrons is peaked very close to the specular direction. As will be seen below, it is possible to distinguish between the two scattering mechanisms by recording the intensity of a particular loss peak as a function of angle around the specularly reflected direction.

There is also the so-called 'normal dipole selection rule'. This is illustrated qualitatively in Fig. 5.6, from Lüth.[7] Any dipole at a surface induces an image

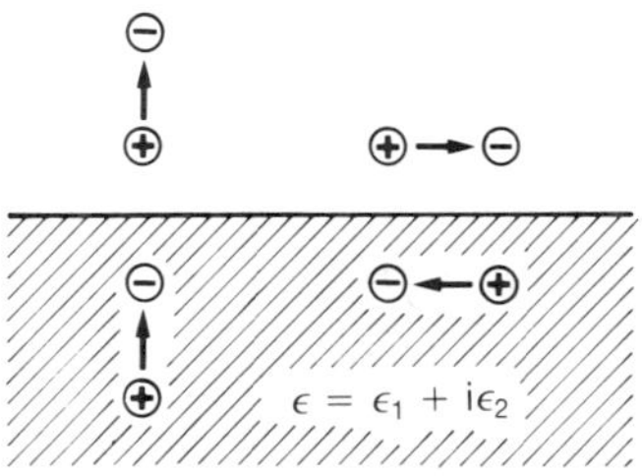

FIG. 5.6. Schematic illustration by Lüth[7] of the operation of the normal dipole selection rule in HREELS. Image dipoles within the surface either enhance (for normal orientation on the left) or negate (for parallel orientation on the right) the effect of dipoles in the vacuum above the surface. There is therefore a selection rule that favours normally oriented dipoles.

charge within the surface; if the dipole orientation is normal to the surface the effect is enhanced by the image dipole, if the orientation is parallel to the surface the effect is annihilated by the image dipole. Thus an orientation quasi-selection rule exists which strongly favours normally oriented dipoles.

The intensity of a loss peak in dipole scattering can be calculated as a proportion of the primary electron current if certain approximations are made. These are, that the energy loss ΔE must be very small compared to the primary energy E_p, that electrons are accepted only in a very narrow angular cone around the specular direction, and that the electron reflectivities before and after a loss are similar. Then the fractional intensity of the loss peak is

$$I(\Delta E)/I(E_p) = C(E_p, \Delta E, \alpha) x |\mu|^2 x n \tag{5.6}$$

where C is a function of kinematic quantities only, i.e. the primary energy E_p, the loss energy ΔE and the half angle of analyser acceptance α, x is the distance normal to the surface, μ is the dipole matrix element, and n is the number of dipoles per unit area. Using this approximation it is possible to calculate the fractional loss intensity for a dipole of charge q and reduced mass M vibrating normal to the surface from the further approximation

$$I(\Delta E)/I(E_p) = C(E_p, \Delta E, \alpha)x(q^2xn/M). \tag{5.7}$$

Two such calculated fractional loss intensities as functions of the primary energy are shown in Fig. 5.7 from Backx *et al.*,[8] for losses of 120 and 375 meV.

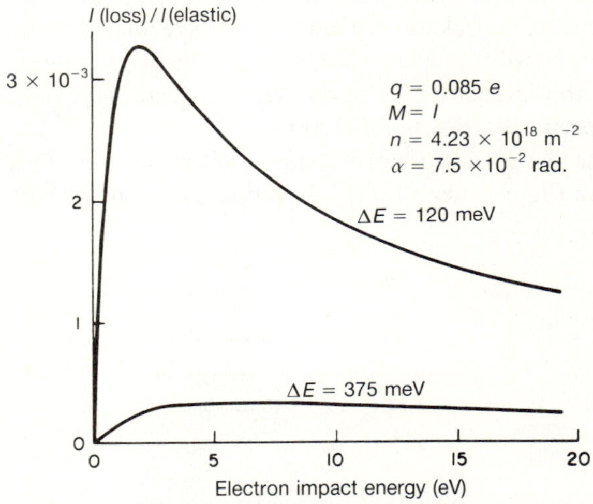

FIG. 5.7. Intensities of loss peaks at 120 meV and 375 meV calculated by Backx *et al.*[8] as a function of impact (primary) energy using the dipole approximation. q and M are the charge and reduced mass of the dipole, α the half angle of analyser acceptance, and n the number of dipoles per unit area. Optimum intensity is obtained when the primary energy is in the range 2–6 eV.

It can be seen that optimum loss intensity occurs when the primary energy is in the range 2–6 eV, which provides a practical reason why in HREELS it is necessary to use very low energy primary electrons.

The discussion so far has been concerned only with the mechanisms of interaction of an incoming electron with a hypothetical vibrating species attached to a surface. It is necessary to consider now what actual modes of vibration are to be expected from a molecule adsorbed on a surface. Taking the simplest case first, that of a homopolar diatomic molecule such as H_2 or N_2, there are in the gas phase, i.e. for the free molecule, 6 degrees of motional freedom—3 *translational*, in each of three mutually orthogonal directions, 2 *rotational*, about two mutually perpendicular axes normal to the bond axis, and 1 *stretching*, along the bond axis. In the gas phase only the last of these, the stretching mode, is detectable, typically by infra-red spectroscopy. Obviously, for larger molecules containing more than two atoms, the stretching mode can be subdivided into many possible sub-modes, and some of these will be mentioned later.

When a molecule is attached to a surface, it can no longer either translate or rotate freely, but is fettered by forces due to the chemical bonds to the surface, which can be visualized as a set of elastic but not easily broken strings. At the surface, then, there is the possibility of observing certain translational or rotational vibrational modes, in a so-called 'hindered' form, not observable in the free molecule, as well as coupling modes between them. For similar reasons there will be stretching modes *additional* to those seen in the free molecule, due for example to interaction between atoms in the molecule and atoms in the substrate, as well as 'hindered stretching' modes in the molecule itself. It will quickly be appreciated that where a complex polyatomic molecule is adsorbed on a surface at several adsorption sites, the resultant spectrum can be very complex.

In spectral identification, then, the first step is a comparison of the observed losses with vibrational frequencies measured by IRS in the gas phase to see if any correlations emerge. For example, in a small molecule such as formaldehyde, the energies corresponding to characteristic frequencies measured by IRS are shown in Table 5.1.

Table 5.1

Energies associated with characteristic stretching frequencies in formaldehyde, meV, as measured by infra-red spectroscopy.

Mode	H_2CO	D_2CO
CH_2, symmetric stretch	345	255
CH_2, asymmetric stretch	352	266
CH_2, scissors	186	137
CH_2, rock	155	123
CH_2, wag	144	122
CO, stretch	216	211

Comparison of normal and deuterated spectra is very common in vibrational spectroscopies of all types, as an aid to the positive identification of modes of vibration. Notice how complete deuteration in formaldehyde has such a drastic effect on those modes involving the C–H bond, i.e. the symmetric and asymmetric stretches, due to the doubling of the mass of the hydrogen. For a pure harmonic oscillator a doubling in mass should lead to a reduction in the characteristic vibrational frequency by a factor $\sqrt{2}$; the factors in Table 5.1 are 1.35 and 1.32 for the symmetric and asymmetric stretches, respectively. Substitution of deuterium for hydrogen has of course very little effect on the CO stretching characteristic energy.

In Table 5.1 the various types of vibrational mode possible in the free formaldehyde molecule are given simple self-descriptive names. However, as remarked earlier, the more complex the molecule the more are the possibilities for different forms of vibration, both individual and collective, and a nomenclature has evolved for the short-hand description of the various possibilities. It is not possible or appropriate here to give a comprehensive list of all the terms that are now used in the vibrational spectroscopies, and the reader is referred to the copious literature if he wishes to know about them all, but Table 5.2 from reference 9 gives some examples of a few of the more commonly appearing terms. Conventionally each type of mode is named with a Greek letter, ν for the stretching modes, and various alliterative ones for the bending modes such as δ for deformation and ρ for rocking. Sub-divisions into symmetric and asymmetric, degenerate, etc., are indicated by subscripts although here there appears to be no general agreement about nomenclature since some authors use letters as in Table 5.2, and others, numbers.

When a simple molecule is adsorbed on a surface it is possible to find in the loss spectrum only those peaks that can be identified immediately with gas-phase vibrational energies. The case of ethylene adsorbed on a silver (110) surface is a good example, studied by Backx *et al.*[8] The loss spectra in specular reflection for ethylene adsorbed on a clean silver surface at $-160°C$, and on an oxygen-covered surface, respectively, are shown in Fig. 5.8 and 5.9. In the former there is only a single peak at 120 meV (90 meV for C_2D_4) but in the latter additional peaks at 164, 194, and 375 meV (124, 182 and 279 meV for C_2D_4). (The peak at 39 meV is due to atomic oxygen.) Table 5.3 compares these values with those observed in the gas phase. Clearly on clean silver the only vibration detectable is the out-of-plane vibration named ν_7, and there must be virtually no hybridization of the molecular orbitals with any surface orbitals since the characteristic energies are almost identical. When oxygen is pre-adsorbed on the silver, very much more ethylene (by a factor of 13) can subsequently be adsorbed, and other molecular vibrations such as the C–H symmetric stretch ν_1, C–C stretch ν_2 and the CH_2 scissors stretch ν_3, become observable. Their energies are still very similar to those of the free molecule, thus the adsorption at $-160°C$ does not involve much electronic interaction with the substrate. Note from Table 5.3 that the symmetry group of a molecule invariably changes when it is adsorbed on a surface.

The case of ethylene adsorbed on oxygen-covered silver also illustrates one of the features of vibrational loss spectroscopy mentioned at the beginning of the section, viz, the possibility of two types of scattering, 'dipole' and 'impact'. It will be recalled that dipole scattering will be strongly peaked in the specular direction by virtue of the selection rule, whereas impact scattering should not

Table 5.2
Nomenclature of vibrational modes.

Local point group	Example	Stretching modes		Bending modes	
C_s	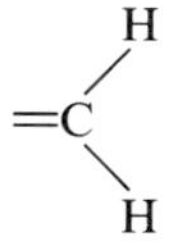		ν stretch	δ in-plane bend	π out-of-plane bend
C_{2v}	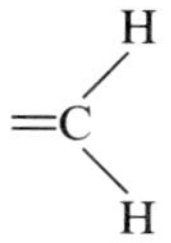	ν_s symm. stretch	ν_a asymm. stretch	δ_s wagging	ρ_r rocking
				ρ_w wagging	ρ_t twisting
C_{3v}	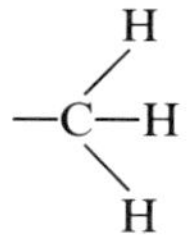	ν_s symm. stretch	ν_d degenerate stretch	ρ_s symm. deform.	ρ_d degenerate rocking
				δ_d degenerate deform	ρ_t twisting

have a strong angular dependence. Now the additional loss peaks seen in Fig. 5.9 and listed in Table 5.3 have the property that their symmetry does not lead to a net change in dipole moment, so that their appearance in the spectrum at all must be explained. If, for C_2H_2 adsorption, the intensities of the single peak at 120 meV found on the clean surface, and of one of the additional peaks, at 375 meV, found on the oxygen-covered surface, are plotted as a function of angle of emergence about the specular direction, the result in Fig. 5.10 is found. The 120 meV peak has maximum intensity in the specular direction, and its intensity drops rapidly on either side of that direction, while the intensity of the 375 meV peaks show hardly any variation with angle. Clearly the 120 meV originates in a dipole scattering mechanism, and the 375 meV in impact scattering. Similar measurements made on the 164 meV and 194 meV loss peaks show that they too originate in impact scattering.

The combination of the use of deuterated molecules and of angular dependence measurements is usually sufficient to allow allocation of vibrational modes to loss peaks in the case of simple molecules. For larger molecules it is often necessary to use other information as well, such as the

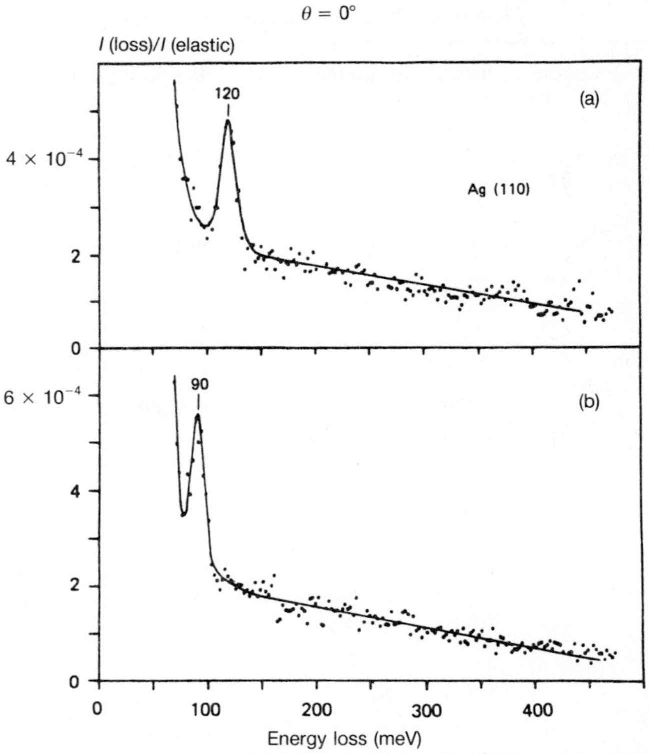

FIG. 5.8. Loss spectra for a primary energy of 3.0 eV for 30 L ethylene in the (a) normal and (b) deuterated forms adsorbed on clean Ag(110) at −160°C, recorded by Backx *et al.*[8] In both spectra only single loss peaks appear, arising from the ethylene out-of-plane vibration (see Table 5.3).

infra-red spectra measured from substituted molecules or from complexes, that either distort the symmetry or simulate the attachment of the molecule to other species, or even from the solid condensed at low temperature. Further supporting information may also be provided by other physical techniques used in parallel with the HREELS measurements, such as Low Energy Electron Diffraction (LEED), Temperature Programmed Desorption (TPD), Electron Stimulated Desorption (ESD), etc. In the course of describing the examples selected to illustrate the applications of HREELS some of the methods used to allocate peaks in the vibrational spectra will be mentioned.

One further type of loss feature found in some spectra must be mentioned. Although the loss spectra from atomically clean metal surfaces are featureless, the same is not true of spectra from semiconductor and insulator surfaces. From the latter surfaces are found inherent loss features due to surface phonon scattering, not possible on metal surfaces. The two outermost layers of the surface have different symmetries due to the termination of the structure, and

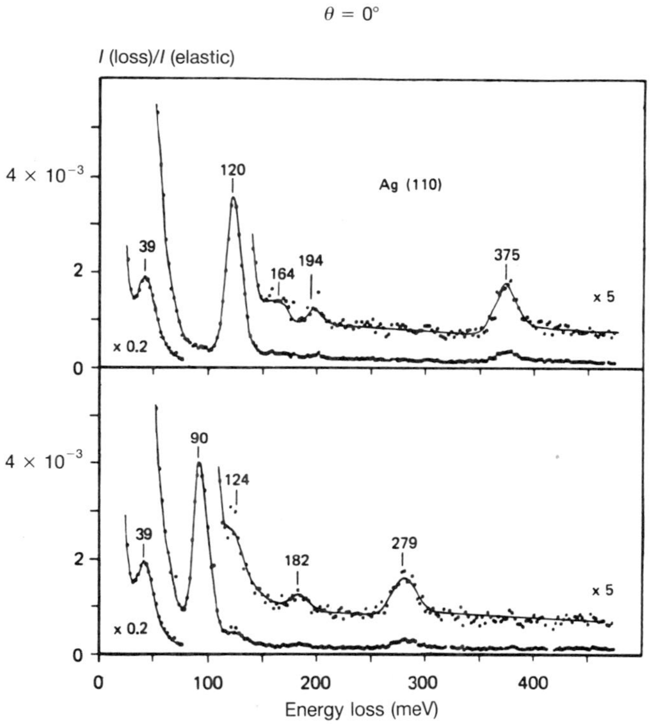

FIG. 5.9. Loss spectra for a primary energy of 2.6 eV for 30 L ethylene in the (a) normal and (b) deuterated forms adsorbed at -160°C on Ag(110) previously exposed to 900 L O_2 at 20°C, recorded by Backx *et al.*[8] $\theta = 0°$ in this figure and in Fig. 5.8 refers to detection in the specular direction. Apart from the peak at 39 meV loss, which is due to atomic oxygen, three additional peaks appear in each spectrum, which can be seen from Table 5.3 to arise from molecular vibrations in the ethylene molecule additional to that seen on clean Ag(110). The amount of ethylene adsorbed was also much greater than on clean Ag(110).

when two two-dimensional sub-lattices vibrate with respect to each other a long-range electric field is set up due to the effective charge in the outer layer. Such a charge would not be present at a metallic surface. The long-range electric field can be considered as a wave-like dipole layer from which the incoming primary electron can lose characteristic amounts of energy by dipole scattering. It is also possible for the incident electron to gain the same characteristic energy, but because the probability of the latter event is low, the intensity in a gain peak is much less than that in the corresponding loss peak.

An example of a feature due to surface phonon scattering is seen in the spectrum from silicon in Fig. 5.11, from Ibach *et al.*,[10] at a loss energy of 56 meV. A small gain peak at -56 meV was also present but is not shown. Note that with increasing oxygen concentration on the silicon surface the surface phonon peak is damped rapidly and it in fact disappears completely at

Table 5.3
Comparison of the observed vibrational frequencies with data of the free molecule.

	D_{2h}		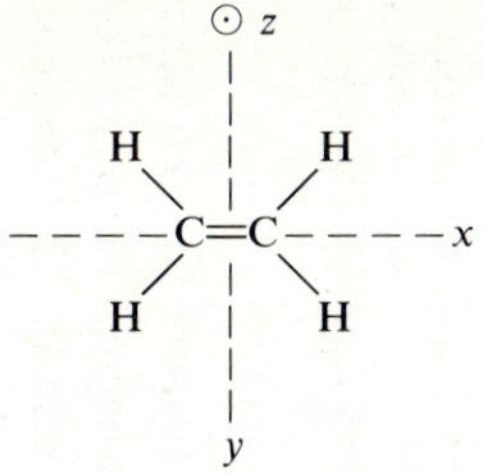					C_{2v}			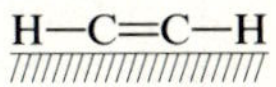
Mode	Symmetry	μ	C_2H_4 (meV)		C_2D_4 (meV)		ν_H/ν_D		Symmetry	μ	Visualized
			Gas	Ads.	Gas	Ads.	Gas	Ads.			
ν_1	Ag	—	374	375	279	279	1.34	1.34	A_1	Z	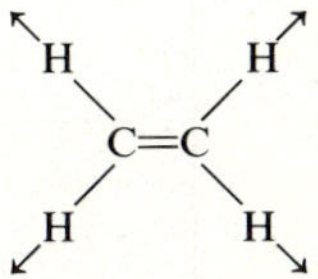
ν_2	Ag	—	201	194	188	182	1.07	1.07	A_1	Z	→C=C→
ν_3	Ag	—	166	164	122	124	1.36	1.32	A_1	Z	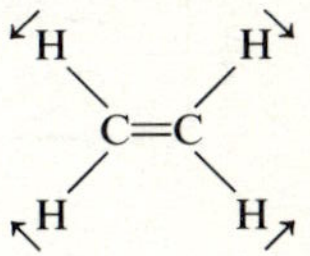
ν_7	B_{1u}	Z	118	120	89	90	1.33	1.32	A_1	Z	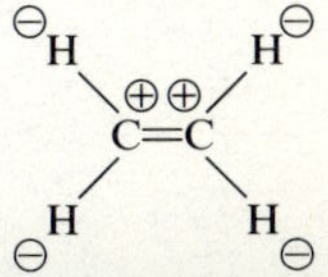

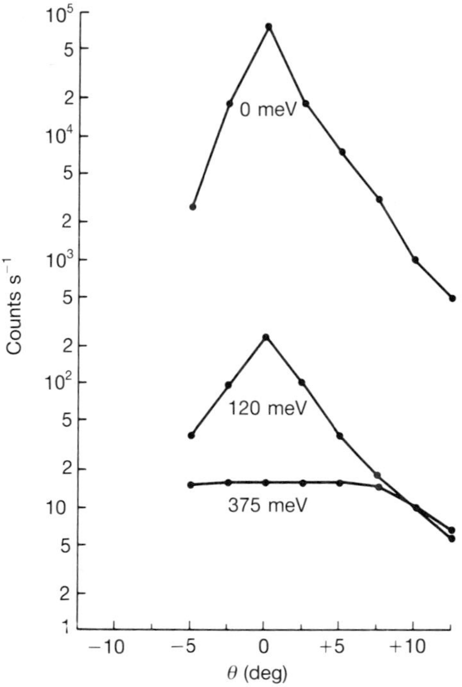

FIG. 5.10. Variations in intensity in the elastic peak and in the intensities of the loss peaks at 120 meV and 375 meV in the upper spectrum of Fig. 5.9, as a function of angle of detection. $\theta = 0°$ corresponds to specular detection. The intensity of the 120 meV peak is maximum in the specular direction, and drops away sharply on either side, indicating a dipole scattering mechanism, whereas the intensity of the 375 meV peak shows hardly any variation with angle, indicating an impact scattering mechanism.

a monolayer coverage of oxygen. This observation along with the fact that the half-width of the phonon loss peak is almost the same as that of the elastically scattered beam confirms the purely surface origin of the feature. Any contribution from bulk phonons would lead to a much broader structure. The strong dependence on angle around the specular direction is also diagnostic of a dipole scattering mechanism.

5.3 Quantification

In general the loss spectroscopies cannot be used directly to provide quantification of the amounts of material present on a surface, but to give chemical and structural information about the ways in which molecules are attached to surfaces. If quantitative information is required, then complementary techniques such as XPS or AES must be used, and often are. There may in some cases be the possibility of the extraction of qualitative information about

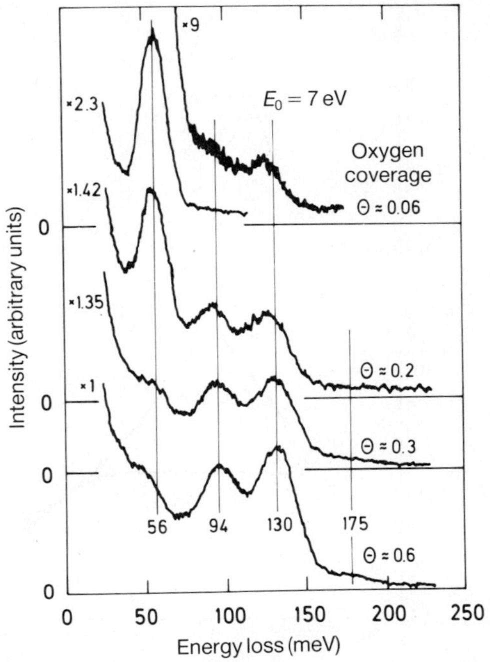

FIG. 5.11. Example of a feature in a HREELS spectrum due to surface phonon scattering, from Ibach *et al.*[10] The peak at 56 meV from the Si(111)-(2 × 1) surface prepared by cleavage in vacuo, at a primary energy of 5 eV, has been ascribed to such scattering. Note the rapid damping of such a peak on exposure to oxygen; $\theta = 1$ corresponds to monolayer coverage. The half-width of the loss feature is the same as that of the elastic beam, which confirms the surface origin.

the concentration of an element, e.g. from the intensities of loss peaks in HREELS, but this has not been pursued and in any case is likely to involve more effort with less certainty than in the more directly quantifiable techniques.

Sometimes quantitative data of a different sort can be derived from loss energy measurements when changes in loss energy are followed as a function of exposure to a reactive gas. For example, Gibson and Thomas[11] observed the change in the surface plasmon energy loss peak of barium as a function of exposure to oxygen, starting with clean barium in the form of an evaporated film. Some of the initial changes that they saw are shown in Fig. 5.12. The oxygen coverage was measured by AES and they also measured the work function of the film by a retarding field method.

Their results are summarized in Fig. 5.13, in which the upper and lower halves are on the same scale of oxygen exposure. As well as the change in work function and the increase in the peak-to-peak height of the oxygen Auger peak, the quantity $(E_{SP}/E_{OSP})^2$ is plotted. Reference to eqns (5.4) and (5.5) shows that the plasmon energy E_{SP} is proportional to $n_s^{1/2}$, n_s being the local free electron

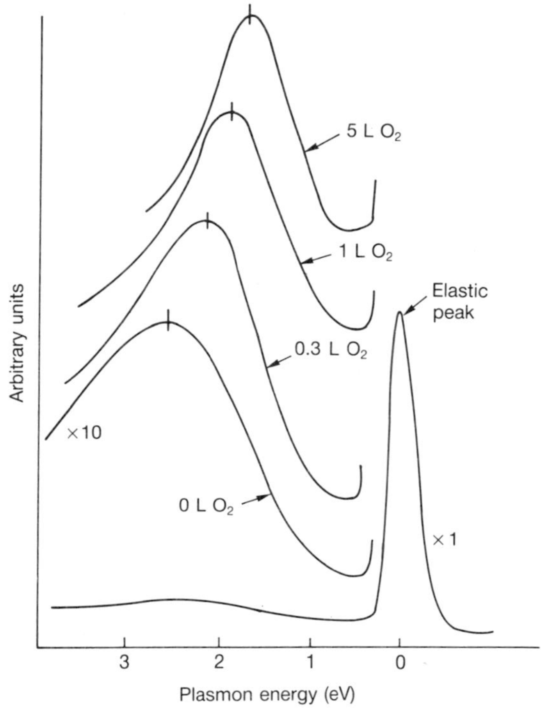

FIG. 5.12. The change in the surface plasmon loss peak observed by Gibson and Thomas[11] from a barium surface during exposure to increasing amounts of oxygen. Oxygen coverage was measured by AES, and the work function was also measured.

density. With E_{OSP} being the value of E_{SP} at zero oxygen exposure, i.e. for the clean barium metal, one can therefore write

$$(E_{SP}/E_{OSP})^2 = n_s/n_{Os} \tag{5.8}$$

where n_{Os} is the local (i.e. surface) free electron concentration before the start of oxidation. The quantity plotted in Fig. 5.13 is therefore a direct measure of the change in the relative free electron concentration at the surface. Thus from the large change in E_{SP} that occurs within 0.5 L exposure, i.e. at a coverage of about a monolayer, it can be deduced that the free electron concentration has been reduced by about 0.5. When this result is taken together with the rather small (≤ 0.1 eV) change in work function that occurs at the same time, and if it is assumed that the Ba–O bonds constitute simple electrostatic dipoles, then it is found that the initially adsorbed oxygen must lie in the plane of the outermost Ba atoms.

The later changes beyond an exposure of ~ 7 L indicate the completion of

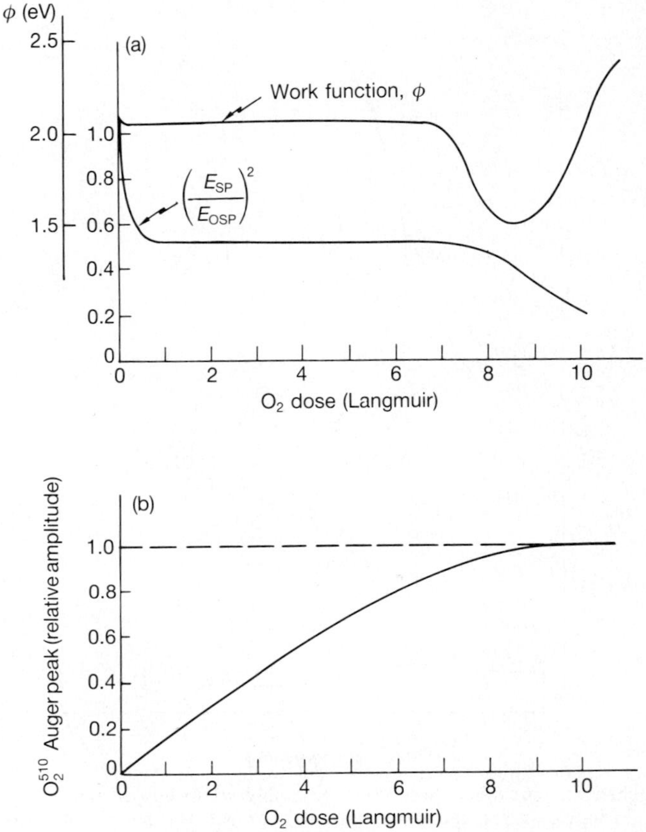

FIG. 5.13. Summary of the observations made by Gibson and Thomas[11] on a barium film as a function of oxygen exposure. (a) Variations in the work function, ϕ, and in the ratio $(E_{SP}/E_{OSP})^2$. E_{SP} is the plasmon energy and E_{OSP} its value for the clean surface. Expression (5.8) shows that the ratio is equal to the relative number of surface free electrons. (b) Variation in the intensity of the O KLL Auger peak at 510 eV. See the text for an interpretation of these results.

oxidation of the barium surface layers, and the appearance eventually of adsorbed oxygen at the external surface. At the minimum in the work function, however, a surface plasmon peak still exists, indicating the presence at that point of some residual metallic barium. This condition is similar to that found on activated BaO cathodes and confirms the presence of free barium in such surfaces.

5.4 Applications

5.4.1 *ELS*

As implied in Section 5.2.1, ELS is not used, or in general usable, as a technique for the analysis of element composition, but as a method of obtaining information complementary to that obtainable from some of the other techniques. For example, since ELS can often be performed with the same equipment as for AES, it is found not infrequently coupled with the latter and with other techniques such as ESD that need primary electron beams. The information provided by ELS includes both the plasmon loss structure (bulk and surface) and the inter- and intra-band transition loss structure, closer to the elastic peak. The two types of information are found in the published applications used in about equal proportions, usually separately, sometimes both together in the same paper. Examples of the use of both types are given below.

In the oxidation of tin neither AES nor XPS, although both are able to measure the growth of the concentration of oxygen on the surface, can give unambiguous information about the oxidation state of the tin, apart from stating that oxidized tin is present. The spectral differences arising from Sn^{2+} and Sn^{4+} respectively are too insignificant to be resolved. Powell[12] was the first to show that the ELS spectra of SnO and of SnO_2 were sufficiently different to be able to use them to distinguish between the oxides, and Stander[13] has followed this up finding by studying the oxidation of tin at room temperature from the originally clean metal surface to saturation. The question to be answered was that of the nature of the oxide formed in the initial stages—was it a layer of SnO over SnO_2, or was it alternatively highly defective and non-stoichiometric SnO_2?

A selection of ELS spectra from Stander's measurements are shown in Fig. 5.14. On the left are the raw spectra at a primary energy of 435 eV, while on the right are the same spectra after deconvoluting the effects of instrumental broadening with a broadening function. In the clean metal spectrum at the bottom the 14.5 eV loss is due to the bulk plasmon, and the 10.0 eV loss to the surface plasmon; the calculated values (see eqns (5.4) and (5.5)) are 14.2 eV and 10.0 eV, respectively. The 26.5 eV loss is a combination loss, i.e. approximately (14.5 + 10.0) eV, or perhaps (2 × 14.5) eV.

Adsorption of oxygen, to saturation by an exposure of 200 torr for 1 h, caused a general decrease in intensity and a *continuous* shift of the surface plasmon peak from 10.0 eV to lower loss values, reaching 6.5 eV loss at saturation. This behaviour is in contrast to that of aluminium and magnesium, in which initial oxidation causes an abrupt disappearance of the original surface plasmon peaks, with the appearance of new loss peaks at lower energies. Now, as mentioned earlier, the surface plasmon loss energy E_{sp} for a clean free electron metal is related to the bulk plasmon E_{bp} by $E_{sp} = E_{bp}/\sqrt{2}$, since for the clean metal/vacuum interface the dielectric constant ϵ in equation (5.5) is equal to unity. As the surface becomes oxidized, however, it is no longer appropriate to equate ϵ with unity, and eqn (5.5) becomes

$$E_{sp}^{0} = E_{bp}/(1+\epsilon)^{1/2} \tag{5.9}$$

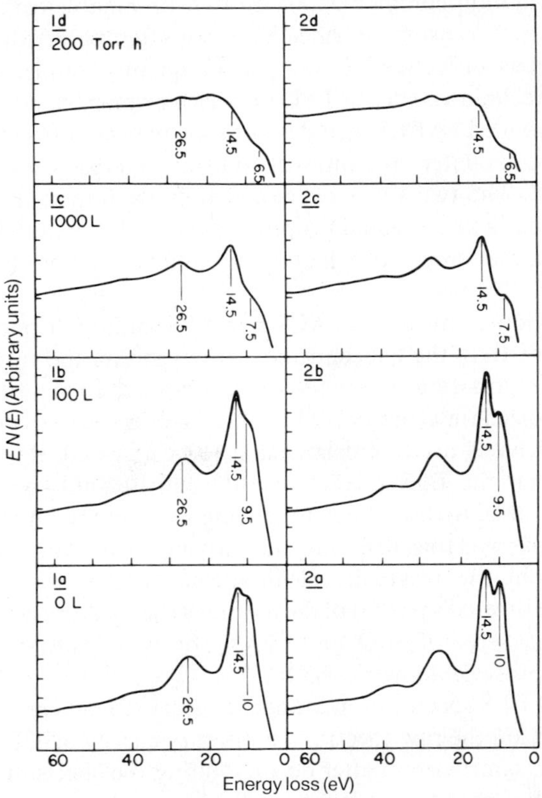

FIG. 5.14. Loss spectra recorded by Stander[13] during the oxidation of tin at room temperature. Left: spectra from the clean surface at the bottom, to saturation with oxygen at the top. Right: results of deconvoluting instrumental broadening from the raw spectra.

where E^0_{sp} is now the surface plasmon loss energy for the oxidized surface, and where ϵ changes from unity for the clean surface to whatever value corresponds to the oxide layer. In the cases of aluminium and magnesium the oxides formed seem to be similar to stoichiometric Al_2O_3 and MgO, respectively, almost immediately oxidation commences, hence the abrupt change in E_{sp}, but for tin the progressive change in E^0_{sp} in a continuous manner indicates a progressive and continuous change in ϵ. The saturation value of E^0_{sp}, 6.5 eV, agrees well with the calculated values of 6.24 eV and 6.79 eV when the dielectric constants of bulk SnO_2[001] and SnO_2[100] are used in (5.9), so clearly the oxide being formed is some form of SnO_2. Although the dielectric constant of SnO is not available, comparison of the properties of SnO and SnO_2 with those of PbO and PbO_2 suggests that SnO should have a dielectric constant of 6 to 7, which would lead to a final value of E^0_{sp} of $\sim$5 eV. Such a low loss was never observed.

Unlike Al_2O_3 and MgO, SnO_2 can exist over a considerable non-stoichiometric range, and its rutile structure allows paths for the ready diffusion of Sn. In the early stages of oxidation, therefore, there will be a gradient in Sn concentration normal to the Sn–SnO_2 interface, and the mean composition will approximate Sn_3O_2, which would have a low dielectric constant (resulting in a value of E_{sp}^0 close to that of the clean metal). As the layer grows thicker, the conductivity drops and the dielectric constant increases, eventually converging to that of bulk SnO_2, with a consequent continuous decrease in E_{sp}^0, as observed. The conclusion is thus that at no stage in the oxidation is SnO formed, but the oxide layer progresses from having a metal-rich semiconductor character, with a high concentration of Sn interstitials, through layers of continually decreasing electrical conductivity and increasing dielectric constant, until SnO_2 is reached.

An example of the application of ELS in which careful measurement of the spectral region very close to the elastic peak, i.e. the interband transition region, has helped in explaining changes in surface electronic properties is that of the caesium–oxygen–metal system. Alkali metal and oxygen coadsorption on metal surfaces is important technologically because by careful adjustment of the conditions it is possible to achieve very low work function values. The technological importance comes from the use of such low work function surfaces as low temperature electron emitters, i.e. as thermionic cathodes and in energy conversion devices, etc. Much of the study has been devoted to the Cs–O–Ag system, which seems to be able to produce the lowest work function values, but other substrates are also important, particularly tungsten and molybdenum. H^- and D^- in fusion devices can be produced at Cs–O–W and Cs–O–Mo surfaces.

The coadsorption of Cs and O on the two metals was studied by Soukiassian *et al.*[14] using ELS. The adsorption was not simultaneous, as is frequently performed when both adsorbents are gases, but sequential, and they compared in separate experiments the results of adsorbing, first caesium followed by oxygen, then caesium on pre-adsorbed oxygen. The incident beam energy used was 45 eV, which ensured maximum surface sensitivity in the measurements, and caesium was deposited from a caesium chromate solid source; the refractory metal surfaces were cleaned before the start of adsorption, and the experiments were carried out at room temperature.

The ELS spectra recorded during the adsorption of oxygen on a monolayer of caesium on tungsten, at exposures of oxygen up to and just beyond 1 L, are shown in Fig. 5.15. On the Cs–W surface two losses are found close to the elastic peak, a principal one P_1 at 1.5 eV and a minor one P_2 at 0.9 eV. On exposure to oxygen, P_1 shifts progressively towards P_2 and eventually merges with it, apparently disappearing. At an oxygen exposure of 0.5 L a new peak P_3 suddenly appears, of much narrower width than P_1, and grows with increasing exposure (note that the sequential spectra in Fig. 5.15 are not all on the same intensity scales). Above an exposure of 4 L, P_3 also disappears. The

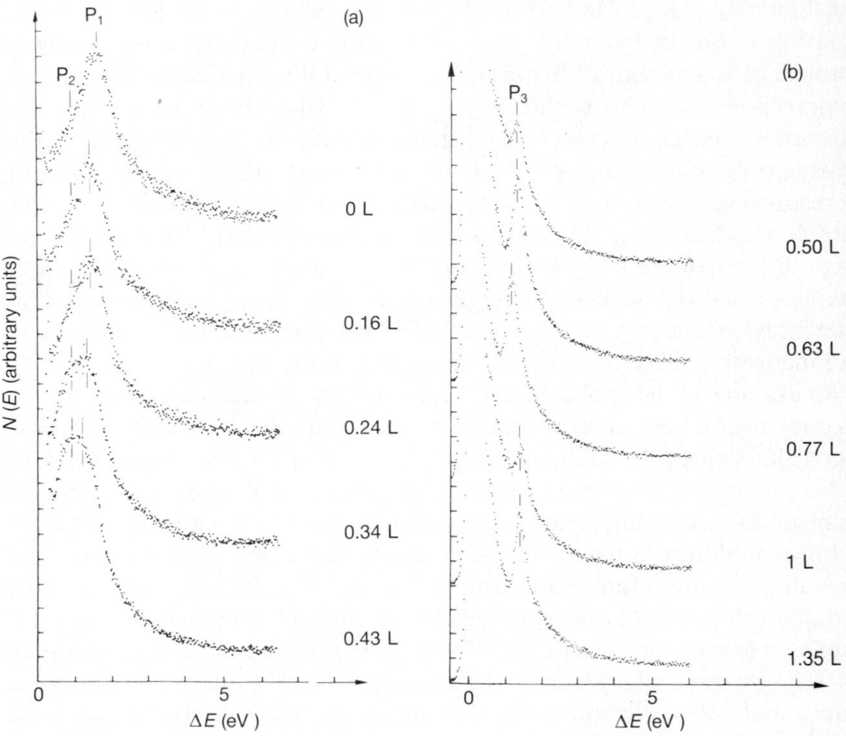

FIG. 5.15. Loss spectra recorded by Soukiassian *et al.*[14] during the adsorption of oxygen on a monolayer of caesium deposited on a W(100) surface. (a) Exposures from 0 to 0.43 L. (b) Exposures from 0.5 to 1.35 L. Primary energy was 45 eV. Peak labelled P_1 shifts towards P_2 and merges with it, then at an exposure of 0.5 L a new peak P_3 appears suddenly. Above an exposure of 4 L, not shown here, P_3 also disappears.

energy shifts of P_1 and P_3 with oxygen exposure are shown in Fig. 5.16. Similar results were found for adsorption of oxygen on caesiated molybdenum.

Such small energy losses must of course be due to the excitation of interband transitions, and it is necessary to establish just how they arise and why they are affected by oxygen. To take the caesiated tungsten surface first, before any exposure to oxygen, it is known that tungsten has a surface state just above the d band, and theoretical calculations[15] predict that there should be strong bonding between the caesium 6s electrons and the tungsten d electrons contributing to the surface state. The result is a hybridized s–d surface state predicted to shift to ~1 eV higher binding energy than the originally unhybridized state near the Fermi level. Since the principal interband transition observed on caesiated tungsten was of energy 1.5 eV (P_1 in Fig. 5.15), the hybridized s–d state must be the initial state for the transition; the final state then lies above the Fermi level by a fraction of an eV, and is probably associated with the caesium 6p states. The width of the loss peak P_3 is

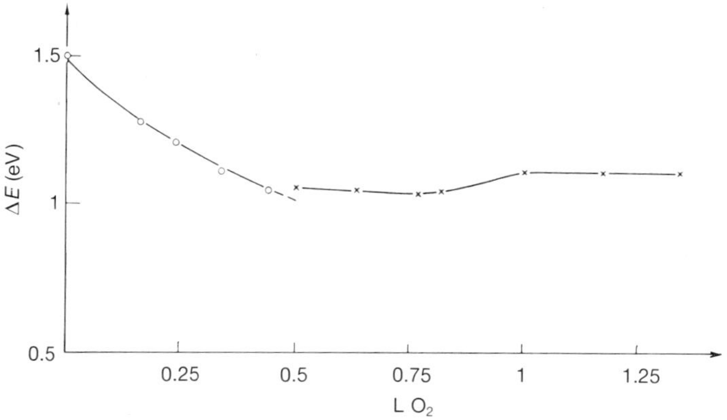

FIG. 5.16. Shifts in the loss peaks P_1 and P_3 in Fig. 5.15 with increasing oxygen exposure. From Soukiassian *et al.*[14]

significantly narrower than that of P_1 and is similar to that of the hybridized s–d state itself, as observed by angle-resolved UPS (ARUPS), so that the implication is that the initial state for P_3 is also the hybridized s–d state, but that the final state is very narrow and lying near the Fermi level. Since the loss energy of P_3 is $\sim$1.1 eV, this implies further that the s–d state must itself shift closer to the Fermi level on oxidation. The proposed energy scheme is shown in Fig. 5.17. On oxidation both the initial and final states for the transition P_1 draw nearer to the Fermi level, and at 0.5 L exposure the final state disappears (probably due to cross-over at the Fermi level). Transition P_3 then becomes possible, and its energy remains almost invariant with exposure until 4 L is reached when it too is extinguished. The variation of the P_1 energy with oxygen exposure, and the eventual extinction of P_3, are due to the progressive destruction of the tungsten surface state by extraction of d electrons to the oxygen.

When the inverse experiments were performed, i.e. when a monolayer of caesium was adsorbed on top of oxygen layers produced by exposures up to 1 L, it was found that adsorption onto an oxygen layer of 0.15 L exposure resulted in exactly the same loss spectrum as that in Fig. 5.15 for adsorption of 0.16 L of oxygen on to a monolayer of caesium. In other words the oxygen can diffuse through the caesium layer at low coverages, allowing bonding between the oxygen and the tungsten leading to reduction of the tungsten surface state. At higher oxygen pre-coverages, the loss spectra started to differ from those in Fig. 5.15, and could not be fully explained. The significance of this finding and of the model for the initial and final states involved, in Fig. 5.17, is that the work function minimum for the Cs–O–W system (and similarly for the Cs–O–Mo system) occurs near L_c, the critical oxygen exposure. This minimum must therefore arise from a combination of an s–d surface state *and* the presence of sub-surface oxygen.

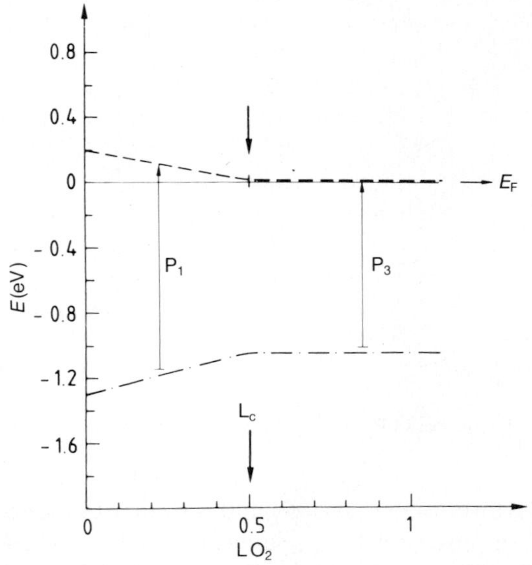

FIG. 5.17. Model proposed by Soukiassian *et al.*[14] to account for their ELS results shown in Figs 5.15 and 5.16. The dot-dashed line shows the variation in the proposed initial state, and the dashed line that in the proposed final state, as a function of oxygen exposure. The peak P_1 is assumed to arise from a transition between an initial state formed from hybridized s and d states, to a final state above the Fermi level. On oxygen exposure the final state shifts down and eventually crosses the Fermi level, making a transition to it impossible. Transition P_3 then suddenly becomes possible to the top of the Fermi level, but it too is extinguished when the oxygen exposure reaches 4 L. This suggests destruction of a tungsten surface state by extraction of d electrons to the oxygen.

5.4.2 *CEELS*

Excitation to localized final states The electronic structure of the lanthanide metals (La–Lu) has fascinated scientists for many years, both because of their intrinsic interest, and because of the wide variety of applications that the lanthanides have, applications that reflect the unusual structure. The fascination arises (a) because of the presence of highly localized 4f electrons, often at such low binding energies that they are at or very close to the valence band but coupled only very weakly to it, and (b) because of the valence fluctuations so often observed in the lanthanides on chemical combination, or even, sometimes, between surface and bulk in the pure metal.

Photoelectron spectroscopy in its various forms (XPS, UPS, SRPS, q.v.) has been very useful in mapping the positions of core levels and of valence band structure, while the newer technique of IPES has mapped the unoccupied densities of states, of the lanthanides, but in many cases a more direct valence monitor is needed. CEELS has been found to be ideally suited to the study of valence changes, and indeed to the basic physics inherent in the lanthanide structure.

Because of the strongly localized 4f states mentioned above, the lanthanide CEELS spectra tend to contain predominantly features due not to direct ionization to continuum states, but to excitation to empty 4f states. This is certainly true of the ionization of the 4d and 3d core levels, but probably not so true of the ionization of the 5p and 4p levels. Such core-level excitations with strongly localized 4f final states are known as 'giant resonances', and they have been investigated comprehensively by Netzer and co-workers. Examples have already been given in Section 5.2.2. These authors have also used the changes observed in the giant resonance loss spectra to monitor the valence changes occurring at the surfaces of some lanthanides during reaction, particularly oxidation.

The most spectacular example is that of ytterbium.[16] In the metallic form ytterbium has a full 4f shell, and there is therefore no giant resonance possible. On oxidation the valence changes from 2 to 3, i.e. the ground state f configuration from $4f^{14}$ to $4f^{13}$, and a channel therefore opens for the giant resonance to be 'switched on'. It can be seen in Fig. 5.18 how dramatically the

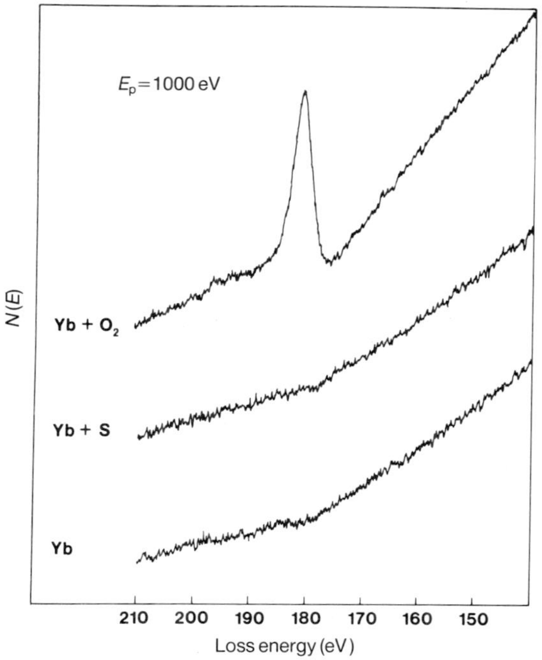

IG. 5.18. Example of the 'switching on' of a giant resonance, that in ytterbium, given by Netzer *et l.*[16] Metallic ytterbium has a full 4f shell, and therefore resonant excitation to the 4f states is not ossible. On oxidation the Yb valency changes from 2 to 3, that is, the ground state configuration om $4f^{14}$ to $4f^{13}$, and the excitation becomes possible. The magnitude of the resonant peak can be sed to follow the oxidation.

giant resonance appears on oxidation; in fact the magnitude of the resonant peak can be used from the earliest stages of exposure to oxygen to follow the surface valence of the ytterbium. Note that exposure to hydrogen sulphide to produce ytterbium sulphide has no such effect since in the sulphide, ytterbium remains divalent.

Another interesting, although not so spectacular, example[16] occurs in the middle of the lanthanide series at europium and gadolinium. In their metallic forms europium and gadolinium are divalent and trivalent, respectively, and being adjacent elements, have therefore the same ground state $4f^7$ configuration. Thus the multiplet structure associated with the giant resonance losses is very similar, and indeed Fig. 5.19 shows that the shapes of the losses are much the same for the two clean metals. On reaction with oxygen or with sulphur gadolinium remains trivalent, and the loss spectra in Fig. 5.19 show that there are insignificant changes. Europium, on the other hand, has stable oxides and sulphides of different valencies, e.g. EuO, Eu_2O_3, and Eu_3O_4. Reaction with sulphur-containing molecules to form a sulphide causes no change in the position of the peak, only a marked narrowing, from which one deduces that EuS has formed, maintaining europium divalent. On oxidation however, the principal loss peak shifts to higher energies by 2.5 eV, and an additional loss peak is resolved to higher energy still. The existence of two loss peaks suggests strongly a shift in the europium valency to a condition of mixed stoichiometry, probably a mixture of EuO and Eu_2O_3. The likelihood of Eu_2O_3 has been confirmed by other techniques, although with measurements made on bulk materials. Here CEELS is monitoring the surface valency in a very thin film of oxide.

Yet another case of valence change on oxidation is that of cerium, studied by Strasser and Netzer.[17] The metal is trivalent and has a ground state of configuration $4f^1$. If on oxidation it becomes tetravalent at any stage then its 4f configuration would change to $4f^0$, whereupon the giant resonance loss spectrum should become similar to that of lanthanum, the next lightest element. In Fig. 5.20 are shown the $4d \rightarrow 4f$ giant resonance spectra for clean cerium, and for cerium after small and large exposures to oxygen, in comparison with the lanthanum spectrum. For both metals the metallic loss spectra consist of a series of small but sharp resonances followed by a broad intense resonance above threshold. For the small exposure, the main resonance is sharpened, but does not basically alter position, while the fine structure is unaltered. The deduction is that at that stage the cerium remains trivalent. On heavy oxidation, *and* on recording the spectrum in an oxygen ambient atmosphere of 10^{-8} torr, the multiplet fine structure reduces to two peaks, and the principal resonance changes shape, sharpening as it does so. There is now more similarity with the lanthanum spectrum, suggesting a valence shift to Ce^{4+} under the conditions of measurement. It was noted by the authors that it was in fact necessary to maintain an oxygen atmosphere

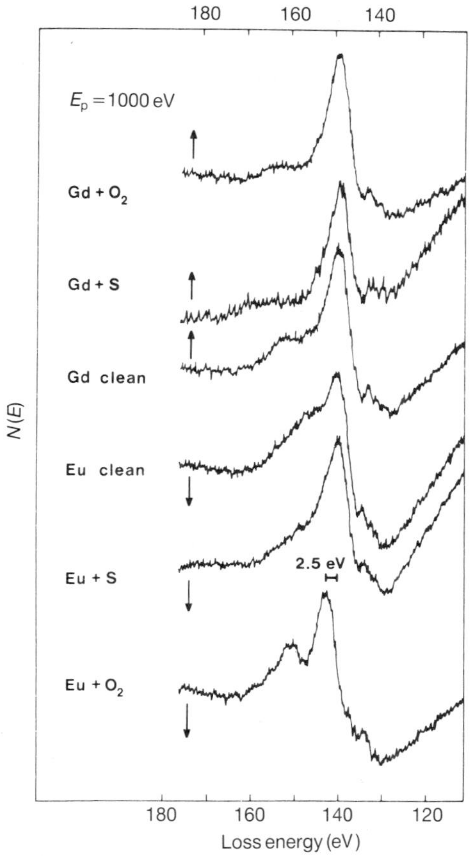

FIG. 5.19. Comparison of the behaviour of the 4d–4f giant resonance peaks for gadolinium and europium during surface reaction, made by Netzer *et al.*[16] Clean Gd and Eu have similar loss structure, having the same ground state configuration $4f^7$. Gd retains the same valency on oxidation or sulphidation, thus its loss structure stays the same, but Eu has oxides of different valencies, so that the principal peak shifts and an additional loss appears. The suggestion is that the surface condition is then one of mixed stoichiometry.

around the cerium in order to retain that spectrum. If this was not done, the spectrum changed back quickly to the upper ones, indicating a reduction in valency by diffusion of oxygen into the bulk.

Direct ionization Direct ionization transitions in CEELS have not had much application in, for example, monitoring surface reactions or providing elemental analysis, because both these tasks can be performed much better by either XPS or AES. The reason is that in general the loss edges do not correspond exactly with the binding energies determinable from XPS, nor do the changes in edge position necessarily correspond to the chemical shifts

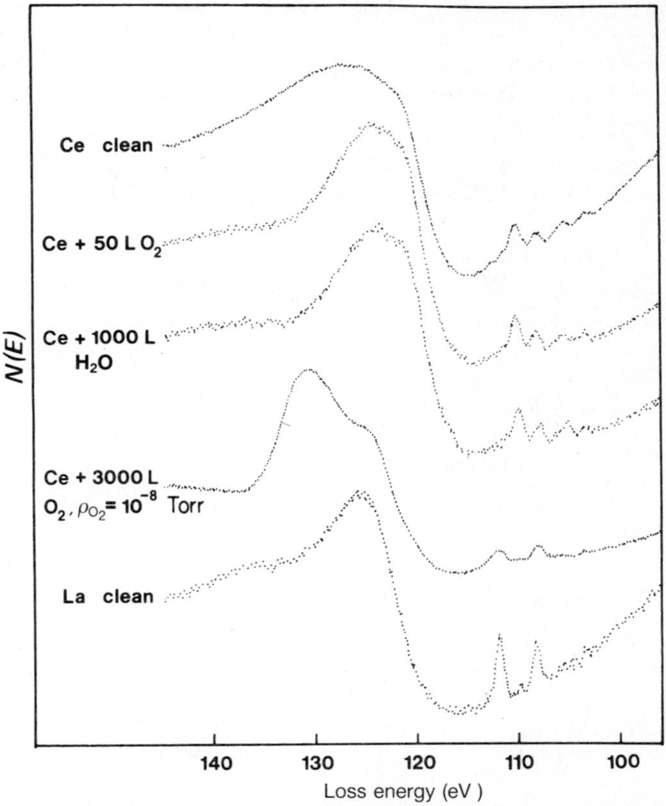

FIG. 5.20. Valence changes in cerium during oxidation followed by ELS, from Strasser and Netzer.[17] At the top the 4d–4f giant resonance spectrum from clean cerium, followed by spectra taken after successively increasing doses of oxygen or water. Up to 1000 L exposure the fine structure on the high energy side of the threshold is sharpened but otherwise unaltered, and the main resonance is also much sharper but basically unaltered in position. At that stage, then, cerium remains trivalent. At higher exposures, and when the spectra are recorded *in* an ambient oxygen pressure of 10^{-8} torr, the fine structure alters character to two peaks only, and the resonance peak changes shape. The similarities with the spectrum from clean lanthanum, at the bottom, suggest a change in cerium valency to tetravalent. Removal of the oxygen caused a rapid change back to the spectrum characteristic of trivalency.

found in the latter technique. The positions of the loss edges will depend in most cases on the structure of the densities of unoccupied states above the Fermi level, which structure is only now beginning to be elucidated accurately by inverse photoemission (IPES). However, there is often fine structure near the principal ionization edge even in materials that cannot produce the giant resonances described above for the lanthanides, and this fine structure can itself be used to derive physical information.

A good example of one of the types of information to be derived comes from the observations of Koma and Ludeke[18] on silicon. They recorded the Si 2p and 2s core-level loss spectra at primary energies of 200 eV and 175 eV respectively, from the Si(111)-7 × 7 annealed surface and the Si(100)-2 × 1 cleaved surface, in an attempt to locate the so-called 'dangling bond' empty surface states. These arise from the population of unsatisfied silicon bonds in the surface layer of silicon atoms, on atomically clean silicon, and they create what is effectively a two-dimensional unoccupied band of states lying just above the Fermi level.

Their spectra are shown in Fig. 5.21, the 2p loss region on the left and that of the 2s core level on the right. Excitation of 2s electrons results in single peaks from both surfaces, near 152 eV. Since the 2s–2p binding energy separation is 51.0 eV, the structure at 100.6 eV in the 2p loss spectra must arise from

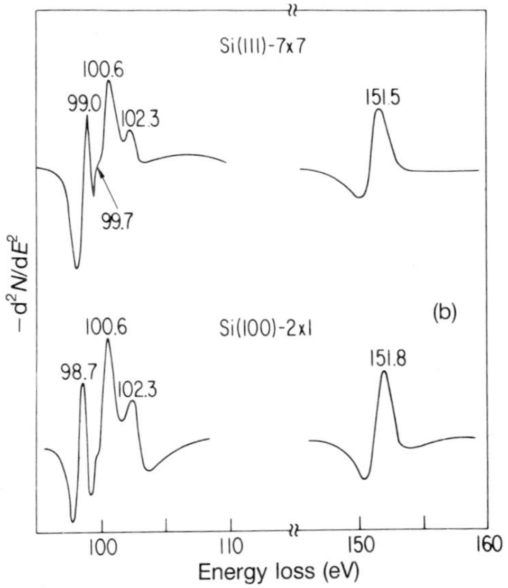

FIG. 5.21. Energy loss spectra arising from direct ionization of the Si 2p and 2s levels recorded by Koma and Ludeke[18] from the Si(111)-7 × 7 (upper) and Si(100)-2 × 1 (lower) surfaces. The 2p loss region on the left was recorded at 200 eV primary energy, and the 2s loss region on the right at 175 eV primary energy. Excitation of 2s electrons results in single peaks near 152 eV. With a 2s–2p binding energy difference of 51 eV, the peak at 100.6 eV in the 2p loss spectrum must therefore correspond to the ionization edge. Thus the peaks at 102.3 eV correspond to transitions into regions of the conduction band at higher energies, while the peaks at 99.0 eV and 98.7 eV must lie near the Fermi level, and are attributed to surface states.

transitions to the same set of final states, and since the peaks at 100.6 eV are the most prominent in the spectra, these states can only be the lower conduction band states. The features seen at 102.3 eV, i.e. at a greater loss energy, must therefore correspond to transitions into parts of the conduction band of high symmetry lying at higher energies. Thus the peaks observed at

99.0 eV for Si(111)-7 × 7 and at 98.7 eV for Si (100)-2 × 1 must lie nearer the Fermi level, and the only possibility for their attribution is that of the surface states mentioned above.

The authors suggest that the reason for the non-observance of fine structure in the Si 2s loss spectrum lies in selection rule effects. Theoretical considerations of the symmetry of the states at higher energy in the conduction band show that such states are not accessible to dipole transitions from s-like states. Since the transition probability from p-like states is also low, the low intensity of the 102.3 eV loss peak compared to that of the peak at 100.6 eV is explained. The observation that the transition into the 'dangling bond' surface state is seen in the 2p loss spectrum but not in the 2s, suggests strongly that the dangling bond states have almost pure s-like character. This is not in agreement with theoretical models proposed for the (100) and (111) surfaces, which indicate p-like character for the surface states, but the differences could well be due to the changes induced by reconstruction. The theory applies mostly to the ideal, unreconstructed, surfaces, whereas the (111) and (100) surfaces studied by Koma and Ludeke were annealed and therefore reconstructed.

Extended energy loss fine structure (EELFS) Although not strictly speaking a method of elemental or chemical analysis in the same sense as the others described here, the technique of EELFS should be mentioned briefly since it is really an extension of CEELS. EELFS is a surface *structural* technique in that it measures the interatomic distances of the various nearest neighbours in the surface, and is therefore directly comparable with Extended X-ray Absorption Fine Structure (EXAFS), not otherwise mentioned in this volume.

EELFS has been pioneered by De Crescenzi.[19] It uses the fine structure that can be observed extending for several hundred eV above the ionization edge in CEELS for many materials, and since the primary electron energies used are generally <2000 eV and the core levels involved are therefore shallow, it is surface sensitive. An example of the type of extended fine structure to be found is seen in Fig. 5.22 for the region above the $L_{2,3}$ ionization edge of titanium; for clarity the contribution from the L_1 edge has been subtracted. For comparison an EXAFS spectrum for the same region using synchrotron radiation is also shown and it can be seen that there is excellent agreement.

Data processing follows the same lines as in EXAFS. The experimental spectrum $X(E) = -\mathrm{d}^2N(E)/\mathrm{d}E^2$, as in Fig. 5.22, is transformed into an $X(K)$ spectrum, where K is the wave vector of the excited electron above the edge, and $X(K)$ is then Fourier integrated according to

$$F(R) = \int_{K_{\min}}^{K_{\max}} X(K)\mathrm{e}^{-2\mathrm{i}KR}\,\mathrm{d}K. \tag{5.10}$$

$F(R)$ is the radial distribution function, that is, it gives in real space the

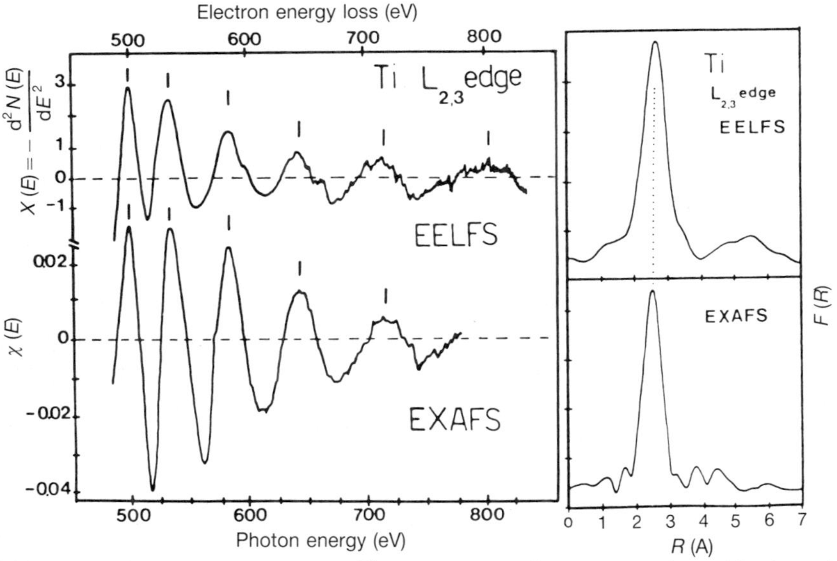

FIG. 5.22. Comparison by De Crescenzi[19] of the extended fine structure observable above the Ti $L_{2,3}$ ionization edge in both EELFS (electron excited) and EXAFS (synchrotron radiation excited). The structure can be seen to extend in an oscillatory manner for several hundred eV above the edge. The L_1 contribution has been subtracted. The right-hand panel shows the result of Fourier transformation of the two spectra. The positions of the major peaks, i.e. the first-neighbour distance in Ti, agree within 0.02 Å between the two techniques.

distances of first, second, third, etc., neighbours around the atom. The right-hand side of Fig. 5.22 shows the plot of $F(R)$ against R, for the Ti $L_{2,3}$ edge, again with the EELFS result compared with the EXAFS. The major peaks resulting from the two techniques agree in position within 0.02 Å. There is also a suggestion of a second peak at greater distance, ~5.5 Å, in the $F(R)$ derived from EELFS, perhaps corresponding to the second-nearest neighbour distance in the surface. There are still many questions to be answered over EELFS, particularly the treatment of inelastic cross-sections at such low incident energies, but on the results seen to date it seems a promising method of obtaining additional information about a surface structure without going to the expense of having to use a synchrotron.

5.4.3 *HREELS*

Since HREELS gives such valuable information about the chemical bonding of adsorbates and about the site and structural specificity of adsorbate-adsorbent systems, its application to a basic study of catalysis is obvious. The emphasis is on basic studies, since it seems unlikely that the technique would ever be suitable for the study of real catalysts; proper interpretation depends on starting with a very well characterized surface such as a clean single crystal. Despite that, the extensive application to model catalyst systems has provided information not easily obtainable in any other way, allowing deeper insight into some aspects of simple catalytic reactions.

One catalytic reaction simple at least in principle is that of ammonia synthesis over iron catalysts. At some stage in the reaction, as ammonia starts to form, knowledge both of the manner in which the ammonia molecule is adsorbed on the iron surface and of the existence of various postulated intermediate species at the surface becomes important. Accordingly the adsorption of ammonia on an Fe(110) single crystal at various temperatures was studied by Erley and Ibach,[20] using HREELS. Admittedly the study of a single crystal, in ultra-high vacuum, and at low temperatures, might seem far removed from the actual operating conditions of a technological catalyst, but the argument is that one has to make a start somewhere. The same argument applies, of course, to the application of most of the surface-specific techniques to the study of technologically important reactions such as corrosion, adhesion, etc.

The sequence of spectra recorded as a function of increasing coverage of ammonia at 110 K are shown in Fig. 5.23, up to the point where no further spectral changes occurred. The clean metallic spectrum is of course featureless, as mentioned earlier. With increasing take-up of ammonia, the loss spectrum becomes increasingly complex, with additional features appearing and shifts occurring in peaks already present (e.g. 1170 to 1105 cm^{-1} from 0.05 L to 0.3 L, and 340 to 410 cm^{-1} from 2 L to 6 L). Interpretation of the spectra was carried out by comparison of the observed vibrational frequencies with those found in the infra-red spectra of gaseous and of solid ammonia, and of the complex hexamine salt $Fe(NH_3)_6Cl_2$. As is normal practice in HREELS, the experiments were repeated using deuterated ammonia, whose spectra were compared with the corresponding deuterated infra-red spectra. The comparison for normal ammonia is shown in Table 5.4.

The interpretation goes as follows. At very low coverage the adsorbed species may be either molecular ammonia, or NH_2, or NH, according to whether or not dissociation has occurred. Only three losses are observed, and angular dependent measurements show that they all arise from dipole scattering. Dipole selection rules can therefore be applied, together with symmetry considerations, on the basis of which the NH_2 species can be excluded at once since the HNH scissor vibrational mode at ~ 1580 cm^{-1} was not observed. The symmetry possibilities for the NH species should have given rise to either two or four vibrational modes, whereas three were observed. Preference is therefore given to molecular NH_3 as the adsorbed species, in C_{3v} symmetry, and the losses are assigned to the symmetric Fe–N stretch, symmetric NH_3 deformation, and symmetric NH_3 stretch. Such a symmetry implies no contribution from the symmetry of the Fe(110) surface, and this can only be so if the NH_3 molecules are singly coordinated, i.e. each sitting on top of one surface Fe atom. Confirmation of the weak bonding situation comes

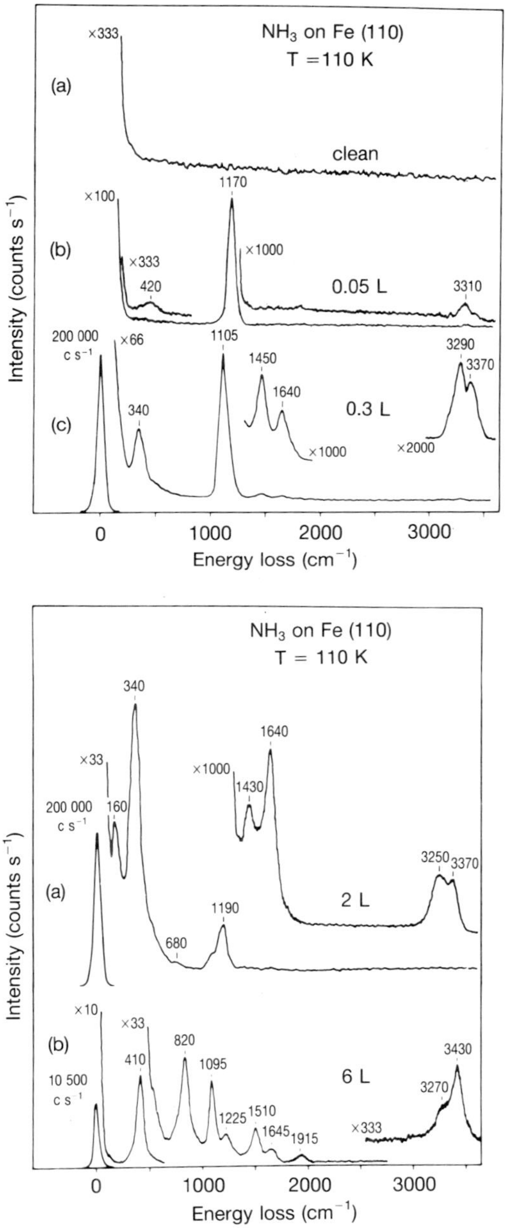

FIG. 5.23. HREELS spectra recorded by Erley and Ibach[20] during the adsorption of ammonia on Fe(110) at 110 K, up to saturation at 6 L. The clean metal spectrum is featureless. The increasingly complex vibrational loss spectrum is interpreted by comparison with the infra-red spectra of gaseous and solid ammonia, as discussed in the text with the help of Table 5.4. Primary energy was 2.4 eV.

Table 5.4

Approximate type of mode		$NH_3/Fe(110)$ First layer (singly coordinated)	$NH_3/Fe(110)$ First layer (multi-coordinated)	$NH_3/Fe(110)$ Second layer	$NH_3/Fe(110)$ Condensed phase	NH_3 (gas) Ref. 7	NH_3 (solid) Ref. 8	$[Fe(NH_3)_6]Cl_2$ Ref. 10
Hindered translation	ν_t	—	—	160	—	—	141	170
Libration	ν_1	—	340	340	410	—	360	—
Fe–N symmetric stretching	ν_{Fe-N}	420	Obscured by 340 band	—	—	—	—	315
$2\nu_1$		—	—	680	820	—	—	—
NH_3 rocking	ρ_r	—	Not observed	—	—	—	—	633
Symmetric NH_3 deformation	δ_s	1170	1105	1190	1095	950	1060	1156
$3\nu_1$		—	—	—	1225	—	—	—
$\nu_1 + \delta_s$		—	1450	1430	1510	—	—	—
Degenerate NH_3 deformation	δ_d	—	1640	1640	1645	1627	1646	1596
$2\nu_1 + \delta_s$		—	—	—	1915	—	—	—
Symmetric NH_3 stretching	ν_s	3310	3290	3250	Shoulder ~3270	3337	3223	3175
Degenerate NH_3 stretching	ν_d	—	3370	3370	3430	3444	3378	3335

from the closeness of the observed NH_3 symmetric stretch vibration at 3310 cm^{-1} to that in the gas phase at 3337 cm^{-1}, and of the observed NH_3 symmetric deformation vibration at 1170 cm^{-1} to that in the complex hexamine salt. Thus no significant molecular distortion occurred on adsorption.

On further adsorption (up to ~1 L), four new loss peaks appear, and the 1170 cm^{-1} loss shifts to 1105 cm^{-1}. The losses at 1640 and 3370 cm^{-1} can be identified readily with degenerate NH_3 deformation and stretch vibrations, respectively, by reference to the gas-phase spectra, and this was confirmed also in the deuterated spectra. That at 1450 cm^{-1} is a combined loss, as indicated in the table. The 340 cm^{-1} loss, which at higher coverage becomes intense, can be associated with the start of the second ammonia layer, since it is close to the 360 cm^{-1} loss in solid ammonia and to the 315 cm^{-1} loss in the hexamine salt; it is attributed therefore to hindered rotation of the ammonia molecule. The progressive shift of the 1170 cm^{-1} loss to 1105 cm^{-1} indicates a continuous change in the degree of charge transfer from the adsorbed ammonia to the metal.

Symmetry considerations indicate that as the coverage increases beyond the very low coverage range, the ammonia molecules become multi-coordinated to the iron surface, i.e. are attached to more than one surface iron atom. The geometry could be either that of a 'bridged' adsorption, to two iron atoms, or at a quasi-triangular site, but there is no way of deciding between them.

After an exposure of 2 L of ammonia, the 160 cm^{-1} and 340 cm^{-1} losses become intense, and additional losses at 680 cm^{-1} and 1190 cm^{-1} appear (the latter does not result from a reverse shift of the 1105 cm^{-1} loss observed earlier). Table 5.4 shows by analogy with the IR spectra of solid NH_3 and the hexamine salt that both the intense losses have their counterparts in the solid materials. The 680 cm^{-1} loss is simply a multiple excitation of the 340 cm^{-1} loss, while the 1190 cm^{-1} loss also has its counterpart in the salt spectrum; the 1105 cm^{-1} loss is still present as a shoulder on the 1190 cm^{-1} peak. At ~3 L exposure the second layer of ammonia is complete and there are suggestions that some multilayer condensation has started.

Above 3 L exposure, the spectrum is characteristic of condensed ammonia. The additional losses at 820 cm^{-1}, 1225 cm^{-1} and 1510 cm^{-1} are interpreted as multiple or combination excitations, as indicated in the table.

Thus the sequence of events on adsorption at 110 K is, at very low coverage, singly coordinated ammonia *molecules*; as the coverage increases to a monolayer, multi-coordinated molecules, beyond a monolayer the formation of a second layer hydrogen-bonded to the first; and, at saturation exposures, formation of many layers of condensed ammonia.

The effects of heating an iron surface already saturated with ammonia at 110 K, to successively higher temperatures, are seen in Fig. 5.24. The spectra were recorded after returning to 110 K. Heating to 155 K converts the loss

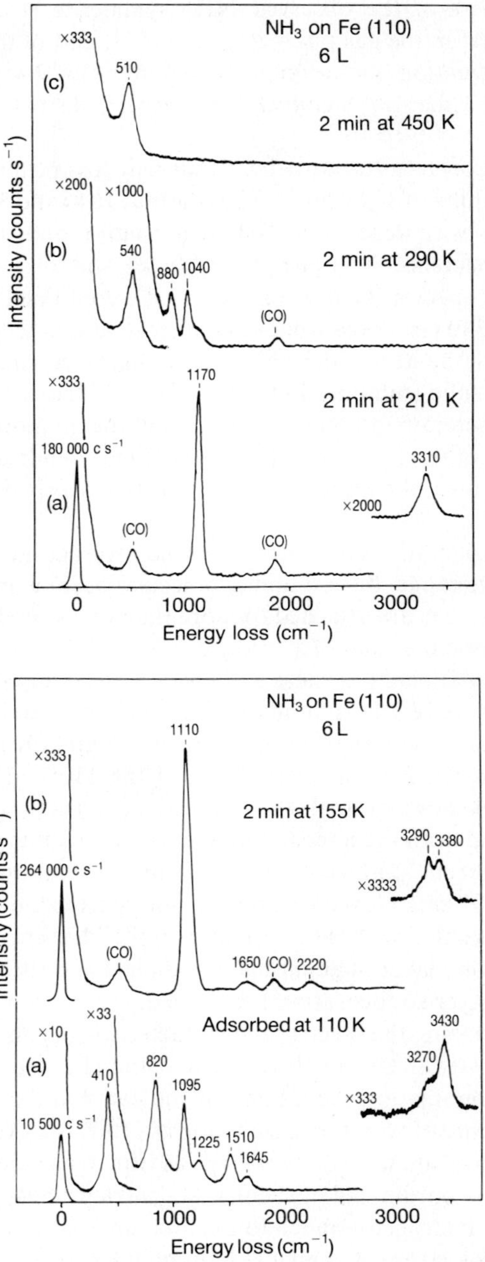

FIG. 5.24. HREELS spectra recorded by Erley and Ibach[20] after heating an Fe(110) surface saturated with ammonia at 110 K (see Fig. 5.23) to successively higher temperatures. The spectra were recorded in each case after return to 110 K. The interpretation of the changes seen as a result of heating is discussed in the text.

spectrum into that previously observed after exposure of 0.3 L ammonia at 110 K, that is, both the condensed phase and the second layer have desorbed completely. Further heating to 210 K causes further desorption, in this case of the multi-coordinated ammonia, leaving only the singly coordinated adsorbed molecules, as at an exposure of 0.05 L in Fig. 5.23. At still higher temperatures, 290 K and 450 K, the spectrum degenerates first into losses only for adsorbed hydrogen and nitrogen and then for nitrogen only at 510 cm^{-1}. The ammonia has dissociated at the elevated temperatures leaving finally only the Fe–N species with a single characteristic stretching vibration.

At no stage in adsorption or desorption was any evidence found for the

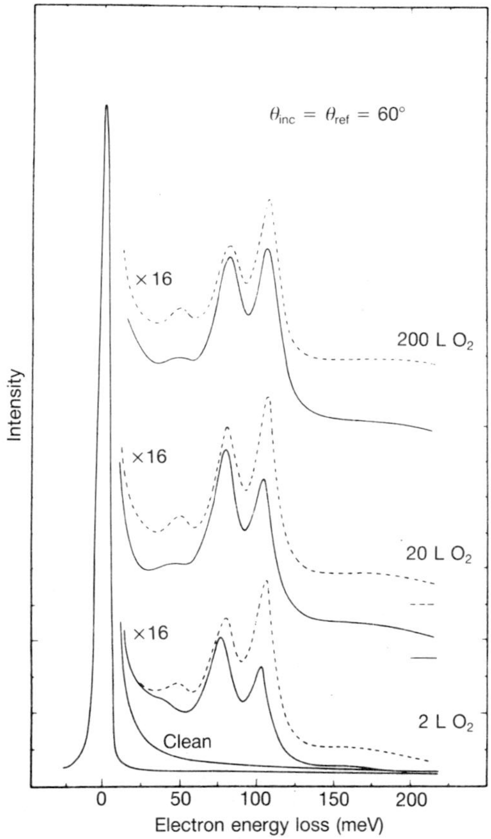

FIG. 5.25. HREELS spectra during the oxidation of Al(111) at 300 K, 2×10^{-7} torr (full curve) and then after heating following each exposure, to 500 K (dotted curve) measured by Erskine and Strong.[21] The primary energy was 4.76 eV, and the spectra were recorded in specular reflection. The relative intensities of the two major loss peaks at 80 meV and 105 meV, respectively, vary with oxygen exposure and with heat treatment. The 80 meV peak is attributed to oxygen adsorbed at the external surface, and the 105 meV to oxygen adsorbed below the surface; at high exposures and after an increase in temperature the sub-surface peak dominates, indicating increased formation of the sub-surface adsorbed phase.

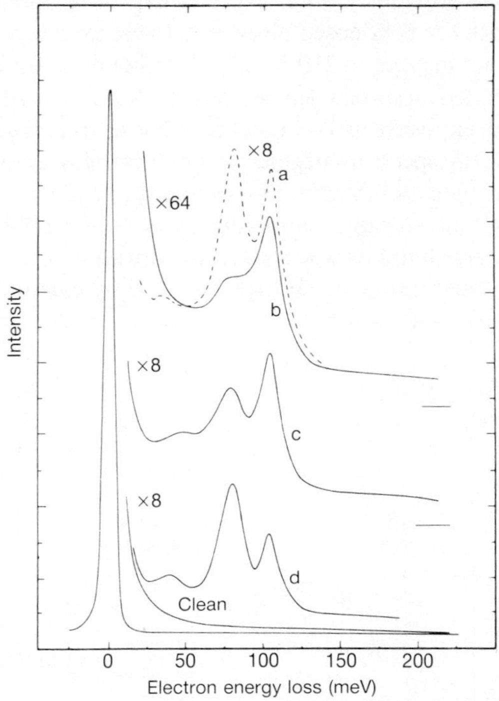

FIG. 5.26. HREELS spectra recorded by Erskine and Strong[21] during various treatments of an oxygen-exposed Al(111) surface. Same analysis conditions as for the spectra in Fig. 5.25. The two lower spectra are from the Al(111) surface, (d) immediately after exposure to 20 L O_2, 1×10^{-6} torr, and then (c) 2 h after that exposure. Allowing the adsorbed layer to stand at 300 K for that time converts much of the externally adsorbed oxygen into sub-surface oxygen. The upper spectra (a) and (b) show the result of gentle argon ion bombardment of the adsorbed layer, followed by heating to 500 K (b). The oxygen is now mostly in the sub-surface sites.

intermediate adsorbed species NH_2 or NH. This is in contradiction to the postulation of an NH intermediate using the more indirect and destructive, respectively, techniques of UPS and SSIMS.

HREELS has also been applied to the study of oxidation, and has been able to resolve difficulties of interpretation found when other techniques have been used. For example, the oxidation of aluminium has been the subject of a variety of studies using, variously, photoemission, extended X-ray absorption fine structure (EXAFS), LEED, and work function change, with differences in opinion as to the sequence of stages of oxidation, the position of the oxygen, and the degree to which a *sub*-surface phase of oxygen occurs in addition to the surface phase. The HREELS experiments of Erskine and Strong[21] were able to provide some new information about the reaction.

A single crystal of Al(111) was used and the oxidation was carried out at room temperature. Figure 5.25 shows a selection from the spectra recorded during increasing oxygen exposure at 300 K (solid lines), and after heating to 500 K in each case (dashed lines). It can be seen that the spectra are all dominated by two loss peaks, at 80 meV and 105 meV, respectively, but that the relative intensities of the peaks vary both with the degree of oxygen exposure, and the heat treatment. The 80 meV peak is attributed to an Al–O stretch mode perpendicular to the surface, where the oxygen is lying in a threefold hollow surface site, in other words adsorbed at the *external* surface. The 105 meV peak, on the other hand, is attributed to the same mode from an Al–O species in which the oxygen is in a similar threefold site *below* the surface. These attributions are strongly suggested by the relative behaviour of the peaks in Fig. 5.25, i.e. at high exposures the sub-surface peak starts to dominate, while an increase of temperature at any stage always accelerates the formation of the sub-surface phase. Further confirmation comes from the results seen in Fig. 5.26. Allowing a surface exposed to 20 L O_2 at 300 K to stand at that temperature for 2 hours causes much of the surface oxygen to convert into the sub-surface phase by penetration into the crystal; and very gentle argon ion sputtering accentuates the 105 meV peak associated with the sub-surface phase, relative to the other. Formation of a layer of approximately Al_2O_3 by heavy oxidation at 500 K and 3000 L O_2 gave a rather different spectrum, as seen in comparison in Fig. 5.27. The major peak at 110 meV agrees closely with the position of the reflectance minimum at 118 meV from bulk Al_2O_3 and there is little resemblance to the 300 K spectra at lower exposures.

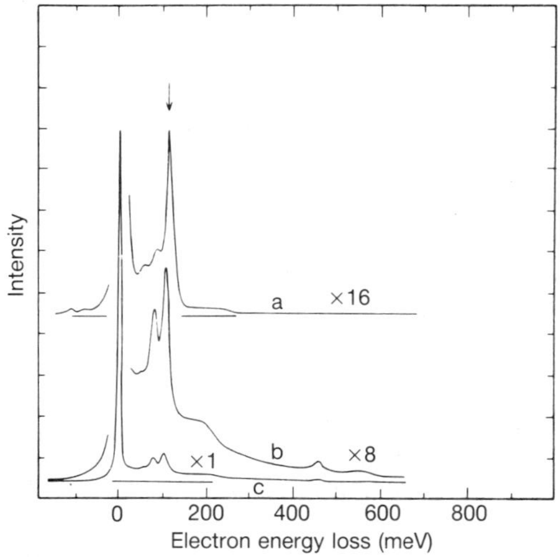

FIG. 5.27. HREELS spectra from an Al(111) surface (a) given heavy oxidation treatment by heating to 500 K during an exposure of 3000 L O_2 measured by Erskine and Strong.[21] $E_{inc} = 17.0$ eV. Same analysis conditions as in Figs 5.25 and 5.26. The spectrum after heavy oxidation does not resemble closely the spectrum from the adsorbed layer, being more typical of the formation of Al_2O_3, as judged by the agreement of the position of the peak at 110 meV with the reflectance minimum of Al_2O_3 at 118 meV. (b), (c) 20 L O_2, $E_{inc} = 4.76$ eV.

Although on the basis of what has been said earlier about the scattering mechanisms in HREELS it might be thought to be difficult for the technique to probe sub-surface species, the authors point out that it is not unreasonable for it to be able to do so. Scattering of low-energy electrons can in principle take place from a sub-surface dipole suitably oriented, although obviously 'sub-surface' in this context implies that the dipole is buried at a depth accessible to the very low energy primary electrons. In the case being considered both theoretical predictions and the results of EXAFS suggest that the depth of burial is only a fraction of an angström. Thus the surface and sub-surface dipoles should have both similar formation energies and similar scattering cross-sections; the vibrational energies (80 and 105 meV) are close together, which implies that the former is true, while varying the primary energy between 5 and 17 eV made no significant difference to the relative magnitude of the loss peaks, implying that the latter is true also.

References

1. Protheroe, A. R., Steinbrunn, A., and Gallon, T. E. (1983). *Surf. Sci.* **126**, 534.
2. Netzer, F. P., Matthew, J. A. D., and Bertel, E. (1988). *Spectroscopy of surfaces*, (eds. R. J. H. Clark and R. E. Hester), Ch. 6. John Wiley, New York.
3. Bargeron, C. B. and Nall, B. H. (1982). *Surf. Sci.* **119**, L319.
4. Pines, D. (1955). *Advances in solid state physics*, (eds. F. Seitz and D. Turnbull). Academic Press, New York, **1**, 373.
5. Curelaru, I. M., Strid, K-G, Suoninen, E., Minni, E., and Rönnhult, T. (1981). *Phys. Rev. B*, **23**, 3700.
6. Strasser, G., Rosina, G., Matthew, J. A. D., and Netzer, F. P. (1985). *J. Phys. F: Met. Phys.* **15**, 739.
7. Lüth, H. (1981). *Festkörperprobleme*, **XXI**, 117.
8. Backx, C., de Groot, C. P. M., and Biloen, P. (1980). *Applic. Surf. Sci.* **6**, 256.
9. Ibach, H. and Mills, D. L. (1982). *Electron energy loss spectroscopy and surface vibrations*, Ch. 4. Academic Press, New York.
10. Ibach, H., Horn, K., Dorn, R., and Lüth, H. (1973). *Surf. Sci.* **38**, 433.
11. Gibson, J. W. and Thomas, R. E. (1983). *Applic. Surf.* **16**, 163.
12. Powell, R. A. (1979). *Applic. Surf. Sci.* **2**, 397.
13. Stander, C. M. (1983). *Applic. Surf. Sci.* **16**, 463.
14. Soukiassian, P., Riwan, R, and Lecante, J. (1985). *Surf. Sci.* **152/153**, 522.
15. Wimmer, E., Freeman, A. J., Weinert, M., Krakauer, H., Hiskeo, J. R., and Karo, A. M. (1982). *Phys. Rev. Lett.* **48**, 1128.
16. Netzer, F. P., Strasser, G., and Matthew, J. A. D. (1983). *Solid State Commun.* **45**, 171.
17. Strasser, G. and Netzer, F. P. (1984). *J. Vac. Sci. Technol.* **A2**, 826.
18. Koma, A. and Ludeke, R. (1976). *Surf. Sci.* **55**, 735.
19. De Crescenzi, M. (1985). *Surf. Sci.* **162**, 838.
20. Erley, W. and Ibach, H. (1983). *J. Electron Spectrosc.* **31**, 161.
21. Erskine, J. L. and Strong, R. L. (1982). *Phys. Rev. B*, **25**, 5547.

6

Electron excitation: soft X-ray appearance potential spectroscopy (SXAPS), Auger electron appearance potential spectroscopy (AEAPS), and dis-appearance potential spectroscopy (DAPS)

6.1 Operation

6.1.1 *SXAPS*

Many different experimental geometries have been used for SXAPS, probably as many as there are experimenters in the field, but the basic configuration is common to all, and is summarized in the simplified diagram of Fig. 6.1, from

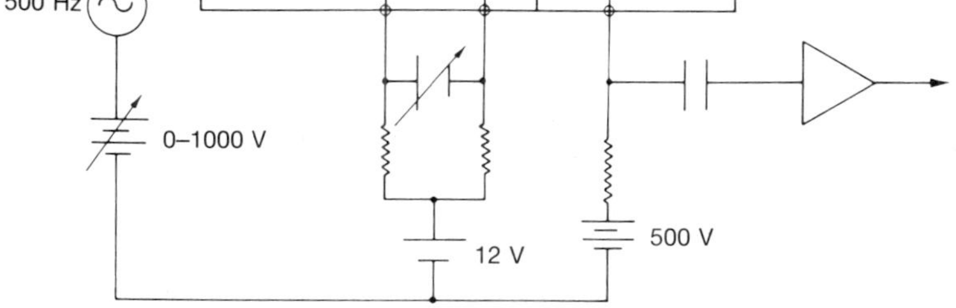

FIG. 6.1. Schematic diagram of a soft X-ray appearance potential spectrometer from Harte *et al.*[1] The target T is bombarded with electrons from F. The accelerating potential between T and F is ramped between 0 and 1000 V via a stabilized ramped power supply. The soft X-ray flux from T passes through a screen S, which retards any electrons, and strikes a photosensitive surface P. Photoelectrons liberated from P are then collected on a biased collector C. The signal-to-noise ratio in the signal can be improved by synchronous detection if a small oscillatory potential is superimposed on the ramp potential.

Harte *et al.*[1] Electrons from a filament F are accelerated through a ramped accelerating potential on to the surface of the material under investigation, which forms the target T. Soft X-rays produced in T pass through a screen S, designed to block passage of electrons, and impinge on a photosensitive surface P, from which photoelectrons are ejected. The latter are collected on a biased collector C. Thus the photocurrent reaching C is proportional to the soft X-ray flux leaving T. As the accelerating potential passes through a threshold for ionization of a core level in the surface material of T, a sharp increase occurs in the total soft X-ray flux. The edge of the increase, or step, in the plot of photocurrent against ramp voltage therefore corresponds to the excitation energy, or binding energy, of the core level, although account has to be taken of the voltage drop across the filament, and of work function terms, in order to arrive at an accurate value of the excitation energy.

The threshold features in the total yield plot are very weak, if sharp, and it is normal practice to accentuate the signals either by electronic differentiation using synchronous detection, or, as is possible these days, by mathematical differentiation after computer acquisition. The former method is still very common and consists in imposing a small AC potential on the ramp voltage at a particular frequency, and then synchronously detecting and amplifying either at that frequency or at a harmonic. Detection at the modulation frequency gives the first derivative, and at the first harmonic the second derivative; both spectra will be found in the literature, according to how authors wish to present their results. The effects of single and double differentiation on a yield curve are shown in Fig. 6.2, from Park and Houston,[2] for a uranium surface.

6.1.2 *AEAPS*

The experimental arrangement needed for AEAPS is even simpler in basic concept than in SXAPS and is shown schematically in Fig. 6.3, from Fukuda *et al.*[3] Electrons from a filament are accelerated to an anode that has an aperture immediately opposite the sample surface. Some electrons pass through the aperture and strike the sample. The primary current I_p to the sample is kept constant, and the incident electron energy is progressively increased, causing emission of secondary electrons from the sample, that are collected on the anode whose potential is held higher than that of the sample. In other arrangements[4] a separate collector is used. When the incident energy is just sufficient to excite electrons from a core state, i.e. when it reaches a threshold, a change ΔI_s takes place in the secondary electron current I_s. This change is made up of two components, $\Delta\delta$, the increase in the secondary electron coefficient, and $\Delta\eta$, the decrease in the elastic backscattering coefficient. That is, according to Pavluch and Eckertova,[5]

$$\Delta I_s = I_p(\Delta\delta - \Delta\eta). \tag{6.1}$$

Ideally, it is the quantity $I_p\Delta\delta$ that should be measured in AEAPS, but with a typically simple non-dispersive arrangement such as that in Fig. 6.3 it is the *total* yield variation ΔI_s that is recorded. Under these conditions, i.e. where all secondary electrons are collected, it is found that ΔI_s is positive, from which i

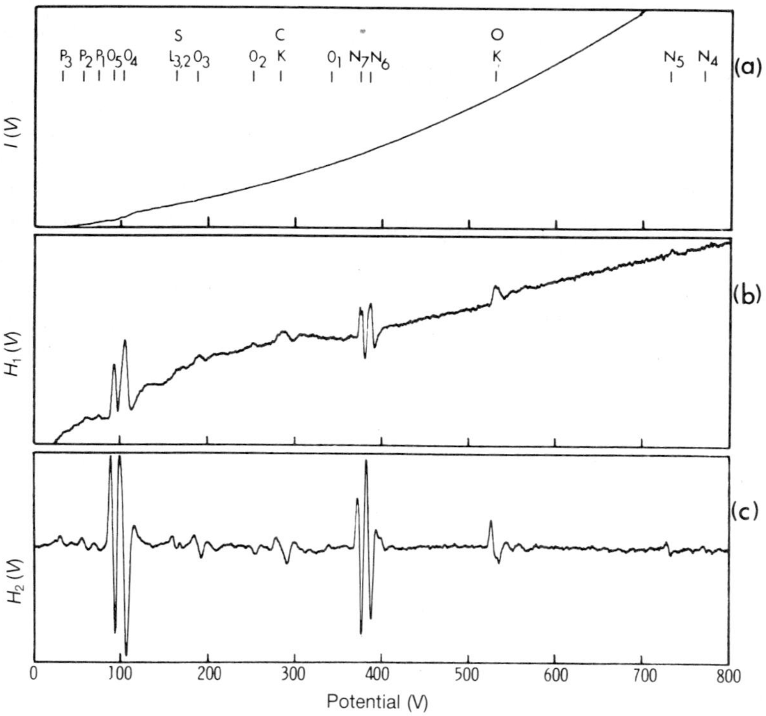

FIG. 6.2. The soft X-ray appearance potential spectrum of uranium, fromPark and Houston.[2] Spectrum (a) is the total yield, spectrum (b) the first derivative of the total yield, and spectrum (c) the second derivative of the total yield. The binding energies of the uranium energy levels and of those of some common contaminants are indicated at the top of the figure.

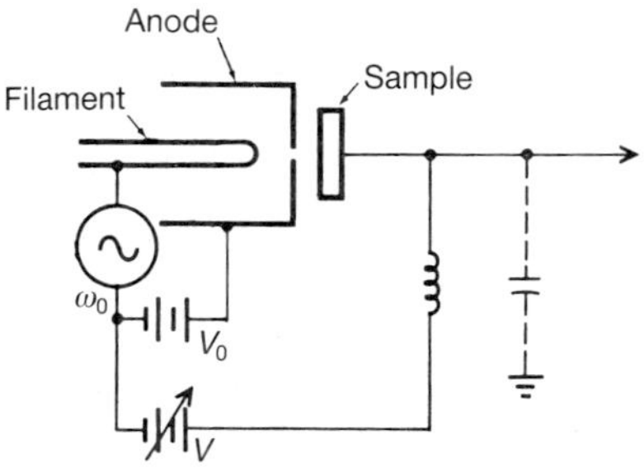

FIG. 6.3. Experimental arrangement used by Fukuda *et al.*[3] for Auger electron appearance potential spectroscopy. Electrons from the filament are accelerated through an aperture in an anode on to the sample surface. Secondary electrons emitted from the sample are collected on the anode, since the sample potential is high than that of the anode. The incident electron energy is increased progressively via a ramped power supply and, using the usual synchronous detection methods, the second derivative of the sample current is recorded as a function of energy.

follows that $\Delta\delta > \Delta\eta$. Both den Boer *et al.*[6] and Eckertova and Pavluch[7] have shown, by using retarding potential analysers, that the bulk of the electrons contributing to the increase in I_s are of energies below 30 eV, that is, true secondary electrons and not Auger electrons. The extra secondary current at threshold arises from inelastic scattering of Auger electrons produced at greater depths within the surface. To a first approximation, therefore, when electrons of all energies are collected, $\Delta I_s \approx I_p \Delta\delta$; this was confirmed rather nicely by den Boer *et al.*,[6] who recorded the secondary electron distribution from a titanium surface in two parts, one of 'loss' features only, with energy correlated with the primary beam, and the other of 'emission' features only, with energy *un*correlated with the primary beam. Their measured distribution is shown in Fig. 6.4, in which it can be seen that the current of 'emission'

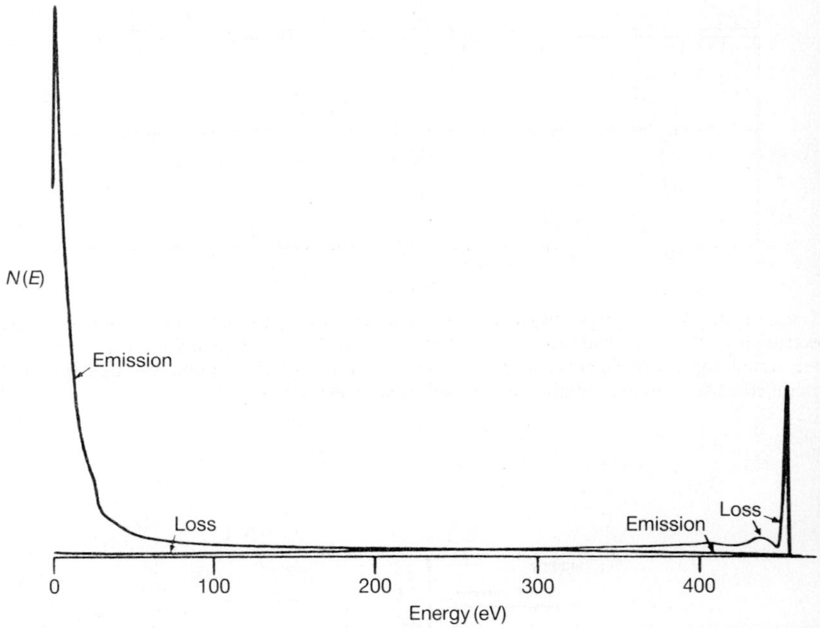

FIG. 6.4. Secondary electron distribution from a titanium surface at a primary energy of 450 eV, from den Boer *et al.*[6] By modulating the primary energy only features associated with primary energy are detected, as shown in the spectrum marked 'loss'. When the sample potential is modulated, on the other hand, only features *un*correlated with the primary energy are recorded demonstrated in the spectrum marked 'emission'. The latter arise from decay of excited states, and include Auger electrons and the so-called 'true' secondary electron peak.

electrons below $\approx$30 eV is much greater than the 'loss' current at the primary energy of 450 eV. If electrons of energies below 30 eV were retarded from reaching the collector, then the spectrum inverted from positive to negative ΔI_s.

6.1.3 *DAPS*

Referring again to eqn (6.1), it is the quantity $-I_p\Delta\eta$ that is measured in DAPS, i.e. the decrease in the elastic backscattering coefficient as the ramped primary energy reaches an ionization threshold. In order to separate the elastically scattered electrons from the rest of the secondary electron distribution some method of energy selection must be used, and the most popular form is that of the retarding-potential analyser (see Section 3.3.3). The originators of the technique, Kirschner and Staib,[8] used a two-grid spherical analyser, but the more recent work of den Boer *et al.*[6] and of Eckertova and Pavluch[7] uses three-grid analysers. Figure 6.5 shows the arrangement of den Boer *et al.*

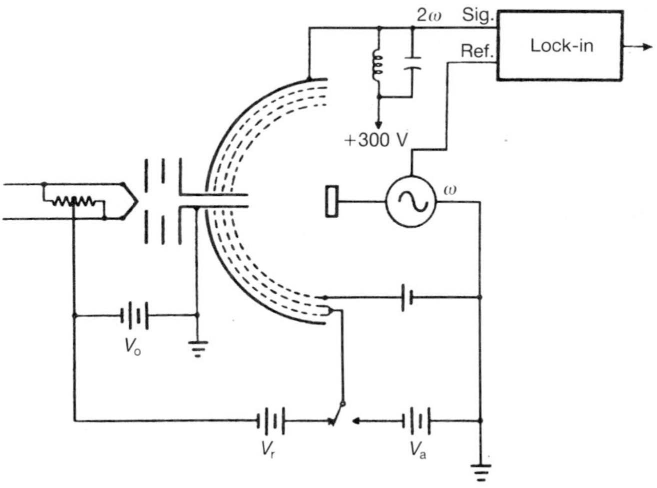

FIG. 6.5. Schematic diagram of the spectrometer used by den Boer *et al.*[6] for disappearance potential spectroscopy, based on LEED optics. Electrons from the LEED gun strike the sample, held at earth potential, at the centre of curvature of a spherical three-grid system. The first grid is also nominally at earth, but in fact has a very small bias to neutralize space charge effects. The second and third grids are connected, and their bias is set to a few volts positive with respect to the electron source, so that only elastically and near-elastically scattered electrons can pass through them. As before, the second derivative of the scattered electron current is measured as a function of incident energy, by modulation techniques.

Electrons from a low energy stabilized electrostatic gun of the type used for LEED impinge on the sample, which is at earth potential. The point of impingement is the centre of curvature of a set of three grids and collector. The first grid is nominally also at earth, but in fact is separated from the sample by a very small positive bias to overcome any space charge effects. The second and third grids are connected together and act as the analysing element; they can be biased either with respect to the electron source (i.e. connected to V_r in the figure) or with respect to the sample (connected to V_a). When in the latter position and $V_a=0$, all secondaries are allowed to pass through the grids and be collected on the collector, corresponding to an AEAPS or total yield measurement. If in the former position, and V_r is set to a few volts positive, then only elastically and near-elastically scattered electrons are collected, allowing a DAPS measurement. By adjusting V_a or V_r, as required, the fraction of total yield collected can be varied in order to study the respective contributions of Auger, true secondaries, and elastically scattered electrons, as described in the previous section.

The AEAPS and DAPS spectra, as the case may be, are obtained by recording the first or second differential of the collector current as a function of the ramped primary energy eV_0, modulated by superimposition of a small sinusoidal signal. It is more usual to record the second differential since in these techniques diffraction effects introduce structure into the secondary electron background that makes the latter poorly behaved. The energy resolution is limited by the energy spread in the primary beam, which, together with slight additional broadening arising from transmission through the grids and from the modulation, amounts[6] to about 0.5 eV.

6.2 Theory and spectral identification

6.2.1 *SXAPS*

When a sample is bombarded by electrons from a thermionic cathode, energy may be conserved either by the emission of a photon or by the excitation of a core-level electron. The two processes are depicted diagrammatically in Fig. 6.6, taken from Park.[9] In the former process the energy of the emitted photon is

$$h\nu = eV + e\phi_c + kT - \epsilon_1 \tag{6.2}$$

where $e\phi_c$ is the cathode work function and ϵ_1 is an unoccupied state above the Fermi level. Variation of eV thus produces a smoothly and continuously varying distribution of photon energies called bremsstrahlung radiation. In the process of excitation of an electron from a core level of binding energy E_B, the state ϵ_2 into which the excited electron goes is given by

$$\epsilon_2 = eV + e\phi_c + kT - \epsilon_1 - E_B. \tag{6.3}$$

The subsequent recombination of the core hole may take place by emission of a characteristic X-ray photon, in which case the total emitted X-ray flux consists of the probabilities for core-level excitations superimposed on the smoothly increasing bremsstrahlung background. Clearly there is a threshold for the excitation of the core electron, at

$$eV_{crit} = E_B - e\phi_c - kT \tag{6.4}$$

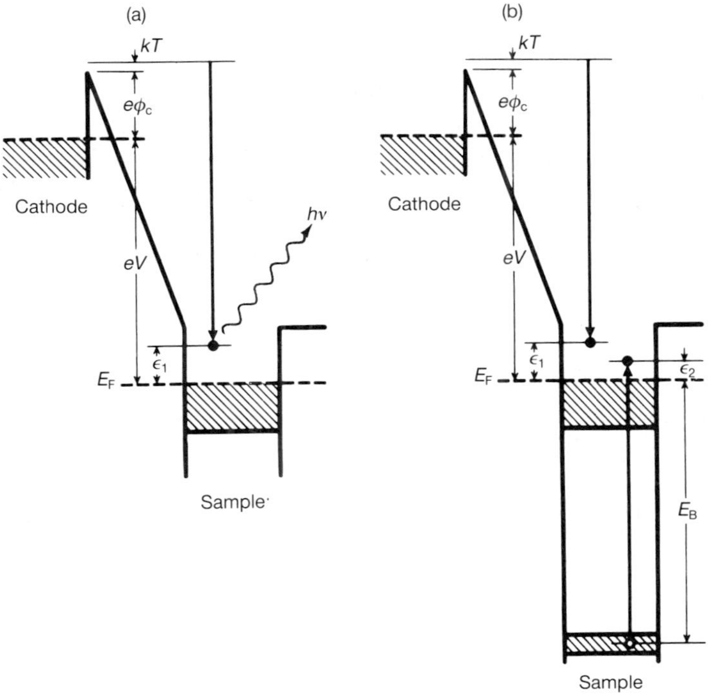

FIG. 6.6. Schematic energy diagrams illustrating the emission of continuum X-rays (a) and of characteristic X-rays (b), from Park.[9] In the former an electron is radiatively captured in a state ϵ_1 above the Fermi level, in which case the emitted photon has energy given by equation (2). Variation of electron energy eV thus provides a distribution of photon energies called bremsstrahlung radiation, and is the principle of IPES (q.v.). In (b) the incident electron may excite a core-level electron of binding energy E_B into a state ϵ_2, whose energy is given by eqn (6.3). The core hole may decay by emission of a characteristic X-ray photon (or by an Auger process), but there is a threshold for the excitation given by eqn (6.4). Above the threshold the excitation probabilities are functions of the densities of unoccupied allowed states.

and above this threshold the probabilities are dominated by the densities of unoccupied available states.

In the earlier days of SXAPS a simple model proposed by Park and Houston[10] met with considerable success in explaining the spectra observed from the first row transition elements Sc to Ni. This model is called the one-electron rigid-band model. In it the excitation probabilities are proportional to the densities of both initial and final states. The initial states, of course, are simply the filled core levels from which electrons are excited. The final states, however, must accommodate both the incident and the excited electrons, and since these two are indistinguishable the state occupancy must include all possible combinations allowed by conservation of energy and Fermi rules. The final state density of the two electrons is then simply the self-convolution of the one-electron empty state density $N(E)$, that is

$$N_2(E) = \int_0^E N(W)\, N(E-W)\, \mathrm{d}W. \tag{6.5}$$

Here the energy zero is at the Fermi level. Expression (6.5) assumes that all states have the same weighting, that is, that across the band of unoccupied states in question the transition matrix elements are constant. The excitation probability P will also depend on the energy width of the initial state, the core level, which is a function of the lifetime of the core hole created in it. A further assumption of the model is that this lifetime is sufficiently long that the energy width is infinitely narrow, in which case one can write

$$P = N \int_0^E N(W)\, N(E-W)\, \mathrm{d}W \tag{6.6}$$

where N is the core level density of states, and effectively constant. The excitation probability is thus directly proportional to the self-convolution of the density of unoccupied available states above the Fermi level.

Park and Houston[10] have illustrated qualitatively what this model means for a 3d transition metal, in Fig. 6.7 taken from their paper. The solid curve (a)

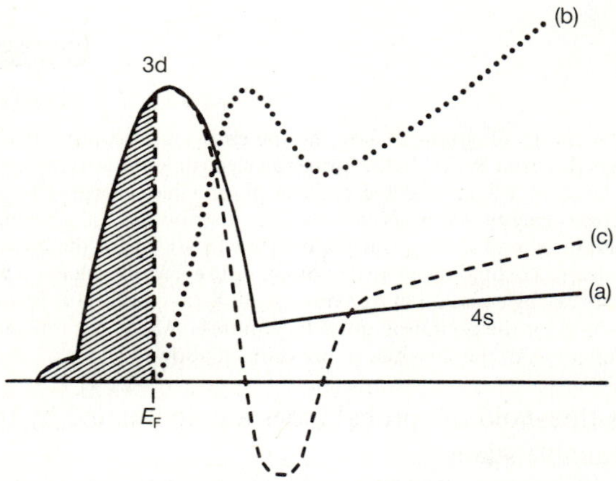

FIG. 6.7. Schematic of the densities of states in a 3d transition metal, and of the origin of the appearance potential spectrum, as given by Park and Houston.[10] Solid curve (a) is the schematic density of states for a typical 3d metal, with a d-band filled up to E_F, on top of a broad free-electron-like 4s band. Curve (b) is the self-convolution of the *unoccupied* part of the density of states, and curve (c) is the first derivative of this self-convolution. Curve (c) is the shape to be expected in the SXAPS spectrum if the simple model is correct.

represents a schematic density of states for a typical 3d metal, with a d-band, filled up to the Fermi level E_F, superimposed on a broad free-electron-like 4s band. As the transition series is ascended, E_F moves upwards as the states are filled, but the shape of the bands is assumed to remain the same, hence the name 'rigid-band'. The dotted curve (b) is the self-convolution $N_2(E)$ of the density of *un*occupied states above the Fermi level, while the dashed curve (c) is the derivative of the self-convolution. It is curve (c) that should approximate to the observed appearance potential spectrum. In other words, the model predicts a positive peak at the threshold whose width should be roughly equal to the width of the unfilled d-band, followed by a negative peak dependent on the 4s contribution. If the latter were absent, there would be equal positive and negative peaks.

The measured[10] SXAPS L-shell spectra from Ti to Mn are shown together in Fig. 6.8. It can be seen that the general trends are correct. The positive peak width decreases progressively as the d-band fills up, and the negative peak decreases in magnitude as the 4s contribution increases along the series. Clearly the Park and Houston model explains those spectra very well, even if there are minor unexplained features here and there. A corollary of the model is that if the d-band is full, then no positive-going peak should be observable, and for Cu this has indeed been found[11] to be the case, the L-shell spectrum being very weak compared to those of the other 3d metals, and exhibiting only ill-defined steps.

When SXAPS measurements were extended to other metals, however, it became apparent that the one-electron rigid band model of Park and Houston could not explain all the observations. In particular severe discrepancies were found by many workers[12] when the lanthanide series was studied. Many unexpected additional peaks were found associated with the 3d and 4d spectra, and when present as the minor elements in dilute alloys the lanthanide SXAPS line shape was essentially the same as that in the pure lanthanide, not that of the host metal, which is contrary to the prediction of the one-electron model. Sometimes, too, the intensities of the lanthanide spectra from lanthanides in very dilute alloys were seen to be far greater than would correspond to their concentration. An example is shown in Fig. 6.9 from Harte *et al.*[1] for the case of La, Ce, Pr and Nd present at less than 1 at.% in a NiCr alloy. When adjustment is made for relative sensitivities the lanthanide spectra are some two orders of magnitude greater than the Ni and Cr spectra.

The reason for the considerable discrepancies observed for the lanthanides is the presence of the highly localized 4f levels in these metals, already referred to in the section on core electron energy loss spectroscopy (q.v.). Instead of scattering into continuous states just above the Fermi level, as shown schematically in Fig. 6.6, the 3d and 4d electrons can take part in resonant scattering mechanisms in which both the d electron and the incident electron scatter into empty 4f levels. This can be represented by

$$(3d)^{10}\,(4f)^{n}+e\rightarrow(3d)^{9}\,(4f)^{n+2}. \tag{6.7}$$

Strong exchange interactions between d core holes and 4f electrons result in multiplet splittings that give rise to much of the additional structure seen in and around the 3d and 4d SXAPS spectra. (Note, however, that because of the

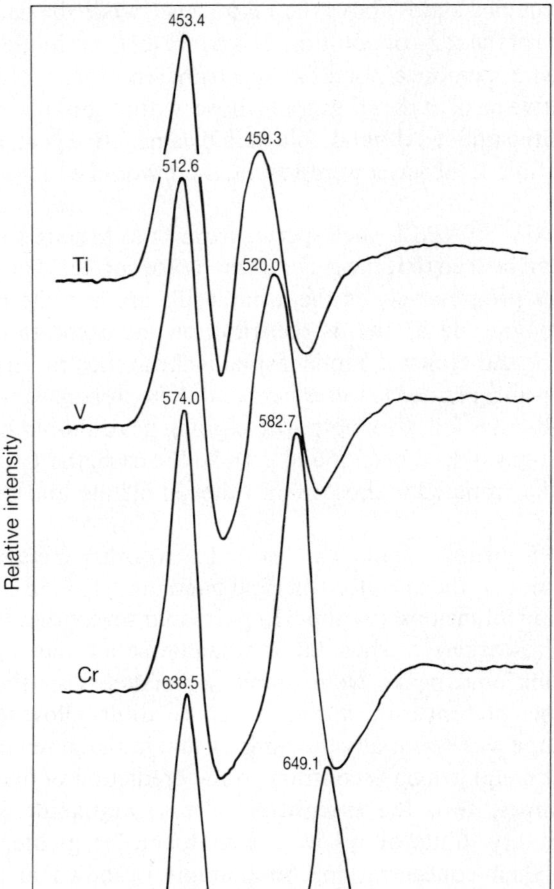

FIG. 6.8. $L_3(2p_{3/2})$ and $L_2(2p_{1/2})$ peaks in the SXAPS spectra of the transition metals from Ti to Mn, measured by Park and Houston.[10] The general trends predicted by the model of Fig. 6.7 are obeyed, in that the positive peak becomes narrower as the d-band fills, and the negative peak decreases as the 4s contribution increases.

additional electron the structure will be different from that seen near the 3d and 4d photoelectron peaks in XPS.) The resonant scattering model has been tested by recording[1] the SXAPS spectrum for Yb_2O_3. Yb metal has a full 4f band, and could not sustain a resonant process in any case, but Yb_2O_3 has a single 4f vacancy. The resonant mechanism is thus still forbidden, but there is the possibility of a non-resonant transition of the type

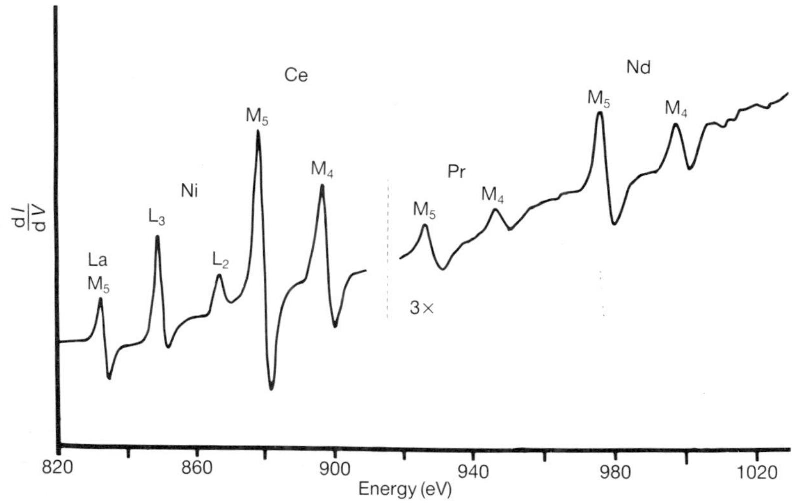

FIG. 6.9. SXAPS spectrum from an NiCr alloy containing the rare earth elements La, Ce, Pr and Nd in concentrations less than 1 at.%, as measured by Harte *et al.*[1] The $L_{2,3}$ lines of Ni can be seen along with the $M_{4,5}$ lines of the rare earths, but the relative intensities of the rare earth lines to those of nickel are anomalously large compared to the relative concentrations.

$$(3d)^{10}\,(4f)^n + e \rightarrow (3d)^9\,(4f)^{n+1} + e. \tag{6.8}$$

It was found that the spectrum was in fact very weak, indicating that the non-resonant transitions are not major contributors to the SXAPS spectra of the lanthanides. Transitions involving interaction between the core hole and the two electrons involved are the dominant ones.

A typical SXAPS spectrum will normally be shown in the doubly differentiated mode, for the suppression of the bremsstrahlung background, and will contain a series of sharp peaks corresponding to the core-level excitation thresholds, with more or less associated fine structure. A good example of a multi-element SXAPS spectrum is that from a stainless steel shown in Fig. 6.10, due to Park.[9] The L_3 and L_2 edges of the constituent elements appearing in the surface are clearly visible, and some L_1 edges as well. Interestingly, vanadium is easily recognizable in the spectrum as being present on the surface, even though it is rarely picked up in AES because of interference by chromium and oxygen. The total yield curve is also shown.

Despite the possibility inherent in expressions (6.5) and (6.6) for the extraction of information about the densities of empty electronic states above the Fermi level, by self-deconvolution of SXAPS spectra, at least for those metals for which the one-electron model applies, such a procedure has in fact seldom been attempted. The mathematical complexity has generally been regarded as too formidable, but modern computing facilities have made

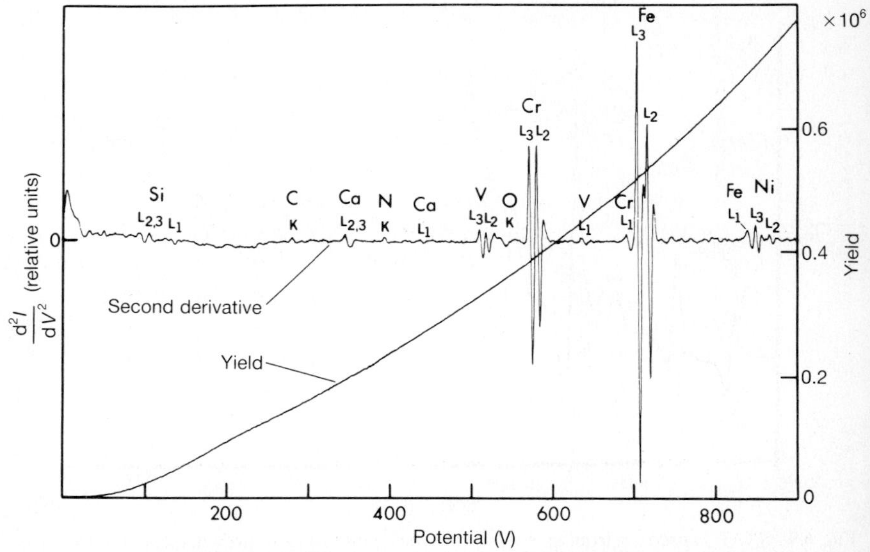

FIG. 6.10. A typical SXAPS spectrum in the doubly differentiated mode, from 304 stainless steel, measured by Park.[9] The total yield curve is also shown. The $L_{2,3}$ thresholds of the alloy constituents are visible, together with the K thresholds of the contaminants C, N, and O, and the $L_{2,3}$ thresholds of S.

inversion of the self-convolution function more routine than hitherto. It is not appropriate here to go into the mathematical operations in detail, since that can be found in the original papers, but the most successful approaches so far have involved the use of spline functions for the self-deconvolution procedure. These have been used, for example, by Dose and co-workers[13] in unravelling the SXAPS spectra of several transition metals.[14] Their application[15] of the method to copper and to two of its metallic glasses with zirconium and titanium gave the results shown in Fig. 6.11, showing (a) the SXAPS spectra of the Cu L_3 edge in the three materials and (b) the resultant densities of unoccupied states. The oscillating structure seen in pure Cu is not visible in the glassy spectra since those features are characteristic of the Cu crystalline structure. The enhancement of empty d-state density in the glassy alloys is quite marked and was explained by the authors on the basis of a d-state hybridization model, e.g. in the case of $Cu_{60}Zr_{40}$ a mixing of the Cu 3d and the Zr 4d wave functions.

A rather more sophisticated approach to the procedure of self-deconvolution of SXAPS spectra has been used by Schleicher *et al.*[16] Before carrying out the self-deconvolution they eliminated the effects of lifetime broadening due to the core hole, using a de-broadening function of Lorentzian shape and of width taken from published values. After integration of the spectra and de-broadening, self-deconvolution was performed by a linear least-squares

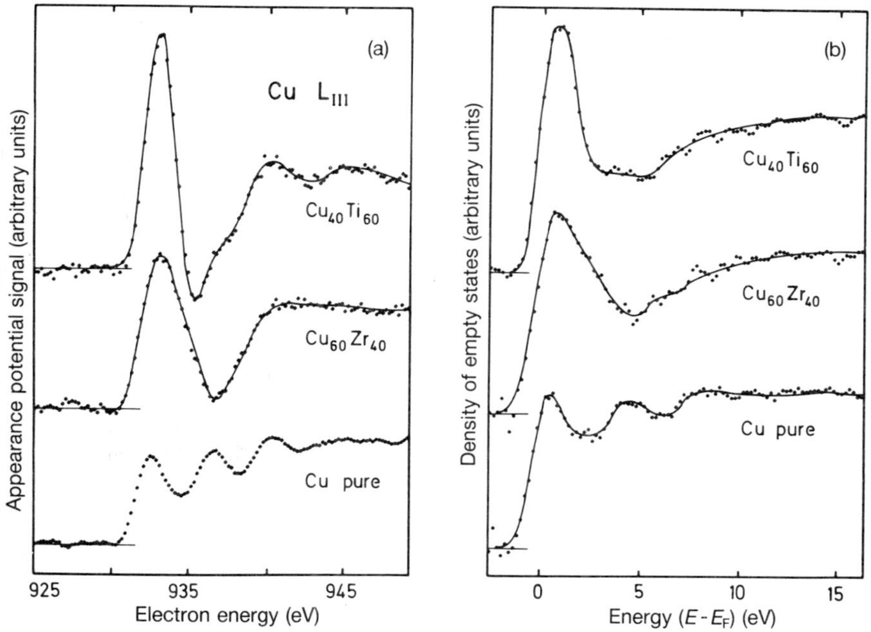

FIG. 6.11. Experimental SXAPS spectra from pure copper and from two glassy CuZr alloys, on the left, and the densities of unoccupied electronic states derived from them, on the right. The measurements and the self-deconvolution procedures were carried out by Dose *et al.*[15] The enhancement of empty d-state density just above the Fermi level in the alloys is very marked, and is attributed to a mixing of Cu 3d and Zr 4d wave functions.

method using quadratic spline functions. The reader is referred to their paper for the computational details.

6.2.2 *AEAPS and DAPS*

As will be clear from the foregoing sections, AEAPS and DAPS provide much the same type of information as is obtainable from SXAPS, the principal differences being in the nature of the secondary radiation used to signal the onset of a core-level excitation. Since AEAPS measures the change in the *total* secondary yield at the threshold, which generally increases, whereas DAPS measures the change in the elastic yield only, it follows that in most cases the spectra will be inverted mirror images of each other, i.e. 180° out of phase. This can be seen clearly in Fig. 6.12 for the 2p and 2s thresholds of titanium recorded by the two techniques, from den Boer *et al.*[6] The reason why the changes in the total yield at threshold are usually positive have already been discussed.

Since both the AEAPS and SXAPS spectral features are positive-going, and reflect events following the same core-level ionization, they should appear

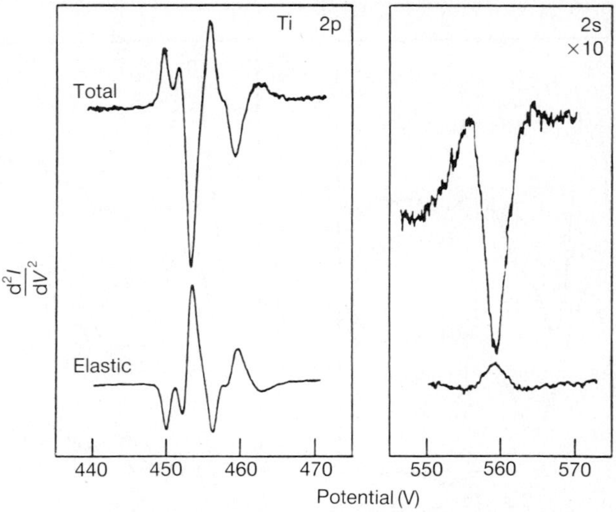

FIG. 6.12. Comparison by den Boer *et al.*[6] of the total yield (i.e. AEAPS) and elastic yield (i.e. DAPS) spectra of the Ti 2p and 2s levels. Since the total yield increases at threshold, while the elastic yield decreases, the two spectra are mirror images of each other.

similar, and indeed generally do, as can be seen in Fig. 6.13 in which the two types of spectra are compared for the $N_{4,5}$(4d) level of cerium in both metallic and oxidized forms, from Hatwar and Chopra.[17] Although there are differences in the magnitude of individual peaks, particularly in the fine structure associated with 4d–4f resonances, there is a close overall resemblance. It is pertinent to ask, then, why SXAPS should be used at all when AEAPS gives the same information from rather similar spectra and when the incident current density required in both AEAPS and DAPS to give the same signal strength as in SXAPS is an order of magnitude lower. The latter experimental fact is due to the very low fluorescence yields of soft X-rays compared to the probability of non-radiative, i.e. Auger, decay of the core hole. The reason for the preference for SXAPS lies in the nature of the spectral background observed in AEAPS and DAPS, which is poorly behaved, that is, exhibits much additional structure, due to diffraction effects on the secondary electron yield. These effects will of course depend on the degree of crystallinity near the surface, but they can extend over several hundred eV, as can be seen in the AEAPS spectrum from titanium in Fig. 6.14, from Fukuda *et al.*[3] When the crystallinity is destroyed by oxidation the structure disappears, as in the lower spectrum after exposure to 30 L of oxygen. Such structure cannot, of course, appear in SXAPS.

In deciding which of these three techniques to use in, say, chemisorption studies there is thus a trade-off. SXAPS gives a cleaner spectrum but requires high current densities to achieve acceptable signal strengths; AEAPS and DAPS can be operated satisfactorily with primary currents of a few μA in a

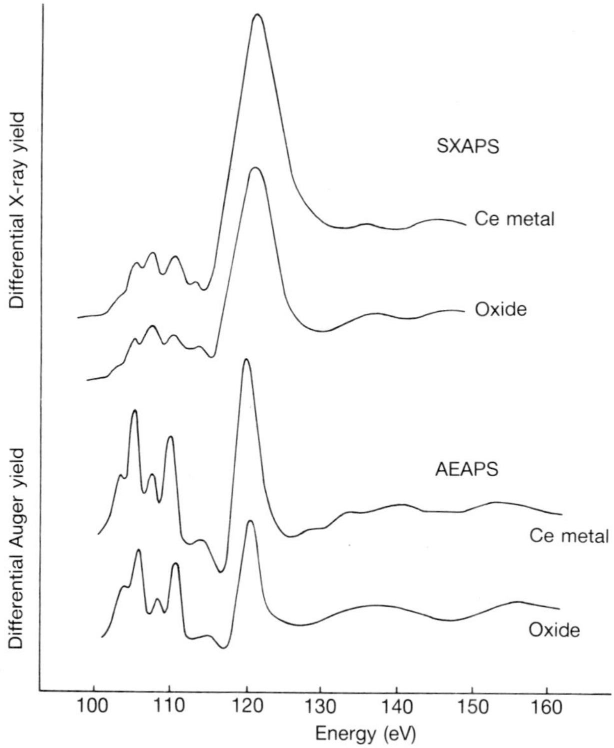

FIG. 6.13. Comparison of SXAPS and AEAPS spectra from cerium metal and cerium oxide, from Hatwar and Chopra.[17] When normalized to constant current and signal-to-background ratio, the intensities in AEAPS are about an order of magnitude greater, due to the low probability of X-ray fluorescent yield in the soft X-ray region compared to the probability of non-radiative, i.e. Auger, recombination.

defocused beam, but diffraction effects cause undesirable and sometimes severe distortions in the background. The answer is probably to have all three available in the same system so that their complementary nature can be exploited fully.

6.3 Applications

6.3.1 *SXAPS*

Amongst the promising alternative sources of energy being studied in view of the eventual exhaustion of fossil fuels is that of hydrogen, in the so-called 'hydrogen economy'. One of the likely problems in connection with this economy is the way in which large quantities of hydrogen can be safely stored and transported without having to resort to very bulky and heavy pressurized

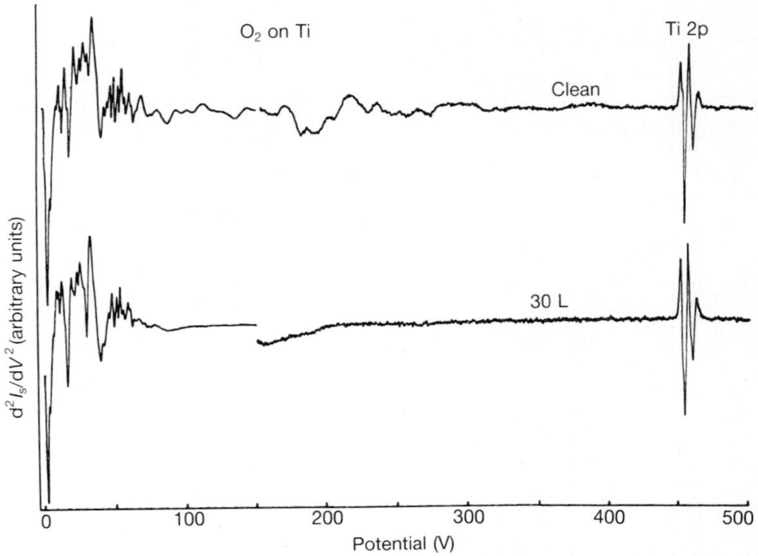

FIG. 6.14. The AEAPS spectrum from clean and oxidized titanium measured by Fukuda *et al.*[3] On the clean surface additional structure due to diffraction effects can be seen in the low-energy end of the spectrum, extending over several hundred eV. Most of this structure disappears on exposure to oxygen, since the surface crystallographic order is destroyed.

containers. In addition there is a requirement for rapid availability of the hydrogen from the storage medium when power consumption increases rapidly. A much-investigated method of storing hydrogen in a small volume is that of dissolution in certain binary intermetallic compounds, of which one component is invariably a transition metal, while the other may also be a transition metal, or may be a lanthanide or an alkaline earth metal. The archetype of these intermetallics is TiFe, although it is now known that that particular one is not necessarily the most efficient for the requirement.

All hydrogen storage intermetallics have to be 'activated' in order to be able to take up hydrogen. The activation process varies a good deal from one to another, and from one experimenter to another, but consists basically of high temperature treatment in hydrogen. The question then arises of the nature of the surface and of any surface film caused by the activation, and of the role of that film in the dissociation and dissolution processes of hydrogen. This question was investigated for TiFe by Sicking and Jungblut[18] using AES and SXAPS.

All surface treatments, including the initial cleaning of the polycrystalline TiFe and the subsequent hydrogen treatment, were carried out in vacuum to avoid interference from ambient gases, principally oxygen. Pure Ti and Fe reference materials were mounted alongside the TiFe and given the same treatments. The two techniques could be used sequentially in the same chamber.

After a sequence of annealing and ion bombardment treatments the TiFe was almost atomically clean as determined by AES; only very small O and C peaks remained. Exposure of this surface to increasing pressures of hydrogen resulted in the SXAPS $L_{2,3}$ spectra for Ti and Fe shown in Fig. 6.15. The most

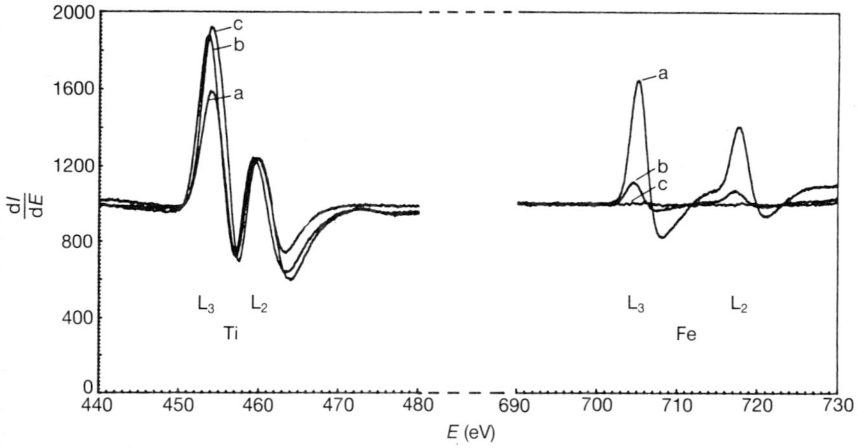

FIG. 6.15. Effect on the Ti and Fe $L_{2,3}$ SXAPS spectra of exposing TiFe to increasing hydrogen treatment. The letters refer to: (a) 6.5×10^{-5} mbar H_2 at 300°C for 30 min. (b) 6.5×10^{-5} mbar H_2 at 800°C for 30 min. (c) 4×10^{-4} mbar H_2 at 800°C for 120 min. Notice the disappearance of Fe from the surface after treatment (c). The nature of the Ti spectrum is identified in Fig. 6.16. From Sicking and Jungblut.[18]

immediately noticeable result is that after the last, prolonged, hydrogen treatment Fe had disappeared from the surface completely. Profiling with ion bombardment and AES established that the surface layer contained no Fe to a depth of about 100 monolayers. The interesting question is then the nature of this titanium-rich layer, particularly since the Auger spectrum showed that the O and C concentrations at the surface had increased many times compared to what they were before hydrogen treatment. Now Park and Houston[2] have compared the Ti $L_{2,3}$ SXAPS spectra before and after complete hydriding and oxidizing, respectively, with the result seen in Fig. 6.16. Firstly, the oxide spectrum shows a distinctive double-peak structure on both L_3 and L_2 threshold peaks, a structure that is definitely not present in the spectra of Fig. 6.15. Secondly, on hydriding the negative-going part of the L_3 signal disappears; such an effect is not apparent in Fig. 6.15. Thus the titanium in the surface layer is neither completely oxidized or completely hydrided, that is, the presence of either or both stoichiometric TiO_2 or TiH_2 seems to be excluded. The authors conclude that the surface layer is probably not of compound nature at all, but consists of a solution of hydrogen in Ti containing also some

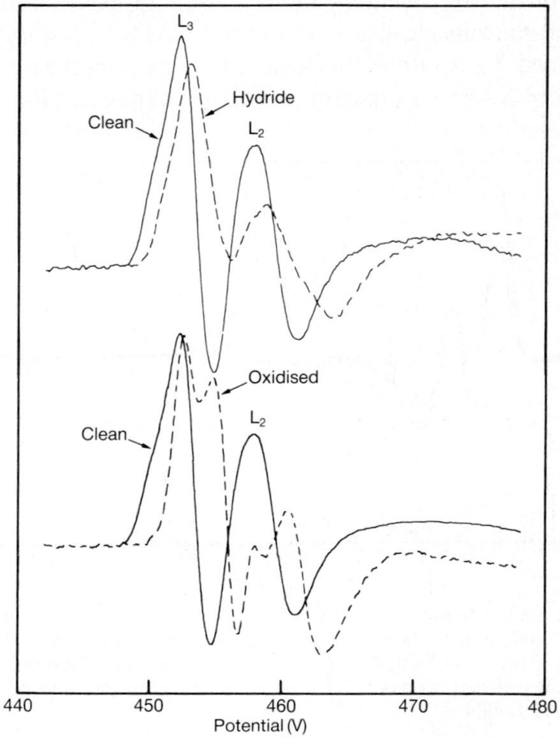

FIG. 6.16. Comparison by Park and Houston[2] of the Ti $L_{2,3}$ SXAPS spectra from clean, hydrided and oxidized Ti. The shapes of the oxidized and hydrided spectra are quite different. The hydrided spectrum is similar to that seen in Fig. 6.15 after hydridation of TiFe, apart from the absence of the negative excursion.

dissolved oxygen, and suggest that the chemical state can be described as TiH_n, with n much less than 2. The oxygen is postulated as being in solution since there is no indication of oxide-like structure in the Ti SXAPS spectra. From their finding of a TiH_n-like surface layer the authors were able to show on thermodynamic grounds that such a surface should be catalytically very active in the dissociation of hydrogen followed by incorporation into the bulk, i.e. mass storage. The effect of the dissolved oxygen was not discussed, but clearly the presence of oxygen can only be deleterious, and for an efficient storage medium some form of activation should be established that minimizes or eliminates oxygen from the surface layer.

Despite the high primary currents that are used generally, SXAPS has been used to study surface reactions, the most comprehensive range of experiments in this field of application being that of Nyberg.[19] A selection of his spectra recorded[19a] during oxidation of various transition metals is shown in Fig. 6.17, for Ti and Cr. By using a low temperature silicon surface barrier detector to detect the soft X-ray flux directly he was able to reduce the noise level so much that he could operate at a primary current of only 150 μA. This, with a defocused beam, resulted in a sufficiently low current density that no electron beam effects were apparent.

In the spectra in the figure, two regimes can be distinguished. The first, attributed to the chemisorption stage, occurs in the range of oxygen exposures up to 10 L, and is characterized by a single-peaked O 1s spectrum and an unchanged metal 2p spectrum. Beyond that exposure, multiple peaks start to appear in the O 1s spectrum, and additional structure in the form of a shoulder on the high energy side of the metal 2p threshold peak. Both these changes are attributed to the formation of incipient oxide, and reflect changes taking place in the unoccupied band structure, as well as the shift in energy of the 2p level due to chemical combination. The appearance at very low exposures of an O 1s threshold peak, while the oxygen is still in the chemisorbed state, implies the introduction of an increase in the local density of states near the oxygen atom. The multiple peaks seen on incipient oxidation are supposed to be characteristic of the densities of unoccupied states in the oxide band structure, but this is not yet proven.

The oxygen peaks when plotted as a function of exposure can be used as monitors of the course of oxidation. In Fig. 6.18 the magnitude of the first O 1s peak at threshold is plotted as a solid line, and the second O 1s peak as a dashed line, vs. oxygen exposure of clean surfaces of Ti, Cr, Fe, and Ni. In some cases the magnitudes of other oxygen features from AES and soft X-ray emission (SXE) are also plotted against exposure, for comparison. All exposures were at room temperature. In the 'first-peak' plots there is a kink visible (for Ti only just visible) at exposures of 10–20 L, which is now generally accepted to be the point at which nucleation of the chemisorbed oxygen–metal system is taking place to form the first oxide islands. Confirmation of this mechanism is seen from the fact that the 'second-peak' plot starts at just these points for each metal. As exposure to oxygen continues, both peaks follow the same trends and eventually saturate at the same exposure. Thus for these metals at least the SXAPS spectra provide a simple way of determining when an oxide is formed.

6.3.2 *AEAPS*

There have not been many applications of AEAPS published in the literature (and none of DAPS), but the few that have appeared have been interesting.

One particularly fascinating study is that made by Dose *et al.*[20] on the comparison of the empty electronic state structure between solid and liquid nickel. As they rightly point out, there were no other studies at the time of the

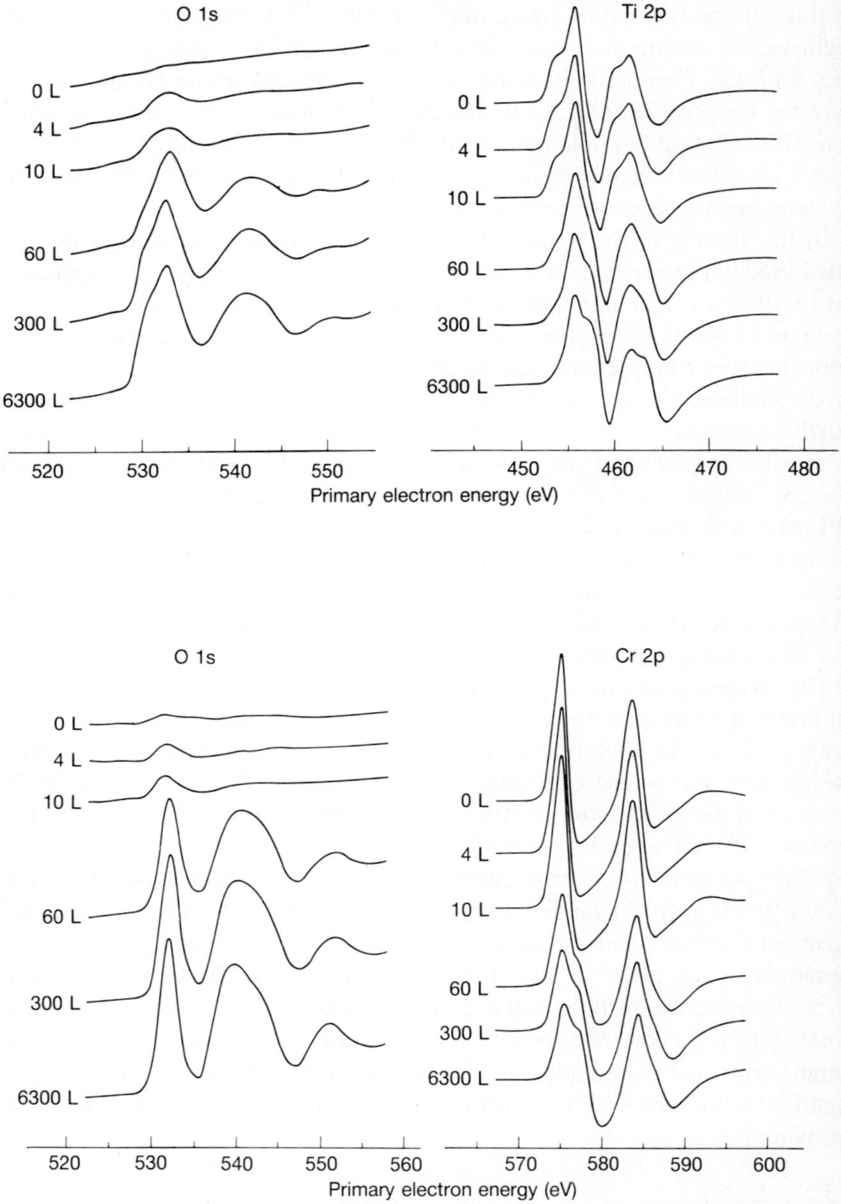

FIG. 6.17. Oxygen K, and Ti and Cr $L_{2,3}$, SXAPS spectra recorded by Nyberg[19a] as a function of oxygen exposure at room temperature. A low primary electron current in a defocused beam was used in order to minimize possible electron beam effects. Up to 10 L exposure the oxygen structure is single-peaked and the metal $L_{2,3}$ spectra are unchanged. Beyond that exposure both the oxygen and metal spectra change rapidly, with much additional structure and changes in peak shape, attributed to the formation of incipient oxide with consequential changes in unoccupied band structure.

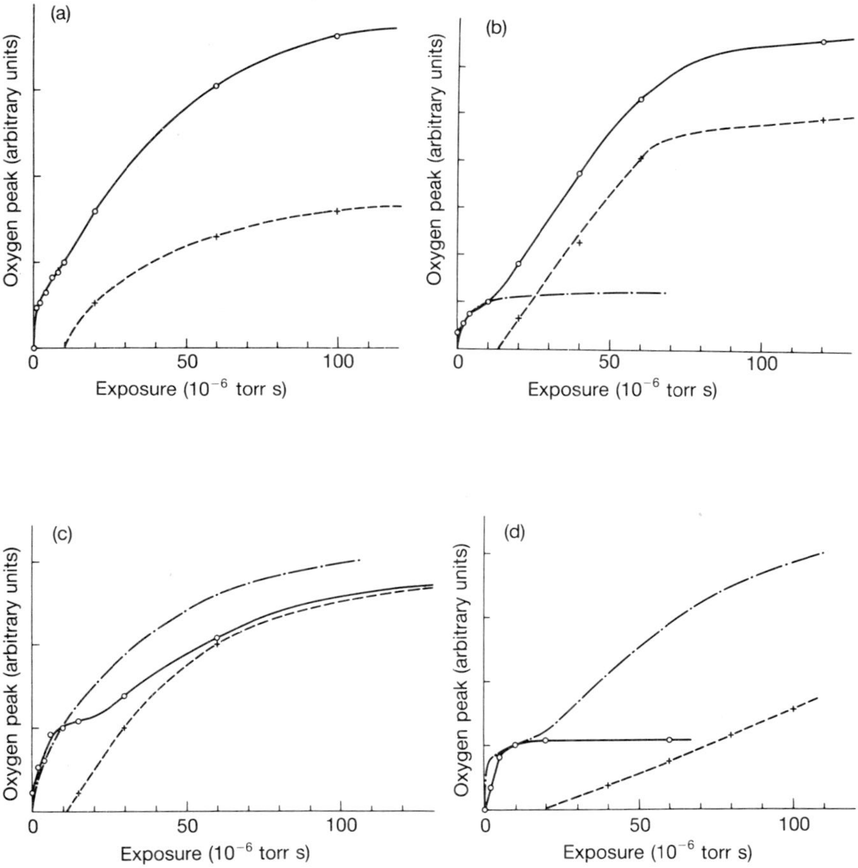

FIG. 6.18. Plot by Nyberg[19a] of the heights of the oxygen SXAPS peaks in Fig. 6.17, above threshold, as functions of the oxygen exposure. Solid line corresponds to the first peak, the dashed line to the second peak, and the dot-dashed line to the oxygen concentration as measured by AES. (a) Ti, (b) Cr, (c) Fe, (d) Ni. The kink in the 'first-peak' plot corresponds to the exposure at which nucleation of oxide to form islands takes place. The second peak appears at that point in confirmation. The differences between the SXAPS and the AES curves is due to the different probing depths in the two techniques, that for AES being smaller.

empty electronic states of a liquid metal; the experimental difficulties in dealing with liquid metals of significant vapour pressure are great.

Their results are given in Fig. 6.19. The Ni L_3 AEAPS spectra for the solid and the liquid are the two lower dotted traces, and they are compared with an SXAPS spectrum from the solid, and with an appearance potential calculated from the theoretical density of states in the upper panel, suitably broadened. In the theoretical spectrum and in the two experimental spectra from the solid, structure is seen that arises from certain critical points in the Ni fcc crystal structure. However, in the liquid spectrum the structure has been washed out. In addition the d-band related peak at threshold has broadened by $\sim 10\%$ and

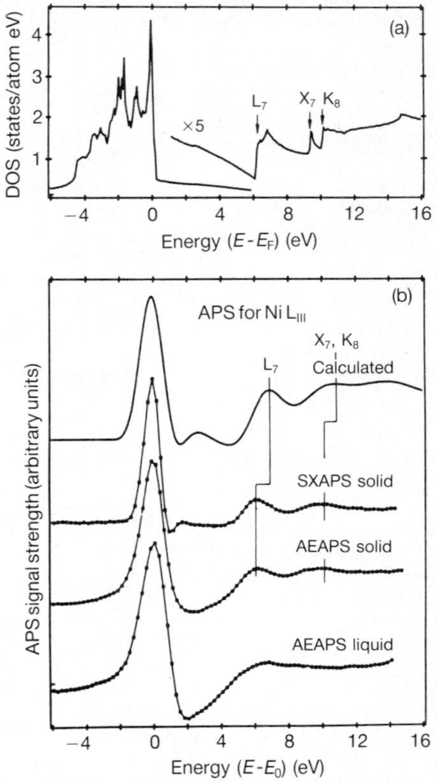

FIG. 6.19. Nickel L_3 AEAPS spectra from solid and liquid nickel (a), compared with (b) the SXAPS spectrum from solid nickel, and with calculated appearance potential spectrum based on the theoretical density of states in the upper panel, from Dose *et al.*[20] The structure seen in the spectra from the solid, and in the theoretical spectrum, is largely washed out in the liquid spectrum, and in addition the threshold peak is broader with a larger negative wing. The observations imply that the densities of empty states in the liquid increase with respect to the sp-contribution.

the negative-going peak is somewhat greater. The latter observations imply that in the liquid the density of empty d-states increases with respect to the sp-contribution. The authors confirmed this conclusion in a very elegant way. First, they carried out self-deconvolutions of the solid and liquid state spectra using their own previously[14] described mathematical method, normalizing the result at 16 eV above the Fermi level, giving the densities of empty electronic states seen in Fig. 6.20. Then they fitted free-electron parabolas to the curves in the region 12–15 eV above the Fermi level, with the origins in the valence band minimum, and subtracted the respective parabolas from the density of state curves. Such a procedure is a first approximation to the isolation of the d-band contribution in each case. If the empty d-band contribution was equated to the known figure of 0.6 d-holes per atom for solid

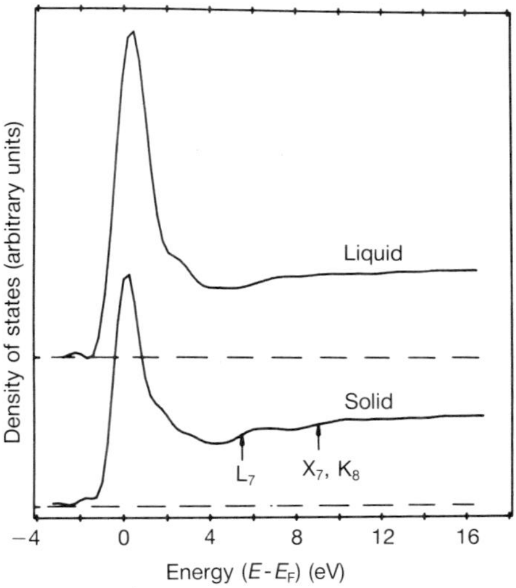

FIG. 6.20. Densities of empty states derived by Dose *et al.*[20] by self-deconvolution of the experimental AEAPS L_3 spectra for solid and liquid nickel. The data were normalized at 16 eV above the Fermi level, based on the assumption that the densities of states of a metal will become free-electron-like sufficiently far from E_F.

Ni, then in the liquid the contribution was found to be exactly 1.0 d-holes. Since the total number of valence electrons per atom remains constant, there must be a d→s, p redistribution of electrons, i.e. a weakening of the d–d interaction in the liquid. This is in line with what is found from electrical resistivity measurements.

In another interesting application of AEAPS, that of the reaction of Ti with N_2, O_2 and CO studied by Fukuda *et al.*,[3] use was made of the aforementioned diffraction effects that are generally a nuisance in AEAPS. Figure 6.14, from their paper, shows the effects associated with a Ti surface, and their disappearance on exposure to oxygen.

In the detailed Ti AEAPS spectrum, in Fig. 6.21 also from their paper, the $2p_{3/2}$ edge shows a characteristic shoulder from the clean surface, due to unfilled bonding orbitals in the 3d band. This shoulder also disappears on oxygen exposure, as illustrated in the figure, and at high exposure another shoulder appears on the other side of the threshold peak. The area of the first shoulder at successive exposures can be estimated by subtracting from the spectra the spectrum in which it has disappeared, as shown by the dotted line. Similar results were found on exposure to nitrogen and to carbon monoxide, and the decreasing shoulder area could be estimated in the same way. Work function measurements by the retarding potential method were made concurrently with the AEAPS observations, and the amplitude of a prominent diffraction feature at ~170 eV was also monitored as a function of exposure.

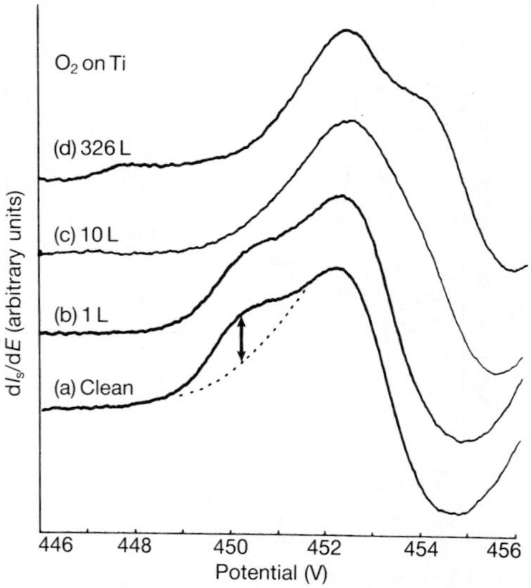

FIG. 6.21. Detailed Ti L_3 AEAPS spectra measured by Fukuda *et al.*[3] during exposure of Ti to oxygen at room temperature. The shoulder on the low energy side of the clean spectrum is characteristic of unfilled bonding orbitals in the 3d-band. On oxygen exposure, this shoulder quickly disappears, and another shoulder appears on the high energy side. The method of estimating the low energy shoulder height is shown.

The changes in shoulder area, diffraction amplitude, and work function for adsorption of the three gases at room temperature are shown in Fig. 6.22, as a function of the logarithm of the exposure. Clearly the loss of the diffraction features is most rapid for oxygen, and only in the case of oxygen is the loss complete, indicating that only for oxygen adsorption is destruction of any Ti surface structure (even for polycrystalline Ti) complete. For adsorption of both oxygen and nitrogen the work function goes through a minimum value *below* that of the clean metal before increasing again, interpreted as arising from initial dissolution of the two gases in the Ti surface, so that the positive poles of the dipoles due to the adsorbed atoms are outwards, thus lowering the work function. Such dissolution does not apparently occur with adsorption of carbon monoxide, since no minimum is observed and the work function increases steadily. In every case the shoulder disappeared at or near an exposure of ~ 10 L, showing that the Ti bonding orbitals are quenched by very small quantities of adsorbed gas. The additional shoulder on the other side of the Ti $2p_{3/2}$ threshold peak after oxygen adsorption could be identified with a peak in the theoretical density of states of TiO, but no such identification with any compound could be made for the reactions with nitrogen and carbon monoxide.

Thus adsorption of oxygen leads in the earliest stages to dissolution of

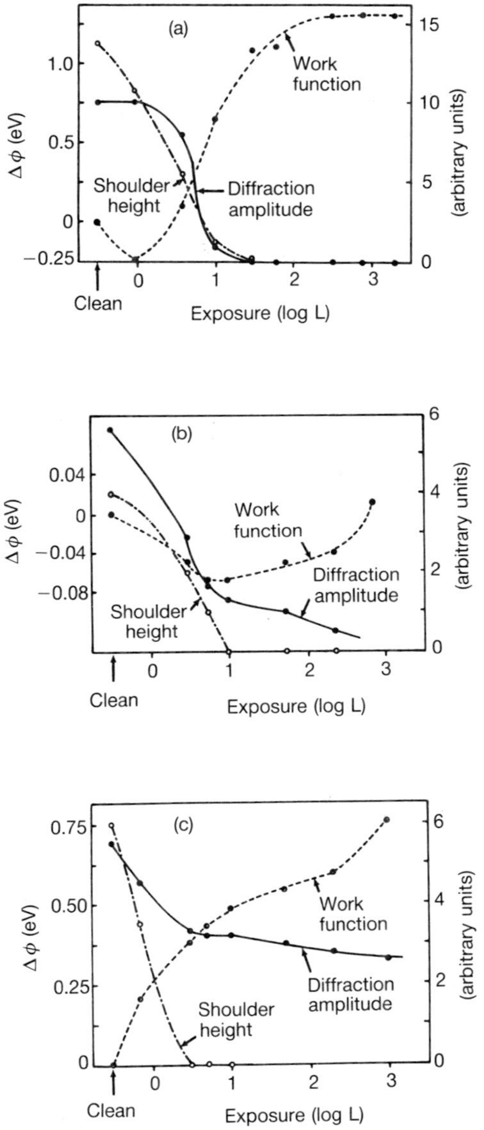

FIG. 6.22. The changes in low energy shoulder height as seen in Fig. 6.21, in work function, and in diffraction structure amplitude (see Fig. 6.14), during the reaction of titanium with (a) oxygen, (b) nitrogen, and (c) carbon monoxide, measured by Fukuda *et al.*[3] The exposures are on logarithmic scales. Only in the case of oxygen reaction at room temperature is the diffraction feature completely lost, indicating that only for oxygen adsorption is the surface structure destroyed. The minima in the work function variations with oxygen and nitrogen exposure indicate initial dissolution in the metal, followed by external adsorption. In every case the unfilled Ti bonding orbitals are quenched by very small quantities of adsorbed gases.

oxygen atoms following dissociation, followed by rapid nucleation to form a surface oxide approximating to TiO, which quenches the bonding orbitals of Ti almost at once, and destroys any surface order. Adsorption of nitrogen also leads to dissolution of nitrogen atoms in the first stages, but not to the same extent as for oxygen, and nucleation is probably not as rapid or as comprehensive since some diffraction effects can still be seen after prolonged exposures. The bonding orbitals are also quenched rapidly, but there is no firm evidence for the formation of any particular compound. Carbon monoxide adsorption is different in that no lowering of the work function occurs, implying no dissolution of oxygen and therefore no dissociation of the adsorbing gas. Nucleation does not seem to take place over the whole surface, since quite pronounced diffraction effects can still be observed at high exposures. However. the Ti bonding orbitals are still quenched rapidly, suggesting that perhaps nucleation does not in fact occur, but that the surface compound formed is the same right through the adsorption regime from the start of exposure.

References

1. Harte, W. E., Szczepanek, P. S., and Leyendecker, A. J. (1983). *J. Less-Common Metals* **93**, 189.
2. Park, R. L. and Houston, J. E. (1974). *J. Vac. Sci. Technol.* **11**, 1.
3. Fukuda, Y., Elam, W. T., and Park, R. L. (1978). *Applic. Surf. Sci.* **1**, 278.
4. Houston, J. E. and Park, R. L. (1972). *Phys. Rev. B* **5**, 3808.
5. Pavluch, J. and Eckertova, L. (1985). *Surf. Sci.* **162**, 896.
6. den Boer, M. L., Cohen, P. I., and Park, R. L. (1978). *Surf. Sci.* **70**, 643.
7. Eckertova, L. and Pavluch, J. (1984). *Czech. J. Phys.* **B34**, 622.
8. Kirschner, J. and Staib, P. (1973). *Phys. Lett.* **42A**, 335.
9. Park, R. L. (1975). *Surf. Sci.* **48**, 80.
10. Park, R. L. and Houston, J. E. (1972). *Phys. Rev. B* **6**, 1073.
11. Park, R. L. and Houston, J. E. (1973). *J. Vac. Sci. Technol.* **10**, 176.
12. e.g. Chamberlain, M. B. and Baun, W. L. (1974). *J. Vac. Sci. Technol.* **11**, 441; Chopra, D., Babb, H., and Bhalla, R. (1976). *Phys. Rev. B* **14**, 5231; Chopra, D., Martin, G., Naraghi, H., and Martinez, L. (1981). *J. Vac. Sci. Technol.* **18**, 44. Chopra, D. and Chung, H. K. (1982). *Applic. Surf. Sci.* **11/12**, 30; and references contained therein.
13. Dose, V., Fauster, Th., and Gossmann, H-J. (1981). *J. Comp. Phys.* **41**, 34.
14. Dose, V., Fauster, Th., and Scheidt, H. (1981). *J. Phys. F.* **11**, 1801.
15. Dose, V., Reusing, G., and Güntherodt, H. J. (1984). *Solid State Commun.* **49**, 1081.
16. Schleicher, K-Th., Schulz, S. W., Gmeiner, R., and Chun, H-U. (1983). *J. Electron Spectrosc.* **31**, 33.
17. Hatwar, T. K. and Chopra, D. R. (1985). *Applic. Surf. Sci.* **22/23**, 267.
18. Sicking, G. and Jungblut, B. (1983). *Surf. Sci.* **127**, 255.
19. Nyberg, C. (1975). (a) *Surf. Sci.* **52**, 1; (1977). (b) *Surf. Sci.* **65**, 389; (1979). (c) *Surf. Sci.* **82**, 165.
20. Dose, V., Drube, R., and Härtl, A. (1986). *Solid State Commun.* **57**, 273.

7

Electron excitation: inverse photoemission spectroscopy (IPES)

7.1 Operation

The technique of Inverse Photoemission Spectroscopy, or IPES, was, and still is by a few workers, called Bremsstrahlung Isochromat Spectroscopy (BIS). Both names give a clue to the mode of operation, which in principle, and to a certain extent in practice, is quite simple. The surface is irradiated with monoenergetic electrons and the bremsstrahlung photons so created are detected at a fixed wavelength, while the energy of the primary electrons is scanned.

If no angular resolution is required in the measurements, then it is adequate to provide primary electrons simply from a hot filament in front of the surface being studied. However, much additional information can be obtained from angularly resolved studies, which require a focused beam, and in any case most workers already have available on their equipment one or other of the various designs of electron gun described under Section 3.2.1. Primary currents on the sample need to be of the order 1–10 μA to produce an adequate emitted photon intensity. The energy spread in a typical gun would be about $\sim$0.5 eV, which in fact is the limitation on the resolution achievable in the spectra, although some workers using specially constructed guns have obtained better energy resolution. The penalty for improved resolution is often lower primary current.

In the experiments concerned with spin polarization, very special electron sources are needed, such as that of the negative-electron-affinity surface of GaAs,[1] which at photoemission threshold produces an emission of which about 40% is spin-polarized. This source also has a low energy spread ($\sim$0.1 eV) and is capable of providing high currents.

Although an incident electron source is common to all IPES instruments, the kinetic energy at which the emitted bremsstrahlung photon flux is detected is the point of division of the technique into its present two unequal parts. If the chosen kinetic energy is in the soft X-ray region, as used by Lang and Baer[2] and by Hillebrecht *et al.*,[3] it is very convenient to use the type of crystal monochromator designed for the monochromatization of the soft X-rays (e.g. Al K_α, $h\nu = 1486.6$ eV) used in XPS, only, of course, in reverse. That is, a detector is placed where the electron bombarded anode would be for XPS. The arrangement is shown in Fig. 7.1 (courtesy VG Scientific). Although IPES and

BIS are really one and the same, the former apellation being the new one, it seems that the acronym BIS is being reserved for *X-ray* IPES, to judge by the usage.

If the chosen kinetic energy, on the other hand, is in the ultra-violet region, as developed by the pioneering work of Dose and his group, then some other type of detector able to select a narrow energy range must be employed. The currently most popular type was introduced by Dose[4,5] and consists basically of a Geiger–Müller counter. The counter is simply a stainless steel cylinder filled with about 500 mbar of helium and containing a few crystals of iodine; there is an axial probe that acts as a collector. The entrance window is made of a CaF_2 single crystal. Figure 7.2 shows Dose's[6] experimental arrangement. This version of the Geiger–Müller counter acts as a band-pass filter. The window is effectively a low-pass filter since the transmission of CaF_2 cuts off around 10 eV. The molecular photoionization yield of iodine, on the other hand, has a threshold at 9.23 eV, and the yield dependence thus provides a high-pass filter. The combined action of the window transmission and the photoionization dependence therefore gives a band-pass filter with a mean energy of 9.71 eV and an energy spread of ± 0.23 eV. The effect of the combination is illustrated in Fig. 7.3, from Dose.[6] The counter window is placed close to the sample surface and collects a very high solid angle of bremsstrahlung radiation.

The much more common variation of the technique using the above counter-type detector for the detection of a selected portion of the radiation in the ultra-violet is called variously ultra-violet bremsstrahlung spectroscopy,[6] inverse UV photoemission,[7] ultra-violet inverse photoemission,[8] or, most commonly, simply inverse photoemission. Like some other techniques in their early stages, a variety of apellations is current until general usage or acceptance of one or another singles it out as the official name, so the reader of the literature must be prepared until that time comes to recognize the technique under several possible acronyms.

As Woodruff *et al.*[9] have pointed out, it is rather limiting to any technique to have to work at a fixed detection energy, be it photon or electron, and of course more information is available if the energy of the UV photons in IPES can be varied continuously. It would also be desirable to have better energy resolution. (The same limitations apply to XPS and UPS vis-à-vis SRPS, q.v.) The obvious step is to a grating spectrograph for selecting the photon energy, and preliminary steps in this direction have already been taken by Fauster and Himpsel,[10] who used a monochromator with two interchangeable spherical gratings, one for the energy region 7–22 eV, and the other for 1–40 eV. The grating resolution was 0.16 eV at 20 eV photon energy, and the overall spectrometer resolution 0.3 eV at 20 eV. They extended the technique further by using a position-sensitive microchannel plate detection system to detect and amplify photons of different energies simultaneously. Since by that

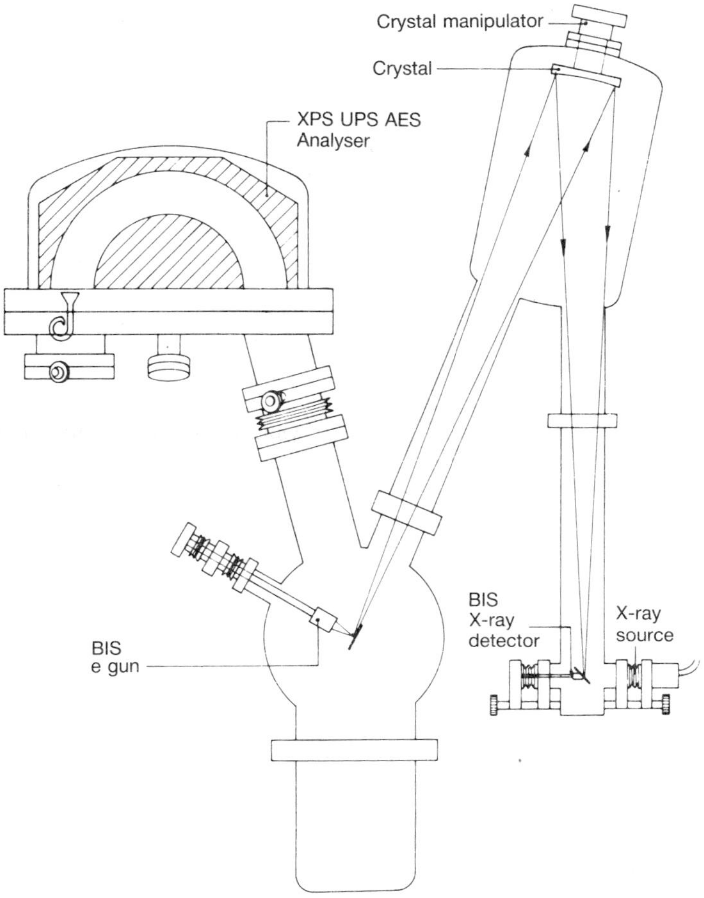

FIG. 7.1. Experimental arrangement for BIS using a commercial X-ray monochromator (courtesy VG Scientific). An electron gun irradiates the sample with a focused beam, the point of impingement being positioned accurately on the Rowland sphere (see Fig. 3.9). X-rays produced in the sample are then focused via the quartz crystal of the monochromator on to an X-ray detector placed where the X-ray source for the monochromator would normally be. The X-ray flux is recorded as the primary electron energy is scanned.

extension a large part of the complete bremsstrahlung photon spectrum is sampled simultaneously, it then becomes possible to keep the incident electron energy constant, rather than ramp it, and to observe transitions from a fixed state into final states just above the Fermi level. Clearly, it is in this direction of development that IPES is likely to go. The major objection to the grating spectrograph approach at the moment is the loss of some two orders of magnitude of sensitivity compared to the use of the fixed-energy counter detector.

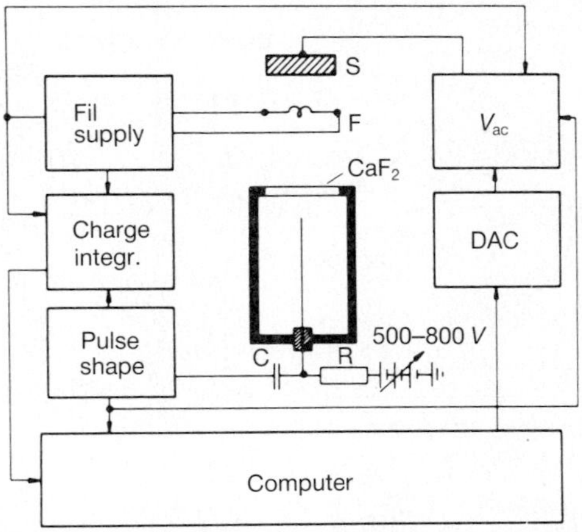

FIG. 7.2. Schematic of the experimental arrangement and of the IPES ultra-violet detector, designed by Denninger *et al.*[5] The detector is basically a Geiger–Müller counter. It is filled with about 500 mbar of helium and contains a few crystals of iodine; the window accepting radiation from the sample S is made of CaF_2. The outer container is stainless steel and there is an axial collector. Since the window is close to the sample a high solid angle of radiation is collected. The detector acts as a band-pass filter since the transmission of CaF_2 cuts off near 10 eV, while the molecular photoionization yield of iodine has a threshold at 9.23 eV. See Fig. 7.3 for more details of the characteristics of the detector.

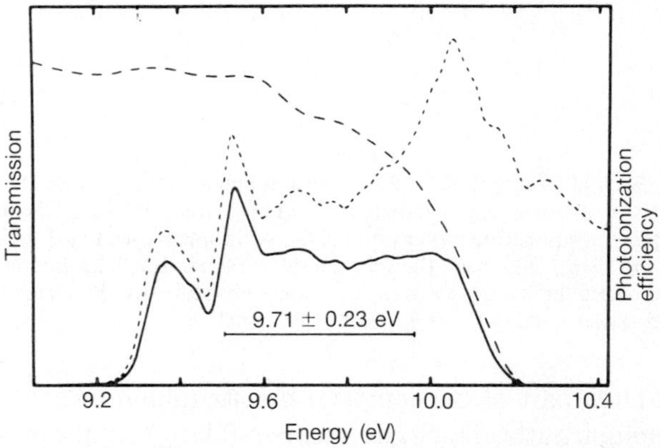

FIG. 7.3. Characteristics of the band-pass detector shown in Fig. 7.2, from Dose.[6] The ultra-violet transmission of the CaF_2 window (long dashes) starts to dip at around 10 eV, and represents a low pass to ultra-violet radiation. The photoionization of iodine (short dashes) has a threshold at 9.23 eV, and provides a high pass to ultra-violet detection. The combination of these two responses is shown by the solid line, giving a band-pass filter with a mean energy of 9.71 eV and a spread of 0.23 eV.

7.2 Theory and spectral identification

The physical mechanism for the production of an inverse photoemission spectrum is shown schematically in Fig. 7.4, from Woodruff *et al.*[9] In the lower half is an indication of the experimental arrangement using the Geiger–Müller counter approach. On the left is a notional density of states for a solid with a partially filled valence band, the unoccupied but allowed states extending upwards indefinitely. An electron of energy E impinges on the solid, giving rise to a wide variety of effects such as ionization, plasmon excitation, etc., but it can also be decelerated simply by stopping in an unoccupied state above the Fermi level, whereupon it decays radiatively with the emisison of a bremsstrahlung photon. If now a particular photon energy $\hbar\omega$ is selectively detected in a system such as a Geiger–Müller counter (for UV photons) or as an X-ray monochromator (for X-rays), and the incident energy E is swept, the variation in emitted flux shoud map the empty densities of states, as seen on the right. The inverse relationship to photoemission is illustrated in Fig. 7.5, also from Woodruff *et al.*,[9] in which the ranges of applicability of photoemission and inverse photoemission are emphasized.

The nature of the inverse photoemission spectrum in the angle-integrated mode is illustrated in Fig. 7.6, from Hüfner *et al.*,[11] for the alloy $CePd_3$, at three selected photon energies. Detection at 9.7 eV and 1486.6 eV (Al K_α) has already been discussed; detection at 72.7 eV is performed by a slightly different method, that of interposing a thin aluminium foil between the electron source and the sample surface, and then of modulating the incident electron energy. Absorption at the L_3 edge of aluminium thus acts as a low-pass filter, and differentiation of the detected signal is achieved by phase-sensitive detection at the modulation frequency.The method has similarities to that used originally in AES using a retarding-field analyser.

The point in Fig. 7.6 of comparing the IPES spectra at three quite different photon energies is simply to see what differences might be observed between surface and bulk. The low and high energy spectra would both correspond to depths sampled of ~25 Å, while that at 72.7 eV would correspond to a depth of only ~5 Å. Features common to all three spectra are at 0.5 eV, 4.5–5.0 eV, and ~6.5 eV. The peak at 2.5 eV appears in the spectra at $\hbar\omega = 9.7$ eV and 1.5 keV, but not at $\hbar\omega = 72.7$ eV. The latter peak must therefore be a feature of bulk $CePd_3$, and can be explained by a Ce 5d–Pd 4d hybridization.

It is the presence of *both* peaks at 0.5 eV (i.e. just above the Fermi level) and at 4.5–5.0 eV that is important in the case of the alloy. If the cerium ground state is $4f^0$ only in the alloy, then there should be only one signal in the IPES spectrum, corresponding to $4f^1$, near the Fermi level. If, on the other hand, the cerium configuration in the alloy is a mixture of $4f^0$ and $4f^1$, then there should appear also a signal corresponding to $4f^2$, and the relative intensity of the $4f^1$ and $4f^2$ signals should be a good measure of the $4f^1$ component of the ground

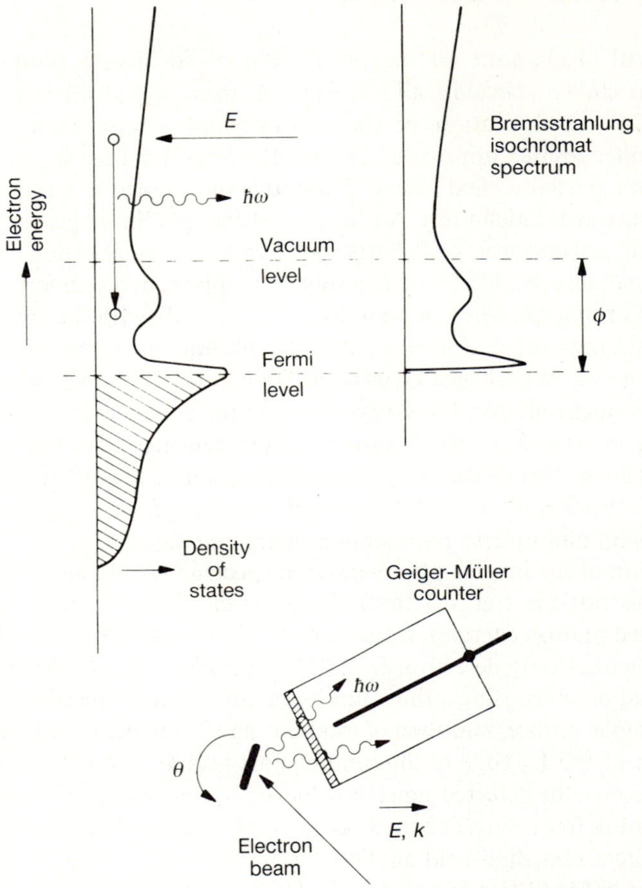

FIG. 7.4. Schematic from Woodruff *et al.*[9] of the energetics of IPES. In the lower part of the figure is shown again an ultra-violet detector of the type shown in Fig. 7.2. On the left is a notional density of states scheme in a solid, with the occupied states shown filled up to the Fermi level, and the unoccupied allowed state densities extending upwards. An electron of energy E can lodge in an unoccupied state above the Fermi level, and decays subsequently with emission of a bremsstrahlung photon. The detector selectively detects a particular photon energy $\hbar\omega$ (e.g. 9.71 eV in Fig. 7.3), and the variation in photon flux at $\hbar\omega$ as E is swept is recorded. The variations in this flux then follow the densities of unoccupied states as indicated on the right.

state. Since both peaks were observed, and that at ~5.0 eV is of greater intensity, it can thus be said that in the alloy the cerium ground state contains a higher proportion of $4f^1$ than of $4f^0$. The fact that the two peaks are still observed in the surface-sensitive spectrum at $\hbar\omega = 72.7$ eV shows that the cerium atoms in the surface are in the same ground state as in the bulk.

Although much useful information can be obtained from angle-integrated IPES measurements, very much more is obtainable from angle-resolved

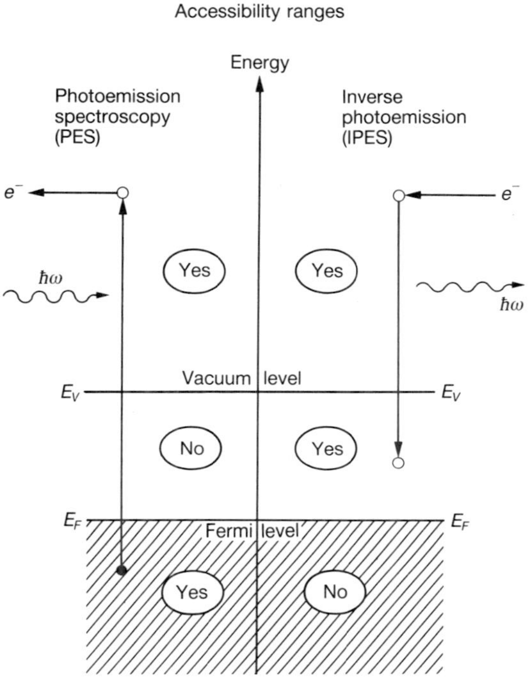

FIG. 7.5. Illustration by Woodruff *et al.*[9] of the relationship between normal photoemission spectroscopy and inverse photoemission spectroscopy. The ranges of accessibility of information in each technique are compared, and it can be seen that they do indeed have an inverse relationship. For this reason it is common to use UPS and IPES together in order to map both occupied and unoccupied densities of states from the same sample at the same time.

experiments. The angular resolution in this case should not be confused with that used in conjunction with photoemission (XPS, UPS, etc.) where it is the angle of *emission* that is important. Here it is the angle of electron *incidence* that is the important experimental parameter. When the angle of incidence at a given initial energy is varied, there is a resultant change in the component $\boldsymbol{k}_{\parallel}$ of the electron momentum parallel to the surface. Since $\boldsymbol{k}_{\parallel}$ is conserved at the surface (as far as a surface reciprocal lattice vector), any change in the structure of an IPES spectrum on varying the angle of incidence reflects the angular dispersion of the momentum in the solid. If the solid is a single crystal it is thus possible to map the three-dimensional *unoccupied* band structure of the crystal (just as ARUPS does for the *occupied* part). Since variation of angle of incidence is effectively varying $\boldsymbol{k}_{\parallel}$, the technique has been called[12] *k*-Resolved Inverse Photoemission Spectroscopy, or KRIPES.

Discussion of the interpretation of KRIPES cannot be satisfactorily entered into without mentioning further theoretical considerations. The theoretical equivalence of photoemission and of bremsstrahlung emission has been

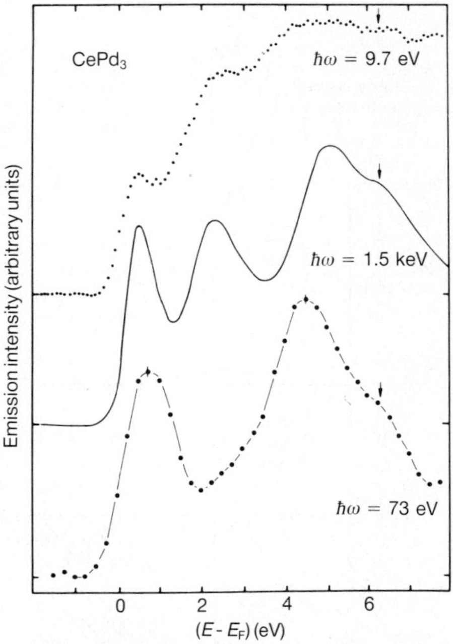

FIG. 7.6. Comparison of the inverse photoemission spectrum from the alloy $CePd_3$ obtained by detection at three different photon energies, made by Hüfner *et al.*[11] Detection at 9.71 eV and at 1486.6 eV (i.e. Al K_α) has already been described; detection at 72.7 eV represents yet another method, described in the text. Sampling depths at 9.71 eV and 1486.6 eV are of the order of 2.5 nm, i.e. the analysis is probably more bulk-like in character, whereas the sampling depth at 72.7 eV is probably only 0.5 nm, more characteristic of the surface. The peak at $\approx$2.5 eV must therefore be a feature of the bulk density of states of $CePd_3$, since it does not appear in the 72.7 eV spectrum. The presence of peaks at $\approx$0.5 eV *and* at $\approx$4.5–5.0 eV is an indication that the Ce ground state configuration in the alloy is a mixture of $4f^0$ and $4f^1$, as discussed in the text.

emphasized by Pendry.[13] Thus it is perfectly possible to derive theoretical models for the latter from the former. A popular model for photoemission first proposed by Berglund and Spicer[14] is the so-called 'three-step' model; it is popular because it gives immediate physical insight into the photoemission process, although in fact it is artificial, and Pendry's 'one-step' model is physically more rigorous. Inverting the three-step model in order to apply it to bremstrahlung emission leads to the following steps:

(a) Penetration of an electron of energy E into the surface region, where it decays into an empty state by an optical transition to an energy $E - \hbar\omega_0$. If it is assumed that the momentum $\boldsymbol{k}$ is conserved, then the initial and final states can be written $E_i(\boldsymbol{k})$ and $E_f(\boldsymbol{k})$.

(b) Transport of the photon so produced through the solid, involving the possibility of inelastic interactions.

(c) Escape of the photon through the surface leading to emission into vacuum.

In the three-step model the three processes are decoupled, and the contribution from electrons that have not suffered inelastic collisions before the radiation decay can be written[6] as

$$I_p(E, \omega) = D(\omega)\, T(E, \omega)\, P(E, \omega). \tag{7.1}$$

In the isochromat mode $D(\omega)$ is constant, while $T(E, \omega)$ depends on the respective photon and electron attenuation lengths; it turns out that in the ultra-violet region T is close to unity since $\lambda_p \gg \lambda_e$. Thus

$$I_p(E, \omega) \propto P(E, \omega) \tag{7.2}$$

and, taking over some standard expressions from photoemission,[12] one may then write

$$P(E, \omega) = \sum_{f,i} \int_{\Omega} d^3k |\boldsymbol{M}_{fi}|^2 c_i \delta\{E_i(\boldsymbol{k}) - E_f(\boldsymbol{k}) - \hbar\omega\} \times \delta\{E_f(\boldsymbol{k}) - E\} \tag{7.3}$$

where $E_i(\boldsymbol{k})$ and $E_f(\boldsymbol{k})$ are band energies above E_F in $\boldsymbol{k}$-space for initial and final state bands i and f, respectively; $|\boldsymbol{M}_{fi}|^2$ is the square of the momentum matrix element for the radiative transition; c_i is the probability that the incoming electron of energy E can be accepted into the initial state i; and Ω is the volume of integration in $\boldsymbol{k}$-space. In an angle-resolved experiment $\boldsymbol{k}_\parallel$ is related to the electron energy E and angle of incidence Θ by

$$\boldsymbol{k}_\parallel = \sin \Theta \{2m(E - E_F)/h^2\}^{1/2}. \tag{7.4}$$

By achieving k-resolution according to equation (7.4), KRIPES is therefore capable of mapping unoccupied band structure dispersion. An example of how this has been carried out for nickel, and the results compared with theory, demonstrates the application of KRIPES. Desinger *et al.*[14] have measured the isochromat spectra at $\hbar\omega = 9.7$ eV for the nickel (001) surface as a function of angles of incidence from normal to 25°, and they can be seen in the left-hand panel of Fig. 7.7. Just above the Fermi level is an inflection point at normal incidence, which does not shift with angle but becomes a well resolved peak at higher angles of incidence. Being so close to the Fermi level it is of course due to transitions into the empty parts of the d-band. The other obvious feature at normal incidence is a broad structure centred at about 2 eV. This structure disperses rapidly with angle, becoming weaker as it does so. At the highest angles used a third feature, also apparently dispersive, appears between 1 and 2 eV.

The peak positions of the dispersing structures are plotted in Fig. 7.8, also from Desinger *et al.*,[14] as a function of $\boldsymbol{k}_\parallel$, as the open circles. Some other experimental results, from Woodruff *et al.*,[12] are also included, as diamonds.

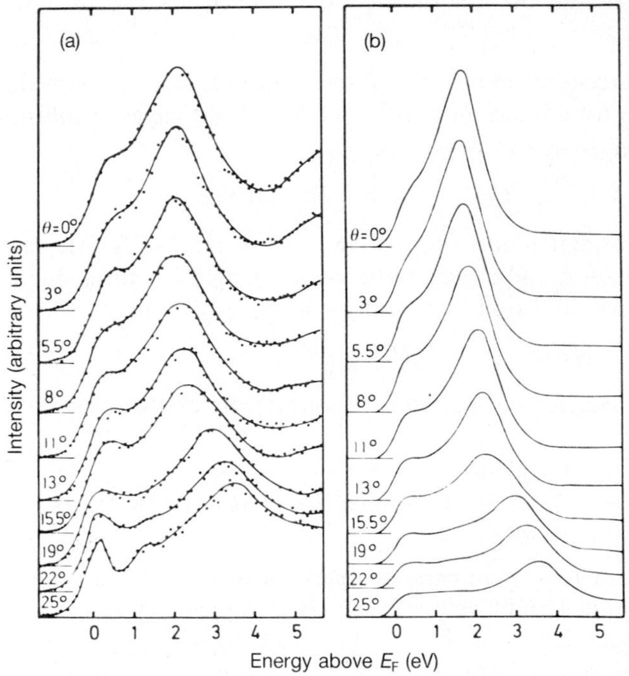

FIG. 7.7. Comparison by Desinger *et al.*[14] of the (a) experimental isochromat spectra ($\hbar\omega = 9.71$ eV) from the nickel (001) [110] surface as a function of angle of incidence of the primary electron beam, with (b) spectra calculated from the one-step photoemission theory. Note the dispersion with angle of the major feature at about 2.5 eV at normal incidence ($\theta = 0°$), and the non-dispersion of the feature just above the Fermi level, that becomes a recognizable peak at $\theta = 25°$.

Three comparisons with theory have been made. The full and dashed lines are from two different band structure calculations based on the three-step model, while the filled circles are from such a calculation[15] based on the one-step model, of inverse photoemission. The agreement of the experimental data with the prediction of one-step theory is particularly good; the branching of band features near $\boldsymbol{k}_{\parallel} = 0.5\ \pi/a$ is well reproduced. When the calculated spectra were convoluted with a Gaussian of appropriate width to simulate experimental broadening the result was the spectra contained in the right-hand panel of Fig. 7.7, which are seen to duplicate the experimental spectra well.

7.3 Applications

Study of the adsorption of gases on metals probably accounts for more of the published literature in surface science than does any other topic. Techniques such as XPS, UPS, and ELS, often in the angle-resolved forms, have

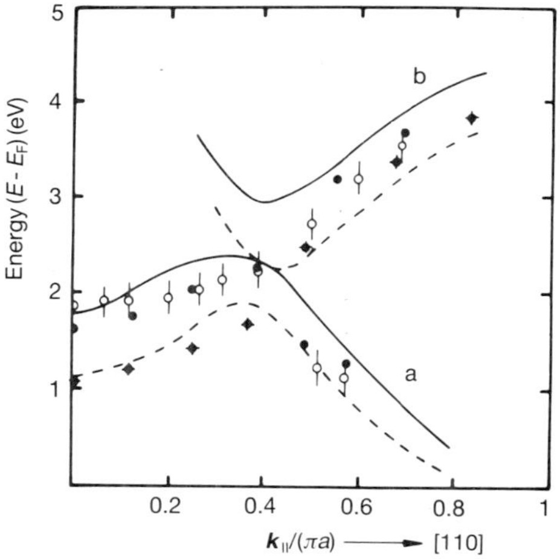

FIG. 7.8. Plot of the peak positions (open circles) in Fig. 7.7 as a function of $k_{\parallel}$ (see expression (7.4)), performed by Desinger *et al.*[14] The diamond-shaped points are additional data from Woodruff *et al.*,[12] and the solid circles are theoretical values calculated by Thörner and Borstel[15] from the one-step photoemission model. The full and the dashed lines are theoretical predictions from two different band structure calculations based on the three-step model. The predictions from the one-step model are in good agreement with experiment. In particular the separation of the band features into two branches a and b near $k_{\parallel}=0.5(\pi/a)$ is well reproduced.

provided an immense amount of information about substrate–adsorbate interactions, but restricted to knowledge about the binding energies distributions, and orientation of *occupied* electronic bands and chemisorption bands. During surface reaction there must be changes in the densities of *unoccupied* substrate states just as there are observed changes in the occupied densities of states, and there must be unoccupied adsorbate states superimposed on the substrate state distribution just as occupied adsorbate states appear in the spectra of techniques such as UPS. These predicted changes have been confirmed in many chemisorption experiments by the group working with Dose.

A particularly interesting example of the effects of chemisorption reflected in the IPES spectra has been recorded[16] by that group for the adsorption of CO on the Cu(100) surface at 120 K. This adsorption leads to successive ordered overlayers, labelled in the LEED notation as $(\sqrt{2}\times\sqrt{2})$R45° and $c(7\sqrt{2}\times\sqrt{2})$R45°, respectively, whose IPES spectra for normal incidence are compared to the spectrum for clean Cu(100) in Fig. 7.9. The additional bands produced by the adsorbed layer in each case are clearly doublets, and in Fig. 7.10 the spectra from the $(\sqrt{2}\times\sqrt{2})$R45° structure taken at several different angles of incidence are resolved into the individual components. The arrows mark the positions of the direct transitions into empty substrate

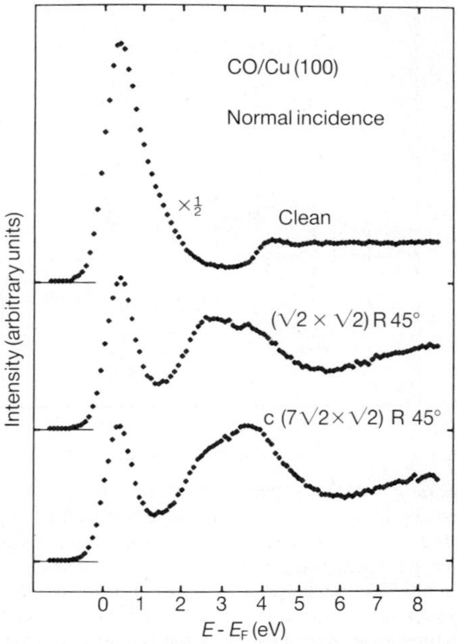

FIG. 7.9. IPES spectra recorded by Rogozik *et al.*[16] from a Cu(100) surface during the adsorption of CO at 120 K. The spectrum from clean Cu(100) is at the top, and the spectra from successive ordered overlayers characterized structurally as $(\sqrt{2}\times\sqrt{2})$ R45° and $c(7\sqrt{2}\times\sqrt{2})$R45°, respectively, below. The additional bands produced by the adsorbed layers are obviously doublets, and that for the $(\sqrt{2}\times\sqrt{2})$R45° structure is analysed in more detail in Fig. 7.10.

d-states, seen clearly in the spectrum from the clean metal. CO adsorption causes considerable attenuation of the peak near E_F due to these transitions.

As the authors remark, the first explanation of the doublet structure that occurs is naturally the presence of two different CO states, i.e. adsorbing forms, on the surface. However, there is a very strong weight of evidence against this explanation from other, complementary, techniques. They suggest instead that the twin peaks derive from distinct *electronic* states of CO on the surface. They substantiate their suggestion by reference to theoretical calculations of the electronic structure of Cu + CO clusters such as CuCO, Cu_5Co, and $Cu_{13}CO$, in which the CO is supposed to be arranged end-on to the Cu, with bonding occurring through the C atom. The calculations show that hybridization occurs between the CO 2π orbitals with the empty metal 4p orbitals leading to two combinations, bonding and anti-bonding. In one (at lower energy) the mixture is 3:1 2π to 4p, in the other (at higher energy) it is 1:2 2π to 4p; in the lower (bonding) state the incident electron is thus captured principally by the CO portion of the surface complex, while in the higher (anti-bonding) state it is principally located on the metal atom. Although the

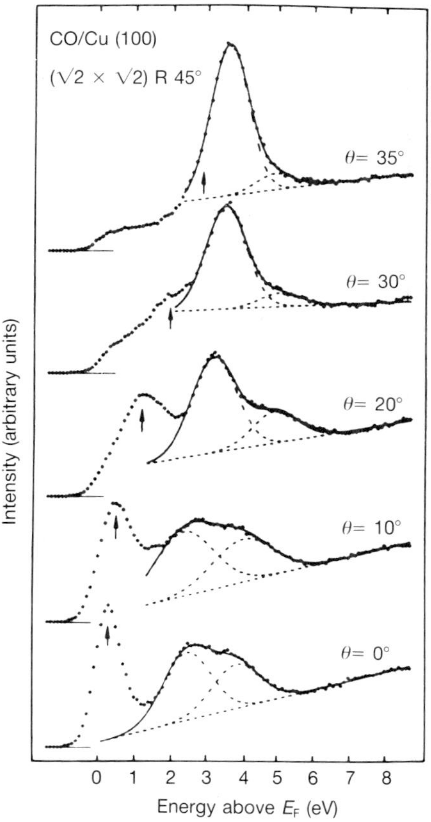

FIG. 7.10. The IPES spectrum from the $(\sqrt{2}\times\sqrt{2})$R45° structure of CO adsorbed on Cu(100), shown in the middle of Fig. 7.9 for normal incidence, analysed as a function of angle of incidence by Rogozik *et al.*[16] The arrows mark the positions of direct transitions into empty Cu d-states. The doublet structure in the additional band due to adsorption is shown separated into its individual components for each angle. The interpretation of the two components is discussed in the text.

electronic condition in the cluster models will not necessarily reflect very accurately the situation in the adsorption complex, it seems that the qualitative explanation fits the observations.

Further evidence for the 2π-4p hybridization comes from consideration of the dispersion of which there is plenty of evidence in Fig. 7.10. When translated into $\boldsymbol{k}$-space as before, the plot of E against $\boldsymbol{k}_{\parallel}$ is shown as the filled circles with error bars in Fig. 7.11. It is not theoretically possible for $p\pi$-derived bands to show an upwards dispersion, as revealed in the plot, and it is clear that hybridization with states from the Cu substrate must have occurred. When the Cu band structure is projected on to the (100) surface, a

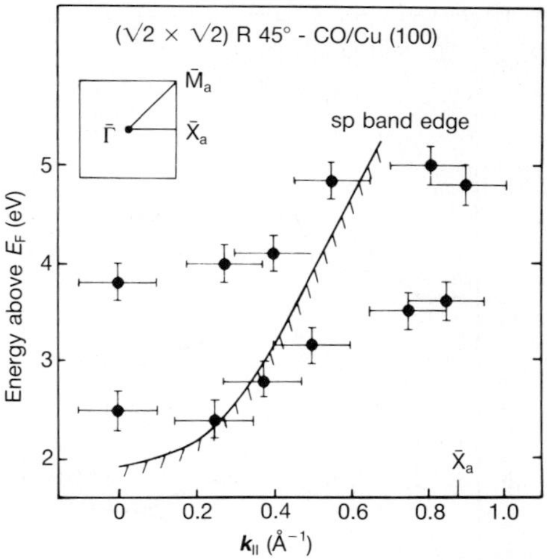

FIG. 7.11. Plot of energetic positions (filled circles with error bars) of the two component peaks seen in Fig. 7.10, as a result of the adsorption of CO on Cu(100), as a function of $k_{\parallel}$, performed by Rogozik *et al.*[16] Also shown is the theoretical position of the Cu 4sp band edge. As discussed in the text, the interpretation of the dispersion observed is in terms of hybridization of the CO 2π orbital with the Cu 4p band.

4sp band edge in fact crosses this particular part of ***k***-space, as depicted in the figure, and the authors point out that where this edge crosses the 2π-4p derived bands, the dispersion is strongest. Thus IPES has been able to produce strong qualitative evidence for chemical bonding effects in adsorption, analogous to the evidence provided by other techniques.

Another most important contribution of IPES to surface physics has been in the identification of the presence of unoccupied electronic surface states on some metal and semiconductor surfaces, analogous to the occupied surface states already established by techniques such as angle-resolved UPS and SRPS (q.v.). A good example of such an identification comes from the work of Reihl and Frank[17] on the Cu(110) surface, using KRIPES. Figure 7.12 shows their angle-resolved IPES spectra at the standard photon energy $\hbar\omega = 9.7$ eV for the atomically clean Cu(110) surface; the positions of the Fermi and vacuum levels are marked. As before, the energetically fixed feature just above the Fermi level is due to bulk transitions into parts of the Cu d-band. The other two features in the spectra, a prominent one labelled A that increases sharply with angle, and another labelled IS, are surface states.

The surface states corresponding to the feature labelled IS constitute a special class of unoccupied electronic states at certain solid surfaces. They were first postulated[18] as a result of some LEED observations, but have now been directly observed by IPES on several metal surfaces. These states arise from the long-range nature of the potential that binds an electron to a surface,

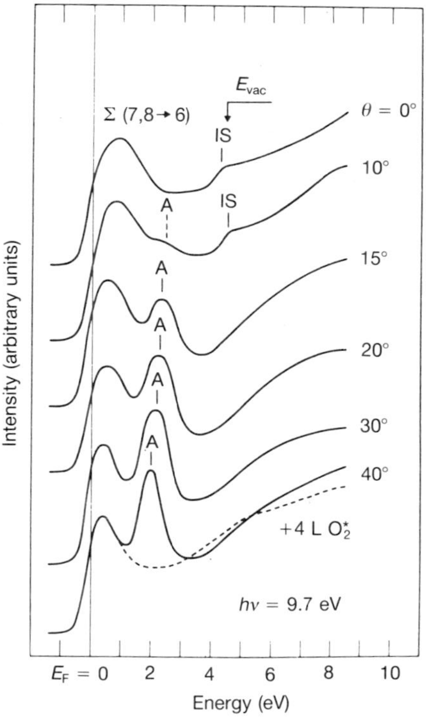

FIG. 7.12. IPES spectra recorded by Reihl and Frank[17] from the Cu(110) surface as a function of angle of incidence along the [100] azimuth, at a photon energy of 9.71 eV. The positions of the Fermi and vacuum levels are indicated. The non-dispersing peak just above the Fermi level is due, as before, to direct transitions into the Cu d-states. The other interesting features are marked A and IS. The former is interpreted as arising from surface states and the latter from image states. See the text for a description of these two types of state.

i.e. the response of the surface to the presence of the electron, or the image potential. If the vacuum level happens to fall within an energy gap of the bulk band structure, there is a repulsive barrier that prevents the electron and its image from combining, with the resultant possibility of a virtual bound state, basically a self-trapping state. This is called an image state or IS. Such states will be weakly bound, and the localization of the electron will be restricted to the direction normal to the surface only, with quasi-free movement in the surface planar directions. Since the image potential is Coulombic a short distance from the surface it can be shown that the states are hydrogenic and have energies that can be described by a Rydberg series.

Although IS features will be by definition found on any metal surface in which E_{vac} coincides with a band gap, the features A are intrinsic to the surface on which they are observed, since they arise by the splitting off of states from bulk electronic bands. They are given the general label of surface states, or SS.

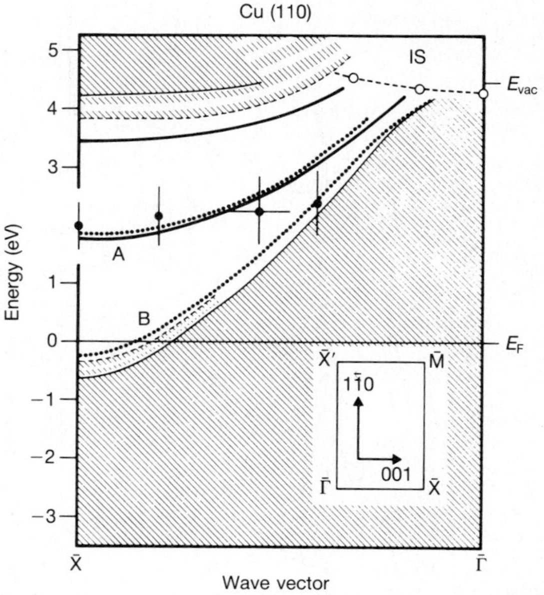

FIG. 7.13. Plot of the energetic positions of the peaks A (full circles) and IS (open circles) in Fig. 7.12, against $\boldsymbol{k}_{\parallel}$, for clean Cu(110) in the [100] azimuth, performed by Reihl and Frank.[17] The bulk band structure projected along the $\bar{\Gamma}\bar{X}$ axis is shown as thin solid and dashed lines, according to the calculation used. The heavy solid and dotted lines are the surface state bands predicted theoretically by Sohn *et al.*[19] and Dempsey and Kleinman,[20] respectively. The agreement between the predicted and observed positions of the surface state is excellent. The closing of the bulk band gap near $\bar{\Gamma}$ allows the formation of weak image states, which have been observed.

For the Cu(110) surface Fig. 7.13 shows the origin of the SS and how they vary with $\boldsymbol{k}_{\parallel}$, as derived from theoretical predictions.[19,20] The full circles show the observed positions of the feature A in Fig. 7.12, and the open circles those of the IS. Clearly the agreement is so close that A can be identified with the predicted SS split off from the bulk d-band. Occupied SS have been observed on many solids, but the presence of unoccupied SS as well could be confirmed only by KRIPES.

References

1. Pierce, D. T., Meier, F., and Zürcher, P. (1978). *Appl. Phys. Lett.* **26**, 670.
2. Lang, J. K. and Baer, Y. (1979). *Rev. Sci. Instrum.* **50**, 221.
3. Hillebrecht, F. V., Fuggle, J. C., Sawatzky, G. A., Campagna, M., Gunnarson, O., and Schönhammer, K. (1984). *Phys. Rev. B* **30**, 1777.
4. Dose, V. (1977). *Appl Phys.* **14**, 117.
5. Denninger, G., Dose, V., and Scheidt, H. (1979). *Appl. Phys.* **18**, 375.
6. Dose, V. (1983). *Prog. Surf. Sci.* **13**, 225.

7. Nilsson, P. L. and Kovacs, A. (1983). *Phys. Scr.* **T4** 61.
8. Dose, V. (1985). *Applic. Surf. Sci.* **22/23**, 338.
9. Woodruff, D. P., Johnson, P. D., and Smith, N. V. (1983). *J. Vac. Sci. Technol.* **A1**(2), 1104.
10. Fauster, T. and Himpsel, F. J. (1983). *J. Vac. Sci. Technol.* **A1**(2), 1111.
11. Hüfner, S., Steiner, P., Dose, V., Straub, D., and Härtl, A. (1983). *Solid State Commun.* **48**, 257.
12. Woodruff, D. P., Smith, N. V., Johnson, P. D., and Royer, W. A. (1982). *Phys. Rev. B* **26**, 2943.
13. Pendry, J. B. (1981). *J. Phys. C: Solid State Phys.* **14**, 1381.
14. Desinger, K., Dose, V., Glöbl, M., and Scheidt, H. (1984). *Solid State Commun.* **49**, 479.
15. Thörner, G. and Borstel, G. (1983). *Solid State Commun.* **47**, 329.
16. Rogozik, J., Dose, V., Prince, K. C., Bradshaw, A. M., Bagus, P. S., Hermann, K., and Avouris, Ph. (1985). *Phys. Rev. B* **32**, 4296.
17. Reihl, B. and Frank, K. H. (1985). *Phys. Rev. B* **31**, 8282.
18. McRae, E. G. and Caldwell, C. W. (1967). *Surf. Sci.* **7**, 41; and McRae, E. G. (1979). *Rev. Mod. Phys.* **51**, 541.
19. Sohn, K. S., Dempsey, D. G., Kleinman, L., and Caruthers, E. (1976). *Phys. Rev. B* **14**, 3193.
20. Dempsey, D. G. and Kleinman, L. (1977). *Phys. Rev. B* **16**, 5356.

8

Electron excitation: cathodoluminescence spectroscopy (CLS)

8.1 Operation

Cathodoluminescence has been known for many years, but its use as a surface analytical technique is only fairly recent. Even now its use in this way is uncommon, but, as will be seen later, it may have some unique contribution to make in the study of interfaces that cannot easily be reached by the more conventional techniques.

The specimen surface is irradiated with electrons from one of the simpler electron guns described under Section 3.2.1, at energies chosen in the range 0.5–3 keV. To avoid electron beam damage as far as possible the beam is defocused and the specimen current kept as low as possible ($<0.1\ \mu A$). Visible light produced at the surface by the electrons can either be collected totally without dispersion, or wavelength-dispersed in a suitable monochromator. An experimental arrangement used by Brillson *et al.*[1] is shown in Fig. 8.1. The light from the bombarded surface is focused by a quartz lens inside the UHV chamber through a quartz window on to the entrance slit of the monochromator. A photomultiplier cooled to liquid nitrogen temperature detects the output signal from the monochromator down to the sensitivity cut-off at about 1 eV photon energy (corresponding to 1.3 μm wavelength). Light reflected from the electron gun filament was eliminated by chopping the beam current and synchronously detecting the photomultiplier signal.

8.2 Theory and spectral identification

Visible light emission from a surface during electron irradiation occurs only with insulators or semiconductors. It is apparently not necessary for the insulator or semiconductor to be a bulk crystal, since Bastasz *et al.*[2] have observed cathodoluminescence from a thorium metal single crystal whose surface had some carbon contamination on it. As the carbon was progressively removed the luminescence diminished. The accepted theory[3] of cathodoluminescence, however, requires within the irradiation depth of the incident electron beam material with a band gap of sufficient size (a few eV) to accommodate electron trapping centres. These centres can be due to

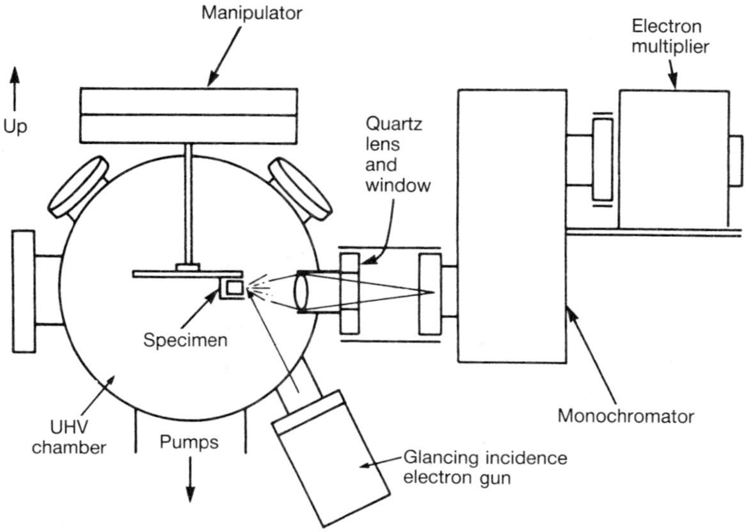

FIG. 8.1. Experimental arrangement used by Brillson *et al.*[1] for cathodoluminescence measurements, under ultra-high vacuum conditions. Electrons from a conventional electron gun, at energies in the range 0.5–3.0 keV impinge on the specimen mounted on a turntable. The resultant luminescent light is focused by a quartz lens inside the system, through a quartz window, onto the entrance slit of a double-prism monochromator. On the output from the monochromator a liquid-nitrogen-cooled photomultiplier detects the light down to the sensitivity cut-off at ≈ 1 eV (i.e. 1.3 μm wavelength). The electron beam current is chopped, and the photomultiplier current amplified synchronously, in order to eliminate light from the filament. The incident electron current was less than 0.1 μA.

impurities (in some cases deliberately added as 'activators') or to defects. Electrons excited into the conduction band by the incident electron beam can be trapped in one of the higher energy levels associated with the centre, from where they can undergo a transition to the ground state of the centre with the emission of visible radiation.

The vast majority of cathodoluminescence studies have not been concerned with surface effects at all, particularly the earlier ones in which incident energies of 50–100 keV were used. Where much lower energies were used, the experiments were aimed at the measurement of luminescent spectra, electron burn (i.e. material degradation), afterglow characteristics, etc., of phosphors designed for fluorescent cathode ray tube screens. The first systematic study of the depth resolution of CLS seems to have been that of Norris *et al.*,[4] who measured the cathodoluminescence spectra from several luminescent crystals as a function of incident electron energy in the range 1–20 keV. In order to maintain the total rate of carrier injection into the crystals linear with electron range, they adjusted the electron beam current I_p at each incident energy E_p to keep the expression $E_p I_p/x_R$ constant, where x_R is the incident electron range. This procedure has the advantage of keeping the deposition of energy in the surface constant so that any surface-related effects could be more easily resolved. Spectra were recorded with a monochromator–photomultiplier system, and the incident beam was chopped electrostatically thus generating a modulated optical signal that was amplified by synchronous detection in a lock-in amplifier.

An example from the observations of Norris *et al.* is given in Fig. 8.2, in which the cathodoluminescence spectra from ZnS at 300 K at several primary energies are compared. The spectra are dominated by a broad peak in the region of blue emission around 4500–5000 Å wavelength, and contain also a weak band in the red at ~7000 Å. As the primary energy, i.e. the penetration depth, increases, the blue peak shifts to shorter wavelengths and becomes narrower, which suggests that it is probably composed of two unresolved components. Although the ZnS crystal used was nominally pure, luminescence due to unknown impurities is a common observation in this material, and indeed the blue band is in the right position for emission due to Cu activation; such activation often leads to two closely spaced emission bands, the so-called Cu blue and Cu green, requiring not only Cu but some unknown co-activator as well. The origin of the weak red band is not fully understood, although it is close to that of another Cu-activated emission, the Cu red resulting from Cu substitution in the ZnS lattice. Note also the increased emission intensity in the 7500–6000 Å region for a primary energy of 1 keV.

The interesting observation is the significant difference between the CLS observations at 1 keV and at 3 keV, corresponding to nominal ranges in this material of 110 Å and 770 Å, respectively. In high quality single crystal phosphors the carrier diffusion length in the bulk should be several μm, but clearly from these experiments it is possible to excite luminescence characteristic of the surface and near-surface regions of depths only a few hundred Å. The increased intensity between 7500 and 6000 Å, with no additional features, and the enhanced red band, in the 1 keV spectrum, suggest some intrinsic surface effect such as lattice disorder, lack of stoichiometry, or unknown surface contamination. The authors state that the crystal was carefully prepared and not intentionally damaged, but it would be difficult to ensure that there was no surface damage within the first 100 Å or so. If there had been some surface disorder then the carrier diffusion length could well have been very much shorter than in the bulk. Whatever the reason, it is clear that 1 keV incident energy for this particular material CLS is significantly surface specific.

8.3 Applications

As with most of the other techniques described here CLS has been used to study oxidation. Its use in this way appears in one case to have been accidental, since Bastasz *et al.*[2] happened to observe not only that 'clean' thorium metal emitted blue cathodoluminescence under 3 keV electron

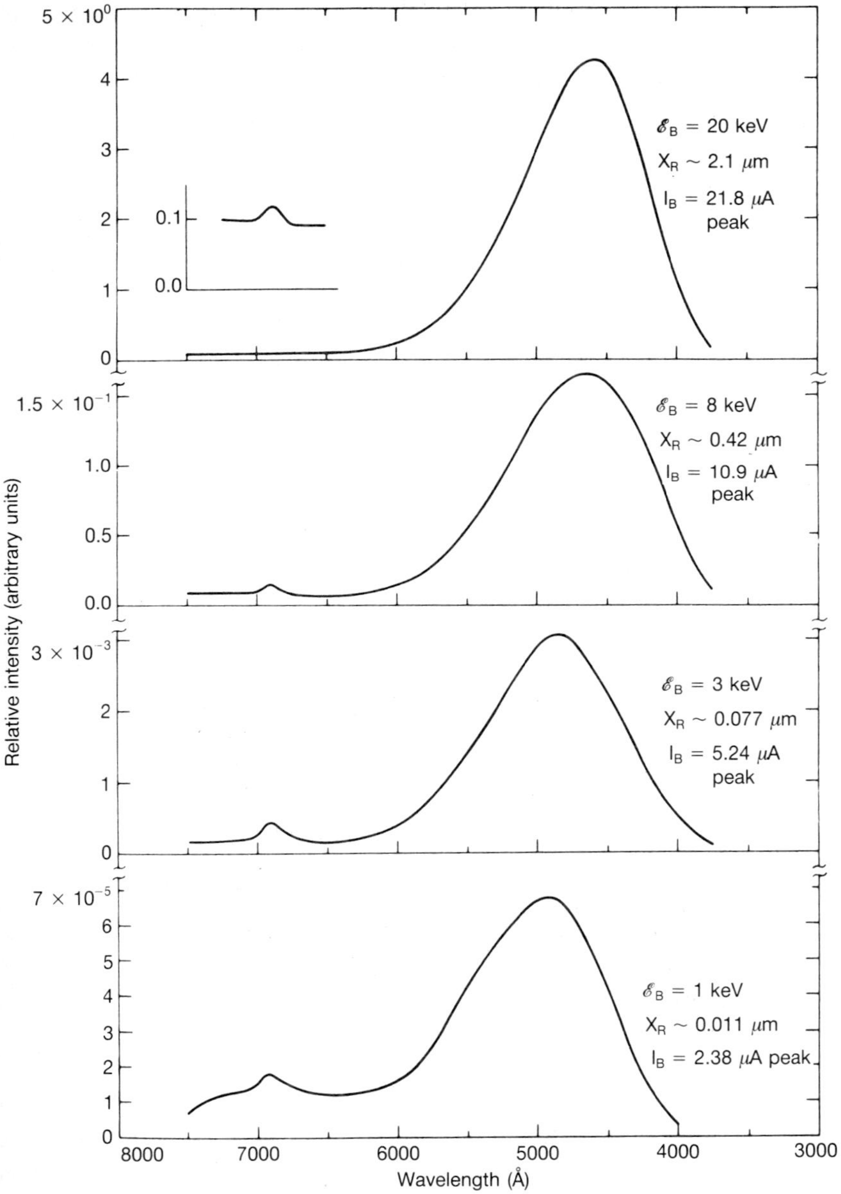

FIG. 8.2. Study of the depth resolution of CLS by Norris *et al.*[4] Cathodoluminescence spectra as a function of primary electron energy from ZnS at 300 K. The deposition of energy into the surface was kept constant by maintaining the expression $E_p\, I_p/x_R$ constant, in which x_R is the incident electron range. The spectra are dominated by a broad blue peak around 4500–5000 Å wavelength, that shifts to shorter wavelengths and narrows, with increasing primary energy, i.e. penetration depth. Between primary energies of 1 and 3 keV (lower two spectra) some additional unresolved structure appears in the 6000–7500 Å region.

bombardment, but that the luminescence was sensitive both to the state of oxidation of the thorium *and* to the pressure of oxygen gas around the specimen. Having seen this dependence, they proceeded to use it as a tool.

They found that the luminescent intensity from the thorium surface increased rapidly when oxygen in the pressure range 7×10^{-6}–1.5×10^{-5} Pa was admitted to the sample, reached a plateau, and then remained constant as long as oxygen was present. Pumping the oxygen away reduced the intensity but not to the original level. The plateau intensity depended on the oxygen pressure, in that changing the pressure from 7×10^{-6} to 1.5×10^{-5} Pa caused the intensity to double. The effect is shown in Fig. 8.3, in which the output

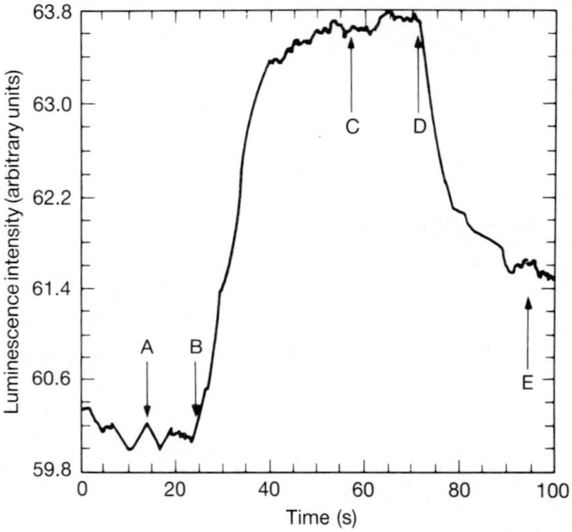

FIG. 8.3. Cathodoluminescent intensity in the 4700 Å emission peak from thorium as a function of exposure to oxygen at $\approx 10^{-5}$ Pa, recorded by Bastasz *et al.*[2] The oxygen is admitted at B and pumped away at D. Letters A–E represent the points in the experiments at which AES analyses were carried out on the surface. Note that removal of oxygen did not reduce the light intensity to the original level, indicating some oxidation of the surface, since thoria gives a higher level of cathodoluminescent intensity than does thorium metal.

from the telephotometer recording the cathodoluminescence is plotted as a function of elapsed time. The emission spectrum from the oxygenated surface was relatively featureless, consisting of a single broad band centred at 4700 Å which did not change shape or position as a result of varying oxygen exposure.

Exposure to argon or to hydrogen in the same pressure range had no effect on the cathodoluminescence, although exposure to carbon monoxide did, but with lower maximum intensity. No effect was observed if gold or stainless steel were substituted for the thorium.

When the authors combined surface analysis by AES with the CLS measurements, some interesting features emerged. It was found that the oxygen concentration on the surface was very similar at the plateau of luminescent intensity and at the intensity level after the oxygen had been pumped away, in fact there was a difference of less than 3% in the Auger peak-to-peak heights, yet the difference in luminescent intensity was nearly 60%! Thus a large part of the luminescence is directly dependent on the presence of gas-phase oxygen near the thorium, and not on the concentration of adsorbed oxygen. The latter has some effect, since the luminescent intensity was higher after oxygen adsorption than before, but not the major effect. The authors concluded that there were two processes contributing to the luminescence during oxygen exposure, one involving the physisorption of molecular oxygen:

$$O_2(\text{gas}) \rightleftarrows O_2(\text{physi}) \xrightarrow{e^-} (O_2^-(\text{physi}))^* \rightarrow O_2^-(\text{physi}) + h\nu_1 \qquad (8.1)$$

and the other involving chemisorbed oxygen atoms:

$$\tfrac{1}{2}O_2(\text{gas}) \rightarrow \tfrac{1}{2}O_2(\text{ads}) \rightarrow O(\text{ads}) \xrightarrow{e^-} (O^-(\text{ads}))^* \rightarrow O^-(\text{ads}) + h\nu_2 \qquad (8.2)$$

Since the emission spectrum is structureless and insensitive to oxygen pressure, the luminescent processes probably do not derive directly from the adsorbed oxygen, but from centres formed by the thorium–oxygen interaction, perhaps surface clusters. It is known that the photoluminescent spectrum of thoria is rather similar to the cathodoluminescent spectrum found by Bastasz *et al.*, suggesting that a thoria lattice is formed immediately on oxygen exposure.

A rather different application of CLS was that of Brillson *et al.*[1,5] They were searching for a technique that would give them some information about 'buried' interfaces, i.e. interfaces at too great a depth below the free surface to be accessible by conventional surface analytical techniques, and whose chemical and electronic characteristics would certainly be altered or destroyed by depth profiling. In particular they were interested in the changes to such interfaces caused by laser annealing, a technique of localized heat treatment becoming popular in semiconductor technology.

Interfaces were formed by evaporating about 50 Å of either Cu or Al on to the cleavage faces of CdS. Subsequent laser annealing was carried out with a pulsed (5 ns) focused excimer laser of emission wavelength 3080 Å. The laser spot of 3000 to 6000 Å diameter was rastered over the surface in such a way that each spot received four pulses. The incident electron energy used to stimulate cathodoluminescence was varied between 0.5 and 2.0 keV to examine electronic states at different depths.

For the Al deposition and the interfaces formed by laser annealing at two energy densities, the CLS results are given in Figs 8.4 and 8.5. The spectra in Fig. 8.4 are all excited by a 2 keV electron beam, and show the progression

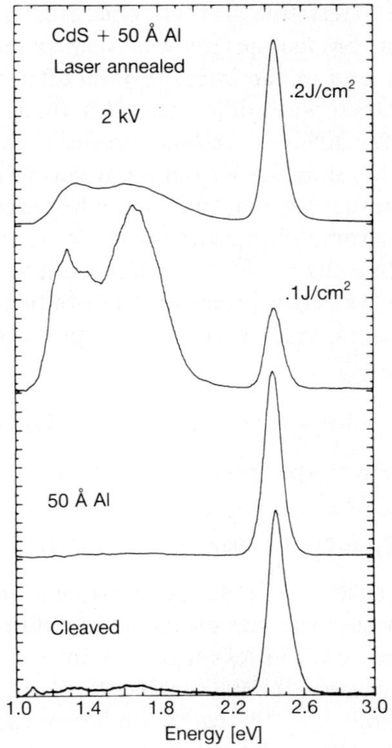

FIG. 8.4. CLS spectra from the CdS(11$\bar{2}$0) surface formed by cleaving in vacuo, recorded by Brillson *et al.*[1] From the bottom, the spectra refer in turn to the clean surface, to the clean surface with a coverage of 50 Å Al, and to the Al-covered surface laser annealed at energy densities of 0.1 J cm^{-2} and 0.2 J cm^{-2}. Excitation was with a 2 keV electron beam. The clean and Al-covered surfaces produce spectra dominated by a single peak at 2.42 eV, although the clean surface also shows very weak features at $\approx$1.3 and $\approx$1.65 eV. On laser annealing at 0.1 J cm^{-2} these weak features are suddenly themselves the dominant ones, but further annealing reduces them again in comparison with the 2.42 eV peak.

from the clean cleaved CdS surface, to coverage with 50 Å Al, to laser annealing at 0.1 J cm^{-2} and at 0.2 J cm^{-2}. The solitary peak at 2.42 eV in the cleaved and as-deposited spectra arises from the so-called band-edge luminescence due to transitions directly from the bottom of the conduction band to the top of the valence band. Very weak features from the cleaved surface are just visible at 1.3 and 1.65 eV, and are probably due to some residual surface damage left behind after cleavage. After laser annealing at 0.1 J cm^{-2} these features suddenly become dominant relative to the band-edge peak, but further annealing at 0.2 J cm^{-2} reduces them again with a corresponding increase in the 2.42 eV peak.

The depth distribution of the CLS features is illustrated in Fig. 8.5 by recording the spectra after 0.1 J cm^{-2} and 0.2 J cm^{-2} laser annealing,

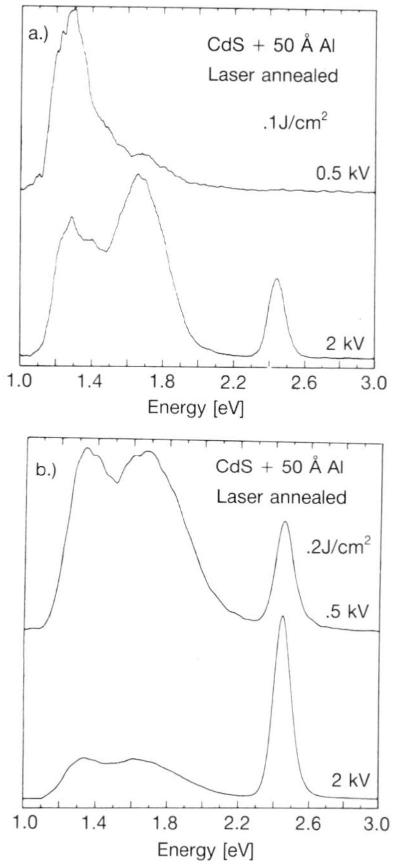

FIG. 8.5. CLS spectra as a function of primary electron energy from the Al-covered CdS surfaces annealed at 0.1 J cm^{-2} (upper) and 0.2 J cm^{-2} (lower), described in Fig. 8.4, from Brillson *et al.*[1] At the lower primary energy, corresponding to greater surface sensitivity, the 0.1 J cm^{-2} annealing results in a dramatic increase in the 1.3 eV peak, with no band-edge luminescence at 2.42 eV visible, while the 0.2 J cm^{-2} annealing shows that band-edge luminescence is still visible, even though the 1.3–1.65 eV band dominates. Clearly the 0.1 J cm^{-2} annealing resulted in a large density of laser-induced defects concentrated close to the surface, similar in nature to those existing at very low density on the cleaved CdS surface. The reduction in the 1.3 eV and 1.65 eV peak intensity on further laser treatment indicates a reduction in the above disorder and a partial return to the CdS structure.

respectively, at 0.5 keV and 2 keV primary energies. After 0.1 J cm^{-2} annealing no band-edge luminescence appears in the 0.5 keV spectrum, and furthermore the 1.3 eV peak increases dramatically relative to that at 1.65 eV. Clearly there is a large change in the density of laser-induced defects between the volumes sampled at the two incident energies, and in addition those defects giving rise to the 1.3 eV transition are located much closer to the surface. The incident electron energy dependence of the CLS spectra after the 0.2 J cm^{-2}

annealing, on the other hand, shows that there is still band-edge emission visible at low energy, and that the 1.3 eV peak does not dominate over the 1.65 eV peak.

The various features observed can be identified with reference to results from related techniques such as photoluminescence. Thus the weak 1.3 eV and 1.65 eV peaks seen after cleavage are consistent with transitions associated with the creation of native defects in CdS, due to mechanical damage, and this would account for their initial increase in intensity after the laser treatment at the lower energy density. The defects are probably related to sulphur deficiency near the surface, since sulphur vapour was detected in a mass spectrometer during annealing. There is possibly also a contribution to the 1.65 eV peak from a complex such as Al associated with a Cd vacancy; in Al-doped CdS a peak in exactly that position has been observed[6] in photoluminescence. The reduction in the 1.3 and 1.65 eV peaks on further laser heating at 0.2 J cm^{-2} suggests a reduction in lattice disorder near the surface with the partial reinstatement of the CdS structure and its characteristic band gap. Thus it seems unlikely that any new compound is formed at the Al/Cds interface (unlike the case of the Cu/CdS interface) and that the effect of the laser annealing is to create both native defects in the CdS and complexes based on defects coupled with Al atoms.

It is clear, particularly from Fig. 8.5, that CLS can be used to identify chemical and structural conditions at 'buried' interfaces without having to resort to the destructive method of ion profiling often employed to attempt to uncover interfaces. The depth resolution of CLS is a function both of the incident primary energy and of the range of inelastically scattered electrons within the solid, and is obviously similar to techniques such as ELS (q.v.) which depend on variation of the incident electron energy.

References

1. Brillson, L. J., Richter, H. W., Slade, M. L., Weinstein, B. A., and Shapira, Y. (1985). *J. Vac. Sci. Technol.* **A3**, 1011.
2. Bastasz, R., Colmenares, C. A., and Somorjai, G. A. (1978). *Surf. Sci.* **71**, 397.
3. Garlick, G. F. J. (1950). *Adv. Electron.* **2**, 151.
4. Norris, C. B., Barnes, C. E., and Beezhold, W. (1973). *J. Appl. Phys.* **44**, 3209.
5. Brillson, L. J. (1986). *Surf. Sci.* **168**, 260 (1986).
6. Susa, N., Watanake, H., and Wada, M. (1975). *Jpn. J. Appl. Phys.* **14**, 1733.

9

Electron excitation: electron stimulated desorption (ESD) and electron stimulated desorption ion angular distribution (ESDIAD)

9.1 Operation

Electron Stimulated Desorption (ESD) and Photon Stimulated Desorption (PSD, q.v.) are probably the most notable examples of the maxim that one man's artefact is another man's technique. Whenever electron, or to a lesser extent, photon, probes are used for surface analysis there is always the worry that the probe itself may be changing the surface composition and/or chemistry during the measurement. In AES the number of recorded examples of electron beam effects is by now considerable. The desorption techniques, however, actually use the probe/surface interaction that is so troublesome elsewhere as a means of characterization in its own right.

Electron irradiation of a surface can give rise to a whole range of secondary particles, including electrons, photons, neutral species, and positive and negative ions. ESD and ESDIAD concentrate on the flux of *positive* ions leaving the surface. ESD as an individual technique is not especially new (refs. 1 and 2 probably represent the beginning of it in much the same basic form as today) but ESDIAD is more recent.[3]

The electron source for ESD and ESDIAD does not require to be finely focused and the conventional guns used for LEED, or for AES such as that shown in Fig. 3.4 are adequate. Electron energies employed are normally a few hundred eV, while the current density on the sample is generally kept at or below 10^{-6} A cm^{-2}, since higher current densities are found to deplete the surface species too quickly.

Detection of the desorbed ions has been performed in several ways. The most popular method uses a quadrupole mass spectrometer, whose mode of operation has been described in Section 3.3.2, to analyse the masses and abundances of the ions; the spectrometer axis is aligned with the point of impingement of the electron beam on the sample surface, and the entrance aperture placed as close to the specimen as practicable. If energy analysis is also required, a set of retarding grids can be added to the front end of the spectrometer. Alternatively energy distributions can be recorded using a Cylindrical Mirror Analyser (CMA), whose characteristics have been discussed under Section 3.3.4, but in this case of course the deflecting potential would be reversed.

For simultaneous determination of desorbed ion mass and kinetic energy the use of a time-of-flight (TOF) ion detector is becoming more common. Such a detection system used by Noell *et al.*[4] is shown schematically in Fig. 9.1. A standard electron gun is operated with the potential on the Wehnelt cylinder just sufficient to prevent emission, and a very fast ($\sim$50 ns) repetitive pulse of 15 V is applied to the cylinder to allow emission only within the 50 ns window. The pulsed beam strikes the sample, and the desorbed ions travel in a field free space into a drift tube, at the end of which is a detector array consisting of two-back-to-back channel electron multiplier plates. From knowledge of the time taken for ions to reach the detector from the sample after desorption, and of the total flight path length, both the mass and the kinetic energy of the ions can be calculated. The repetition rate of the pulse is $\sim$100 kHz, which gives about 10 μs to collect all the ions. In the system used by Noell *et al.* the total path length was 7.6 cm, but other workers have used longer distances; obviously the greater the length the better is the mass separation of adjacent masses since the times of arrival of ions of different masses at the detector are then spaced out. Fast triggering and counting circuitry is needed for TOF, and Fig. 9.1 shows one configuration using a time-to-amplitude converter (TAC).

Following the first observation of anisotropies in the angular distributions of desorbing ions by Czyzewski *et al.*,[3] the variation of ESD called ESDIAD has now been taken up by many laboratories. Two methods of recording the angular distribution 'patterns' have been used. The most direct and frequently employed is the display-type method, rather akin to LEED in reverse. Electrons from a conventional gun strike the sample, usually at 45° incidence, and the desorbed ions travel through a field-free space to an array of hemispherical grids backed by a double micro-channel electron multiplier plate, the sample being at the centre of the grids. The ESDIAD pattern is observed by accelerating the electrons from the multiplier plate output onto a fluorescent screen. By reversing the potentials the LEED pattern from the surface can also be observed. The incident electron beam is not pulsed, and the display is thus a static one.

A typical experimental arrangement combining ESD and ESDIAD is that of Schwalke *et al.*[5] shown schematically in Fig. 9.2. Mass-resolved and threshold ESD measurements can be made if the sample is turned to face the quadrupole mass spectrometer with retarding grids, and ESDIAD or LEED if it is rotated to face the hemispherical optics. For the monitoring of sample cleanliness, or measuring the uptake of adsorbing gases, a standard CMA for Auger analysis is provided, also accessible by sample rotation. As in most ESD systems, the adsorbing gas is piped into the chamber through a tube brought to within a few mm of the sample surface, a more efficient dosing method than that of simply filling the chamber to the required pressure.

The alternative to the display method is the mapping method.[6] Instead of the desorbed ion angular distribution being viewed directly, the ions in a very

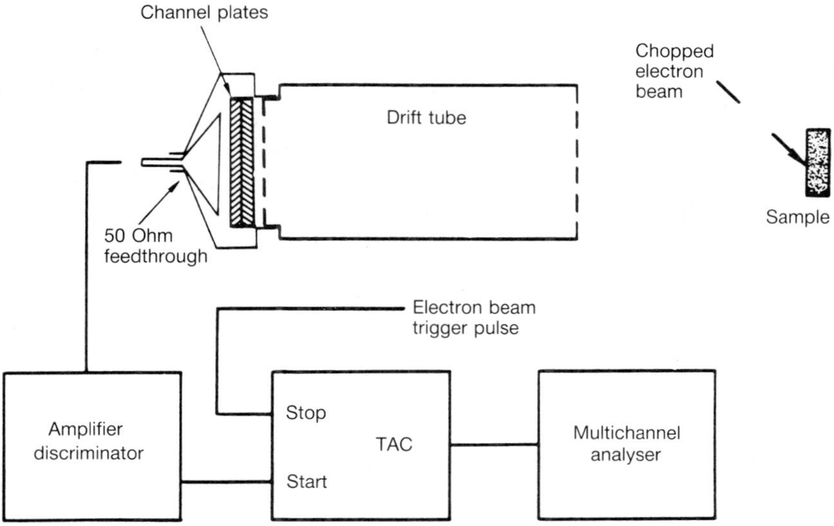

FIG. 9.1. Schematic of the time-of-flight ion detector used for ESD by Noell *et al.*[4] The incident electron beam is chopped by imposing a fast (≈ 50 ns) pulse of 15 V on top of the potential on the Wehnelt cylinder, that is just enough to prevent emission. Emission then occurs only in a 50 ns window. Desorbed ions travel in field-free space down a drift tube, at the end of which is a detector array consisting of two channel plates back-to-back. The total path length is 7.6 cm. The pulse repetition rate is 100 kHz, and fast triggering and counting circuitry is used to measure the flight times.

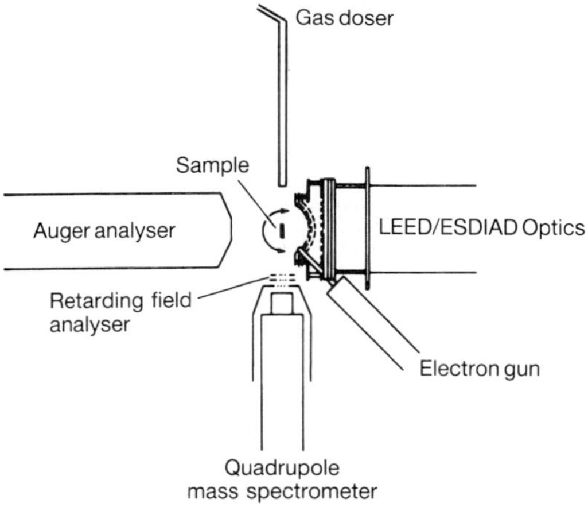

FIG. 9.2. Experimental arrangement for combined ESD, ESDIAD, AES and LEED used by Schwalke *et al.*[5] The sample can be turned to face the quadrupole mass spectrometer for mass-resolved and threshold ESD measurement, or the LEED optics for ESDIAD or LEED. In the latter position, simple inversion of the grid potentials displays either the LEED or the ESDIAD pattern. The analyser for AES is a standard CMA. Angle of incidence for both ESD and ESDIAD is 45°, and the electron current used is 10^{-7} A.

narrow angular core are collected in a detector similar to a Faraday cup collector for electrons. The detector is able to traverse the surface of a sphere of which the sample is the centre, i.e. it is mounted on a goniometer mechanism that allows it to move in both azimuths with respect to the sample. The ion collector inside the cup is a shielded electron multiplier. Acquisition of data is controlled by stepping motors that can step the detector through angular steps of 1° if required, and with the help of a computer a map of ion distribution might take 30 minutes to collect. The arrangement used by Niehus and Krahl–Urban,[6] who originated the mapping variation, is seen in Fig. 9.3. The

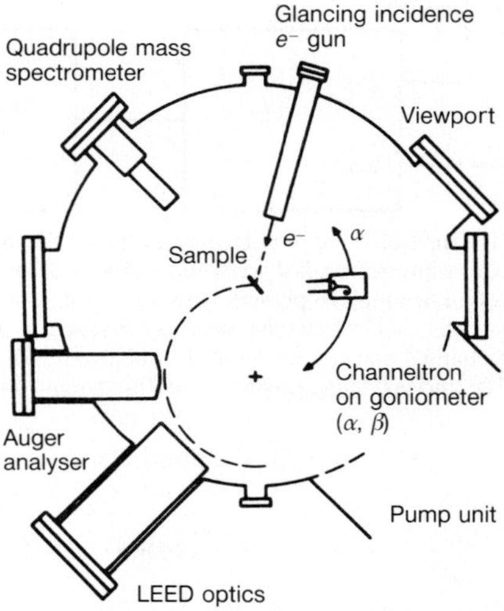

FIG. 9.3. Schematic of the ESDIAD arrangement used by Niehus and Krahl-Urban[6] for mapping the desorbed ion distributions. The channeltron detector is mounted on a goniometer that can step the detector through angular steps of 1° in either azimuth (indicated as α and β). The sample can also be swung to face either an Auger analyser or LEED optics. A quadrupole mass spectrometer is focused on the ESD position for mass-resolved desorption measurements. The electron gun is pulsed at 2 MHz, and with a sample to detector distance of 15 cm the flight time for, e.g., O^+ ions, is 15–20 μs.

electron gun is pulsed at 2 MHz and the flight path to the collector is 15 cm, which allows a flight time for O^+ ions, for example, of 15–20 μs. In their experiments oxygen ions were the only desorbed ions expected, and it was possible therefore to measure the kinetic energies of the ions, which are related to the flight time t, the length of the flight path s, and their mass m by

$$E=(ms^2/2)t^{-2}. \tag{9.1}$$

The mapping technique has the advantage of increased sensitivity and greatly improved angular resolution compared to the display method, but recording times are much longer and there is therefore always the possibility that the adsorbed material may degrade undesirably during the measurement. However, the ability to measure the kinetic energy distributions of the desorbed ions is unique to mapping.

9.2 Theory and spectral interpretation

The first mechanism proposed for the desorption of ions by electron impact was that of Menzel and Gomer[1] and Redhead,[2] independently. In their model it was realized that the small momentum of a low energy electron is insufficient to knock an ion out of a surface directly by inelastic collision, and that an electronic excitation must be the first step in the desorption process. The excitation was envisaged as that of valence electrons from their bonding ground state into an anti-bonding excited state, from which the particle (ion or neutral) could desorb if it was already far enough from the surface during the transition. Since the timescale for vibration of adsorbed species or surface atoms normal to the surface is longer by some orders of magnitude than that for an electronic transition, the probability of desorption by this mechanism is finite. Return to the ground state is a more probable process than desorption, and much more probable for ions than for neutrals, so that the cross-sections for neutral desorption are invariably higher than for ion desorption.

Most of the earlier ESD experiments were performed at low incident energies, < 150 eV, and the thresholds for desorption were generally found to be between 10–30 eV, consistent with the prediction from the model that the threshold should be equal to the energy of the Franck–Condon-type excitation described above. However, there were always difficulties in the Menzel–Gomer–Redhead (MGR) model, and these became even more apparent when higher incident electron energies were used. Desorption thresholds were found that were at higher energies than possible in the MGR model, and that correlated with the onset of electron loss peaks corresponding to core-level ionization losses. In addition the MGR model could not account for the desorption of oxygen as O^+, when oxygen is normally bonded strongly electronegatively as O^{n-}, where $1 \leq n \leq 2$. Clearly a further desorption mechanism was operating, and this was put forward by Knotek and Feibelman,[7] based on their observations on ESD from TiO_2, V_2O_5 and WO_3.

Their curves of O^+ yield vs. electron energy for the three oxides are shown in Fig. 9.4. It can be seen that there are no desorption thresholds corresponding to the valence electron bonding–antibonding transitions, as would be expected in the MGR model, whereas there are thresholds at higher energy correlating with ionization loss thresholds for Ti(3p), W(4f) and V(3p). Based on this evidence they proposed a core-hole Auger decay model to explain both the higher energy thresholds and the large charge transfer involved in the desorption of species such as O^+. They illustrated their (KF) model[8] by reference to TiO_2, whose schematic energy band structure is shown in Fig. 9.5.

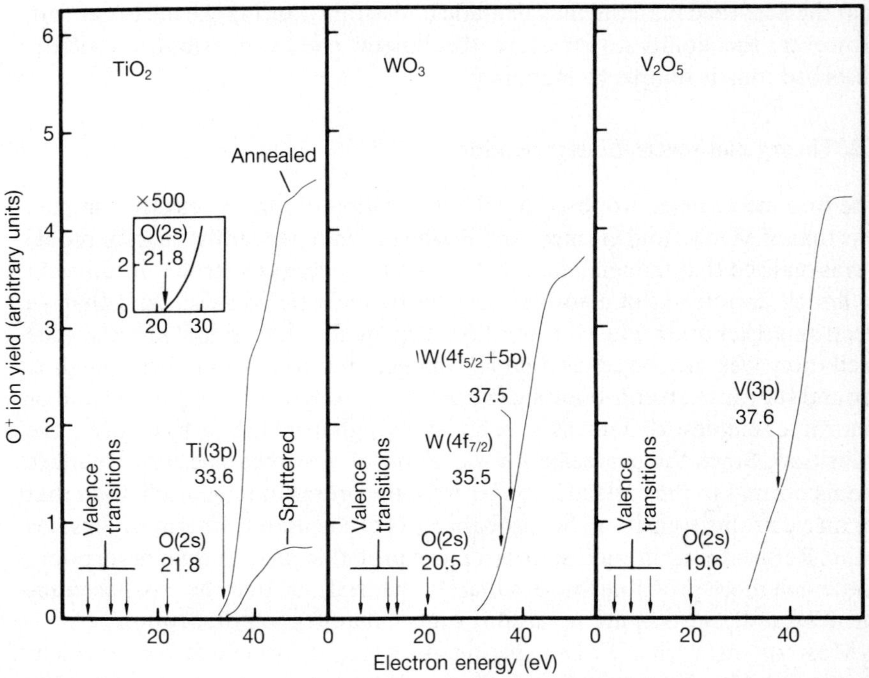

FIG. 9.4. O^+ ion yields as a function of incident electron energy for the three oxides TiO_2, WO_3, and V_2O_5, measured by Knotek and Feibelman.[7] The desorption thresholds are not located anywhere near the valence electron thresholds, but correlate with the ionization thresholds of the Ti 3p, W 4f, and V 3p core levels, respectively.

In the Ti^{4+} ion the highest occupied level will be the Ti(3p) at about 34 eV below the conduction band. Ionization of this level leads to an inter-atomic Auger process as the dominant decay mechanism, in which an O(2p) electron fills the Ti(3p) hole, and a further one or two Auger electrons are ejected from the O(2p) band. The ion that was originally O^{2-} thus becomes O^0 or O^+ after the Auger event. With a change of sign on the anion, it then finds itself in a strongly repulsive Coulomb potential, and will desorb if at the surface (or shift interstitially if in the bulk). The same is not true of the Ti^{4+} cation which always has some negatively charged neighbours, and therefore never finds itself in a totally repulsive potential. Cations are seldom seen in the desorption spectra from maximal valency oxides such as TiO_2.

Although the KF model was proposed originally to explain the ESD observations from oxides, it has now been shown to operate for chemisorbed atoms. This is illustrated in Fig. 9.6 from Schwalke *et al.*,[5] where the dependence of O^+ yield on incident energy is plotted for the adsorption of NO on Pt(100) (solid line) and for the adsorption of CO on W(100) (dotted line, data from Franchy and Menzel[9]). It can be seen there that for NO/Pt(100) the

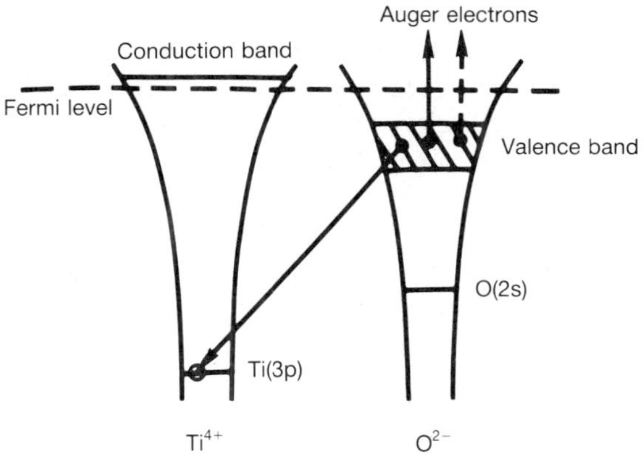

FIG. 9.5. Schematic energy band structure of TiO_2 illustrating the desorption mechanism that forms the basis of the Knotek and Feibelman[8] model. In the Ti^{4+} ion in TiO_2 the highest occupied level will be the Ti 3p at about 34 eV below the Fermi level. Ionization of this level leads to an interatomic Auger process in which an O 2p electron from the valence band fills the Ti 3p hole, and a second Auger electron is then ejected from the valence band. The result is to leave the original O^{2-} ion in either an O^0 or even O^+ state after the Auger process. The ion then finds itself in a repulsive Coulomb potential, and is expelled from the surface. The Ti^{4+} cation always has some negatively charged neighbours and cannot itself be ejected.

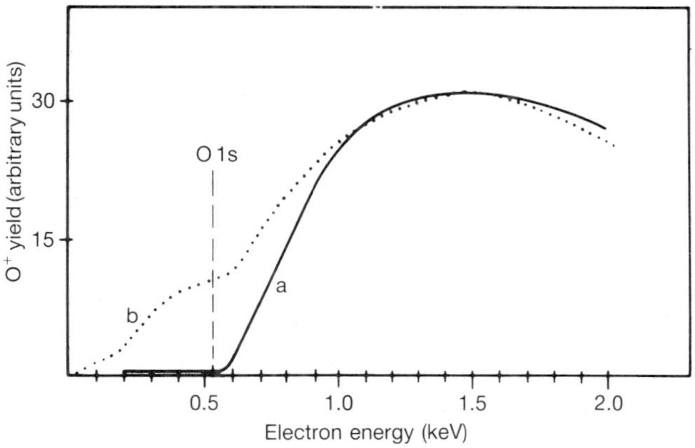

FIG. 9.6. Dependence on incident electron energy of the O^+ yield from (a) NO adsorbed on Pt(100), Schwalke *et al.*,[5] and from (b) CO adsorbed on W(100), Franchy and Menzel.[9] For the NO adsorption the desorption threshold is at the O 1s ionization threshold, which indicates operation of the Knotek and Feibelman[8] mechanism only. For the CO adsorption, there is again a threshold at the O 1s ionization, but it is a secondary threshold, the primary one being at much lower energies (<50 eV). Thus in the latter case both the Knotek–Feibelman and the Menzel–Gomer–Redhead[1,2] mechanisms are operating.

only ESD mechanism operating is the high energy KF one, whereas for CO/W(100) both the MGR and the KF mechanisms are operating. However, in this case the simple KF mechanism as described above is not completely adequate, since the ESD threshold for NO/Pt(100) is at 565 eV, some 34 eV higher than the O 1 s binding energy and a modification to the mechanism is necessary. One suggestion[10] is that in addition to the initial core ionization additional valence electron excitation (i.e. shake-off) is necessary.

Results of ESD can be found in the literature plotted in a variety of ways, according to the information required. Desorption yield curves, i.e. yield of a selected ion as a function of incident electron energy, are important in the determination of desorption thresholds. When combined with ion kinetic distribution curves for the same chosen ion, plotted either for different incident energies at a fixed adsorbate coverage, or for a fixed incident energy at increasing coverages, it is often possible to deduce a desorption mechanism for that particular ion. Another way of presenting results is to plot various selected ion yields at a fixed incident energy as a function of exposure of the adsorbing gas. All these ways will be described in the next section on applications.

In ESDIAD experimental results appear in the form of patterns on a fluorescent screen, somewhat reminiscent of LEED patterns, but with relatively much less well focused and more diffuse features, or 'spots'. Some typical examples are shown in Fig. 9.7 from Madey *et al.*,[11] where temperature effects on the angular distribution of O^+ from oxygen adsorbed on W(100) are recorded. Note that the ESDIAD pattern has the same symmetry as the substrate, an universal observation whenever a regular pattern becomes discernible. The other, currently less common, method of presenting ion desorption distributions is that mentioned already above, the mapping method. Figure 9.8 gives some results of mapping the angular distributions of desorbing O^+ ions from oxygen adsorbed on Mo(111), as a function of annealing temperature, from Niehus and Krahl–Urban.[6] The ion energies measured in the time-of-flight detection are also shown. For comparison patterns obtained at the same time by the display method are included.

Since ESDIAD patterns appear to have the same structural symmetry as the substrate, it seems likely that as a surface technique ESDIAD is providing surface structural information, and is therefore out of place in this volume, but since it is an offshoot of ESD it merits some discussion.

In desorption the initial state is that of an atom or molecule adsorbed on a surface; the adsorption is characterized by the position of the adsorbed species with respect to the substrate, the bond length and strength, the bond angle, the vibrational frequency, and so on. The final state is that of the ion, or the neutral or metastable species at the surface that result from interaction with the incident electron. The principal question to be answered is then 'To what extent does the ion angular distribution reflect the directionality of the initial state adsorbate–substrate bond?' It is clear that the observed ESDIAD patterns are final state effects because of the experimental observation that the mean kinetic energies of the desorbed ions are rather too high to be explained simply by momentum transfer from the incident electron. If the patterns are to

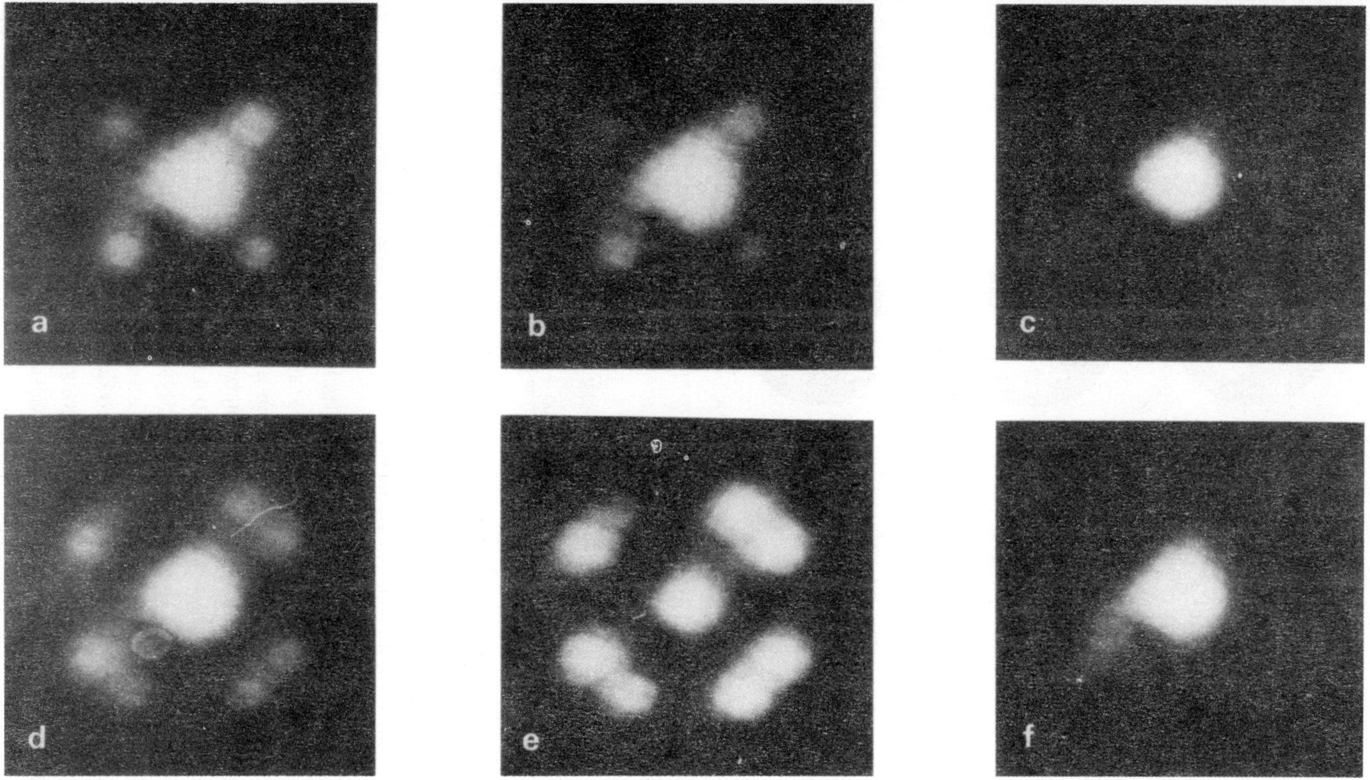

FIG. 9.7. Examples of ion angular distributions in the technique of ESDIAD, from Madey *et al.*[11] Oxygen is adsorbed to high coverage on W(100) at $\approx$100 K. The crystal is then heated to successively higher temperatures for 10 s before return to 100 K for the analysis. Electron energy 190 eV. Temperatures are in (a) 100 K, (b) 300 K, (c) 600 K, (d) 700 K, (e) 870 K, and (f) 950 K. The desorption pattern has the same symmetry as the substrate.

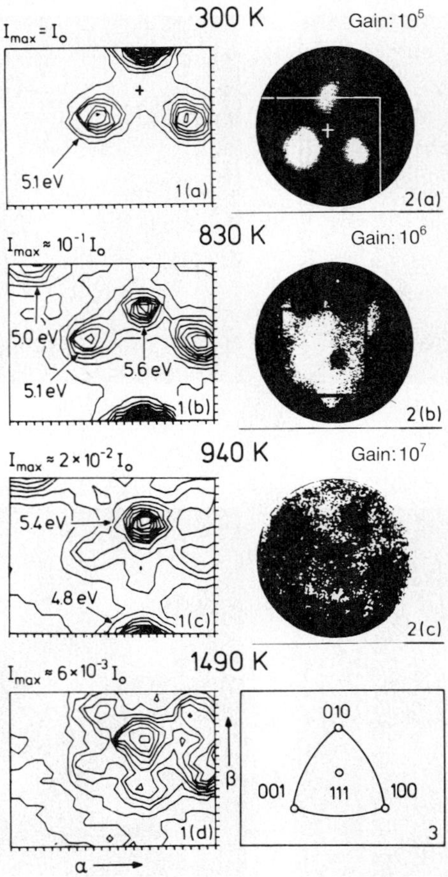

FIG. 9.8. Examples of ion angular distributions recorded by Niehus and Krahl-Urban[6] in the mapping variation of ESDIAD. The yields are of O^+ ions ejected from Mo(111) after adsorption at 300 K, as a function of the annealing temperatures indicated. The portions of the ESDIAD patterns used in the mapping are shown to the right. The energies marked on the maps are those of ions originating at those points, as determined from time-of-flight measurements.

relate directly to the *initial* state bond directions, it is then necessary that the configuration of the excited adsorbate + substrate couple does not change before desorption occurs. Since for species adsorbed on metals the lifetime of an excited ion is much less than the characteristic molecular vibration time, it is very unlikely that any rearrangement to a new configuration will take place prior to desorption. In the gas phase, electron and photon impact dissociation gives rise in general to products localized along the original (initial state) bond direction, and the same seems to apply to desorption from adsorbed species. To put it another way, the repulsive forces giving rise to desorption may well act along the same directions as the attractive forces holding the configuration together in the initial state. The likelihood of the correlation of the ESDIAD pattern with the initial state bond directionality is reinforced by the appearance of spots in the pattern, akin to LEED.

Like LEED patterns, not all ESDIAD patterns are easy to understand in terms of the surface structure, but simple correlations sometimes suggest themselves, such as the positions of adsorbed oxygen atoms on a W(100) surface shown schematically in Fig. 9.9 with the fourfold symmetry of the

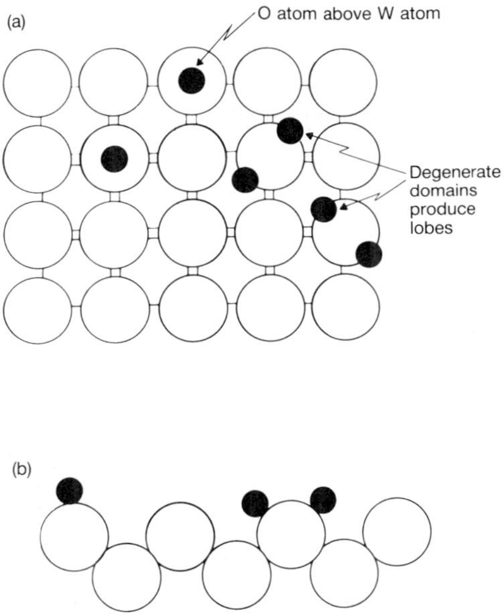

FIG. 9.9. Model of the chemisorption of oxygen on W(100) at high coverage proposed by Madey *et al.*[11] to explain the observations of Fig. 9.7. (a) Plan view. (b) Section view along 110 direction. The central spot in the patterns in Fig. 9.7 is identified with desorption of O^+ from the 'on-top' site, in which oxygen is adsorbed directly above a substrate tungsten atom. The four lobes, seen especially in Fig. 9.7(d) and (e), are attributed to O^+ desorption from each of the two possible orientations arising from adsorption of oxygen into an O–W–O complex.

patterns in Fig. 9.7. Madey *et al.*[11] identify the central spot with desorption of O^+ from the 'on-top' site and the four lobes with the two possible orientations of symmetric O–W–O complexes. Where diffuse or hazy patterns occur, either the desorption layer is disordered or there are many possible bonding configurations leading to overlap of the directions of desorption.

9.3 Applications

One cannot have desorption without first having adsorption, and all applications of ESD and ESDIAD have involved studies of the interactions of simple molecules with (usually) single crystal surfaces, nearly always in conjunction with one or more of the other surface characterization techniques. The adsorption of water seems to have been one of the most popular subjects for study. For example, Ding and Williams[12] combined ESD and AES in their experiments on the adsorption of water on Al(100), using a time-of-flight technique to detect the desorbed ion masses and to measure their energy distributions. Incident energies used were 500 eV for ESD and 2.4 keV for AES; the electron gun was pulsed at 10 kHz for pulses of 1 μs duration, and the flight path was 23 cm.

A typical flight-time distribution found during desorption after exposure of the Al(100) surface to 150 L of water vapour is shown in Fig. 9.10. The

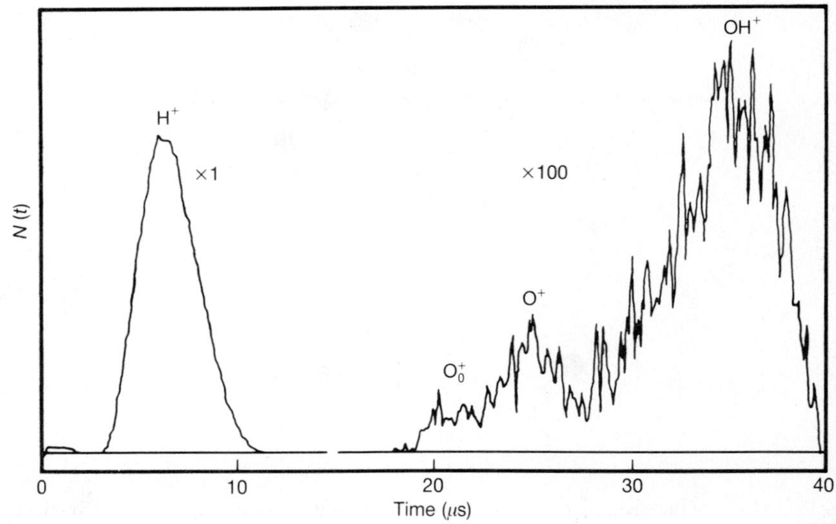

FIG. 9.10. Time-of-flight distribution of desorbing species from an Al(100) surface after adsorption of 150 L of water vapour at 300 K, from the experiments of Ding and Williams.[12] The types of ion giving rise to each peak are indicated. Note the two types of oxygen ion, denoted O^+ and $O_0{}^+$, respectively, associated with chemisorption and with oxide formation, respectively.

principal desorbing species is H^+, with OH^+ and two types of oxygen ion at some two orders of magnitude lower intensity. The energy distributions of these four ions are illustrated in Fig. 9.11, with the values of the most probable energy indicated in each case. With increasing water vapour exposure the distributions shifted progressively toward higher energies as the work function of the aluminium surface decreased, but did not change shape. When the ion yields were plotted against oxygen exposure up to saturation at 150 L the responses in Fig. 9.12 were found. Note that in the low exposure range (right-hand panel) no signal from the oxide ion $O_0{}^+$ was found, and only a very weak signal from O^+. The oxygen Auger signal in the same region was found to have

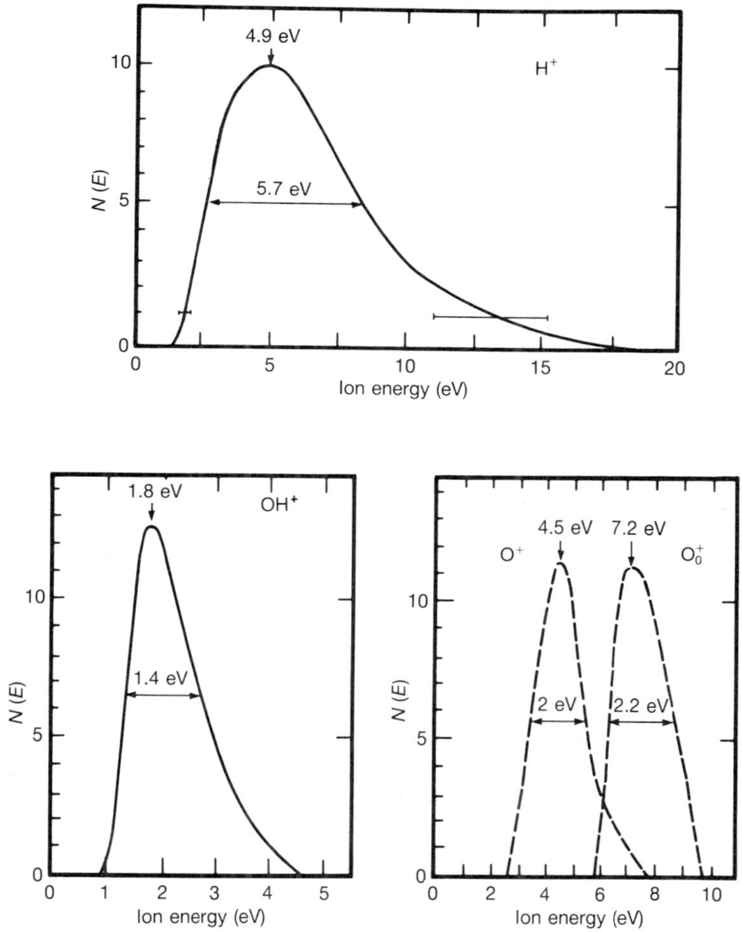

FIG. 9.11. The energy distribution of the four types of ion observed in the time-of-flight spectrum of Fig. 9.10, by Ding and Williams.[12] The values of the most probable energies are indicated. The shapes remained unaltered with increasing exposure to water vapour, although the peaks shifted towards higher energies as the work function of the Al(100) surface increased.

the same form of increase up to an exposure of ~5 L, suggesting a common source for it and the H^+ and OH^+ ESD signals. Since OH^+ is being desorbed, the most likely source is in fact OH groups on the surface.

With increasing exposure beyond 5 L, the oxygen Auger signal growth goes through a point of inflection before increasing again, behaviour that has always been identified with nucleation of the adsorbed oxygen into oxide and the start of oxide growth in the form of islands. At just about that point two other changes occur, the sudden decrease in the OH^+ yield seen in Fig. 9.12 and the appearance in the Auger spectrum of an Al LVV feature shifted to the

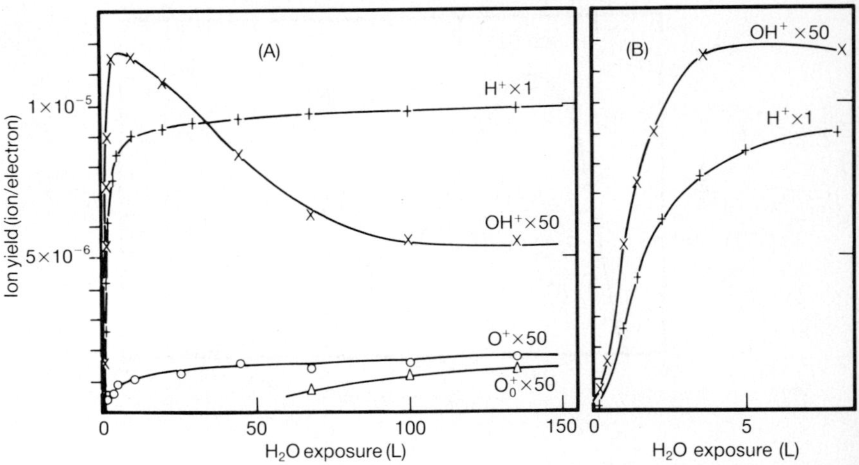

FIG. 9.12. The variation in the yields of the four types of ion observed in Figs. 9.10 and 9.11, with exposure to water vapour, in two different regimes, (A) 0–150 L, and (B) 0–8 L. In the low exposure range the dominant desorbing species are OH^+ and H^+, with a weak signal from O^+ ions, but none from $O_0{}^+$ ions. The latter do not appear until an exposure of about 50 L is reached. Note the plateau in yield reached by the OH^+ ions at about 5 L, followed by a progressive decrease, although the H^+ yield remains constant. From Ding and Williams.[12]

oxide position. These changes are brought out more dramatically in Fig. 9.13 by plotting various signals as a function of the oxygen Auger signal. Clearly in the 5–8 L exposure region the metal is transforming into oxide, with a concomitant modification in the surface coverage of OH groups; however, the yield of H^+ ions hardly alters (Fig. 9.12). At an exposure of $\sim$45 L there is a change in slope of the OH^+ signal in Fig. 9.13, and it is at that point that the first $O_0{}^+$ signal appears.

The model proposed by the authors to take account of all the observations is shown in Fig. 9.14. In the earliest stages of adsorption the OH groups are restricted solely to the outer surface (sites X), from where both H^+ and OH^+ ions could be expected to be desorbed under electron impact. In the exposure range 5–8 L the OH groups start to become incorporated in the surface of the Al, as in site Y, from which it is still just as easy to desorb H^+ but not so easy to desorb OH^+, hence the abrupt decrease in the OH^+ yield. With further exposure the OH^+ groups are converted by dissociation progressively into O^{2-} ions, with the appearance of the Al (oxide) Auger peak position, and finally, beyond 45 L, the oxide ions appear at the surface, and desorbed $O_0{}^+$ is seen.

Because the most probable energies of H^+ and OH^+ desorbing ions are high, they must originate in an Auger decay process of the type described above due to Knotek and Feibelman. The most likely is ionization of the oxygen 2s shell in the OH group, followed by de-excitation involving transfer

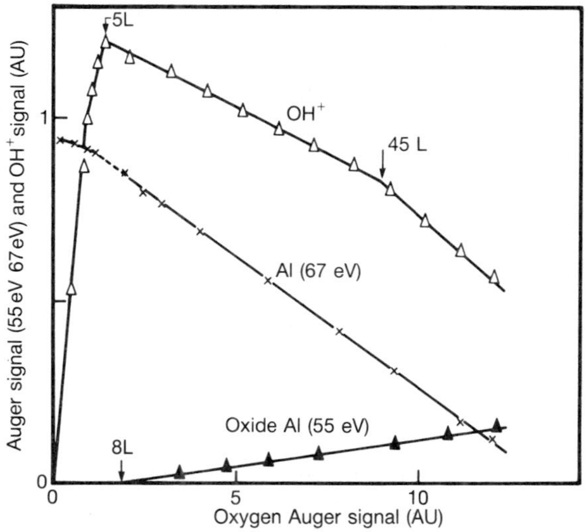

FIG. 9.13. Plot of the peak-to-peak heights of the Al Auger peaks corresponding to metallic (67 eV) and to oxidized (55 eV) Al, and of the OH^+ ion yield, against the peak-to-peak height of the O Auger peak, from the experiments of Ding and Williams.[12] In the 5–8 L range of exposure incipient oxide nucleation is taking place, with a consequent drop in the number of surface OH groups. At 45 L there is a change in slope of the OH^+ ion yield, coincident with the appearance of the oxide O_0^+ ions in the yield spectrum.

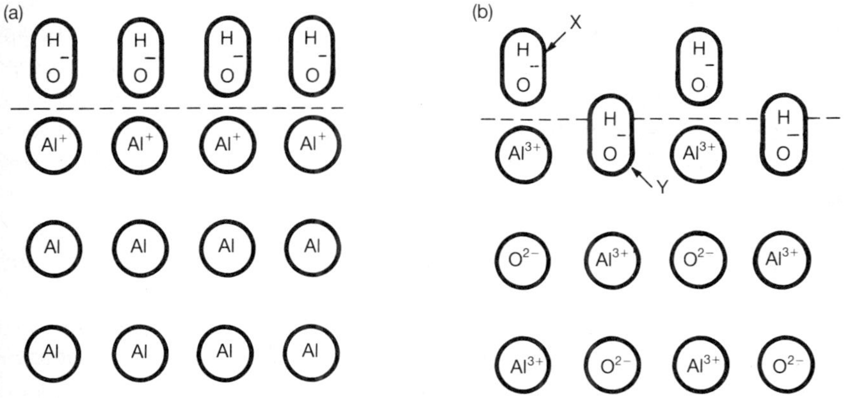

FIG. 9.14. (a) Chemisorption. (b) Oxidation. Model proposed by Ding and Williams[12] to take account of the various observations made during the interaction of water vapour with Al(100), such as those in Figs 9.10 to 9.13. In the chemisorption stages (0–5 L) the OH groups are restricted entirely to the external surface, and consequently both H^+ and OH^+ are desorbed under electron impact. Between 5 and 8 L exposure the OH groups begin to become incorporated in the surface, as indicated by Y, from which H^+ can still be desorbed easily, but OH^+ not so easily, hence a decrease in the OH^+ yield. With further exposure the incorporated OH groups are converted progressively into O^{2-} ions, signalled by the appearance of the Al Auger peak in the oxide position, and finally at 45 L, oxide ions appear at the surface and are desorbed as O_0^+.

of charge from hydrogen to oxygen so that the hydrogen experiences a repulsive force within the OH group and is desorbed. For desorption of OH^+, it is suggested that ionization of the oxygen followed by de-excitation to leave two core holes on the oxygen can also lead to removal of an electron involved in the bonding of the OH to the surface. This would then create a positively charged OH species, resulting in desorption.

One conclusion from the above experiments is that ESD is more surface-sensitive than AES, since changes due to initial formation of the 'incorporated OH' state were visible in ESD before there was any indication of oxide formation in AES.

ESDIAD has been shown to be capable of providing striking evidence for the effects of small amounts of impurities on the ordering of certain adsorbed molecules in a way which could not have been established otherwise. An excellent example is the study by Netzer and Madey[13] of the effects of very small doses of pre-adsorbed oxygen on the adsorption of NH_3 on Ni(111). When NH_3 is adsorbed on a clean Ni(111) surface at 150 K to a coverage of about 0.25 monolayer, the ESDIAD pattern of H^+ desorption appears as a continuous ring or halo, as seen in Fig. 9.15b. In 9.15a is the LEED pattern at 110 eV from clean Ni(111), showing the sharp hexagonal pattern typical of the surface structure of this crystal face. The H^+ ESDIAD halo suggests adsorption of NH_3 via the N atoms, with no preferred orientation of the H atoms which point away from the surface. If, on the other hand, the Ni(111) surface is pre-dosed with a very small amount of oxygen, to an estimated coverage of ~0.05 monolayer, cooled to 80 K, exposed to NH_3 as before, and then allowed to warm to ~140 K, the H^+ desorption pattern now shows significant threefold symmetry, as seen in Fig. 9.15c. The ESDIAD pattern appears as an intermittent halo, with intensity maxima in the $[\bar{1}\bar{1}2]$ directions. When the NH_3 exposure is much greater, and the crystal is warmed to 610 K, a sharp hexagonal ESDIAD pattern is observed, as in Fig. 9.16D; the other panels in this figure show the effect of warming to 150 K on the ESDIAD patterns at different NH_3 coverages.

The difference in the ESDIAD patterns between adsorption of NH_3 on clean and on oxygen pre-dosed, Ni(111), is clear evidence for chemical interaction between the adsorbed NH_3 and O. The authors suggest that the interaction is via a weak hydrogen bond, which would then remove some H atoms from participation in the desorption process, those that do not participate then contributing to the pattern. In their model each adsorbed O atom can interact via H-bonds with three NH_3 molecules in neighbouring sites, thereby inducing short-range preferred azimuthal ordering. The remaining adsorbed NH_3 molecules are still randomly ordered and desorption of H^+ from them contributes to the background of the halo in Fig. 9.15c, and in Figs 9.16A, B and C. At high NH_3 coverages (Fig. 9.16D) the principal ion normal to the surface is in fact O^+, suggesting that the oxygen has caused

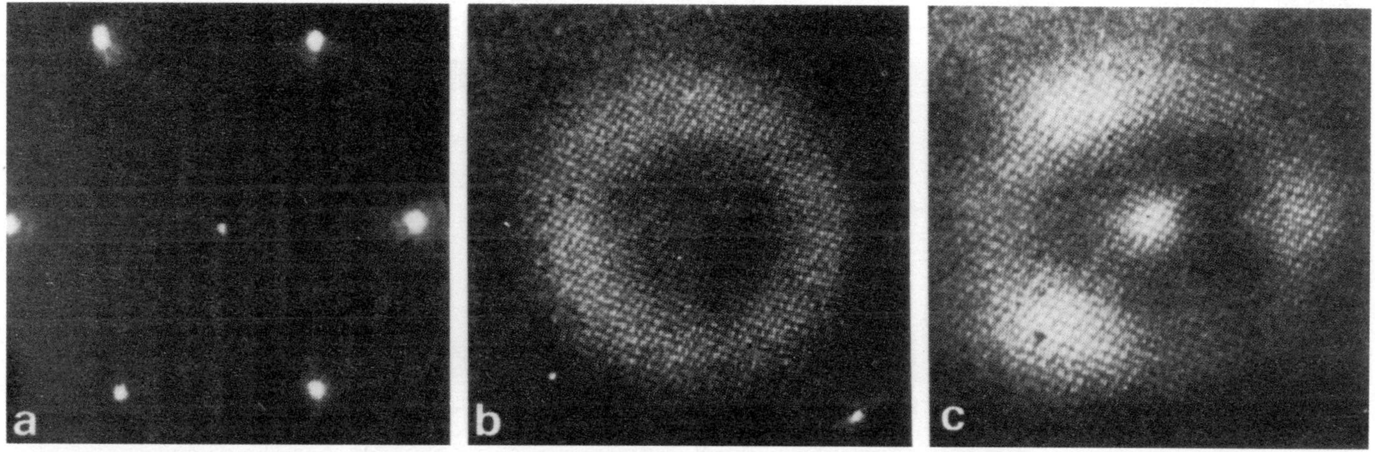

FIG. 9.15. Observations made by Netzer and Madey[13] on the adsorption of NH_3 at 150 K on clean and on oxygen-dosed Ni(111). (a) LEED pattern from the clean surface at a primary energy of 110 eV; (b) ESDIAD pattern of H^+ desorption from NH_3 on clean Ni(111) at a primary energy of 300 eV: (c) ESDIAD pattern of H^+ desorption from NH_3 adsorbed on Ni(111) after pre-adsorbing oxygen to a coverage of $\theta = 0.05$. The continuous halo in (b) indicates an arrange of NH_3 molecules bonded via the N atom, with no preferred orientation of the H atoms. In (c) some clear threefold symmetry in the H^+ desorption is observed, with intensity maxima in the $[\bar{1}\bar{1}2]$ directions.

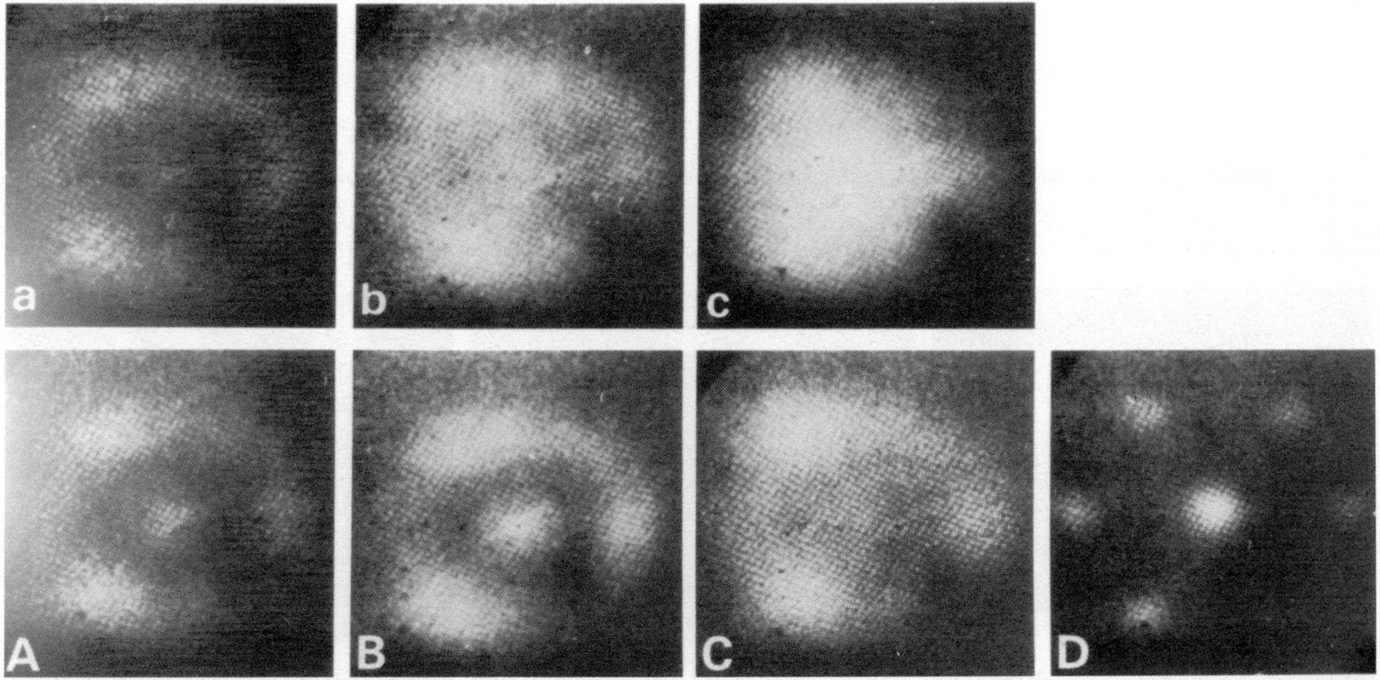

FIG. 9.16. Series of ESDIAD patterns observed by Netzer and Madey[13] during the adsorption of NH_3 at 80 K in increasing amounts on Ni(111) pre-dosed with oxygen to a coverage of $\theta \approx 0.05$, followed by heating to 150 K. Upper row (a, b, c) adsorption at 80 K; lower row (A, B, C) heating to 150 K. Exposures: (a, A) θ $(NH_3) \approx 0.08$; (b, B) θ $(NH_3) \approx 0.12$; (c, C) θ $(NH_3) \approx 0.25$. The pattern in D was observed after repeated exposures of NH_3 to the oxygen-dosed surface, followed by heating to 610 K. At that stage the principal ion normal to the surface was O^+. Evidently the presence of the oxygen on the Ni is promoting dissociation of the adsorbed NH_3.

dissociation of the NH_3 into a mixed layer of N and O, with the remaining NH_3 molecules being left in a highly ordered array.

References

1. Menzel, D. and Gomer, R. (1964). *J. Chem. Phys.* **41**, 3311.
2. Redhead, P. A. (1964). *Can. J. Phys.* **42**, 886.
3. Czyzewski, J. J., Madey, T. E., and Yates, J. T. (1974). *Phys. Rev. Lett.* **32**, 777.
4. Noell, J. O., Melius, C. F., and Stulen, R. H. (1985). *Surf. Sci.* **157**, 119.
5. Schwalke, V., Niehus, H., and Comsa, G. (1984). *Surf. Sci.* **137**, 23.
6. Niehus, H. and Krahl-Urban, B. (1981). *Rev. Sci. Instrum.* **52**, 56.
7. Knotek, M. L. and Feibelman, P. J. (1978). *Phys. Rev. Lett.* **40**, 964.
8. Knotek, M. L. and Feibelman, P. J. (1979). *Surf. Sci.* **90**, 78.
9. Franchy, R. and Menzel, D. (1979). *Phys. Rev. Lett.* **43**, 865.
10. Jaeger, R., Stöhr, J., Treichler, R., and Baberschke, K. (1981). *Phys. Rev. Lett.* **47**, 1300.
11. Madey, T. E., Czyzewski, J. J., and Yates, J. T. (1975). *Surf. Sci.* **49**, 465.
12. Ding, M. Q. and Williams, E. M. (1985). *Surf. Sci.* **160**, 189.
13. Netzer, F. P. and Madey, T. E. (1982). *Surf. Sci.* **119**, 422.

10

Photon excitation: X-ray photoelectron spectroscopy (XPS) and X-ray excited Auger electron spectroscopy (XAES)

10.1 Operation

10.1.1 *XPS*

In conventional XPS, X-radiation from either an Al or an Mg unmonochromatized source of a type similar to that described under section 3.2.2 and illustrated in Fig. 3.7 falls on the surface of the sample to be analysed, in a chamber at ultra-high vacuum. Less frequently, and usually for special purposes, monochromatic radiation from an Al source produced according to the principle of Fig. 3.9 is used. The energies of photoelectrons ejected from the sample are measured in a Concentric Hemispherical Analyser (CHA), whose principles of operation are described under section 3.3.4, and which is shown schematically in Fig. 3.22. In most modern spectrometers the CHA has an input lens system that allows a greater solid angle of photoelectrons to be collected and in which pre-retardation of electrons to the entrance slit of the CHA is performed. A schematic diagram of a typical configuration in a spectrometer is shown in Fig. 10.1 (courtesy of Leybold Ltd.). In that case the sample for analysis is placed in a slot cut in a shaft that slides in and out of the analysis chamber, which is just one of the many different methods of rapid sample insertion that are now available on current equipment.

Choice of analysing parameters depends on what needs to be known about individual samples. If simple elemental quantification is all that is needed, it is customary to record a spectrum over a sufficiently wide energy range to include photoelectron peaks from all the elements that might be present. Such a spectrum is often termed a 'wide' or 'survey' scan, and it might, for example, when using Al K_α (1486.6 eV) radiation, run from 400 eV to 1500 eV. Since detailed chemical information is not being sought in that case, it is usually adequate to record at 1.0 eV or 0.5 eV steps and to operate the CHA at moderate resolution, so that sensitivity is maintained and the spectrum acquisition does not take too long. For most analyses, chemical information about selected elements will also be required, in which case more detailed spectral recording will have to be performed, involving choice of an energy range that covers the photoelectron features of a particular element, acquisition at much smaller energy intervals, typically 0.1 to 0.05 eV and

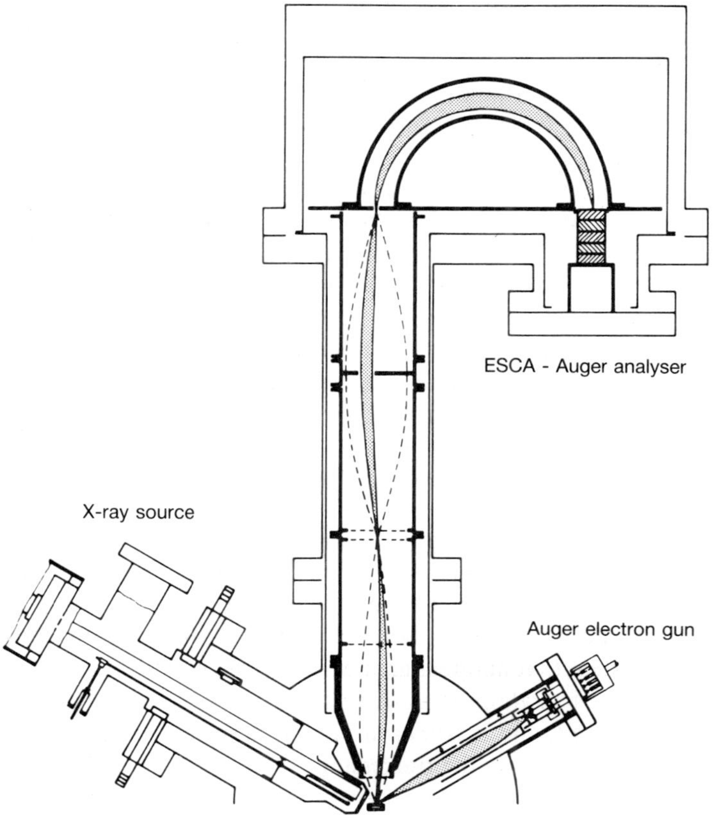

FIG. 10.1. Typical configuration in a modern XPS spectrometer. The specimen sits at the centre of a spherical chamber, and can be irradiated either with soft X-rays (on the left) or electrons (on the right). Ejected photo- or Auger electrons are energy analysed in the concentric hemispherical analyser above. An input lens system focuses the emitting area on to the entrance slit to the analyser, and introduces some magnification as well. (Courtesy of Leybold Ltd.)

adjusting the CHA to operate at as high an energy resolution as is compatible with adequate signal-to-noise ratio. When the signal is weak this may involve long counting times, although eventually instabilities inherent in the whole system will impose an upper limit on the length of time over which it is worthwhile continuing to count. With progressively increasing sensitivity in XPS from the use of such improvements as multi-detectors, the recording of weak signals at adequate resolution will become less of a problem.

The great majority of XPS systems, and certainly all modern ones, are controlled by dedicated microcomputers that store the spectral data on hard or soft discs. Every manufacturer sells a data system to go with his instrument, and this aspect of the technology changes so rapidly that any one such system is liable to be out of date with respect to capability, sophistication, and speed of operation within a couple of years. The fundamental requirements of spectrometer control and of data processing are, however, common to them all, and it is only in the details of the way in which these requirements are implemented that they differ.

The basic parameters that need to be set up in spectrum acquisition are the kinetic energy at start, the kinetic energy range, the number of channels in the range (i.e. the energy step), the counting time per channel, the number of scans to achieve the required statistics, the energy resolution (either the pass energy or the retard ratio), and the X-ray source voltage and current settings. In some systems it is also possible to ask the computer to switch ion guns on and off, to rotate a specimen carousel for automatic sequential analysis of specimens, and to operate other auxiliary devices such as flood guns and heating stages. The control system will also display the spectrum as it is being acquired and, usually, allow previously acquired spectra on the same disc to be viewed while acquisition continues.

Once the data have been acquired and stored they will then need to be processed. Processing operations are far more complex and time consuming than the straightforward acquisition, and the mathematical and physical assumptions made in the setting-up of some of the operations are still matters of controversy. If the throughput of analytical work on an XPS spectrometer is liable to be large, the data processing side of the operation could represent a bottleneck in output, and it is then desirable to have an additional off-line minicomputer dedicated to the processing.

Certain comparatively trivial operations that will be needed to manipulate the display spring to mind at once. These are, expansion of the energy and counts scales, selection of any portion of the energy range, normalization of the maximum count to the maximum of the display scale, and provision of a cursor to identify the energy and count rate at any point in the spectrum. It is also essential, although not so obvious, to have a facility for spike removal. Spikes are merely abnormally high count rates in individual channels or pairs of channels caused by electrical transients affecting the counting circuitry. If spikes are not removed subsequent operations such as smoothing can be rendered useless.

The non-trivial processing operations are those of smoothing, instrumental de-broadening, satellite subtraction, inelastic background removal, peak synthesis and curve fitting, spectrum comparison involving addition or subtraction of spectra, peak area measurement, and quantification. A complete exposition of all the methods used for these operations is not within the scope of this volume, and in any case they have been dealt with fairly exhaustively by both Sherwood[1] and by Carley and Joyner,[2] to which works the reader is referred. However, some brief description and comments are in order here.

Smoothing is used to eliminate small systematic variations in intensity of instrumental origin, by increasing the correlation between data points at the expense of uncorrelated noise. It is valuable where the acquisition statistics are poor, e.g. where very weak signals can only be recorded by taking a long time over the accumulation, or where a material is unstable and a spectrum must therefore be recorded quickly. In general, however, smoothing must be used with great caution since the operation will always introduce some artefacts into the spectrum, and it is safest if possible to employ unsmoothed data for analysis. The most common method of smoothing, and the one that is found in all data processing packages in various forms, is the least squares approach of Savitsky and Golay.[3] It consists of fitting a polynomial of chosen degree to a set of an odd number of data points, the smoothed point being the central one in the interval, minimizing the least squares sum at each successive point. Thus a three-point smooth would consist of applying the operation

$$I_s(N) = \tfrac{1}{4}\{C_{N-1}\, I(N-1) + 2C_N\, I(N) + C_{N+1}\, I(N+1)\} \qquad (10.1)$$

to each channel, $I_s(N)$ being the intensity in channel N after the smoothing, and the C_i being the coefficients to be established by the least squares fit. With this method the safest approach is to use the greatest possible number of points in the smoothing interval, but the lowest possible degree of the polynomial. Provided that the smoothing interval is not large with respect to the peak width, the best (in the sense of least introduction of artefacts) use of this method is to repeat the smoothing operation several times with the smallest possible smoothing interval.

The result of smoothing a noisy spectrum compared with the same spectrum recorded with good statistics is shown in Fig. 10.2, from Sherwood,[1] for an O 1s spectrum. The smoothed spectrum in (c) contains no more information than the unsmoothed 'good' spectrum in (a), but the smoothing has made the overall shape of the peak much more obvious visually than that of the unsmoothed noisy spectrum in (b), revealing, for example, the contribution from the smallest peak at about 535 eV. The energetic positions of the three contributions to the oxygen spectrum resolved from the smoothed spectrum (c) were almost identical to those retrieved from (a), but the measured areas of the second and third peaks had increased by 7% and 18%, respectively, compared to the largest peak.

The other, less well used, approach to smoothing uses a Fourier transform method. Fourier inversion of a spectrum consisting of an array of data points containing signal and noise components results in a distribution of frequencies for the signal and the noise. If a suitable weighting function can be devised, the distribution can be convoluted with the function to suppress the noise contribution, following which re-inversion produces a smoothed spectrum.

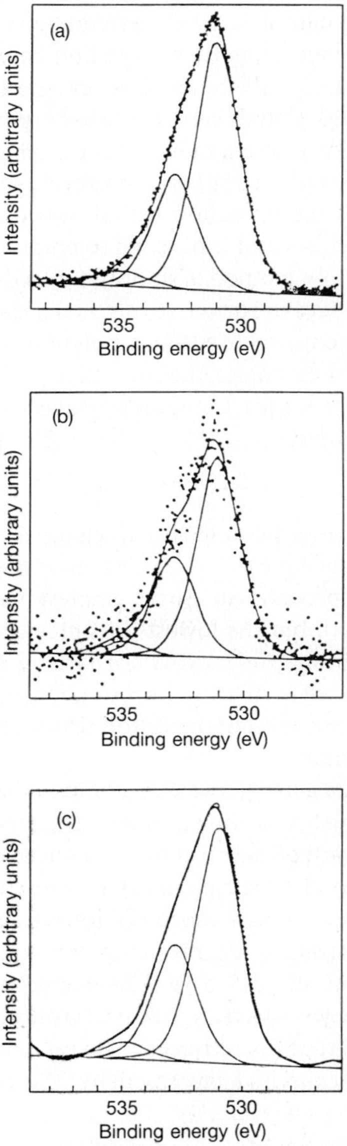

FIG. 10.2. The effect of smoothing on some O 1s spectra, compared with the results of long and short spectral acquisition times. In each panel the spectrum has been fitted with three peaks. (a) Spectrum with good statistics achieved by acquisition over a long period. (b) Spectrum with poor statistics resulting from a short acquisition period. (c) Result of smoothing spectrum (b) 100 times with a 21-point smoothing interval. The energetic positions and half-widths of the fitted peaks were similar in all three treatments, but the smoothing increased the areas of the two smaller peaks slightly with respect to the largest. (From Sherwood.[1])

The difficulty in the method is that of finding a suitable weighting function that will not affect the signal contribution to an undesirable extent, and no universal recipe for doing this has yet been established.

Whether or not the experimental spectrum has been smoothed, it is generally desirable to remove the broadening introduced by the finite width of the incident X-radiation and by the response of the analyser itself. The process is known as instrumental broadening deconvolution. If the observed spectrum as a function of energy is $f_{\mathrm{obs}}(E)$ and the related 'true' spectrum is $f_{\mathrm{t}}(E)$, then one can be obtained from the other by

$$f_{\mathrm{obs}}(E)=\int f_{\mathrm{t}}(y)\, X(E-y)\, A(E-y)\, \mathrm{d}y \tag{10.2}$$

where X is the X-ray source broadening function and A the analyser broadening function. Usually X and A are lumped together into a total instrumental broadening function $B(E)$. If B can be chosen suitably, and it is usual to choose a Gaussian whose width is dependent on the energy resolution, then f_{t} can be extracted either by direct solution of the deconvolution equation, by Fourier transform analysis, or by iteration. Iterative techniques are well suited to computer operation and are the most popular. It is worth noting, however, that if the original data are noisy, attempts to deconvolute instrumental broadening may simply introduce unwanted artefacts into the data since the presence of much high frequency noise could make the finding of an unique solution impossible. Good statistics are in this case an essential starting point.

Where an X-ray monochromator is not being used, it is also sometimes desirable to be able to remove satellite photoelectron peaks. As pointed out in section 3.2.2, unmonochromatized radiation from Al and Mg sources consists not only of the principal $K_{\alpha_{1,2}}$ doublet, but also the $K_{\alpha_{3,4}}$ at $\sim$10 eV higher energy, and the K_β at some 55–70 eV higher energy. The intensity of the $K_{\alpha_{3,4}}$ is about 10% of that of the $K_{\alpha_{1,2}}$, and satellite peaks arising from it can be seen in any spectrum, associated with all the major peaks excited by $K_{\alpha_{1,2}}$. The K_β intensity is much lower, and it does not normally cause any problems, although there are instances where its presence could be mistaken for an impurity element (e.g. Zn with its $2p_{3/2}$ peak at $\sim$1022 eV when there is a large O 1s peak at $\sim$952 eV). Since the intensities and energy separations of the satellites relative to the principal line are well known and fixed, it is easy to construct a programme to remove them. Starting at the low kinetic energy end of a spectrum, proportions of the intensity in any one channel are subtracted from those channels situated at fixed energy distances ahead of that channel. There are inherent problems, however, which are not trivial. Although the satellite subtraction process is in itself straightforward, it is necessary before carrying it out to have subtracted the correct inelastic background from the spectrum, which, as will be seen below, is not simple. Secondly, X-ray excited Auger peaks will have no associated satellite, and they must therefore either be removed or their spectral contribution known exactly. As a result of these factors, satellite subtraction is not generally recommended; it is far better to use monochromatic radiation in which satellites are absent in any case.

In XPS spectra, peaks appear situated not only on a continuously increasing background due to photoemission by bremsstrahlung radiation (when using an unmonochromatized source) but also on the background steps originating from inelastic electron losses from peaks at higher kinetic energies. Figure 10.3, in which a wide XPS scan from an almost clean copper surface is

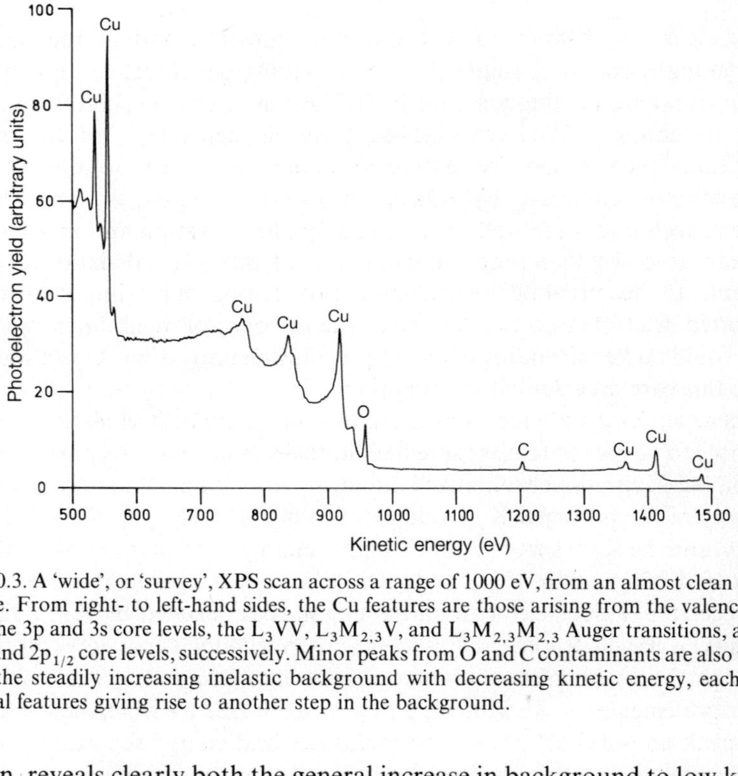

FIG. 10.3. A 'wide', or 'survey', XPS scan across a range of 1000 eV, from an almost clean copper surface. From right- to left-hand sides, the Cu features are those arising from the valence band (3d), the 3p and 3s core levels, the L_3VV, $L_3M_{2,3}V$, and $L_3M_{2,3}M_{2,3}$ Auger transitions, and the $2p_{3/2}$ and $2p_{1/2}$ core levels, successively. Minor peaks from O and C contaminants are also visible. Note the steadily increasing inelastic background with decreasing kinetic energy, each set of spectral features giving rise to another step in the background.

shown, reveals clearly both the general increase in background to low kinetic energy and the inelastic steps that add to it. When attempting to resolve a complex photoelectron peak into its individual components, corresponding to different coordination states of the element giving rise to the peak, it is essential to remove the background. Such peak resolution is the most important aspect of the data processing of XPS spectra since without the possibility of performing it unambiguously, both the inherent chemical information and the ability to quantify are lost.

There are at present two methods in general use for background removal.

The simpler merely draws a straight line between two points chosen on either side of a peak. There is no obvious criterion for the choice of the points, which is entirely subjective, and no physically realistic reason why a straight line should be a reasonable approximation. However, it is easy and quick, and if used in a consistent manner throughout a series of spectra from the same material, will give good *relative* accuracy, although not good *absolute* accuracy.

The other method has slightly more physical basis, and is due to Shirley,[4] whose name it carries. It assumes that the background intensity at any point is proportional to the total spectral area to higher energy. In other words, it assumes that for each photoelectron ejected at a particular energy there is a constant background to lower energies. Thus the true intensity at energy E can be expressed as

$$I_t(E) = I_{obs}(E) - K \int_0^E I_t(y)\,dy \tag{10.3}$$

where $I_{obs}(E)$ is the observed intensity at E (which includes the background) and where the energy zero is taken to be at the point to higher kinetic energy of the peak where the background can be regarded as flat. Since the true intensity appears also in the integral the method of evaluating I_t is iterative. The proportionality constant K is determined by matching the calculated to the observed background at some chosen energy below the peak, and since no rules exist for identifying the matching energy on physical grounds, the match is made by eye and the method is to that extent subjective. Where there is a succession of closely spaced peaks the method may well be prone to severe errors, but at least the background so removed is likely to have rather more physical reality than the use of a straight line. Some idea of the shape of a Shirley background can be had from Fig. 10.4, from the work of Bishop,[5] for the case of a synthesized spectrum corresponding to equal amounts of Fe and Fe_2O_3; two possible positions of linear backgrounds are also shown.

Once the above operations have been performed in the correct sequence, the experimental spectra should then be in a standard form suitable for the final and most important operation, that of curve resolution or synthesis. By this is meant the separation into its constituent contributions of a photoelectron peak from an element existing in two or more separately identifiable chemical states within the volume analysed by XPS. In such a procedure an attempt is made to duplicate the envelope of the experimental spectrum as closely as possible by adding together a series of standard peaks, where the variables are the number, position, intensity, width, shape, and type of standards. For this purpose, the standards must be given analytical mathematical forms that maintain physical reality, in order that the procedure can be performed automatically on a mini-computer. The form found to be most effective and at the same time physically realistic is the so-called mixed Gausso–Lorentzian

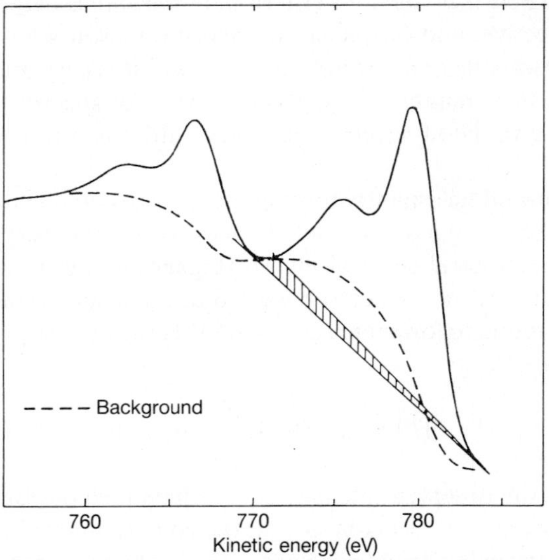

FIG. 10.4. Spectrum synthesized by Bishop,[5] corresponding to equal amounts of metallic Fe and of Fe_2O_3. The dashed line is the shape of the Shirley[4] background calculated according to expression (10.3). The two straight lines indicate two possible positions of a linear background, i.e. tangentially to the onset of the $2p_{1/2}$ peak, and at the minimum between the peaks, respectively. The hatched region is the difference in measured area when using one or other linear background.

function; some workers have used pure Gaussians and others pure Lorentzians, but in fact neither of these forms is sufficient by itself. The Gausso–Lorentzian takes the form

$$f(E)=h/(\{1+M(E-E_0)^2/\beta^2\}\exp[(1-M)\{\ln 2(E-E_0)^2\}/\beta^2])^{-1} \quad (10.4)$$

where h = peak height
M = mixing ratio (1 for pure Lorentzian, 0 for pure Gaussian)
E_0 = energy at peak centre
β = parameter proportional to peak FWHM.

Fitting to the experimental spectrum is of course iterative, and to start the iteration it is necessary to make intelligent guesses at the variables involved. It is also necessary to limit the ranges over which the variables are allowed to roam, otherwise the program will take a long time examining all possibilities and will probably produce an unrealistic answer. The goodness-of-fit of the synthesized to the experimental spectrum is usually decided by evaluating χ^2, and the iteration is repeated, if necessary with adjusted guesses, until χ^2 is reduced to the level required. Much of the mathematical detail to this procedure is given in the article by Sherwood,[1] to which the reader is referred for more information. An example of the fitting of standard peaks of the form of expression (10.4) to an experimental spectrum is given in Fig. 10.5, in which the tungsten 4f spectra from tungsten metal and from tungsten oxide are fitted with four Gausso–Lorentzians.

Other manipulations that may be required in the processing of data are the subtraction and addition of spectra. Simple subtraction and addition are trivial, but in general the manipulations will not be trivial since in most cases spectra will need to be aligned energetically, and also scaled up or down in intensity. Where there are many overlapping peaks, alignment and scaling could be a matter of considerable trial and error before the manipulations give an acceptably realistic result.

10.1.2 *XAES*

Since a core hole is created in the absorption of an X-ray photon in XPS, just as it is by the interaction with an incident electron in AES, relaxation by Auger emission can occur just as readily in XPS and indeed any 'wide scan' XPS spectrum such as that of Fig. 10.3 will contain Auger features. Thus in a sense XAES is inherently a part of XPS, but is normally regarded as a separate technique because of the ways in which it is used. Thus, in some applications it may be necessary to be able to study Auger peak shapes without having to worry about the potentially damaging effect of an incident electron beam, in others the better energy resolution employed in XPS than in AES may be required in order to resolve the fine structure associated with some Auger transitions. An increasing use of XAES is in the identification of chemical state through use of the Auger Parameter. The concept of the Auger Parameter is due to Wagner,[6] and as originally conceived the AP was equal to the difference in energies between the most prominent photoelectron and Auger features from the same element in an XPS spectrum. Being a difference in energies the AP is independent of any charging effects, and is often more sensitive to chemical changes than the positions of either photoelectron or Auger peaks. Unfortunately, the original AP could occasionally result in negative values, and so Wagner[7] defined a 'modified' AP as the sum of the kinetic energy of the most prominent Auger peak and the binding energy of the most prominent photoelectron peak, for a given element. This was equivalent to adding the energy of the X-ray photons to the original AP, resulting in positive values.

The reason for mentioning the Auger Parameter in this section is that for some elements an Auger peak of suitable intensity and energetic position cannot be found within the conventional XPS energetic range. Thus for Si the only Auger peak found in the range up to 1500 eV, i.e. using Al K_{α}, would be the LVV at ≈ 90 eV, which is in an energy region of very high background and therefore almost impossible to identify accurately with respect to position. However, as remarked under Section 3.2.2 and illustrated in Fig. 3.8, there is much useful bremsstrahlung intensity from an unmonochromatized X-ray

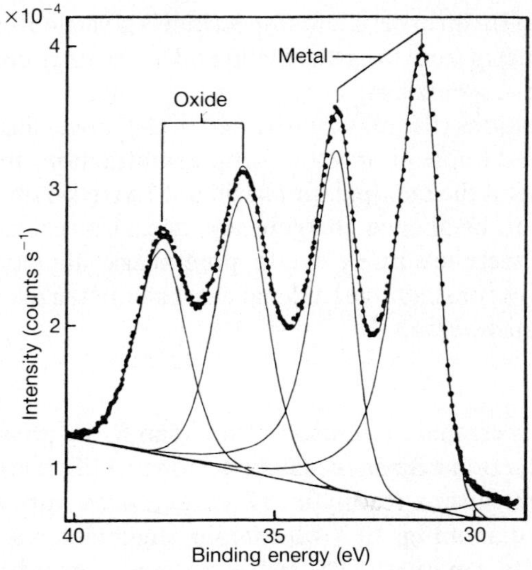

FIG. 10.5. Example given by Sherwood[1] of the fitting of standard Gausso–Lorentzian peak shapes according to expression (10.4), to an experimental spectrum, in this case of the tungsten 4f spectrum from a thin film of oxide on the metal. The curve through the experimental points represents the sum of the fitted standard peaks.

source above the characteristic line radiation, and since all that is required to initiate the Auger process is ionizing radiation at any energy, Auger peaks arising from ionization of core levels at energies higher than the X-ray line can be found if the recording range is extended upwards. For Si, the KLL principal peak in the region of 1610 eV can therefore be used for the calculation of the AP.

Some workers have gone further than this, and developed sources from metals with characteristic lines at higher energies, in order to excite both photoelectron and Auger emission from core levels not accessible by Al or Mg K_α radiation. The pioneers in this field have been Castle and co-workers,[8,9] who used a Zr source with principal line, L_α, at 2042 eV, and Yates and West,[10] who developed a monochromatized Ag source whose L_α line energy is 2984.3 eV. Both these sources have rather greater natural widths (1.6 eV and 1.3 eV, respectively) than the conventional Al and Mg sources, but their particular applications do not in general require high energy resolution.

Control of data acquisition in XAES is exactly the same as in XPS, and data processing, when used, is also very similar. Obviously, when dealing with Auger features, satellite subtraction is irrelevant since the characteristics of the X-ray source are not reflected in the Auger process. With a few exceptions, such as Na, Mg, Zn, Ga, Ge, and As, in which elements either the principal photoelectron peak is in a difficult energy region for XPS, or the Auger peaks are in any case the most prominent features, Auger peaks are not used for quantification in XPS or XAES and so such operations as peak area measurement and quantification would not be used. However, in studying Auger peak shapes and associated fine structure the same instrumental de-broadening, background subtraction, and peak resolution operations used so extensively in XPS are used also in XAES. An example of the processing of the X-ray excited Auger spectrum of cadmium by Aksela and Aksela[11] is shown in Fig. 10.6. The figure also demonstrates the wealth of fine structure in some Auger spectra due to final state effects; this will be discussed in more detail later.

10.1.3 *Improvement of spatial resolution*

XPS and XAES still operate in most laboratories as relatively large analysed area techniques. While in conventional AES, using, say, an electron source of the type illustrated in Fig. 3.2, the analysed area might be $2\text{–}5 \times 10^{-9}$ cm^2, in typical XPS operating conditions the analysed area might be 5×10^{-2} cm^2, i.e. about seven orders of magnitude greater. Since XPS has the advantages over AES that it both provides chemical information much more directly and is non-destructive, the technological pressures to attempt to obtain XPS analyses from significantly smaller areas are great. Some success has already been achieved in this direction, as will be described below for the various methods that have been tried.

Selected area analysis Two methods have been used to select a small area for analysis from a surface flooded with X-rays in the normal manner. The simpler of these is collimation of the beam of photo- and Auger electrons emerging from the surface by interposing a mechanical restriction between the sample and the entrance to the analyser, as has been carried out by Keast and Downing.[12] Their arrangement is shown in Fig. 10.7. A tube of internal diameter 0.7 mm and length 8 mm is positioned over the sample so that its axis is accurately aligned with that of the input lens to the analyser, and so that its lower end is as close as possible ($\sim$2 mm) from the surface compatible with the angle of the X-radiation. With a test specimen consisting of a gold strip 1.4 mm wide on a molybdenum substrate, and by traversing the specimen while recording the Au $4f_{7/2}$ and Mo $3d_{5/2}$ signals at $\sim$200 μm intervals, the results of Fig. 10.8 were obtained. The distance over which the Au signal rose to maximum and the Mo signal fell to zero was 0.74 mm, indicating an area resolution almost exactly that of the tube diameter. The disadvantage of this method is that the sensitivity, for a given X-ray flux, goes down directly proportional to the area selected.

Another method of selecting a small area for analysis from a uniformly irradiated surface is to use a lens with magnifying optics between the sample and the entrance to the analyser. For a magnification *a* and an entrance

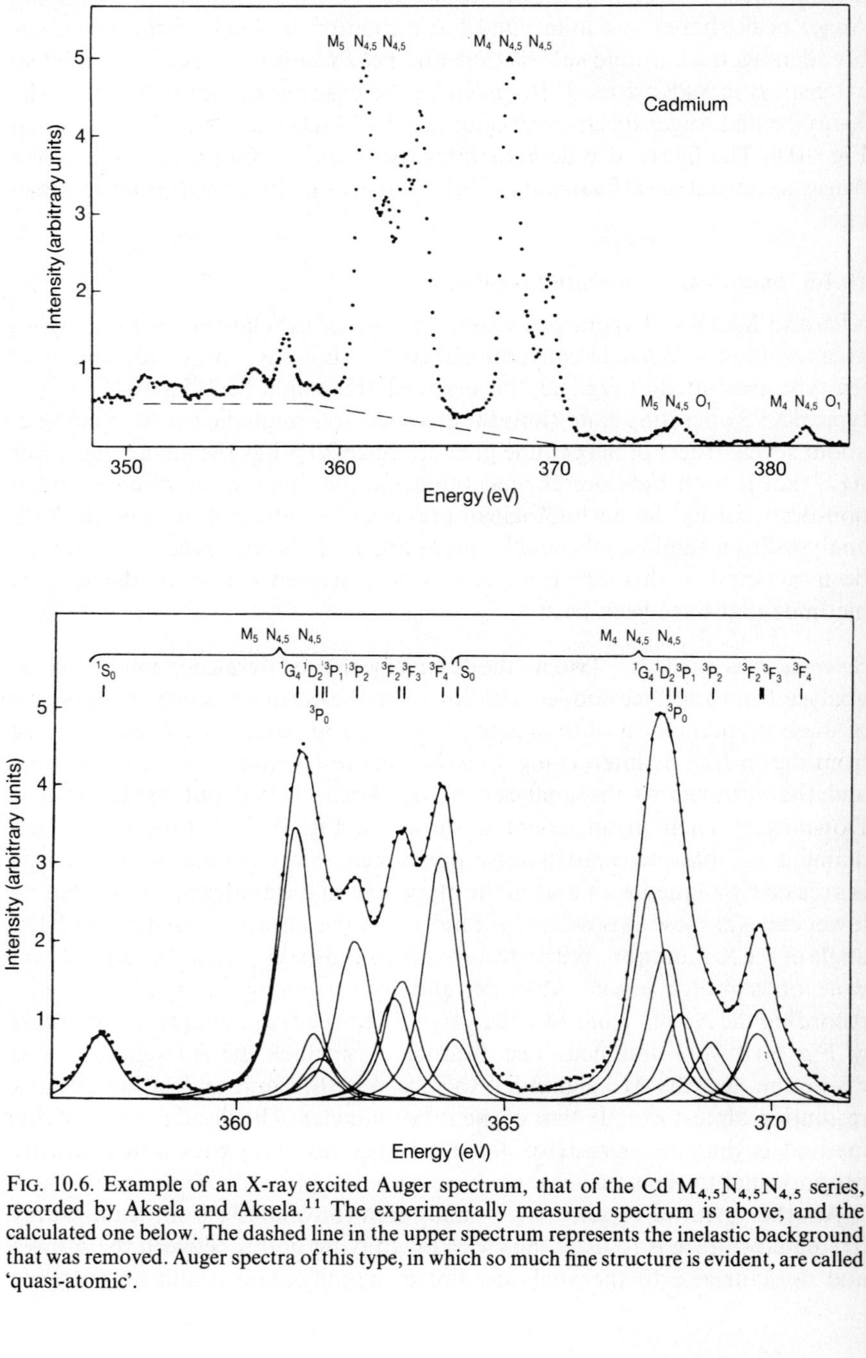

FIG. 10.6. Example of an X-ray excited Auger spectrum, that of the Cd $M_{4,5}N_{4,5}N_{4,5}$ series, recorded by Aksela and Aksela.[11] The experimentally measured spectrum is above, and the calculated one below. The dashed line in the upper spectrum represents the inelastic background that was removed. Auger spectra of this type, in which so much fine structure is evident, are called 'quasi-atomic'.

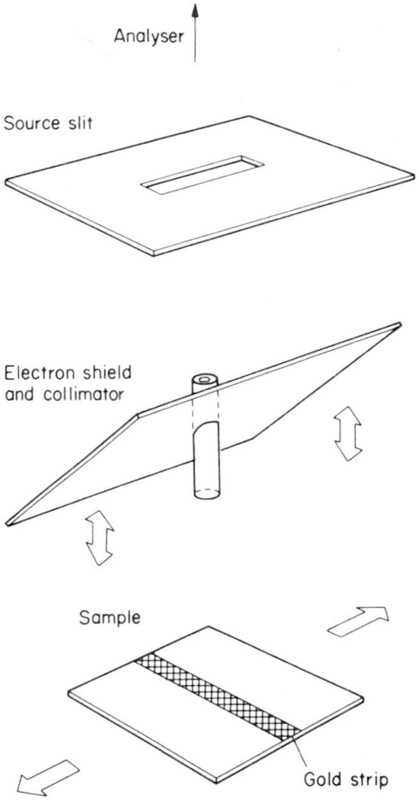

FIG. 10.7. Schematic of the arrangement devised for small area XPS by Keast and Downing.[12] Interposed between the sample and the entrance slit to the analyser is a small tube of internal diameter 0.7 mm and length 8 mm, positioned so that its axis is aligned accurately with that of the input lens, and with its lower end as close as practicable to the sample surface. The tube is fixed in a metal plate angled at 45° to the tube axis. The plate acts as an electron shield, and also as a mirror for alignment of the sample, since the underside of the plate is highly polished. Small area analysis is achieved simply by restriction of the area on the sample by collimation in the tube.

aperture of diameter b, photoelectrons from an area of diameter a/b will be analysed, assuming no optical aberrations within the lens that would distort electron paths and thus increase the effective area. With this system, and with a lens that also has a high acceptance solid angle it has been possible[13] to analyse areas down to diameters of $\sim 150\ \mu m$.

Scanning XPS Although present technology does not allow an incident X-ray beam to be rastered over the surface of a specimen, an ingenious method of obtaining scanning XPS images under certain special conditions has been devised by Cazaux.[14] The sample to be analysed, which must be in the form of a very thin film, is placed or deposited on the surface of an aluminium foil a few μm thick, and faces the entrance to an energy analyser. The back of the foil, i.e.

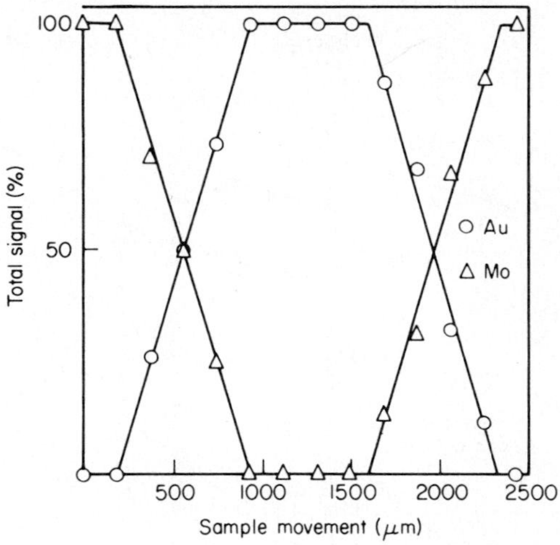

FIG. 10.8. Variation in the Au 4f and Mo 3d signals from a test specimen consisting of a gold strip 1.4 mm wide on a molybdenum substrate, when traversing the specimen underneath the collimator tube seen in Fig. 10.7. The points are about 200 μm apart. The rise and fall distance for the Au and Mo signals, respectively, was 0.74 mm, indicating an area resolution almost exactly that of the tube diameter. From Keast and Downing.[12]

on the other side from the sample, is irradiated with electrons of energy 10–20 keV. Characteristic Al K_α X-rays are generated in the foil and there is sufficient intensity of them at the sample side, if the foil is thin enough (<5 μm), to produce photoelectrons in the sample, which are then analysed. By rastering the focused electron beam across the back of the foil, and tuning the analyser to a photoelectron peak characteristic of a particular element, a two-dimensional XPS map of the distribution of that element may be obtained.

Although suggested by Cazaux, the first successful application of scanning XPS was by Hovland,[15] whose arrangement using a double-pass CMA is shown in Fig. 10.9. With it she was able to obtain images from areas of less than 20 μm in diameter. Cazaux has since pursued the idea in a series of papers, but it seems that the method will always suffer from the drawback that the sample must be in the form of a very thin film. However, it may have possibilities in biological research since the damage by the X-ray photons is minimal compared to what would be caused by an incident electron beam to achieve the same resolution.

Photoelectron spectro-microscopy (*PESM*) The principle of this method depends on the ability of an axially highly symmetric but divergent magnetic field to generate a magnified image of an electron-emitting object while at the same time maintaining the original energy distribution. The originator of the method was Turner, and Fig. 10.10 illustrates the instrumental arrangement

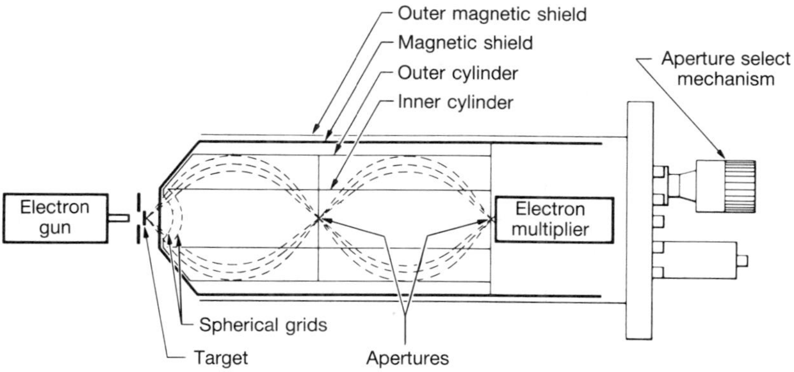

FIG. 10.9. Arrangement used to obtain scanning XPS images by Hovland.[15] The sample is in the form of a very thin film deposited on Al foil of thickness a few μm. An electron gun with a rastered beam irradiates the *reverse* side of the foil, thereby generating characteristic Al K_α photons in the foil, which in turn generate photoelectrons in the film. The energies of the photoelectrons are analysed in the double-pass CMA, opposite the entrance to which the sample is positioned. By selecting a photoelectron energy characteristic of the sample, and by rastering the electron beam, a scanning image can be obtained. The method suffers from the drawback that the sample must be very thin, i.e. ≤ 20 nm, otherwise all photoelectrons will be inelastically scattered before they reach the surface.

first used[16] by him and co-workers. The sample sits in the centre of a magnetic field produced by a superconducting solenoid, the field being variable from 0 to about 8 T, and is surrounded by an electrode structure that establishes a region of uniform electrostatic field. Apertures in the electrode structure allow the introduction of UV or X-ray photons to the sample surface, and the exit of photoelectrons. Away from the solenoid the magnetic field is reduced progressively along a cylindrical magnetic shielding cage down to about 0.001 T at its furthest end. Such a reduction provides a magnification of about 100.

The magnification arises through the laws governing the orbital motion of an electron moving in a divergent magnetic field. The electrons follow helical paths about magnetic flux lines passing through the point of emission on the surface, and as the flux lines diverge, the helical motion is self-adjusting so that orbital momentum is conserved. If the field starts off as B_1 and ends as B_2, then the lateral resolution is $(B_1/B_2)^{1/2}$, and the magnification B_1/B_2. Another important consequence of the laws of motion of an electron in a magnetic field is that, after a sufficiently great change in field, the photoelectron energy becomes concentrated in the axial motion along the flux line. This enables energy analysis to be carried out at some convenient point along the divergent magnetic field; in the original device a crude retarding field analyser was used, but later versions have used trochoidal energy analysers.

The PESM has a lateral resolution that is energy dependent, going

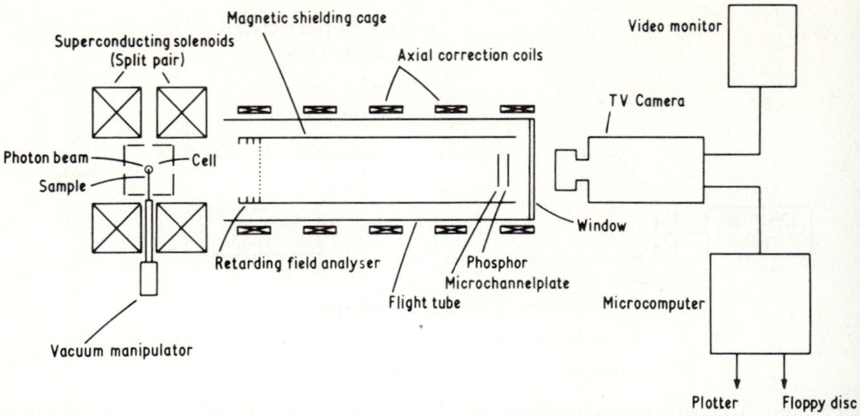

FIG. 10.10. Schematic of the principle of photoelectron spectromicroscopy, from Beamson *et al.*[16] The sample is placed in the centre of a high magnetic field of typically 8 T, produced by a superconducting solenoid, and is surrounded by an electrode structure establishing a uniform electrostatic field. Apertures allow the introduction of photon beams, and the exit of photoelectrons. Beyond the sample volume the magnetic field reduces progressively to 0.001 T along a flight tube of about 1 m in length. Emitted electrons follow helical paths around magnetic flux lines passing through the point of emission, and are constrained to the flux lines through conservation of orbital momentum. Thus a detector at the end of the flight tube will record an image of those photoelectrons entering the tube, with a magnification of 100–1000 times. The effective lateral resolution depends on both the field strength and on the electron kinetic energy.

approximately as $E^{1/2}$, and the theoretical resolution at 500 eV and a field of 8 T is about 1 μm. The resolutions obtained so far in practice have been rather worse than this, of the order of 10 μm at a few hundred eV, but even so the potentiality for greatly improved spatial resolution in XPS and UPS is obvious. The imaging capability is particularly dramatic. Figure 10.11 shows an image of a microscope grid sitting on a mesh, using all the emitted photoelectrons, when flooded with HeI radiation at 21.21 eV. The mesh size of the grid was 18 μm, and is easily resolved; the illustration also shows the depth of focus obtainable by the method.

Electrostatic imaging using an energy analyser The imaging properties of a concentric hemispherical analyser (CHA), whose characteristics were described under Section 3.3.4, can be used to obtain photoelectron images, as shown by Gurker *et al.*[17] Electrons of different energies entering a CHA are dispersed in energy along the dispersive plane of the CHA. Normally a small aperture or slit is placed at the analyser exit in order to select electrons of a particular energy for amplification. If the aperture/electron multiplier combination is replaced with a two-dimensional positional detector such as a microchannel plate, then spatial information can be obtained in the non-dispersive plane and energy information in the dispersive plane. The principle is shown in Fig. 10.12. A change in source position in the y_S direction results in

FIG. 10.11. A microscope grid with the letter M imaged by PESM using all the photoelectrons ejected by HeI radiation at 21.21 eV. The mesh size is 18 μm, and a good depth of focus can be achieved.

a corresponding opposite change $-y_D$ in the image position, while the photoelectron energies are dispersed along x_D. Thus if the sample is scanned mechanically in the x_S direction, two-dimensional energy, in other words compositional, maps may be constructed. It is not clear whether the method has actually been used yet in practice, but computer image simulation based on a kinetic energy of 500 eV and a peak width of 1.5 eV indicates a potential resolution of $\sim$100 μm.

Crystal focusing of the X-ray beam The principles of operation of an X-ray monochromator have already been described under Section 3.2.2. In conventional commercial monochromators, e.g. the popular size involving a Rowland circle of diameter 0.5 m, the area irradiated on the specimen surface is a line of length about 1 cm and width about 100 μm at optimum focus, the shape being set by the crystal geometry and the rather crude focusing

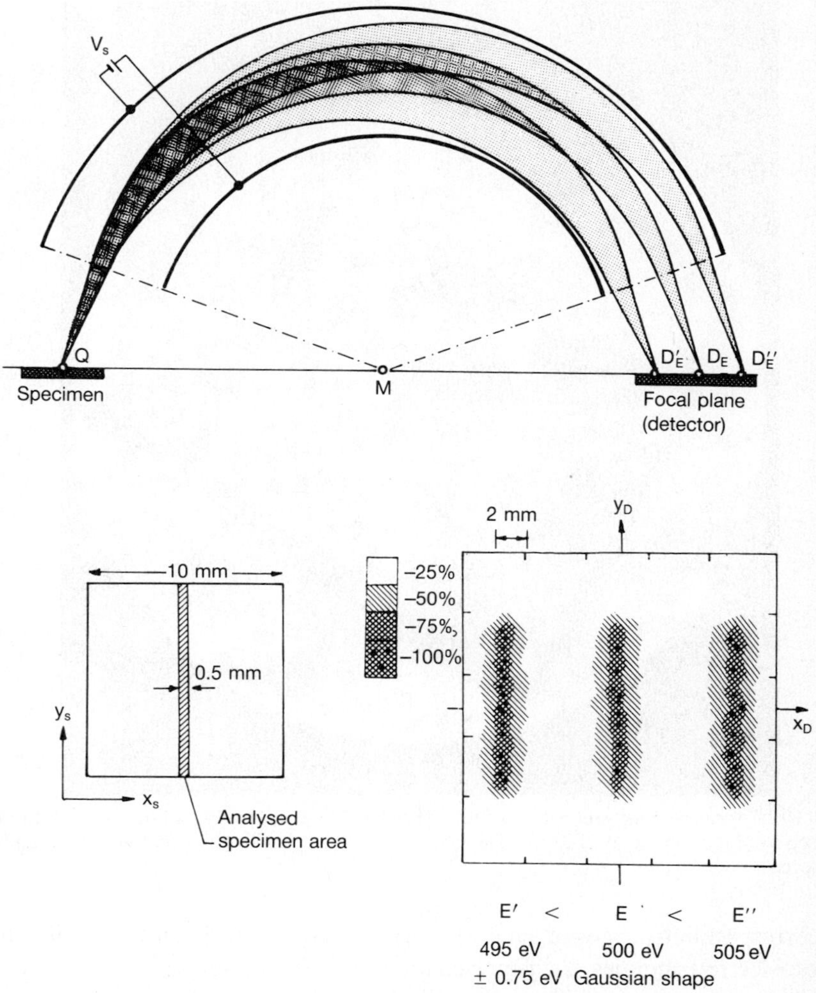

FIG. 10.12. Principle of electrostatic imaging using a concentric hemispherical analyser (CHA), as described by Gurker *et al.*[17] The normal exit aperture/multiplier combination used in a CHA is replaced with a two-dimensional position sensitive detector such as a microchannel plate. Electrons of different energies are dispersed in energy along the dispersal plane, so that spatial information can be obtained in the non-dispersive plane, and energy information in the dispersive plane. Referring to the diagrams in the lower part of the figure, a movement of the source position in the y_S direction results in an image movement in the $-y_D$ direction, while the photoelectron energies are dispersed along x_D. Scanning the sample thus produces a compositional map.

arrangements. If much more attention is paid to producing a proper point focus on the anode surface, and if a quartz crystal is bent and ground according to the Johansson condition, so that the Bragg diffraction condition is met at all points on the crystal, then the monochromator will produce a point focus on the sample. This is the method that has also been used commercially[18] to provide variable spatial resolution in XPS; the spot size on the sample is varied by varying the focus on the anode, between 150 μm and 1 mm. The arrangement is shown schematically in Fig. 10.13, by courtesy of

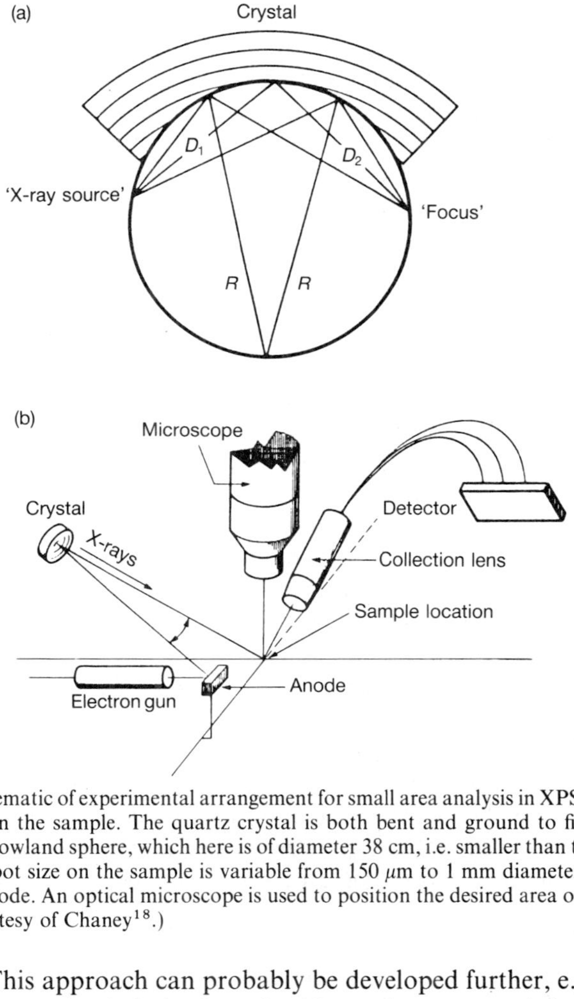

FIG. 10.13. Schematic of experimental arrangement for small area analysis in XPS by focusing the X-ray source on the sample. The quartz crystal is both bent and ground to fit accurately the surface of the Rowland sphere, which here is of diameter 38 cm, i.e. smaller than that described in Fig. 3.9. The spot size on the sample is variable from 150 μm to 1 mm diameter by varying the focus on the anode. An optical microscope is used to position the desired area on the sample for analysis. (Courtesy of Chaney[18].)

Chaney.[18] This approach can probably be developed further, e.g. by using a much larger Rowland circle than the 38 cm diameter used, but because of inherent limitations in the optics and in the X-ray source is unlikely ever to rival the very high spatial resolution achievable in AES. It seems unlikely, too, to be able to match the resolution in the imaging mode available theoretically in PESM, although the latter approach must always suffer from high capital cost.

10.2 Spectral interpretation

10.2.1 *XPS*

A typical 'wide' or 'survey' XPS spectrum has already been shown in Fig. 10.3 for an almost clean copper surface. Such a spectrum will of course show only the major spectral features, but even so there are many of them and they should be described before going on to the more detailed features to be seen in the 'narrow' or 'detail' scans.

The most prominent features in most photoelectron spectra are the peaks arising from direct excitation of core-level electrons according to eqn (2.2). Thus in Fig. 10.3 one can see in ascending order of kinetic energy the $2p_{1/2}$, $2p_{3/2}$, 3s, 3p and 3d photoelectron peaks of copper, with the first two of these being by far the most intense for excitation by Al K_α radiation. The 3d electrons come, of course, from the copper valence band. In general the photoelectron peaks will be the narrowest features in the XPS spectrum, although some Auger peaks are also quite narrow. Auger peaks appear because the Auger de-excitation process can occur following ionization by any means, not just by electron excitation as in AES. The techniques of Auger electron emission following excitation by X-rays or by ions are known respectively as XAES and IAES and will be discussed separately, but it is sufficient to note at this point that Auger peaks will be major features in most XPS spectra, and in a few cases (e.g. Ge, Se) exceed in intensity the photoelectron peaks. In Fig. 10.3 the peaks in the copper LMM Auger series are indicated. (Correlation of the X-ray and spectroscopic notations used in AES and XPS, respectively, is given in Table 4.1.) The combination of the 2p doublet, and the LMM triplet of peaks, is a characteristic fingerprint of the first transition metals from titanium to copper.

Some other minor features (apart from the traces of contaminant O and C) are visible in Fig. 10.3, associated with the major peaks. Those situated to the high kinetic energy side of a peak occur because unmonochromatized radiation has been used, and are the photoelectron peaks excited by the X-ray $K_{\alpha_{3,4}}$ satellite already described under Section 3.2.2. If a monochromator is used, these satellite lines disappear. On the low kinetic energy side of a peak will be found plasmon loss peaks, whose origin has been given in detail under Sections 4.2.6 and 5.2.1. Both bulk and surface plasmon losses may be seen, and multiples and combinations of both; an example was shown in Fig. 4.9 for plasmons associated with the aluminium KLL Auger peak, and another example connected with a photoelectron peak, again that of aluminium, is given in Fig. 10.14 from the work of Barrie.[19]

The other types of spectral feature to be found in XPS spectra will not necessarily appear in all spectra, and in any case will normally be visible only when detailed scans of individual spectral regions are recorded. By far the most important of these, and forming the central strength of XPS, are the so-

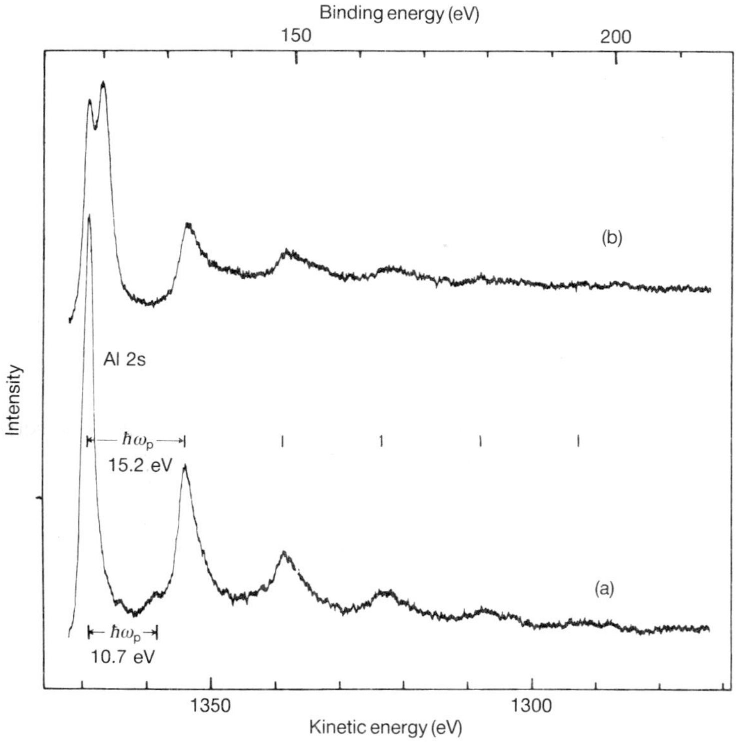

FIG. 10.14. Example from Barrie[19] of successive bulk plasmon losses observable in association with a sharp intense peak. Spectrum (a) clean aluminium, spectrum (b) lightly oxidized aluminium. Note that the surface plasmon, just visible at a loss of 10.7 eV, is quenched on oxidation, and note also the chemical shift on the aluminium 2s photoelectron peak.

called 'chemical shifts'. The theoretical ideas behind their origins will be discussed in more detail below, but briefly, when an atom enters into combination with another atom or group of atoms there will be a change in the valence electron charge cloud density, positive or negative according to whether charge is accepted or donated, which will alter the electrostatic potential 'seen' by a core electron. The binding energy of the core electron will therefore change, which will appear in the XPS spectrum as a shift in a peak position. The simplistic relationship between chemical state and binding energy change has led to the term 'chemical shift', universally used, although shifts can occur for other reasons, and sometimes the observed shift seems anomalous based on purely charge exchange grounds.

A striking example of a chemical shift is shown in Fig. 10.15. In the upper spectra (b) (at two different gain factors) are the 4f doublet and associated satellites from the surface of clean tungsten metal. When the surface is slightly

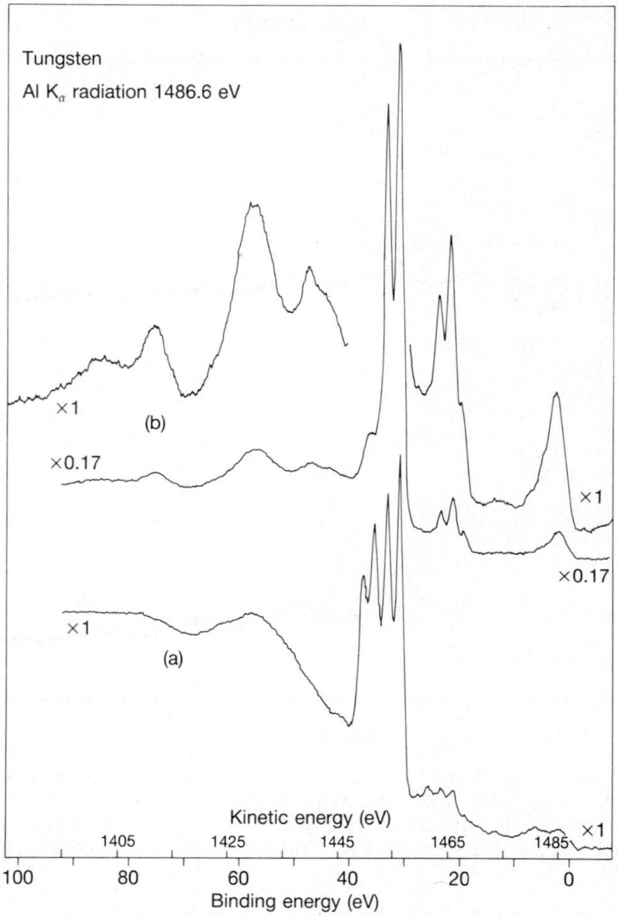

FIG. 10.15. Example of a chemical shift in an XPS spectrum. The tungsten 4f photoelectron spectra from (a) lightly oxidized tungsten, and (b) clean tungsten. In (b) can be seen the sharp 4f doublet and the associated plasmon and $K_{\alpha 3,4}$ satellite peaks. In (a) the oxide film is sufficiently thin that photoelectrons from the metal substrate can still escape, and there are now four 4f peaks rather than two, since the chemical shift is large enough ($\sim$4.5 eV) to move the oxide doublet clear of that from the metal. Note also the increased inelastic background beyond the oxide 4f spectrum, indicating that the metallic contribution is now 'buried' below the oxide.

oxidized so that the oxide film is still thin enough to allow photoelectrons from the unoxidized substrate to escape, one finds the spectrum (a) in the lower half, in which instead of two 4f peaks there are now four. The chemical shift in the 4f peaks in going from the metallic to the oxidized state is great enough ($\sim$4.5 eV) for the doublet to move clear of that from the metal, thus giving rise to four distinct peaks. Chemical shifts are rarely as immediately apparent and measurable as in Fig. 10.15. Much more often the shifts are small, and there may be several chemical states of the one element in a particular material or as a result of a surface reaction so that the spectrum may appear considerably misshapen compared to the shape of a pure single peak. It is in that situation that the operation of peak resolution or curve synthesis described earlier is so necessary. Examples will arise when the applications of XPS are discussed.

Notice also in Fig. 10.15, in which the spectra were recorded with unmonochromatized X-radiation, that the $K_{\alpha 3,4}$-excited peaks just to the high kinetic energy side of the 4f peaks mirror faithfully the 4f structure, but on a smaller scale; that, to the low kinetic energy side of the 4f doublet from the clean surface can be found both bulk and surface plasmon losses (the latter is obliterated rapidly on oxidation); and that, near 1485 eV kinetic energy (i.e. near zero binding energy) is the tungsten valence band emission. On oxidation the shape of the latter changes drastically.

Other, usually but not always, minor, features often seen in an XPS spectrum are the 'shake-up' satellites. They tend to be of significant intensity when there are unpaired electrons in 3d or 4f levels, that is, they are associated almost always with formation of certain compounds. They arise from the perturbation to the atom caused by the photoemission process itself. Loss of a core electron causes a change in the nuclear screening which leads to a temporary rearrangement of atomic states, known as relaxation. The energy associated with the relaxation may be sufficient to excite a valence electron to an unfilled level at higher energy, the so-called 'shake-up' process. Since photoelectron ejection and relaxation occur simultaneously, the energy can come from that of the outgoing photoelectron, which thus suffers a discrete loss in energy. The probability of shake-up varies greatly between different materials and between different oxidation states of the same material, as illustrated (Scrocco[20]) in Fig. 10.16, which gives a classic example. The spectrum from CuO shows strong shake-up satellites, since Cu^{2+} has an open shell configuration, whereas that from Cu_2O does not, since Cu^+ has a closed shell. Thus the presence or absence of the 'shake-up' peaks in a Cu spectrum is diagnostic of the oxidation state.

Passing mention has already been made about valence band features appearing in the XPS spectrum, in discussion of the Cu spectrum in Fig. 10.3. Electrons in degenerate levels that go to make up the valence band of a solid are of course those of lowest binding energy, and hence of highest kinetic energy. In a 'wide scan' spectrum such as that of Fig. 10.3 it is impossible to see any detailed structure in the valence band, but in fact for many metals there is much interesting detail to be seen if the valence band spectral region is recorded with sufficient energy resolution and good statistics. Thus, if monochromatic X-radiation is used, the Au valence band has a wealth of fine structure, as seen in Fig. 10.17, from Wertheim *et al.*[21]

The other class of spectral feature to be found in detailed studies of the XPS spectrum is multiplet splitting. This occurs when an element has unpaired

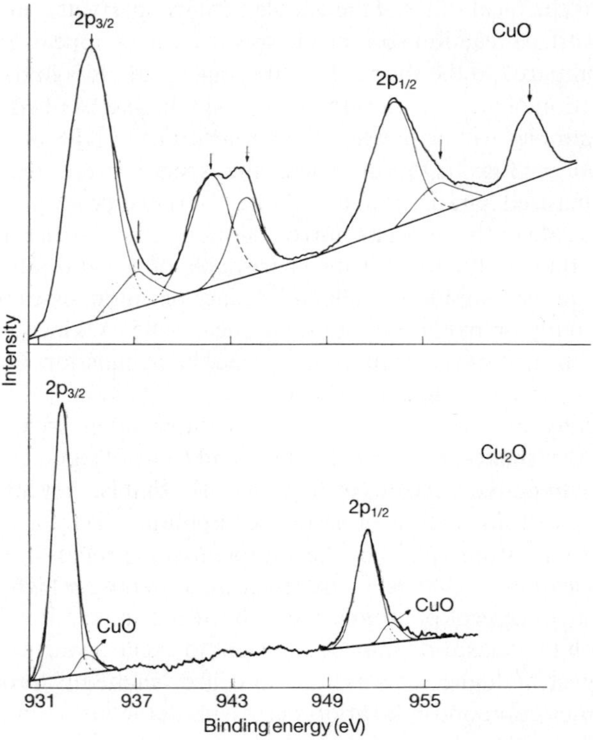

FIG. 10.16. A striking example of the appearance of 'shake-up' peaks associated with photoelectron peaks under certain electronic conditions within the atom. Such features tend to appear when there are unpaired electrons in 3d or 4f levels. Cu^{2+} as in CuO has an open shell configuration whereas Cu^{+} has a closed shell, thus the 2p spectrum from CuO shows intense shake-up structure while that from Cu_2O does not. The shake-up process is described in the text. Note in the figure that line shape analysis of Cu_2O reveals in the 2p spectra minor contributions from CuO, presumably as surface oxidation. (From Scrocco.[20])

electrons in either the valence bands or in core levels very close to the valence band. It can manifest itself either as a result of compound formation or it can be intrinsic to the element itself. The most striking examples occur amongst the rare earths. The 4f electrons of the rare earths are often situated energetically very close to, if not actually coincident with, the valence band, but in fact they are very strongly localized and the hybridization with states in the valence band is weak. Thus correlation between the spin and angular momenta of the 4f electrons can occur in a variety of different ways, giving rise to multiplets at different energies. If a core hole is also present, then other correlations are also possible. When the multiplets lie close together, it is not always possible to separate them with the best resolution available in XPS, but even so there is a great deal of fine structure in some of the spectra. The 4f spectral regions from the heavy rare earths, recorded by Campagna *et al.*,[22]

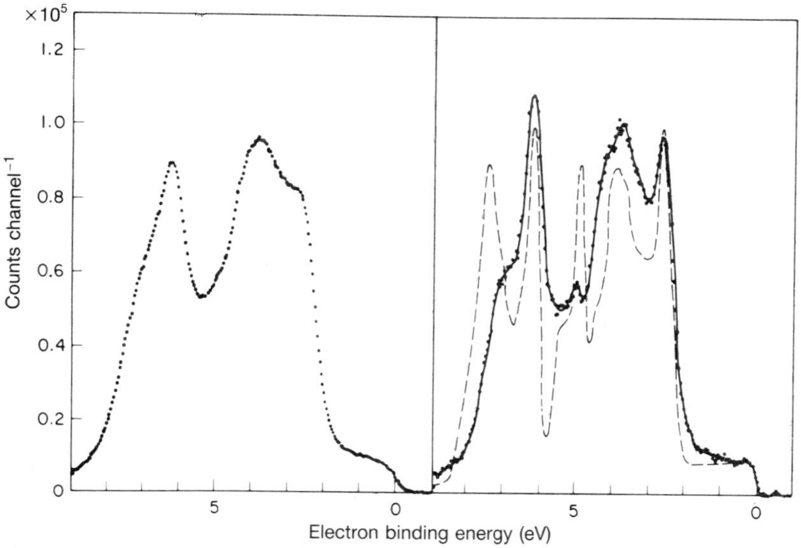

FIG. 10.17. Valence band spectrum of gold recorded using Al K_α radiation, by Wertheim *et al.*[21] In the left-hand panel are the raw data after background subtraction. In the right-hand panel the solid line running through points is the result of deconvoluting the instrumental broadening from the spectrum on the left. The dashed line is the result of a theoretical calculation of the valence band structure. Note the sharpness of the Fermi edge at zero binding energy, after deconvolution.

are shown in Fig. 10.18. The positions of the final state multiplets and their relative intensities, are indicated underneath the experimental envelopes.

10.2.2 *XAES*

An example of the fine structure that can be found in the Auger spectrum when recorded at high energy resolution in the undifferentiated, or $N(E)$, mode, has already been given in Section 10.1.2 for cadmium from the work of Aksela and Aksela.[11] A detailed study of such structure for the various elements that show it has been made over some years by Weightman and co-workers.[23] The structure arises in those elements in which the Coulomb interaction energy between the two holes left behind in the final state of the Auger process is much greater than the width of the valence band, thus preventing the holes from decaying as rapidly as they would if this interaction energy were lower. The result is that the holes become localized near the ionized atom for long enough for electronic correlations between them to occur, leading to a splitting of the Auger spectrum into multiplets due to these final-state effects. Since the line widths are much narrower than expected, certainly narrower than the valence band width, and the situation is formally similar to that in a free atom, the elements showing this behaviour are said to be of 'quasi-atomic' character. It is the narrowness and the wealth of fine structure in these cases that necessitate the use of high energy resolution as used in XAES.

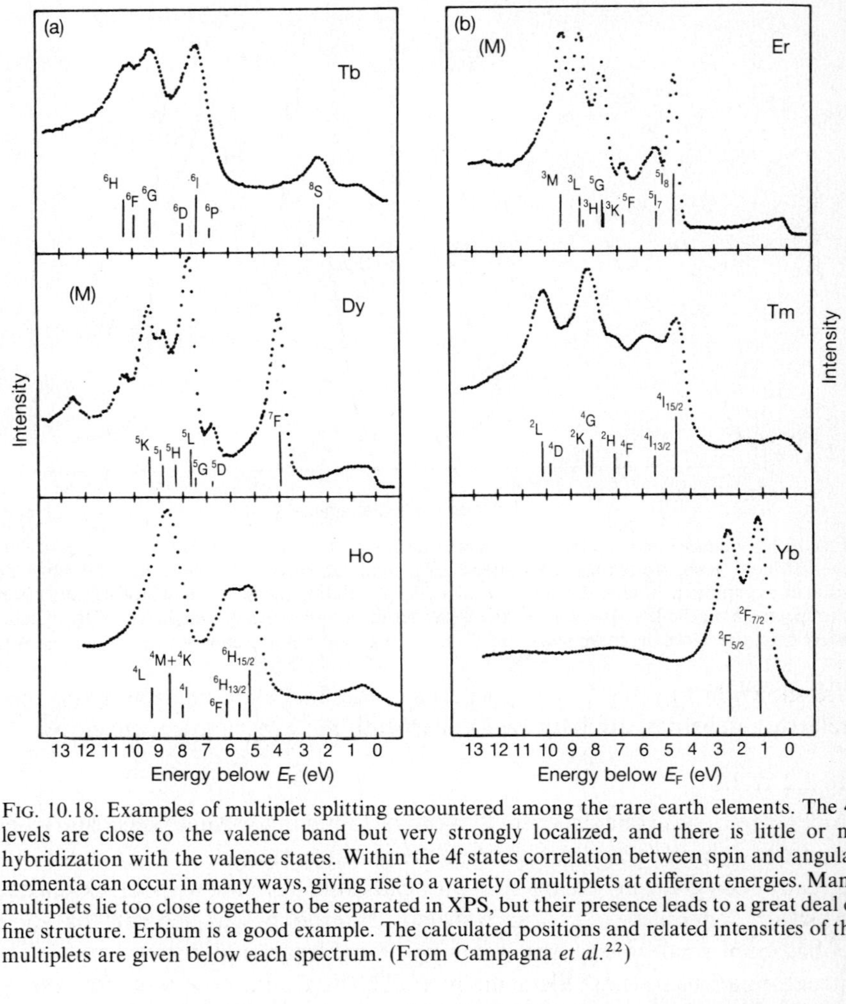

FIG. 10.18. Examples of multiplet splitting encountered among the rare earth elements. The 4f levels are close to the valence band but very strongly localized, and there is little or no hybridization with the valence states. Within the 4f states correlation between spin and angular momenta can occur in many ways, giving rise to a variety of multiplets at different energies. Many multiplets lie too close together to be separated in XPS, but their presence leads to a great deal of fine structure. Erbium is a good example. The calculated positions and related intensities of the multiplets are given below each spectrum. (From Campagna *et al.*[22])

As mentioned earlier in Section 10.1.2, XAES does not depend only on characteristic X-ray line radiation for the excitation of Auger transitions. Both bremsstrahlung radiation from the conventional Al and Mg sources, and radiation from special sources with characteristic emission lines at higher energies than those of Al and Mg, have been used. Thus Wittberg and Wang[24] were able to find Auger peaks at kinetic energies in the 4000 eV region from TiO_2 and $TiH_{1.64}$, excited by the bremsstrahlung radiation from an Mg source, as seen in Fig. 10.19. These Ti KLL Auger spectra are, as far as the author is aware, at the highest kinetic energies yet reported for excitation by bremsstrahlung radiation. An example of excitation using characteristic X-ray lines at higher energies than those of the conventional Al and MgK_α sources is

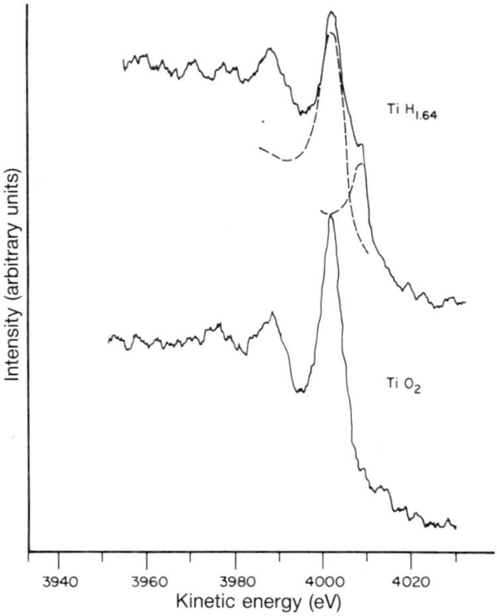

FIG. 10.19. X-ray excited Auger peaks using bremsstrahlung radiation from an Mg K_α X-ray source, measured by Wittberg and Wang.[24] The Ti KLL Auger peaks from, upper, $TiH_{1.64}$ powder, and lower, a TiO_2 film. The most intense peak in each spectrum is at about 4002 eV, corresponding to the KL_2L_3 ($^1D_2$) calculated transition energy. The shoulder on the high kinetic energy side of the $TiH_{1.64}$ spectrum is due to surface TiO_2.

seen in Fig. 10.20 from West and Castle,[8] where the Si KLL Auger line excited by Zr L_α radiation is plotted as a function of depth into a thin film of SiO_2 on Si. Also shown are the changes in the Si 1s photoelectron peak during the same profile; this core level would not have been accessible using Al or Mg $K_\alpha$radiation. Clearly the energetic difference between Si^0 and Si^{4+} in the KLL spectrum is much greater than in the photoelectron spectrum, revealing one of the advantages of XAES for certain elements.

10.3 Theory

10.3.1 *XPS line shapes*

In photoemission following creation of a core hole the shape, width, asymmetry, etc., of the threshold feature, which in the case of XPS is a photoelectron peak, will be decided by the nature of the magnitude of the perturbation to the final state by the hole. This perturbation was visualized by Mahan[25] as a singularity in the scattering of the excited electron at the Fermi energy from the transient and screened Coulomb potential of the core hole, leading to a singularity in the soft X-ray absorption cross-section threshold.

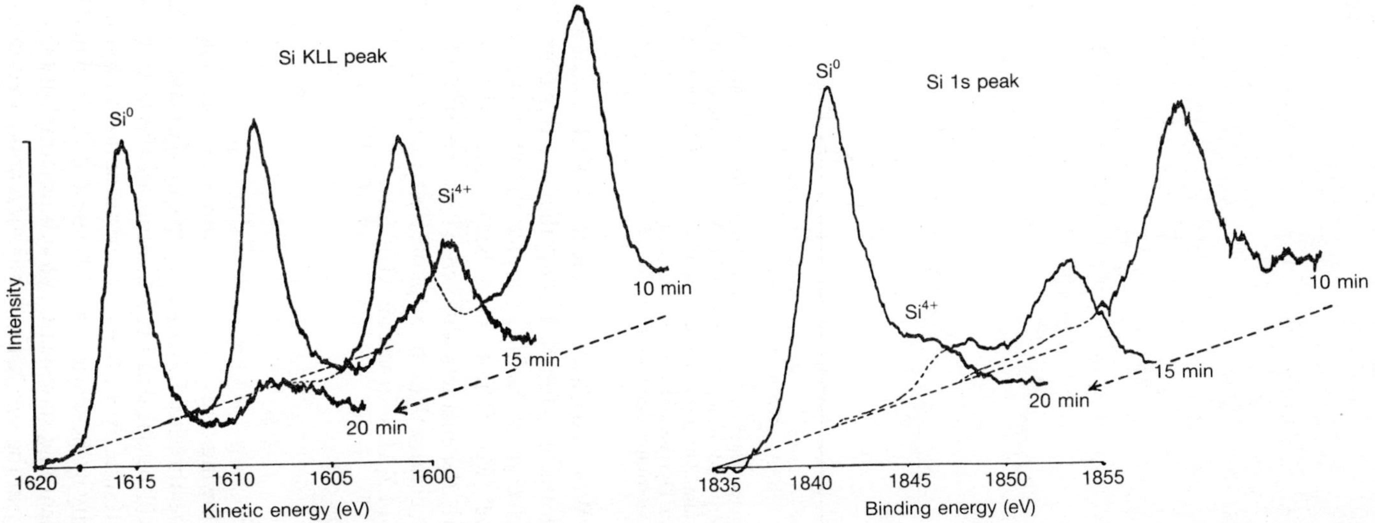

FIG. 10.20. XPS studies using an X-ray source of higher characteristic energy than the conventional Al K_α and Mg K_α sources, made by West and Castle.[8] The Si KLL Auger and Si 1s photoelectron peaks are excited by Zr L_α radiation at 2042.4 eV; although the KLL transition could have been excited by the bremsstrahlung from the conventional Al and Mg K_α sources, the Si 1s level would not have been accessible using the latter sources. Note the much greater shift between Si^0 and Si^{4+} in the KLL spectrum compared to that in the 1s spectrum.

The idea was extended by Nozières and de Dominicis[26] who pointed out that as well as the effect suggested by Mahan there should also be a transient and singular re-adjustment of the ground state of the entire Fermi sea due to the presence of the potential associated with the core hole. Depending on their respective signs and magnitudes, the two effects may reinforce or cancel each other. In fact the latter authors solved the theoretical problem exactly.

Using the Mahan–Nozières–de Dominicis theory, Doniach and Šunjić[27] showed that the final-state effect in core-level photoemission should manifest itself as an asymmetry, or 'tailing', to higher binding energies. The processes described in the MND theory are known as many-body processes, and the results of the deliberations of Doniach and Šunjić was the so-called many-body line shape in XPS, which had the form

$$I(\epsilon)=\frac{\Gamma(1-\alpha)}{(\epsilon^2+\gamma^2)^{(1-\alpha)/2}}\cos\{(\pi\alpha/2)+\theta(\epsilon)\} \tag{5}$$

where $\epsilon=$ energy

$\gamma=$ lifetime width of core hole

$\theta(\epsilon)=(1-\alpha)\tan^{-1}(\epsilon/\gamma)$

$\alpha=2\sum_l (2l+1)(\delta_l/\pi)^2.$

Γ is the gamma function and δ_l are the partial-wave phase shifts. α is an asymmetry parameter, and when $\alpha=0$ expression (10.5) reduces to a Lorentzian shape.

Expression (10.5) has been compared directly with photoemission data, although great care has to be taken to ensure that the instrumental broadening inherent in any spectrum does not make the comparison meaningless. Deconvolution to remove the instrumental broadening, as described earlier, does not necessarily remove all the instrumental width. The approach adopted by Wertheim and Hüfner[28] in their comparisons was the reverse in fact, and that was to convolve the theoretical line shape of a particular core-level peak with the measured instrumental broadening function, the result of which was then compared with the raw spectrum. Their comparison of theory with experiment for the $3d_{5/2}$ and the 4d photoemission spectra of tin is shown in Fig. 10.21. It can be seen that for these peaks the fit is almost exact. The only problems that can arise are those due to the presence of plasmon loss peaks close to the photoelectron peaks, since the MND and the Doniach and Šunjić theories take no account of plasmon losses.

10.3.2 *Nomenclature*

Some mention has already been made in Section 4.2.1 of the spectroscopic nomenclature. It is based on *j–j* coupling, as is the simple X-ray notation for

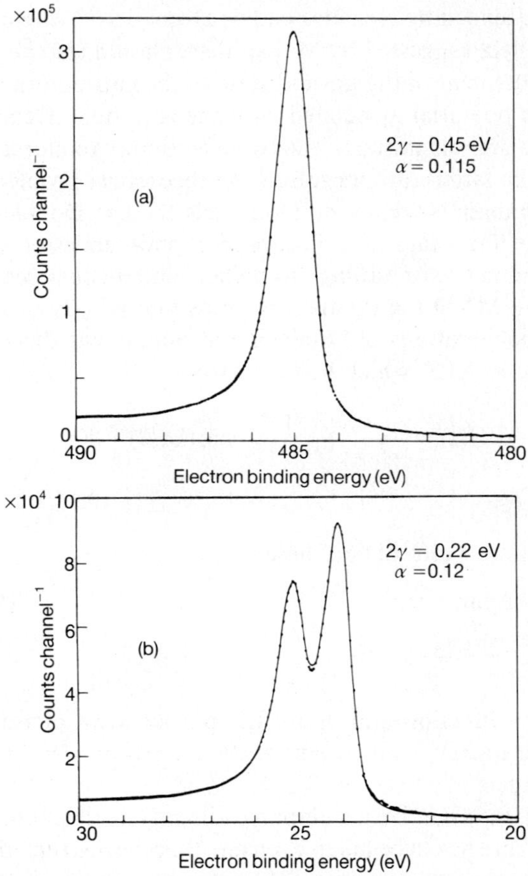

FIG. 10.21. Comparison of theoretical and experimental line shapes in XPS, for the $3d_{5/2}$ (a) and the $4d_{5/2,3/2}$ (b) photoelectron spectra of tin, by Wertheim and Hüfner.[28] The theoretical curves were derived by convolving the theoretical line shape with an instrumental broadening function measured for their spectrometer. The experimental data are represented by the points, the line shape derived from eqn (10.5), by the lines. γ and α are defined below eqn (10.5) in the text. The agreement is almost exact.

Auger transitions, but the labelling of atomic states is different. States with values of the total one-electron orbital angular momentum quantum number $l=0, 1, 2, 3, \ldots$, are designated s, p, d, f, . . . , respectively, and the latter is preceded by the value of the principal quantum number n. The j values (where $j=l+s$, s being *here* the total one-electron spin momentum quantum number) are then appended as subscripts. A state with $n=2$, $l=1$, and $j=\frac{3}{2}$ would thus be written as $2p_{3/2}$. Table 4.1 sets out the relationship between the quantum numbers and the notation for levels up to $3d_{5/2}$.

In published XPS spectra it is conventional to label the photoelectron features as above, and the Auger features in the X-ray notation already described in 4.2.1, so that confusion is avoided. Where one or more of the electrons involved in the Auger process originate in the valence band of the solid, the notation V is introduced, e.g. L_3M_3V and L_3VV.

10.3.3 *Binding energies and chemical shifts*

Since the observed kinetic energies of photoelectrons are related directly to the binding energies of the core levels from which they have been ejected, according to expression (2.2), it is not in general as necessary to consider the detailed calculation of an energy as it is in AES, since reference can be made to published tables of electron binding energies. However, it is advisable to be aware of some of the factors that enter the photoelectron energy expression, particularly when one comes to consider the so-called 'chemical shifts'.

Any system consisting of N electrons can be described by an initial or ground state, with energy $E_i(N)$ and, after interaction with a photon of energy $h\nu$, by a final state with energy $E_f(N-1)$ together with a resultant photoelectron of kinetic energy E_K. Conservation of energy requires that

$$E_i(N)+h\nu=E_f(N-1)+E_K. \tag{10.6}$$

If the photoelectron originated in a level X, then clearly the binding energy of X with respect to the vacuum level is

$$E_B(X)=E_f^X(N-1)-E_i(N). \tag{10.7}$$

The simplest assumption to make regarding the after-effects of the photon interaction, also called Koopman's Theorem, is that nothing is affected apart from level X, i.e. all energies are still in their initial state. The Theorem then states that

$$E_B(X)=-\epsilon_X \tag{10.8}$$

where ϵ_X is the orbital energy of the photoelectron. However, reference to the discussion already given in Section 4.2.3 on Auger energies will indicate immediately that (10.8) cannot be correct. With a core hole present, additional screening of the nucleus is required, which takes place by inward orbital contraction, giving rise to the same relaxation energy terms as before. Furthermore, (10.8) ignores relativistic and correlation processes, both of which will tend to increase the core electron binding energy. Expression (10.8) should thus be written more correctly as

$$E_B(X)=-\epsilon_X-R_X^{in}-R_X^{ex}+\epsilon_{rel}+\epsilon_{cor}. \tag{10.9}$$

Note that the relaxation energies operate to reduce the binding energy, whereas of course in the two-electron Auger process they operated to increase the Auger kinetic energy.

It is of course the existence of the 'chemical shift', a term borrowed from NMR, that provides so much of the power of the XPS technique. By 'chemical shift' is meant merely the difference in binding energy of a particular core level of an element between two different chemical states of the element; the term is also applied to core-level binding energy differences arising from less obviously chemical origins, such as different structures, different lattice sites, and surface vis-à-vis bulk, which are really physical differences. Genuinely chemical shifts can be of magnitude a few eV, but the more subtle physical shifts do not generally amount to more than a few tenths of an eV.

The change in binding energy between two different chemical situations can be written down generally from (10.9) as

$$\Delta E_{\mathrm{B}}(X) = -\Delta\epsilon_X - \Delta R_X^{\mathrm{in}} - \Delta R_X^{\mathrm{ex}} + \Delta\epsilon_{\mathrm{rel}} + \Delta\epsilon_{\mathrm{cor}}. \tag{10.10}$$

The last two terms in (10.10) are usually negligible, so that one is left with the changes in the initial state orbital energies and in the relaxation energies.

Various simplified theories of chemical shifts in XPS have been used in attempts to calculate the shifts and therefore to be able to predict what the shift should be in some untested situation. The first and simplest of these was introduced by Siegbahn and co-workers,[29] and is called the 'charge-potential model'. It ignores relaxation effects, simply because there is no room in the model for them. The atom is considered electrostatically as a hollow sphere of radius r over the surface of which is distributed the charge q due to the valence electrons, from which it follows that the potential inside the sphere is uniformly q/r. Since no charge fine structure due to different binding energies comes into the model, it also follows that all core-level binding energies are assumed to be shifted to the same extent. The binding energy of an electron in a core level i of an atom A surrounded by other atoms B is given in the model by

$$E_{\mathrm{A}}(i) = E^0 + k_i q_{\mathrm{A}} + \sum_{\mathrm{B}} q_{\mathrm{B}}/4\pi\epsilon r_{\mathrm{AB}} \tag{10.11}$$

where E^0 is a constant energy reference term (e.g. to the vacuum or to the Fermi levels), k_i is a proportionality constant, q_{B} is the uniform charge on the neighbouring atoms B, and r_{AB} is the atomic separation. The constant k_i is approximately equal to the average value of the repulsion between an orbital in level i and the orbitals of the valence shell. Thus the chemical shift is

$$\Delta E_{\mathrm{A}}(i) = l + k_i \Delta q_{\mathrm{A}} + \Delta \sum_{\mathrm{B}} q_{\mathrm{B}}/4\pi\epsilon r_{\mathrm{AB}} \tag{10.12}$$

l being the difference in references between the two situations, and normally determined from a standard reference material. For some gaseous compounds containing light elements the model has been successful, as shown in Fig. 10.22 from Gelius[30], in the case of the carbon 1s level. The third term in (10.11) is usually called the Madelung potential, being similar to the lattice potential in a crystal.

A variation on the charge-potential model is the valence-potential model of

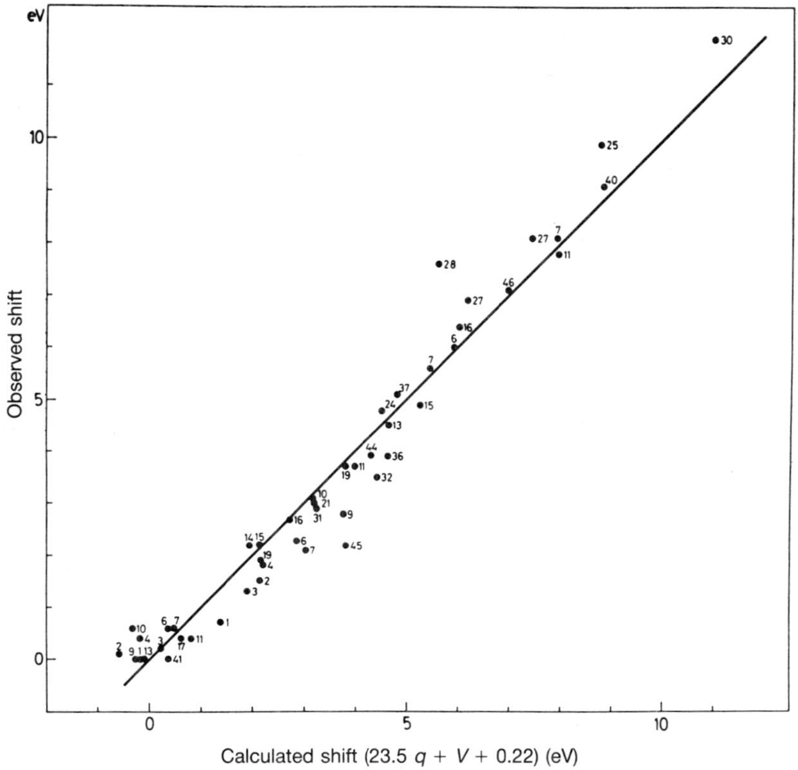

FIG. 10.22. Comparison between experimentally measured chemical shifts in the carbon 1s level, and shifts calculated by Gelius[30] using the charge-potential model described in the text.

Gelius;[30] it is more fundamental but still neglects the important relaxation effects. In it the potential experienced by a core electron in a particular atom is approximated by that due to the cores of neighbouring atoms *plus* that due to the valence electron density. Table 10.1 indicates the extent of agreement with experiment that this model has had for a variety of simple molecules, for three light elements.

A rather different approach to the calculation of chemical shifts is based on thermodynamics, and is termed the 'equivalent cores method'. It uses the assumption that when a core electron is removed from an atom, the valence electrons relax in a way equivalent to the increase of the atomic nuclear charge by one. An example from Clark and Adams[31] will serve to illustrate the method.

Suppose one wishes to calculate the chemical shifts in the N 1s binding energy between N in NH_3 and in N_2, that is $\Delta E_N(NH_3–N_2)$. The shift is then equal to the difference in energy for the process

Table 10.1

Theory. Comparison of experimental and calculated chemical shifts from the valence-potential model, eV.

	Expt	Calc
Carbon (measured from CH_4)		
CH_4	0	0.3
$^*CH_3COH$	0.6	0.8
CH_3OH	1.6	2.3
CS_2	2.4	4.1
CH_3COCH_3	3.1	3.3
CO_2	6.8	6.8
CF_4	11.1	10.2
Nitrogen (measured from N_2)		
NH_3	−4.3	−3.8
N_2	0	−1.6
*NNO	−1.4	−1.7
N*NO	2.6	2.8
Sulphur (measured from CS_2)		
CS_2	0	0.7
H_2S	0.4	0.8
SO_2	5.0	4.3
SF_6	10.6	11.1

$$NH_3 + \overset{*}{N}_2{}^+ \rightarrow \overset{*}{N}_3{}^+ + N_2 \tag{10.13}$$

where $*$ indicates a N 1s core hole. Now the electron deficient core of $\overset{*}{N}_2{}^+$ is equivalent in the model to the ion NO^+, and the reaction

$$\overset{*}{N}H_3{}^+ + NO^+ \rightarrow OH_3{}^+ + \overset{*}{N}_2{}^+ \tag{10.14}$$

has of course $\Delta E = 0$. Combining (10.13) and (10.14) gives

$$NH_3 + NO^+ \rightarrow OH_3{}^+ + N_2 \tag{10.15}$$

for which gaseous thermodynamic data are known. The energy of reaction (10.15) is thus equal to the chemical shift ΔE_N required. Table 10.2 gives some results for another range of simple gaseous molecules in which the equivalent-cores method has been applied to shifts in C, N, and O 1s core levels with some success.

10.3.4 *XPS intensities*

Calculation of photoelectron intensities is based on the calculation of the transition probabilities involved in photoelectron ejection, which are in effect the photoelectric cross-sections. The cross-section is defined as the transition

Table 10.2
Theory. Chemical shifts in XPS. Comparison of experimental and calculated chemical shifts from the equivalent cores method, eV.

	Expt	Calc
Carbon		
HCN	2.0	1.43
CO	4.6	3.81
CO_2	6.2	5.09
CS_2	1.8	1.73
Nitrogen		
*NNO	−1.4	−1.32
N*NO	2.6	2.31
HCN	−3.8	−2.54
Oxygen		
CO	1.3	1.74
NNO	−0.4	−0.57
C_3O_2	−1.1	−0.96

probability per unit time for the excitation from an initial state of wave function ψ_i to a final state of wave function ψ_f, at a unit incident photon flux of 1 cm^{-2} s^{-1}. The article by Fadley[32] sets out the detailed formalism, which will be outlined only briefly here.

Transition probabilities always have the mathematical form of squared matrix elements involving the initial and final state wave functions, and, for an assumed travelling plane wave form for the electromagnetic wave associated with the incoming photon, the transition probability between ψ_i and ψ_f is proportional to

$$|M_{\mathrm{if}}|^2 = h^2 A_0^2 |<\psi_f| \sum_{i=1}^{N} \exp(\mathrm{i}\boldsymbol{k}\cdot\boldsymbol{r})\boldsymbol{e}\cdot\nabla_i|\psi_i>|^2. \quad (10.16)$$

Here A_0 is the amplitude of the travelling wave, N the number of electrons, $\boldsymbol{k}$ the wave vector of propagation, $\boldsymbol{r}$ the radial field vector, $\boldsymbol{e}$ a unit vector parallel to the electric field, and ∇_i has the usual meaning. In the so-called dipole approximation it is assumed that the photon wavelength is much larger than a typical dimension in the system absorbing the photon, in which case (10.16) simplifies to

$$|M_{\mathrm{if}}|^2 = h^2 A_0^2 \sum_{\mathrm{i,f}} |\boldsymbol{e}\cdot <\psi_f| \sum_{i=1}^{N} \nabla_i|\psi_i>|^2. \quad (10.17)$$

Normally it is not the total cross-section that is measured experimentally, but the differential cross-section, since the photoelectron emission is collected over a limited solid angle $d\Omega$, the differential cross-section then being denoted by $d\sigma/d\Omega$. The differential cross-section is related directly to the transition probability expression in (10.17) by

$$\frac{d\sigma}{d\Omega}=\frac{C}{g_i}\left(\frac{1}{h\nu}\right)\times|M_{if}|^2 \tag{10.18}$$

in which C is a combination of fundamental constants, g_i is the degeneracy of the initial state, and A_0 is eliminated by normalization to unit photon flux.

Using a central-potential model in the dipole approximation, a general expression can be devised for the total cross-section for transitions from a given nl subshell in an atom, which includes all possible one-electron events, viz.

$$\sigma_{nl}(E)=\frac{4\pi\alpha_0 a_0^2}{3}(h\nu)\{lR_{l-1}^2(E)+(l+1)R_{l+1}^2(E)\} \tag{10.19}$$

where E is the photoelectron energy, α_0 the fine structure constant, a_0 the Bohr radius, and $R_{l+1}(E)$ are radial dipole length matrix elements. The differential photoelectric cross-section for a given subshell then has the form

$$\frac{d\sigma_{nl}}{d\Omega}(E)=\frac{\sigma_{nl}}{4\pi}\{1+\tfrac{1}{2}\beta_{nl}(E)(\tfrac{3}{2}\sin^2\alpha-1)\} \tag{10.20}$$

where β_{nl} is called the asymmetry parameter and α is the angle between photon incidence and photoelectron emission detection.

Since values of J_{nl} have been computed by Scofield,[33] and of β by Reilman *et al.*,[34] for all elements for the two commonly used X-ray lines Al K_α and Mg K_α, reasonably accurate values of the differential cross-section for any element in any situation should be calculable. It should be clear from the form of eqn (10.20) that inter-comparison of peak intensities in XPS would be meaningless without making allowance for the angular geometry in the experimental arrangement. The asymmetry parameter β can be neglected only for the situation where $(\tfrac{3}{2}\sin^2\alpha-1)=0$, which corresponds to the 'magic-angle' $\alpha=54.74°$; there are no spectrometers that operate in a geometry in which the angle between photon incidence and photoelectron detection is that of the 'magic-angle'.

10.4 Quantification

In a manner analogous to that of eqn (4.3) for the detected Auger current, an expression for the X-ray photoelectron current may be written down. If an incident X-ray photon of energy $h\nu$ ionizes the core level X in atom of type A in a solid, giving rise to a photoelectron to be detected in a spectrometer, then the photoelectron current is

$$I_A(X) = I_0 \rho \sigma(hv: E_X) T(E_X) D(E_X) L(hv: X)$$
$$\int_0^{\infty} N_A(z) \exp\{-z/\lambda_m(E_X) \cos\theta\} \, dz \qquad (10.21)$$

where I_0 is the characteristic X-ray flux on the surface, ρ is a surface roughness factor, $\sigma(hv: E_X)$ is the photoelectric cross-section for ionization of X by a photon of energy hv, T is the analyser transmission for the kinetic energy E_X, $D(E_X)$ is the efficiency of the detector at energy E_X, $L(hv: X)$ is the asymmetry parameter for emission from X at a photon energy hv (given above in eqn (10.20)), $N_A(z)$ is the distribution of atoms of type A with depth z, $\lambda_m(E_X)$ is the inelastic mean free path in the matrix M containing A, of electrons of energy E_X, and θ is the angle of emission to the surface normal.

All the energy- and matrix-dependent terms in (10.21) have already been discussed, either in the section 4.3 on Quantification in AES, or in the above section. As before, if a fixed set of operating conditions are assumed, i.e. constant X-ray flux, and analyser operated in the constant pass energy mode, eqn (10.21) may be simplified and abbreviated to

$$I_A(X) = K\sigma L \int_0^{\infty} N_A(z) \exp(-z/\lambda \cos\theta) \, dz. \qquad (10.22)$$

If the usual additional simplifying assumption is also made that the distribution of element A with depth is uniform, with an average atomic density $\bar{N}_A$, then

$$I_A(X) = K\sigma L \bar{N}_A \lambda \cos\theta, \qquad (10.23)$$

very similar to eqn (4.9).

Values for the total cross-section relative to carbon for Al K_α calculated by Scofield[33] are shown in Fig. 10.23, and for the asymmetry parameter β by Reilman *et al.*[34] in Fig. 10.24, as plotted by Seah.[35] Together with the expressions 4.7 for inelastic mean free path λ, and a knowledge of θ from the experimental arrangement, a linear relationship between the average atom density and the photoelectron current can be derived directly from (10.23), but still containing a constant of proportionality. It is normal practice to eliminate the proportionality constant by putting all quantification on a relative basis, i.e. by choosing one particular elemental peak, usually the F 1s, as a standard, and referring all measurements to it. Thus for measurements performed under fixed conditions one would have

$$\frac{I_A(X)}{I_F(1s)} = \frac{K_\sigma(hv: E_X) L(hv: X) \bar{N}_A \lambda_A(E_X) \cos\theta}{K_\sigma(hv: E_{1s}) L(hv: 1s) \bar{N}_F \lambda_F(E_{1s}) \cos\theta}. \qquad (10.24)$$

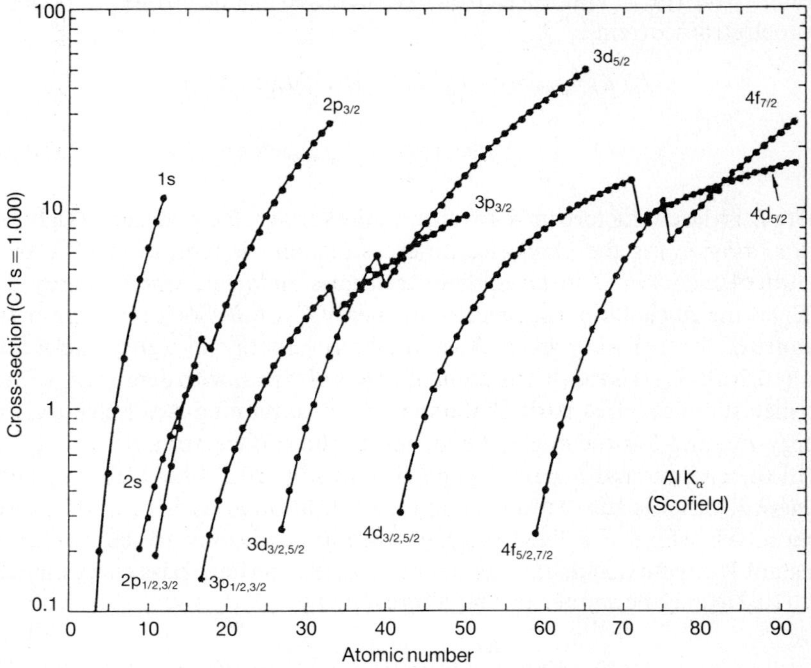

FIG. 10.23. Values of the photoionization cross-sections for Al K_α radiation, relative to that of the C 1s taken as unity, calculated by Scofield.[33] The calculation for each successive shell is taken up the periodic table to the point where the ionization energy becomes greater than the Al $K_{\alpha_{1,2}}$ line energy.

With $I_F(1s)$ defined as unity, one then has

$$I_A(X) = \frac{\sigma(h\nu : E_X)L(h\nu : X)\lambda_A(E_X)}{\sigma(h\nu : E_{1s})L(h\nu : 1s)\bar{N}_F\lambda_F(E_{1s})}\bar{N}_A \tag{10.25}$$

where now $I_A(X)$ is measured from pure A, and a well characterized fluorine standard is used. The ratio that is the multiplier of $\bar{N}_A$ is then the *relative sensitivity factor* of A, and it can either be calculated, or measured empirically with a set of standards. Tables of such relative sensitivity factors for XPS have been produced by various authors over the years, and all the tabulations suffer from one or more defects, but the most generally useful is that of Wagner *et al.*[36] For a very informative assessment of the reference data sets the reader is referred to the paper by Seah.[37]

It will be clear from what has already been said about XAES that it is not in general used for quantification. There are, however, several instances where it is more convenient to use an X-ray excited Auger peak from an element for quantification purposes than a photoelectron peak. Thus, XAES peaks may lie in a more accessible part of the spectrum than XPS peaks (e.g. Na, Zn), or

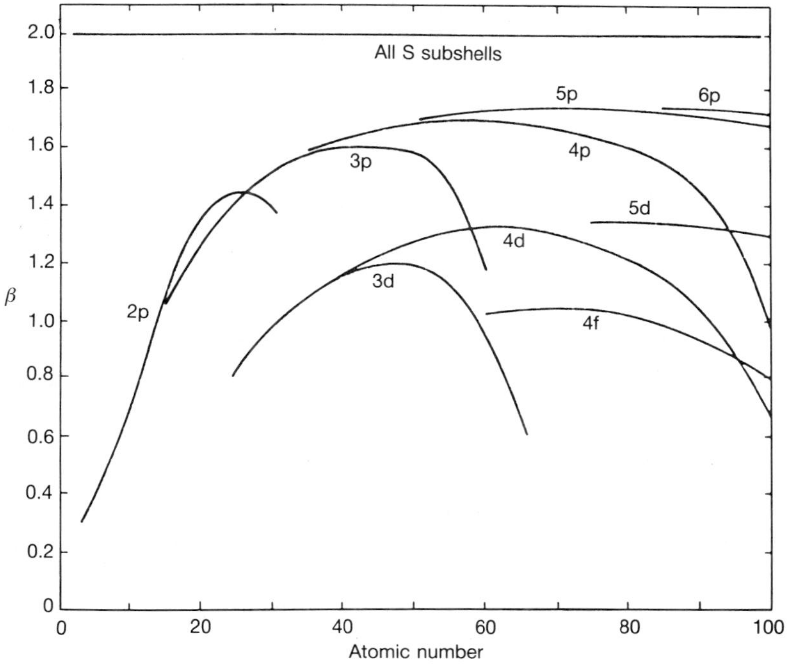

FIG. 10.24. Values of the asymmetry parameter β appearing in eqn (10.20), as calculated by Reilman *et al.*[34] and re-plotted by Seah[25] for Al K_α radiation. Evaluation of eqn (10.20) then requires the further knowledge of the angle between incident photon flux and the direction of photoelectron detection.

the XAES peaks may be the most intense elemental features in any case (e.g. Ge, Se), or spectral overlap in a multi-component sample may force the use of XAES peaks where they might not normally be used. For these purposes the XAES peak or peaks can be allocated sensitivity factors relative to the F 1s peak just as if they were XPS peaks, although not many of the tabulations contain such factors for Auger peaks and it is normally necessary to measure them individually in the laboratory.

Although eqn (10.25) when reduced to a set of relative sensitivity factors should be straightforward in use, the question must be asked as to what exactly is the appropriate measure of I_A and what difficulties arise in the measurement. If photoelectron peaks had zero width then I_A would be a delta-function, and all that would be necessary would be a measurement of peak height above background. However, photoelectron peaks have significant widths, and it is therefore necessary to measure the *area* under the peak to arrive at the total current arising from the transition. Two major problems then arise. Firstly, as seen in Fig. 10.16, there are cases where 'shake-up' satellites appear to the low kinetic energy side of the parent peak. Since both these and the less easily distinguishable 'shake-off' peaks originate by loss of electrons from the parent peak, they must be included in any measurement of area. Often this is not easy to do because of the presence of other closely adjoining peaks, and in general surface analysts tend to try and ignore shake-up peaks in the hope that they will not represent too large a proportion of the correct area. Secondly, the area of a peak to be measured is again that above the underlying background. The currently used methods of background subtraction have already been described in Section 10.1, but their application is far from automatic, indeed a significant amount of subjectivity is involved. The subjectivity arises in the decision as to where to place the end points, both in the simple straight line method, and in the Shirley method. Where a peak is sitting on a nearly flat background, there is no problem, but the steep inelastic backgrounds under some of the photoelectron spectra from the transition metals have given rise to endless discussion and controversy. As remarked by Seah[37] in the handbook 'Practical Surface Analysis', there is not as yet any universally recommendable method of defining I_A uniquely. On the other hand, if a particular method is applied in a completely consistent fashion, it will almost certainly give good relative accuracy in quantification, if not absolute accuracy, provided that it is not being asked to cover a very wide range of materials from metals to wide-gap insulators.

10.5 The Auger Parameter

One of the principal, and expanding, applications of XAES in conjunction with XPS is in the measurement of the Auger Parameter. The reasons why the Auger Parameter is so useful in many situations where chemical information is needed are: (i) it is independent of any surface charging, and (ii) it is often more sensitive to the precise chemical state of an element than the energies of either photoelectron or Auger peaks on their own.

The Auger Parameter was conceived by Wagner,[6] who defined it in its original form as the difference in kinetic energies between that of a prominent and conveniently situated Auger peak and that of the most prominent photoelectron peak, and called it α. Thus

$$\alpha = E_K(\text{Auger}) - E_K(\text{photoelectron}) \tag{10.26}$$

or

$$\alpha = E_K(\text{Auger}) + E_B(\text{photoelectron}) - h\nu. \tag{10.27}$$

There were soon minor difficulties with this definition, since there was not always in the same spectrum an Auger peak of sufficient intensity to be accurately measurable, but yet with a kinetic energy greater than that of a prominent photoelectron peak, and not infrequently negative values of α resulted. Wagner at first[38] tried to overcome the problem by adding 1000 or

2000 eV to α quite arbitrarily, but that was clearly not satisfactory, so he subsequently[7] defined a 'modified' Auger Parameter α^*, which was simply the sum of α and the photon energy $h\nu$, leading to

$$\alpha^* = \alpha + h\nu$$

or

$$\alpha^* = E_K(\text{Auger}) + E_B(\text{photoelectron}) \tag{10.28}$$

α^* is thus independent of photon energy, and is always positive.

The original concept of both α and α^* was empirical, but in fact they are both related directly to the extra-atomic relaxation energy, changes in which are in turn related directly to the changes in the *local* chemical environment of the atom in which the Auger and photoelectron transitions are taking place. The correct expression for the energy terms appearing in an Auger transition XYZ is

$$E(XYZ) = E_B(X) - E_B(Y) - E_B(Z) - \mathscr{F} + R^{\text{ia}} + R^{\text{ea}} \tag{10.29}$$

where $\mathscr{F}$ is the energy involved in the multiplet coupling between two holes in the final state of the Auger process, and R^{ia}, R^{ea} are the intra-atomic and extra-atomic relaxation energies, respectively. Now the relation between the photoelectron kinetic energy and the binding energy of the level from which it came is of course

$$E_K(\text{photoelectron}) = h\nu - E_B$$

so that one can write for α and α^*

$$\alpha = 2E_B(X) - E_B(Y) - E_B(Z) - \mathscr{F} + R^{\text{ia}} + R^{\text{ea}} - h\nu \tag{10.30}$$

and

$$\alpha^* = 2E_B(X) - E_B(Y) - E_B(Z) - \mathscr{F} + R^{\text{ia}} + R^{\text{ea}}. \tag{10.31}$$

Normally the most intense, i.e. those most likely to be easily measurable, Auger transitions are those for which $Y = Z$. In addition, to a first approximation one can take the chemical shifts in X and Y as equal, i.e. $\Delta E_B(X) = \Delta E_B(Y)$. Also, the terms $\mathscr{F}$ and R^{ia} are purely atomic and will be independent of chemical state, so that finally one can write that

$$\Delta\alpha = \Delta\alpha^* \approx \Delta R^{\text{ea}}. \tag{10.32}$$

The derivation of eqns (10.30) and (10.31), and hence (10.32), assumes that $E_K(XYZ)$ and $E_B(X)$ can be measured in the same spectrum. Even with the conventional X-ray line sources Al K_α and Mg K_α that are used every day in XPS it is possible in most cases to measure $E_K(XYZ)$ and $E_B(X)$ together, but there are nevertheless several important situations where the binding energy $E_B(X)$ is too high. In these cases it is often possible[39–41] to measure $E_K(XYZ)$

by means of excitation via the bremsstrahlung continuum beyond the characteristic X-ray line, and then to compromise by using another prominent photoelectron line, rather in the spirit of Wagner's original conception. More correctly, one should turn to X-ray line sources of higher energy, as has been done by several workers.[38,39,42] A particularly useful such source is that of Zr, with an L_α line at 2042.4 eV, and an L_β, at half the intensity, at 2124.0 eV.

Examples of the application of the Auger Parameter in identifying chemical state will be given in the section on applications.

10.6 Fine structure in Auger spectra studied by XAES

When electron spectrometers of sufficiently good energy resolution became available in the early 1970's, several workers made the same observation at much the same time that certain Auger features were remarkably narrow, too narrow in fact to be characteristic of an atom in a solid. The first of these observations was made for the LVV peaks of copper[43] and of zinc,[44] and the MVV peaks from silver.[45,46] Not only were the principal features very narrow, but there was also a wealth of fine structure in the form of additional peaks, shoulders, and inflections in the spectra. An example of such a spectrum, that of the MNN Auger transitions in cadmium, has already been shown in Fig. 4.7 for *electron* excitation, and another example, using *X-ray* excitation of the LVV Auger spectra of copper, is given in Fig. 10.25, from the work of Roberts *et al.*[47] Using an energy resolution of the order of 1 eV, and operating in the pulse-counting mode with long acquisition times allowed the latter authors to establish the presence of many minor fine structural features associated with the copper Auger peak, that in the simple *j–j* coupling notation would be labelled $L_3M_{4,5}M_{4,5}$. The principal peak can be seen to have an intrinsic width of about 1.5 eV; if the measured width of the $M_{4,5}$ level, which forms the main contribution to the copper valence band, were to govern the width of the Auger peak, then the latter would instead be several eV wide.

Auger spectra of the above type are, as already mentioned under Section 4.2.5, called 'quasi-atomic' since they are more characteristic of transitions taking place in a free atom than in the same atom in a solid. It was the realization that there was such similarity between the atomic and solid-state Auger spectra of elements like copper and zinc that led to the physical explanation for the occurrence of quasi-atomicity. In the final state of the Auger process the atom is left with two holes in outer levels, i.e. it is doubly ionized. If the ionized atom is free, the holes are strictly localized because clearly there is no internal mechanism for their decay which depends on interaction with the surroundings. With the two holes together on the atom for so long (relatively), significant electronic correlation between them can occur, giving rise to many possible final-state configurations and therefore to much fine structure in the Auger spectrum. A direct comparison in the same

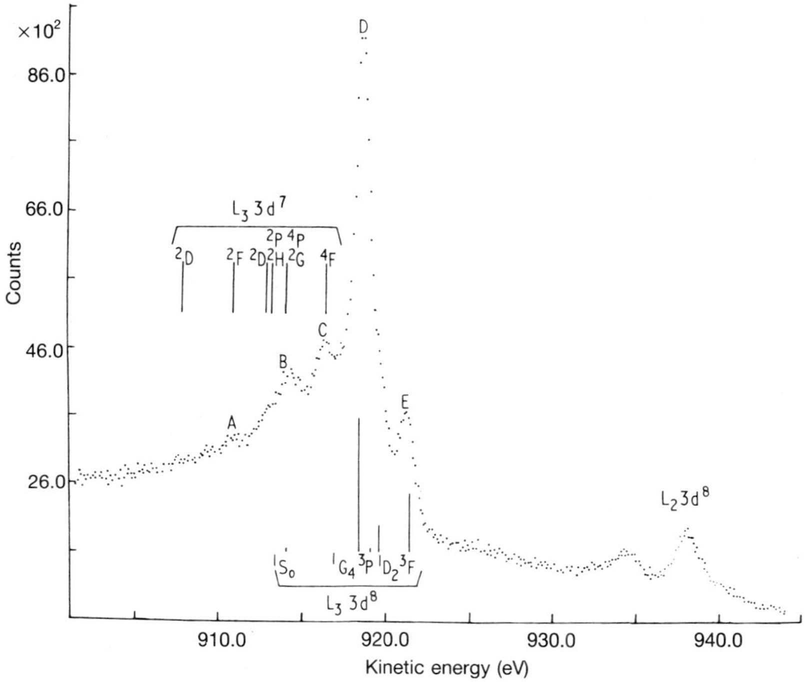

FIG. 10.25. The X-ray excited $L_{2,3}M_{4,5}M_{4,5}$ Auger spectrum of copper measured by Roberts *et al.*[47] using Mg K_α and high energy resolution with long acquisition times. In the L_3-derived part of the spectrum there are five clearly discernible peaks, A, B, C, D, and E. Shown above and below the spectrum are the line positions calculated for the configurations $L_3 3d^7$ and $L_3 3d^8$, respectively, and in the latter case the lengths of the lines are proportional to calculated intensities.

instrument of the Auger spectra of free cadmium atoms and of solid cadmium has been made by Aksela *et al.*,[48] and their spectra are reproduced in Fig. 10.26. Apart from better resolution in the free atom case, due to solid state broadening, the two are almost identical. The final-state spectroscopic terms arising from the hole–hole correlation, in intermediate coupling notation (see Section 4.2.1) are indicated, together with the expected magnitudes based on calculated intensities.

The observation that atoms in certain solids produced Auger spectra very similar to those of free atoms, i.e. the two holes in the final state of the ionized atom in the solid are also localized for a sufficient length of time, led naturally to consideration of the reasons for such localization, and thus to the criteria to be applied. As both the 3d and 4d transition series are ascended, the d bands become progressively narrower, and the sp bands fill with electrons. Quasiatomic behaviour of the type described is most prominent in those metals with virtually full d bands, e.g. copper and silver and for those metals models have been proposed by Cini[49,50] and by Sawatzky[51] to explain the

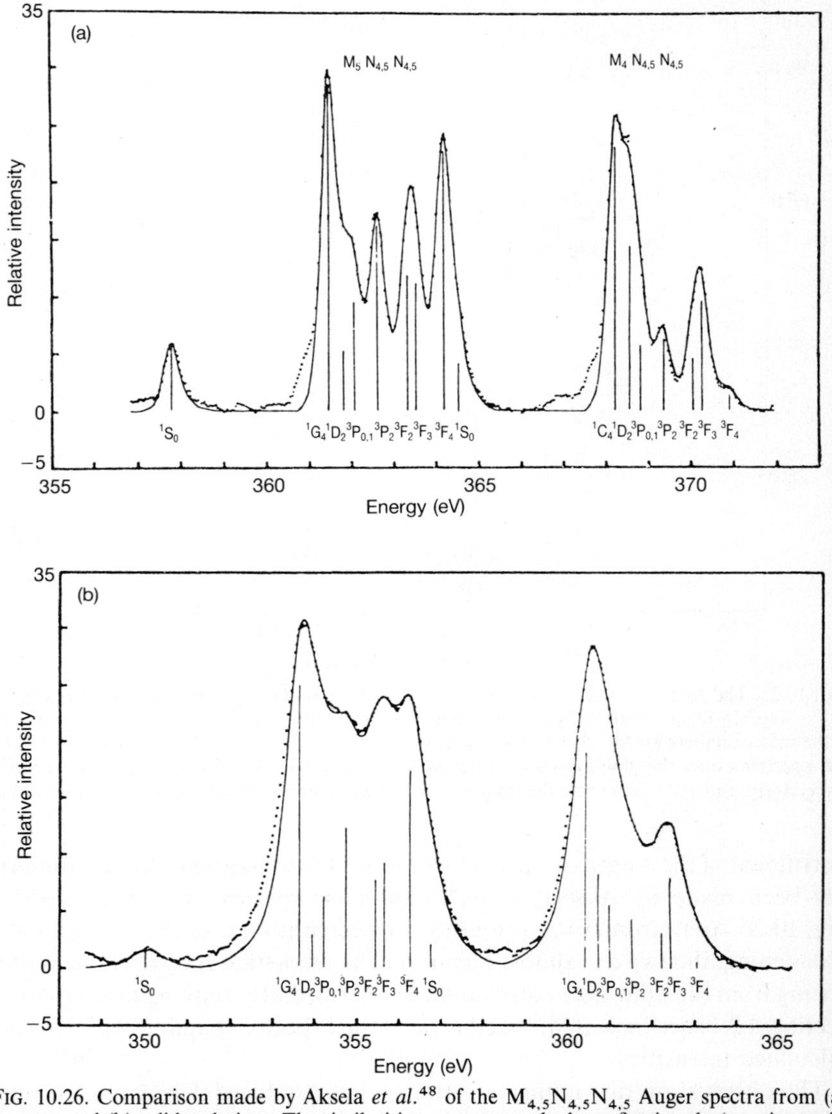

FIG. 10.26. Comparison made by Aksela *et al.*[48] of the $M_{4,5}N_{4,5}N_{4,5}$ Auger spectra from (a) gaseous and (b) solid cadmium. The similarities are very marked, confirming the 'quasi-atomic' nature of the Auger spectrum from solid cadmium. In each case the experimental data are represented by points and the calculated spectral envelopes by the solid lines. Underneath are shown the calculated line positions and relative intensities, derived from multiplet splittings arising in the intermediate coupling scheme.

behaviour. The prediction from their models, which are very similar, differing only in the mathematical treatment, is that the Auger spectra will be quasi-atomic if the effective Coulomb interaction U_{eff} between two holes on any pair of levels in an atom is at least twice the d, or valence, band width. The interaction energy U_{eff} for the Auger transition XYZ can be derived from the expression

$$U_{\text{eff}}(XYZ) = E_{\text{B}}(X) - E_{\text{B}}(Y) - E_{\text{B}}(Z) - E_{\text{K}}(XYZ). \tag{33}$$

Values of U_{eff} measured by Parry-Jones *et al.*[52] for the principal component ($^1G_4$) in the X-ray excited MNN Auger spectra for the metals silver, cadmium, indium, and tin, are compared in Table 10.3 with the measured widths W of the 4d bands in these metals.

Table 10.3

Comparison of the effective Coulomb interaction U_{eff} with the observed 4d band width, for Ag, Cd, In, and Sn

Metal	$U_{\text{eff}}(^1G_4)$	W
Ag	5.1 ± 0.7	3.6 ± 0.2
Cd	6.0 ± 0.6	1.9 ± 0.1
In	6.8 ± 0.7	1.9 ± 0.1
Sn	6.9 ± 0.6	2.0 ± 0.1

Clearly the criterion that $U_{\text{eff}} \geq 2W$ holds for cadmium, indium, and tin, but silver can be seen to be a borderline case. Indeed, multiplet structure calculations and calculated transition rates for silver generated spectral envelopes that showed discrepancies between theory and experiment; these are obvious from Fig. 10.27 from the paper by Parry-Jones *et al.* Not all the spectral components show the same quasi-atomic behaviour, in particular for the $^3F_4$ component, another prominent one, U_{eff} is only 2.8 eV, so that its behaviour is said to be more 'band-like' than quasi-atomic. Thus it is possible for a spectrum to exhibit some components that are quasi-atomic and others that are the opposite, band-like.

Both Cini and Sawatzky have developed and extended their theories further, to include, for example, coupling between holes and plasmons[53] in transitions involving only one valence electron, incompletely filled valence bands,[54] and the use[55] of a band structure of realistic shape. Detailed consideration of these developments is beyond the scope of this book, and the reader is referred to the above and subsequent papers from those authors for information about the state of that aspect of Auger theory. The basic criterion remains unaltered.

10.7 Applications

In Section 4.4 it was remarked that the range of applications of AES were so diverse, and the number of publications now so great, that it was impossible to

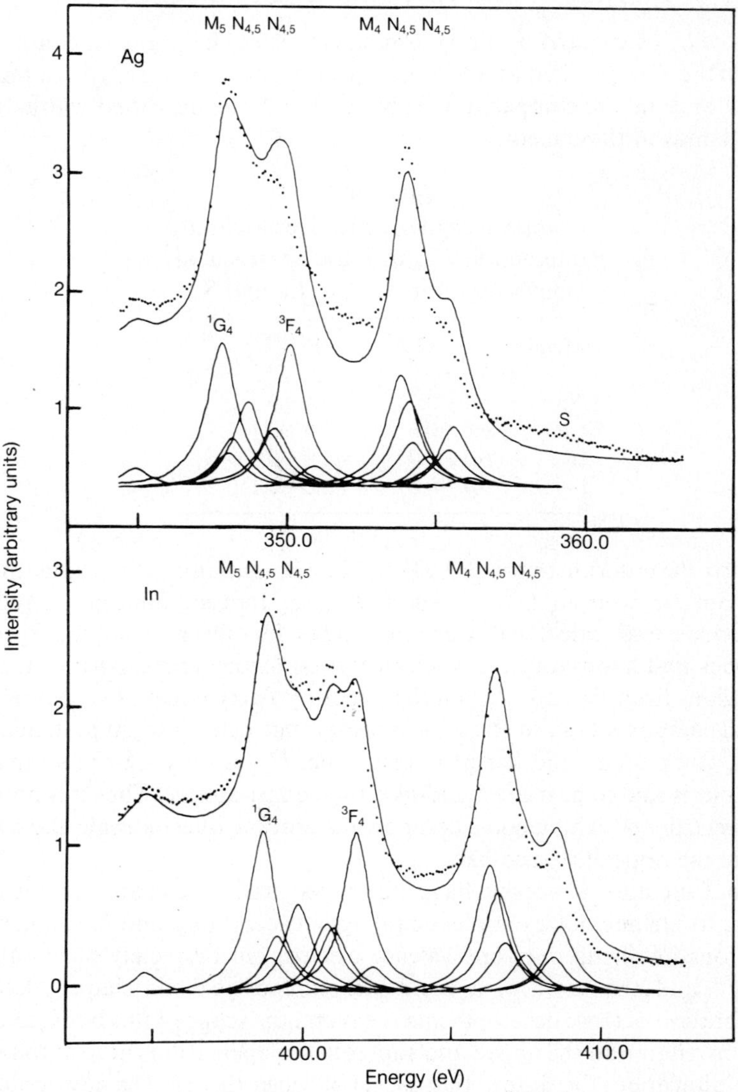

FIG. 10.27. Comparison of the experimentally measured $M_{4,5}N_{4,5}N_{4,5}$ Auger spectra for solid Ag and In, represented by points, with theoretical spectra calculated on a purely 'quasi-atomic' basis, by Parry-Jones *et al.*[52] Discrepancies can be seen between experiment and theory, particularly for Ag. In the latter case, the $^3F_4$ component is overestimated theoretically, there is too little theoretical intensity near 352 eV, and theory does not account for the broad feature S. In the In spectrum, agreement is much better. The reason, as discussed in the text, is that Ag is a metal intermediate in characteristics between 'quasi-atomic' and 'band-like'.

cover all of them in a volume of this size, and that some selection had therefore to be imposed. If that is true of AES, it is even more so of XPS, which has become the workhorse of surface analysis. The following chosen examples must thus be regarded as a very minimal selection from the vast literature of the last twenty years or so.

10.7.1 *Corrosion and passivation*

The ability of XPS to establish the nature of the chemical state(s) of an element in any particular situation has probably been of greater immediate use in those applications involving surface reactions than in any others. Amongst such reactions are not only those of direct technological origin, such as corrosion, passivation, catalysis, adhesion, etc., but also the enormous range of basic studies in which the starting point is always that of an atomically clean, well characterized surface, often that of a single crystal. The first examples of the application of XPS in this general area of surface reaction will be taken from the field of corrosion and passivation.

In a nuclear reactor, the radioactive fuel must be contained in a material chosen so that at the extreme conditions of operation there is compatibility (i.e. lack of reaction) between the fuel and the containment, resistance to corrosion of the containment by the coolant gas flowing past it, and long-term mechanical stability under temperature cycling and irradiation. The technical name for the containing material is 'cladding'. Operating conditions are generally severe; in the advanced gas-cooled reactor (AGR), for instance, the cladding is in contact with the fuel UO_2 on one side, and with a CO_2/1% CO mixture at 40 atm on the other, at temperatures up to 1123 K. In addition thermal cycling over a range of 600 K is not uncommon. The material chosen as cladding for the AGR is a steel containing 20 wt.% Cr, 25 wt% Ni and about 1 wt% Nb, the latter acting as a metallurgically stabilizing addition.

The cladding is in the form of tube fabricated by drawing, which introduces about 25% cold work, so it is annealed in hydrogen at 1203 K for an hour before use. An interesting observation is that the annealed and unannealed material exhibits different oxidation behaviour, the latter showing a greater degree of oxidation under the same conditions. Such differences could be very important if at any time annealing were not carried out thoroughly, and have been investigated by Tyler[56] using XPS (and AES). He was concerned with the relatively early stages of oxidation, up to scale thicknesses of $\approx 0.25\ \mu m$, and was of course not able to reproduce the reactor operating pressure of 40 atm, instead working at 50 torr CO_2. Photoelectron spectra, both survey and detailed, were recorded at intervals during oxidation, up to 100 h; clearly the oxidizing atmosphere would have to be removed at each interval to enable analysis to be performed. Quantification in terms of at.% was achieved using the relative sensitivity tables of Wagner *et al.*[33] already mentioned in Section 10.4.

It was evident even after the very first period of oxidation, 2 min, that there were significant differences between the oxide compositions on the annealed and on the unannealed surfaces, respectively. Figures 10.28 and 10.29 show the quantified surface compositions of the two surfaces as a function of oxidation time; although the absolute atomic percentages are probably not accurate, on a relative basis a detailed comparison can be made. On the annealed surface Ni and Cr virtually disappeared, and the oxide was entirely Fe oxide up to the duration of the experiment. On the unannealed surface, however, Fe dropped to a low level at once, whereas Cr rose to and stayed at $\approx$20 at.%. In the latter case the oxide film contained an appreciable Mn concentration as well. Notice that the oxide film on the unannealed surface also showed some structural variation in that after about 2 h oxidation a peak in the Mn concentration appeared, with a corresponding dip in the Cr concentration. Subsequently the relative concentrations became effectively constant at $\approx$21 at.% Cr and $\approx$8 at.% Mn, with small quantities of Fe and Ni.

Detailed analysis of the various metal component photoelectron spectra also revealed differences between the behaviours of the annealed and unannealed surfaces. For the unannealed surface the Fe and Cr 2p peaks were in energetic positions characteristic of the III oxidation states at all stages of oxidation, while the Mn 2p peak showed initially a mixture of II and III before shifting to the II position for the bulk of the oxidation period. On the annealed surface, the behaviour was slightly more complex, as shown in Fig. 10.30, in which the Fe 2p and O 1s spectra are given for selected stages of oxidation. At first the Fe 2p spectra showed a mixture of Fe(II) and Fe(III) (binding energies 709.7 eV and 711 2 eV, respectively), but after 5 h oxidation the proportion of Fe(III) increased substantially as indicated by a narrowing of the 2p peaks, the appearance of characteristic shake-up satellites at 719.6 eV and 732.7 eV, and a slight shift in the O 1s peak position. Between 5 and 30 h oxidation the Fe(II) contribution increased again, that is, the features diagnostic of Fe(III) decreased in magnitude, but then after 100 h the situation reverted once more so that the spectra were very similar to those observed at the start of oxidation.

The corrosion resistance of steels such as the 20/25/Nb steel studied here depends on the formation of a chromia layer at an early stage, which then acts as a barrier to further oxidation by hindering the outward diffusion of metal cations. However, although a Cr-rich oxide was formed on the unannealed surface, on the annealed surface there was only Fe_3O_4, shown by the variable mixture of Fe(II) and Fe(III) seen by XPS, and by X-ray diffraction (XRD). The presence of some Mn and Fe in the film on the unannealed surface was shown by XRD to be consistent with the formation of an $MnCr_{2-x}Fe_xO_4$ spinel, the lattice parameter indicating a spinel with much more Cr then Fe, in agreement with XPS. Thus a chromia film could not form on the annealed surface. The principal difference between the two states of the steel lay in the

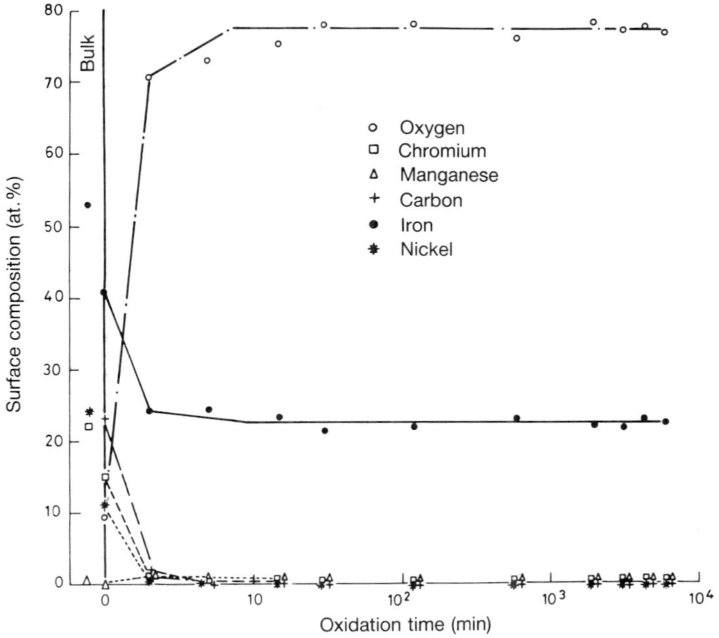

FIG. 10.28. Surface composition determined by XPS of a 20Cr–25Ni–Nb steel oxidized in 50 torr CO_2 at 973 K, as a function of oxidation time, from the measurements of Tyler.[56] The steel had been first annealed in hydrogen at 1203 K for 1 h. Atomic percentages were derived using the relative intensity factors of Wagner *et al.*[33]

average grain size. The cold worked, i.e. unannealed, material had been amorphised to a depth of $\sim 5\ \mu m$, that is, the original grain boundaries had been destroyed, leaving a very high concentration of defects and dislocations in the surface region, providing a large number of short-circuit diffusion paths for the movement of Cr to the surface. In the annealed material, on the other hand, the average surface grain size after the annealing treatment was of the order of 9 μm, with the consequence that the number of available short-circuit diffusion paths was much lower, and Cr could not reach the outer surface fast enough to be included in the oxide. The differing Mn contents observed in the two cases are attributed to the different rates of diffusion of Mn through Fe_3O_4 and through chromia. The variations in Mn content seen on the unannealed surface as a function of oxidation time were probably due to the mode of growth of the film, which is envisaged as the lateral spread of spinel patches over the initial chromia layer. If the diffusion rates of Mn in chromia and in the MnCrFe spinel were different, then at the point where the patches coalesce some such variations might be expected.

Passivation is complementary to corrosion since a surface that is completely passivated will by definition not corrode. The goal of all those studying

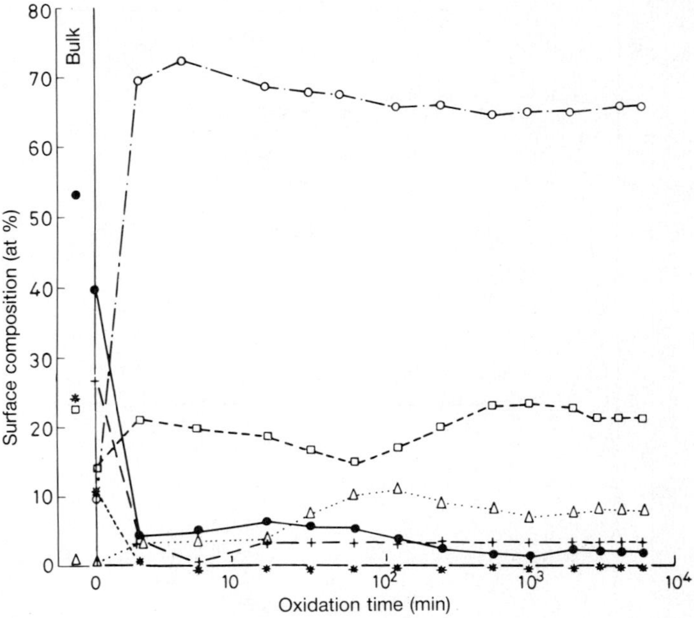

FIG. 10.29. Same as in Fig. 10.28, but with the steel in the previously *un*annealed condition. Note that on the annealed steel in Fig. 10.28 the oxide formed was entirely iron oxide. On the unannealed surface, however, Fe dropped to a low level at the start of oxidation, whereas Cr and Mn became the principal metallic components in the oxide. Some structural variation in the oxide at about 10^2 h of oxidation is also apparent. From Tyler.[56]

corrosion, whether in gaseous or aqueous media, is to find ways of slowing it down or, better still, of stopping it, by the formation of passive layers. Thus the detailed chemistry of such layers, both during their formation and during subsequent exposure to corrosive environments, is a vital piece of information. XPS is one of the variety of techniques that has been used to attempt to establish the chemistry, with some success, although it should be emphasized that XPS on its own is never sufficient and that information from techniques used in parallel with it must always be available.

A typical technological area in which the breakdown of a passive film could lead eventually to catastrophic failure is that of turbine blades in the low-pressure steam section of a generator, where condensation can occur at temperatures in the range 50–100°C. The blades are normally made of a martensitic stainless steel containing $\approx 12\%$ Cr, and the aqueous medium on condensation approximates to a de-aerated dilute chloride solution.

The nature of the passive film on several commercial alloys, including that normally used for blade fabrication, has been studied by Brüesch *et al.*[57] Alloy compositions ranged from 7 to 29% Cr, with pure Fe and pure Cr added at

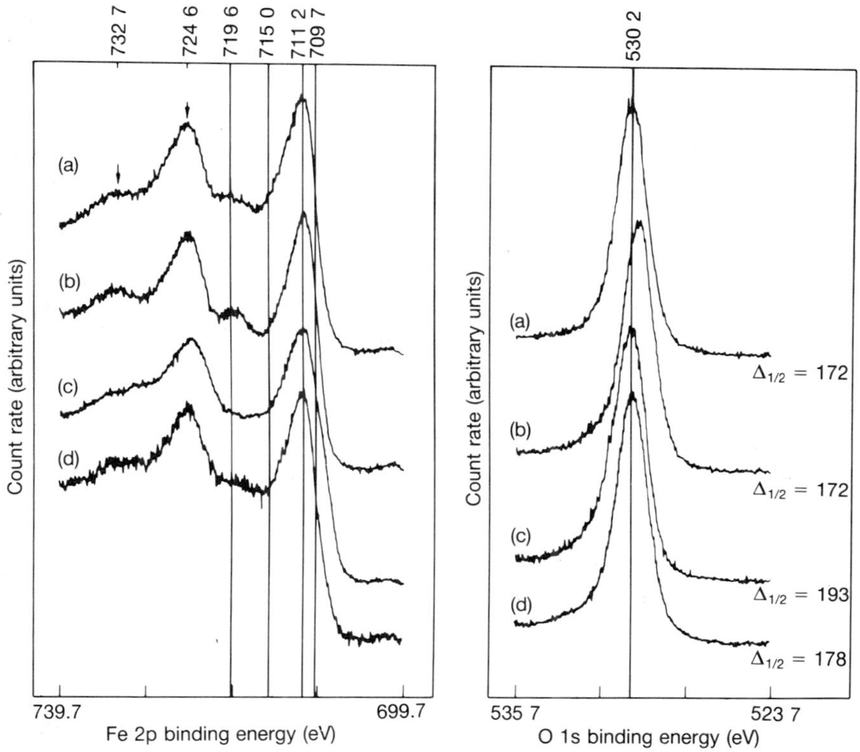

FIG. 10.30. The Fe 2p and O 1s photoelectron spectra from a 20Cr–25Ni–Nb steel oxidized in 50 torr CO_2 for (a) 2 min, (b) 5 h, (c) 10 h, and (d) 100 h. The steel was in the annealed condition (see caption to Fig. 10.28). Note the shift from a mixture of Fe(II) and Fe(III) (binding energies 709.7 and 711.2, respectively) after 2 min to principally Fe(III) after 5 h, confirmed by the appearance of shake-up peaks at 719.6 and 732.7 eV. With further oxidation, the proportion of Fe(II) increased again, until at 100 h the spectrum was very similar to that at the beginning. Corresponding slight shifts can be seen in the O 1s spectra. From Tyler.[56]

each end of the range. The authors were rigorous in their approach in that the passivation treatment was carried out in a cell attached to the spectrometer. With the cell full of argon gas the sample surface was brought close to the open end of a glass tube containing the solution so that the liquid meniscus came into contact with the surface. The open-circuit potential between the sample and a reference electrode (saturated calomel) was then recorded as a function of time up to a duration of 10^4 s, following which the electrolyte was removed from the surface and the latter rinsed with de-aerated water, still under argon gas. The sample was then transferred into the preparation chamber of the spectrometer, where residual water was pumped away, and finally into the analysis chamber.

The behaviour of the open-circuit potential with time of exposure to the 0.1 M de-aerated NaCl solution, for each of the six alloys, and for the pure metals, is shown in Fig. 10.31. Passivation is indicated by approach to zero potential with respect to the reference electrode, and as expected the alloys with increasing Cr content show an increasingly rapid approach to zero, with the exception of Cr itself, which is slower. By contrast, pure Fe and the low-Cr alloy show opposite behaviour, with the open-circuit potential moving in the other direction with time, and they are said to remain in the 'active' region. Alloy 2, which is the one used in the turbine blades, passivates after an induction period of about 400 s.

Elemental concentrations through the various films formed, passive or otherwise, were obtained by depth profiling with argon ions at 4 keV with XPS analyses at selected intervals. Since it is well known[58] that ion bombardment can cause reduction of many compounds, and especially the oxides of the transition metals, the authors plotted either total elemental (i.e. oxide plus metallic), or total metallic and total oxide (irrespective of state), concentrations against bombardment time. To have attempted to plot, for example, the Fe^{3+} and Fe^{2+} individual concentrations would have been meaningless. The profile of the total Cr concentration as a function of sputter time (i.e. depth) for each of the materials is given in Fig. 10.32. For Cr bulk concentrations above $\approx 15\%$ a sharp peak in Cr atomic percentage appeared at a depth between 10 and 15 Å, followed by a region depleted in Cr with a minimum near 40 Å. The low-Cr alloy showed no such peak or depletion, while the 12.6% Cr alloy had a broad Cr concentration peak at ≈ 30 Å, but again with no depleted region. The Cr concentration at the sharp maximum increased with increasing Cr in the bulk, but started to show saturation above 30 at.%; the Cr concentration in pure Cr_2O_3 is 40 at.%.

Profiles for Fe and Cr broken down into the metallic and total oxide contributions are shown in Figs 10.33 and 10.34 for the low and high Cr containing materials, respectively. On pure Fe there was a relatively thick oxide film of order 50 Å, while on pure Cr the oxide was only about 10 Å thick. In between the oxide film became progressively thinner with increasing Cr content in the alloy, and the proportion of Fe oxide in the film also decreased with increasing Cr. Not shown in Figs 10.33 and 10.34 is the profile for O. At the surface itself before profiling it was the major atomic constituent, and the O 1s spectrum could be decomposed easily into contributions from oxide, hydroxide, and water. By increasing the surface specificity by reducing the photoelectron take-off angle, as described in Fig. 3.27, the contributions from OH^- and H_2O were found to be concentrated very much near the external surface. The smooth variation indicated that no sharp interface existed between the hydroxide + water outer region and the oxide inner region. It was also obvious that the alloy with the greatest Cr content had the highest concentration of surface OH^- and H_2O.

Clearly, the passive films were enriched in Cr, and it was found that at a critical bulk Cr concentration near 13%, the Cr concentration in the film

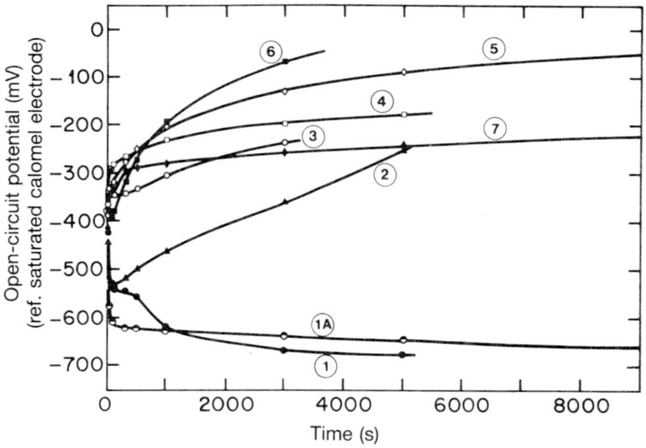

FIG. 10.31. Behaviour of the open-circuit potential as a function of time of exposure to 0.1 M deaerated NaCl solution, of pure Fe and Cr, and of six ferrous alloys, recorded by Brüesch *et al.*[57] Passivation is indicated by an approach to zero potential with respect to the reference saturated calomel electrode. Alloys with increasing amounts of Cr show an increasingly rapid approach to zero, except for Cr itself. Pure Fe and the Fe–7Cr alloy show opposite behaviour, and are said to remain in the 'active' region. (1) Fe; (1A) 7Cr; (2) 12Cr 1Mo; (3) 15Cr 6Ni; (4) 17Cr 13Ni; (5) 26Cr 6Mn; (6) 29Cr 4Mo; (7) Cr.

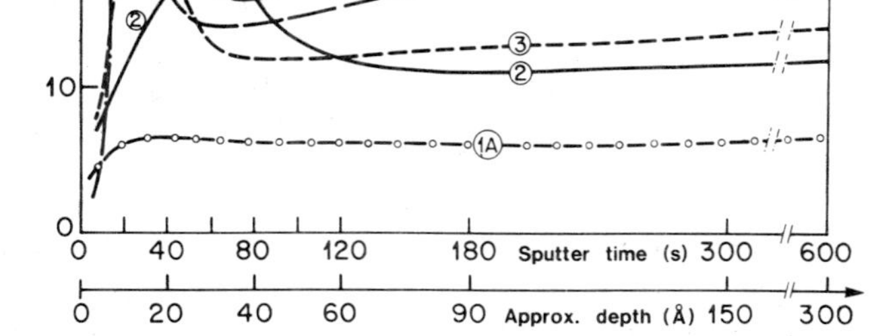

FIG. 10.32. Profile of the total Cr (i.e. $Cr^{ox} + Cr^{m}$) atomic concentrations through the passive and other oxide films formed on the alloys shown in Fig. 10.31, achieved by bombardment with 4 keV Ar^+ ions, by Brüesch *et al.*[57] For concentrations of Cr in the bulk alloy above about 15% there is a sharp peak in Cr concentration at a depth between 10 and 15 Å, followed by a region depleted in Cr. The concentration at the peak maximum increased with increasing Cr concentration in the bulk, saturating beyond 30 at.%.

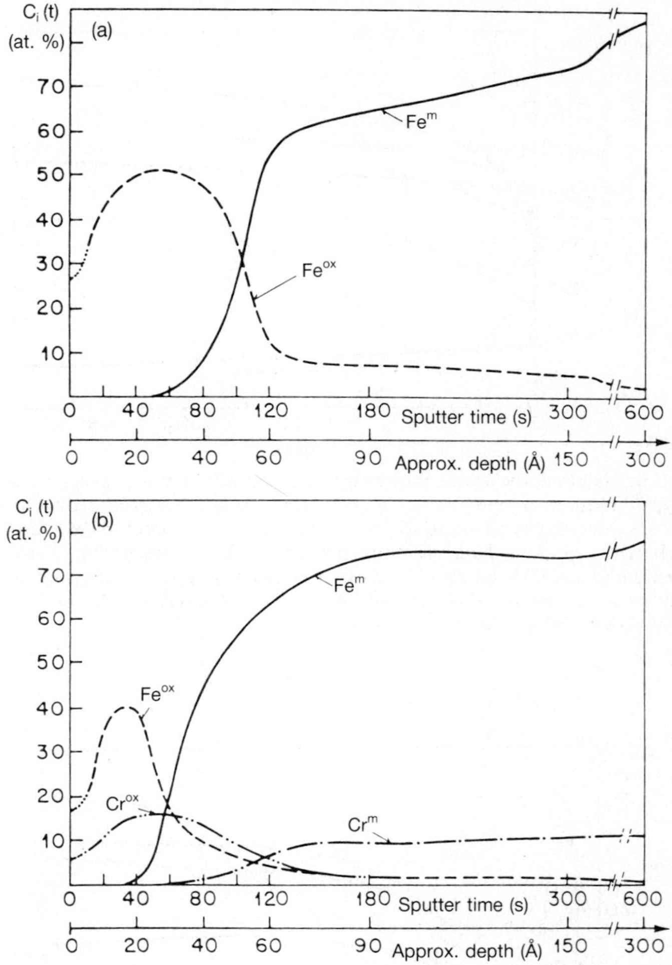

FIG. 10.33. Profiles of the Fe and Cr metallic and oxide contributions through oxide films on the low-Cr containing materials (a) pure Fe, (b) sample 2 (12 Cr 1Mo). The oxide contributions are $Fe^{ox} = Fe^{2+} + Fe^{3+}$, and $Cr^{ox} = Cr^{3+}$. The oxide film on Fe is relatively thick, ≈ 50 Å, while on the alloy it is much thinner, and contains Cr. From Brüesch *et al.*[57]

increased non-linearly with the bulk concentration. At the same point the mean thickness of the film decreased sharply. It is also known that below a critical bulk Cr concentration of about 12% the corrosion rate of Fe–Cr steels increases dramatically, consistent with the results of Fig. 10.31. This critical concentration suggests that the transition from corrosive to passive behaviour is accompanied by a phase transition, and indeed the stoichiometries estimated from XPS in the central region of the films suggest a change from a predominantly Fe–Cr spinel to a predominantly Cr_2O_3 structure at that

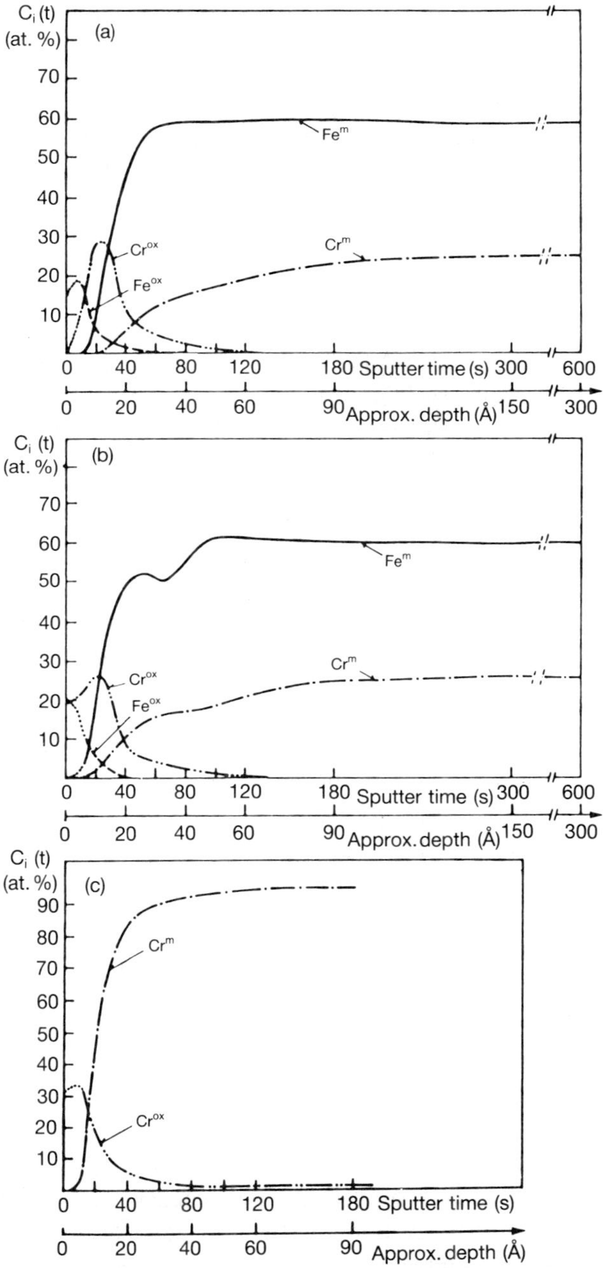

FIG. 10.34. Profiles of the Fe and Cr metallic and oxide contributions through oxide films on the high-Cr containing materials (a) sample 5 (26Cr–6Mn), (b) sample 6 (29Cr–4Mo), and (c) pure Cr. Note the decreasing thickness of the oxide films with increasing Cr; on pure Cr the oxide is only about 10 Å thick. Again $Fe^{ox} = Fe^{2+} + Fe^{3+}$, and $Cr^{ox} = Cr^{3+}$. From Brüesch *et al.*[57]

point. The stable passive films on the alloys containing more than the critical amount of Cr are therefore substantially independent of alloy composition, and consist of a sequential layered structure of a hydrated layer containing OH^- and H_2O, a very thin mixed Fe–Cr oxide layer, a thicker layer mostly of Cr_2O_3, and a metallic layer depleted in Cr. On the above basis the alloy used for the fabrication of the turbine blades would appear to be in the undesirable transition region between the active and passive behaviours.

10.7.2 *Catalysis*

The understanding of the basic mechanisms of heterogeneous catalysis was from the very beginnings of XPS one of the prime motivations for the application of the technique. This motivation has driven research workers in this field in two principal directions—one in which 'real' catalyst systems, i.e. those approximating closely to the catalyst materials used technologically, have been studied, and the other in which the number of operational parameters involved is reduced to a minimum using 'model' catalysts in the form of well-characterized single crystal surfaces. There is still considerable argument as to whether the two approaches are ever likely to converge. The two examples discussed below are typical of these approaches; both use XPS as well as other techniques in parallel but by the very nature of the different forms of the materials studied the information sought in the two cases is also different.

One of the technological fields which is a major user of catalysts is of course the oil industry; this includes both the treatment of petroleum products from raw oil feedstock and of oily by-products from coal. An important step in these treatments involves the removal of sulphur, which is an atmospheric pollutant, with simultaneous hydrogenation. The process is called hydrodesulphurization (HDS) and the catalyst used is molybdenum (in the form of an oxide) supported on alumina with promotion by cobalt, also in the form of an oxide. Like all catalysts, it requires an activation procedure before use to establish the catalytically active condition, and the activation often includes reduction, usually in hydrogen. The effect of reduction on the chemical state of molybdenum in a cobalt–molybdenum–alumina commercial catalyst was studied by Patterson *et al.*;[59] the catalyst composition was, by wt.%, 3 CoO, 14.5 MoO_3, and 82.5 Al_2O_3.

A succession of Mo 3d photoelectron spectra taken after each of several stages of reduction is shown in Fig. 10.35. The vertical lines at binding energies of 233.0 eV and 236.1 eV correspond to the positions of the 3d doublet for the Mo(VI) state, while that marked 229.9 eV corresponds to the $3d_{5/2}$ binding energy in the Mo(IV) state. It can be seen that an Mo(IV) contribution is just discernible after 50 min reduction (spectrum C), but in fact the spectra during reduction are more complex than would correspond to a mixture of only two

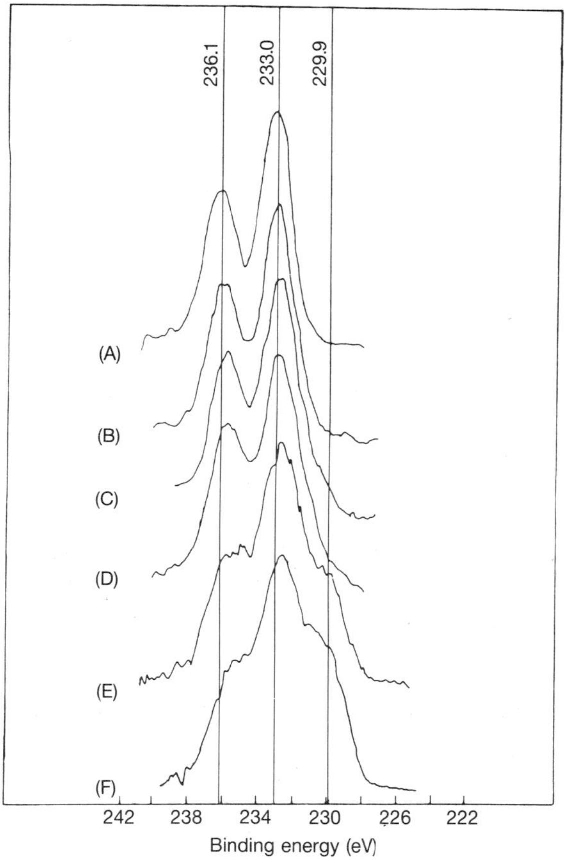

FIG. 10.35. Mo 3d photoelectron spectra from the catalyst 3CoO–14.5MoO$_3$–82.5Al$_2$O$_3$ as a function of various reduction treatments, recorded by Patterson *et al.*[59] (A) catalyst fired in air at 500°C for 4 h. (B) 15 min reduction in hydrogen at 500°C. (C) 50 min reduction. (D) 60 min reduction. (E) 120 min reduction. (F) 200 min reduction. The vertical lines at 233.0 eV and 236.1 eV correspond to the positions of the 3d doublet components for the Mo(VI) state, and that at 229.9 eV to the 3d$_{5/2}$ component for the Mo(IV) state. The latter state is just discernible after 50 min reduction, but curve resolution (see Fig. 10.36) indicates more than two states of Mo.

states. The authors carried out careful curve resolution of both the Mo 3d and 3p spectral envelopes, and an example of such resolution is shown in Fig. 10.36, which corresponds to stage E in Fig. 10.35. Clearly, there are not two, but three chemical states contributing to the spectrum, and the peak positions for the intermediate state are of course those for Mo(V). Using the curve resolution procedure at all stages of reduction, and taking the areas under the resolved peaks as proportional to the concentrations of the individual state, the mole fraction of each state as a function of reduction time

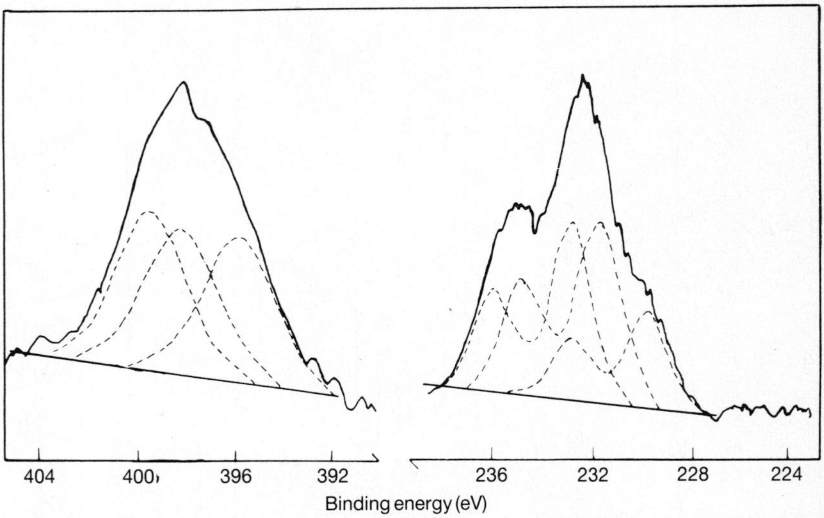

FIG. 10.36. Curve resolution carried out by Patterson *et al.*[59] on the Mo 3p (left) and 3d (right) spectral envelopes from the catalyst $3CoO–14.5MoO_3–82.5Al_2O_3$ after reduction in hydrogen for 120 min at 500°C. The 3d spectrum is similar to that labelled E in Fig. 10.35. The straight base lines indicate the linear backgrounds that were subtracted. Three states of Mo can be resolved in each of the envelopes, corresponding to Mo(IV), Mo(V), and Mo(VI).

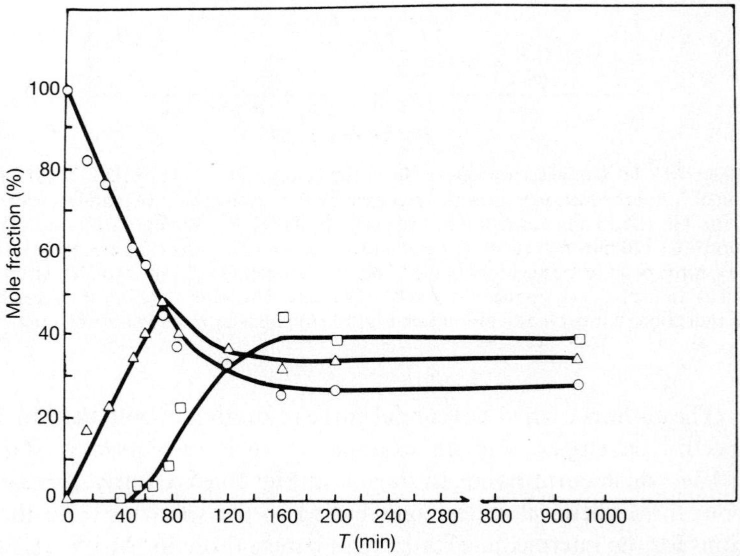

FIG. 10.37. Plot of the areas under each resolved component from the result of curve resolution of all envelopes, as a function of time of reduction in hydrogen at 500°C, carried out by Patterson *et al.*[59] ○, Mo(VI); △, Mo(V); □, Mo(IV). It can be seen that after about 3 h reduction an equilibrium state is reached in which the proportions of the three states are roughly equal.

could be plotted, and the result is shown in Fig. 10.37. After about 3 h of reduction in H_2 at 500°C an equilibrium state is reached in which the proportions of the three states are roughly equal. Tests with thiophene on the catalyst at various stages of treatment showed that firing alone (in air at 500°C for 4 h) did not activate the catalyst for desulphurization, whereas in the reduced condition it was able to desulphurize the thiophene, with the formation of MoS_2.

The material studied by Patterson *et al.* was of one particular composition, although there have been suggestions that the chemical states of both Co and Mo in the catalyst can be altered by varying the Co concentration. In order to examine this possibility, Makovsky *et al.*[60] prepared compositions in the range (0–8)% CoO, 15% MoO_3/Al_2O_3, and also reduced them in H_2 at 500°C. The resultant Co $2p_{3/2}$ and Mo 3d photoelectron spectra are shown in Fig. 10.38 for increasing loadings of CoO from 2 to 8%, after 5 h reduction.

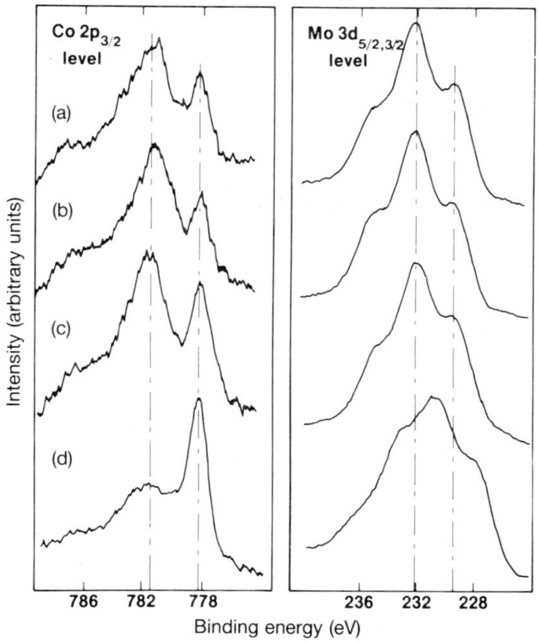

FIG. 10.38. Co 2p (left) and Mo 3d (right) photoelectron spectra from a series of $CoO–MoO_3–Al_2O_3$ catalysts after reduction in hydrogen for 5 h at 500°C, as a function of the CoO concentration, from Makovsky *et al.*[60] (a) 2% CoO, (b) 3% CoO, (c) 5% CoO, and (d) 8% CoO. Compare the Mo 3d spectrum for composition (b) with that at F in Fig. 10.35. In the Co $2p_{3/2}$ spectra the vertical lines correspond to the Co^{2+} and Co^0 states, respectively, while in the Mo 3d spectra the lines correspond to the Mo(IV) binding energies in the 3d doublet. There is little difference between the spectra until a CoO concentration of 8% is reached, when for both Co and Mo contributions from the metallic state appear.

The Mo 3d spectrum for the 3% CoO content can be seen to be identical with the bottom spectrum of Fig. 10.35. The vertical lines in the Mo 3d spectra correspond to the Mo(IV) binding energies; in the Co $2p_{3/2}$ spectra they correspond to Co^{2+} and Co metal, respectively. There is little significant difference between any of the spectra until the CoO loading is increased to 8%, whereupon in both spectra contributions from the metallic state appear, particularly prominently for Co. It is known that both Co_3O_4 and $CoMoO_4$ contain Co reducible to metal, and that $CoMO_4$ contains Mo reducible to metal, and evidence from both Raman spectroscopy and X-ray diffraction confirmed that at CoO loadings $\leq 5\%$ there was $CoMoO_4$ present on the surface, while for CoO at 8% Co_3O_4 was present as well. Measurements of the catalytic activity for desulphurization and for conversion of liquefied coal extract showed the synergistic effects of cobalt, since the optimum CoO concentration was $\approx 3\%$; thus the addition of a higher proportion of CoO is not advantageous since the surface regions are then reduced partly to metallic species which reduce the activity.

It is often possible to change the selectivity of a particular catalyst in terms of the nature of the reaction products or their relative proportions by the addition of a so-called 'promoter'. Good examples are the iron and cobalt used in the Fischer–Tropsch type synthesis, where potassium is always used as a promoter to increase the yield of high molecular weight hydrocarbons. The promotion mechanism is said to involve the enhancement of electron transfer from the iron or cobalt to adsorbed CO, which strengthens the metal–CO bond, but weakens the C–O bond, allowing both longer residence time of CO and a greater probability of CO dissociation. The mechanism has been studied on well characterized metal surfaces in a series of papers by Bonzel and co-workers. Their observations[61] on potassium-promoted iron surfaces are worth describing.

The iron surface was that of a polycrystalline foil, and it was cleaned by the usual procedure of cycles of ion bombardment followed by heating to a high temperature for annealing. Introduction of the promoter was achieved by removing the pre-cleaned specimen to the introduction chamber, covering it immediately with a drop of a solution of either KOH, KNO_2 or K_2CO_3, transferring the specimen to the adjacent turbo-molecularly pumped chamber where the water was pumped away, and then replacing it in the UHV chamber where it was heated first to 420 K and then to 690 K. XPS spectra were recorded at each stage, and those for the K_2CO_3-covered surface are shown in Fig. 10.39. In the as-deposited state (spectra a) the C 1s peak at ≈ 290 eV and the shoulder in the O 1s spectrum at about 532 eV are characteristic of a carbonate. Heating to 420 K does little to decompose the carbonate, but after heating to 690 K both the carbonate-associated features have disappeared. Subsequent reduction in a special microreactor system attached to the spectrometer, in 1 bar H_2 at 570 K completely reduced the Fe to the metallic

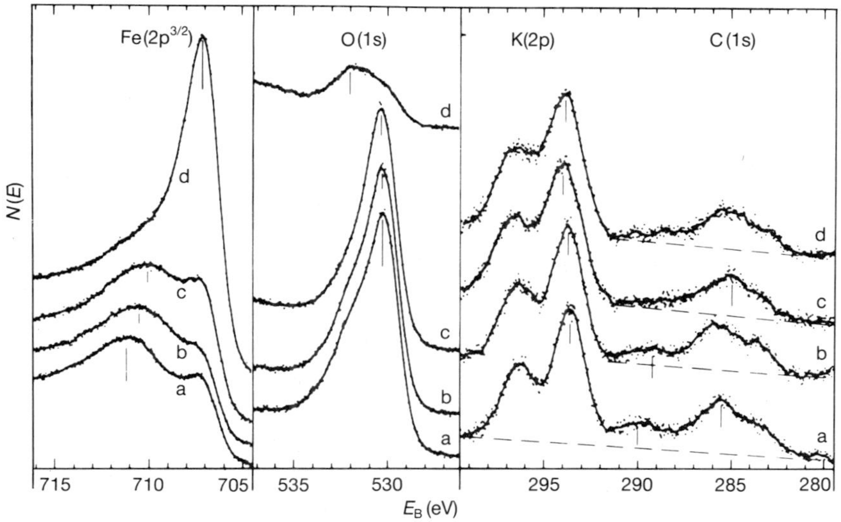

FIG. 10.39. Fe $2p_{3/2}$, O 1s, K 2p and C 1s photoelectron spectra recorded by Bonzel *et al.*[61] from K_2CO_3 deposited on iron foil, as a function of heat treatment. $h\nu = 1253.6$ eV. (a) as prepared; (b) heated in UHV to 420 K for 1 min; (c) heated in UHV to 690 K for 10 s; (d) reduced in H_2. In the as-deposited state a shoulder on the O 1s peak, and the carbon peak at ≈ 290 eV are characteristic of the carbonate. The carbonate has to be heated to 690 K to be decomposed. Reduction in hydrogen reduces the underlying Fe to the metallic state, and leaves about a monolayer of K on the surface. Residual O and C remain.

state, and left about 1 monolayer of K on the surface as judged from the relative intensities of the Fe and K signals.

When the K-promoted Fe foil prepared in the above way was then heated in the microreactor in a 1:20 mixture of CO and H_2 at 1 bar pressure and at 520–570 K, the resultant K 2p and C 1s spectra were those in Fig. 10.40a. For comparison, Fig. 10.40b gives the C 1s spectra for an unpromoted pure Fe foil. Several interesting features of the spectra in Fig. 10.40a should be noted; (a) the intensity of the K 2p region is almost independent of the carbon concentrations; (b) three separate C 1s peaks can be distinguished at 520 K, compared to one in the same condition from unpromoted Fe; (c) at the highest reaction temperature there is only one C 1s peak, in a position corresponding to graphitic carbon. If the surface after the 520 K treatment was flashed briefly to 720 K in vacuum the high binding energy components were removed.

Thus at the lower reaction temperature three distinct forms of adsorbed carbon are present, and the peak positions indicate that the forms are adsorbed carbonate, adsorbed CO, and atomic carbon. The proposed structural surface models are shown in Fig. 10.41. Because the higher binding energy components associated with carbonate and CO do not appear on the unpromoted Fe it is clear that the promotion by K has increased the adsorption energy of CO, leading to high CO coverages. The latter then allow

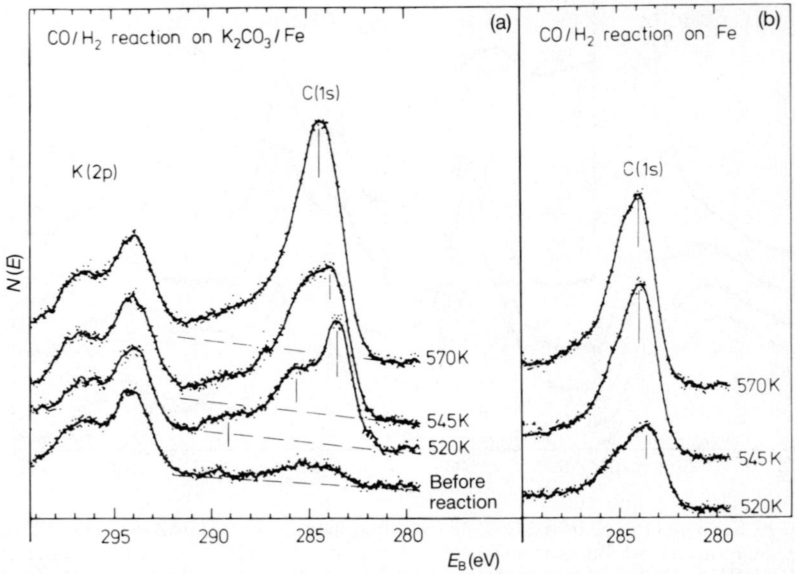

FIG. 10.40. (a) K 2p and C 1s photoelectron spectra from the K-promoted Fe foil, and (b) C 1s photoelectron spectra from unpromoted pure Fe foil, recorded by Bonzel *et al.*[61] The temperatures refer to heating in a 1:20 mixture of CO and H_2 at 1 bar pressure for 10 min. A specially constructed microreactor was used for the high pressure. The intensity of the K peak is almost constant with treatment. At 520 K three C 1s peaks can be identified from the promoted surface compared to only one on the unpromoted.

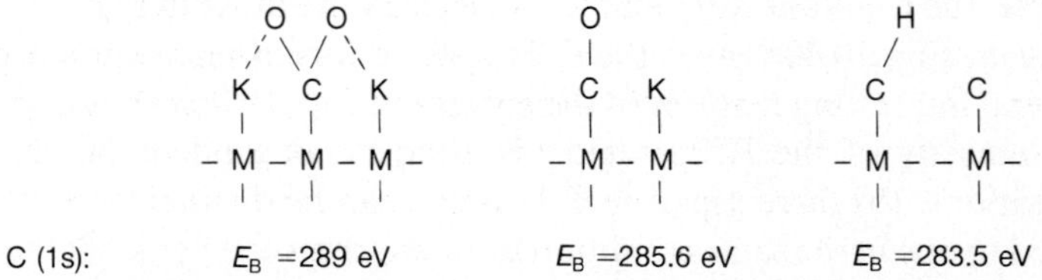

FIG. 10.41. Structural models proposed by Bonzel *et al.*[61] to account for the three C 1s peaks observed in Fig. 10.40 at 520 K from the promoted Fe surface. The corresponding binding energies are shown.

formation of an adsorbed carbonate, which decomposes above 700 K as expected. The independence of the K concentration on the carbon overlayer thickness (up to ≈ 50 monolayers by varying the CO/H_2 ratio) indicates that the K promoter remains always on top of the carbon, i.e. it 'floats', probably by diffusion through the graphitic layer via a transient state. Very similar results were obtained for the KOH and KNO_2 covered surfaces.

10.7.3 *Adhesion*

Increasingly in many industries, e.g. automobile and aircraft manufacture, are materials bonded together not by welding but by adhesion. The adhesive materials are polymeric and are usually complex mixtures of several chemical compounds including a variety of additives intended to increase joint strength and durability; a particular additive or group of additives is chosen on a selective basis for a particular type of joint. Clearly the characteristics of an adhesive joint depend largely on the chemistry of the adhesive/metal interface, and the initial contact and interaction of the two materials is very much a surface reaction. However, it is impossible for any of the surface specific techniques being described here to analyse such an interface without a method of exposure, and so it is normal practice for the joint to be ruptured in a small test machine, followed by analysis of each side of the fracture path. Fracture would occur along the weakest plane in the joint. It is very rare for fracture to be entirely within the adhesive material itself, and it normally follows a path close to the metal surface.

Adhesive joints formed by polysiloxane-based compounds with the metals Ag, Ni and Cr were studied by Torrisi *et al.*[62] The metal surfaces were prepared electrolytically, but no other special treatment was given, so that they did in fact represent 'real' surfaces, i.e. covered with oxide films to a greater or lesser degree before any contact with the adhesive. Commercial polysiloxane resins were used, being moulded to the metals in normal ambient conditions; no environmental tests were carried out at that stage. As stated above, the joints were pulled apart in a pull test machine, which of course measured the strength of the joint. To the naked eye the metal side of the interface after fracture appeared to be bare metal, but XPS analysis showed that it was in every case covered with a very thin polymeric layer.

Examination of each side of the interfaces produced on the three adhesive/metal systems showed marked chemical differences. An easy-to-visualize picture of the situation on each side of the Ni–resin interface is given in Fig. 10.42. The comparison shows that on the metal side (a) the shake-up satellite near the C 1s peak associated with the phenyl ring had disappeared, (b) there were additional peaks at higher binding energy near the C 1s peak, associated with –COO– and –CO– species, (c) a pronounced Pb 4f doublet was present, (d) there was more Ca and less Si, and (e) a small Ni signal was present. The presence of the Ni indicates that the film thickness on the metal was only 1–2 nm. The decrease in Si concentration coupled with the disappearance of the phenyl group shake-up feature suggest that the layer of resin immediately adjacent to the metal contained very little polysiloxane, but some SiO_2, which is added to the raw resin as a filler. Similarly, the oxidized C species, not present in the spectrum from the bulk resin itself, indicate a breakdown of the polysiloxane structure at the metal surface followed by

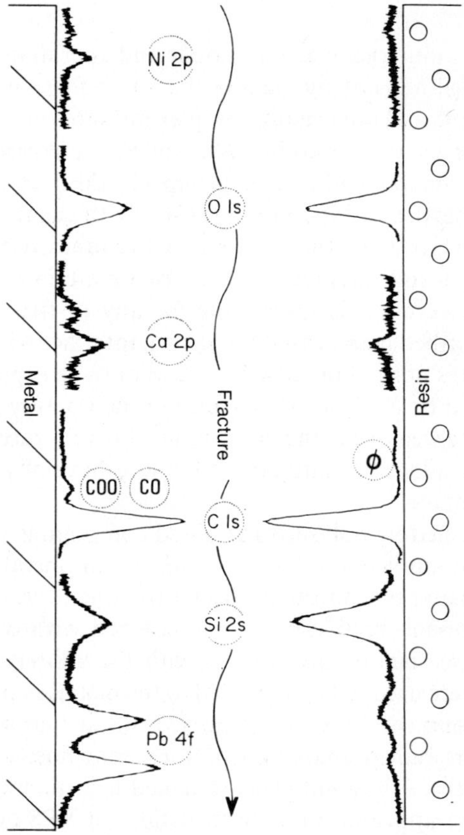

FIG. 10.42. Illustration of the chemical differences established by XPS on the two sides of an adhesive nickel–resin joint fractured by pulling, according to Torrisi *et al.*[62] Comparison of the two interfaces shows that on the metal side (a) a small amount of nickel is present, (b) rather more calcium is present than on the resin side, (c) there are C 1s peaks at binding energies corresponding to –COO– and –CO–, and (d) there is a pronounced Pb 4f doublet present. On the resin side there are (a) a shake-up satellite due to a phenyl ring, and (b) more Si than on the metal side.

interaction of C with the oxide species originally present before moulding. Ca and Pb come from a releasing agent and a polymerization catalyst, respectively. It seems that Pb is particularly important in deciding the strength of the joint. If the Si/Pb intensity ratio is plotted against the measured joint strengths, it can be seen from Fig. 10.43 that a relatively high Pb level at the interface favoured high joint strengths. However, that was not the only factor, since the degree of segregation of Pb and Ca to the interfaces on the three metals was markedly different, as was the thickness of the residual resin film on the metal. Based on the XPS observations, the authors were able to rank the behaviour of the metals in the pictorial manner of Fig. 10.44, where Cu has

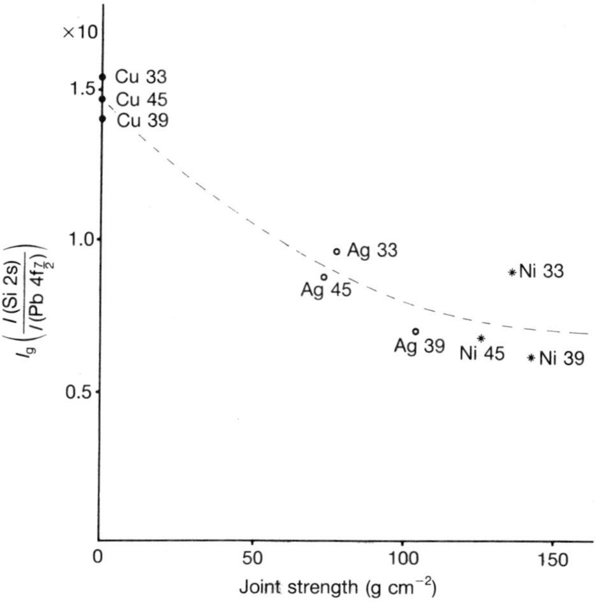

FIG. 10.43. Plot of the Si/Pb intensities in XPS against the metal–resin joint strength, by Torrisi *et al.*[62] The numbers against each of the metals used in the joints are the spiral flows, which are measures of the viscosity, i.e. the lower the flow the higher the viscosity. The plot shows that a level of Pb high relative to that of Si favours high joint strengths. The presence of the Pb is due to a releasing agent incorporated in the adhesive formulation.

relatively high resin coverage, but little additive segregation, through Ni as an intermediate situation, to Ag where the coverage is slight but segregation high.

10.7.4 *Surface reactions*

As remarked earlier XPS has been used extensively in basic studies of reactions and interactions on well characterized surfaces. In fact the general impression formed from the published literature is that it is in this area that the major application of the technique has been made, despite the conviction among the early workers that it would make its greatest impact in catalysis. The choice of examples is here therefore even more difficult; those below are selected merely to show some typical approaches used in the basic studies.

Amongst the many hundreds of publications on surface reactions probably the largest section has been concerned with the oxidation mechanisms of the first row transition elements from scandium to copper. Again, of these most effort has understandably been devoted to iron, chromium and nickel. A typical approach using XPS, UPS and work function measurement was that of Chesters and Rivière[63] in studying the oxidation of an Fe(110) surface at room temperature. After the usual considerable length of time spent in producing an

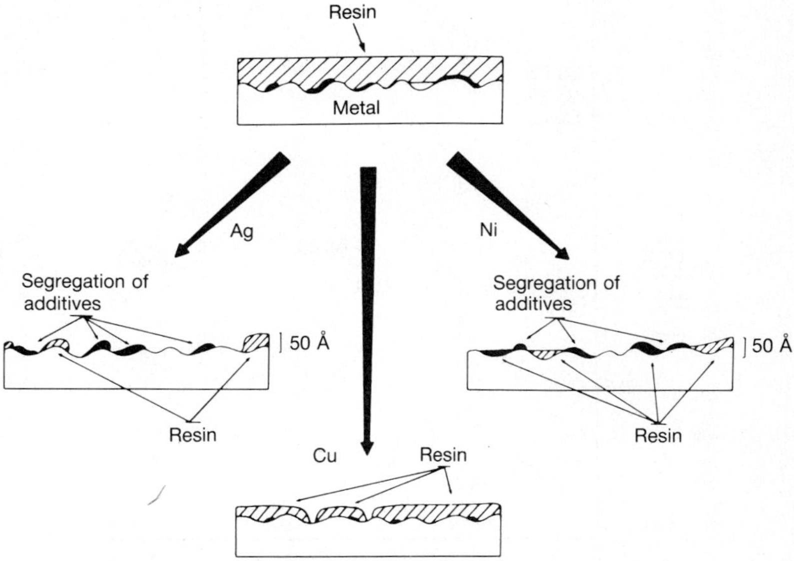

FIG. 10.44. Schematic of ranking of behaviour of the three metals Ag, Ni, and Cu, based on XPS analysis of the metal–resin adhesive joint fracture surfaces. Additives refer to inorganic additives and resin to polysiloxanes and carbonaceous residues. Cu has high resin coverage but little additive segregation, while Ag and Ni both show a high level of additional migration. From Torrisi *et al.*[62]

atomically clean surface on a transition metal by many cycles of ion bombardment and annealing (below the α–γ transition temperature!), oxygen was admitted to the crystal in short carefully controlled doses of 0.5 L at a time. The course of the oxidation was followed by monitoring the O 1s, Fe 2p, and Fe 3p spectra as a function of exposure as far as the point at which no further changes appeared to be taking place, i.e. up to 750 L. Figure 10.45 shows the changes in the O 1s photoelectron peak as a function of exposure in the earliest stages; there was a slight change in peak position to 955.0 eV at exposures beyond 300 L. The parallel changes in the Fe $2p_{3/2}$ spectrum are shown in Fig. 10.46. Two principal effects occurred, (a) a gradual reduction, but not complete disappearance, of the peak in the metallic position at 779.2 eV kinetic energy, and (b) the emergence of a broad peak centred around 775 eV. It was found that the leading edge of the metallic contribution to the spectrum at each exposure had exactly the same shape as that of the clean spectrum, which meant that the latter could be scaled down exactly and then subtracted to produce a difference spectrum. The difference spectra so produced showed only a single shifted peak near 775 eV, the energetic position corresponding to Fe^{3+}.

Since the O 1s peaks were always situated on a flat background, it was an easy

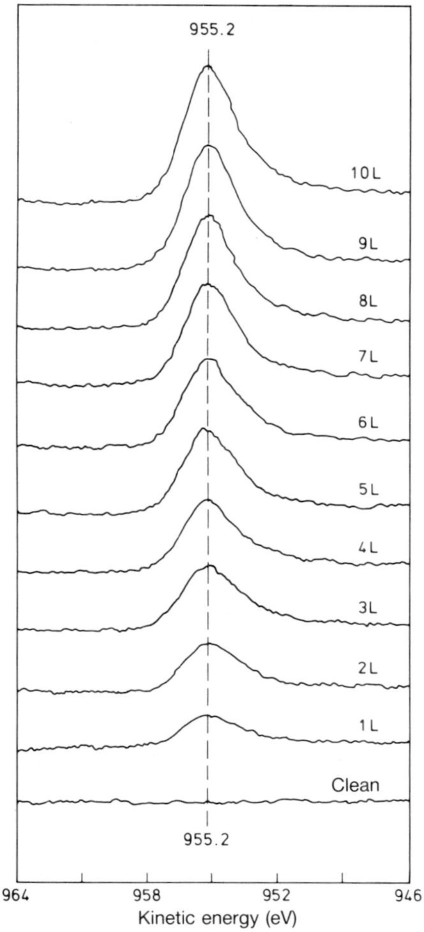

FIG. 10.45. Changes in the O 1s spectrum as a function of exposure to oxygen of Fe(110) at room temperature, recorded by Chesters and Rivière.[63] A single peak appears and grows, eventually saturating in area at exposures greater than 750 L. At low exposures, as in the figure, no change in peak position was observed, but eventually a slight shift of 0.2 eV to 955.0 eV was noted.

matter to measure the area under the peak as a function of exposure, the result of which is shown in the lower half of Fig. 10.47 as far as an exposure of 50 L. Also shown in that figure in the upper half is the variation of work function of the Fe(110) surface during the same exposure. The O 1s peak area vs. exposure plot shows a behaviour which has been found in several other cases of oxidation of transition metals, either in single crystal or in polycrystalline forms. There is an initial straight section, up to an exposure of ≈ 4 L, followed by a kink, and then another straight section from ≈ 10 L to ≈ 20 L before

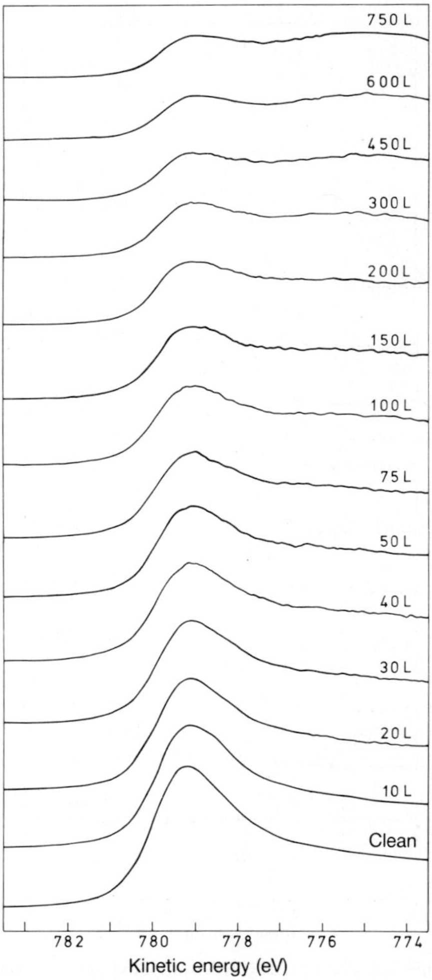

FIG. 10.46. Changes in the Fe $2p_{3/2}$ spectrum as a function of exposure to oxygen of Fe(110) at room temperature, recorded by Chesters and Rivière.[63] With increasing exposure the peak in the metallic position at 779.2 eV decreased progressively in size, but never disappeared, and another peak centred at ≈ 775 eV appeared and grew. The peak corresponding to the metallic state retained the same line shape throughout the exposure. The peak at ≈ 775 eV corresponds to the Fe^{3+} state of iron.

levelling off to saturation. The work function variation follows this behaviour by at first increasing up to ≈ 6 L, i.e. along the initial linear oxygen uptake, to the kink in the uptake plot, then decreasing again along the second linear section up to ~ 20 L, and finally going below the initial value of the clean surface, to saturate at an exposure of ≈ 35 L.

The mechanisms put forward to explain the above observations are as

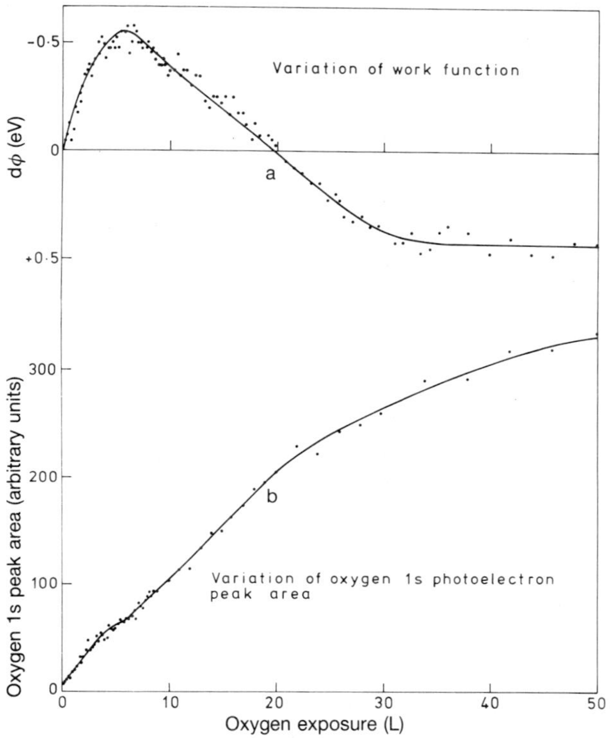

FIG. 10.47. a: Variation of area under the O 1s photoelectron peak as a function of exposure of Fe(110) to oxygen at room temperature, from the work of Chesters and Rivière.[63] b: Variation in work function of the Fe(110) surface under the same conditions, where the zero line corresponds to the work function of the clean surface. Negative values indicate an increase in work function, positive a decrease.

The kink in the oxygen uptake curve is due to nucleation of oxide islands from the initially chemisorbed oxygen layer. During chemisorption on the external surface the work function increases, but during incorporation of oxygen as a result of nucleation and oxide growth, the work function decreases since the oxygen is then situated internally.

follows. The initial strictly linear uptake is due to chemisorption following dissociation of O_2, on the external surface, leading to an increase in work function since oxygen is electronegative. When the concentration of chemisorbed oxygen is high enough, probably half a monolayer, the surface layer becomes unstable with respect to the formation of oxide, and nucleation of the oxide takes place, giving rise to the kink in the uptake curve at ≈ 5 L. This nucleation involves a place exchange at the surface in which the oxygen then finds itself in sites just *below* the surface; therefore the work function starts to decrease since the electronegative species below the surface induces an electropositive double layer at the outer surface. Beyond 10 L the oxygen that is taken up continues to pass through the outer surface into the sub-surface layers, with the result that the work function continues to decrease. Eventually the oxygen O 1s peak area saturates. This could be due to either of two effects, genuine cessation of uptake, or a continuation of uptake that puts the inwardly diffusing oxygen beyond the escape depth of the O 1s electrons, so that the additional oxygen would not be detected. The fact that there is even at saturation a metallic contribution to the Fe $2p_{3/2}$ spectrum indicates that the underlying unoxidized metal is still visible and therefore that the oxide thickness has indeed saturated. The UPS spectra taken with HeII radiation confirmed that there was still a remnant of the Fe 3d valence band remaining after 750 L exposure. The single unshifting position of the O 1s peak and the appearance of only one shifted peak in the Fe $2p_{3/2}$ spectrum indicates that the reaction at room temperature between oxygen and the Fe(110) surface leads at once to the Fe^{3+} state and that the oxide formed after nucleation is Fe_2O_3.

Other interesting surface reactions have been studied on semiconductor surfaces. Compound semiconductors are invariably grown as single-crystals by molecular beam epitaxy, indeed it is necessary in general for the material to

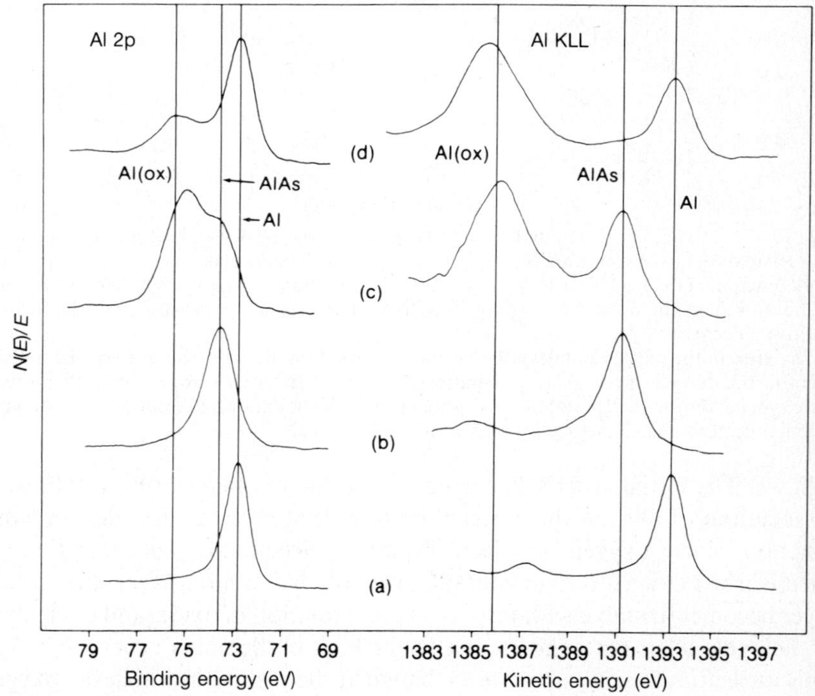

FIG. 10.48. The Al 2p photoelectron and Al KLL Auger spectra from (a) clean metallic Al, (b) clean AlAs, (c) thin oxide layer on AlAs, and (d) thin oxide layer on Al, measured by Taylor.[64] The film of AlAs had been passivated by formation of a thin oxide layer for transfer from one system to another for analysis. The oxide layer was then removed by ion bombardment. The spectrum of the sputtered AlAs film in (b) shows no features due to metallic Al, and no trace of residual oxide, indicating that the passivation treatment had been successful.

be of high crystalline quality for its particular application, so that semiconductor technology is probably the only field in which basic experiments on single crystals can be taken as directly relevant to what happens in practice. Manufacture of devices nearly always requires the production or deposition of an insulating layer at some stage, but on some semiconductors it has been found quite difficult to produce the required layers by direct oxidation. This is true, for example, of GaAs, and much effort has been expended over the years in attempts to form a stable insulating layer on the compound. One approach that has been tried is to grow a layer of AlAs on top of the GaAs, which is then thermally oxidized. The nature of the resultant oxide has been studied by Taylor,[64] using XPS.

On top of a GaAs substrate previously grown by MBE was grown a further layer of AlAs about 150–300 nm thick, also by MBE. Since the MBE and surface analysis systems were separate, and no interference from atmospheric contamination could be tolerated, the AlAs film was passivated after growth by heating in dry O_2 at 200°C. This treatment forms a very thin film of Al_2O_3 which is stable for long enough to transfer the sample from one system to another; once inside the spectrometer the thin oxide film was removed by ion

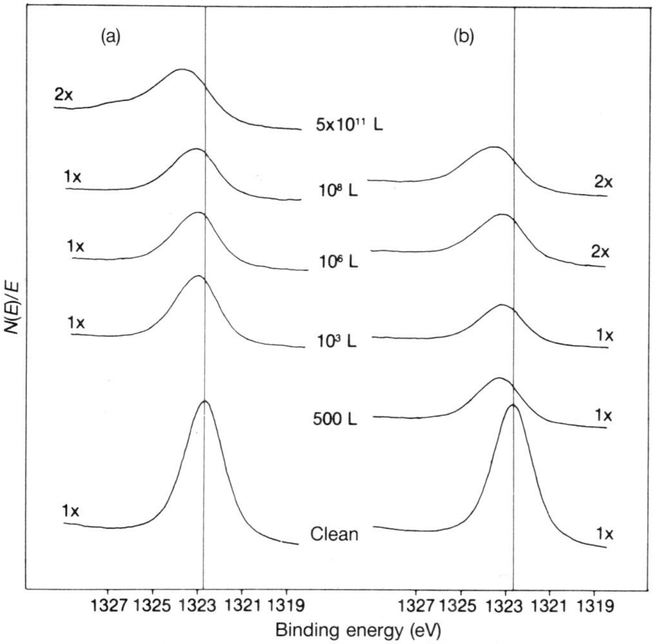

FIG. 10.49. The As 2p photoelectron spectra from an AlAs film exposed to (a) oxygen and (b) water vapour at room temperature, recorded by Taylor.[64] No new peak appears, but a progressive shift to higher binding energy occurs, together with a loss in intensity. The gradual shift is interpreted as being due to a change from the chemical state of As in AlAs to that in free As. This arises from the oxidation of the Al in the AlAs (see Fig. 10.50).

bombardment. Oxidation was then carried out in either O_2 or H_2O in a separate reaction chamber in which high pressures could be used.

Before oxidation was commenced, it had first to be established that the bombardment to remove the passive oxide film had not altered the surface composition or chemistry appreciably. The Al 2p and KLL Auger spectra in Fig. 10.48 confirm that no such alteration had occurred. The peaks for clean AlAs (i.e. after sputtering) at (b) show no shoulders due to metallic Al, since the Al positions in the element and in AlAs are sufficiently far apart (0.8 eV) for such shoulders to have been obvious if present. The oxide peak positions are shifted even further.

The effects of oxidation in O_2 and in H_2O on the As and Al 2p spectra are shown in Figs 10.49 and 10.50, respectively. In the case of As, the 2p line shows no appearance of an additional peak, but merely a progressive shift to higher binding energy of ~1 keV, coupled with a marked loss in intensity. Only at the very highest exposure, 5×10^{11} L of O_2, is there a suggestion of an extra peak at higher energy. On the other hand, the Al spectrum develops an additional peak at some 1.4 eV higher binding energy with a contribution from Al in AlAs

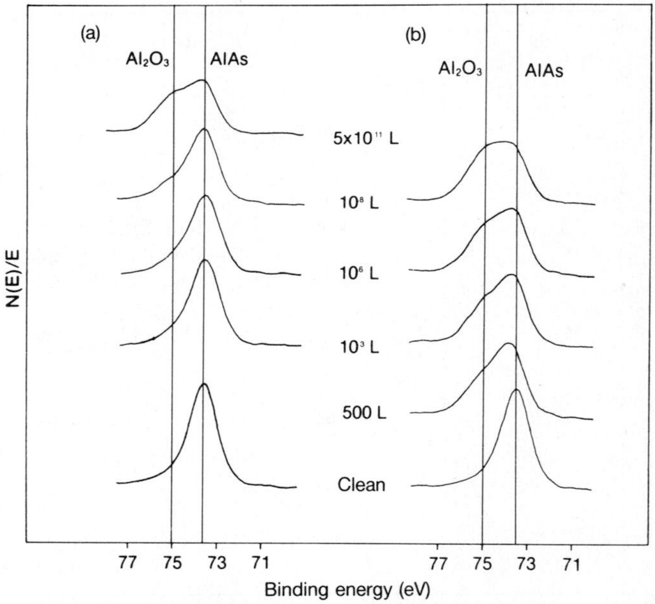

FIG. 10.50. The Al 2p photoelectron spectra from an AlAs film exposed to (a) oxygen and (b) water vapour at room temperature, recorded by Taylor.[64] A new peak develops at higher binding energy, the shift being the same for O_2 or H_2O exposure. Even at saturation of O_2 or H_2O the contribution from Al in AlAs remained a major component. The measured Auger Parameter was identical to that for an oxidized Al foil, although lower than in Al_2O_3. Thus the oxidation of AlAs produces predominantly oxidized Al, the oxide being defective since its state is not quite that of Al_2O_3, with As being released as free As.

remaining even at the highest exposures. The shifts were the same in O_2 or in H_2O. The measured Auger Parameter was identical to that in oxidized Al foil, but lower than from Al_2O_3 in the forms either of sapphire or corundum. For As the Auger Parameters indicated As as being partly in a state close to that in As_2O_3 and partly similar to that in As^0. The progressive shift seen in Fig. 10.49 is due to the change from As in AlAs to free As.

Thus the findings from XPS are that the oxide film is predominantly oxidized Al, in a state not quite that of Al in crystalline Al_2O_3, so that the film is certainly defective. Due to oxidation of the Al, free As is formed, which eventually at very high exposures also oxidizes; the free As, and subsequently the small amount of oxidized As, are mixed with the oxidized Al in the film.

References

1. Sherwood, P. M. A. (1983). *Practical surface analysis* (eds. D. Briggs and M. P. Seah), p. 445. John Wiley & Sons (New York).
2. Carley, A. F. and Joyner, R. W. (1979). *J. Electron Spectrosc.* **16**, 1.
3. Savitsky, A. and Golay, M. J. E. (1964). *Anal. Chem.* **36**, 1627.
4. Shirley, D. A. (1972). *Phys. Rev. B* **5**, 4709.
5. Bishop, H. E. (1981). *Surf. Interface Anal.* **3**, 272.
6. Wagner, C. D. (1975). *Faraday Discuss. Chem. Soc.* **60**, 291.
7. Wagner, C. D., Gale, L. H., and Raymond, R. H. (1979). *Anal. Chem.* **51**, 466.
8. West, R. H. and Castle, J. E. (1982). *Surf. Interface Anal.* **4**, 68.
9. Edgell, M. J., Paynter, R. W., and Castle, J. E. (1985). *J. Electron Spectrosc.* **37**, 241.
10. Yates, K. and West, R. H. (1983). *Surf. Interface Anal.* **5**, 133.
11. Aksela, H. and Aksela, S. (1974). *J. Phys. B: Atom. Mol. Phys.* **7**, 1262.
12. Keast, D. J. and Downing, K. S. (1981). *Surf. Interface Anal.* **3**, 99.
13. Yates, K. and West, R. H. (1985). *Spectrochim. Acta* **40B**.
14. Cazaux, J. (1973). *Rev. Phys. Appl.* **8**, 371. Cazaux, J. (1975). *Rev. Phys. Appl.* **10**, 263.
15. Hovland, C. J. (1977). *Appl. Phys. Lett.* **30**, 274.
16. Beamson, G., Porter, H. Q., and Turner, D. W. (1981). *Nature* **290**, 556.
17. Gurker, N., Ebel, M. F., and Ebel, H. (1983). *Surf. Interface Anal.* **5**, 13.
18. Chaney, R. L. (1987), *Surf. Interface Anal.* **10**, 36.
19. Barrie, A. (1973). *Chem. Phys. Lett.* **19**, 109.
20. Scrocco, M. (1979). *Chem. Phys. Lett.* **63**, 52.
21. Wertheim, G. K., Buchanan, D. N. E., Smith, N. V., and Traum, M. M. (1974). *Phys. Lett.* **49A**, 191.
22. Campagna, M., Wertheim, G. K., and Baer, Y. (1979). *Photoemission in Solids II*, (eds. L. Ley and M. Cardona), Chapter 4, p. 234. Springer-Verlag, Berlin.
23. e.g. Weightman, P. and Andrews, P. T. (1979). *J. Phys. C: Solid State Phys.* **12**, 943.
24. Wittberg, T. N. and Wang, P. S. (1983). *J. Electron Spectrosc.* **31**, 81.
25. Mahan, G. D. (1967). *Phys. Rev.* **163**, 612.

26. Nozières, P. and de Dominicis, C. T. (1969). *Phys. Rev.* **178**, 1097.
27. Doniach, S. and Šunjić, M. (1970). *J. Phys. C: Solid State Phys.* **3**, 285.
28. Wertheim, G. K. and Hüfner, S. (1975). *Phys. Rev. Lett.* **34**, 53.
29. Siegbahn, K., Nordling, C., Johansson, G., Hedman, J., Heden, P. F., Hamrin, K., Gelius, U., Bergmark, T., Werne, L. O., Manne, R., and Baer, Y. (1969). *ESCA applied to free molecules.* North-Holland, Amsterdam.
30. Gelius, U. (1974). *Phys. Scr.* **9**, 133.
31. Clark, D. T. and Adams, D. B. (1972). *J. Chem. Soc., Faraday Trans. II*, **68**, 1819.
32. Fadley, C. S. (1978). *Electron spectroscopy: theory, techniques and applications*, Vol. 2, Chapter 1 (eds. C. R. Bundle and A. D. Baker). Academic Press, London.
33. Scofield, J. H. (1976). *J. Electron Spectrosc.* **8**, 129.
34. Reilman, R. F., Msezane, A., and Manson, S. T. (1976). *J. Electron Spectrosc.* **8**, 389.
35. Seah, M. P. (1983). *Practical surface analysis* (eds. D. Briggs and M. P. Seah), p. 207. John Wiley & Sons, Chichester.
36. Wagner, C. D., Davis, L. E., Zeller, M. V., Taylor, J. A., Raymond, R. M., and Gale, L. H. (1981). *Surface interface analysis*, **3**, 211; reproduced in *Practical surface analysis* (eds. D. Briggs and M. P. Seah), p. 511. John Wiley & Sons, Chichester.
37. Seah, M. P. (1980). *Surf. Interface Anal.* **2**, 222.
38. Wagner, C. D. (1978). *J. Vac. Sci. Technol.* **15**, 518.
39. Castle, J. E. and West, R. H. (1979). *J. Electron Spectrosc.* **16**, 195.
40. Wagner, C. D. and Taylor, J. A. (1980). *J. Electron Spectrosc.* **20**, 83.
41. Taylor, J. A. (1981). *Applications Surf. Sci.* **7**, 168.
42. Castle, J. E., Hazell, L. B., and Whitehead, R. D. (1976). *J. Electron Spectrosc.* **9**, 247.
43. Yin, L., Adler, I., Tsang, T., Chen, M. H., and Crasemann, B. (1973). *Phys. Lett.* **46A**, 113.
44. Rojo, J. M. and Baro, A. M. (1976). *J. Phys. C: Solid State Phys.* **9**, L543.
45. Bassett, P. J., Gallon, T. E., Matthew, J. A. D., and Prutton, M. (1973). *Surf. Sci.* **35**, 63.
46. Powell, C. J. (1973). *Phys. Rev. Lett.* **30**, 1179.
47. Roberts, E. D., Weightman, P., and Johnson, C. E. (1975). *J. Phys. C: Solid State Phys.* **8**, L301.
48. Aksela, S., Aksela, H., Vuontisjärvi, M., Väyrynen, J., and Lähteenkorva, E. (1977). *J. Electron Spectrosc.* **11**, 137.
49. Cini, M. (1976). *Solid State Commun.* **20**, 605.
50. Cini, M. (1977). *Solid State Commun.* **24**, 681.
51. Sawatzky, G. A. (1977). *Phys. Rev. Lett.* **39**, 504.
52. Parry-Jones, A. C., Weightman, P., and Andrews, P. T. (1979). *J. Phys. C: Solid State Phys.* **12**, 1587.
53. Cini, M. (1978). *Phys. Rev. B* **17**, 2486.
54. Cini, M. (1979). *Surf. Sci.* **87**, 483.
55. Sawatzky, G. A. and Lenselink, A. (1980). *Phys. Rev. B* **21**, 1790.
56. Tyler, J. W. (1985). *Oxidation of metals* **24**, 149.
57. Brüesch, P. Müller, K., Atrens, A., and Neff, H. (1985). *Appl. Phys. A* **38**, 1.
58. Kelly, R. (1980). *Surf. Sci.* **100**, 85.
59. Patterson, T. A., Carver, J. C., Leyden, D. E., and Hercules, D. M. (1976). *J. Phys. Chem.* **80**, 1700.

60. Makovsky, L. E., Stencel, J. M., Brown, F. R., Tischer, R. E., and Pollack, S. S. (1984). *J. Catalysis* **89**, 334.
61. Bonzel, H. P., Brodén, G., and Krebs, H. J. (1983). *Applic. Surf. Sci.* **16**, 373.
62. Torrisi, A., Marletta, G., Puglisi, O., and Pignataro, S. (1983). *Surf. Interface Anal.* **5**, 161.
63. Chesters, M. A. and Rivière, J. C. (1977). *Proc. 7th Int. Vac. Cong., 3rd ICSS*, p. 873. Vienna.
64. Taylor, J. A. (1982). *J. Vac. Sci. Technol.* **20**, 751.

11

Photon excitation: ultra-violet photoelectron spectroscopy (UPS) and synchrotron radiation photoelectron spectroscopy (SRPS)

11.1 Operation

11.1.1 *UPS*

Ultra-violet radiation from a source of the type described under section 3.2.2 and illustrated in Fig. 3.6 illuminates the surface of the sample. The most commonly used line sources are those of HeI at 21.21 eV and of HeII at 40.82 eV, produced by a discharge in helium gas, but NeI (16.8 eV) and NeII (26.9 eV) are also found occasionally. Just as with XPS, the energies of the photoelectrons ejected from the sample are measured in a CHA. Since the photoelectron energies in UPS are so low and span such a narrow range compared to XPS, elemental quantification is impossible, and there is no such thing as a survey scan at only moderate energy resolution for purposes of quantification. All scans cover the whole kinetic energy range (i.e. from $\sim$22 eV to zero for HeI, and from $\sim$42 to $\sim$22 eV for HeII) and are recorded at the best possible energy resolution compatible with good signal-to-noise ratio in the spectrum. For similar reasons it it usual to acquire the spectrum in steps of no more than 0.1 eV.

The same dedicated micro computer that drives the XPS acquisition will also drive that of UPS, since it is merely a matter of setting different operating parameters in the control unit for the two techniques. However, it is not usual for the UV lamp operating conditions to be set by the computer since operation of the discharge depends not only on the voltage applied to the discharge tube, but also on the gas pressure, which is not so easy to control automatically. Once a lamp has 'struck', it is normal for it to be left switched on throughout the course of a series of experiments, for many hours if necessary. As pointed out under 3.2.2, the partial pressures of helium in the system during operation of the lamp are too low to interfere with any gas adsorption studies.

Compared to XPS, UPS does not require much in the way of data processing facilities. The line widths of the resonant lines from the He or Ne discharges are so narrow ($\sim$20 meV) that there is negligible contribution to instrumental broadening from that direction. In addition, any satellites that do occur are only a very small proportion of the intensity of the resonant line.

Thus the instrumental de-broadening and satellite subtraction operations described for XPS are not needed here. Similarly, since quantification is not appropriate there is very little call for inelastic background subtraction or peak area measurement. Occasionally smoothing is used, but generally speaking is unnecessary since the high photon flux from the gaseous discharge sources invariably gives good statistics in the spectra.

The operation that is used very widely, probably more so than in XPS, is that of spectrum subtraction, resulting in the so-called difference curves. These are very useful when a small coverage of some adsorbate is adsorbed on a surface, for the additional features or bands due to the adsorbate can be accentuated and if required amplified for comparison, for example, with energy levels in the gaseous adsorbate. Where there is extra intensity due to adsorption, the difference spectrum will show a positive-going feature, and where intensity has been lost from the original or base spectrum, by removal of electrons from certain bands of the adsorbent, the difference spectrum will show a negative-going portion. An example of the use of difference spectra is given in Fig 11.1 for the adsorption of oxygen on Fe(110) at room temperature, in which the 'clean' spectrum at the bottom of the left-hand panel has been subtracted from each of the others. The resultant difference spectra on the right reveal both the extraction of electrons from the metal d-band (after an exposure of 1 L) and the additional band at $\sim$35 eV kinetic energy or $\sim$6 eV binding energy, taking E_F as the zero of energy.

A most important variation of UPS is angle-resolved UPS, or ARUPS. The reasons why it is important will be discussed in the next section on theory and spectral interpretation, but suffice it to say here that it is one of the principal techniques for mapping band structure in the surface region. The most popular method of recording angle-resolved spectra is to mount a CHA (of smaller radius than those used normally for UPS and XPS) on a mechanical system similar in operation to a goniometer, that allows the analyser to traverse the space around the sample in both polar and azimuthal directions, while the angle of incidence of radiation on the sample surface is kept fixed. Rotation of the sample on its manipulator (e.g. Fig. 3.25) allows the angle of incidence to be changed. A schematic of an angle-resolved system is shown in Fig. 11.2. With such a system it is perfectly possible to perform angle-resolved XPS and AES as well if the appropriate sources are positioned in the right places.

11.1.2 *SRPS*

Since SRPS is basically an extension of XPS and UPS using a photon source of variable energy, its operation as far as data collection is concerned is identical to that of the other two techniques. Indeed the spectrometers attached to synchrotron rings for this purpose are the same as those used as stand-alone instruments for UPS and ARUPS. The major difference in operation is

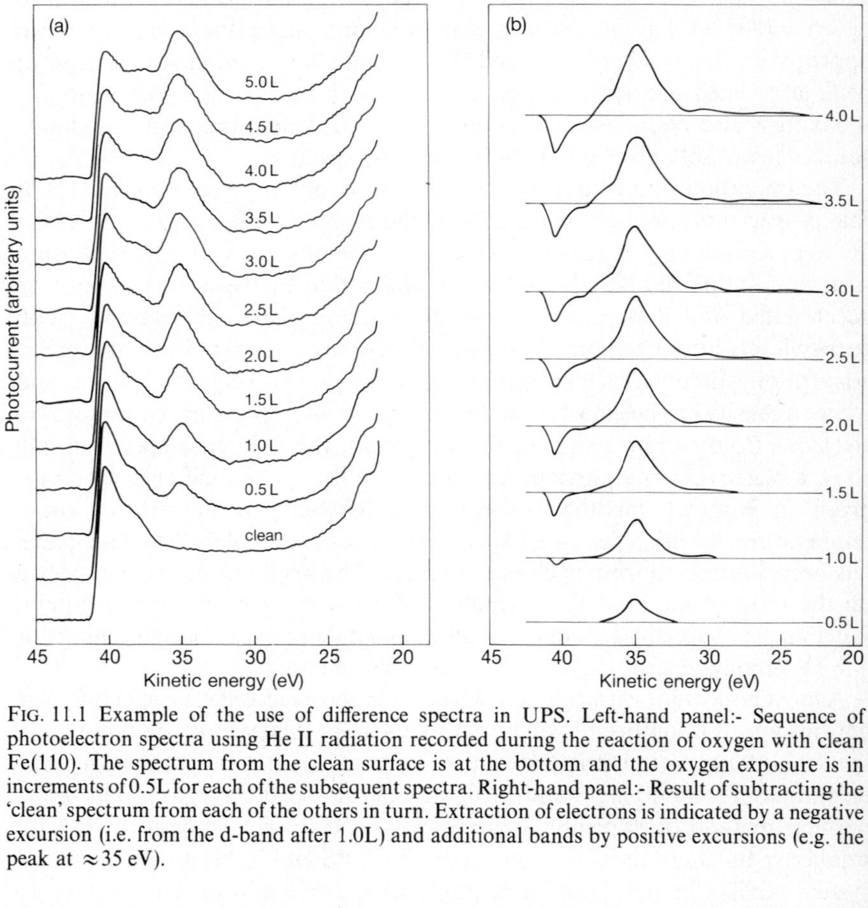

FIG. 11.1 Example of the use of difference spectra in UPS. Left-hand panel:- Sequence of photoelectron spectra using He II radiation recorded during the reaction of oxygen with clean Fe(110). The spectrum from the clean surface is at the bottom and the oxygen exposure is in increments of 0.5L for each of the subsequent spectra. Right-hand panel:- Result of subtracting the 'clean' spectrum from each of the others in turn. Extraction of electrons is indicated by a negative excursion (i.e. from the d-band after 1.0L) and additional bands by positive excursions (e.g. the peak at $\approx$35 eV).

obviously that the spectroscopist does not 'switch on' the radiation in the same sense that he would an X-ray line source or a UV resonant lamp. When the synchrotron is running properly, the synchrotron radiation is present all the time, and to use it means opening a shutter in a flight tube to allow it to pass out to the experiment. On the way to the electron spectrometer the photon energy required must be selected from the continuous spectrum of energies (see Fig. 3.10) by passage through a monochromator. Extension of the energy range available from monochromators is currently one of the lines of rapid development of the technique; description of the various types of monochromator is outside the scope of this volume, and the reader is referred to the by now voluminous literature on the subject.

It should be mentioned in passing that there are two properties of synchrotron radiation that make it unique (apart from the continuous tunability) and these are that it is pulsed and that it is 100% plane polarized.

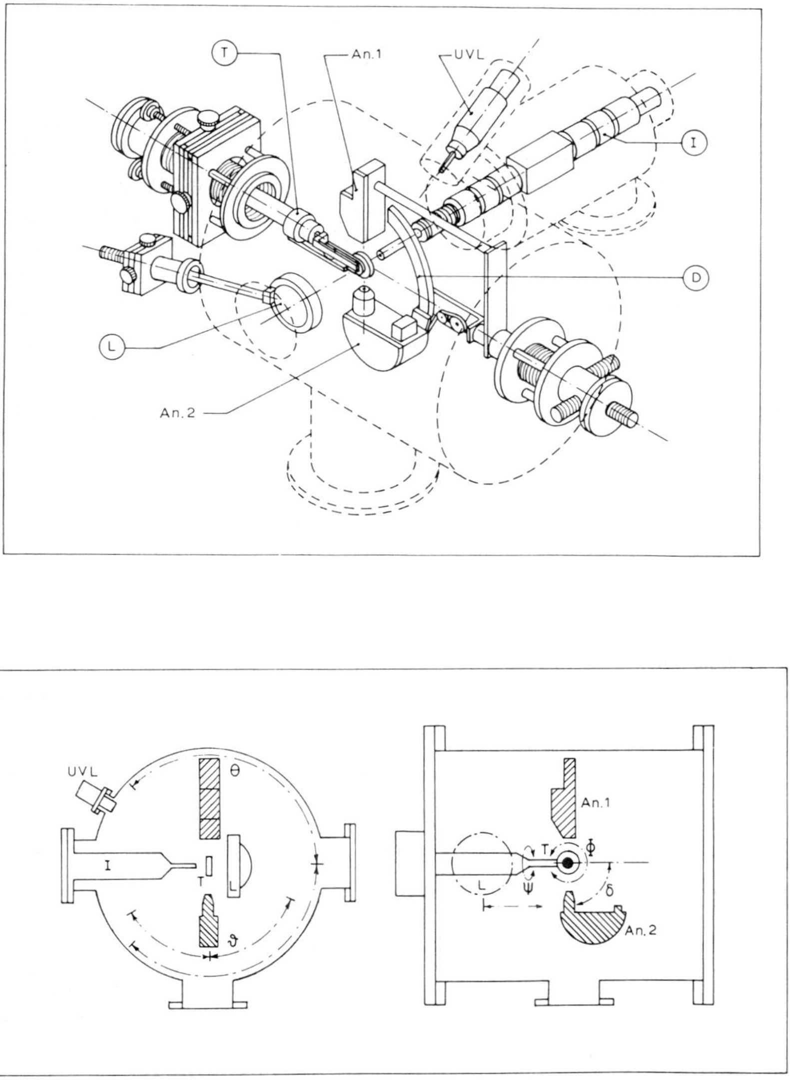

FIG. 11.2 Schematic of an experimental arrangement for angle-resolved UPS. The sample is positioned on the end of the target manipulator T, and can be irradiated either with UV photons from the lamp UVL or from the ion gun I. The detection system D consists of two analysers, a double-parallel-plate analyser An. 1, and a hemispherical analyser An.2. Also included is a LEED system L. The lower panel shows the ranges of angles traversed by the analysers, centred on the sample. On the left is a cross-section through the plane of UVL and I, and on the right a longitudinal cross-section. An.1 can travel in an arc θ of 260° with an angular reproducibility of $\pm 0.4°$, while An.2 can travel either in the arc δ of 90° or be rotated through ν of 125°, each with a reproducibility of $\pm 0.2°$. In addition the sample can be rotated around two mutually orthogonal axes, through angles ψ or Φ. (From P. A. A. F. Wouters, P. A. Zeijlmans van Emmichoven, J. M. Fluit, and A. Niehaus, J. Phys. E: Sci. Inst., accepted for publication).

The pulsed nature allows some rather clever time-resolved experiments to be conducted, while the plane polarization property leads to other experiments that study the detailed electronic structure of matter.

Synchrotrons are expensive to build and expensive to operate, and they tend to be out of action for maintenance or repair rather more frequently than would an electron spectrometer for conventional XPS or UPS in one's own laboratory. As a result those who wish to use synchrotron radiation are allocated a block of time starting on a fixed date, and thus have little control over exactly when they would like to carry out their experiments. What is more, that allocated time is continuous as long as the beam in the ring is available, which means that the experimenters must be prepared to run the experiment night as well as day, and it is normal to arrange shift working to make maximum use of the time. Despite these apparent discomforts the advantages to be gained by using a variable photon energy source are very considerable, and there are always more research workers wishing to use time on a synchrotron than there is time available.

11.2 Theory and spectral interpretation

11.2.1 *UPS*

Angle-integrated UPS Photoemission by excitation with ultra-violet photons from typical gas discharge sources such as He and Ne lamps obeys the same physical relation as does photoemission in XPS, i.e.

$$E_{\text{kin}} = h\nu - E_{\text{B}} - e\phi\,, \tag{11.1}$$

already given in chapter 2. Here, however, the E_{B} refer not to core levels, but to the very shallow valence band levels. Since the binding energies of electrons in solids are referred to the Fermi level E_{F}, it follows that at E_{F}, $E_{\text{B}} = \text{O}$, so that photoemission commences near $E_{\text{kin}} = h\nu$. Onset of photoemission will not occur at exactly $h\nu$ because the electron energies have to be measured in an energy analyser, which will have an effective work function barrier that the electrons must surmount. Thus a work function term appears on the right-hand side of expression (11.1), but it remains sensibly constant during any one set of experiments. Photoemission can continue at all energies down to that at which electrons can no longer leave the sample surface by virtue of the work function barrier at *that* surface, which means, taking into account the analyser work function, that an emission cut-off occurs at the difference in work functions between sample and analyser. A typical UPS spectrum using HeI (21.21 eV) radiation would therefore appear as in Fig. 11.3a, for a clean d-band metal such as iron. For HeII (40.82 eV) radiation, however, the UPS spectrum from the same metal appears as in Fig. 11.3b, in which there is no secondary emission cut-off since HeI radiation cannot be eliminated except

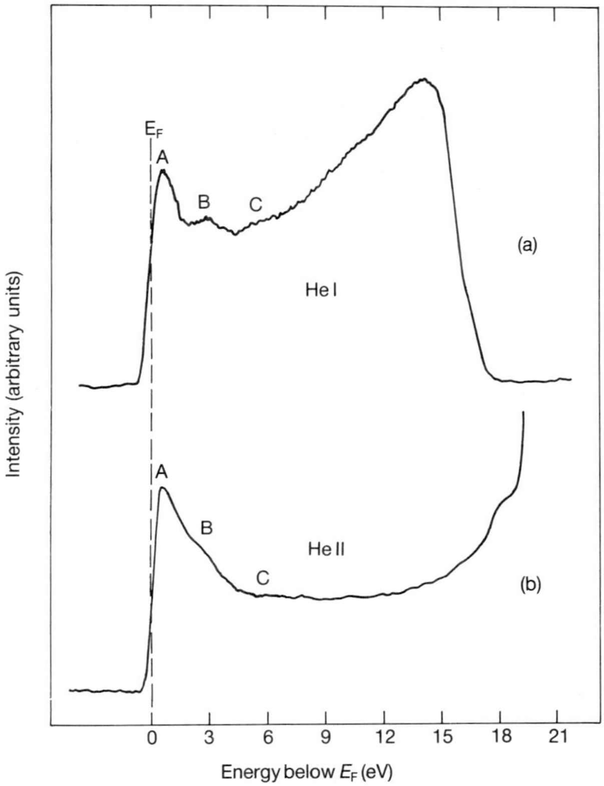

FIG. 11.3 Comparison of UPS spectra from a clean Fe(110) surface. (a) HeI (21.21 eV) spectrum; (b) HeII (40.82 eV) spectrum. There is no secondary emission cut-off in the HeII spectrum as there is in the HeI because of course the spectrum using HeII leads into that of HeI. The features A, B, and C in the density of states of iron appear in both spectra, but with different relative intensities and widths because of different ionization cross-sections.

with an appropriate filter, and at the high binding energy end the HeII-excited spectrum simply continues into the HeI-excited region.

In both HeI- and HeII-excited spectra from clean iron there are three features labelled A, B and C, at respective distances from the Fermi edge of 0.6, 3.0 and $\approx$5.8 eV. Since these features arise from the photoemission of electrons of lowest binding energies, i.e. from the valence band, it is necessary to ask in which way the observed spectrum actually reproduces the density of states (DOS) in the valence band, and whether it refers to bulk or surface DOS.

The essential feature of photoemission is contained in eqn (11.1), and that is conservation of momentum. Since photon momentum is so small compared to the momentum of a photoelectron of energy E_{kin}, the nucleus of the photoionized atom can absorb the momentum recoil due to energy exchange, and the photoelectron energy is unaffected. This applies, of course, to a localized electron in a core level. Electrons in valence bands are, however, regarded as non-localized on any particular atom, and the only characterizing electron momentum is the so-called *reduced* momentum or wave vector. In the reduced momentum scheme conservation of energy and momentum can be achieved by the entire solid absorbing the recoil momentum of photoemission by just one reciprocal lattice vector. Since a change of momentum by any unit reciprocal lattice vector cannot be distinguished from no change at all, the electronic transition between allowed bands is a purely *vertical* one with no change in the reduced wave vector. For a chosen photon energy and for a particular direction in the surface Brillouin zone, such a vertical transition should give rise to a fairly sharp peak in the photoelectron spectrum. An example of what might be expected to be seen in a valence band photoelectron spectrum for a fixed photon energy, based on 'only vertical transitions allowed', is given in Fig. 11.4 for Cu in the Γ-X direction in the Brillouin zone,

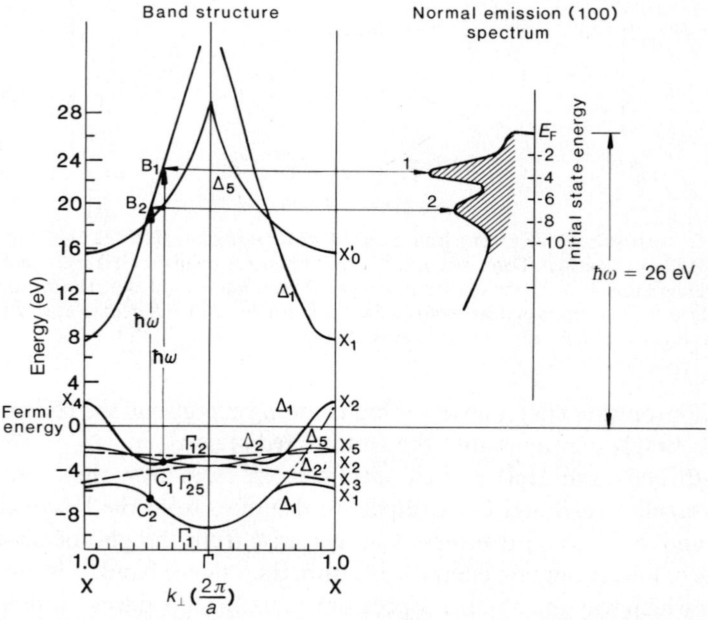

FIG. 11.4 Example adapted from Plummer and Eberhardt[1] of the origin of features in an UPS spectrum based on only vertical transitions within the band structure being allowed. On the left is the band structure of Cu along the Γ-X direction in the Brillouin zone, with vertical transitions B_1C_1 and B_2C_2 shown. On the right is a hypothetical photoemission spectrum at a photon energy of 26 eV which would arise from the two transitions.

for $h\nu = 26$ eV (adapted from Plummer and Eberhardt,[1] based on the theory of Smith and Mattheiss[2]). For a full exposition of the theory, the reader is referred to the articles by Plummer and Eberhardt,[1] and to chapter 4 in the volume by Ertl and Küppers.[3] It will be seen later when considering angle-resolved UPS (ARUPS) that such peaks and their dispersion are used to map the band structure. However, when photoelectrons emerging from a three-dimensional band structure are collected over a large solid angle, the resultant spectrum is related to an averaged joint DOS. That is, the spectrum is a convolution of the initial and final DOS, always provided that there is no localization.

In the iron valence band spectra of Fig. 11.3a and 11.3b, therefore, the features A, B and C should be interpreted by comparison with a theoretical calculation of the joint DOS, and just such a comparison has been carried out by Pessa *et al.*[4] in Fig. 11.5. The calculated joint DOS came from the

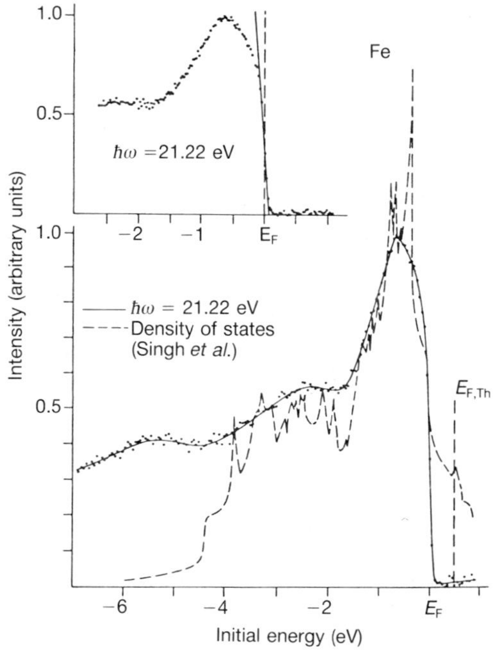

FIG. 11.5 Comparison by Pessa *et al.*,[4] of the HeI spectrum of iron with the theoretical density of states calculated by Singh *et al.*[5] The major peaks in the vicinity of -1 eV have been aligned for ease of comparison, which imposed a shift of 0.5 eV on the theoretical spectrum; hence the theoretical Fermi level $E_{F,Th}$ does not quite coincide with the experimental one. The peaks at ≈ 1 eV, ≈ 2.2 eV, and ≈ 5 eV correspond to those labelled A, B, and C in Figure 11.3. Peak C has no apparent counterpart in the theoretical DOS, and may be a satellite of similar origin to that seen prominently in nickel.

theoretical work of Singh *et al.*[5] In the figure the maxima in the experimental and in the DOS curves have been aligned and normalized, which involves a slight shift of 0.5 eV in the position of the Fermi level. Feature A can then be seen to coincide with the maximum in the densities of majority-spin states in the iron d-band, while feature B consists of a superposition of majority-spin states from the sp band and a small contribution of minority-spin d states. On the other hand, feature C has no apparent counterpart in the theoretical DOS, and its origin must be sought elsewhere. By analogy with nickel, whose valence band spectrum has a pronounced satellite at binding energies between 5 and 6 eV, feature C is believed to originate from a d-state with two holes rather than one. If so, then there should certainly be no prediction from theory of any maximum in the DOS at that energy.

Referring again to Fig. 11.3a, the large peak between 14 and 15 eV is due to the background of 'true' secondary electrons produced in the sample by a cascade process. This background decreases sharply due to the work function of the sample, causing the secondary distribution to peak at some energy; with perfect energy resolution and at absolute zero of temperature, the work function cut-off should be vertical. The additional shoulder near the bottom of the cut-off orginates from secondaries produced inside the spectrometer, and is thus an artefact. The method used to measure work function accurately will be described later.

In general, interpretation of the angle-integrated valence band photoelectron spectra from pure clean metals or alloys will be based on similar comparisons as described above for iron. The large amount of additional information obtainable from angle-resolved measurements will be discussed below. When attempting to interpret the often complex spectra resulting from reaction of a clean surface with a particular molecule, it is always necessary to start by comparing the resultant adsorbate spectra with those from the molecule in the gas phase. The gas-phase spectra might be acquired in the same experiments by admitting the gas to a sufficiently high pressure, or results from other spectroscopies might be used. Thus, Ehlers *et al.*,[6] in studying the low-temperature adsorption of methanol on the Pt(111) surface, used a gas-phase methanol spectrum generated previously in the same equipment. Their comparison of the 100 K adsorbed layer spectra using both HeI and HeII excitation, with the HeII gas-phase spectrum, is shown in Fig. 11.6. The adsorbate spectra are presented as *difference* spectra, that is, the spectrum from the clean surface has been subtracted, leaving just the additional features due to the methanol on the surface. It is clear that after an exposure of only 1.5 L the methanol is present as an undissociated molecule in the adsorbed layer, and after 5 L the spectrum resembles that of the gas-phase molecule even more closely. At the higher exposure the two upper orbitals of the methanol undergo a slight bonding shift. On annealing to temperatures

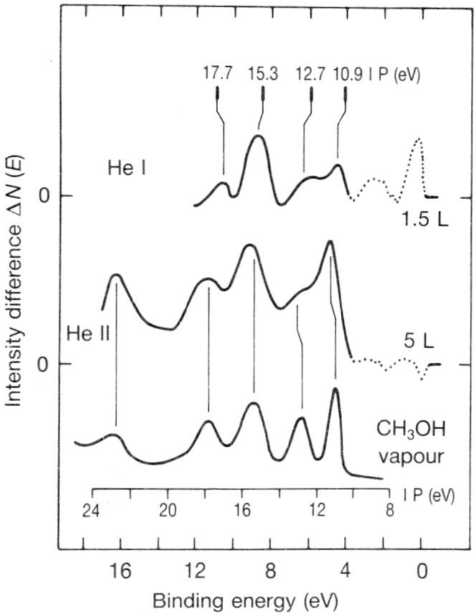

FIG. 11.6 Comparison by Ehlers *et al.*,[6] of the HeI and HeII difference spectra of methanol adsorbed on Pt(111) with the spectrum of methanol in the gas phase. The gas phase spectrum has been shifted to align the respective bands for better comparison. After 5L exposure the UPS spectrum shows marked similarities to the gas phase spectrum, with slight bonding shifts in the outer orbitals.

above 100 K, the spectrum changed radically, indicating dissociation of the methanol and chemical bonding.

Even the adsorption spectra of relatively large organic molecules can be interpreted in terms of the molecular orbital assignments made for the free molecular spectra. A good example is that published by Netzer and Mack[7] for the adsorption of pyridine on a Pd(111) surface, shown in Fig. 11.7. They had at first recorded a set of HeI-excited spectra for the saturation coverage ($\approx$5 L) for pyridine, and found five recognizable adsorption bands, those labelled A-E in Fig. 11.7. Previous workers had made assignments of these bands according to their energy sequence in the gas phase, but unambiguous assignment did not seem possible without establishing the presence and position of the orbital $4a_1$, which is at slightly higher ionization energy in the gas phase. Clearly part of the problem was the steeply sloping secondary electron background in the HeI spectra. In a HeII spectra there is a nearly flat region extending over a greater energy range than in the HeI excited spectrum, although to counterbalance that advantage the sensitivity is lower due to a lower photon flux, and when the authors recorded the HeII spectrum a further band labelled F was found. However, the position of F was in fact too high in binding energy for the $4a_1$ orbital of pyridine, and could be assigned only to

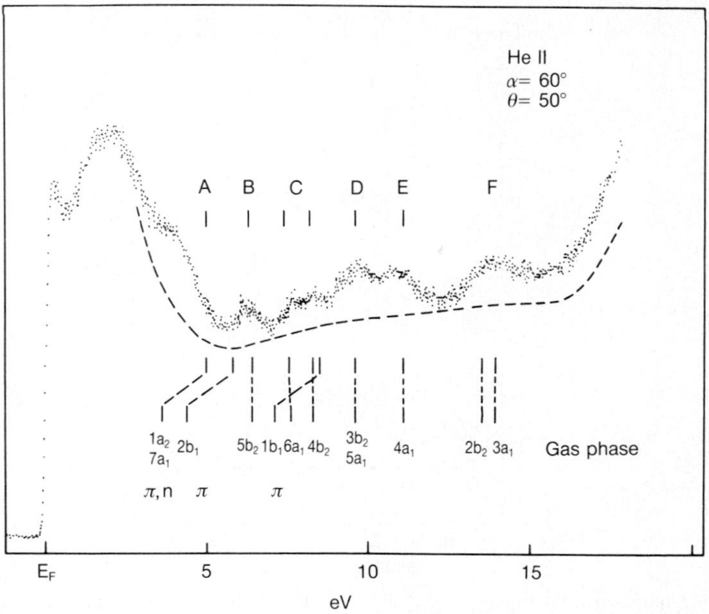

FIG. 11.7 Molecular orbital assignments made by Netzer and Mack[7] in the HeII spectrum of pyridine adsorbed on Pd(111), based on the orbital energies of the molecule in the gas phase. The dashed line is a portion of the spectrum from clean Pd(111).

the even deeper lying orbitals $2b_2$ and $3a_1$. To accommodate $4a_1$, therefore, needed a reconsideration of all the other assignments, and that could be done only by a detailed angle-resolved study of the system. As a result the various absorption bands were finally allocated the assignments shown in Fig. 11.7, in which the previous attributions of bands D and E have been changed. An interesting additional result to emerge from the work of Netzer and Mack was that at saturation coverage the pyridine molecules were in fact adsorbed neither parallel to nor vertical to the Pd surface, but tilted at an angle to it.

Angle-resolved UPS (*ARUPS*) The important additional information to be gained by recording UPS spectra as a function of angle has been mentioned several times, and the nature of this information as well as how it is obtained should now be discussed in more detail. In so doing some of the ground covered above briefly will be revisited slightly more extensively.

It has already been pointed out that in considering transitions involving non-localized electrons, e.g. in the valence band of a solid, it is the reduced wave vector $\boldsymbol{k}$ that must be conserved, and that this leads to the rule that in such a case only vertical, i.e. optical, transitions are allowed. If the transition is from a state in a band i of initial states to a state in a band f of final states, and if the dispersions with energy of these bands are $E_i(\boldsymbol{k})$ and $E_f(\boldsymbol{k})$, respectively, then all vertical transitions of energy $h\nu$ are confined to a surface in k-space whose equation is

$$E_f(\boldsymbol{k}) - E_i(\boldsymbol{k}) - h\nu = 0\,. \tag{11.2}$$

Defining the zero of energy in the usual way as being at the Fermi level, the kinetic energy of the photoelectron is then

$$E_{\text{kin}} = E_f(\boldsymbol{k}) - e\phi \tag{11.3}$$

where ϕ is the work function. With this definition the energy of the initial state $E_i(\boldsymbol{k})$ is always negative, being zero at the Fermi level.

Photoemission is normally described in terms of the so-called 'three-step' model, in which the steps consist sequentially of (i) excitation of an electron through the vertical transition, (ii) transport through the solid which might involve inelastic scattering, and (iii) escape through the surface into the surrounding vacuum. Thus the energy distribution actually observed externally in a spectrometer is the sum of two components, a distribution $I_p(E, h\nu)$ of electrons that have *not* suffered any inelastic collisions, and a distribution $I_s(E, h\nu)$ of electrons that have suffered one or more such collisions, i.e.

$$I(E, h\nu) = I_p(E, h\nu) + I_s(E, h\nu)\,. \tag{11.4}$$

Clearly, it will be the no-loss electrons in I_p whose energy distribution contains the fine structure details relating to the band structure of the solid. Because of the three-step model assumption, I_p can be broken down into the product of three separate functions, the energy distribution $N(E, h\nu)$ of photoexcited electrons produced at the atomic site, a transport function $T(E)$, and an escape function $D(E)$, the two latter functions depending of course only on the kinetic energy E of the photoelectron. Thus

$$I_p(E, h\nu) = N(E, h\nu)\; T(E)\; D(E)\,. \tag{11.5}$$

$T(E)$ will depend on the inelastic scattering probability, which will in turn be a function both of the inelastic mean free path $\lambda_e(E)$ of the photoelectron of energy E, and the attenuation length $\lambda_p(h\nu)$ of the photon. $T(E)$ can be written down as

$$T(E) = \frac{\{\lambda_e(E)/\lambda_p(h\nu)\}}{1 + \{\lambda_e(E)/\lambda_p(h\nu)\}}\,. \tag{11.6}$$

$T(E)$ varies only slowly with E and can be considered constant over the narrow range of kinetic energies used in UPS; it will not introduce any structure into I_p.

An electron can escape from a solid surface only if it has sufficient kinetic energy to surmount the work function barrier, or, more accurately, if its component of energy normal to the surface is sufficient. If the distribution of photoexcited electrons inside the solid is considered isotropic, and if they spread out as a plane wave, then the escape function $D(E)$ has the following conditions, based on the above definition of the zero of energy at the Fermi level:

$$D(E)=\tfrac{1}{2}\{1-(e\phi/E)^{1/2}\} \qquad \text{if } E>e\phi$$
$$=0 \qquad \text{if } E\leq e\phi\,. \qquad (11.7)$$

Beyond the work function cut-off $D(E)$ is a smooth function of E, and again will not introduce any additional fine structure into I_p, although the combination of $T(E)$ and $D(E)$ will cause a smooth distortion of $N(E, h\nu)$ compared to the actually observed I_p.

Since I_p will contain the fine structure in $N(E, h\nu)$, it is pertinent to examine the form of the energy distribution of photoemitted electrons. The close analogy with optical absorption is very useful. The imaginary part $\epsilon_2(h\nu)$ of the dielectric constant can be written as

$$(h\nu)^2\epsilon_2(h\nu)\propto \sum_{\mathrm{i,f}} \int \mathrm{d}^3\boldsymbol{k}|\langle\, \psi_\mathrm{f}|\nabla|\psi_\mathrm{i}\rangle|^2\, \delta\,\{E_\mathrm{f}(\boldsymbol{k})-E_\mathrm{i}(\boldsymbol{k})-h\nu\}\,. \qquad (11.8)$$

Integration has to be caried out over all $\boldsymbol{k}$ states in the initial and final bands separated by $h\nu$, and over all pairs of bands i and f. However, if the dipole matrix element in (11.8) is assumed to be constant, then there results a quantity J called the joint DOS:

$$J(h\nu)=(8\pi)^{-3}\sum_{\mathrm{i,f}} \int \mathrm{d}^3\boldsymbol{k}\cdot\delta\{E_\mathrm{f}(\boldsymbol{k})-E_\mathrm{i}(\boldsymbol{k})-h\nu\}\,. \qquad (11.9)$$

J is the total number of transitions possible for a photon energy $h\nu$ taking into account energy and $\boldsymbol{k}$ conservation. The measured distribution $I_p(E, h\nu)$, on the other hand, selects those transitions in the total J in which the final state energy $E_\mathrm{f}(\boldsymbol{k})$ coincides with E seen by the spectrometer, so that in the final expression for N a further δ function must appear. Also, of course, integration can cover only occupied initial states, which means within the Brillouin zone when considering a three-dimensional solid. Thus the photoelectron distribution can be written.

$$N(E, h\nu)\propto \sum_{\mathrm{i,f}} \int_{\mathrm{BZ}} \mathrm{d}^3\boldsymbol{k}\cdot\delta\{E_\mathrm{f}(\boldsymbol{k})-E_\mathrm{i}(\boldsymbol{k})-h\nu\}\;\delta\,\{E_\mathrm{f}(\boldsymbol{k})-E\}\,. \qquad (11.10)$$

From the assumptions made in arriving at expression (11.10) it can be seen that only for constant matrix elements and structureless final states will the measured distribution I_p reflect the densities of occupied initial states.

It is now necessary to consider what happens when electrons setting out with an energy distribution given by (11.10) arrive at the surface. As electrons cross the surface, momentum is not conserved. In the presence of a surface, periodicity may be assumed only parallel to the surface, which implies that the only conservation will be that of the wave vector parallel to the surface. This can be written as

$$\boldsymbol{k}_{\parallel}(\text{outside}) = (\boldsymbol{k}_{\parallel}(\text{inside}) + G_{\parallel} \tag{11.11}$$

where $G_{\parallel}$ is the parallel component of any reciprocal lattice vector, in other words a surface reciprocal lattice vector. The perpendicular component $\boldsymbol{k}_{\perp}$ of the photoelectron wave vector is not conserved, and cannot be determined experimentally; in fact for a chosen $\boldsymbol{k}_{\parallel}$, $\boldsymbol{k}_{\perp}$ could have any value along a rod in $\boldsymbol{k}$-space extending from the two-dimensional surface Brillouin zone into the three-dimensional bulk Brillouin zone. Thus, *in the general case*, it is not possible to determine the full three-dimensional band structure of a solid from angle-resolved UPS. Having said that, however, there are now so many examples of the successful elucidation of band structure by ARUPS that it is clear that some of the assumptions made in the method must be valid. In particular, if it is assumed that the solid is free-electron-like or even nearly-free-electron-like, then the $E(\boldsymbol{k})$ relationships are basically the same on either side of the surface, and the $\boldsymbol{k}$ values can be directly measured away from the Brillouin zone boundaries. On the other hand, not all metals approximate to free-electron-like, and other ways of circumventing the problem are to study layer compounds, in which the band dispersion as a function of $\boldsymbol{k}_{\perp}$ should be small, i.e. the bands in $\boldsymbol{k}_{\perp}$ are assumed flat, or to use normal emission, for which $\boldsymbol{k}_{\perp}(\text{outside}) = 0$, and vary the photon energy $h\nu$ using synchrotron radiation. The latter will be discussed below under SRPS.

Assuming free-electron-like behaviour and an abrupt potential barrier at the surface due to the work function, there is a simple relationship between $\boldsymbol{k}_{\parallel}$, the kinetic energy E, and the polar angle of emission θ, namely

$$\boldsymbol{k}_{\parallel} = (2mE/\hbar^2)^{1/2} \sin\theta\,. \tag{11.12}$$

The polar angle θ is defined in Fig. 11.8, which shows a typical arrangement[8] for an angle-resolved UPS experiment. The angle of incidence of the photon beam to the surface normal is ψ and the azimuthal angle is ϕ. $\delta\theta$ is the angular acceptance into the detector. By measuring energy distributions as a function of θ, it is then possible to construct dispersion curves $E(\boldsymbol{k}_{\parallel})$, since as discussed above, in the expression for E

$$E_{\text{kin}} = (\hbar^2/2m)\,(\boldsymbol{k}_{\parallel}^2 + k_{\perp}^2) \tag{11.13}$$

It is assumed that $\boldsymbol{k}_{\perp}$ is effectively invariant.

An example of the application of the ARUPS method to the elucidation of band structure is that of Tokailin *et al.*[9] for rhombohedral arsenic. Figure 11.9 shows the bulk Brillouin zone of rhombohedral arsenic, with the axes and the symmetry points labelled. Measurements were performed along three planes of high symmetry, namely Γ-L-X, Γ-X′-L′, and Γ-W-K. It can be seen from the

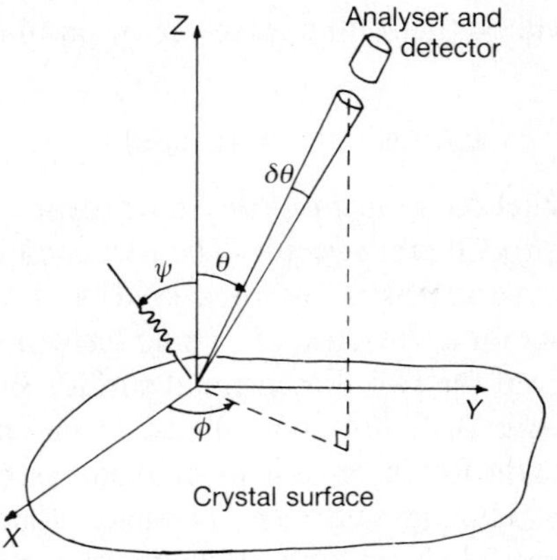

FIG. 11.8 Schematic of experimental arrangement for angle-resolved UPS. The photon beam makes an angle of incidence ψ with the surface normal Z. The polar angle, i.e. the angle made by the axis of the detector with Z is θ, and the angular acceptance of the detector is $\delta\theta$. The azimuthal angle is ϕ, i.e. the angle made by the projection of the detector axis on the XY plane with the plane containing the polar angle ψ. From Leckey[8]. Compare this diagram with the experimental equipment shown in Fig. 11.2.

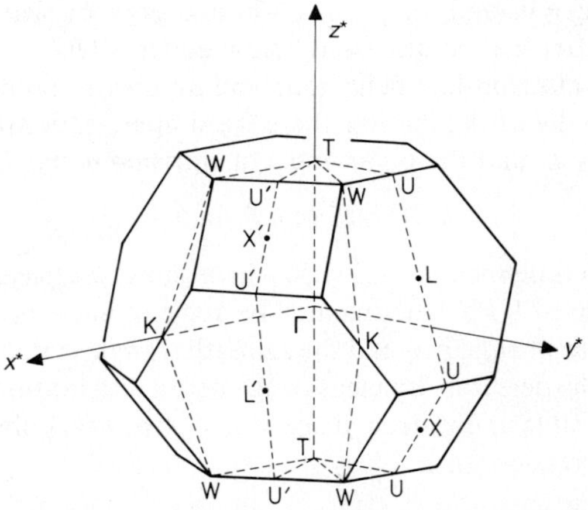

FIG. 11.9 The bulk Brillouin zone of rhombohedral arsenic, from Tokailin *et al.*[9] Axes and symmetry points are labelled.

figure that Γ-L-X and Γ-X′-L′ are equivalent through inversion symmetry, but should not be equivalent in ARUPS since the three-dimensional symmetry breaks down at the surface. Figure 11.10 shows the results for the latter two planes for polar angles from 0 to 34° using HeI radiation. Various spectral

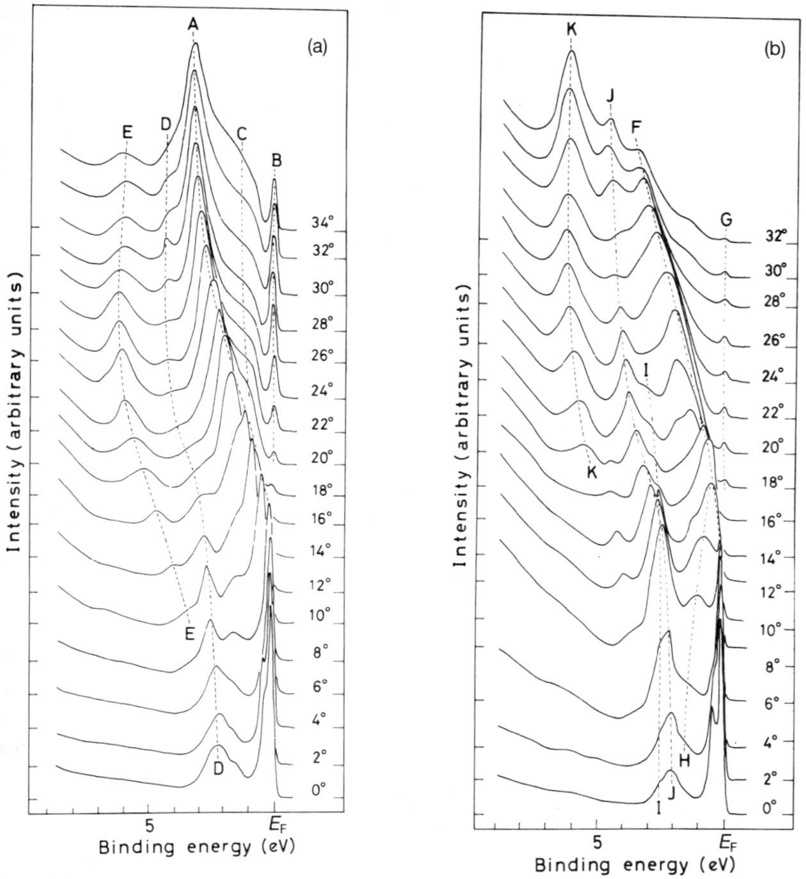

FIG. 11.10 Angle-resolved HeI UPS spectra from rhombohedral arsenic in the Γ-L-X (left-hand-side) and Γ-X′-L′ (right-hand-side) planes. The polar angle for each spectrum is indicated. Various features are labelled A to K in the two panels, and can be seen to disperse in energy with angle. From Tokailin *et al.*[9]

features are labelled A–K, and are seen to disperse in energy with changing angle.

The spectra for the two planes seem at first sight quite different, due principally to the different relative intensities of the features as they disperse, but when they are each plotted in *k*-space according to expression (11.12), the resultant bands in fact behave very similarly. Indeed each reproduces quite well the band structure calculated along the T-U, Γ-L, and Γ-X directions by a self-consistent pseudopotential method. The interesting point is that for the two planes the derived band structures are so similar. If at the surface there was a maintenance of momentum conservation in a direction perpendicular to the surface as well as parallel to it, then photoelectrons originating on the Γ-L-X and Γ-X′-L′ planes with the same E and at the same θ should have different initial states since the points on the planes defined by $\boldsymbol{k}_{\parallel}$ and $\boldsymbol{k}_{\perp}$ would not be equivalent in general. Since they have apparently the same initial states there must clearly be either or both a loss and a mixing of momentum, leading to loss of momentum conservation in one or both directions. This is termed momentum broadening and is a consequence of lack of approximation to free-electron-like behaviour.

The above example was provided firstly to demonstrate the usual way in which ARUPS is applied, leading to quite satisfactory agreement with theory, but secondly to show that ambiguities are not uncommon and have to be explained if possible. For metals with behaviour close to free-electron-like there is less chance of ambiguity.

A method for the precise location of the initial state $E_i(\boldsymbol{k})$ was devised by Kane[10] and applied to the band structure of gold by Heimann *et al.*,[11] It circumvents the disadvantage inherent in the normal methods that it is difficult to map the initial state bands without making theoretical assumptions about the final state bands. This is implicit in the form of expression(11.10). Kane's method depends on energy-coincidence measurements. Two sets of ARUPS data are obtained for two different orientations of a single crystal, chosen so that the angle between them is small and in a convenient range, without changing the azimuthal angle. An attempt is then made to identify the same transition on the two different surfaces by visual inspection of the spectra, and if such an identification can be made, then the two values of $\boldsymbol{k}_{\parallel}$ so found will define uniquely the value of $\boldsymbol{k}$ itself. Knowing the polar angles and the energies enables a simple triangulation construction to be used.

The application by Heimann *et al.*, to gold is a good illustration of the method. They started by recording the UPS spectrum at a photon energy of 16.85 eV from the Au(111) surface at *normal* emission. Under that condition $\boldsymbol{k}_{\parallel}$ is approximately zero and if the Au(111) is assumed to be an ideal surface then conservation of $\boldsymbol{k}_{\parallel}$ means that it is also zero inside the crystal. The spectrum from Au(111) is shown at the top of Fig. 11.11. Angle-resolved photoemission was then measured from another gold surface making a fairly small angle with the (111), the Au(112), over a range of polar angles from 0° to 75°, the resultant spectra being displayed also in Fig. 11.11. Inspection of the data shows at once that near polar angles of 25°, emission from the (112) face resembles very closely that at normal emission from the (111), except for a feature near the Fermi energy which should not appear from the (112) in any case.

The triangulation method is illustrated in Fig. 11.12 for the transition marked f′ in Fig. 11.11. Figure 11.12 shows a section through the repeated Brillouin zone, containing the [111], [112] and [110] directions. Knowing the angle, 25°, to the [112] direction, the final state energy, and the work function of the [112] surface, the free-electron (or vacuum) wave vector $\boldsymbol{k}_{v,112}$ can be calculated and hence the parallel component $\boldsymbol{k}_{\parallel,112}$ by triangulation as

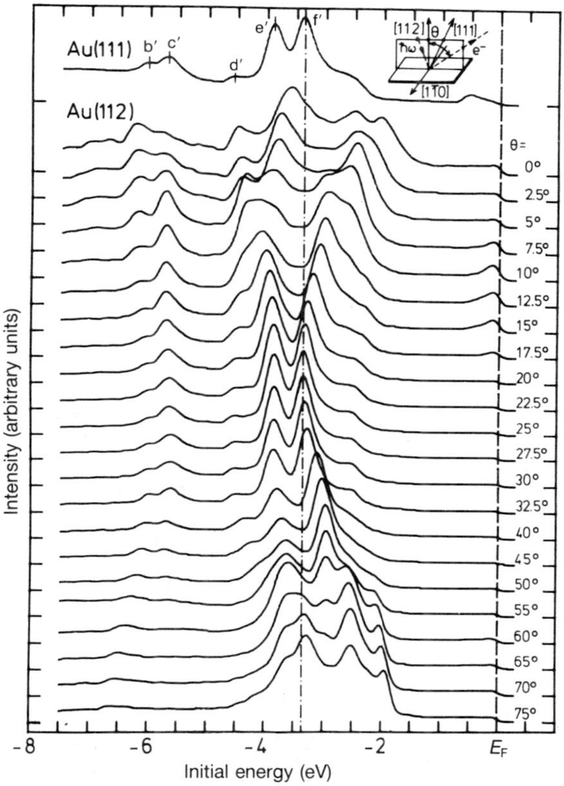

FIG. 11.11 UPS spectra at a photon energy of 16.85 eV from the Au(111) surface at *normal* emission (top) and from the Au(112) surface as a function of polar angle, measured by Heimann *et al.*[11] Near a polar angle of 25° the photoemission spectrum from the Au(112) surface closely resembles that from the Au(111) at normal emission. This observation is used in the triangulation method shown in Fig. 11.12.

indicated. $\boldsymbol{k}_{\parallel,112}$ is then projected onto the ΓL direction as shown in order to obtain the transition vector giving the position of the final state, $\boldsymbol{k}_{b,111}$. The calculated length of this vector then puts the origin of the transition f′ at A in the second zone. When the process was repeated by measuring angle-resolved photoemission from the Au(110) surface, the triangulation method, also shown in Fig. 11.12, gave a very similar result to that using the Au(112) surface.

11.2.2 *SRPS*

The same theoretical considerations apply to photoemission using a synchrotron source as to that using a conventional He or Ne discharge source, but obviously the ability to vary the photon energy provides an extra degree of

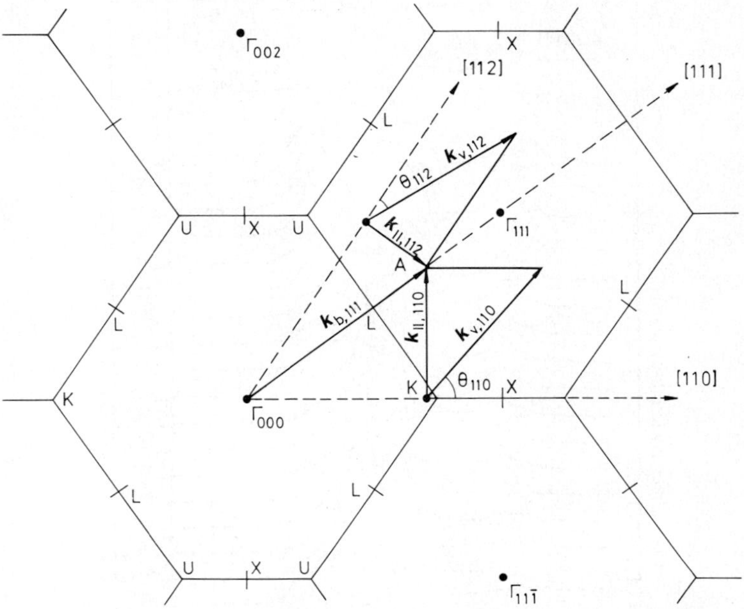

FIG. 11.12 Description of the triangulation method of Kane[10], as applied by Heimann *et al.*,[11] to the feature marked f′ in Fig. 11.11. The diagram shows a section through the repeated Brillouin zone, containing the [111], [112] and [110] directions. From the results of Fig. 11.11, the angle, 25°, between the [111] and [112] directions, is known. Using the final state energy of the photoelectrons, and the work function, the wave vector $\boldsymbol{k}_{v,112}$ can be calculated and thence the parallel component $\boldsymbol{k}_{\parallel,112}$ by triangulation. The latter vector is then projected onto the ΓL direction as shown to derive the transition vector $\boldsymbol{k}_{b,111}$ for the final state, whose calculated magnitude then puts the origin of the feature f′ at the point A.

freedom in the elucidation of band structure. In the direction normal to the surface, expression (11.12) can be written as

$$\boldsymbol{k}_{\perp} = \{2m(h\nu - E_B)/h^2\}^{1/2}. \tag{11.14}$$

Variation of $h\nu$ thus gives the final-state momentum normal to the surface.

Energy-resolved SRPS based on expression (11.14) as well as angle-resolved SRPS have been used by Kevan *et al.*,[12] to confirm the existence of surface states on Al(111) and (001) surfaces. In Fig. 11.13 are shown the spectra in normal emission from the A1(111) surface as a function of photon energy from 45 to 57 eV. In this range initial states along Λ near the L point in the bulk band structure are sampled, L corresponding closely to $h\nu = 53$ eV. Above and below $h\nu = 53$ eV two rather broad bands can be seen to disperse in energy with varying $h\nu$; in addition there is a sharp, non-dispersing peak at $E_B = 4.56$ eV which is at maximum intensity near $h\nu = 53$ eV. The two broad dispersing bands are free-electron initial state bands split by about 0.2 eV, but

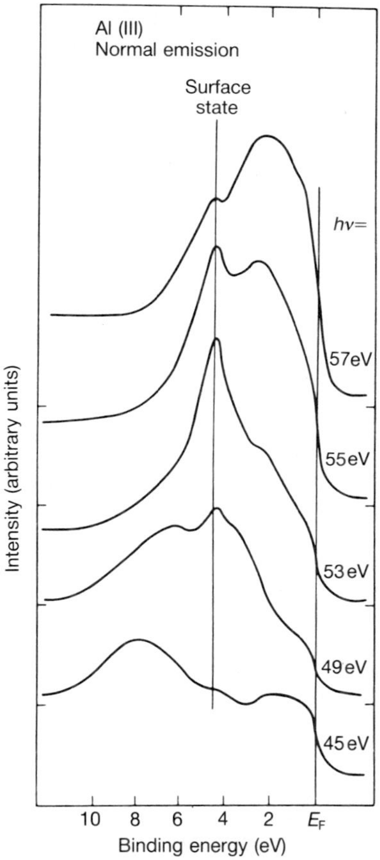

FIG. 11.13 Photoemission spectra from the Al(111) suface at normal emission, as a function of photon energy, measured by Kevan *et al.*,[12] using synchrotron radiation. At 53 eV, corresponding to the L point along Λ in the bulk band structure, a sharp, non-dispersing peak at 4.56 eV has maximum intensity. This peak is attributed to a surface state, since it does not disperse and can be quenched rapidly by surface reaction.

the sharp peak can only be a surface state peak since it does not disperse, and is quenched rapidly by surface reaction. The angular dispersion, that is, the mapping of the dispersion of the surface state *parallel* to the surface, is revealed by the spectra in Fig. 11.14. Application of expressions (11.12) and (11.14) then enabled the experimental dispersion relation to be fitted to the projection of the three-dimensional bulk continuum on to the surface two-dimensional Brillouin zone, as seen in Fig. 11.15. The effective electronic mass needed for fitting was 1.03 m_e, i.e. very close to genuinely free-electron-like. The surface state remains in the band gap up to about $\boldsymbol{k}_{\parallel}=0.90$ Å^{-1}, at which point it merges with a bulk transition and cannot be separately distinguished.

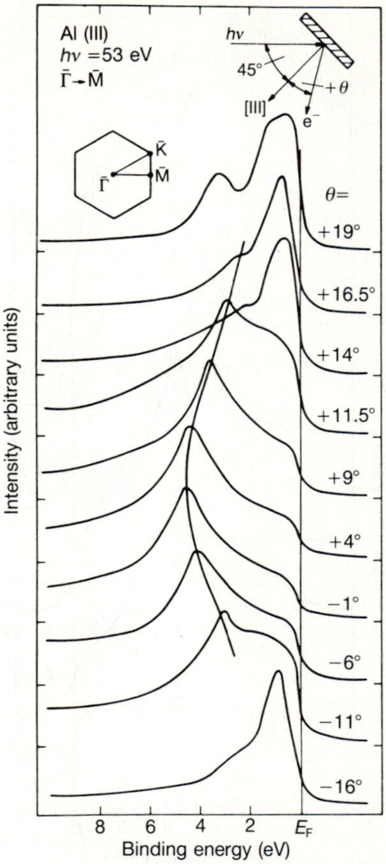

FIG. 11.14 Photoemission spectra from the Al(111) surface at a photon energy of 53 eV (see Fig. 11.13) as a function of polar angle, from Kevan *et al.*[12] The dispersion of the surface state parallel to the surface, i.e. in the $\bar{\Gamma}$-$\bar{M}$ azimuth, is mapped. Near $\theta = \pm 15°$ the state merges with a bulk feature.

As far as photoemission is concerned, the other major advantage of being able to vary the photon energy through use of synchrotron radiation is that the variation with photon energy of the photoionization cross-section can be exploited. In general the cross-section reaches a maximum close to the threshold of ionization (i.e. the core-level binding energy) and then decreases smoothly with increasing photon energy. However, where the orbital of the initial state has a radial node, which happens especially for the 2s, 3p, 4d and 5d core levels, there is a destructive interference in the transition matrix element at certain energies, leading to a sharp reduction in the photoionization cross-section. Thus for those levels the cross-section at first increases, goes through a maximum, then decreases rapidly again and passes through a

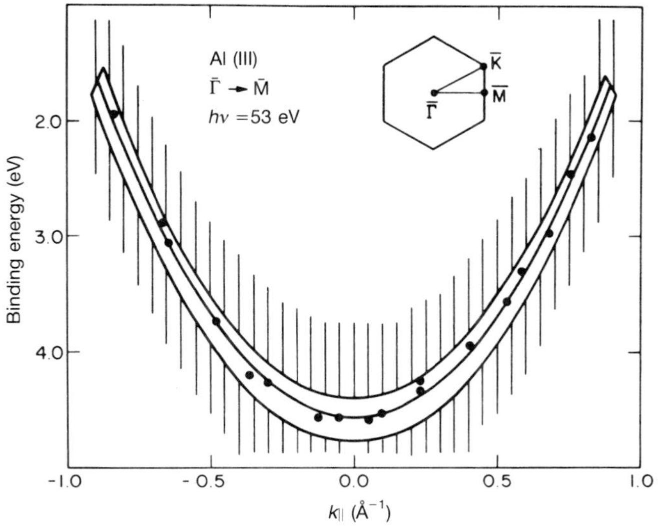

FIG. 11.15 Fitting by Kevan *et al.*,[12] of the experimentally observed surface state dispersion to the projection of the bulk structure onto the two-dimensional surface Brillouin zone. The hatched region is the bulk continuum. The points are the experimental observations that have been matched to a parabola with an effective mass of 1.03 m_e, close to free-electron-like. The surface state remains in the band gap up to about $k_{\|} = \pm 0.9$ Å^{-1}, when it becomes indistinguishable from the bulk.

minimum before a final smooth decline. The effect was first explained by Cooper[13] and the minimum is called the 'Cooper minimum'. Such a minimum can be seen[14] in Fig. 11.16 near 125 eV for the Sb 4d photoionization cross-section; compare the shape of that cross-section with that for the Ga 3d level in the same figure, which shows no such minimum.

The existence of the Cooper minimum has been put to use increasingly often in the study of surface reactions, particularly in the important field of metal–semiconductor interfaces arising out of semiconductor technology. As an example of such use, Rossi *et al.*,[15] wanted to establish for the Ag–Ge system whether any reaction between the deposited Ag and the Ge surface had taken place, and if so whether the interface had become diffuse as a result. No stable bulk silver germanide is known to exist, and thus the interface would normally be regarded as abrupt, with implications for the normal Schottky barrier model for such systems, which assumes an abrupt interface.

Now it is known[16] that the photoionization cross-section of the Ag 4d states goes through a deep Cooper minimum at a photon energy of $h\nu = 130$ eV. By comparing spectra taken at the minimum, and at another photon energy where the 4d cross-section is much higher, it should be possible to emphasize the contributions from the Ge and the Ag, respectively. Figure 11.17 shows this comparison, between valence band spectra at 80 eV and at 130 eV recorded as a function of increasing coverage of Ag, at room temperature. In

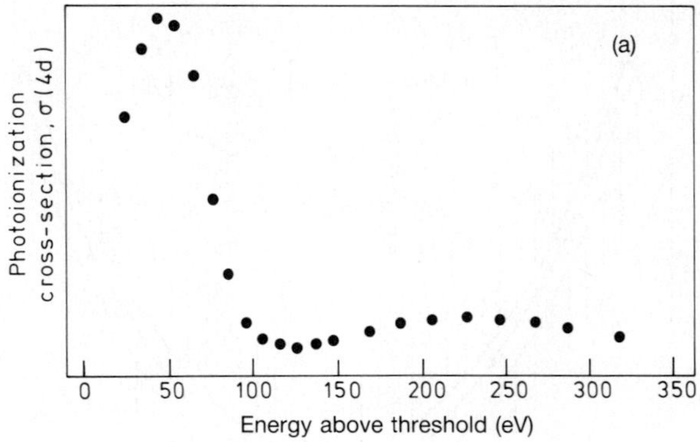

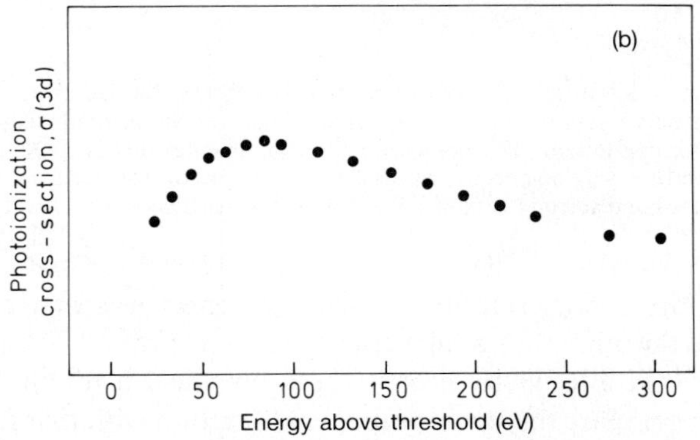

FIG. 11.16 Photon energy dependence of the partial photoionization cross-sections of the Sb 4d(upper) and Ga 3d (lower) levels in GaSb, measured by Lindau *et al.*[14] The minimum in the Sb cross-section is termed the Cooper[13] minimum. No such minimum appears for the Ga 3d cross-section since the 3d wavefunctions do not have any nodes, and there is thus no possibility of destructive interference in the transition matrix element at certain energies.

the left-hand box the spectra at 80 eV show Ag features dominant over the Ge features, and it is interesting that at very low coverages the Ag 4d feature is apparently a single sharp peak, almost symmetrical in shape, situated at 6 eV below the Fermi level. This is indicative of the presence of the Ag on the Ge surface as isolated atoms or very small clusters of atoms, since the 4d peak is clearly more atomic-like than band-like. Apart from a growth in intensity, there is no change in the position or shape of the Ag 4d peak up to about 0.3 monolayers of deposit. Within this coverage range, the right-hand box shows that there is no change in the valence band structure of the Ge. Beyond 0.3 monolayers of Ag, however, the situation begins to change. Not only does the Ag 4d band begin to broaden and shift towards lower binding energies, i.e. become more bandlike, but a shift in the Ge s states, lying at about 10.5 eV

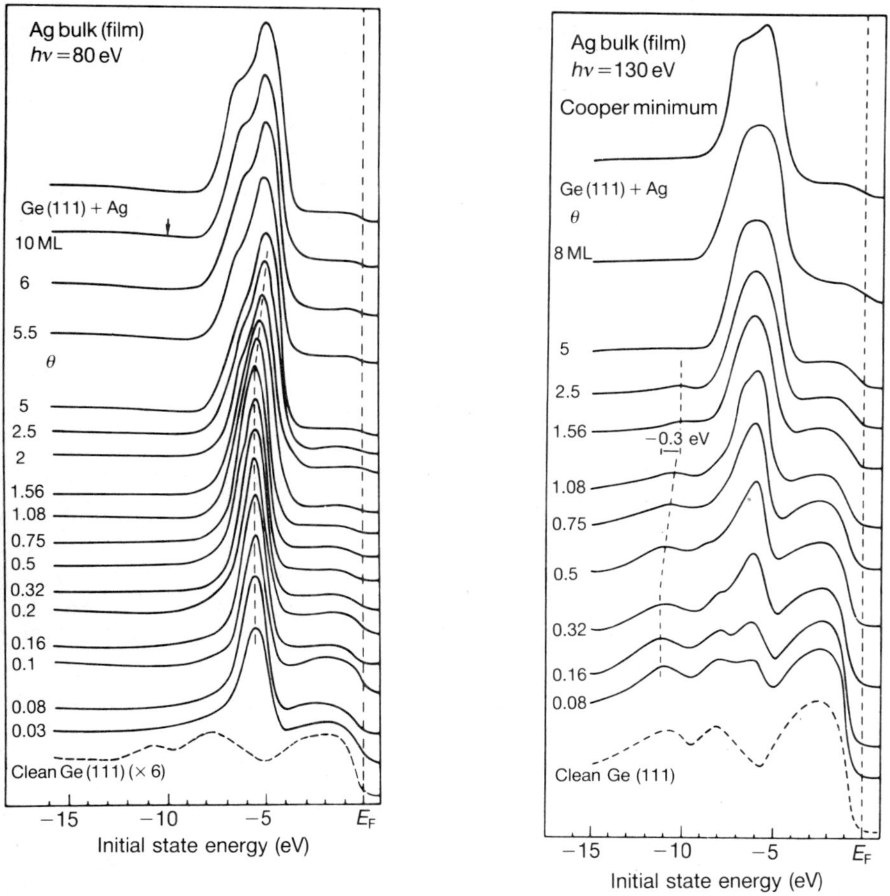

FIG. 11.17 Photoemission spectra from Ge(111) as a function of coverage by a silver film, recorded by Rossi *et al.*[15] The top spectra in each panel are for bulk silver. The spectra in the lef-hand panel were recorded at a photon energy of 80 eV, and those on the right at a photon energy of 130 eV, corresponding to the Cooper minimum in the photoionization cross-section for the Ag 4d states. Thus the Ge contribution should be proportionately greater on the right, and in fact the Ge s states are discernible and can be followed up to a coverage of silver of 2.5L.

below the Fermi level, takes place. The shift is not large, 0.3 eV, but is easily discernible since the s states are well clear of the Ag feature, and eventually of course the Ge features disappear under the coverage of Ag anyway. Since in other elemental semiconductor–metal systems that show reactivity the same features shift in the same direction, although by differing amounts, as a result of modification of the sp^3 configuration, clearly the same happens in the Ag–Ge system. This is thus the first evidence of a chemical interaction between Ag and Ge at the interface, even though no bulk compound between them exists.

Further information about the extent of the reaction was obtained from the variation in the intensities of the core levels Ag 3d and Ge 3d with Ag coverage, shown in Fig. 11.18. The dashed line in the Ag 3d plot shows the expected variation if the Ag were forming a uniform layer; clearly some of the photoemission intensity is lost as the coverage increases. Furthermore, the Ge 3d plot shows that even after 10 monolayers of Ag the intensity is still $\approx 30\%$ of that of pure Ge. The inescapable conclusion is that intermixing is taking place, probably near a depth of 2–3 layers, with consequent chemical interaction to form, presumably, a quasi-silver germanide.

11.3 Quantification

Neither UPS nor SRPS can be used as elemental analytical techniques in the same sense that can, say, XPS and AES; the information they provide is not of that nature, as should have been obvious from the foregoing descriptions. However, the changes in certain features and characteristics of UPS spectra as a function of surface reaction have been used successfully by several authors as a measure of the uptake of various species on the surface. The most comprehensive use of such features seems to have been that of Brodén *et al.*,[17] in their study of the chemisorption of various hydrocarbons on the Ir(110) surface. Their HeI spectra, in total (upper panel) and difference (lower panel) forms, are displayed in Fig. 11.19 for various exposures of Ir(110) to acetylene at room temperature. Also shown in the upper panel is the method of measuring work function change from the change in the width of the spectrum. Adsorbate induced features in the difference curves are labelled G–N; notice also the increasingly negative excursion near the Fermi level as adsorption removes electrons from the Ir d band.

Brodén *et al.*, used three features of the curves in Fig. 11.19, namely the area under peak L (the σ peak), the variation in work function, and the depth of the negative peak at the Fermi level. When each of these measured values was normalized to unity at saturation coverage, and plotted against ethylene exposure, they found the result of Fig. 11.20. All the measurements agreed within experimental error, and fell on a curve corresponding to a first order Langmuir type adsorption. The self-consistency and the agreement with theory indicate that in this case at least several features in a UPS spectrum can be used in a qualitative manner as a measure of reactant uptake.

Another interesting use of the intensities of UPS spectral features was that of Ruckman and Strongin,[18] who were studying the adsorption of CO on very thin Pd layers on Ta(110). Owing to modification of the electronic properties of the overlayer by hybridization with the underlying metal, the nature of the adsorption can be drastically altered. Fig. 11.21 shows some of their UPS spectra taken after exposure to 21 L of CO of Ta(110) covered by increasing amounts of Pd up to 3 monolayers. The appearance of the O 2p and C 2p

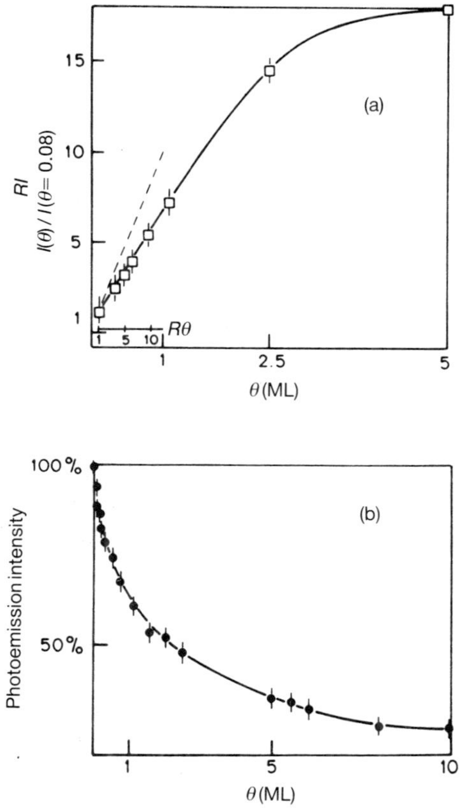

FIG. 11.18 Plots of intensity of the Ag 3d (upper) and Ge 3d (lower) core level photoelectron lines, as a function of silver coverage on Ge(111), at selected photon energies, from Rossi *et al.*[15] The dashed line in the upper panel is the behaviour expected if the silver were forming a uniform layer. The loss of intensity there, and the observation in the lower panel of some Ge 3d intensity even after 10 monolayers of silver, indicate intermixing at the surface.

features is characteristic of dissociation of the CO, while molecular adsorption is characterised by the CO 4σ and 1π–5σ features. After background subtraction, the intensities of the O 2p and the CO 1π–5σ features (in terms of peak height) were plotted against the thickness of Pd, as in Fig. 11.22, for two exposures of CO. Clearly, at a Pd coverage of about 1.2 monolayers no adsorption at all of CO occurs. Below that coverage the adsorption is dissociative, with the CO coverage decreasing linearly with Pd coverage, and above 1.2 monolayers Pd the adsorption is non-dissociative. The increase in chemisorption of non-dissociated CO as the Pd thickness was increased above 1.2 monolayers could be correlated with the appearance of d states near the Fermi level, characteristic of bulk Pd. The conclusion is that the single layer condition behaves very much like an inert metal like Cu or Ag, because of hybridization with the substrate.

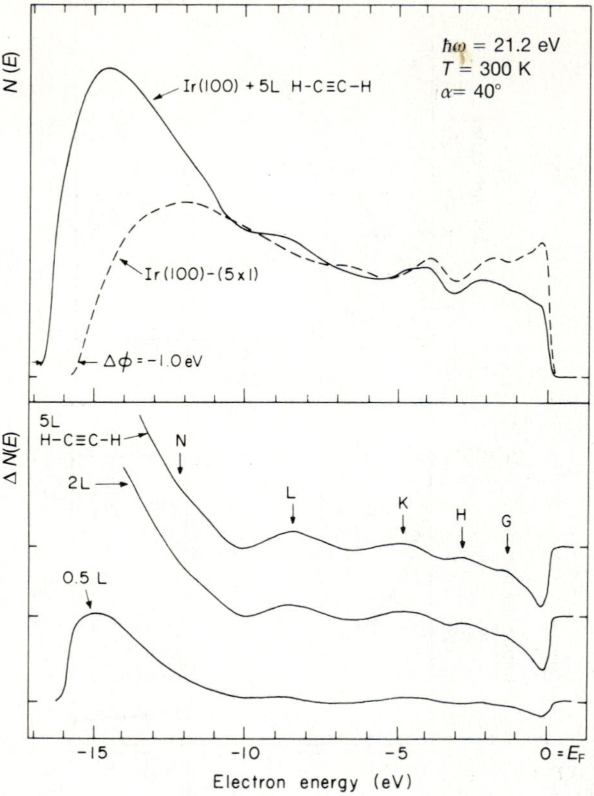

FIG. 11.19 HeI UPS spectra recorded by Brodén *et al.*,[17] for the adsorption of acetylene on the Ir(100)–(5 × 1) surface. Upper panel, spectra from clean iridium and from the adsorbed surface. Lower panel, difference spectra for three exposures of acetylene. Additional features due to acetylene are labelled G-N. The adsorption saturated at 5L.

11.4 Applications

11.4.1 *UPS*

Surface reactions The major use of UPS in both angle-integrated and angle-resolved forms has been in basic studies of the reactions of pure gases with well characterized surfaces, usually of metal. Since characterization in terms of surface cleanliness and structure is always necessary, which is not possible with UPS, the technique is therefore always used in conjunction with one or more complementary techniques such as XPS, AES and LEED.

Very typical of the sets of angle-integrated UPS spectra as a function of coverage that are common in the literature are those in Fig. 11.23 from the paper by Ya-po Hsu *et al.*[19] They were interested to know whether at very

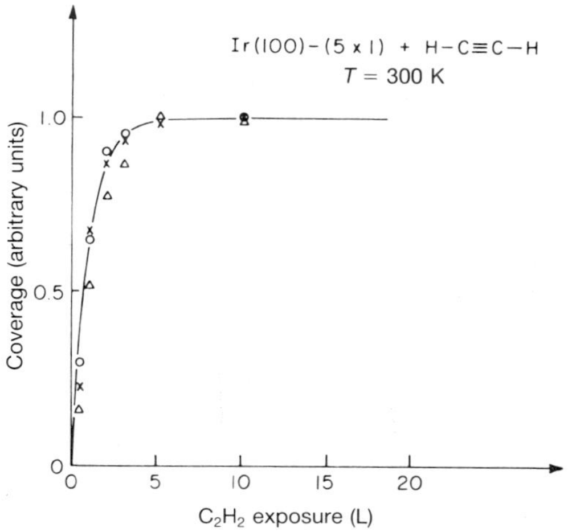

FIG. 11.20 Calibration by Brodén *et al.*,[17] of the uptake of acetylene on Ir(100)$-(5\times 1)$. They used the area under the peak L, the decrease in height of the d band at the Fermi edge, and the change in work function deduced from the change in total width of the photoemission curve, normalized to unity at saturation, to provide a self-consistent calibration. The measurements agreed with each other, and obeyed a first-order Langmuir type adsorption.

low temperature, c. 20 K, any evidence of chemical interaction between the Ni(110) surface, which is known to be highly reactive at room temperature, and the common gases O_2, CO, and N_2 could be found. The cleanliness of the crystal was checked with AES after the usual cycles of ion bombardment and annealing, and when clean it was cooled to 20 ± 5 K in a rotatable cryostat. Exposure of the cold crystal to each of the three gases to high coverage, 4–5 L, produced in every case a spectrum containing features characteristic of the condensed phase. Thus the four peaks for condensed O_2 in the top left panel correspond to $3\sigma_u$, $3\sigma_g$, $1\pi_u$ and $1\pi_g$ orbitals in molecular oxygen, while in the top right panel the three peaks for condensed CO correspond respectively to the 4σ, 1π and 5σ orbitals in molecular carbon monoxide. Before the condensed phases were reached, however, other changes had occurred. First, for both O_2 and CO in the earlier stages broad peaks developed in the 6–8 eV region, while there was a progressive shift in the secondary emission cut-off to lower binding energies, indicating a steady increase in work function. The positions and shapes of the peaks are known from previous work to be due to dissociated O and to chemisorbed CO, respectively, and the work function changes also are what would be expected from extraction of electrons from the Ni due to chemical reaction. Note from Fig. 11.23 that the Ni d-band near the Fermi level also decreased during the early stages, confirming the interaction. Therefore even at 20 K there was chemisorption of O_2 and CO. Secondly, for N_2 there were again features at saturation that could be associated with

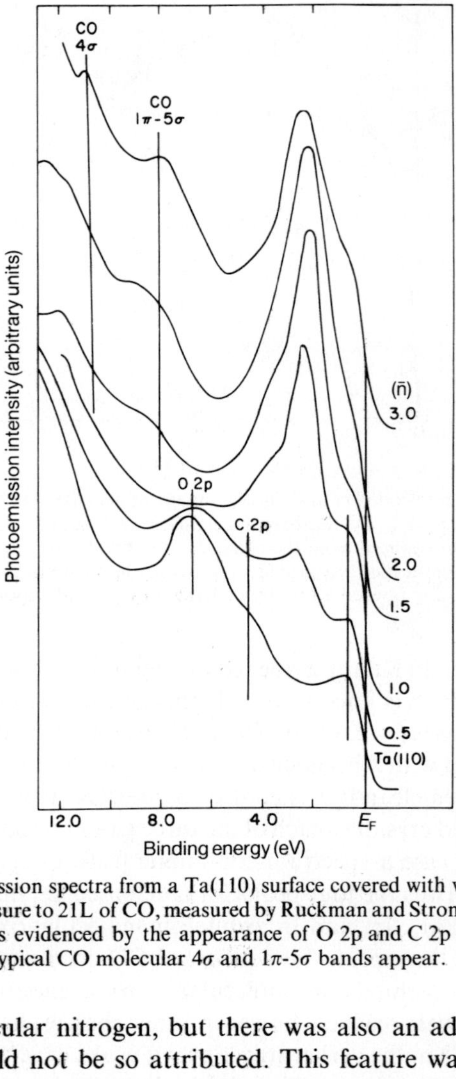

FIG. 11.21 Photoemission spectra from a Ta(110) surface covered with various thicknesses n̄ of palladium, after exposure to 21L of CO, measured by Ruckman and Strongin.[18] At low coverages the CO dissociates as evidenced by the appearance of O 2p and C 2p features, but at higher coverages of Pd the typical CO molecular 4σ and 1π-5σ bands appear.

orbitals in molecular nitrogen, but there was also an additional feature at 10.5 eV that could not be so attributed. This feature was visible at earlier stages of adsorption as well. In addition, there was *no* change in the work function during the exposure to N_2, although, as for O_2 and CO, the Ni d-band reduced progressively. The behaviour of nitrogen was therefore somewhat different from those of the other two gases. By carrying out angle-resolved measurements on the spectra following various coverages of N_2, and again by comparison with other work, the authors were able to come to the conclusion that N_2 does indeed chemisorb on Ni at 20 K, but molecularly as does CO, and that the N_2 molecules are probably adsorbed at some angle to the surface. However, the distinction between chemisorption and physisorp-

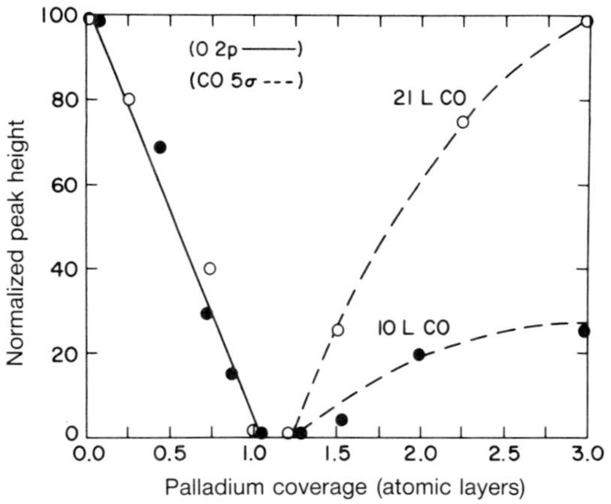

FIG. 11.22 Use of features in the spectra of Fig. 11.21 to estimate the coverage of CO on Ta(110) as a function of palladium coverage, by Ruckman and Strongin.[18] For Pd coverages below one monolayer the height of the O 2p feature was used, and above one monolayer the height of the CO molecular 1π-5σ feature. It is clear that at just over one monolayer coverage of Pd no chemisorption of CO occurs at all.

tion in the case of N_2 must be very blurred in view of the lack of measurable change in work function, even though the Ni d-band was depleted. Some chemical reaction must have occurred, but of a very weak nature. The overall result from UPS is thus that the three gases do indeed react chemically with Ni(100) surfaces at very low temperatures, implying negligible activation energies for chemisorption, with a question mark against N_2 as to the exact extent of genuine chemical interaction.

The usefulness of UPS in helping to identify the nature of different states of the same molecule on a surface has been demonstrated by Fisher[20] in a study of ammonia adsorption on the Pt(111) surface. According to the coverage of ammonia at low temperature three different states can be identified by thermal desorption spectroscopy (TDS) as shown in Fig. 11.24. At low coverages (spectra a–d), there was no evolution of mass 17 (NH_3) until the temperature reached 150–200 K, and there was then a maximum desorption rate at about 360 K. The state giving rise to this desorption peak is called the α state. At higher coverages, beyond spectrum e, another desorption peak appeared, narrow and well defined, at about 160 K, designated β; it grew rapidly in intensity as the coverage of NH_3 was progressively increased. For the highest exposures of NH_3, a third peak labelled γ, was observed, and since it was extremely sharp and its maximum desorption rate was at 110K, was attributed to the desorption from solid or multi-layer NH_3. When the TDS experiment was repeated for other masses, it was found that no H_2 or N_2 were desorbed over the same temperature range, indicating that the NH_3 must have been molecularly adsorbed. The question, therefore, was 'What is the nature of the α and β states?'

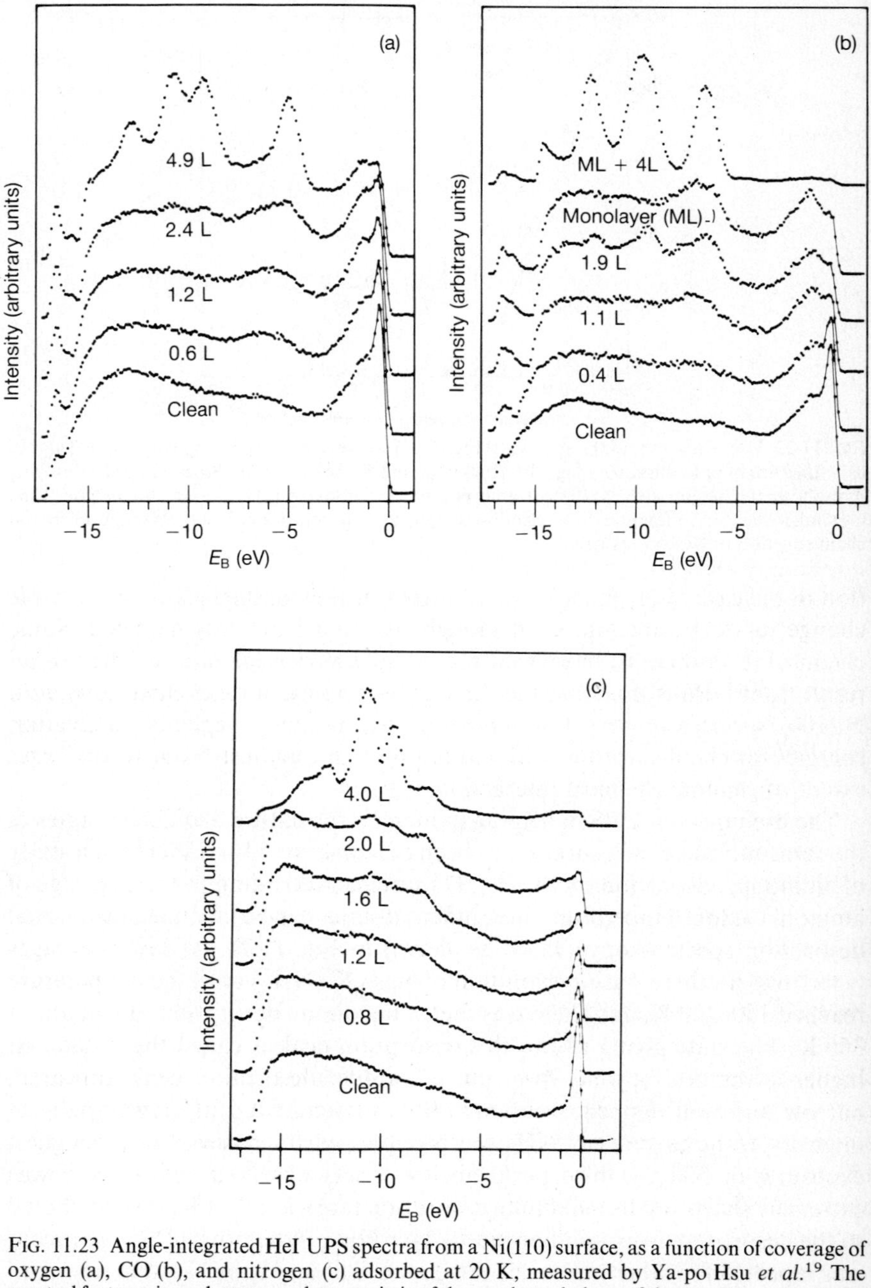

FIG. 11.23 Angle-integrated HeI UPS spectra from a Ni(110) surface, as a function of coverage of oxygen (a), CO (b), and nitrogen (c) adsorbed at 20 K, measured by Ya-po Hsu *et al.*[19] The spectral features in each case are characteristic of the condensed phase of the gas. For example, for condensed oxygen at the top of the upper panel, the four peaks correspond to the $3\sigma_u$, $3\sigma_g$, $1\pi_u$, and $1\pi_g$ molecular orbitals in oxygen. Note the progressive decrease in the nickel d-band intensity, indicating extraction of electrons due to reaction, even at 20 K.

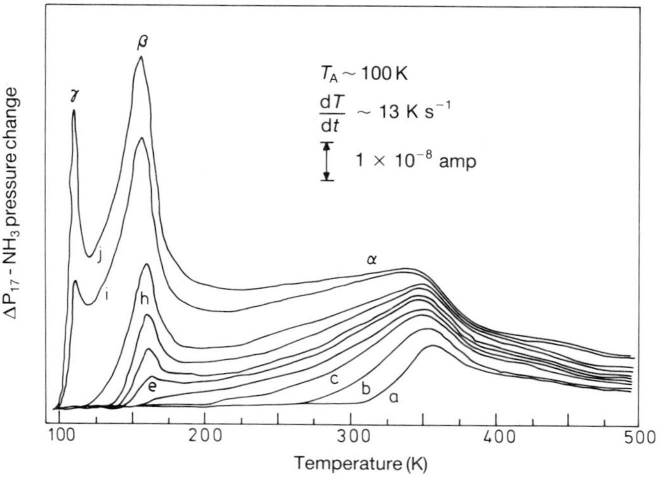

FIG. 11.24 Thermal desorption spectra (see Chapter 23) of ammonia adsorbed on Pt(111) at 100 K to various coverages, measured by Fisher.[20] The α and β states are labelled, the γ state being attributed to desorption from solid, i.e. multilayer, ammonia. The α state had a peak desorption at ≈ 360 K and the β state at ≈ 160 K. No H_2 or N_2 were found to desorb in other experiments.

Figure 11.25 shows the HeII-excited UPS spectra that have been used to provide part of the answer to the above question. At the bottom is the spectrum from clean Pt(111), and at the top the NH_3 gas-phase spectrum, shifted in energy as appropriate to match the adsorbate peaks. Spectra b–d are *difference* spectra. Spectrum d corresponds to the high-coverage thermal desorption spectra in Fig. 11.24, without peak γ since the crystal was warmed to 120 K to remove any solid NH_3. Spectrum b corresponds to the low-coverage regime of Fig. 11.24, after warming to 250 K to remove the β state. Thus spectrum d contains α and β together, but b contains α only, and to arrive at the spectrum corresponding to β only, a further subtraction of spectra b and d was carried out to produce spectrum c.

Comparison of the gas-phase spectrum with spectrum c shows that there is clearly a close correspondence between the two. Such close correspondence can only mean that the adsorbed NH_3 is rather loosely bound to the surface, and indeed that is confirmed by the low desorption temperature of about 160 K as well as by the low heat of desorption of 9 kcal mol^{-1}. It is therefore unlikely that the bonding to the surface is via the nitrogen atom, but through the hydrogen atoms; hydrogen bonding is usually weak.

The spectrum b for the low-coverage α state is different. There are two major peaks at 11.2 and 8.1 eV, which must be NH_3-related since this is a difference spectrum, and must refer to a molecular state. The only reasonable assignment is again to the same orbitals 1e and 3a$_1$ as in spectrum c, but this

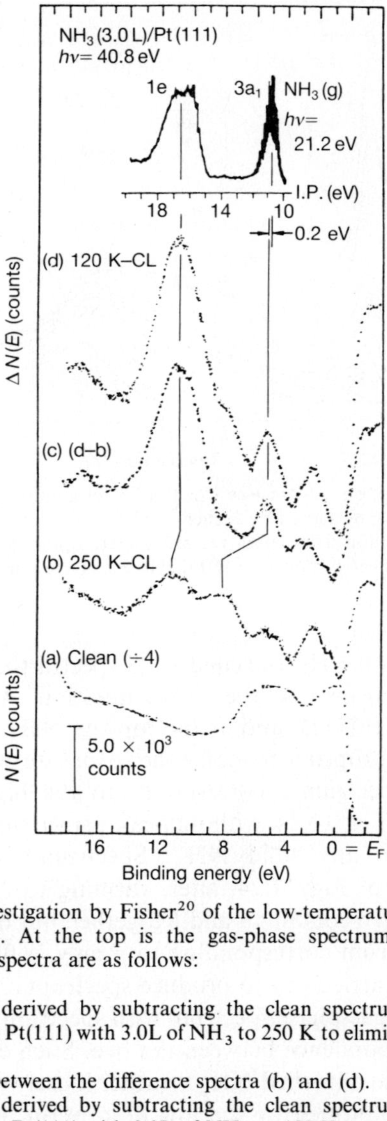

FIG. 11.25 Results of investigation by Fisher[20] of the low-temperature adsorption of NH_3 on Pt(111) using HeII UPS. At the top is the gas-phase spectrum of NH_3 recorded with HeI radiation. The other spectra are as follows:
(a) Clean Pt(111).
(b) Difference spectrum derived by subtracting the clean spectrum (a) from the spectrum recorded after heating Pt(111) with 3.0L of NH_3 to 250 K to eliminate the β and γ states (see Fig. 11.24).
(c) Difference spectrum between the difference spectra (b) and (d).
(d) Difference spectrum derived by subtracting the clean spectrum (a) from the spectrum recorded after heating Pt(111) with 3.0L of NH_3 to 120 K to remove the γ state only (see Fig. 11.24).
Thus curve (b) refers to the α state alone, and curve (c) to the β state alone.

time with a significant shift in the 1e–$3a_1$ splitting compared with that in the β state, of some 2.3 eV. The major shift is in the $3a_1$ orbital, which increases in binding energy. The most plausible explanation is that there is a substantial transfer of charge from this orbital, and this explanation is supported strongly by the change in the work function of about 3.0 eV during the time that the α state is building up. Such a large work function change implies a large induced dipole at the surface, which could occur only if the nitrogen atom of the NH_3 were close to the surface, and if the threefold axis of the molecule were normal to the surface. In other words, in the α state the NH_3 is adsorbed rather like an inverted umbrella.

Since the maximum coverage of the NH_3 in the α state was determined from the magnitude of the N 1s peak in XPS to be $\frac{1}{4}$ monolayer, and since some of the higher coverage desorption spectra show both α and β states, the two can clearly co-exist on the Pt surface. Consideration of the ways in which the two states could co-exist suggests that the α state is adsorbed in a threefold symmetric hollow in the (111) surface. The orientation of the molecule normal to the surface in the form of an inverted umbrella with the hydrogen atoms spread out would block any further adsorption in such sites, and the only subsequent possibilities for adsorption would be either on top of the surface metal atoms still exposed, or on top of the already adsorbed ammonia, via hydrogen bonding. The latter possibilities constitute the weakly bound β state.

Isolated atoms and clusters Catalysts in the form of dispersions on support material behave as if the catalytically active material were present as very small clusters of atoms or even individually isolated atoms. There has therefore been much study of both experimental and theoretical natures of the electronic structure of such small entities in an attempt to predict the catalytic behaviour as a function of cluster size. Clearly, if a particular size can be shown to be optimum for the efficiency or selectivity of a desired catalytic reaction then the preparation and subsequent treatment of the catalyst must be aimed at maintaining that size.

UPS has been able to contribute in this field in, for example, the interesting series of experiments conducted by Schmeißer *et al.*[21] They were able to produce Cu in the forms of both isolated atoms and clusters of different sizes, by the so-called matrix isolation technique. In this technique a substrate surface, in their case Ga chosen because of its featureless UPS spectrum, is cooled to 7 K, and then Cu is incorporated into an isolating matrix by co-condensation from a Knudsen source in the presence of one of the noble gases Ar, Kr or Xe. If the noble gas pressure is varied during deposition of the copper, then the metal/gas ratio can be changed controllably over the range 10^{-3}–10^{-2}, allowing either atoms or clusters to be embedded in the solid noble gas matrix.

The HeI and HeII-excited UPS spectra for pure Kr on Ga, for Cu atoms in Kr, and for Cu clusters in Kr, are shown in Fig. 11.26. Energies are referred to the Ga Fermi energy, which although very weak, is detectable by amplifying the HeI spectra as indicated. The major feature is the 4p

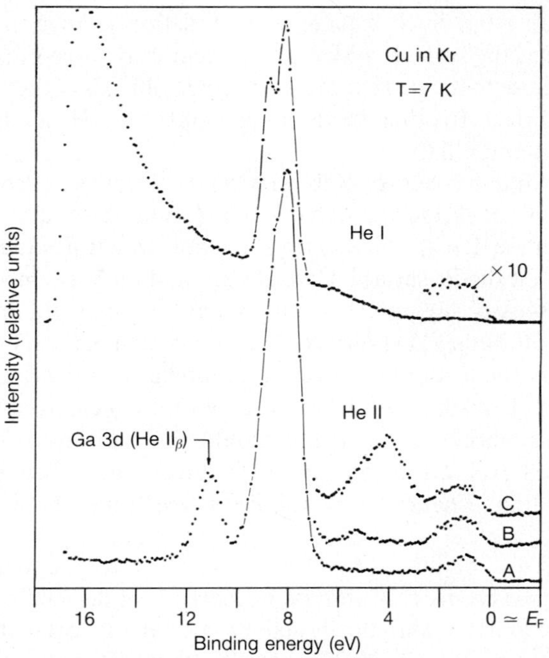

FIG. 11.26 HeI and HeII (curve A) UPS spectra of krypton condensed on a gallium substrate at 7 K, and HeII UPS spectra for copper atoms (curve B) and copper clusters (curve C) in solid krypton, measured by Schmeisser *et al.*[21] In the HeI spectrum the enlarged section demonstrates the Fermi edge for gallium. The spin-orbit splitting of the Kr 4p state is seen in the HeI spectrum but not in the HeII because of lower energy resolution used in the latter.

photoemission peak from Kr, in which the spin-orbit splitting of about 0.7 eV is visible in the HeI spectra since it was recorded at better energy resolution. When the solid Kr is doped with Cu, emission peaks appear in the energy region between 2 and 6 eV; that near 2 eV is in fact overlapping with another at $\approx$1.0 eV due to Kr 4p excitation by the HeII satellite line HeIIβ. The low energy parts of the HeII spectra in Fig. 11.26 are shown amplified in Fig. 11.27, and compared with the spectrum of bulk Cu taken under the same conditions. Spectrum b corresponds to an isolation of Cu as atoms almost exclusively, while c corresponds to clusters. It can be seen that a weak single peak at $\approx$5.1 eV is characteristic of the isolated atoms and a stronger broad band at $\approx$4 eV of the clusters. Similar results were found when Cu was isolated in solid Ar and Xe.

Because of its relatively high energy position the weak peak at 5.1 eV for the isolated atoms was assigned by the authors to the $3d^9 4s^1$ final state, so that the ground state would be $3d^{10}4s^1$, i.e. the same as in the gas phase. Comparison with gas-phase data produces a relaxation shift between isolated atoms in the

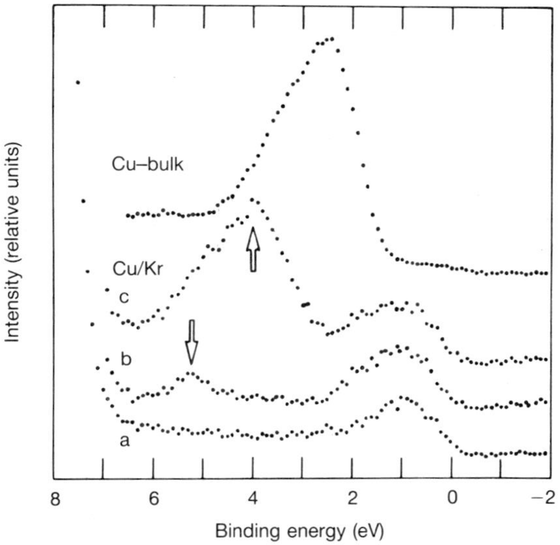

FIG. 11.27 Amplification of the low-energy portion of spectra A, B, and C in Fig. 11.26, with the spectrum from bulk copper for comparison. Isolation of Cu as atoms gives rise to a weak peak at ≈5.1 eV, and as clusters to a stronger broad band at ≈4 eV. From Schmeisser *et al.*[21]

gas and condensed phases of 1.3 eV, which was independent of the noble gas matrix. The independence on matrix also indicated that the relaxation shift was entirely intra-atomic, that is, governed by the Cu 4s electron.

As the concentration of condensed Cu is increased, the isolated atoms start first to 'feel' the proximity of other isolated atoms, and then to interact electronically to form clusters of increasing size. As this happens the atomic peak originally at 5.1 eV moves progressively to lower binding energy and broadens; spectrum c in Fig. 11.27 corresponds to a stage in this progression between the isolated atom and bulk copper. Since the metal/gas ratios used to produce the conditions for spectra b and c were 10^{-3} and 10^{-2}, respectively, i.e. it is likely that for spectrum c at least 10 Cu atoms were on average in the clusters, it is interesting to enquire as to how many Cu atoms are needed in a cluster for the band structure to resemble that from the bulk. The answer from theoretical cluster calculations[22,23] is that for a cluster of only six atoms the bands should already resemble those of the bulk. Clearly that is not the case based on these experiments, and the results give a clear indication that the theoretical ideas should be modified, and that the band structure of Cu clusters is markedly different from that of the bulk up to cluster sizes much larger than previously considered. If true, this has considerable implications for dispersed catalysts in consideration of the electronic reactions between catalyst particles (i.e. clusters) and reactant gases, since the particles will have electronic structure varying substantially across the range of particle size, with consequent variation in the nature of the reactivity. By using the basically simple technique of the above experiments, the electronic structure for any chosen size of cluster can be established.

11.4.2 *SRPS*

Interfaces In the last few years the electronic properties of the more or less abrupt interfaces between semiconductor surfaces and metal overlayers have assumed great importance because of the formation either of Schottky barriers or of ohmic contacts. Since the thickness of these overlayers varies between a few monolayers and a few score monolayers, they are clearly candidates for the application of surface analytical techniques. The essential information required is the nature of the relationship, if any, between the surface chemistry and the electronic properties.

In the example described earlier of the application of SRPS to the examination of an interface, that of Ag on Ge, the unique tunability of the photon energy from a synchrotron was used in conjunction with the Cooper minimum in the photoionization cross-section to emphasize or suppress photoelectron features as required. Another way of using the tunability is to choose that photon energy that will lead to ejected photoelectron energies corresponding to the minimum inelastic mean free path, or escape depth, i.e. to the maximum surface sensitivity. Reference to Fig. 2.7 shows that on average the minimum imfp occurs at a kinetic energy of 50–70 eV. This was the approach adopted by Petro *et al.*,[24] in their study of the Au–GaAs(110) interface. Ga and As have 3d binding energies at $\approx$20 eV and $\approx$45 eV, respectively, for which a photon energy of 100 eV was used, and Au has 4f binding energies at $\approx$85 eV, for which a photon energy of 130 eV was used. The surface sensitivity in each case was therefore of the order of 5 Å.

Au was deposited at room temperature on freshly vacuum-cleaved GaAs(110) surfaces to coverages ranging from 0.3 monolayers to 80 monolayers, while the Ga and As 3d and the Au 4f photoelectron spectra were monitored. Figure 11.28 shows a selection of the spectra at various chosen coverages. Several features may be noted. After deposition of the lowest coverage, 0.33 monolayers, of Au, both the Au and Ga peaks shift upwards in kinetic energy (downwards in binding energy) by $\approx$0.9 eV due to bending of the GaAs valence bands at the surface induced by the thin layer of Au. However, in both cases the core levels continue to shift, the As 3d only very slightly, but the Ga 3d by a further $\approx$0.4 eV at the highest coverage. The Ga peak shape remains the same throughout the coverage, with no evidence of any feature arising from metallic Ga, but at high coverages the As peak develops a pronounced shoulder at $\approx$0.6 eV higher binding energy, which probably indicates the formation of elemental As. Another interesting observation is that while the Ga peak intensity decreases monotonically as the Au coverage increases (but does not disappear!), the As peak intensity

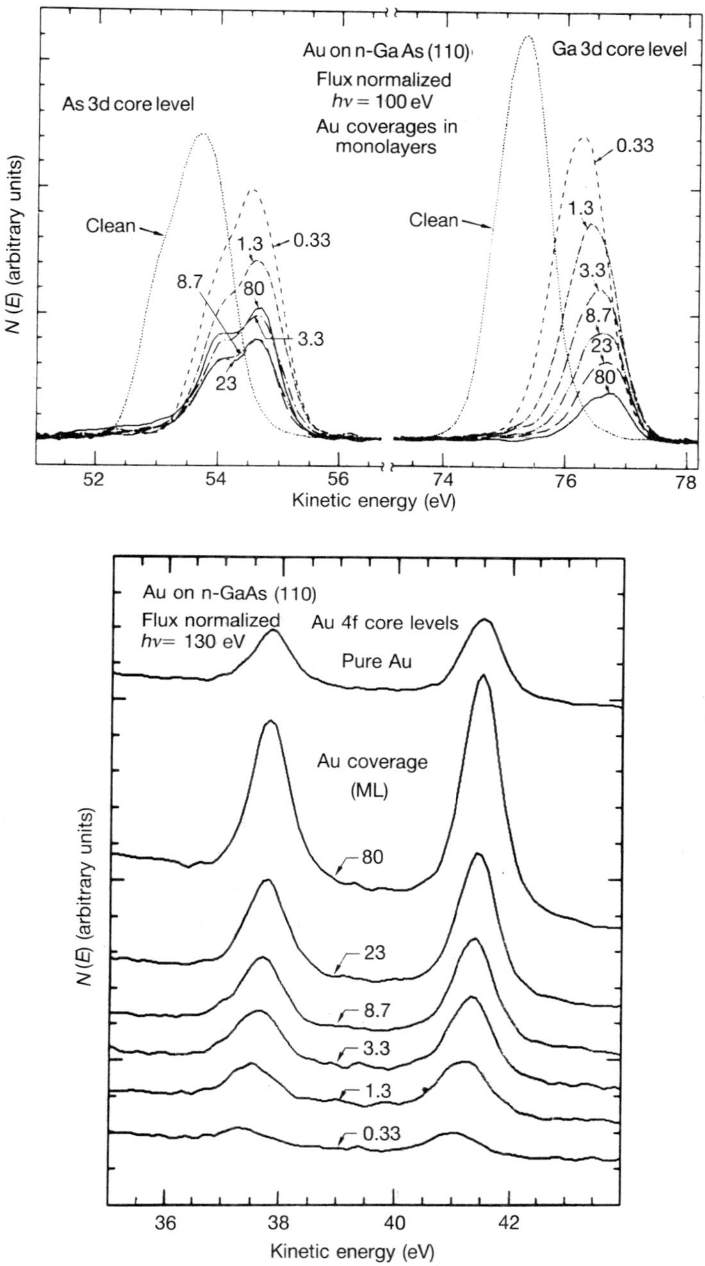

FIG. 11.28 Photoemission spectra of the As 3d and Ga 3d core levels (upper) and of the Au 4f core levels (lower) recorded as a function of the coverage of Au on GaAs(110) using synchrotron radiation, by Petro *et al.*[24] The top spectrum in the lower panel is from pure Au recorded under the same conditions. The photon energies of 100 eV and 130 eV, respectively, were chosen to restrict the depth of analysis to ≈5 Å in each case.

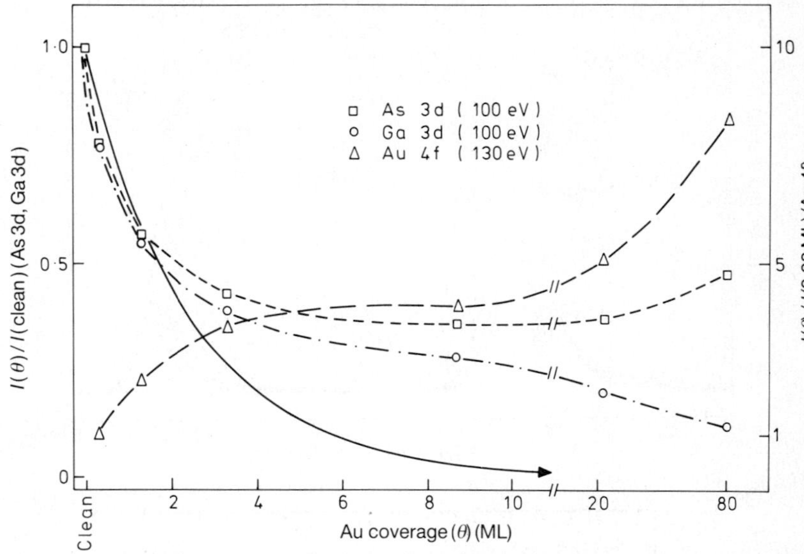

FIG. 11.29 Intensity profiles as a function of the coverage of gold on GaAs(110), measured from the As 3d, Ga 3d, and Au 4f peak areas in Fig. 11.28, by Petro *et al.*[24] The solid curve is the behaviour to be expected for the attenuation of the As and Ga areas if a uniform layer of Au were to be deposited, without any intermixing. The observed behaviour indicates an intermixed Au-Ga-As layer with the As/Ga ratio increasing with coverage.

appears to go through a minimum between 8 and 20 monolayers of Au. If the areas of the Ga 3d, As 3d and Au 4f peaks are plotted as a function of Au coverage, the result in Fig. 11.29 is obtained. The solid line is the expected attenuation of the Ga and As peaks if the Au overlayer were going down on the GaAs surface simply as an abrupt non-interacting uniform film. The lower half of Figure 11.28 shows a slight shift towards higher kinetic energy (lower binding energy) of ≈ 0.5 eV for the Au 4f peaks with increasing Au coverage. This shift is in the wrong direction to be due to any chemical interaction involving the Au and must be attributed to a change in screening and relaxation around the Au atoms.

On the above evidence there was quite clearly a strong intermixing of the Au and the GaAs in the surface layers, occurring even at room temperature. Neither the Ga or the As signals disappear, in fact at high coverages the As actually increases again and a contribution from elemental As appears. Thus, although there seems to be no actual chemical interaction, the Au is breaking down the GaAs structure at the surface, allowing the formation of free As, which segregates to the surface, leaving underneath a mixed layer of Au–Ga–As and Ga–As. The implications for the electronic properties of Schottky barriers or contacts formed by Au films are therefore considerable.

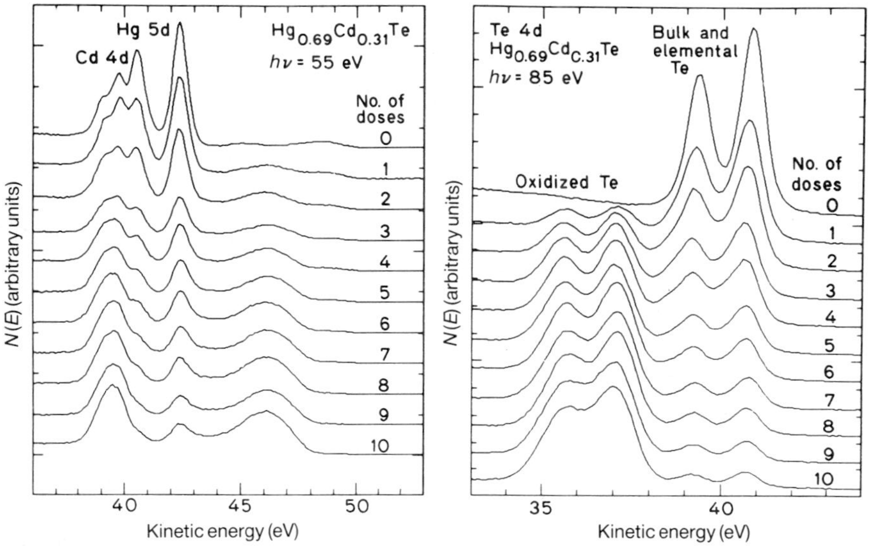

FIG. 11.30 Photoemission spectra of the Cd 4d and Hg 5d core levels (left) and the Te 4d core levels (right) as a function of exposure to oxygen, using synchrotron radiation, measured by Silberman *et al.*[25] The photon energies of 55 eV and 85 eV, respectively, were chosen to maximize the surface sensitivity to about 1 atom layer in each case and to ensure that the same inelastic mean free path applied to each element. The material was $Hg_{0.69}Cd_{0.31}Te$ whose surface was prepared by cleaving in vacuum. The Hg 5d spectrum attenuated progressively but did not show any other changes, there was little apparent effect on the Cd 4d spectrum, but the Te 4d spectrum showed very dramatic changes, from the first exposure.

Activated oxidation of semi-conductors Many semiconductors are rather difficult to oxidize in pure oxygen at room temperature, and recourse has to be made to the use of more extreme conditions such as elevated temperatures in steam. One of the obstacles to oxidation in molecular oxygen is the inability of semiconductor surfaces to dissociate the oxygen, which is the step that must precede all oxide formation. It has been found that if the oxygen is pre-dissociated or activated before it reaches the surface then much faster oxidation can be induced to occur. The activation can be achieved either with a hot filament placed near the sample surface or with laser irradiation.

Among the semiconductor materials whose oxidation is normally very slow are HgTe, CdTe and HgCdTe. Their enhanced oxidation in the presence of a hot filament has been studied by Silberman *et al.*[25] As in the example quoted above, photon energies were chosen to maximize the surface sensitivity, since the resultant oxide films were not expected to be very thick. Thus $h\nu = 55$ eV and $h\nu = 85$ eV were selected for the Cd 4d and Hg 5d, and the Te 4d, spectra, respectively, in order to place the kinetic energies in the 35–50 eV region. Oxygen exposure was carried out at a fixed pressure, 5×10^{-5} torr, and the doses were also fixed at 500 s each; thus each dose corresponded to 2500 L exposure.

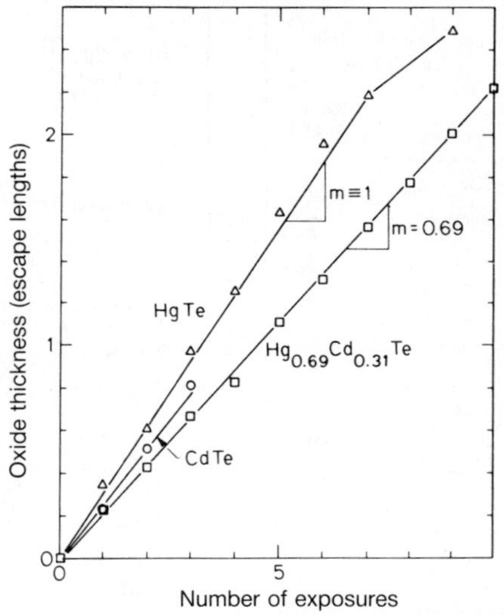

FIG. 11.31 Thickness of the oxide films produced on HgTe, CdTe, and on $Hg_{0.69}Cd_{0.31}Te$ as a function of oxygen exposure, derived from spectra similar to those in Fig. 11.30, by Silberman *et al.*[25] The assumptions made were an abrupt interface, a homogeneous distribution of Te atoms, and the same values of the inelastic mean free path in all materials. Oxidation proceeds linearly with exposure. The slope of the plot for $Hg_{0.69}Cd_{0.31}Te$ is just 69% of that for HgTe, suggesting that the rate of oxidation depends on the number of Hg-Te bonds, and that the important step is therefore the breaking of the Hg-Te bond.

The changes in the Cd, Hg and Te photoelectron spectra from $Hg_{0.69}$ $Cd_{0.31}$ Te as a function of the number of doses in the presence of the hot filament are shown in Fig. 11.30. With increasing dose a peak arises near 46 eV kinetic energy which can be attributed to the oxygen 2p level, and the Hg $5d_{5/2}$, $5d_{3/2}$ doublet becomes progressively attenuated. On the other hand, there is little apparent effect on the Cd 4d peak. The most striking changes occur in the Te 4d spectra in the right-hand panel of the figure, for with the first dose another doublet appears, shifted by 3.0 eV with respect to Te in the compound; with increasing dose the new doublet intensity increases as the original decreases. The shift is of course due to the formation of oxidized Te. The reduction in the Hg peak and the oxidation of Te were found also for HgTe; however, for CdTe the Cd was found as a constituent in the oxide.

By assuming an abrupt oxide/semiconductor interface, a homogeneous distribution of Te atoms, and the same inelastic mean free path in both media, the thicknesses of the growing oxidized Te films could be estimated. These are shown plotted as a function of oxygen dose in Fig. 11.31. Oxidation is seen to proceed in proportion to the exposure, with little evidence of saturation as far as the experiment went. Interestingly, the rate of oxidation of the $Hg_{0.69}Cd_{0.31}Te$, based on the slope of the plot, is found to be just 69% of the rate of HgTe. That is, the rate of oxidation scales directly with the number of

Hg–Te bonds. Taking into account the progressive disappearance of Hg, as evidenced in Fig. 11.30, this suggests that the important step in the oxidation is the breaking of the Hg–Te bond, following which free Hg is released and is not incorporated in the growing oxide. Figure 11.31 also shows a few data for CdTe. If HgCdTe were to oxidize merely as a mixture of HgTe and CdTe, its rate would be expected to fall between those of the latter two compounds. Since it does not, there is clearly a modification of the Cd–Te bond in $Hg_{0.69}Cd_{0.31}Te$ compared to the binary.

The above oxidation data could not have been obtained in any reasonable time without activation of the oxygen by the hot filament, nor would such unambiguous results have been obtainable without the ability to tune synchrotron radiation to maximize surface sensitivity. Enhancement of reactivity to oxygen through activation has also been studied by Petro *et al.*[26] for GaAs(110) using laser irradiation.

References

1. Plummer, E. W. and Eberhardt, W. (1982). *Adv. Chem. Phys.*, **49**, 533.
2. Smith, N. V. and Mattheiss, L. F. (1974). *Phys. Rev. B* **9**, 1341.
3. Ertl, G. and Küppers, J. (1985). *Low energy electrons and surface chemistry* VCH Verlag, Weinheim.
4. Pessa, M., Heimann, P., and Neddermeyer, H. (1976). *Phys. Rev. B* **14**, 3488.
5. Singh, M., Wang, C. S., and Callaway, J. (1975). *Phys. Rev. B* **11**, 287.
6. Ehlers, D. H., Spitzer, A., and Lüth, H. (1985). *Surf. Sci.* **160**, 57.
7. Netzer, F. P., Mack, J.-U. (1983). *Chem. Phys. Lett.* **95**, 492.
8. Leckey, R. C. G. (1982). *Applic. Surf. Sci.* **13**, 125.
9. Tokailin, H., Takahashi, T., Sagawa, T., and Shindo, K. (1984). *Phys. Rev. B* **30**, 1765.
10. Kane, E. O. (1964). *Phys. Rev. Lett.* **12**, 97.
11. Heimann, P., Miosga, H., and Neddermeyer, H. (1979). *Solid State Commun.* **29**, 463.
12. Kevan, S. D., Stoffel, N. G., and Smith, N. V. (1985). *Phys. Rev. B* **31**, 1788.
13. Cooper, J. W. (1964). *Phys. Rev. Lett.* **13**, 762.
14. Lindau, I., Pianetta, P., and Spicer, W. E. (1976). *Phys. Lett.* **57A**, 225.
15. Rossi, G., Abbati, I., Braicovich, L., Lindau, I., and Spicer, W. E. (1982). *Phys. Rev. B* **25**, 3619.
16. Johansson, L. I., Lindau, I., Hecht, M. M. and Källne, E. (1980). *Solid State Commun.* **34**, 83.
17. Brodén, G., Rhodin, T., and Capehart, W. (1976). *Surf. Sci.* **61**, 143.
18. Ruckman, M. W. and Strongin, M. (1984). *Phys. Rev. B* **29**, 7105.
19. Ya-po Hsu, Jacobi, K., and Rotermund, H. H. (1982). *Surf. Sci.* **117**, 581.
20. Fisher, G. B. (1981). *Chem. Phys. Lett.* **79**, 452.
21. Schmeißer, D., Jacobi, K., and Kolb, D. M. (1981). *J. Chem. Phys.* **75**, 5300.
22. Rösch, N. and Menzel, D. (1976). *Chem. Phys.* **13**, 243.
23. Messmer, R. P., Knudson, S. K., Johnson, K. H., Diamond, J. B., and Yang, C. Y. (1976). *Phys. Rev. B* **13**, 1396.

24. Petro, W. G., Babalola, I. A., Kendelewicz, T., Lindau, I., and Spicer, W. E. (1983). *J. Vac. Sci. Technol.* **A1**, 1181.
25. Silberman, J. A., Laser, D., Lindau, I., and Spicer, W. E. (1985). *J. Vac. Sci. Technol.* **A3**, 222.
26. Petro, W. G., Hino, I., Eglash, S., Lindau, I., Su, C. Y., and Spicer, W. E. (1982). *J. Vac. Sci. Technol.* **21**, 405.

12

Photon excitation: reflection–absorption infra-red spectroscopy (RAIRS) and surface-enhanced Raman spectroscopy (SERS)

12.1 Reflection–Absorption Infra-Red Spectroscopy

This technique is known also as Infra-Red Reflection Absorption Spectroscopy (IRAS), by a rearrangement of title words, but RAIRS seems now to be more generally accepted. As will be described below, the modern version of infra-red spectroscopy employing Fourier transform techniques has also been applied here, leading to the acronym variations FT-RAIRS and FT-IRAS.

12.1.1 *Operation*

In conventional transmission infra-red spectroscopy the spectra before and after passage of infra-red radiation through a specimen are compared, which gives rise to a set of absorption bands at frequencies characteristic of the material of the specimen. Changes in the absorption spectrum as a consequence of surface reaction then give information about the nature of the reaction products at the surface. However, the application to general surface studies is severely restricted since the substrate material must clearly be both infra-red transparent and very thin (e.g. $<200\ \mu m$), while the changes that occur at the surface are only a very small proportion of the total absorption. Since the majority of surface studies are concerned with metal surfaces and with very thin layers of adsorbed species, conventional transmission techniques are obviously inapplicable. Instead reflectance techniques must be used.

In RAIRS infra-red radiation is reflected from a plane surface through an adsorbed layer, in which the reflected light loses intensity due to absorption in the layer. The intensity losses occur at those frequencies at which the light frequency matches that of one of the vibrational modes at the surface, corresponding to vibrations either within the adsorbed species itself or as a result of interaction with surface atoms of the substrate. RAIRS is thus very closely related to HREELS (q.v.).

To be applicable at all, there are several requirements that must be met in RAIRS. First, and most obviously, the substrate surface must be an extremely good reflector. It is normal to use single crystal faces of metals that can be polished to a mirror finish, such as those of Cu, Ag, Au, Pt, Al, and Ni.

Secondly, the absorption cross-section of the adsorbed layer at the characteristic frequencies must be sufficiently large that the resultant signal-to-noise ratio allows adequate energy resolution to be used. Thirdly, an important factor, that will be discussed below under theory, is the use of plane polarized radiation. Fourthly, also discussed below, the angle of incidence of the infra-red radiation must be close to grazing; this requirement also imposes conditions of near-atomic flatness, i.e. high reflectivity.

These requirements have led workers to the type of experimental arrangement shown in Fig. 12.1, from Erley.[1] His apparatus consisted of a three-level vacuum system in which HREELS could be performed at the lowest level, AES and LEED at the middle level and FT-RAIRS at the upper level. Infra-red light from the FT interferometer is focused by a KBr lens through a KBr window onto the sample surface at an incident angle of 80°. After reflection the light passes through a wire grid polarizer and leaves the UHV chamber by another KBr window. KBr has to be used as window material since it is transparent to the infra-red radiation. Outside the chamber the light is collimated by another KBr lens and is then collected by an elliptic mirror which focuses it onto a liquid nitrogen cooled HgCdTe detector. Both interferometer and detector housings are evacuated to $\approx 10^{-2}$ torr to reduce absorption due to atmospheric molecules. The incidence angle of 80° was chosen as a compromise between the theoretically optimum angle of 88°, and reduction of signal due to the illuminated area becoming much larger than the sample area near the optimum. The wire polarizer passes only radiation polarized parallel to the plane of incidence.

To date most RAIRS has been performed with the use of dispersive spectrometers. Various specialized modifications of the technique have been devised, all with the aim of improving signal-to-noise ratios, but it seems that they may be superseded with the advent of Fourier transform methods. The theory of FT spectroscopy has been outlined by Bell,[2] and may be illustrated here with the help of some of his diagrams.

Figure 12.2 shows schematically the optical arrangement in an FT spectrometer. A parallel infra-red beam enters a Michelson interferometer in which it is split by a beam splitter at 45° to the axis of the beam, half being reflected to a fixed mirror and the other half to a movable mirror. After reflection, the beams recombine at the splitter. The path lengths of the two parts of the beam after reflection will not be exactly equal, and therefore after recombination the two beams will interfere either constructively or destructively. If the movable mirror is then displaced backwards and forwards along the axis the recombined beam leaving the interferometer will be modulated. For a displacement frequency v the modulation frequency at a wave number $\bar{v}$ will be given by $2v\bar{v}$. The modulated beam is then reflected out of the interferometer and focused on the specimen as indicated in Fig. 12.1.

Because of the modulation of the incident radiation the signal observed by

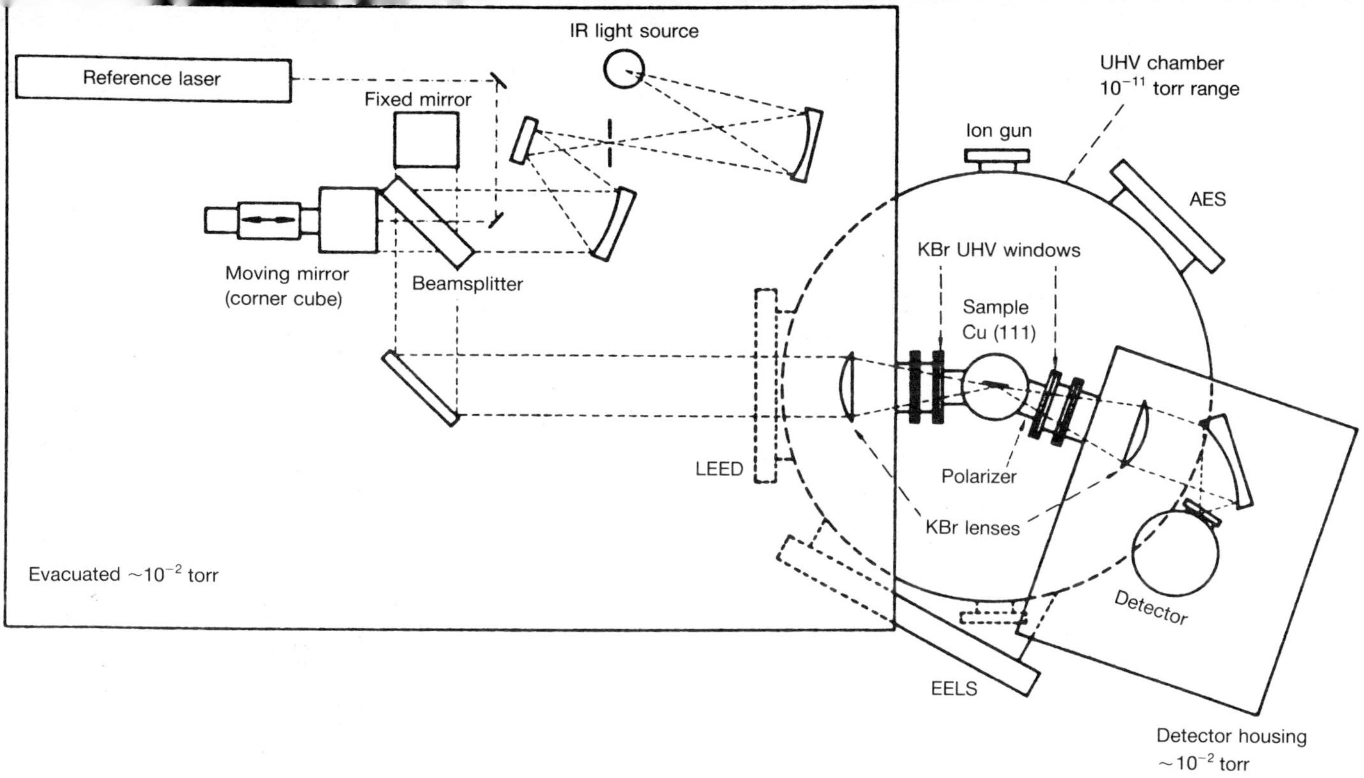

FIG. 12.1 Multi-technique system developed by Erley[1] for combined HREELS, AES, LEED, and FT-RAIRS, at three different levels. For the FT-RAIRS measurements light from the FT interferometer (Mattson Sirius 100) on the left is focused by a KBr lens through a window of KBr onto the sample surface at an angle of 80°. After reflection from the sample surface the light passes through a wire grid polarizer and passes out to a liquid nitrogen cooled HgCdTe detector, again via KBr window and lens. An elliptic mirror focuses the light onto the detector. The wire polarizer passes only light polarized parallel to the plane of incidence.

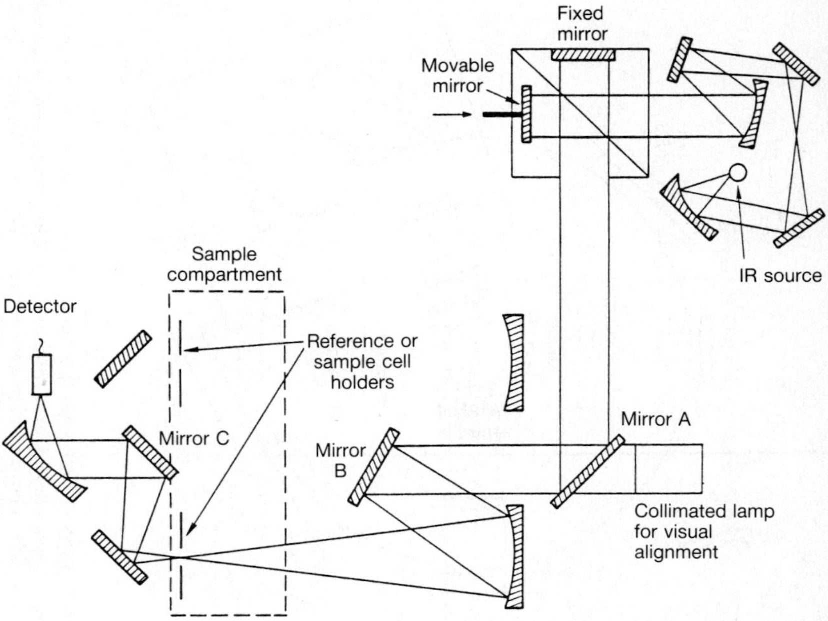

FIG. 12.2. Schematic diagram of the typical optical arrangement in an FT spectrometer, from Bell.[2] A series of mirrors picks up the light from the infra-red source, and passes it as a strictly parallel beam into a Michelson interferometer. There it is split by a 45° beam splitter, half the beam being reflected to a fixed mirror and half to a movable mirror. After reflection the beams recombine at the splitter. Their path lengths at that stage will not be equal, and therefore constructive or destructive interference will occur. If the movable mirror is then displaced back and forth along its axis the recombined beam will be intensity modulated. The beam is then reflected out of the interferometer and focused on the specimen. Because of the modulation on the incident beam, the detected signal will also be modulated, giving rise to an interferogram. To recover the spectrum from the interferogram it is necessary to take the Fourier transform of the fluctuating contribution.

the detector after reflection from the specimen will be fluctuating, and is known as an interferogram. If the displacement of the movable mirror is s then the interferogram intensity is a function $I(s)$ of s. To recover a spectrum $(I\bar{v})$ from the interferogram it is necessary to take the Fourier transform of the fluctuating contribution to $I(s)$, which is $\bar{I}(s)=\{I(s)-0.5I(0)\}$. Thus the spectrum recovered is

$$I(\bar{v})=\int_{-\infty}^{\infty}\bar{I}(s)\cos(2\pi\bar{v}s)\,\mathrm{d}s \qquad (12.1)$$

In Fig. 12.3, from Bell,[3] are shown examples of spectra recovered from $\bar{I}(s)$. In (a) the natural relationship between $\bar{I}(s)$ and $I(\bar{v})$ is given for a spectrum consisting of three lines of infinitely narrow widths, for integration over s from $-\infty$ to $+\infty$. Clearly, however, s cannot be varied over an infinite range, and

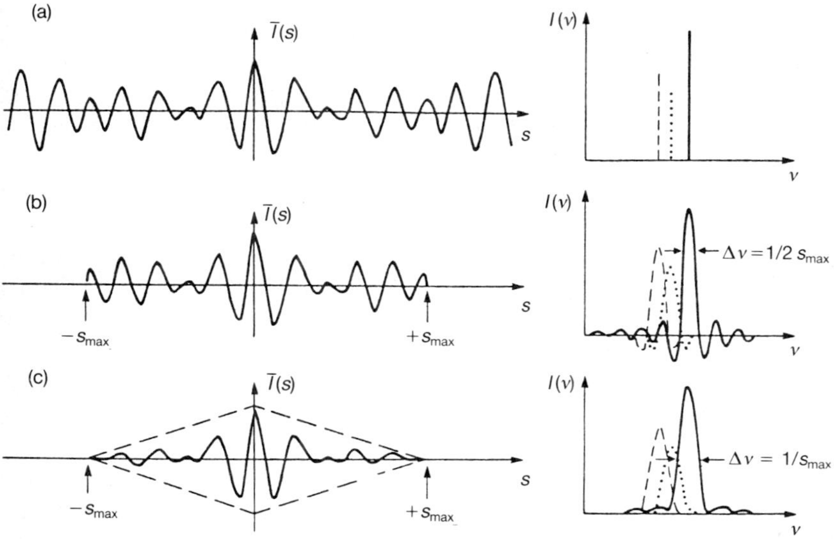

FIG. 12.3. Examples from Bell[2] of the spectra recovered from interferograms produced in various ways. In each case $\bar{I}(s)$ is the fluctuating contribution to the detected intensity, and $I(\nu)$ is the corresponding infra-red spectrum.
(a) The relationship between $\bar{I}(s)$ and $I(\nu)$ for a hypothetical spectrum consisting of three lines of infinitely narrow width. s extends from $-\infty$ to $+\infty$.
(b) Consequence of determination of s over a finite range, from $-s_{max}$ to $+s_{max}$. The resultant spectrum contains both broadened peaks and secondary maxima to each side of the main peaks, arising from the mathematical transformation.
(c) Example of the use of triangular apodization, in which $\bar{I}(s)$ is multiplied by a triangular function which causes it to approach zero at $s = \pm s_{max}$. The secondary maxima are thereby much reduced, but the peaks are broadened further.

an interferogram can be determined experimentally only over a finite range, leading to truncation of the integration limits. The consequence of variation of s from $-s_{max}$ to $+s_{max}$ is shown in (b). Broadening of the spectral lines results, and also the appearance of 'wings' or secondary maxima to each side of the main peak. The latter represent a nuisance since they can mask a weak spectral feature close to a strong one, and to minimize this interference the technique known as triangular apodization is used, in which the interferogram is multiplied by a function which causes it to approach zero in a continuous fashion at $s = \pm s_{max}$. This apodization is indicated schematically in (c), where the resultant spectrum has much weaker 'wings'.

Since truncation of the integration limits causes broadening of the spectral features, it is clear that the energy resolution available in the technique must be a function of the maximum displacement of the movable mirror, and indeed it can be shown that the resolution $\Delta\bar{\nu}$ is given by

$$\Delta\bar{\nu} = 1/s_{max} \tag{12.2}$$

where triangular apodization is used.

The advantages of FT techniques over dispersive techniques are twofold. Firstly, the high optical transmission in FT, which does not need defining slits, allows either much improved signal-to-noise ratios for the same data collection time, or more rapid data collection for the same signal-to-noise. It is possible to record an interferogram of the full mid-infra-red range (i.e. 400–4000 cm^{-1}) in less than a second. Secondly, in FT spectroscopy all spectral features are recorded simultaneously, whereas a dispersive spectrometer has to scan across the spectral range with a narrow energy window, recording features in succession. This is called the multiplex advantage. If only a very narrow spectral range is of interest, the latter advantage diminishes rapidly.

It is instructive to compare the advantages and disadvantages of FT-RAIRS and HREELS, since they are so complementary. The principal advantages of FT-RAIRS over HREELS are those of energy resolution, speed, and ability to operate in non-vacuum environments, e.g. at high pressures. Thus, while a typical resolution used in FT-RAIRS would be 2 cm^{-1}, that in HREELS would be in the range 20–60 cm^{-1}. On the other hand, owing to the characteristics of thermal sources in the infra-red, and to the transmission characteristics of optical components such as the KBr windows and lenses, the effective lower end of the spectral region attainable in FT-RAIRS is at about 400 cm^{-1}, whereas HREELS can cover the whole range almost to 0 cm^{-1}. It is therefore desirable to combine both techniques in the same instrument, which is in fact what has been done by Erley in the system shown in Fig. 12.1. At the moment there are very few systems allowing such a combination of techniques.

12.1.2 *Theory and spectral interpretation*

The theory of external reflectance infra-red spectroscopy was developed by Francis and Ellison[3] and by Greenler,[4] the latter having put the technique on a quantitative footing by showing how to calculate the absorption intensity from an isotropic model. The important qualitative consequences of the theory can be understood with reference to Fig. 12.4, which shows the electric fields associated with incident and reflected radiation at a bare metal surface, for s-(i.e. perpendicular)-polarized and p-(i.e. parallel)-polarized incidence. The plane of incidence is the xz plane, and the surface fields arising from the radiation have been resolved into their respective vector components perpendicular and parallel to the surface. On reflection the s-polarized component reverses phase by almost 180°, i.e. from E_s to E'_s, at all angles of incidence, and with a reflection coefficient near unity in the infra-red region for metal surfaces, the vector sum is close to zero. On the other hand, the p-polarized component undergoes a phase change which is strongly dependent on the angle of incidence, from E_p to E'_p, so that the vector sum, or

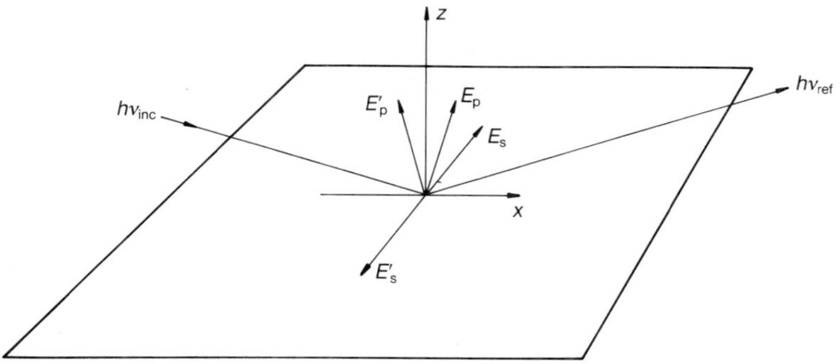

FIG. 12.4. Schematic of the electric fields associated with incident and reflected radiation at a bare metal surface. E_s and E_p are the s-polarized and p-polarized, respectively, vector components of the incident radiation, and E'_s, E'_p the corresponding components of the reflected radiation. The plane of incidence is xz. The s-component is almost reversed in phase on reflection at all angles, whereas the p-component undergoes a phase change strongly dependent on the angle of incidence. With a reflection coefficient near unity for metal surfaces in the infra-red, the s-vector sum is close to zero, while the p-vector sum varies with angle.

surface field varies with the angle. The practical consequence of the latter is illustrated in Fig. 12.5, in which the perpendicular and tangential vector components of E_p, as well as E_s, are plotted as a function of angle of incidence. It can be seen that at grazing incidence the perpendicular electric vector $E_{p\perp}$ is almost doubled, while the tangential electric vector $E_{p\parallel}$ remains insignificant at all angles. The $E_{p\perp}$ component can therefore couple strongly to any vibrational modes associated with adsorbed species that have dipole derivative perpendicular to the surface, but the sensitivity to any parallel modes of vibration is far lower. This constitutes an important selection rule in RAIRS.

Another important practical factor is the enhancement of the absorption intensity as the angle of incidence is increased, by virtue of an increase in the number of adsorbed species sampled with an extension in the area irradiated. This area varies as sec θ, where θ is the angle of incidence, so that the total absorption intensity is proportional to E^2 sec θ, E being the amplitude of the resultant electric field component. The angular dependences of each component relative to E_0, the amplitude of the incident beam, are shown in Fig. 12.6, where it can be seen that there is a sharp maximum near grazing incidence. The optimum angle is in fact 88°.

The mathematical analysis of the conditions for obtaining an infra-red absorption spectrum from molecules adsorbed on a metal surface was first given by Greenler.[4] Subsequently the analytical expressions were simplified by McIntyre and Aspnes[5] using linear approximations as described below, for the case where the thickness of the adsorbed layer is very much less than the wavelength of the infra-red radiation, which is of course always the case.

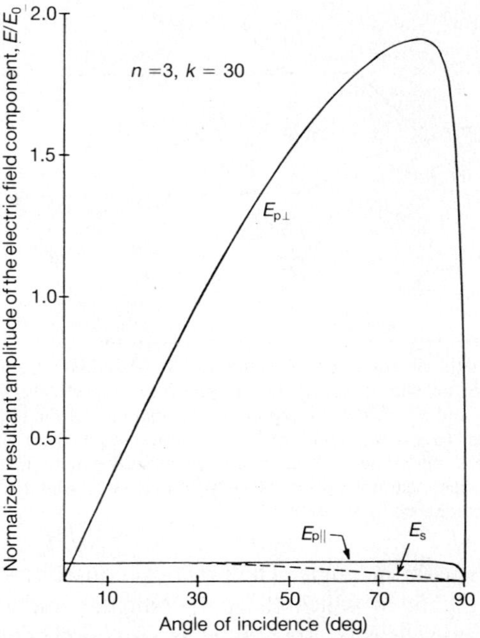

FIG. 12.5. Dependence on angle of incidence of the amplitudes of electric field vectorial components reflected from a metal surface, calculated by Greenler[4] for a metal with refractive index $n=3$ and absorption index $k=30$. Near grazing incidence ($\theta \approx 90$), the perpendicular component of the p-polarized contribution can be seen to be nearly doubled relative to E_0, the amplitude of the incident beam. The tangential component is much weaker, while E_s falls to zero. The perpendicular component of the p-polarized contribution can therefore couple strongly to any vibrational modes having dipoles perpendicular to the surface.

Consider a three-phase system consisting of an ambient phase (in this case vacuum), a thin film of thickness d, and a metallic substrate, labelled 1, 2, and 3, respectively, whose complex refractive indices are $\tilde{n}_1$, $\tilde{n}_2$, and $\tilde{n}_3$. In terms of the real refractive index n_i and absorption coefficient k_i, $\tilde{n}_i = n_i - ik_i$. Since the complex refractive indices are related to the complex dielectric constants $\tilde{\epsilon}$ by $\tilde{n}_i = (\mu_i \tilde{\epsilon}_i)^{1/2}$, and since the magnetic permeability μ_i can be taken as unity in the optical region, the real refractive indices and real dielectric constants are related by $\epsilon_i = n_i^2 - k_i^2$.

For reflection of radiation of wavelength λ (in vacuum) from such a three-phase system, the total electric field amplitude reflected back into the ambient is given by the Fresnel reflection coefficient

$$r = \frac{r_{12} + r_{23} e^{-2i\beta}}{1 + r_{12} r_{23} e^{-2i\beta}} \tag{12.3}$$

where r_{12} refers to the phase 1/phase 2 interface between ambient and film and

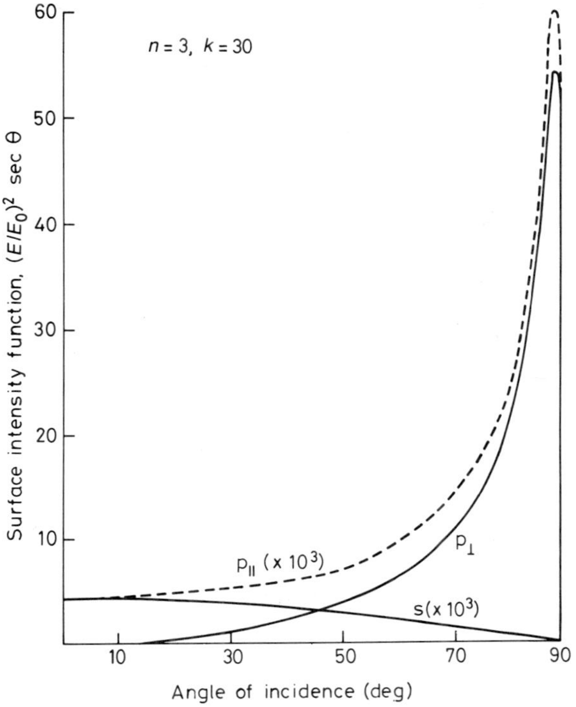

FIG. 12.6. Enhancement in absorption intensity as the incident angle is increased, due to an increase in the number of adsorbed species sampled as the irradiated area increases. The area varies as sec θ, where θ is the angle of incidence, so that the total absorption intensity is E^2 sec θ, E being the amplitude of the electric field component. Calculated by Greenler[4] for the same metal optical constants as in Fig. 12.5. There are sharp maxima in the angular dependences of the p-polarized components, relative to E_0. The optimum angle is 88°.

r_{23} to the phase 2/phase 3 interface between film and substrate. β is the change in phase of the light beam caused by one traversal of phase 2, the film, and is given by

$$\beta = (2\pi \tilde{n}_2 d \cos \phi_2)/\lambda \tag{12.4}$$

where ϕ_2 is the phase thickness of the film and, for an absorbing film, is complex. The corresponding quantities for phases 1 and 3 are ϕ_1 and ϕ_3. Note, however, that because phase 1, the vacuum ambient, is completely transparent, $k_1 = 0$ and therefore, n_1, ϵ_1, and ϕ_1 are all real. Thus ϕ_1 is the real angle of incidence of the radiation.

The linear approximation comes in the expansion of eqn (12.3). For $d \ll \lambda$, the first-order expansion of r is

$$r = \frac{r_{12} + r_{23}(1 - 2i\beta)}{1 + r_{12}r_{23}(1 - 2i\beta)}. \tag{12.5}$$

Expression (12.5) refers to the three-phase system. If the amplitude in the absence of the film, i.e. for a two-phase system, is r_0, then it can be shown that

$$r/r_0 = \frac{1-2i\beta\{r_{23}/(r_{13}+r_{23})\}}{1-2i\beta\{r_{12}r_{23}/(1+r_{12}r_{23})\}} \tag{12.6}$$

which can again be expanded neglecting terms of higher order than β to give

$$r/r_0 = 1+2i\beta\left\{\frac{r_{23}(r_{12}^2-1)}{(r_{12}+r_{23})(1+r_{12}r_{23})}\right\}. \tag{12.7}$$

Now the intensity of the reflected beam, i.e. the reflectivity, is given by $R=rr^*$, where r^* is the complex conjugate of r. If (12.7) is therefore multiplied by its complex conjugate, and as before only first order terms in (d/λ) are retained, one has for perpendicularly polarized light

$$(R/R_0)_\perp = 1+\frac{(8\pi d n_1 \cos\phi_1)}{\lambda}\,\mathrm{Im}\left(\frac{\tilde{\epsilon}_2\cos^2\phi_2-\tilde{\epsilon}_3\cos^2\phi_3}{\epsilon_1\cos^2\phi_1-\tilde{\epsilon}_3\cos^2\phi_3}\right) \tag{12.8}$$

and for parallel polarized light

$$(R/R_0)_\parallel = 1+\frac{(8\pi d n_1 \cos\phi_1)}{\lambda}\,\mathrm{Im}\left(\frac{\tilde{\epsilon}_3\cos^2\phi_2-\tilde{\epsilon}_2\cos^2\phi_3}{\epsilon_3\cos^2\phi_1-\epsilon_1\cos^2\phi_3}\right). \tag{12.9}$$

In order to put these expressions in forms similar to the experimentally measured quantities, the change in reflectivity can be defined as

$$\Delta R/R = (R-R_0)/R_0 \tag{12.10}$$

from which, for the optical frequency range, one has

$$(\Delta R/R)_\perp = \frac{8\pi d n_1\cos\phi_1}{\lambda}\,\mathrm{Im}\left(\frac{\tilde{\epsilon}_2-\tilde{\epsilon}_3}{\epsilon_1-\tilde{\epsilon}_3}\right) \tag{12.11}$$

and

$$(\Delta R/R)_\parallel = \frac{8\pi d n_1\cos\phi_1}{\lambda}\,\mathrm{Im}\left[\left(\frac{\tilde{\epsilon}_2-\tilde{\epsilon}_3}{\epsilon_1-\tilde{\epsilon}_3}\right)\left\{\frac{1-(\epsilon_1/\tilde{\epsilon}_2\tilde{\epsilon}_3)(\tilde{\epsilon}_2+\tilde{\epsilon}_3)\sin^2\phi_1}{1-(1/\tilde{\epsilon}_3)(\epsilon_1+\tilde{\epsilon}_3)\sin^2\phi_1}\right\}\right] \tag{12.12}$$

When the changes $(\Delta R/R)$ for the two directions of polarization are computed from (12.11) and (12.12), for real situations using infra-red radiation, it is found that $(\Delta R/R)_\perp$ is negligible for all angles of incidence. This confirms the qualitative conclusions reached earlier, that only parallel polarized radiation need be considered.

For typical experimental conditions normally encountered in RAIRS, in which the metal substrate is highly reflecting, the film is relatively weakly absorbing, and the angle of incidence is not too close to the optimum 88°, further approximations lead to

$$(\Delta R/R)_{\parallel} = -(4 \sin \phi_1 \tan \phi_1/n_2^3)(4\pi k_2 \, d/\lambda). \tag{12.13}$$

The first term on the right corresponds to the electric field enhancement, and the second to the attenuation factor in the Beer–Lambert law, where the transmitted intensity I is given in terms of the incident intensity I_0 by

$$I = I_0 \exp(-4\pi k_2 \, d/\lambda). \tag{12.14}$$

Extensions of the simple isotropic model to considerations of surface roughness and of a non-isotropic uniaxial film have been made by Dignam[6–8] and co-workers, but their analysis is outside the scope of this volume, and the reader is referred to their publications.

As with the technique of HREELS, the first step in spectral identification is a comparison of the observed absorption bands from a film on a surface with those of the film material in the gas phase. Because of the attachment to the surface, some of the vibrational modes that are possible in the free molecule will either be absent or will be altered in nature due to the 'hindering' or damping effect of the surface bond. By the same token, modes that were not possible in the free molecule will become possible due to the presence of the surface. For example, translational and rotational modes are not of course detectable in a free molecule in the gas phase, but directly one or more parts of the molecule become fixed, as in a surface bond, then such modes can occur in the so-called 'hindered' form.

In RAIRS the same selection rule applies as in HREELS. Both grazing incidence infra-red photons and specularly reflected very low energy incident electrons experience the same long-range surface interaction, and in both cases only vibrations providing dipole changes resolved perpendicularly to the surface will couple with the incident radiation. The only difference between the techniques in this respect is that in HREELS there is the possibility of 'impact' scattering as well as 'dipole' scattering, for off-specular reflection. See chapter 5 for a full description of the processes operative in HREELS.

12.1.3 *Quantification*

Expression (12.13) for the change in reflectivity contains the thickness d of the adsorbed film. If the monolayer thickness is d_0 then the fractional coverage θ is given by

$$\theta = d/d_0 \tag{12.15}$$

and expression (12.13) can then be written

$$(\Delta R/R) = -(4 \sin \phi_1 \tan \phi_1/n_2^3)(4\pi k_2 \, d_0\theta/\lambda). \tag{12.16}$$

For a fixed angle of incidence and a fixed infra-red wavelength, then, the change in reflectivity is proportional to the coverage of the adsorbed film, *provided* that neither the optical nor chemical characteristics of the film change as a function of coverage. For proportionality of reflectivity and coverage it is obvious that n_2 and k_2 should remain constant, but there are other considerations too. As the coverage increases, changes in the nature of the chemical interactions either between the film and the substrate or between molecules in the film itself can cause broadening or narrowing of band widths, shifting of bands, and appearance of additional bands. Since experimentally the changes in reflectivity would be followed by monitoring the changes in a selected absorption peak, rather than the integrated changes over the whole spectrum, it is clear that the finding of a direct proportionality must be the exception rather than the rule. One example of such an exception is that of the adsorption of N_2 on Ni(110) at 81 K from the work of Trenary and Brubaker.[9] Figure 12.7 shows the dependence of absorption peak height on coverage from

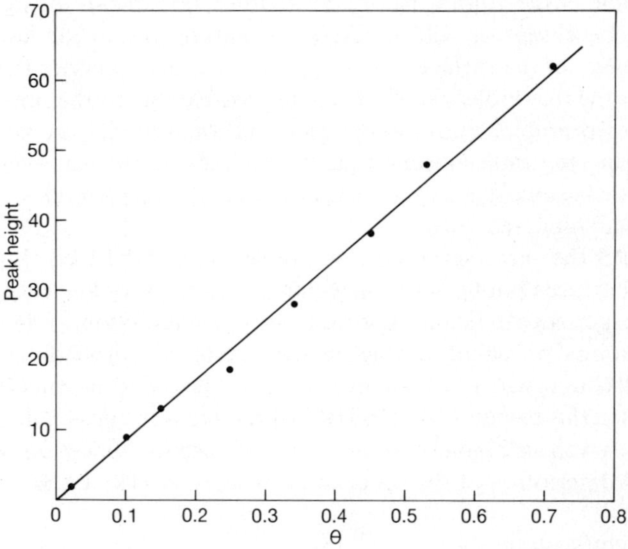

FIG. 12.7. Linear correlation between infra-red absorption peak height and coverage, during the adsorption of N_2 on Ni(110) at 81 K. Derived from the measurement of Brubaker and Trenary.[9] The maximum coverage reached in their experiments was $\theta=0.71$.

Fig. 1 of their paper, and it can be seen that there is a close linear relationship up to the maximum coverage used of $\theta=0.71$. Strictly speaking it should be the area under the peak rather than the height which is plotted, but in this case the half-width remained constant during the adsorption, indicating that the peak height was proportional to area.

In general, then, the behaviour seen in Fig. 12.7 is not found, and the estimation of coverage is made from other techniques operated in parallel with RAIRS, especially Thermal Desorption Spectroscopy (TDS) and LEED, but occasionally also AES and XPS. Similarly to HREELS, RAIRS is not used in the type of work that requires the technique itself to provide a quantification of surface composition, its application being in the realm of chemical speciation, as will be shown below.

12.1.4 *Applications*

Just as for HREELS, the great majority of RAIRS experiments and analyses have been carried out with single-crystal metal surfaces as the substrates on which films have been adsorbed or deposited. Many of these experiments have been of a very basic nature and apparently unconnected with any technological processes, but some on the other hand have been directed more deliberately towards a better understanding of a surface reaction of industrial origin, albeit still on well characterized surfaces.

A good example of the latter type of experiment comes from the work of Trenary *et al.*,[10] who set out to understand more about the well known effects of promotion by alkali metals and of poisoning by sulphur, in the synthesis of hydrocarbons from CO and H_2 over nickel catalysts. In particular they wanted to address the question as to whether the effect of adsorbed sulphur on subsequent CO adsorption was of a short range or long range nature, that is, whether the electronic perturbation due to the sulphur is restricted to the atomic sites in its immediate vicinity, or whether it is delocalized over a large area thus involving many sites.

Trenary *et al.* made use of one of the properties of RAIRS unusual in surface analytical techniques, that is, the ability to operate under relatively high pressure conditions as well as in UHV, by comparing the characteristics of CO adsorption over a range of pressures from 2×10^{-10} torr to 2 torr. Before studying the adsorption on sulphided Ni(111) surfaces they first established the behaviour of CO on clean Ni, to have a basis for comparison. Figure 12.8 shows the infra-red spectra in the CO absorption region after adsorption at 80 K, as a function of thermal treatment in (a) and of coverage in (b). There are two peaks in the spectrum, although they are not always both present; at saturation coverage of CO ($\theta = 0.57$) they appear at 1926 cm^{-1} and 2057 cm^{-1}, the former corresponding to the so-called 'bridge' adsorption. As the names suggest, the bridge position is that of CO adsorbed so that both ends of the molecule are bonded to the surface, and the on-top, or terminal, position that of CO with only one end bonded to the surface. The on-top position can be visualized as having the CO molecular axis vertical to the surface and directly over the centre of a surface Ni atom. Figure 12.8(a) indicates that the population of the on-top CO is maximized by a brief anneal at 240 K before return to 80 K, while after another brief anneal at 300 K no on-top CO remained, and the peak associated with the bridge adsorption had sharpened and also shifted slightly to 1980 cm^{-1}. The LEED structures to the right of the spectra are based on the parallel LEED observations, and confirm that after

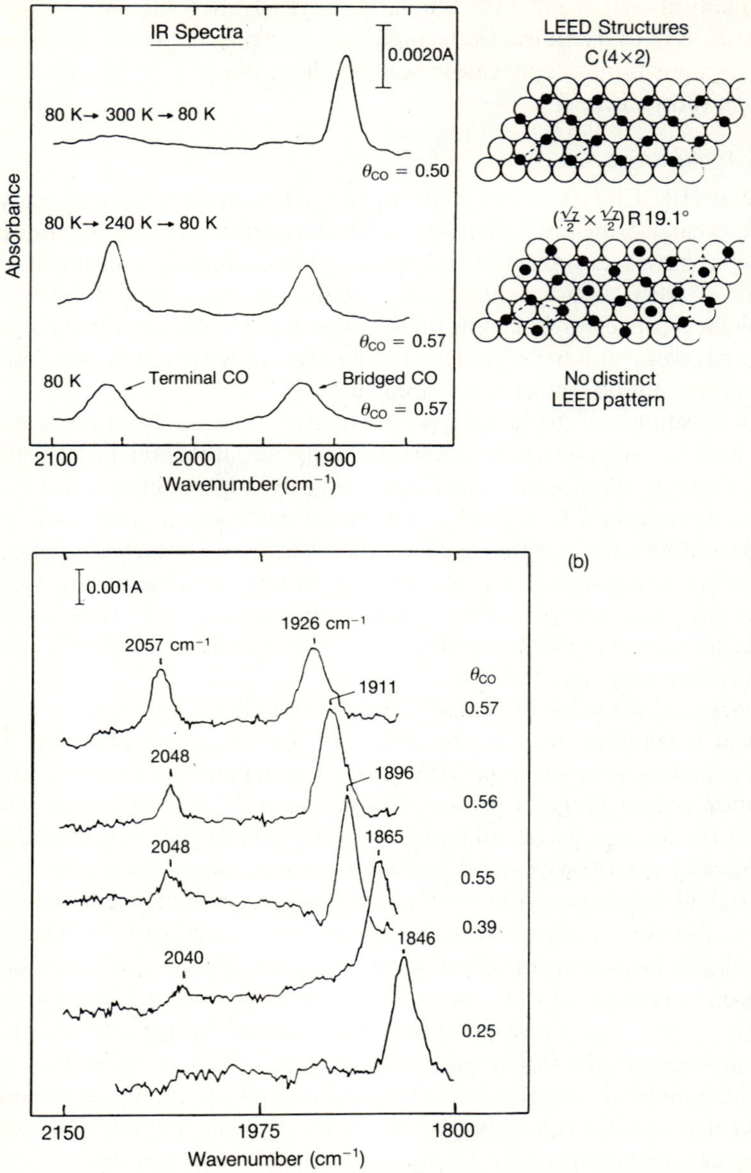

FIG. 12.8. Infra-red spectra in the CO absorption region as a function in (a) of heat treatment, and in (b) of coverage, recorded by Trenary *et al.*[10] during the adsorption of CO on clean Ni(111), at 80 K. Saturation coverage is $\theta = 0.57$.
In the upper panel, the absorption peaks at 1926 cm^{-1} and 2057 cm^{-1} correspond to CO adsorbed in the 'bridged' and 'on-top' or terminal, orientation, respectively. The 'on-top' population is maximized by a brief anneal at 240 K, but after heating to 300 K only the 'bridged' CO remains. The lower panel shows that the 'bridged' sites are occupied first on adsorption at low temperature, followed by occupation of the 'on-top' sites as the former are filled.

the room temperature excursion no on-top CO remained; they also show that the CO coverage reduced from 0.57 to 0.50 as a result of the anneal and the surface re-structuring. The information from Fig. 12.8(b) is that the bridge sites are occupied first by CO, followed by occupation of the on-top sites as the former fill.

When the authors came to study the adsorption of CO at high pressures and on surfaces containing small amounts of sulphur, some striking differences in behaviour were found. Figure 12.9 shows the absorption spectra recorded at 300 K as a function of coverage of sulphur, both at a relatively high CO pressure of 1 torr, and then after returning to UHV. The sulphur-free surface (bottom spectra) shows the peaks attributable to bridge and on-top CO appearing at the high pressure, and persisting, even though reduced in intensity, after removing the CO. The similarity of the high pressure spectrum to that at 80 K in Fig. 12.8(a) suggests that at 1 torr the CO coverage is 0.57. Small fractions of a monolayer of sulphur make a significant difference to the spectra. Firstly, a new adsorption state, labelled CO*, at 2108 cm^{-1}, makes its appearance, but is active in CO adsorption only at high pressures, since on pumping away the CO it disappears completely. Secondly, concentrations of sulphur greater than about 4×10^{14} cm^{-2} allow no CO adsorption at all at low pressures, since neither the bridge-type or the on-top-type peaks remain once the pressure is reduced. Also, at that sulphur coverage, equivalent to less than half a monolayer, the new CO* site is the only one occupied at high pressure, as seen in the top spectrum on the left. A further observation, not obvious from Fig. 12.9, is that there were virtually no shifts in the position of either the CO* peak or the on-top peak as the sulphur coverage increased.

The observations that the sulphur-induced CO* band is positioned only 35 cm^{-1} below the gas phase frequency, that it is very narrow (only 8 cm^{-1} in width), and that the sites giving rise to it are populated only at high pressures indicate the CO molecule is weakly adsorbed. Indeed temperature-dependent measurements at 1 torr CO pressure produced a heat of adsorption of only 19 kcal mol^{-2} compared to 26.5 kcal mol^{-1} on a sulphur-free surface. This is all consistent with a model in which the sulphur, being electronegative, withdraws electronic charge from the adjacent nickel atom(s) so that less charge is available for the normal 'back donation' to the CO orbitals, and so the CO is much less tightly bound. (Electropositive atoms such as the promoter potassium have, of course, the opposite effect.) Furthermore, the virtual invariance with sulphur coverage of the CO* bond frequency, coupled with a similar invariance in the on-top and bridge CO bonds, indicate that the perturbing effect of the sulphur is very local in nature; indeed calculation shows that only about 6 Ni sites are involved for each sulphur atom. Thus the poisoning effect of the sulphur is that of reducing the capacity of the Ni surface to hold CO by adsorption, through reduction in adsorption energy, and the effect is local and not long range in character.

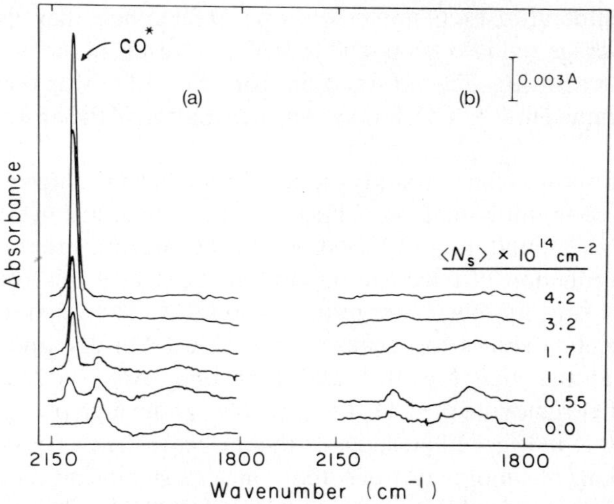

FIG. 12.9. Infra-red spectra recorded by Trenary *et al.*[10] during the adsorption of CO at high pressure on a Ni(111) surface containing variable amounts of sulphur. The coverage of sulphur is shown on the right against each spectrum corresponding to it. The spectra in (a) were measured at 1 torr pressure of CO, in (b) after the CO had been removed.
The presence of sulphur causes the appearance of a new state, labelled CO*, at 2108 cm^{-1}, active only at high pressures, since it disappears on pumping away the CO. At a sulphur coverage of $\approx 4 \times 10^{14}$ atoms cm^{-2} no CO adsorption can occur at low pressure, and the new state is the only one occupied at high pressure. The results are discussed in the text.

A rather interesting series of experiments were carried out by Erley[1] using the apparatus shown in Fig. 12.1 in which the RAIRS spectra of various organic molecules adsorbed on copper surfaces were compared with the HREELS spectra from the same surfaces recorded in the same system consecutively. His results provide an excellent comparison of the relative strengths and weaknesses of the two techniques, and an example from his work is given in Fig. 12.10, for the adsorption of acetonitrile, CH_3CN, on Cu(111). The film thicknesses for HREELS and for RAIRS measurements were approximately one and two monolayers, respectively.

The first point to note is that the RAIRS spectrum cuts off at about 800 cm^{-1}, corresponding to the frequency cut-off of the mercury cadmium telluride detector, whereas HREELS can be recorded right down to zero loss. On the other hand, and secondly, the resolution available in RAIRS is far better than in HREELS; in Fig. 12.10 a resolution of 2 cm^{-1} was used for RAIRS, but only of ≈ 60 cm^{-1} for HREELS. Thus, while the bands labelled ν_4 and ν_7, separated by ~ 120 cm^{-1}, can be resolved in HREELS, those labelled ν_3 and ν_6, separated by 78 cm^{-1}, cannot, although there is a suggestion of a shoulder on the high energy side of the broad peak at ~ 1400 cm^{-1}. The assignment of the various RAIRS absorption bands was made by comparison with the Raman spectra of isolated CH_3CN, and is given in Table 12.1. The

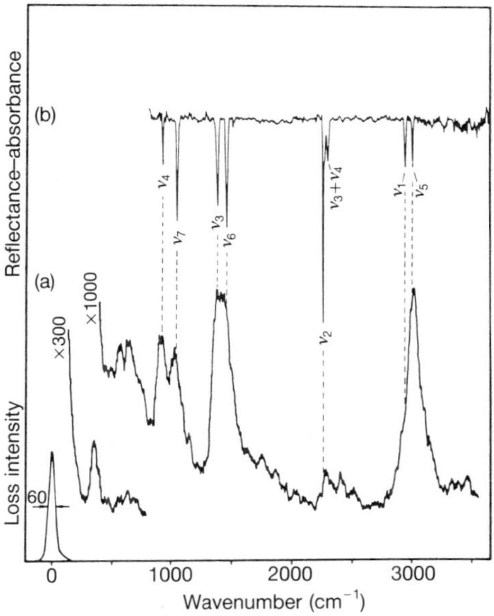

FIG. 12.10. Comparison by Erley[1] of the RAIRS and HREELS vibrational spectra of acetonitrile, CH_3CN, on Cu(111) at 100 K. The RAIRS spectrum (b) cuts off at ≈ 800 cm^{-1}, due to the frequency cut-off in the detector, whereas HREELS can be recorded to zero loss. However, the resolution in RAIRS is of the order of 2 cm^{-1} as against ≈ 60 cm^{-1} in HREELS. Thus bands ν_3 and ν_6, ν_1 and ν_5, can be resolved by RAIRS but not by HREELS. These are the reasons why the two techniques are often used together.

author suggests that the reason for the much greater intensity of the CN stretching vibration in RAIRS was due to the thicker film used for the latter technique.

The three loss peaks in the HREELS spectrum, not accessible by RAIRS, appeared at 385, 600, and 670 cm^{-1}, and were assigned respectively to the C–C–N bending mode, and to the symmetric and asymmetric Cu–CH_3CN stretching modes.

12.2 Surface-enhanced Raman spectroscopy

Until recently conventional or unenhanced Raman spectroscopy had not been applied to the study of surface films simply because there was insufficient intensity in the scattered signal. Thus Campion[11] estimated that in a typical experiment involving Raman scattering from a monolayer on a surface, a count rate on a peak of a few counts per second could be expected using a conventional scanning double monochromator as the detector. This would require a spectral acquisition time of many hours for an acceptable signal-to-noise ratio, which is too long for an experiment in surface chemistry, even in UHV. That situation is now changing, with the advent and increased use of

Table 12.1
Assignment of vibrational modes in RAIRS spectra from CH_3CN adsorbed on Cu(111) at 100 K

Observed energy, (cm^{-1})	Assignment	Nomenclature
921	C–C stretch	ν_4
1040	CH_3 rocking, deg.	ν_7
1377	CH_3 deform., symm.	ν_3
1455	CH_3 deform., deg.	ν_6
2253	C–N stretch	ν_2
2945	C–H stretch, symm.	ν_1
3003	C–H stretch, deg.	ν_5

multichannel detectors, which will reduce acquisition time to a few minutes, but even so the number of published instances in which 'normal' Raman spectroscopy has been used is still very small. On the other hand, the discovery of SERS in 1974 by Fleischmann *et al.*[12] has given enormous impetus to the application of the Raman technique to the study of surfaces even though SERS is still restricted mainly to surfaces of the noble metals silver, copper, and gold, and though the reasons for the enhancement are still not fully understood. The fact remains that an intensity enhancement by factors between 10^5–10^6 over observed gas-phase scattering intensities is so enormous that Raman scattering from thin films immediately becomes very useful, despite the restrictions. This section will therefore deal with SERS only, although it should be borne in mind that the unenhanced version is making rapid strides.

12.2.1 *Operation*

An experimental arrangement typical for the application of SERS to surface studies in conjunction with other surface characterization techniques is that of Pockrand and Otto[13] in Fig. 12.11. Light from an argon laser operated at 250 mW and producing a wavelength of 514.5 nm passes through a filter monochromator to remove the plasma lines of the laser, and through a polarizer to polarize the light in the plane of incidence. It is reflected into the chamber through a window and focused on to the sample surface via a cylindrical lens and another mirror. The angle of incidence is 75°. Scattered light is collected over a large solid angle by a strong (high *f*-number) lens and focused onto the entrance slit of a spectrometer, in which detection is by conventional photon counting. Specularly reflected light can be collected if required through another port with an axis at 15° to the sample surface. The sample was mounted so that its surface was on a circle that brought it by rotation in front of either LEED optics, or an Auger analyser, or an ion gun.

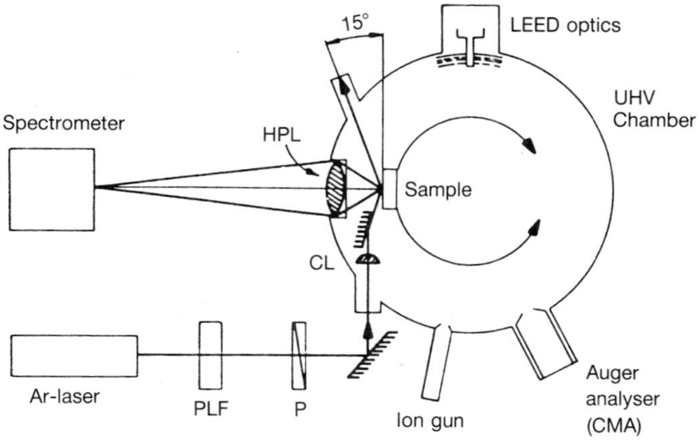

FIG. 12.11. Experimental arrangement used by Pockrand and Otto[13] in the application of SERS in conjunction with LEED and AES. Light from an argon laser of wavelength 514.5 nm is passed through a filter monochromator PLF to remove the laser plasma lines and then through a polarizer P to polarize the light in the plane of incidence (i.e. p-polarized). After reflection into the UHV chamber it is focused on the sample at an angle of incidence of 75° via a cylindrical lens CL and a mirror. Scattered light is collected over a large solid angle by a strong lens HPL and focused onto the entrance slit of a spectrometer.

The original surface enhancement effect was discovered as occurring from surfaces of silver that had been electrochemically roughened in solution, and the Raman scattering measurements were performed at the metal/electrolyte interface. When similar measurements were attempted at the metal/vacuum interface, there was at first considerable disagreement as to whether SERS was actually being observed or not, since some of the early experiments were made on smooth single crystal surfaces. However, it is now realized that indeed such surfaces are too smooth, for reasons that are still not fully understood, and that SERS can be observed in vacuum if the surface is prepared or treated in the right way. Thus enhancement has been found to occur in vacuum by adsorption on surfaces prepared by sputtering or evaporating on to cold substrates at 120–150 K, by mechanically polishing polycrystalline material, by colloidal deposition, or by decomposition of salts on to substrates. In most cases silver has been used, but copper and gold can be almost as effective. The preparation of a SERS-active surface is still subject to considerable variability, and seems from the literature to be largely a matter of luck even now. For this reason no hard and fast recipe can be given for the preparation, and the reader must consult the literature for hints.

12.2.2 *Theory and spectral interpretation*

As remarked at the beginning of this section, the exact reasons for the SERS effect are still not established with certainty, but considerable progress has been made in the last few years towards reaching an understanding. In his review article, Campion[11] points out that two classes of mechanism of enhancement have survived the requirement of being able to account for all or most of the observed characteristics of SERS, namely electromagnetic and chemical. In any one adsorption situation both are likely to be operative.

In the electromagnetic mechanism the incident laser radiation interacts with the roughness features on the metal surface to produce conduction electron resonances in the form of surface plasmons, which in turn produce enormous local electric fields at the surface. It is the increased amplitude of these fields that provides the electromagnetic contribution to the enhancement. In other words, as Campion puts it, the rough surface is acting as an amplifier for both the inward laser radiation and the outward Raman radiation. A very simple model, based on the electron microscopic observation that an electrolytically roughened surface has quasi-spherical protrusions of average size ≈ 50 nm, provides qualitative demonstration of the electromagnetic mechanism. If the surface features are assumed to be isolated spheres, then, since the size of the features is very much less than the wavelength of the incident laser radiation, the local field at the sphere surface E_{out} is related to the uniform external electromagnetic field E_{in} of the laser by

$$E_{\text{in}} = \left(\frac{\epsilon_1(\omega) - \epsilon_2}{\epsilon_1(\omega) + 2\epsilon_2}\right) E_{\text{out}} \tag{12.17}$$

where $\epsilon_1(\omega)$ is the complex frequency-dependent dielectric function of the metal, and ϵ_2 is the dielectric function of the surrounding medium. The form of expression (12.17) shows that resonant frequencies occur when the real part of $\epsilon_1(\omega)$, $\text{Re}(\epsilon_1)$, is equal to $-2\epsilon_2$, and clearly at and very near those frequencies extremely large fields can be produced on the surface of the sphere. This result in itself is enough to explain the enhancing properties of metals such as silver, copper, and gold, since the resonant condition for them is actually in the visible region of the spectrum, and since the imaginary part of the dielectric function happens to be small in the resonant region.

There is further field amplification to consider. If one calculates the dipole field induced in the model sphere by the Raman dipole, the magnitude of the induced dipole has the same form as (12.17) but now with the laser frequency replaced by the Raman or Stokes-shifted frequency. With the dipole located at the centre of the sphere, the radial distance dependence is given by incorporating the dipole decay law twice, and finally, since intensities are proportional to the squares of the fields, the enhancement factor is given by

$$G = \left(\frac{\epsilon_1(\omega_{\text{L}}) - \epsilon_2}{\epsilon_1(\omega_{\text{L}}) + 2\epsilon_2}\right)^2 \left(\frac{\epsilon_1(\omega_{\text{s}}) - \epsilon_2}{\epsilon_1(\omega_{\text{s}}) + 2\epsilon_2}\right)^2 \left(\frac{r}{r+d}\right)^{12} \tag{12.18}$$

where ω_{L} and ω_{s} are the laser and Stokes frequencies, respectively, r is the radius of the sphere, and d is the distance of the molecule above the sphere surface. Many of the properties of SERS can be explained by reference to expression (12.18). Thus the above-mentioned resonance condition, at $\text{Re}(\epsilon_1) = -2\epsilon_2$, is satisfied for either ω_L or ω_s, the largest enhancement occurring for very small Stokes shifts when both resonances are satisfied simultaneously. The reduction in intensity at higher frequencies is also explained, since then one of the frequencies must be off resonance. Again, since the enhancement at resonance is proportional to $\{\text{Im}(\epsilon_2)\}^{-4}$, it is clear that metals of high reflectivity will be the most efficient enhancers. Finally, the form of the dipole decay law indicates that long-range enhancement can occur at relatively large separations from the surface of the model spherical protrusions; Campion has estimated that at a distance of 5 nm from a sphere of radius 25 nm, molecules still experience 10% of the enhancement of those in the first layer next to the surface.

Despite the qualitative success of the above model in explaining many features of SERS, it is known that it cannot explain all aspects of the enhancement completely. Other mechanisms must be contributing. The evidence for this is that the enhancement effect of molecules adsorbed in the first layer is often much larger, by as much as a couple of orders of magnitude, than would be expected based on the distance dependence, in the electromagnetic mechanisms, and that very small fractions (in some cases as little as 3% of a monolayer) of some metals on the surface can act as poisons and quench the enhancement completely. The latter observation seems to suggest that only a small proportion of surface sites are active in SERS.

To account for these anomalies a chemical enhancement mechanism has been proposed as likely to be operative either in parallel with, or sometimes instead of, the electromagnetic mechanism. It has been known for some years that Raman scattering is enhanced, sometimes by many orders of magnitude, when the laser frequency is close to that of an electronic absorption frequency in a molecule. However, it so happens that those molecules that have shown SERS activity do not have electronic transitions in the visible part of the spectrum in which the usual laser frequencies reside. The resonant enhancement within the molecule must therefore occur via resonant states not present in the free molecule but present when it is adsorbed on a surface. These new states have been proposed by Demuth and co-workers[14,15] as being charge transfer states, in which a molecular adsorbate energy level is shifted and broadened as a result of transfer of electronic charge from the metallic substrate to the molecule. As the adsorbate levels shift due to this charge transfer, so additional transitions become possible, and if these new transitions have frequencies close to the exciting laser frequency then resonant Raman scattering occurs, and therefore enhancement.

Such a charge transfer mechanism would of course have the greatest effect on the first layer of adsorbed molecules, which would account for the observation that the enhancement is proportionately so much higher in the first layer. It would also account for the poisoning effect of certain metallic atoms, if their electronic structure was such that transfer was impossible, and if they also occupied the sites at which transfer would be possible in their absence. An extension of the charge transfer scheme involves the concentration of atomic defects at the surface. Since the band width of the electron donating densities of states near the surface will be sensitive to the nature and number of atomic defects, and since this band width is an important factor in the case of charge transfer, it is very likely that defective surfaces will show a greater enhancement than more perfect ones. Some experiments to explore this idea have been performed, but as yet are not conclusive. However, it is to be expected that some definitive experimental work will soon decide the extent of the importance of the defective structure.

Just as in HREELS and RAIRS, the first interpretation of a SERS spectrum from an adsorbed molecule on a surface is made by comparison with the vibrational spectrum from the same molecule in the gaseous state. Again, as in the other techniques, vibrational modes either additional to, or modified from, those in the gas phase as a result of attachment to the surface will appear in the SERS spectrum as new bands. In general, interpretation of new bands, particularly if found well separated from identifiable bands, must be made by a combination of physical possibilities, information from other techniques or from known chemistry, theoretical estimation, and intuition.

12.2.3 *Quantification*

Since one of the properties of the SERS effect is that the scattered intensity does not increase uniformly or in a predictable manner with concentration, it does not seem possible to use the technique in a quantifiable way. The principal interest is to use the intensities to obtain enough wavelength resolution to identify peak positions, i.e. Stokes shifts, as accurately as possible in order to extract chemical information.

12.2.4 *Applications*

A considerable advantage of all Raman techniques is that they can be applied as readily in aqueous media as in air or in vacuum, so it is hardly surprising that the great majority of applications have been to processes occurring at electrode surfaces while operating in electrolytes. The initial discovery[12] of SERS was made during the adsorption of pyridine from solution on to a silver electrode surface, and many of the subsequent studies of SERS have also dealt with the adsorption of pyridine and its derivatives, since the enhancement seems to be particularly great for such molecules. Parallel studies of enhancement on vacuum-prepared surfaces have not been nearly so common, but are increasing in number as the applicability of the technique extends and becomes appreciated.

An interesting study in which pyridine adsorption in vacuum has been investigated by SERS is that of Gao and Lopez-Rios;[16] they were interested in the reasons why very small amounts of contamination in the form of adsorbed gases or metals have such a profound effect on the SERS intensities. As remarked earlier, in some cases as little as 3% of a monolayer of some metal impurities can quench the enhanced signal completely. Gao and Lopez-Rios used aluminium as the impurity metal in this particular study, with silver films deposited by evaporation at 120 K as the 'SERS-active' surface.

Before discussing the results of co-adsorption of aluminium and pyridine it is of interest, as a starting-point, to compare the SERS spectra from coldly deposited pure silver and aluminium films with adsorbed pyridine on each. These are shown in Fig. 12.12. The metal films were both 1000 Å thick, with

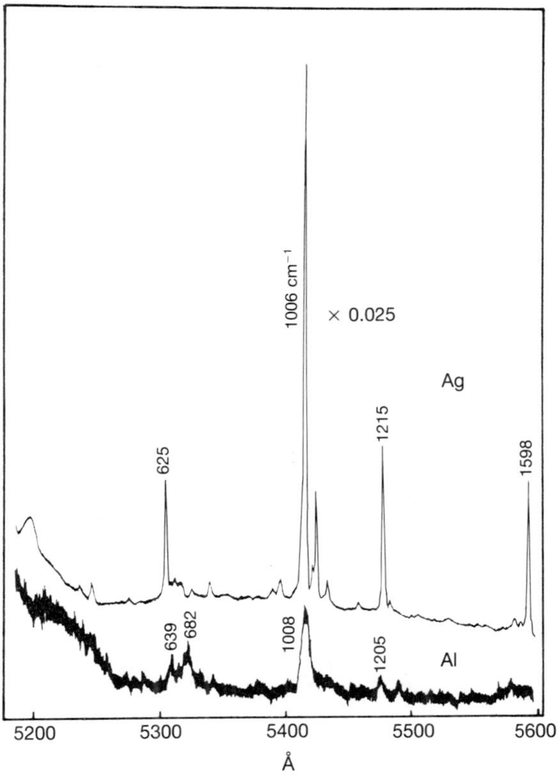

FIG. 12.12. Comparison of SERS spectra from coldly deposited films of Ag and of Al, about 100 nm thick, each with adsorbed pyridine. The enhancement on Ag is obviously much greater than on Al, by a factor of $\approx 10^3$. The absorption frequencies are also slightly different, due to slight differences in the adsorption sites. From Gao and Lopez-Rios.[16]

the silver exposed to 3 L, and the aluminium to 12 L, of pyridine vapour. It can be seen at once that the enhancement on silver is much greater than that on aluminium, the difference being a factor of about 10^3. Also of importance are the slightly different absorption frequencies of the pyridine on the two substrates; this differentiation is useful in being able to identify the adsorption site of pyridine in the subsequent experiments.

Two types of co-adsorption experiment were performed. In the first the cold silver film was first exposed to 3 L of pyridine, corresponding on a flat surface to about 0.3 monolayers of pyridine, and then increasing thicknesses of aluminium were deposited on top. The resultant SERS spectra are given in Fig. 12.13. At the top, (a), is the spectrum from the pure silver with 3 L of

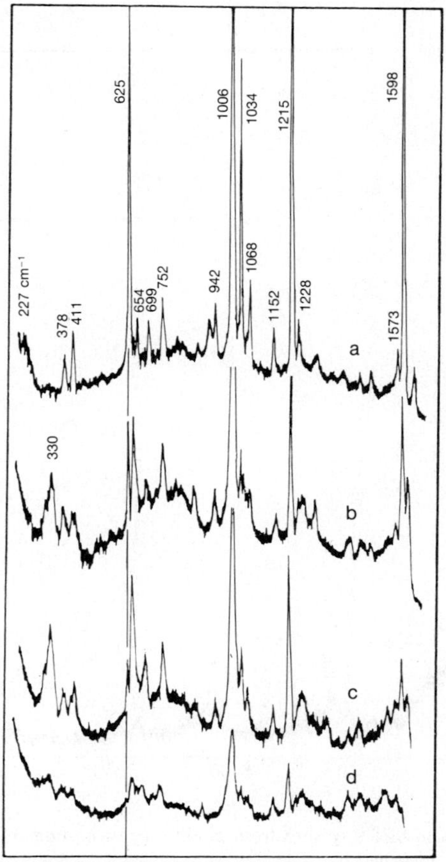

FIG. 12.13. SERS spectra recorded from a cold Ag film after various treatments, by Gao and Lopez-Rios.[16]
(a) Exposure to 3 L of pyridine (similar to upper spectrum of Fig. 12.12).
(b) Deposition of 0.6 Å of Al on the pyridine-covered surface.
(c) Deposition of 1.7 Å of Al.
(d) Deposition of 3.0 Å of Al.
The ν_3 (625 cm^{-1}) band is indicated by a vertical line.

pyridine, while (b), (c), and (d) were recorded after thicknesses of aluminium of 0.6 Å, 1.7 Å, and 3 Å, respectively, were laid down on top of the pyridine. The vertical line passes through the position of the 625 cm^{-1} peak. Notice the rapid quenching of the principal lines as the concentration of aluminium increases. These lines are all associated with various in-plane vibrating modes; the lesser lines that are not significantly affected are out-of-plane vibrations, which the authors think may indicate that the aluminium atoms are inducing a re-orientation of the adsorbed pyridine. As the original major lines reduce, so additional lines close to them appear and grow, the shift from one to the other being only a few cm^{-1}. The effect associated with the so-called 'breathing' mode of the pyridine ring is demonstrated on an expanded scale in Fig. 12.14.

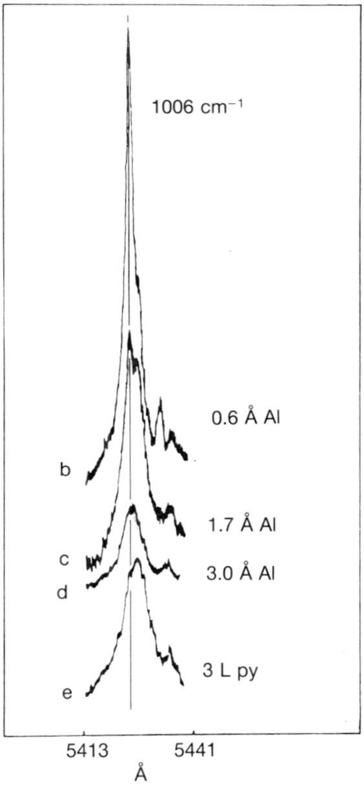

FIG. 12.14. Curves (b), (c), and (d) are expanded sections of the correspondingly labelled spectra in Fig. 12.13. The peak at 1006 cm^{-1} is the ν_1 mode, the so-called 'breathing' mode of the pyridine ring. Spectrum (e) was recorded after further exposure to 3 L of pyridine on top of the deposited Al. The ν_1 line in spectrum (a) in Fig. 12.13 was too intense to be accommodated on the same scale. From Gao and Lopez-Rios.[16]

By reference to the lines associated in Fig. 12.12 with adsorption on pure aluminium, the new shifted lines in Fig. 12.13 and 12.14 are due to pyridine bonded to aluminium atoms and not to silver. Further exposure to 3 L of pyridine after the aluminium deposition, as in spectrum (e) of Fig. 12.14 confirms that the effect is due to aluminium and not due to possible removal of the adsorbed pyridine by the arriving aluminium atoms.

One of the interesting results from a series of experiments such as those of Fig. 12.13 and 12.14 is the rapidity with which quite small coverages of aluminium quench the enhancement. Figure 12.15 shows the decrease in the

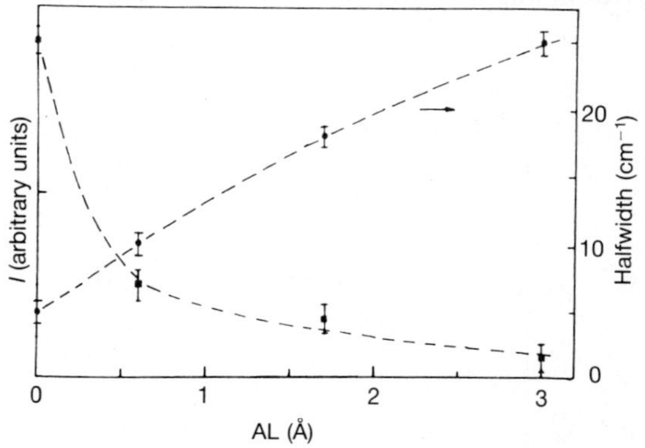

FIG. 12.15. Plot by Gao and Lopez-Rios[16] of the intensity and the half-width of the ν_1 line seen in Fig. 12.13 and 12.14, as a function of the thickness of the Al deposit on top of adsorbed pyridine. The Al deposit quickly quenches the pyridine ring breathing mode, most of the reduction occurring within the first Å thickness. The line broadening may be due to an increasing tail from the new shoulder separated from ν_1 by only a few cm^{-1}.

intensity of the ring breathing mode as a function of aluminium thickness; the overall reduction at a coverage of 3 Å is by a factor of about twenty, but most of that occurs within the first Å of deposition. The line broadens at the same time, as seen in the figure, probably due to the increasing tail from the new line alongside it.

The other type of co-adsorption performed was to deposit aluminium to small thicknesses first, and then adsorb pyridine afterwards, i.e. the reverse of the above procedure. In Fig. 12.16 three spectral regions corresponding to the positions expected for three of the principal Raman lines are shown as a function of increasing pyridine coverage at 120 K on aluminium of 1.5 Å thickness on top of a coldly deposited silver film. The lines are detectable only for pyridine coverages greater than about 1.2 L, and their shapes, intensities, and positions change with increasing coverage. Below a coverage of about 2 L only one peak appears in each of the regions, and reference to Fig. 12.12 shows that the positions correspond to pyridine bonded to aluminium; they are also

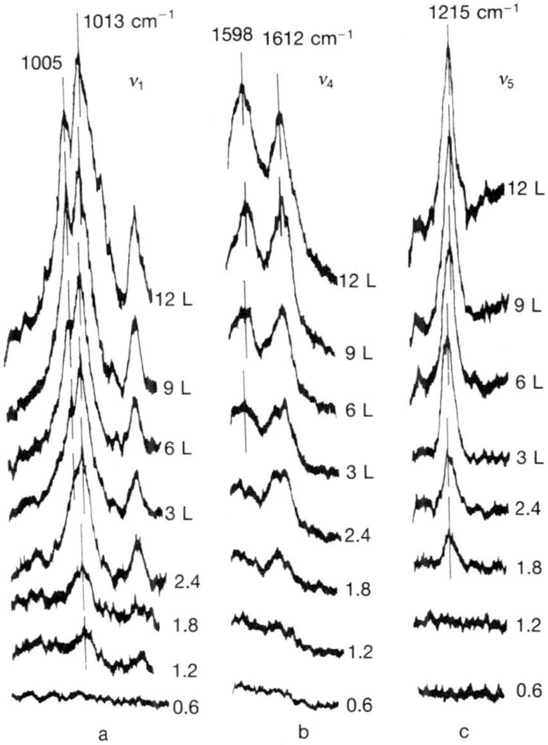

FIG. 12.16. SERS spectra from a coldly deposited Ag film as a function of pyridine exposure up to 12 L, after deposition of an Al film of thickness 1.5 Å. The sequence of adsorption is therefore the reverse of that in Figs 12.13 and 12.14. The selected regions are those of the ν_1, ν_4, and ν_5, modes of pyridine. Below 2 L only one peak appears, characteristic of pyridine bonded to Al (see Fig. 12.12). Beyond 3 L additional peaks appear and grow in positions corresponding to pyridine bonded to Ag (see Fig. 12.12 again). From Gao and Lopez-Rios.[16]

in similar positions to those in Fig. 12.13 following the deposition of aluminium on top of pre-adsorbed pyridine. Beyond a pyridine coverage of ≈ 3 L additional peaks appear and grow in intensity, and can be identified as corresponding to pyridine bonded to silver atoms. Thus, irrespective of whether the deposition–adsorption sequence is aluminium-on-pyridine, or pyridine-on-aluminium, for low coverages of pyridine the Raman spectra indicate bonding to aluminium and not to silver. In the same low coverage region of pyridine, i.e. less than 2 L, the enhancement quenching is greatest as shown in Fig. 12.17. The integrated intensity of the ν_1 band in Fig. 12.16 is plotted there as a function of pyridine coverage in comparison with the same plot in the absence of the aluminium film. The ratio of these intensities drops very sharply from ≈ 20 to ≈ 5 within the first two L of pyridine exposure.

It is clear from these experiments that the presence of very small amounts of

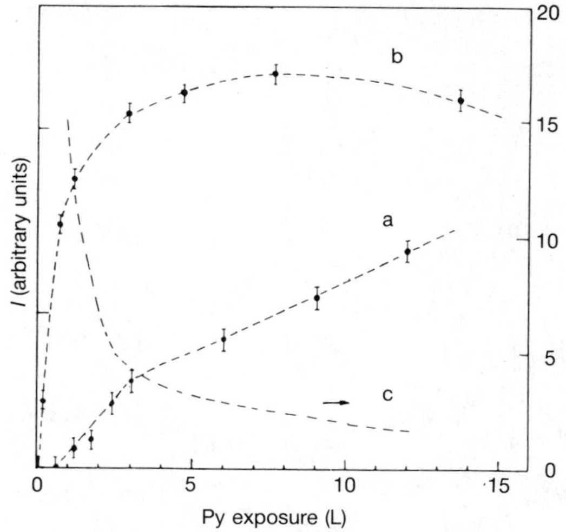

FIG. 12.17. Integrated intensity of the v_1 band (a) in Fig. 12.16, and (b) for an uncoated cold Ag film, plotted as a function of pyridine exposure. Curve (c) is the ratio of the intensities. The intensity ratio drops sharply from ≈ 20 to ≈ 5 within the first 2 L of pyridine exposure. The interpretation is discussed in the text. From Gao and Lopez-Rios.[16]

aluminium has a profound effect on the enhancement, and that the effect is due to the transfer of bonding from silver to aluminium in the earliest stages of pyridine adsorption. Figure 12.16 shows that for more extensive adsorption, bonding is again to the silver, from which one deduces that the adsorption is then taking place on sites not blocked by the aluminium. From these experiments and from other experiments in which the films were annealed to ≈ 245 K, Gao and Lopez-Rios concluded that in fact the deposited aluminium atoms migrated into the micro-pores of the cold silver films, and that the enhancement seen in SERS from silver in particular has at least part of its origin from the situation of adsorbed molecules actually in the pores. Any impurity that can migrate into the pores will thus lower the enhancement.

References

1. Erley, W. (1987). *J. Electron Spec.* **44**, 65.
2. Bell, A. T. (1984). *Chemistry and physics of solid surfaces IV* (eds. R. Vanselow and R. Howe), Ch. 2, p. 25. Springer-Verlag, Berlin.
3. Francis, S. A. and Ellison, A. H. (1959). *J. Opt. Soc. Am.* **49**, 131.
4. Greenler, R. G. (1966). *J. Chem. Phys.* **44**, 310.
5. McIntyre, J. D. E. and Aspnes, D. E. (1971). *Surf. Sci.* **24**, 417.
6. Dignam, M. J. and Moskovits, M. (1973). *J. Chem. Soc. Faraday Trans. II* **69**, 56.
7. Dignam, M. J. and Moskovits, M. (1973). *J. Chem. Soc. Faraday Trans. II* **69**, 65.
8. Dignam, M. J., Moskovits, M., and Stobie, R. W. (1971). *Trans. Faraday Soc.* **67**, 3306.

9. Brubaker, M. E. and Trenary, M. (1987). *J. Electron Spectrosc.* **44**, 47
10. Trenary, M., Uram, K. J., and Yates, J. T. (1985). *Surf. Sci.* **157**, 512.
11. Campion, A. (1986). *Chemistry and physics of solid surfaces VI* (1986). (eds. R. Vanselow and R. Howe), p. 261. Springer-Verlag, Berlin.
12. Fleischmann, M., Hendra, P. J., and McQuillan, A. J. (1974). *Chem. Phys. Lett.* **26**, 163.
13. Pockrand, I. and Otto, A. (1980). *Applic. Surf. Sci.* **6**, 362 and *Vibrational spectroscopy of molecules at surfaces* (1987). (eds. J. T. Yates and T. E. Madey), Ch. 8, p. 345.
14. Demuth, J. E. and Sandra, P. N. (1981). *Phys. Rev. Lett.* **47**, 57.
15. Schmeisser, D., Demuth, J. E., and Avouris, Ph. (1982). *Chem. Phys. Lett.* **87**, 324.
16. Gao, Y. and Lopez-Rios, T. (1988). *Surf. Sci.* **198**, 509.

13

Ion excitation: ion-excited Auger electron spectroscopy (IAES) and proton-excited Auger electron spectroscopy (PAES)

13.1 Operation

13.1.1 *Low energy ions (2–10 keV)*

For operation in the low energy region, ions are usually produced in a conventional gun of the type also used for sputter cleaning and depth profiling, such as the duoplasmatron described under Section 3.2.3 and show in Fig. 3.13. The general experimental arrangement would be very similar to that used for electron-excited AES, as seen in Fig. 13.1 from the work of Kempf and

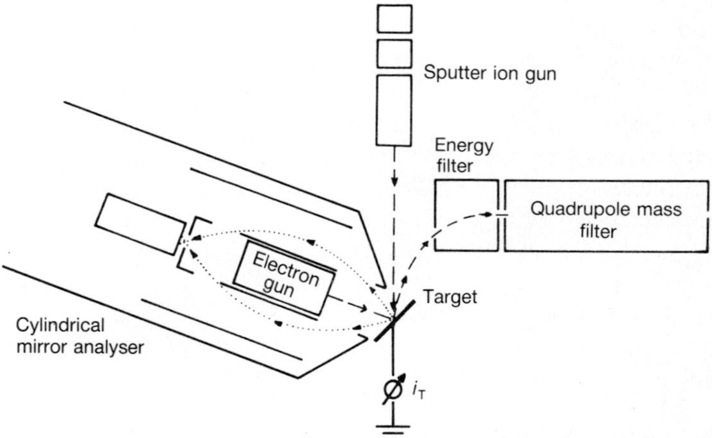

FIG. 13.1. Experimental arrangement used by Kempf and Kaus[1] for ion-excited AES. Ion energies were variable between 2 and 15 keV, and currents used were in the range 0.5–10 μA into an irradiated area about 1.5 mm in diameter. Auger electrons were energy analysed in the same single-pass CMA (see chapter 3) as for electron-excited AES. By tilting the specimen the angles of incidence of the primary beams could be made the same, which assisted inter-comparison of the spectra.

Kaus.[1] With such an arrangement the techniques of AES, IAES, SSIMS, and ISS could all be performed, although the conditions for the two latter would not be ideal. For electron energy analysis a cylindrical mirror analyser (CMA) is commonly used, since high energy resolution is not normally required. The comparison between electron-excited and ion-excited Auger spectra is important for the understanding of the latter, and both incident electron and incident ion energies are varied over the range 2–10 keV, incident currents being of the order of 1 μA. The spectra tend to be recorded in the integral or undifferentiated mode more frequently than in conventional AES, where the differential spectrum is still the more common.

Specially constructed ion sources are used in some laboratories, where the standard commercial guns are not adequate for the type of research envisaged. Such a source has been described already under Section 3.2.3, and in Fig. 3.15; it is in fact capable of operating over a wide range of ion energies, covering both the low and medium ion energy regions.

13.1.2 *Medium energy ions (10–100 keV)*

The ion source mentioned above can be used to cover the lower end of this energy range, up to $\approx$20 keV, but for the much higher energies either more sophisticated dedicated sources must be built or the beam from an ion implanter employed. An example of the custom-built type of source is shown in Fig. 13.2, from Benazeth *et al.*;[2] this particular source has been the principal one used so extensively over the years by the Toulouse group in their major contribution to the physics of IAES. Ions of either rare gases or of hydrogen (protons) are produced in a high frequency (100 MHz) discharge (4) at the top of the column, and are extracted into a magnetic separator (5) which deflects them through a small angle, essentially to separate H^+ from $H_2{}^+$ and $H_3{}^+$. The selected masses are passed through a valve (6) into the UHV chamber (7) pumped by ion and cryogenic pumps. The differential pumping system (1), (2), and (3), necessary between the ion source and the magnetic separator is shown separately on the left. The energy analysis arrangement used recently[3] by the same group is shown in Fig. 13.3, where a 127° electrostatic analyser is employed. The arrangement allows angular electronic distributions to be measured as well. With the source in Fig. 13.2 ion energies up to 100 keV are available.

If the particle beams from accelerators or implanters are to be used for excitation, the principal problem is one of vacuum engineering since the operating vacua in such devices are normally much worse than UHV, i.e. typically in the range 10^{-5}–10^{-6} mbar. To maintain the required UHV in the region of the specimen thus requires several stages of differential pumping around a succession of limiting apertures. Since the problem is the same for the major facilities providing sources of high energy ions, it will be discussed in more detail in the next section. A schematic of the experimental arrangement in the analysis chamber used by Metz *et al.*,[4] seen in Fig. 13.4, shows that the CMA is still the favoured analyser for the Auger electrons.

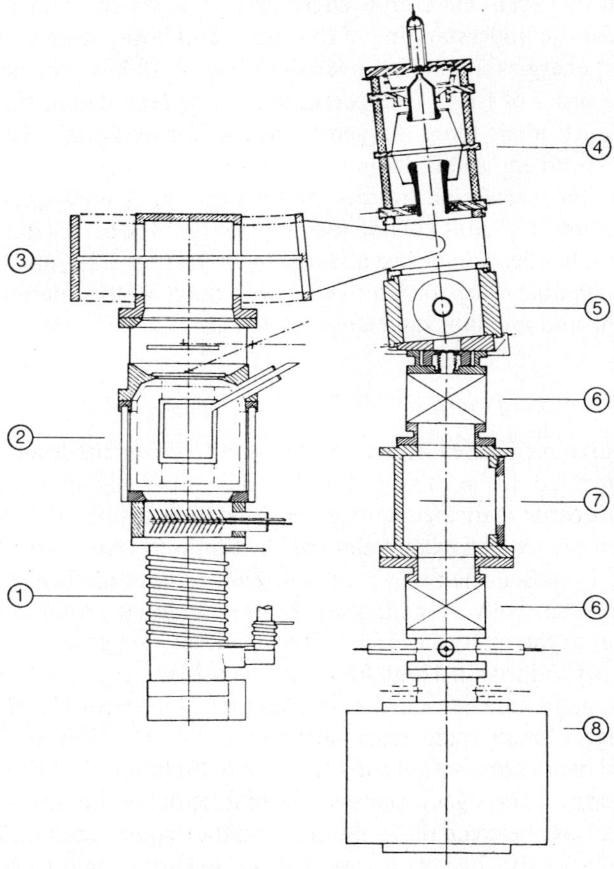

FIG. 13.2. Purpose-built system for IAES and PAES used by the Toulouse group, as described by Benazeth *et al.*[2] It consists of a column at the top of which is a chamber (4) in which a high frequency discharge (100 MHz) is initiated. Ions of either the rare gases or of hydrogen (protons) are extracted into a magnetic separator (5), which deflects them through a small angle, chiefly to separate H^+ from H_2^+ and H_3^+. The filtered beam then passes through a valve (6) into the measuring chamber (7). Differential pumping is necessary between source (4) and separator (5), which is provided by the diffusion pump/trap stack shown on the left; pumping manifold (3), liquid nitrogen trap (2), and diffusion pump (1). The analysis chamber is pumped by combined ion and cryogenic pumps (8) via another valve (6).

13.1.3 *High energy ions (100 keV–5 MeV) and protons*

For the very high energy region nearly all workers have used Van de Graaf accelerators, usually in the MeV range but occasionally at lower energies. As remarked above, for surface studies the principal problem is that of bringing the fast ion beam into the experimental chamber without disturbing the UHV environment of the surface being studied. A clear example of how this can be done by a careful arrangement of successive differential pumping stages has

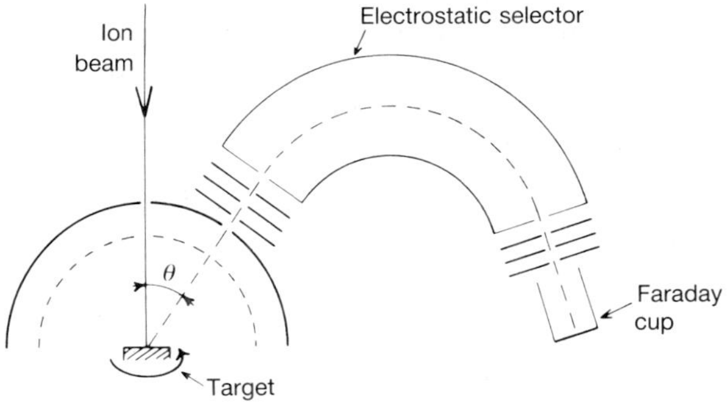

FIG. 13.3. Schematic of experiment arrangement used by Nègre *et al.*[3] for measuring the angular distribution of ion-excited Auger electrons. The electrostatic analyser used was a 127° sector analyser with a Faraday cup detector, and it could be rotated around the specimen in the plane of the page through the polar angle θ. Change in azimuthal angle was effected by rotating the sample.

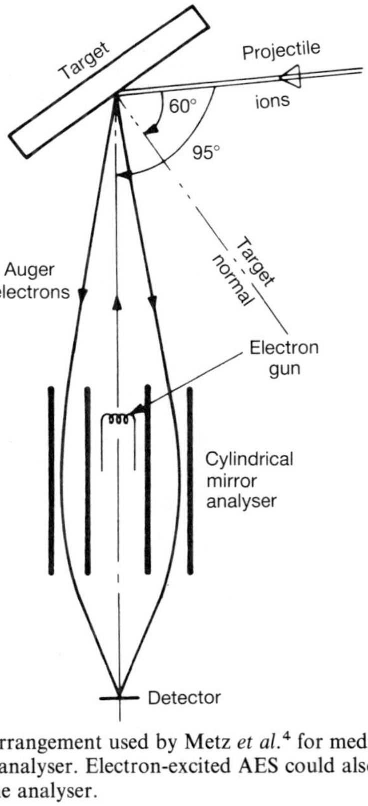

FIG. 13.4. Fixed geometry arrangement used by Metz *et al.*[4] for medium energy IAES, using a CMA (see chapter 3) as the analyser. Electron-excited AES could also be carried out using the electron gun integral with the analyser.

been given by Sitter *et al.*[5] They wished to add a UHV chamber to the end of a 2.5 MeV Van de Graaf, in which the normal operating pressure was 1–5 × 10^{-4} Pa (≈ 10^{-6} mbar), and a schematic diagram of their arrangement is shown in Fig. 13.5. The Van de Graaff beam line is attached to two UHV

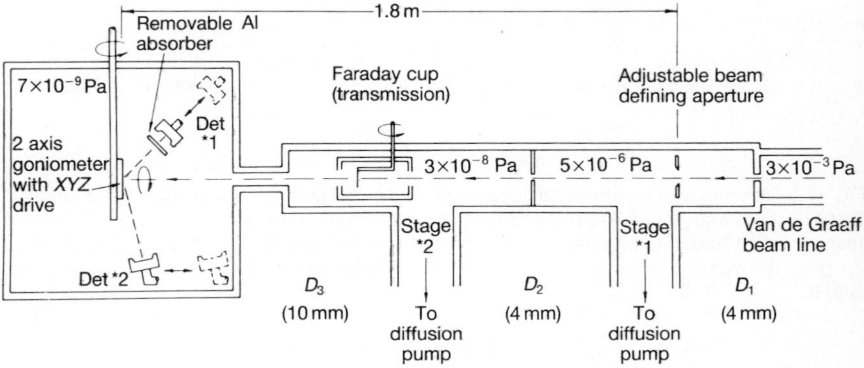

FIG. 13.5. Use of the beam from a Van de Graff generator for high energy IAES in ultra-high vacuum. Since the vacuum in a generator is normally no better than ≈5 × 10^{-4} Pa, a series of differential pumping chambers was required in order to allow experimentation in UHV. Between the various stages there were placed apertures to restrict gas flow from one stage to the next, of diameters shown in the figure. The solid angle subtended at the target from aperture D_1 was < 10^{-5} sr. From Sitter *et al.*[5]

chambers in series (Stages 1 and 2), each pumped by a large (1300 l s^{-1}) trapped diffusion pump. Between the beamline exit, stage 1, stage 2, and the experimental chamber are apertures of diameters 2 mm, 4 mm, and 25 mm, respectively, and there is also another aperture in stage 1 adjustable from 1–3 mm diameter. The vacua achievable as a result of the substantial pumping speeds of the pumps, and of the restrictions imposed by the apertures, are indicated in the figure. There is in fact another problem that has to be overcome when using accelerators, and that is the one of molecular entrainment in the beam itself, basically a gathering-up and sweeping along of contaminant molecules near the beam position. This cannot necessarily be overcome completely, except by designing the accelerator itself to operate at near-UHV pressures, but can be minimized by making the drift distances as long as practicable so that the solid angle subtended at the specimen is as small as possible. In the arrangement of Sitter *et al.* the distance between the first aperture (D_1) and the sample was about 2 m, and the solid angle subtended at the sample from that aperture less than 10^{-5} sr.

The differential pumping problem was solved in a slightly different way by Schmidt,[6] who used not apertures but narrow-bore tubes of sufficient lengths to reproduce the low conductance of apertures, and not diffusion pumps but a mixture of ion getter and electrostatic pumps. The overall effect was the same.

Schmidt's experimental chamber is shown diagrammatically in Fig. 13.6, and confirms once more that the universally used analyser for those measurements is the CMA.

As far as the requirements for surface analysis are concerned, the use of protons and α-particles as the exciting particles involves exactly the same arrangement for differential pumping and aperturing as for high energy ions. Production of H^+, He^+, or $H_2{}^+$ ions is of course carried out by charge stripping or foil neutralization, as the case may be, in the accelerator beamline immediately after extraction of ions from the Van der Graaff. If necessary beams of H^0 particles can also be provided by magnetic deflection of charged particles. Such a system has already been described under Section 3.2.3 and Fig. 3.17.

13.2 Theory and spectral interpretation

Comparison of the Auger spectra produced as a result of bombardment by various ion species at various energies with those produced by electron bombardment reveals some striking differences. In Fig. 13.7, for example, from Thomas,[7] are shown the Auger spectra from Mg and Si induced by 190 keV Ar^+ ions and 1 keV electrons, respectively. Clearly the ion-induced spectra bear some resemblance to their electron-induced counterparts, in that there is a similar background contribution from inelastic decay processes in the solid in each case, but equally clearly there are several features in the former not present in the latter. Peaks appear on top of the background distribution, of a width that is too narrow to be ascribed to any process taking place in the solid, there are additional subsidiary shoulders and peaks (visible especially in the Si spectrum) and a peak also appears at higher energy than that of the principal electron-excited peak. The latter appears only with ion bombardment, not with electron or proton bombardment, as indicated in Fig. 13.8, from MacDonald *et al.*;[8] note that the Auger spectra in this figure are in the differentiated form, unlike most of the spectra published for ion excitation.

These differences, and their interpretation, have given rise to a great deal of research, both experimental and theoretical. Since the physical mechanisms that are thought to be operating are different according to the energy regime and the type of exciting particles, they will be discussed separately here.

13.2.1 *Proton excitation*

During electron irradiation the mechanism of excitation is by direct Coulomb interaction with the core-level electron and insufficient momentum is transferred to the ionized atom, so that the atom is not displaced. The same is true during proton irradiation. Proton impact can cause excitation only by Coulomb interaction and normally little momentum is transferred to matrix

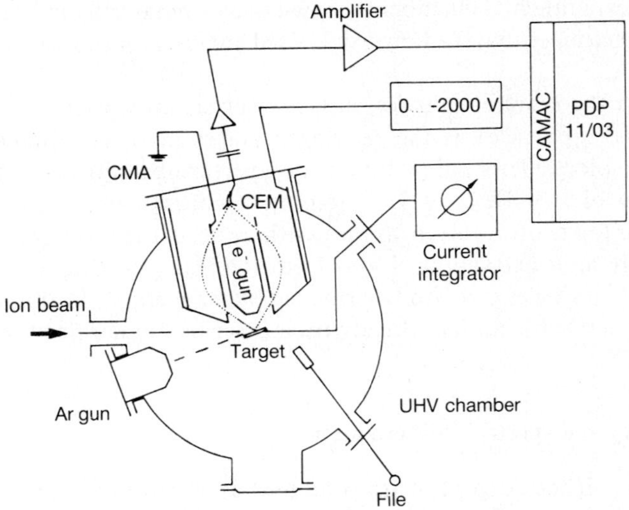

FIG. 13.6. Experimental chamber for high-energy IAES used by Schmidt.[6] Instead of using apertures with differential pumping to allow use of UHV in the experimental chamber, as in Fig. 13.5, he employed narrow-bore tubes of length sufficient to act as pumping resistances. These did not impede the beam from the Tandem Van der Graaff generator. Final energy from the generator was of the order of 4 MeV. Notice from this figure that one of the methods used to clean specimens was by abrasion, i.e. using a file on a bellows in this case.

atoms. The line shapes, positions, and fine structure will therefore be similar between the two excitations, since the initial ionized state is the same. There will, however, be a difference in the shape of the background, since primary electrons can suffer inelastic losses by a variety of mechanisms, whereas fast protons will suffer negligible loss. The depth distribution of the ionized atoms will also be different, since backscattered inelastic electrons can cause additional ionizations in the surface region over and above those caused by the primaries.

13.2.2 *High energy ions*

Auger spectra produced as a result of high energy ($\approx$MeV) ion excitation generally show marked differences between the gaseous and solid states. Ion–atom collisions in this energy region lead to very complex Auger spectra with many additional lines, as seen in Fig. 13.9 from Maeda *et al.*,[9] in which the effects of electron, proton, and α-particle excitation of gaseous argon are compared. In normal electron-excited Auger emission from argon the principal transition is based on ionization of the $L_{2,3}$ shell; owing to very rapid Coster–Kronig transitions any ionization of L_1 is converted immediately to that of $L_{2,3}$. M shell ionization can occur as well but transitions based on the M shell hole are normally weak. Simultaneous ionization of $L_{2,3}$ and M shells is never observed.

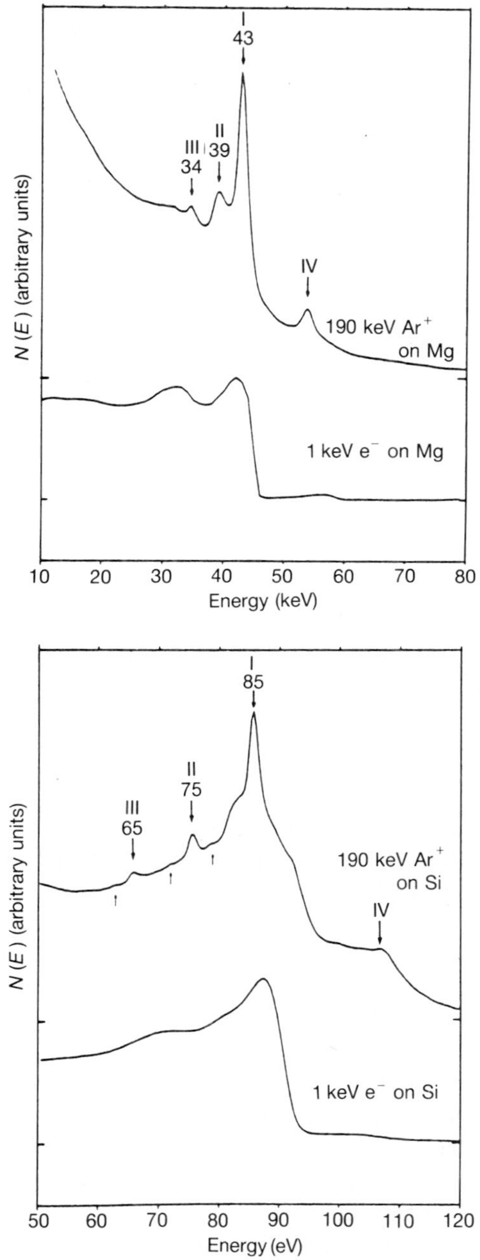

FIG. 13.7. Comparison by Thomas[7] of the electron-excited and ion-excited Auger spectra from Mg and Si. Electron energy 1 keV, Ar^+ ion energy 190 keV. There is a superficial resemblance between the spectra, but there are several features of the ion-excited spectra that do not appear in AES. The principal peaks are narrower, and have additional shoulders, there are additional peaks, and also a peak at higher energy (IV) than any seen in the AES spectrum.

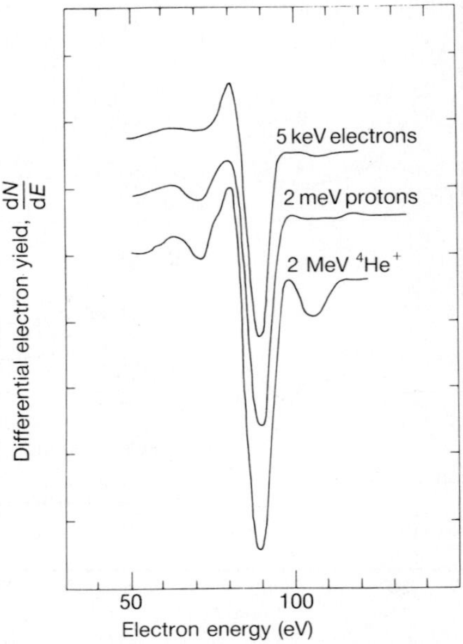

FIG. 13.8. Demonstration by MacDonald *et al.*[8] that the peak at higher energy than the principal Auger peak (i.e. peak labelled IV in Fig. 13.7) arises only during ion excitation and not during electron or proton excitation. The spectra are those of the Si LVV and associated Auger transitions, and are in the differentiated form, although most IAES spectra are presented in the undifferentiated form.

With both proton and α-particle excitation, however, there is an increased probability of simultaneous ionization of atomic levels; in the case of argon using protons this probability is quite low and hence the proton-excited spectrum in Fig. 13.9 looks very similar to the electron-excited one. For α-particles and heavier ions the probability is much greater, and it is thus possible that Auger transitions with initial states containing several holes will occur. The energies of such transitions have been calculated by Larkins[10] and those of highest probability are indicated in the figure. Using the electron-excited spectrum as a standard for the single-hole transitions $L_{2,3}MM$, the proportionate contributions of the multiple-hole transitions could be estimated by simple spectrum scaling and subtraction. This results in estimates of 40% additional single ionization in the M-shell for proton excitation, but of 140% single, 80% double, and 30% triple additional ionizations in the M shell for α-particle excitation.

When irradiating solids with high energy ions the resultant spectra appear quite different and much closer in character to the electron and proton-excited spectra. This can be seen in the Auger spectra in Fig. 13.10 from Saemann-Ischenko and Schmidt,[11] for irradiation of Al and Si with various very high

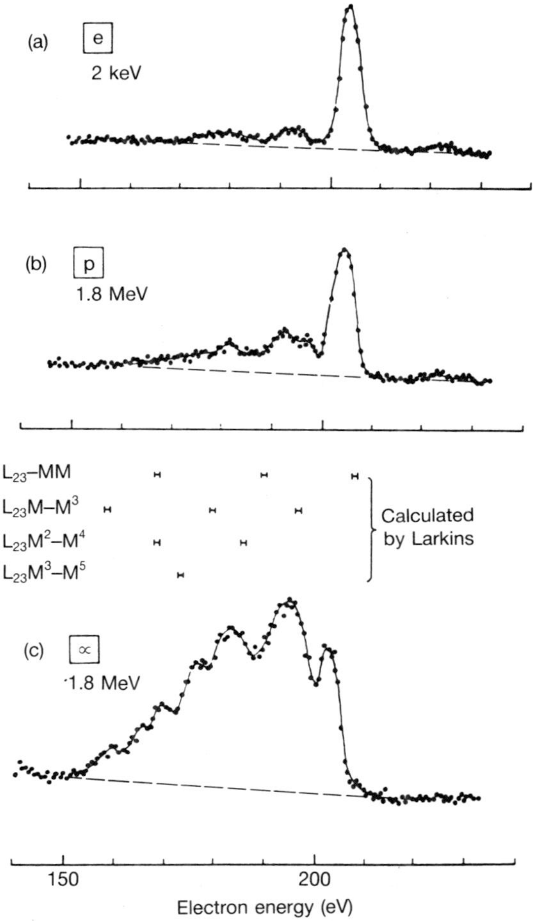

FIG. 13.9. Comparison by Maeda *et al.*[9] of (a) electron, (b) proton, and (c) α-particle-excited Auger spectra of gaseous argon. The proton-excited spectrum looks similar to the electron-excited spectrum since the probability in argon of simultaneous excitation of several atomic levels is low. The principal transition is based on ionization of the $L_{2,3}$ shell. For α-particle ionization the probability of simultaneous ionization is much higher, leading to the possibility of Auger transitions based on initial states containing several holes. The energies of some of these states with highest probability have been calculated by Larkins[10] and are indicated.

energy heavy ions, with protons, and with electrons, respectively. Although the higher energy satellites L*VV and L**VV arising from multiple ionization of $L_{2,3}$ are present in the heavy ion excited spectra, there are virtually no additional features due to *simultaneous* excitation of the M shell. The reason of course is that in solid Al and Si the M shell is part of the valence band, and is almost completely de-localized, in contrast to the atomic situation. Thus any

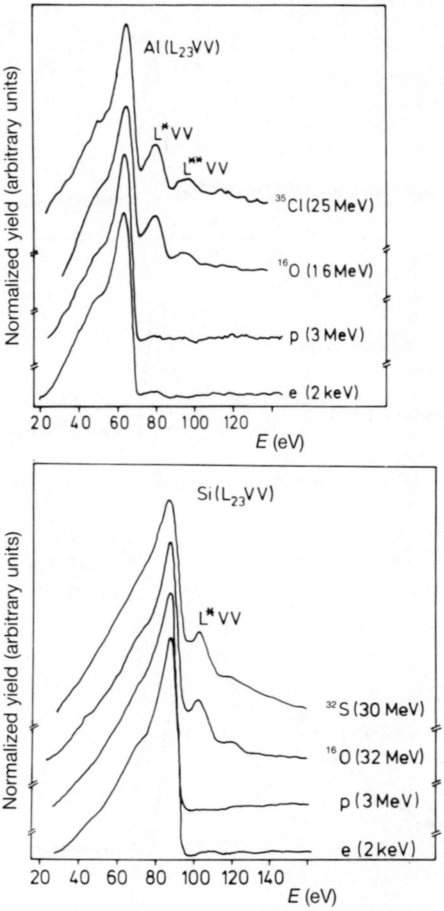

FIG. 13.10. Comparison by Saemann-Ischenko *et al.*[11] of electron, proton, and very high energy ion excitation of the L_3VV Auger spectra of Al (upper) and Si (lower). Apart from the presence of the high-energy peaks marked L*VV and L**VV, the spectra excited by the various projectiles all look similar. No features due to simultaneous excitation of the M-shell appear, as in Fig. 13.9 for argon, because for Al and Si the M-shell is in the valence band and thus de-localized.

M-shell vacancies created by the fast heavy ion are annihilated in times shorter than the Auger transition time by relaxation of the valence band through plasmon oscillations. If this argument is extended to the cases of metals such as Ni and Cu, a rather different situation should be predicted. For Cu in particular it is known that the Auger transitions involving the valence band should be narrow and atomic-like since the two final state holes have a Coulomb interaction energy which is similar to the width of the valence band, and they are therefore highly localized. For Ni the localization is less rigid. Thus one might expect for these metals some contribution from satellites due to multiple ionization during heavy ion excitation. However, in the same paper by Saemann-Ischenko and Schmidt the Auger spectra of Ni and Cu excited in this way do *not* show any such satellites conclusively, only a line broadening without any additional fine structure. Clearly more theoretical considerations are needed here.

13.2.3 *Low to medium energy ions*

Excitation mechanism The excitations already described for high energy ions and protons can be regarded as falling into the class of weak perturbations of the core electrons during interaction since the latter is between an incident particle and an electron. As the energy is reduced into the low and medium ranges this description is no longer adequate and it is necessary to think instead in terms of the transient 'quasi-molecule' formed by the ion and the target atom or molecule as their distance of approach reaches some critical value. The original formulation for the excitation mechanism was proposed by Barat and Lichten.[12] In this scheme the energy levels of the separated atoms are correlated with calculated or estimated energy levels of the projectile-target quasi-molecule, in the so-called 'correlation diagrams'. Such a diagram for the impact of 100 keV Ar^+ ions on Fe is shown[13] in Fig. 13.11. At the right are indicated on a logarithmic energy scale the various atomic energy levels for the separate Ar and Fe atoms. Internuclear distance decreases from right to left. At approximately zero separation the coalescence of the Ar–Fe electron systems is equivalent to the electron orbital system of Ru, which is drawn on the left. At intermediate separations there must be quasi-molecular orbitals giving rise to energy levels connecting the two extreme situations, and the connections from right to left as the two entities approach each other are made following the rules laid down by Barat and Lichten. These are that (a) free atom and coalesced atom energy levels can be correlated if their wave functions have the same number of nodes (n–l–1), where n is the principal quantum number and l the orbital angular momentum; (b) the component of the orbital angular momentum projected along the inter-nuclear axis, Λ, is a good quantum number, and is conserved, so that free atom levels with orbital angular momentum l produce molecular orbitals in the coalesced atom with $\Lambda = 0, 1, \ldots l$. It is then possible to draw correlation lines between states of the same symmetry, starting with the deepest level, i.e. the 1s level, and proceeding to the valence band.

As a result of the construction of a correlation diagram such as that in Fig. 13.11, it can be seen at once that as the internuclear separation decreases there is the likelihood of one or more of the correlation lines crossing several others. The obvious example in the figure is that of the Ar 2p atomic level which rises through the molecular orbital 4fσ to the coalesced atom 4d. In this case the Ar 2p electron is said to be 'promoted', and the phenomenon is called

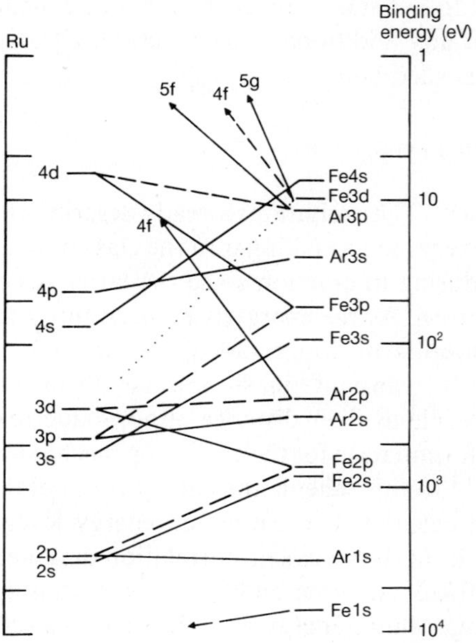

FIG. 13.11. Correlation diagram based on the Barat and Lichten[12] mechanism for ion excitation, drawn for the case of Ar^+ excitation of Fe by Legg *et al.*[13] On a logarithmic scale on the right are the energy levels for separate Ar and Fe atoms. The internuclear distance decreases from right to left. At zero distance the atoms coalesce, giving an Ru atom, whose energy levels are drawn on the left. At intermediate distances there are quasi-molecules of Ar–Fe, whose orbitals give rise to energy levels connecting the two extremes right and left. The rules of Barat and Lichten[12] are mentioned in the text. As the separation decreases there is increasing likelihood that one or more lines will cross several others, e.g. the Ar 2p atomic level which rises to contribute to the coalesced atom 4d via the molecular 4fσ. The Ar 2p electron is said to be 'promoted'. If a promoted electron crosses a normally empty state, electron transfer can occur, and on separation the consequent vacancy may be retained as a hole, giving rise to strong Auger transitions.

electron promotion. When this happens, it is always possible that the promoted electron will cross a state normally empty or with a vacancy, and that electron transfer can occur. When the two atoms separate again after such interaction the vacancy so created may be retained as a hole in the Ar 2p shell, so giving rise to strong Auger peaks in the secondary electron spectrum. Note in Fig. 13.11 that there is also some promotion of Fe 3p electrons, and the holes thus created do indeed produce Fe $M_{2,3}VV$ Auger peaks as well.

One consequence of the above picture of the ion excitation mechanism is that, since the promotion due to molecular orbital formation occurs at small internuclear distances (a fraction of an Å) where there is strong nuclear repulsion, there must be an incident ion energy threshold for excitation. In addition there will be a rapid rise in the excitation probability above threshold, leading to constant ionization cross-section at sufficiently high energies. All these consequences have been confirmed experimentally.

Auger decay mechanisms A further consequence of the molecular orbital picture of ion excitation, just described, is that because of the rather large repulsion energy involved during the approach to the distance at which molecular orbital crossings can occur, significant momentum may be transferred to a target atom. If sufficient, then the atom will be displaced, either recoiling into the solid or being ejected as a free atom from the target surface. The recoiled atoms will have essentially the same Auger spectra as those of stationary atoms, since the valence band screening will be very fast, but if the ejected atom is already in an excited or ionized state then the resultant Auger spectrum will resemble that from a free atom. It is for precisely this reason that low to medium energy ion-excited Auger spectra have the appearance typified by Fig. 13.7, that is, with a series of very sharp peaks superimposed on a much broader structure. The sharp Auger peaks are in fact quasi-atomic in nature.

Despite this recognition that the most prominent features are quasi-atomic in origin, there are still difficulties in interpretation since the observed ion-excited solid state Auger spectra do not coincide in energy exactly with those obtained from gas-phase excitation, and there are also additional lines not present in the gas-phase observations. The problem has been discussed at length by Thomas.[7] Thus the spectrum in Fig. 13.7 from Mg excited by 190 keV Ar^+ contains four prominent and sharp peaks labelled I–IV, of which only III can be identified as arising from an $L_{2,3}MM$ transition in a free atom with a 2p ($L_{2,3}$) vacancy. Thomas has pointed out that there is always the possibility that the ejected species might be that of an atom which is in fact electronically neutral but nevertheless has vacancies in core levels. He gives for Mg the example of the state $2p^5 3s^2 3p$, which has an electron removed from the 2p shell, thus providing a vacancy, but another added to 3p to preserve neutrality. Auger transitions based on such an initial state can be calculated, and indeed the energies for the transitions $2p^5 3s^2 3p \rightarrow 2p^6 3s$ and $2p^5 3s^2 3p \rightarrow 2p^6 3p$, which might also be written $L_{2,3}M_1M_{2,3}$ and $L_{2,3}M_1M_1$, agree well with the positions of peaks I and II, respectively, in Fig. 13.7. The calculated relative probabilities of these transitions are also in rough agreement with the relative intensities of peaks I and II. In addition the weak high energy feature labelled IV can be identified on this basis as arising from an ion singly charged but with a double vacancy, i.e. from the transition $2p^4 3s^2 3p \rightarrow 2p^5 3s$, whose calculated energy again agrees well with the position of IV.

Confirmation of the above ideas about interpretation of ion-excited Auger spectra, at least for Mg and Al, has been provided[14] by a spectrum simulation using the various suggested contributions to the spectrum. Thus, Fig. 13.7 shows a secondary electron background increasing progressively to lower kinetic energies that can be fitted to a polynomial function at each end of the spectrum where no features are present. Then there is a contribution from Auger electrons excited in the bulk Mg only which is given by the electron-excited spectrum shown in the lower half of the figure. These two contributions are added together and give rise to the curve A in Fig. 13.12, which therefore

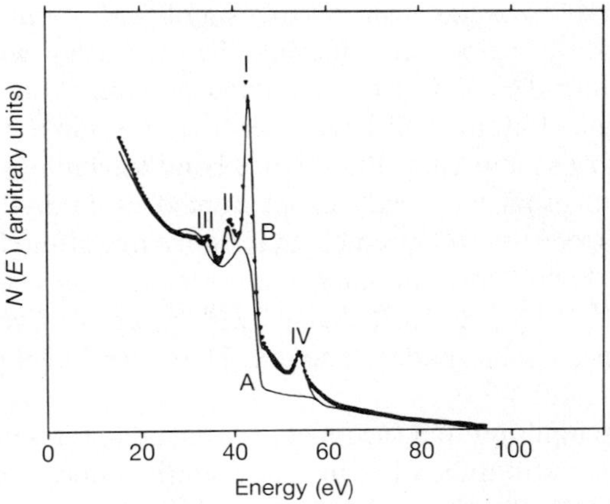

FIG. 13.12. Spectrum simulation by Whaley and Thomas[14] of the 190 keV Ar^+ excited $L_{2,3}VV$ Auger spectrum of Mg. Curve A is the sum of a secondary background fitted by a polynomial function and the contribution from Auger electrons excited in *bulk* Mg. Curve B is the sum of curve A and Auger energies calculated for excited atom transitions, broadened appropriately with Gaussian functions. In fitting to the experimental results, the Auger line separations were kept constant, but the relative intensities and the Gaussian widths were allowed to vary. Curve B matches the experimental data very well.

simulates the purely bulk contribution to the spectrum. Auger lines at energies calculated as described above, broadened by Gaussian functions, are then added to curve A by a curve-fitting routine that allows the relative intensities and the Gaussian widths to vary, but keeps the Auger line separations constant. The result is curve B, which can be seen to match the experimental points very well. The widths of the quasi-atomic features necessary for the fit are all consistent with Doppler broadening of particles with energies around 50 eV.

When the same procedure is applied to the ion-excited Si Auger spectra, it is found that the sum of the background and calculated Auger contributions does not match the observed spectrum very well, as seen in Fig. 13.13. However, for Si, and presumably for other elements as well, a further consideration needs to be taken into account, and that is the distance from the surface within which the Auger decay is likely to occur. In the cases of Mg and Al it can be estimated from the Doppler shift that decay occurs a long way outside the extension of the surface potential ($\approx$5 Å), so that the atoms are essentially in free space. For Si it is clear that a substantial proportion of the

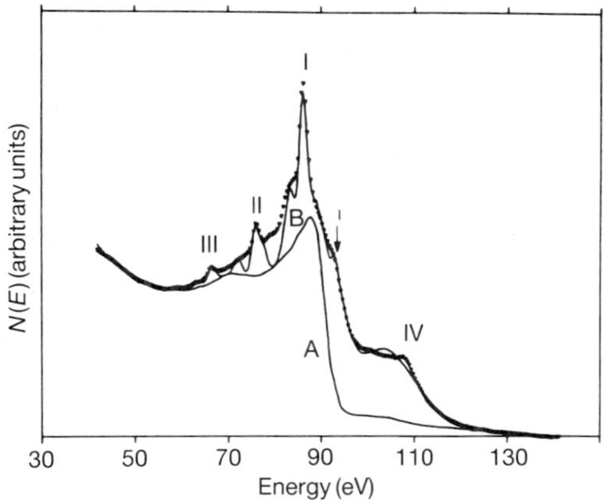

FIG. 13.13. Same procedure[14] as in Fig. 13.12 for Mg used on the ion-excited L_3VV Auger spectrum of silicon. Again curve A is the bulk contribution to the spectrum, and curve B is the sum of A plus broadened calculated Auger energies. The agreement is not as good as for Mg, due, according to Whaley and Thomas,[14] to the fact that the Auger decay in excited Si atoms takes place much closer to the surface, while the atoms are still within the influence of the surface. The outer valence electrons will not have made the full adjustment from their configuration in an atom in the solid to that in a free atom. This will give rise to additional features, as seen in the figure.

decay will happen while the atom is still within the influence of the surface potential. In this case the outer valence electrons will not quite have made the configuration transition from an atom in the solid to a free atom, and will be in some intermediate situation. This will give rise not only to a broadening of the Auger peaks characteristic of decay in the free atom, but to additional features, and both effects can be seen in the Si spectra of Fig. 13.13. On the other hand, at the moment it is difficult to make accurate estimates of the peak positions and widths to be expected.

13.3 Quantification and applications

It will have become clear from the foregoing that IAES and PAES are not techniques that can either easily be made quantitative or that are likely to have wide application. Although the Auger spectra produced by electron excitation and proton excitation are similar both in appearance and in physical origin, to achieve proton velocities equivalent to those of the energies of electrons normally used in AES would require proton beams of some MeV, which rules out their routine use. It is also difficult to see how excitation by rare-gas ions at low to medium energies can ever be useful in a general analytical sense because of the many additional free-atom contributed features in the spectra; such features do not provide any clear insight into either the composition or the chemistry of the surface.

It will also be clear from the previous section that for the foreseeable future the principal use of IAES is in the realm of basic physical research into the properties of quasi-free atoms whose electronic structure is perturbed by proximity to a surface. This is such an esoteric area that there are only a few groups actively involved in it internationally, but it is nevertheless a very interesting subject, and it may well be that with much further work and improved physical insight it will be able to make an increased contribution to the knowledge about the surface itself.

References

1. Kempf, J. and Kaus, G. (1977). *Appl. Phys.* **13**, 261.
2. Benazeth, C., Benazeth, N., and Viel, L. (1978). *Surf. Sci.* **78**, 625.
3. Nègre, M., Mischler, J., and Benazeth, N. (1985). *Surf. Sci.* **157**, 436.
4. Metz, W. A., Legg, K. O., and Thomas, E. W. (1980). *J. Appl. Phys.* **51**, 2888.
5. Sitter, C., Davies, J. A., Jackman, T. E., and Norton, P. R. (1982). *Rev. Sci. Instrum.* **53**, 797.
6. Schmidt, W. (1984). *Ph.D. Thesis*, Universität Erlangen-Nürnberg; and Schmidt, W., Müller, P., Brückner, V., Löffler, F., Saemann-Ischenko, G., and Schubert, W. (1981). *Phys. Rev. A* **24**, 2420.
7. Thomas, E. W. (1984). *Vacuum* **34**, 1031.
8. MacDonald, J. R., Feldman, L. C., Silverman, P. J., Davies, J. A., Griffiths, K., Jackman, T. E., Norton, P. R., and Unertl, W. N. (1983). *Nucl. Instrum. Meth.* **218**, 765.
9. Maeda, N., Kobayashi, K., Hori, H., and Sakisaka, M. (1976). *J. Phys. Soc. Jpn.* **40**, 1430.
10. Larkins, F. P. (1971). *J. Phys. B: Atom. Mol. Phys.* **4**, 1.
11. Saemann-Ischenko, G. and Schmidt, W. (1983). *Nucl. Instrum. Meth.* **218**, 757.
12. Barat, M. and Lichten, W. (1972). *Phys. Rev. A* **6**, 211.
13. Legg, K. O., Metz, W. A., and Thomas, E. W. (1980). *J. Appl. Phys.* **51**, 4437.
14. Whaley, R. and Thomas, E. W. (1984). *J. Appl. Phys.* **56**, 1505, and Thomas, E. W. and Whaley, R. (1984). *Nucl. Instrum. Meth.* **B2**, 516.

14

Ion excitation: ion neutralization spectroscopy (INS) and metastable quenching spectroscopy (MQS)

14.1 Ion neutralization spectroscopy

Strictly speaking INS should be included with IAES in the previous chapter, but it is treated separately here for several reasons. Firstly, the incident ion energies are very much lower (5–10 eV) than any discussed previously, and the production and transport of the ions therefore require special experimental design. Secondly, for reasons that will become apparent, the only ion type used is helium. Thirdly, the processing of the experimental data leads necessarily into complex but fascinating mathematical manipulations involving, principally, deconvolution. And fourthly, it is still almost a one-man technique; nearly all the published work on INS has come from Hagstrum (and co-workers), who was both the originator and the developer of the technique. What follows, therefore, is taken almost entirely from his publications.

14.1.1 *Operation*

The apparatus built by Hagstrum for INS experiments has been described in several publications,[1–3] and the most recent version is that reproduced here in Fig. 14.1. The original design[1] was at the time the most complex ever built for surface studies, and in addition the level of vacuum sought and achieved, $\approx 5\times10^{-11}$ torr, was significantly better than the current norm.

The experimental chamber is a three-dimensional symmetric cross, with the experimental facilities on the four horizontal ports, specimen manipulator on the top port, and pumping on the bottom port. The manipulator allows rotation of the specimen (or target T) around the vertical axis A–A so that the surface of the specimen can be presented in turn to any of the four experimental ports, and also (not shown in Fig. 14.1) allows a further limited rotation around another vertical axis passing through the specimen surface. In the LEED position (port 3) the latter rotation is useful for operation at normal incidence of the electron beam. Port 2 is used for sample processing, containing a retractable electrode system for sputtering the specimen in order to clean it, and an evaporator for coating the surface with whatever material is required. The LEED optics are conventional apart from the use of a transparent phosphor-coated glass screen rather than the hemispherical metal screen normally used, in order to be able to view the LEED pattern from

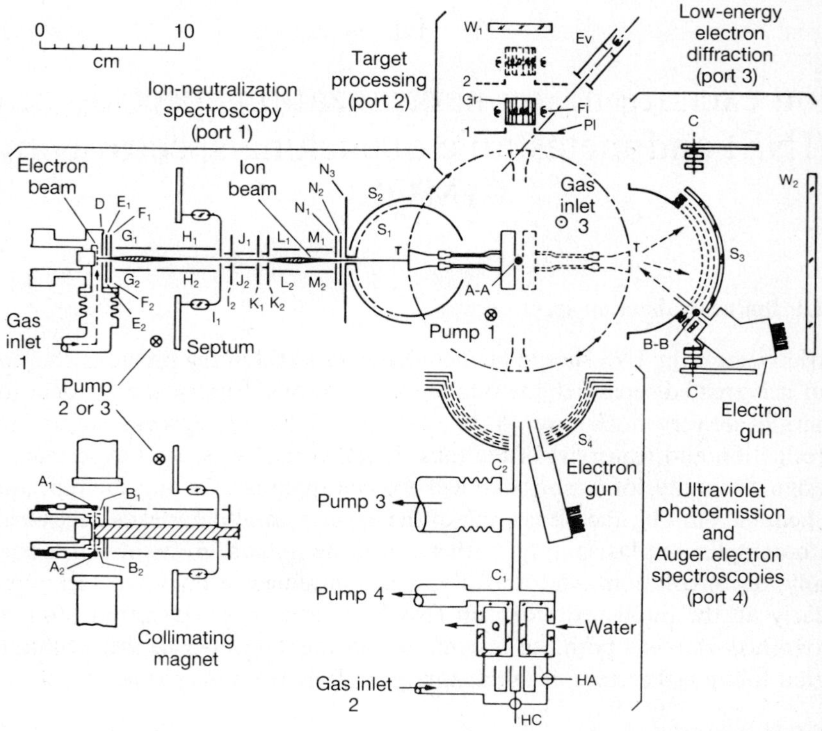

FIG. 14.1. The complex multi-technique experimental system built by Hagstrum[3] for INS; LEED, UPS, and AES can also be performed. The diagram is a plan view, that is, the plane of the page is the horizontal plane. The target T is mounted on a manipulator that allows its surface to be swung around for presentation to any one of four experimental ports. Port 2 for sample processing contains a primitive ion gun (filament Fi, grid Gr, and plate P1) for sputtering, and an evaporator Ev for coating, the target. Port 3 carries a 3-grid LEED system with a phosphor-coated glass screen S_3 for rear viewing, and an electron gun that can be scanned over the axes B-B and C-C if required. Port 4 is the analysis station, with a 4-grid retarding field analyser S_4 and electron gun for AES, and a differentially pumped capillary discharge lamp for UPS. Energy analysis in UPS is also effected with the retarding-field analyser. Port 1 carries the source and analyser for INS. Ionization takes place in the chamber D by acceleration of electrons from filament A_1–A_2 through a slit B_1–B_2 into helium gas. Extraction and beam formation then takes place via electrodes E_1–E_2 and F_1–F_2, and the lens system G_1–G_2, H_1–H_2, J_1–J_2, L_1–L_2, and M_1–M_2. Collimation is imposed by electrodes I_1–I_2, K_1–K_2, and N_1, N_2, N_3. Beam deflection can take place at any pair of symmetrical electrodes. The analysing arrangement is shown in greater detail in Fig. 14.2.

behind through the window. Port 4 is an analysis station, containing a four-grid retarding-field analyser (see Section 3.3.3 for a description of the analyser) with an electron gun for AES measurements and a capillary discharge lamp (see under Section 3.2.2).

The attachment of particular relevance to this chapter is that on Port 1 for INS measurements. On the main diagram of Fig. 14.1 the ion source and the focusing and collimating electrodes are shown in side view for better understanding, with a smaller separate top view inserted below. Electrons from the filament $A_{1,2}$ are accelerated through a slit in the electrode $B_{1,2}$ into the ionization chamber D through which helium gas is passed. Helium ions so formed are extracted through electrodes $E_{1,2}$ and $F_{1,2}$. Focusing then takes place electrostatically along the lens system from $G_{1,2}$ to $N_{1,2,3}$, with collimation by apertures along the way in I_1, K_1, and K_2. Beam deflection can be imposed by varying the opposing voltages on any pair of symmetrical electrodes.

Ions of the required energy in the range 5–10 eV strike the surface of the specimen T, the point of impact being arranged to be at the centre of curvature of a simple retarding-field analyser formed by the hemispherical grid S_1 and collector S_2. A more detailed diagram of the specimen and analyser region is shown in Fig. 14.2, from Hagstrum and Becker.[4] In order to separate currents

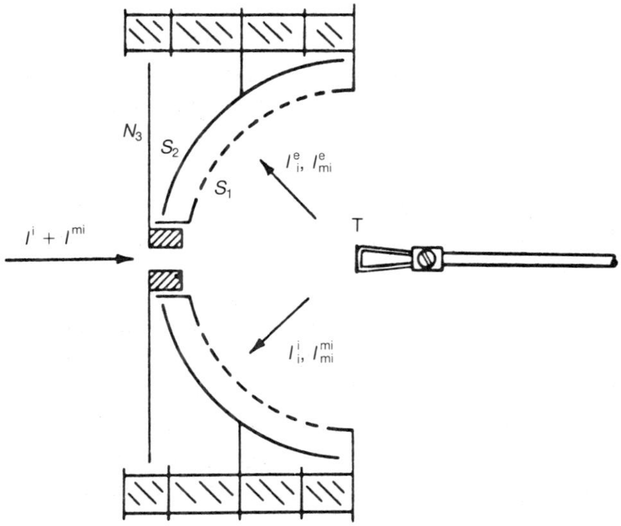

FIG. 14.2. The retarding-field energy analyser used by Hagstrum and Becker [4] to analyse the very low energy secondary electrons emitted in INS. It consists of a single spherical grid S_1 and a spherical collector S_2, both centred on the point of impact of the primary ions on the target T. The primary beam is of energy 5–10 eV. Secondary electrons produced at grid S_1 by reflected ions are separated by applying accelerating potentials of ≈ 5 eV between N_3, S_1, and the target. The primary beam will contain currents I^i and I^{mi} of ions and of metastable ions, respectively. The notation of the secondary currents is: I^e_i and I^e_{mi}, secondary electrons ejected by primary ions and metastable ions respectively; I^i_i, secondary ions ejected by primary ions, and I^{mi}_{mi}, secondary metastable ions ejected by primary metastable ions.

of secondary electrons produced at the grid S_1 by reflected ions, accelerating potentials of about 5 eV are applied between N_3, S_1 and the specimen surface T.

For INS, as will be discussed in the next section, it is necessary to work with an ion beam of pure He^+ (1s), and this can be obtained when the energy of the ionizing electron beam is kept below 65.4 eV, which is the onset potential for the production of He^+(2s). If an extension of INS into the technique of Metastable Quenching Spectroscopy (MQS) is required then the metastable excited states He* $^3S_1$ and He* $^1S_1$ are needed. MQS will be discussed briefly later.

14.1.2 *Theory and data processing*

As pointed out in the discussion in the previous chapter on IAES, when an ionized or excited atom directed at a surface approaches sufficiently close ($\lesssim 1$ nm) to that surface, it can be considered as forming a quasi-molecule with one or more surface atoms. If the energy (velocity) of the approaching species is very low, as in INS, any transient changes in the electronic structure of the interacting surface atom(s) have time to be screened by the surrounding atoms in the solid, and the transient species formed is not so much a quasi-molecule as an ionized (excited) atom-solid system. In INS it is the mechanisms of de-excitation of the resultant ion-solid system that form the basis of the technique.

Consider in Fig. 14.3 (from Hagstrum[5]) the approach to a metal surface of either an He^+ or an He^{++} ion. The schematic energy level diagram of the metal is shown at the left; the vacuum level is at V, the Fermi level at F, and the work function is ϕ. The valence band extends from F down to B (lowest filled level ζ_0), and ζ_1 and ζ_2 are two valence band levels separated by Δ from their mean position ζ.

In the middle and at the right of Fig. 14.3 are shown the schematic energy level diagrams of two atomic wells corresponding to the ions He^+ and He^{++} approaching the metal surface. The binding energies with respect to the vacuum level of He^+ are of course those of the neutral atom He^0, and of He^{++} those of He^+, since, for example, the binding energy of He^0 ($1s^2$) is the energy required to ionize that level. Actually, owing to interaction with the solid at very close distances, the effective ionization energy is slightly less than the free-atom ionization energy.

According to whether the atomic levels in the approaching ion do or do not fall within the range of allowed levels in the solid, so different processes of de-excitation are possible. Thus, where, as for He^{++}, the levels (He^+(2s) and He^+(3s)) lie *within* the allowed range of the solid, tunnelling of electrons through the surface potential barrier can occur from the valence band into the empty levels, so that the latter are effectively resonance states and the process is known as resonance neutralization. Only if the energy of the hole so left in the valence band is more than twice the work function does electron emission occur, and the electron energies are not informative about surface valence band structure. If, on the other hand, as for He^+, the atomic levels lie *outside* the allowed range in the solid, then the tunnelling electron from the valence band can drop into the hole available in the ground state, which, being below

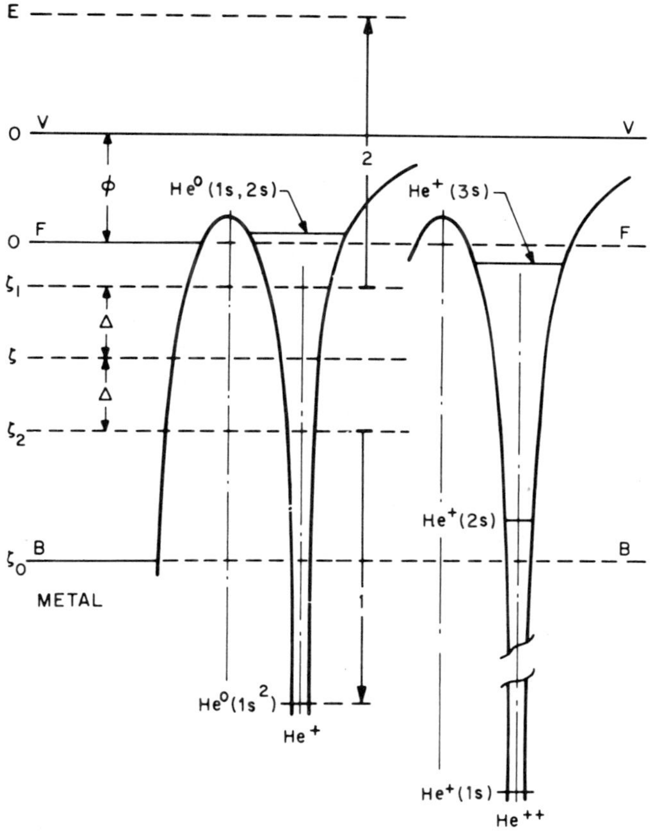

FIG. 14.3. Schematic energy diagram by Hagstrum[5] of the process of Auger neutralization that forms the basis of INS. On the left, notional energy levels in a metal; the vacuum level is at V, the work function is ϕ, and the valence band extends from F down to B. ζ_1 and ζ_2 are two valence band levels separated by Δ from their mean position ζ. In the middle and to the right of the diagram are the schematic energy levels associated with the atomic wells of the ions He^+ and He^{++}. Where, as for He^{++}, the atomic levels lie within the allowed range of valence levels of the solid, tunnelling of electrons can occur from the valence band into the empty states by a process of resonance neutralization. If, as for He^+, however, the atomic levels lie *outside* the valence band range, then a tunnelling electron can drop into a hole in the ground state, making excess energy available which can then be given to another valence electron, which is ejected into vacuum by an Auger process, if its energy is great enough. The transitions labelled 1 and 2 in the figure give an example of such a process.

the bottom of the band, increases the binding energy of the electron with respect to the vacuum level. There is thus excess energy available and this can be given up to another electron in the valence band, which is ejected into the vacuum if its energy is high enough, in effect greater than twice the work function. This is an Auger process, and is called Auger neutralization; it forms the basis of INS.

An example of such an Auger neutralization process is given in Fig. 14.3. An electron from level ζ_2 in the valence band drops into the level $He^0(1s^2)$ via the transition 1, giving up an amount of energy $E_i-\zeta_2-\phi$, which is transferred to an electron in level ζ_1, E_i being the energy of the initial state in the He^+ ion. The energy of the ejected electron, E_f, i.e. in the final state, is then given by

$$E_f=E_i-(\zeta_2+\phi)-(\zeta_1+\phi). \tag{14.1}$$

Since ζ_1 and ζ_2 can each lie anywhere within the entire valence band, it is clear that E_f is not a single energy but a broad band of energies, and, importantly, that contained in E_f is information about the distribution with energy of the ζ, in other words the surface density of states (DOS). However, since both ζ_1 and ζ_2 have the same distribution, the observed E_f is in fact a self-convolution of the DOS, and of course contains as well self-convolutions of the transition probabilities from each state and of any energy broadening processes. It is around the mathematical problems inherent in the self-deconvolution, or 'unfolding', that the major contribution by Hagstrum[6] has been made.

If one starts with the assumption that the transition probabilities are constant, and independent of the position of ζ within the band, then the probability of the neutralization process involving valence band electrons is simply the product $N(\zeta_1)N(\zeta_2)$, where $N(\zeta)$ is the DOS. Since all pairs of electrons symmetrically distanced in energy from ζ will contribute, and ζ lies halfway between E_f and E_i, the total probability of production of an excited electron at E_f is the integral of the state density product, which can be written

$$F_c(\zeta)=\int_{-\zeta}^{\zeta} N(\zeta-\Delta)\, N(\zeta+\Delta)\, d\Delta \tag{14.2}$$

where $F_c(\zeta)$ is the pair distribution function based on constant transition probability. If now the restriction on constant transition probabilities is relaxed, the more general function $F(\zeta)$ is obtained, equal to

$$F(\zeta)=\int_{-\zeta}^{\zeta} |H_{fi}|^2\, N(\zeta-\Delta)\, N(\zeta+\Delta)\, d\Delta \tag{14.3}$$

where $|H_{fi}|^2$ is the square of the transition matrix element between the initial and final states. H_{fi} has the form of a Coulomb interaction integral, and involves the wave functions of the atomic ground state of the approaching ion, of the two electrons in the valence band, and the excited electron at E_f. In general the H factors will not be the same for ζ_1 amd ζ_2, but if they can be assumed to be equal, then they can be subsumed with the state density function H to give the more general function U, namely

$$F(\zeta)=\int_{-\zeta}^{\zeta} U(\zeta-\Delta)\, U(\zeta+\Delta)\, d\Delta. \tag{14.4}$$

The function $U(\zeta)$ is termed the transition density function. Note that the internal excited electron energy distribution $F(E_f)$ can be obtained from $F(\zeta)$ by changing the variable from ζ to E_f. This can be done from eqn (14.1); since $\zeta_2 = \zeta + \Delta$ and $\zeta_1 = \zeta - \Delta$, it follows that

$$E_f = E_i - 2(\zeta + \phi). \tag{14.5}$$

Thus

$$F(E_f) = \int_{E_i + 2(\zeta - \phi)}^{E_i - 2(\zeta + \phi)} U(\zeta - \Delta)\, U(\zeta + \Delta)\, d\Delta. \tag{14.6}$$

It can be seen that the integrals on the right-hand sides of expressions (14.4) and (14.6) are simply the self-convolution of U, also called the convolution square, or the fold. Thus

$$F(\zeta) = F(E_f) = U * U \tag{14.7}$$

Self-deconvolution, or unfolding, of $F(\zeta)$ therefore gives the function $U(\zeta)$, which is an average of the surface density of states, modified by the presence of the ion, and weighted with both an energy broadening function and an energy-dependent transition matrix element. This unfolding is the basis of the INS technique.

The distribution $F(E_f)$ is not of course that observed externally. The latter distribution, $X(E)$, is related to $F(E_f)$ by

$$X(E) = F(E_f)\, P(E) \tag{14.8}$$

where $P(E)$ is the probability of escape of an electron of kinetic energy E from the surface into the vacuum, i.e. over the potential energy barrier. $P(E)$ is a smooth function related to the variation of inelastic mean free path with kinetic energy. The $F(E_f)$ appearing in expression (14.8) will also contain some instrumental and other broadening that must be removed, and Hagstrum has indicated[5,6] how that can be achieved experimentally. $P(E)$ can also be obtained experimentally[2,5] or its theoretical form used. If the de-broadened experimentally measured distribution is $X_0(E)$, then

$$F(E_f) = X_0(E)/P(E). \tag{14.9}$$

A change of variable of E_f back to ζ then gives the function $F(\zeta)$ to be self-deconvoluted. Thus

$$F(E_f) = X_0(E)/P(E) = F(\zeta) = U * U. \tag{14.10}$$

The self-deconvolution of $F(\zeta)$ to arrive at the transition density function U is an 'ill-posed' mathematical problem[7] that has been discussed in detail by Hagstrum and Becker.[6,8] This is not the place to discuss the mathematical complexities, but it is sufficient to say that most of the difficulties arise in the use of the rigorous eqn (14.3), in which the transition probability matrix element between the initial and final states appears. If $H(\zeta_1) \neq H(\zeta_2)$, then $F(\zeta)$ cannot be self-deconvoluted analytically. In fact the authors developed a step-wise numerical sequential deconvolution procedure that had to depend on certain assumptions about distribution shape, and on information from other experimental sources about the starting-point of the process. It was self-consistent in that there were built-in mathematical tests for the genuineness or otherwise of peaks in the distribution so produced.

14.1.3 *Spectral interpretation*

Typical raw kinetic energy distribution data, i.e. the $X(E)$ of eqn (14.8), are shown in Figs 14.4 and 14.5, in the former[9] as a function of surface condition

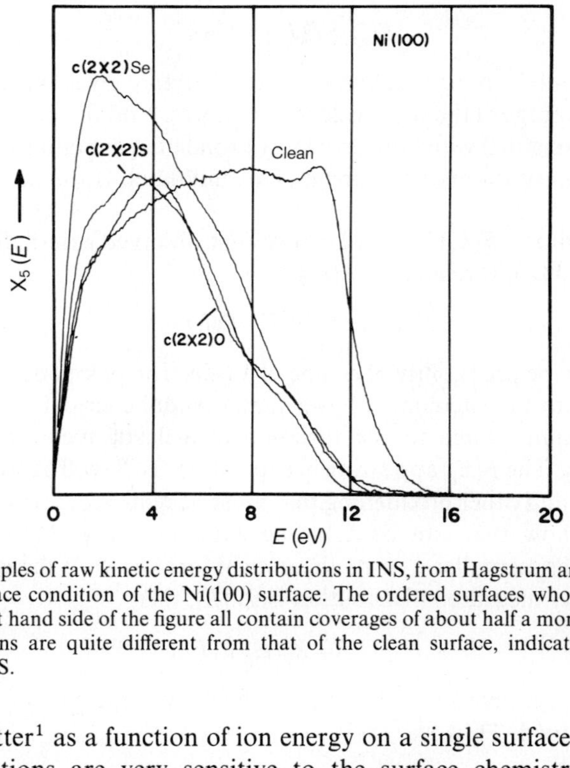

FIG. 14.4. Examples of raw kinetic energy distributions in INS, from Hagstrum and Becker,[9] as a function of surface condition of the Ni(100) surface. The ordered surfaces whose distributions appear to the left hand side of the figure all contain coverages of about half a monolayer, and yet their distributions are quite different from that of the clean surface, indicating the surface sensitivity of INS.

and in the latter[1] as a function of ion energy on a single surface. Clearly the INS distributions are very sensitive to the surface chemistry, since the distributions peaking to the left-hand side of Fig. 14.4 arose from surfaces containing chemisorbed material to coverages of less than a monolayer. Figure 14.5 shows the result, in the form of the distribution $X_0(E)$, of the de-broadening procedure; it is basically derived from an extrapolation from two distributions at different ion energies, in this case 4 eV and 10 eV. The assumption is made that energy broadening varies linearly with ion velocity.

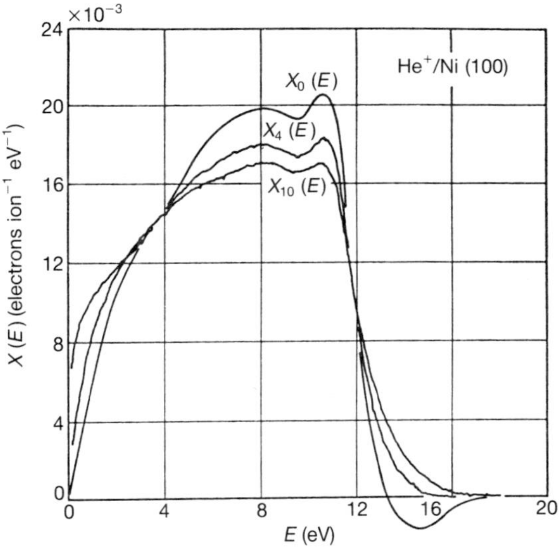

FIG. 14.5. Kinetic energy distributions from a Ni(100) surface; $X_4(E)$ and $X_{10}(E)$ are the raw distributions at ion energies of 4 eV and 10 eV, respectively. $X_0(E)$ is the distribution resulting from $X_4(E)$ and $X_{10}(E)$ following a de-broadening procedure that assumes energy broadening varies linearly with ion energy. From Hagstrum.[1]

The F functions derived from X_0 by division by the function $P(E)$, the form of $P(E)$ used, and the consequent U functions, are shown in Fig. 14.6 for the clean surfaces of Cu(111) and Ni(100), from Hagstrum and Becker.[2] In this case the function $P(E)$ included not only the escape probability but also other E dependences, such as final-state density.

Since $U(\zeta)$ is in a sense the end-product of the self-deconvolution, or unfolding procedure in INS, its significance needs to be examined. It was stated earlier that $U(\zeta)$, the transition density function, is a weighted average of the surface DOS. Accordingly, $U(\zeta)$ should bear some resemblance to both theoretically calculated DOS and the DOS, probably weighted in some different way, obtained from other experimental measurements such as those of UPS. A comparison of the U functions obtained from the low-index faces of copper, with the theoretical DOS calculated by Mueller,[10] has been made by Hagstrum and Becker,[2].and is shown in Fig. 14.7. It can be seen that INS does indeed reproduce the maximum in the d band DOS, but without revealing any of the fine structure apparent in the theoretical DOS. There are several reasons for the latter disagreement. Firstly, of course, the energy resolution in INS is relatively coarse owing to the nine-point data smoothing that has to be carried out before the self-deconvolution, but that is only part of the story. Secondly, the DOS distributions determined by INS and by theory are different functions, in that the theoretical DOS is an average over the whole Brillouin

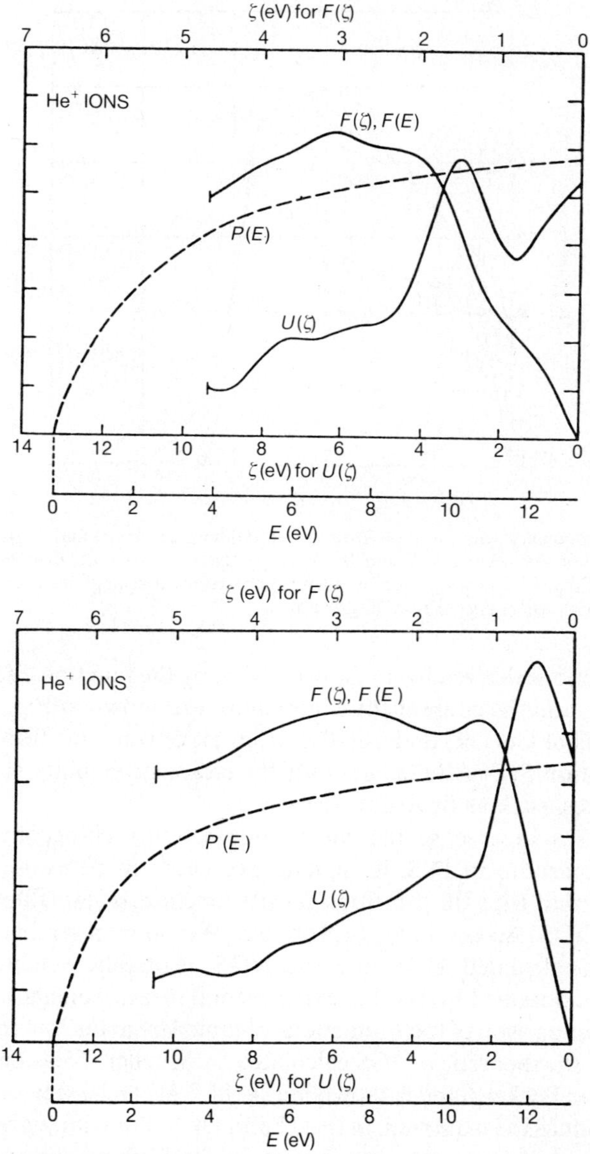

FIG. 14.6. Forms of the F and P functions for clean (a) Cu(111) and (b) Ni(100) surfaces used by Hagstrum and Becker[2] in arriving at the *U* functions, i.e. the transition density functions. The *F* functions are derived from the raw *X* functions (see Figs. 14.4 and 14.5) by division by the *P* function, which is the probability of escape of an electron from the surface into the vacuum. As can be seen, *P* is a smoothly varying function of kinetic energy and is related to the inelastic mean free path. The *U* function, which is the one sought in INS, is then obtained from *F* by a process of iterative numerical self-deconvolution.

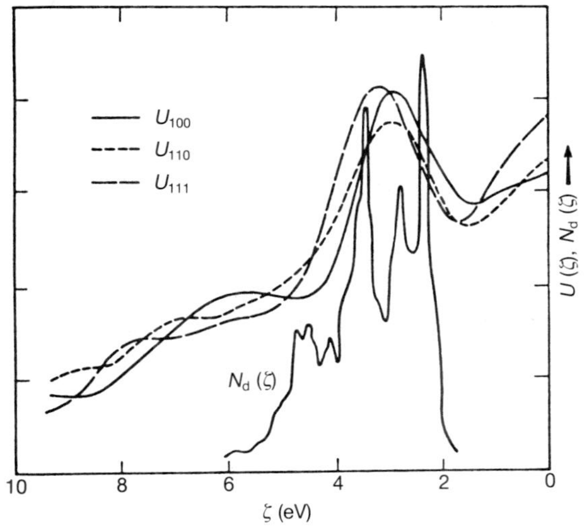

FIG. 14.7. Comparison of the U functions for the Cu(100), (110), and (111) surfaces, obtained by Hagstrum and Becker,[2] with a theoretically derived density of states for the copper d bands (Mueller[10]). INS reproduces the maxima in the d band DOS, but does not show the fine structure. The reasons for this are discussed in the text.

zone, whereas $U(\zeta)$ has been determined for each crystal face and therefore includes momentum selection. Thus the DOS peaks associated with different points in the Brillouin zone will be weighted differently in the individual face $U(\zeta)$ functions, which accounts not only for the differences between the three $U(\zeta)$ distributions in Fig. 14.7, but also for the appearance of structure in U_{100} at about 6 eV. Thus, although one should expect some agreement with the theoretical DOS, it is not likely to be more than qualitative, and that is indeed what is found.

A comparison of INS-derived $U(\zeta)$ functions with experimental DOS obtained from UPS has been carried out by Hagstrum[3] himself, since the apparatus described in Fig. 14.1 is capable of performing UPS measurements. In Fig. 14.8, taken from his paper, are examples of two such comparisons, in (a) for a clean Ni(100) surface, and in (b) for the same surface on which both C and Hg have been adsorbed. The UPS spectra are labelled L in each case. For clean Ni(100) the two DOS distributions appear quite similar, although there are a couple of minor differences that should be noted.[7] One is the more intense tail of inelastically scattered electrons seen in UPS compared to INS, due to the fact that UPS samples more of the bulk than does INS. The other is the proportionately lower contribution of the d electrons in INS compared to UPS, due to the localized nature of d orbitals which lowers their tunnelling probability in INS compared to that of s and p electrons.

In the case of the C and Hg contaminated Ni(110) surface, on the other

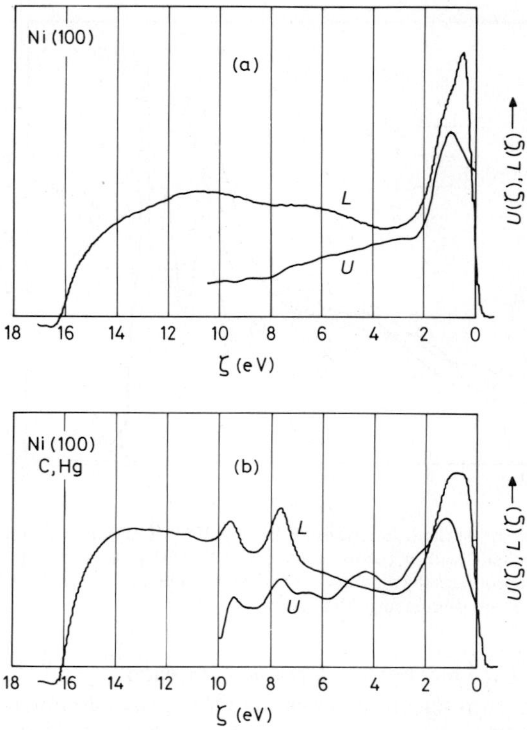

FIG. 14.8. Comparison of the INS-derived U functions from (a) clean Ni(100) and (b) Ni(100) with C and Hg adsorbed on it, with the spectra labelled L from the same surfaces obtained by UPS in the same system (Fig. 14.1), by Hagstrum.[3] Note that for the clean surface there is a longer tail of inelastically scattered secondary electrons in the UPS spectrum, since the kinetic energies involved in the latter are higher than in INS. Also there is a smaller contribution from the nickel d electrons in INS, since the localized nature of the d orbitals reduces their tunnelling probability compared with that of the s and p electrons. On the C and Hg contaminated surface, peaks at 7.8 eV and 9.7 eV from the Hg $5d_{5/2}$ and $5d_{3/2}$ levels can be seen clearly in both distributions. However, the INS U-distribution shows additional structure at 4.2 eV and ≈6.5 eV not present in UPS, which may be due to surface resonances originating from the presence of C, since in the absence of C the additional structure was also absent.

hand, there are both clear similarities and clear dissimilarities. The peaks at 7.8 eV and 9.7 eV arise from the $5d_{5/2}$ and $5d_{3/2}$ levels of the mercury atom and are reproduced in both UPS and INS. However, there is much additional structure in the INS $U(\zeta)$ distribution that has no counterpart in UPS, i.e. the peaks at 4.2 eV and ≈6.5 eV, and the shoulder at ≈2.5 eV. Hagstrum does not show INS distributions from surfaces with C contamination in any other publication, and gives no explanation for the additional structure, so its precise origin is not clear but it must originate in surface resonances due to the presence of C, that only INS can detect. When Hg was present on the Ni(100) surface alone the additional structure was absent.

It would be surprising if there were not differences in the INS and UPS distributions. The description given earlier of the physical processes taking place in INS as the slow ion approaches the surface indicates that the maximum resonance interaction will be with those orbitals with wave functions projecting furthest from the surface into vacuum. On the other hand, the optimum interaction in UPS will be with those orbitals where wave functions are oriented along the direction of the electric vector of the incident light (assuming no selection rule exclusion). Sometimes these respective interactions coincide, and at other times they do not; angularly resolved UPS measurements can occasionally resolve interpretational difficulties.

As with the theoretical comparison, then, the experimental comparison reveals some agreement and some disagreement. The reasons for the latter are largely understood, but on the whole insufficient INS experimental data exist as yet for a complete understanding.

14.1.4 *Applications*

Because of the complexities in both the equipment required and in the mathematical treatment, INS has not been popular and there are very few published applications. The best known are again from the work of Hagstrum himself, in papers[9,11] with Becker in which the effects on the INS distributions from various Ni low-index crystal faces, of the absorption of chalcogens, were studied. The most quoted example[11] is that shown in Fig. 14.9 for the chemisorption of O, S, and Se atoms on the Ni(100) surface; the solid line distributions in this figure should be compared with the kinetic energy distributions in Fig. 14.4 from which they have been unfolded. According to the length of exposure of the Ni(100) surface to O_2, H_2S, and H_2Se, and to the heat treatment, structure corresponding to LEED patterns indexed either $c(2 \times 2)$ or $p(2 \times 2)$ could be obtained.

In Fig. 14.9 the vertical lines marked 'p' are in the positions of the orbital energies of the p-electrons in free O, S, or Se *atoms*, while those marked '$1b_1$', '$3a_1$', and '$1b_2$' are in the positions of molecular orbital energies of free O_2, H_2S, and H_2Se *molecules*. Several features of the $U(\zeta)$ distributions in Fig. 14.9 should be noted. Firstly, the peak at about 1 eV in the clean Ni distribution due to the d-band is much reduced on adsorption; such a reduction can also be observed in UPS. Secondly, a number of additional peaks can be seen in the distributions following adsorption, particularly when the $c(2 \times 2)$ surface structure is present. These correspond to molecular orbitals in the pseudo-molecules formed at the surface between an adsorbed chalcogen atom and a nickel atom. For S and Se adsorption the energies of electrons in these orbitals in the $c(2 \times 2)$ structures are close to those in the free molecule orbitals, indicating adsorption leading to bonds of S or Se with two metal atoms, i.e. a surface pseudo-molecule of composition $M_2X(X = S, Se)$, in the so-called bridge-bonding configuration. For O adsorption, however, there is only a

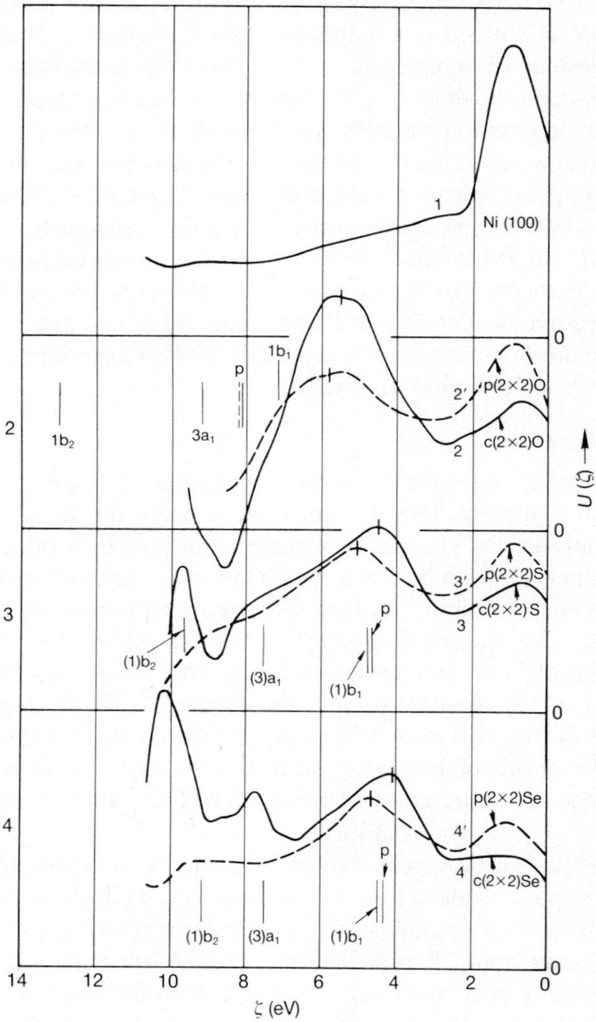

FIG. 14.9. Transition density U functions derived by INS for the chemisorption of O, S, and Se atoms into ordered structures on a Ni(100) surface, by Becker and Hagstrum.[11] Top curve, clean Ni(100). The energy positions labelled p are the orbital energies of the p-electrons in free O, S, or Se atoms, adjusted by the work function. The energies labelled $1b_1$, $3a_1$, and $1b_2$ in panel 2, and $(1)b_1$, $(3)a_1$, $(1)b_2$ in panels 3 and 4 are the molecular orbital energies of free H_2O, H_2S, and H_2Se molecules, again adjusted for the work function.
Note the reduction in the d band of nickel on chemisorption; this can also be observed in UPS. Note also the additional structure in the distributions that arises on chemisorption, particularly for the $c(2 \times 2)$ surface structure. The new energetic structure can be seen to correlate well with the free molecular orbital energies for S and Se adsorption, indicating formation of a pseudo-molecule at the surface. For O adsorption, however, there is no such correlation, and the conclusion was that the pseudo-molecule is not M_2X but M_4X. See the text for further discussion.

single peak below the d-band, shifted towards the Fermi level by much greater energies than for either S or Se. This implies more negative charge on the oxygen atom than on either the sulphur or selenium atoms. The lack of additional structure also implies that the pseudo-molecule in this case is bonded to more metal atoms, and the conclusion was that the configuration was M_4X, i.e. O bonded to four nickel atoms in an in-plane arrangement in which the O atom is positioned in the middle of the (100) surface unit mesh.

When the population of adsorbed species is halved in going from the $c(2 \times 2)$ structure to $p(2 \times 2)$, there is little change in the structure due to O adsorption, apart from intensity, but for S and Se absorption the distribution changes drastically to that of a single peak below the d-band. These new peaks lie very close to the positions of the energies of the p electrons in the free chalcogen atom, indicating a complete alteration in the nature of the surface bonding. This change also indicates that in fact the latter configuration in the $p(2 \times 2)$ structures is energetically more favourable than that in the $c(2 \times 2)$ structures since otherwise a mixture of both distributions would be observed. From parallel work function measurements the pseudo-molecule in the $p(2 \times 2)$ structure was assigned again to M_4X, but in a pyramidal arrangement in which the S or Se atoms were positioned *above* the centre of the (100) unit mesh.

14.2 Metastable quenching spectroscopy

An atom is in a metastable state if it is excited but electrically neutral, carries its excitation in the form of potential energy, and cannot return to the ground state by photon emission due to quantum mechanical selection rules. When such excited metastables are incident on a surface, electron emission is observed just as in INS. The various emission channels available according to the relative juxtaposition of the metastable atomic level and the surface DOS will be discussed below.

It transpires that the light noble gases, in particular He, have metastable properties that make them suitable for MQS, and these properties have been listed by Ertl and Küppers.[12] There is a singlet level He* 2^1S_1 and a triplet level He* 2^3S_1, at 20.62 eV and 19.82 eV, respectively, above the ground state, with respective lifetimes of 2×10^{-2} s and 4.2×10^{-3} s. These lifetimes are certainly long enough for the metastable species to be usable in the technique.

An example of the way in which metastables can be produced and used is shown in the diagram in Fig. 14.10, from Bozso *et al.*[13] Helium at 1.4 bar pressure emerges from a nozzle source with an orifice of 50 μm diameter, and passes through a beam skimmer with an orifice of 1.25 mm into a cold cathode discharge region. The latter contains a ring electrode surrounded by grounded plates, with an electric field of 4000 V cm^{-1} perpendicular to the direction of the beam and in the same direction as an applied magnetic field. The combined

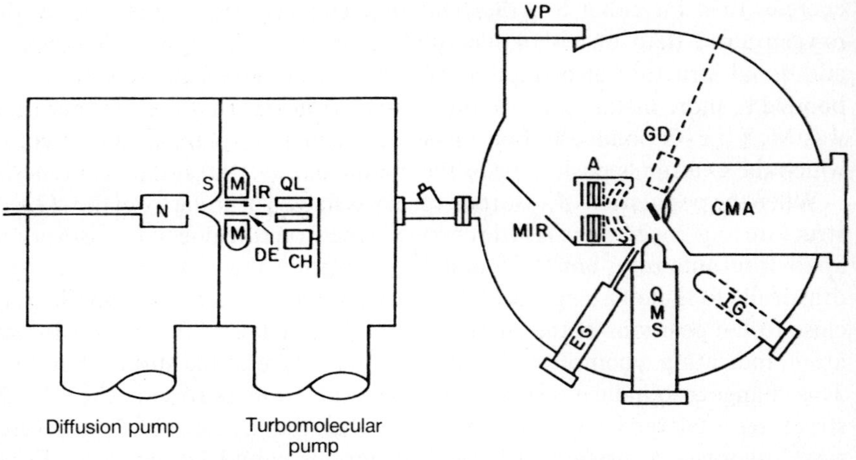

FIG. 14.10. Experimental arrangement devised by Bozso *et al.*,[13] for Metastable Quenching Spectroscopy (MQS). Helium at high pressure (1.4 bar) emerges from a 50 μm aperture in a nozzle source N and passes through a beam skimmer S which has an orifice of 1.25 mm, into a discharge region. The electrodes DE consist of a central ring with grounded plates on either side, the applied electric field being about 4000 V cm^{-1}. A permanent magnetic field is imposed in the same direction. As a result the electrons created in the discharge have high excitation efficiency due to their long spiral paths. Fast neutrals cannot escape from the discharge region, and the high field also removes highly excited Rydberg states from the beam, so that the emergent beam contains only ground state and metastable atoms, and photons. The beam then passes through the radiation from a He quench lamp QL that returns metastables from the 2^1S state to the ground state via excitation to the 2^1P state.
The beam is chopped by a chopper CH at several kHz, and then passes into a chamber where it strikes the target. Energy analysis is performed in the analyser A fitted with a multi-channel plate, with synchronous detection using a lock-in amplifier. The other components consist of EG, an electron gun, QMS, a quadrupole mass spectrometer, IG, an ion gun for sputtering, a CMA, and GD, a calibrated gas doser.

fields cause spiral oscillations of electrons between the plates, causing high excitation efficiency. Ions created in the discharge also move perpendicular to the beam direction, which ensures that fast neutrals formed by charge exchange do not emerge from the discharge region. The high electric field also removes the ions and any highly excited Rydberg states from the beam, so that as it emerges it should contain only ground state and metastable atoms, and photons. After emergence from the discharge region the beam passes through a helium quench lamp and thence via a valve into the UHV chamber. The quench lamp is used to return the 2^1S metastables to the ground state via excitation to the 2^1P state with He 2.06 μm radiation. Since the beam after quenching produced an emission spectrum that was the same shape as that produced by the unquenched beam, but reduced in intensity by $\approx 90\%$, it is clear that the beam contained mostly He* 2^1S, with a small amount of He* 2^3S, but negligible photon intensity. If there had been any appreciable proportion of photons, the spectrum produced by the quenched beam would have had the appearance of a UPS spectrum rather than that of an MQS spectrum. The components shown in Fig. 14.10 in the analyser chamber are self-explanatory.

The emission mechanisms that can occur when a metastable atom approaches a surface are now well understood, and the differences between them have been exploited in several elegant experiments. The two basic possibilities (there are some variations on them) are set out in Fig. 14.11, from

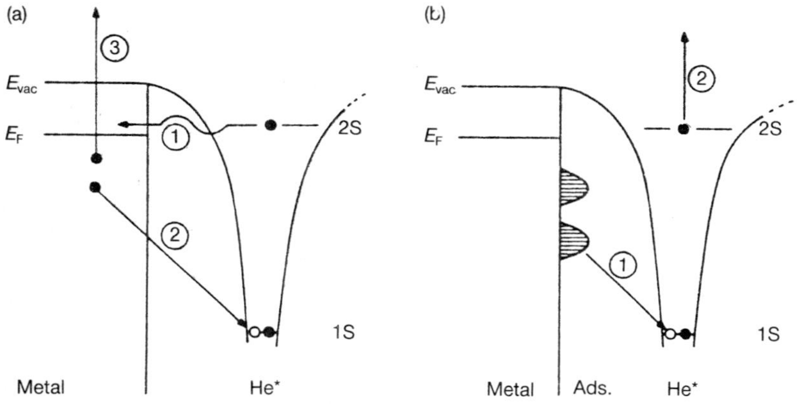

FIG. 14.11. The two possible basic mechanisms of de-excitation of a metastable He^*2^1S atom at a metal surface, as depicted by Bozso *et al.*[14]
(a) If the excited atomic level can resonate with empty states just above the Fermi level, then an electron can transfer (transition 1) from the excited state to the surface, causing resonance ionization (RI) at some critical distance from the surface. The resulting ion moves further towards the surface and can then become neutralized by Auger neutralization (AN) (transitions 2 and 3) just as in INS. In this case the transition density function observed is very similar to that in INS.
(b) If the excited atomic level cannot resonate with empty states because the electrons in it are 'opposite' filled states in the metal surface, then Auger de-excitation occurs. The hole in the inner atomic shell is filled from a surface state, and the excess energy is given to an electron in the excited state, which is then ejected. This is also known as Penning ionization. In this mechanism only *one* electron originates in the surface. If there are adsorbed species with an individual set of molecular orbitals, then the neutralized electron will come from them and not from the metal, indicating very high surface specificity.

the paper by Bozso *et al.*[14] If the excited atomic level can resonate with empty states above the Fermi level of the surface, which assumes that there is sufficient overlap of the respective wave functions, then the mechanism on the left operates. In this mechanism an electron is transferred from the excited atomic level into the surface, causing resonant ionization (RI) of the metastable. The resultant ion continues moving towards the surface, and then becomes neutralized by the same process of Auger neutralization (AN) as occurs in INS. Thus the transition density function observed by RI + AN is essentially the same as in INS.

If the ionization potential of the metastable is greater than the work function of the surface, in other words if the electrons in the excited atomic level are 'opposite' filled and not empty levels in the surface, or if there is some other reason why the surface and metastable wave functions cannot overlap sufficiently, then the de-excitation occurs principally by the mechanism on the right of Fig. 14.11. In this mechanism the hole in the inner shell of the metastable is filled by an electron from a surface state, followed by ejection of the excited electrons; in other words an Auger process, and the mechanism is called Auger De-excitation (AD). It is often also referred to as Penning ionization. Since one of the reasons why resonance ionization might be blocked is the presence of adsorbed species on the surface, with its own molecular orbitals, as indicated schematically in the right hand diagram of Fig. 14.11, the emitted electron spectrum in Penning ionization is characteristic *only* of the adsorbed species. With such high surface sensitivity the technique has been labelled[14] SPIES (Surface Penning Ionization Electron Spectroscopy). Of course, the RI + AN mechanism has equally high surface sensitivity, but has not received the accolade of a separate label since the spectrum is so close to that of INS in any case.

There is one important difference to note between the two mechanisms, which should be obvious from Fig. 14.11. Where RI + AN is operative, *two* electrons are emitted from the surface, and the resultant spectrum is a convolution of their individual local DOS, as in INS. Where Penning ionization is operative, however, only *one* electron originates in the surface, and the spectrum is thus a much more direct reflection of the local one-electron DOS, bearing therefore resemblances to a UPS spectrum.

It should also be obvious from the foregoing that there will be situations in which the two processes compete. After all, if RI + AN is observed from a clean metal surface, and Penning ionization from the same surface covered with an absorbed species, there must be intermediate conditions for low adsorbate coverage in which both types can find the appropriate local electronic situations. Such competition has been demonstrated by Bozso *et al.*[14] during the adsorption of CO and NO on Ni(111) at 270 K and 90 K, respectively. Figure 14.12 from their paper shows the sequence of MQS spectra recorded from the clean state of the Ni surface to saturation at coverages of 3.2 L in each case. The spectrum from clean Ni(111) bears a striking resemblance to the INS transition density function from clean Ni (cf. Fig. 14.6), and clearly the RI + AN mechanism is operating in that case. At the other extreme, saturation coverages of either CO or NO, the whole character of the spectrum has changed, with a high secondary electron background at low kinetic energies and very well developed emission from the 4σ and $(1\pi + 5\sigma)$ orbitals of the adsorbed molecules, i.e. very similar to a UPS spectrum. Also visible is another, smaller, peak at ≈ 13 eV attributed to emission from the $2\pi^*$ orbitals. Obviously at saturation coverage the Penning ionization mechanism has

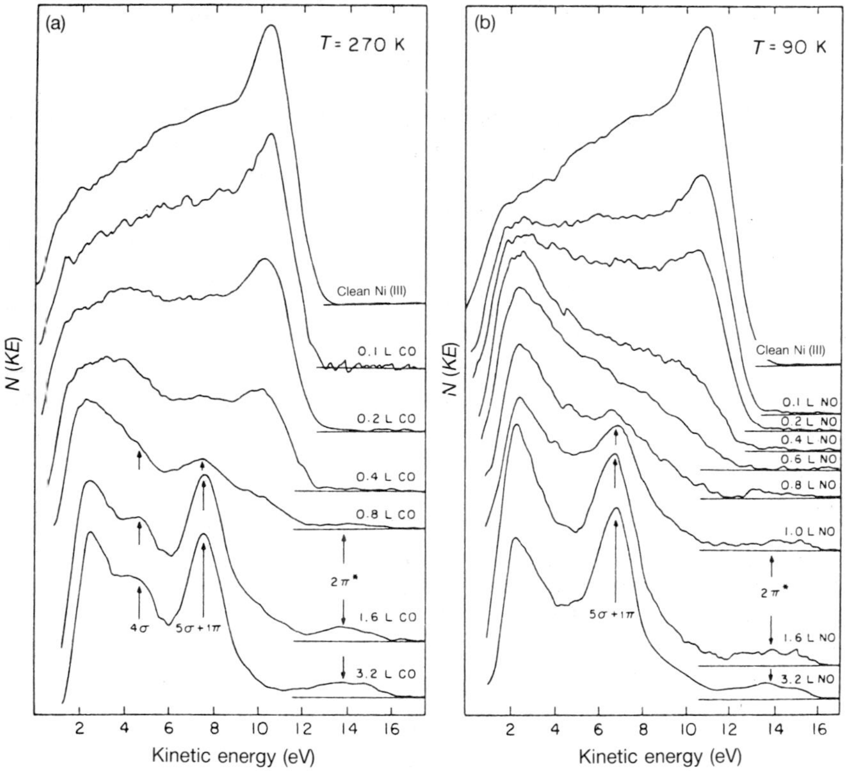

FIG. 14.12. He*2^1S MQS spectra from clean and from CO- and NO-covered surfaces of Ni(111), measured by Bozso *et al.*[14] The CO adsorption was performed at 270 K, and the NO at 90 K, saturation coverage being at 3.2 L in each case.
Note that the spectrum from clean Ni(111) bears a striking resemblance to the INS transition density function of Fig. 14.6, indicating that the RI + AN mechanism in Fig. 14.11a is operating. At saturation, however, the spectral character has changed completely, with well-developed emission from the 4σ and the $(1\pi + 5\sigma)$ orbitals of the adsorbed molecules, rather as in UPS. Thus at saturation the Auger de-excitation mechanism has taken over (Fig. 14.11b). The adsorbed CO and NO molecular orbitals are shielding the underlying Ni valence levels from electronic interaction. In the intermediate coverage region there is a mixture of the two mechanisms in operation; see the text for a discussion of the interpretation.

taken over, and the CO and NO molecules are shielding the underlying Ni completely from electronic interrogation by the approaching metastable.

The coverage region up to about 0.8 L in each case is most interesting. Other techniques indicate that even at low coverages both CO and NO are adsorbed on the surface, but in the so-called bridge-bonded configuration, i.e. attached to a Ni atom at each end, and more or less parallel to the surface. Yet MQS shows no appearance of the characteristic CO and NO orbital emissions, and there is still a substantial Ni 3d contribution. The interpretation is that in the bridge-bonded configuration there is insufficient external projection of the molecular orbitals outwards from the surface to interact with the excited orbital in the metastable, which still 'sees' the metal occupied levels at the surface. However, beyond about 0.8 L, CO and NO start to adsorb in an 'on-top' configuration, i.e. more or less vertical to the surface, in which case the molecular orbitals do project sufficiently far into vacuum, and at once the Penning ionization mechanism comes into operation. Thus the extent of competition between the two processes depends critically on the degree of projection of the adsorbate molecular orbitals from the surface, and not just on the mere presence of the adsorbate.

Many other adsorption experiments similar to that described above have been carried out,[15–18] and a great deal of additional very surface specific information can be extracted by variation of metastable species, of temperature, of angle of acceptance, and of work function by alkali metal pre-adsorption.

References

1. Hagstrum, H. D. (1966). *Phys. Rev.* **150**, 495.
2. Hagstrum, H. D. and Becker, G. E. (1967). *Phys. Rev.* **159**, 572.
3. Hagstrum, H. D. (1972). *Science* **178**, 275.
4. Hagstrum, H. D. and Becker, G. E. (1973). *Phys. Rev. B* **8**, 107.
5. Hagstrum, H. D. (1970). *J. Res. NBS* **74A**, 433.
6. Hagstrum, H. D. and Becker, G. E. (1971). *Phys. Rev. B* **4**, 4187.
7. Baragiola, R. A. (1982). *Radiat. Eff.* **61**, 47.
8. Hagstrum, H. D. and Becker, G. E. (1973). *Phys. Rev. B* **8**, 1592.
9. Hagstrum, H. D. and Becker, G. E. (1971). *J. Chem. Phys.* **54**, 1015.
10. Mueller, F. M., (1967). *Phys. Rev.* **153**, 659.
11. Becker, G. E. and Hagstrum, H. D. (1972). *Surf. Sci.* **30**, 505.
12. Ertl, G. and Küppers, J. (1985). *Low energy electrons and surface chemistry*, p. 147. VCH Verlagsgesellschaft, Weinheim, West Germany.
13. Bozso, F., Yates, J. T., Arias, J., Metiu, H., and Martin, R. M. (1983). *J. Chem. Phys.* **78**, 4256.
14. Bozso, F., Arias, J., Hanrahan, C. P., Martin, R. M., Yates, J. T., and Metiu, H. (1984). *Surf. Sci.* **136**, 257.
15. Bozso, F., Arias, J. M., Hanrahan, C. P., Yates, J. T., Metiu, H., and Martin, R. M. (1984). *Surf. Sci.* **138**, 488.
16. Sesselmann, W., Woratschek, B., Ertl, G., Küppers, J., and Haberland, H. (1984). *Surf. Sci.* **146**, 17.
17. Mucchielli, F., Boiziau, C., Nurolone, R., and Roussel, J. (1980). *J. Phys. C: Solid St. Phys.* **13**, 2441.
18. Conrad, H., Ertl, G., Küppers, J., Sesselmann, W., and Haberland, H. (1980). *Surf. Sci.* **100**, L461.

15

Ion excitation: ion beam spectrochemical analysis (IBSCA) and glow discharge optical spectroscopy (GDOS)

Although these two techniques depend on rather different physical mechanisms, they are discussed together here since the one thing that they have uniquely in common is identification of sputtered species by spectroscopic analysis of light in the visible region. They bear approximately the same relationship to each other as do SSIMS and GDMS, to be the subjects of following chapters.

15.1 Ion beam spectro-chemical analysis

Note that the technique probably has more alternative acronyms than any other surface analytical technique. They include SCANIIR (Surface Composition by Analysis of Neutral and Ion Impact Radiation), SIPS (Sputter-induced Photon Spectroscopy), BLE (Bombardment-induced Light Emission), and ILSA (Ion Luminescence Surface Analysis). This list may not be exclusive, but at least each of the various titles is reasonably self-explanatory.

15.1.1 *Operation*

Positive ions, usually Ar^+, in the energy range 2–50 keV, and at current densities of 10–100 μA cm^{-2}, bombard a surface, producing a great variety of secondary particles, amongst which are atoms in highly excited states. The decay of these excited atoms to their ground states is accompanied by emission of light of wavelengths characteristic of the particular atom, thus providing a means of analysis.

Ion sources used to produce beams of ions in the different energy regions are very similar to those already described in Chapter 13 for the operation of IAES, and range from the conventional duoplasmatron type to ion accelerators. In the more sophisticated experimental arrangements care is taken to produce a very pure beam of primary ions by first mass analysing the beam to remove doubly charged and contaminant ions, and then deflecting the beam slightly to remove the component of neutrals. Such an arrangement is shown in Fig. 15.1 from Loxton and Tsong,[1] who used ions in the energy range 2 to 7 keV, which were mass analysed by passage through a filter of crossed electric and magnetic fields and then deflected through 5° before being focused on to the target surface. As can be seen from the figure, their system

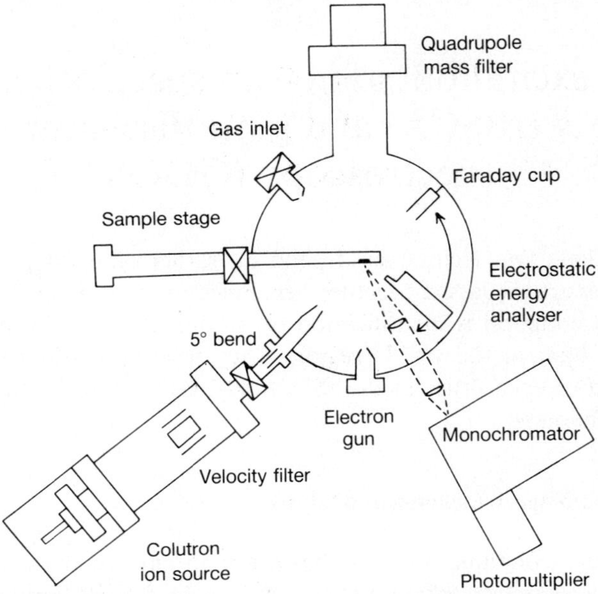

FIG. 15.1. Schematic of the experimental arrangement used by Loxton and Tsong[1] for IBSCA, SSIMS and ISS. Ions produced in a discharge source are mass analysed in a Wien filter and then deflected through 5° to remove the neutral component before entering a chamber and impinging on the target surface. The spot size on the target is of the order of 70 μm diameter, with a beam current of 0.4 μA, and the beam is rastered over an area of 0.5 mm^2. Photons emitted from sputtered particles are focused by two quartz lenses into a scanning monochromator. For SSIMS secondary ions are energy selected and then mass analysed in a quadrupole mass spectrometer. Scattered primary ions are energy analysed for ISS in an electrostatic analyser rotatable about the point of ion impact on the target.

included both mass and energy analysers as well as a light emission analyser, so that the techniques of SSIMS and ISS (q.v.) could be performed as well. The emitted photons were focused by two quartz lenses into a scanning monochromator, and detected at the monochromator output with a photomultiplier tube. All operations were computer controlled.

Another arrangement in which two ion sources were available, according to which incident ion energy range was being studied, is illustrated in Fig. 15.2, from the paper by Wright and Gruen.[2] For the low energy region, up to 3 keV, a standard ion gun of the type used universally for sputter cleaning was used, without mass analysis, while for the higher energy region up to 18 keV a mass analysed beam from a duoplasmatron could be brought to bear on the target. Obviously the target would have to be rotated to face one source or the other. Light passed through an LiF window and was focused by a quartz lens into the input of a scanning monochromator, whose internal design is shown schematically. Wavelength analysis in a scanning monochromator was

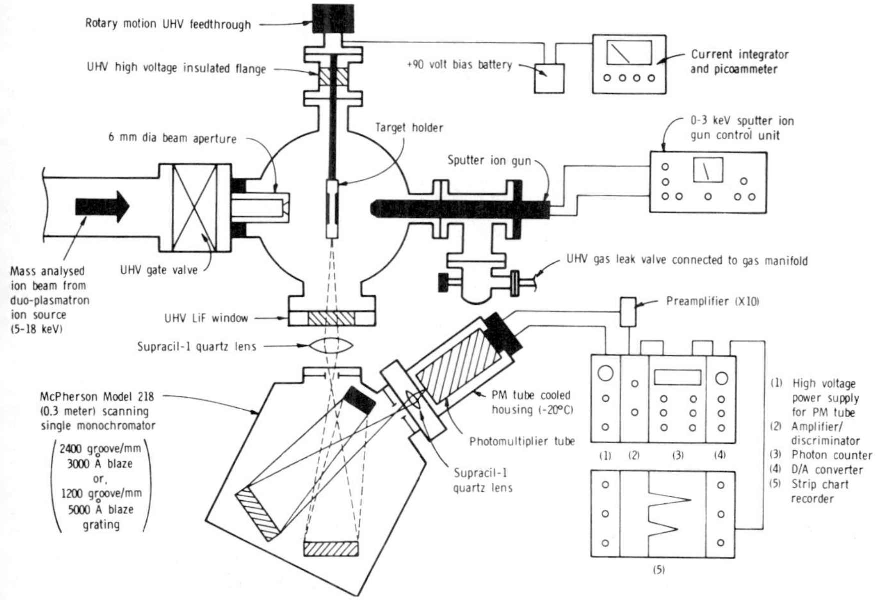

FIG. 15.2. Experimental apparatus used by Wright and Gruen[2] for the study of photon emission from ion-bombarded beryllium. For ion energies up to 3 keV, a standard ion gun of the type used for sputter cleaning was employed. For higher ion energies, up to 18 keV, a beam from a mass analysed duoplasmatron source was used. The target could be rotated to face one or other source. Light from the bombarded region passed through a LiF window and was focused via a quartz lens into a scanning monochromator.

carried out by rotation of a ruled grating while monitoring the output of the photomultiplier tube; the response must previously have been calibrated with light of accurately known wavelength.

A typical intensity vs. wavelength plot that might be recorded is that seen in Fig. 15.3, from Hamilton and Anderson,[3] from the surface of an Fe–Cr–Mo single crystal. The positions of the spectral lines for each element can be obtained readily from various published tabulations.

15.1.2 *Theory and quantification*

It is commonly observed in IBSCA that those elements, particularly the transition metals, that emit intense, sharp, spectra on bombardment, show emission characteristics very similar to those of an arc discharge, or plasma. If this analogy is correct, then it might be possible to extract analytical information using standard spectrochemical methods, as has been discussed by Martin and MacDonald.[4] Should the surface compositions so obtained agree with those found by other surface analytical techniques, then the presumed analogy with conditions in an arc discharge is confirmed.

Since the model relating sputtered ion intensity to elemental concentration,

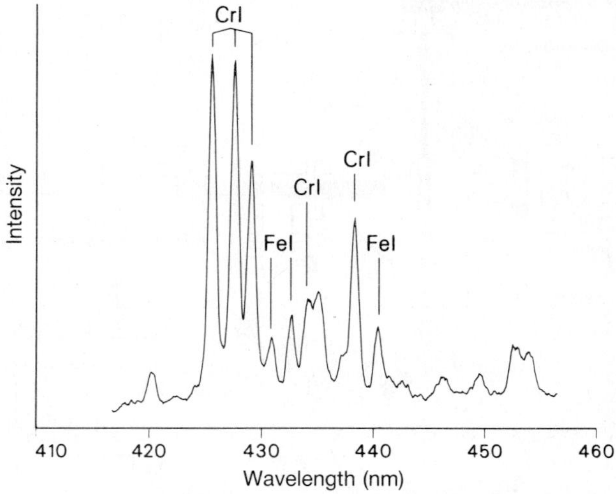

FIG. 15.3. Example of an IBSCA spectrum, from an Fe–18Cr–3Mo(100) surface under bombardment by Ar^+ at 4 keV. The spectral elemental lines are indicated; the allocations of such lines can be made by reference to published tables. From Hamilton and Anderson.[3]

of Andersen and Hinthorne,[5] has been marginally more successful than other models, and is based on the formation of a plasma in local thermal equilibrium (LTE) in the target surface region, Martin and MacDonald took that model as their starting point. For a plasma in LTE one can calculate the relative concentrations of ions, electrons, and neutrals from the Saha–Eggert equation:

$$\frac{N_i^+ N_e}{N_i^0} = \frac{Z_i^+(T)}{Z_i^0(T)}\, 2\, \frac{(2\pi mkT)^{3/2}}{h^3} \exp\{-(I_i - \Delta I_i)/kT\} \tag{15.1}$$

where N_i^+ = concentration of singly charged ions of element i
N_e = free electron concentration
N_i^0 = concentration of neutral atoms of element i
$Z_i^+(T)$, $Z_i^0(T)$ = respective partition functions of ions and neutrals of element i
T = temperature (effectively the equilibrium arc temperature)
I_i = ionization potential of element i
and Δ_i^I = correction to ionization potential in plasma conditions.

The factor 2 arises from the electronic partition function.

In eqn (15.1) the only unknown parameters are the effective temperature T and the electron concentration N_e. However, if well characterized standards are used in which the concentrations of at least two elements are known accurately, then T and N_e can be calculated from the Debye-Hückel expression:

$$\Delta I = 2 \times 10^{-8}(N_e/T)^{1/2}. \tag{15.2}$$

In principle, then, knowing T and N_e for a given set of analysing conditions, it should be possible to analyse unknown materials containing the same elements. The philosophical problem is that the values of T extracted in this way are unusually high, in the range 4000 to 11000 K, i.e. higher than the melting point of the material. In addition the energy distributions of the secondary particles are found to be non-Maxwellian. In the case of photon emission, however, the experiments of Martin and MacDonald appeared to show that the photon source was one whose excited level population did in fact obey Maxwell–Boltzmann statistics, that is, it resembled a plasma in LTE. On the other hand, the values of T that they obtained were still high, in the range 3300 K to 5500 K, and they tended to regard T as merely an adjustable parameter without any clear physical meaning.

Now the intensity of an emission line arising from a transition from an upper level q to a lower level p can be written as

$$I_{qp} = A_{qp}\, h\nu_{qp}\, N g_q Z(T)^{-1} \exp(-E_q/kT) \tag{15.3}$$

where A_{qp} = transition probability from q to p
$h\nu_{qp}$ = energy of transition
N = density of excited particles
g_q = statistical weight of level q
$Z(T)$ = partition function
E_q = energy of excitation
and T = excitation temperature.

Thus the relative intensities of chosen lines from two elements A and B would be:

$$I_A/I_B = \frac{A_A}{A_B}\frac{\lambda_B}{\lambda_A}\frac{N_A}{N_B}\frac{g_A}{g_B}\frac{Z_B}{Z_A}\exp\{-(E_A - E_B)/kT\} \tag{15.4}$$

where $h\nu$ has now been replaced by λ, the wavelength of the emission line. Since T can be measured, the λ values are chosen, and the other terms can be found from spectroscopic data, it is possible to find N_A/N_B, the ratio of the elemental concentrations. Using this method when applied to analysis of a commercial stainless steel, Martin and MacDonald found a value for the concentration of Cr of 15%, which agreed quite well with the known bulk figure of 17.4%.

A much more empirical method, and one similar to that used in other surface analytical techniques, is simply to measure the intensity of emission from a pure element and assume that the intensity of emission of the same line from the same element in an alloy or compound is proportional to the concentration of that element. This was used by MacDonald[6] in quantifying Nb and V concentrations in a series of Nb–V alloys. Both the alloy and the elemental standard surfaces were given the same polishing treatment, and then mounted together in the vacuum system so that photon intensities could always be measured consecutively without the need for removal from vacuum. The relative intensity values used in the analysis were always those reached after stabilization, i.e. after any transient effects had disappeared.

MacDonald's results, and indeed also those of Martin and MacDonald,[4] emphasized one of the considerable problems to be overcome in any attempt to use IBSCA quantitatively, a problem closely related to a similar one encountered in SSIMS. Under UHV conditions the quantification was quite good, in that for all alloys, ranging from 2%V to 93.5%V, the sum of the measured Nb and V concentrations was $99.1 \pm 1.7\%$, although the deviation from the known Nb/V ratios averaged about 30%. However, when the same analysis was attempted in the presence of a high background of oxygen, the sum of the measured Nb and V concentrations was now $98.5 \pm 24.5\%$, and the ratio deviation averaged about 70%. A similar devastating effect of oxygen had been noticed in the earlier work; Fig. 15.4 from Martin and MacDonald

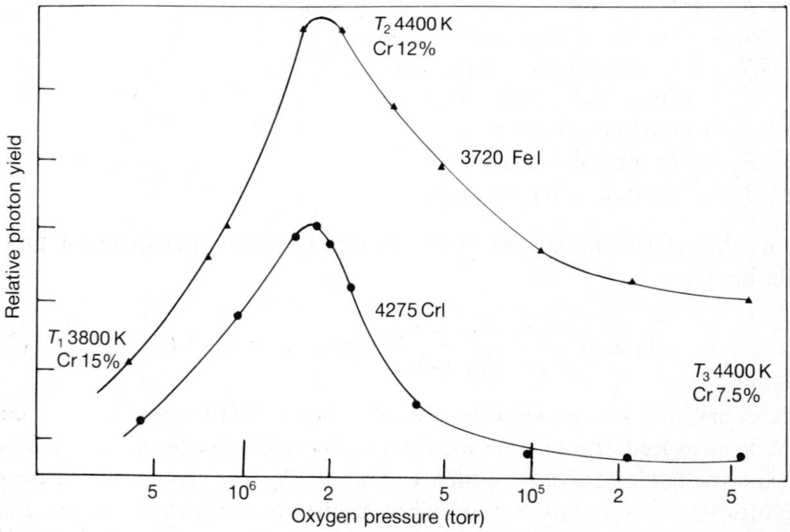

FIG. 15.4. Variations in the photon yields in the emission lines FeI and CrI, at 372 nm and 427.5 nm, respectively, from bombarded stainless steel, as a function of oxygen pressure, recorded by Martin and MacDonald.[4] T_1, T_2, and T_3 refer to the effective excitation temperatures at the indicated points, calculated in the Anderson and Hinthorne[5] LTE model. The calculated Cr concentrations at the surface of the Fe–17Cr–7Ni steel are also shown. It can be seen that the photon yields change in a non-uniform fashion, and in ways different between the two lines.

shows how the yields from two emission lines characteristic of Fe and Cr, as the stainless steel surface was bombarded in an increasing pressure of oxygen, increased by factors of 3 and 2, respectively, at their maxima, before decreasing again. The effective excitation temperatures and Cr concentrations were calculated at three points, and it can be seen that according to the oxygen pressure the apparent concentration of Cr could vary by 100%.

The reasons for the oxygen effect are still not well understood. One theory suggests that the relative positions of the excited levels in the sputtered neutral and the band structure at the surface are important, and that as the work function of the surface changes during oxygen adsorption, so electron exchange leading to de-excitation becomes less likely with consequent enhancement of radiative de-excitation. At higher oxygen pressures the band structure changes to that of the oxide, and apparently electron exchange again becomes possible, with a resultant decrease in photon emission. However, this mechanism cannot account for the differences in behaviour of individual emission lines from the same element as a function of oxygen pressure. Another idea is that the local chemical environment in the surface before sputtering determines the degree of enhancement or otherwise. Hamilton and Anderson suggest that in a Cr-containing alloy the stable oxide at high temperatures (i.e. in an arc discharge) is Cr_2O_3, and therefore that in the presence of sufficient oxygen quasi-molecular species containing Cr and O are formed preferentially, leading to an enhancement of Cr emission on subsequent dissociation of the molecule into excited neutrals. However, it is clear that, irrespective of the precise mechanism, the presence of oxygen during IBSCA analysis complicates quantification, which makes the study of oxidation processes rather difficult with the technique.

It will have become obvious from the above description that the surface specificity of IBSCA must be identical to that of SSIMS (q.v.). The particles giving rise to the characteristic photon emission are merely a part of the many species produced by ion bombardment such as is used in SSIMS, and they originate in the outer atomic layer of the surface by a cascade process just as do the secondary ions that are used in SSIMS. Thus, like SSIMS, IBSCA is in principle one of the more highly surface specific of the surface analytical techniques; just as in SSIMS, on the other hand, there may well be questions that cannot easily be answered about the relationship between the emission intensities and the surface elemental concentrations, because the particle giving rise to the emission used for analysis is already in the gas phase, albeit still within the influence of the surface. Much more basic work is still required to establish the physical mechanisms with more certainty.

15.1.3 *Applications*

Not a large number of applications have appeared in the literature, many of those that have being concerned with profiling of insulating materials. In particular Baucke and co-workers have applied the technique extensively to studies of leaching and corrosion of glasses, of ionic transfer in glasses and across glass-solution interfaces, and of the protonation of glasses. An example[7] of the latter study was that of the field-driven protonation of lithium silicate glasses; knowledge of the mechanisms of the migration of protons within glasses is required when considering conductivity and mobility. In this study the displacement of lithium ions by protons was studied. A thin membrane of lithium silicate glass was sealed on to the open end of a tube of glass of much higher resistance, by melting, and then formed the diffusion barrier in an electrolytic cell as shown in Fig. 15.5. Application of a high

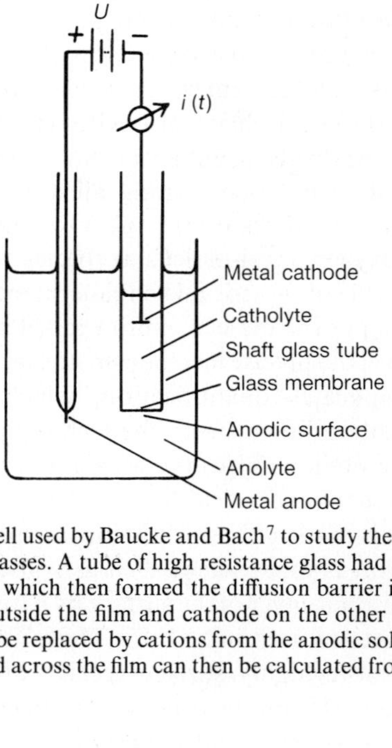

FIG. 15.5. The electrolytic cell used by Baucke and Bach[7] to study the displacement of lithium by protons in lithium silicate glasses. A tube of high resistance glass had sealed on to one end a thin film of lithium silicate glass, which then formed the diffusion barrier in the cell. Application of a voltage V between anode outside the film and cathode on the other side causes cations to drift towards the cathode and to be replaced by cations from the anodic solution. The total number of cations that have thus passed across the film can then be calculated from the total charge that has passed through the cell.

voltage of the correct polarity across the anode outside the membrane, and the cathode inside it causes cations to drift towards the cathode and to be replaced by cations from the anodic solution. The total number of cations that have passed across the membrane could be calculated from the total charge q that passed through the cell during each electrolysis.

Following electrolysis for different durations, under different applied voltages, and using different solutions, the membrane was removed and examined by a variety of techniques, including depth profiling with IBSCA. A set of such profiles, for a multi-component lithium silicate glass that had been electrolysed for progressively longer periods, is shown in Fig. 15.6, for lithium itself. It should be noted in passing that the very capability of analysing for lithium so easily and without interferences is not easy in some surface analytical techniques and impossible in others, and that such capability is a significant advantage in IBSCA. The profiles show that with increasing electrolysis time the lithium ions in the glass are displaced progressively towards the cathodic surface, and indeed, after a sufficiently long time, they can be removed completely. The lithium must be replaced by something in the glass network, and by using infra-red spectroscopy, which revealed an increase in intensity of the OH stretching vibration as electrolysis proceeded, coupled with the use of non-aqueous but proton-containing anodic solutions, it was established beyond doubt that the lithium was being replaced with protons. As the authors point out, this therefore represents a rare example of proton migration in a solid.

Having established the one-to-one replacement of lithium by protons by this method, the authors were able to produce glasses with any desired proton and lithium concentrations and distribution, by choosing anodic solutions containing the appropriate ratios of lithium to hydrogen ion activity, and varying the electrolysis conditions. Thus the pure lithium silicate glass, which is a lithium ion conductor, could be changed controllably to a mixed lithium ion-proton conductor, or even to a pure proton conductor. On the other hand, IBSCA depth profiles showed that when other alkali ions were attempted to be transferred into the glass, steric hindrance prevented complete substitution. Thus only 60% of Na ions could be transferred, and no ions at all of K, Rb, or Cs. From the study the authors were also able to derive information about the stability conditions associated with migrating boundaries, i.e. elemental concentration edges, in glasses.

15.2 Glow discharge optical spectroscopy

Unlike IBSCA, which still remains more of a research laboratory technique than of a routine analytical tool, GDOS is widely used in many industries and there are commercial versions on the market. However, it must be emphasized that in the great majority of applications GDOS is not being used as a surface-specific analytical technique at all, but as a very rapid and reasonably well quantifiable technique for providing elemental concentration profiles through oxide and other films many μm in thickness. Nevertheless it has been shown by some workers to be capable of surface analysis with a depth resolution approaching that of some of the other surface analytical techniques described in this volume. In what follows, therefore, the description of the equipment, mode of operation, and theory will be mostly that appropriate to the technique as it is used universally, but, particularly when discussing

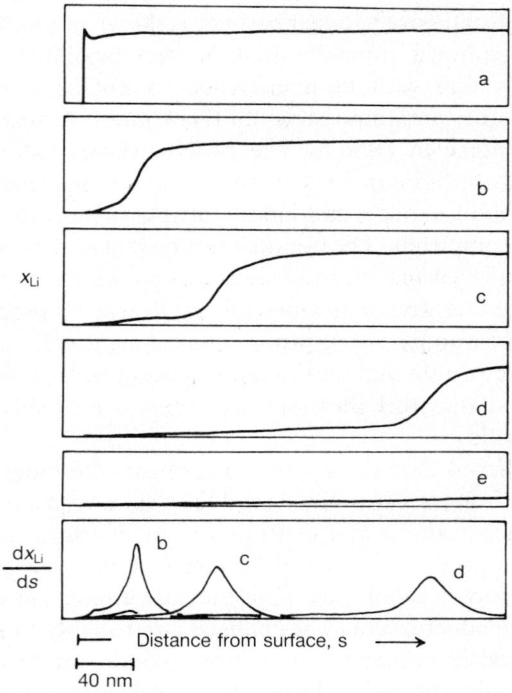

FIG. 15.6. Profiles using IBSCA of a multi-component lithium silicate glass that had been electrolysed in the cell shown in Fig. 15.5 for progressively longer periods, measured by Baucke and Bach.[7] The vertical axis is the concentration of Li itself, except in the lowest panel. With increasing electrolysis time the lithium ions in the glass are depleted by progressive diffusion towards the cathode, and eventually (panel e) can be virtually eliminated. The lithium is being replaced in the glass by protons from the anodic solution. The differential spectra in the lowest panel show clearly the movement of the profile edge towards the cathode.

applications, the more surface specific aspects of GDOS will be brought out as much as possible.

15.2.1 *Operation*

The heart of virtually all GDOS systems is the glow discharge lamp designed by Grimm.[8] This is basically a very simple device, and has been modified only slightly by the many builders of GDOS equipment. A diagram of a typical Grimm-type source is shown in Fig. 15.7, from the paper by Ohashi *et al.*[9] The sample (1) is clamped against a cathode block (2) via an O ring (3), and actually forms the cathode during operation. Argon is admitted to the lamp via inlet port (8) to a pressure between 1 and 10 torr, the presssure being maintained constant dynamically by balancing the incoming argon flow with the pumping speed of mechanical pumps via evacuation ports (6) and (7).

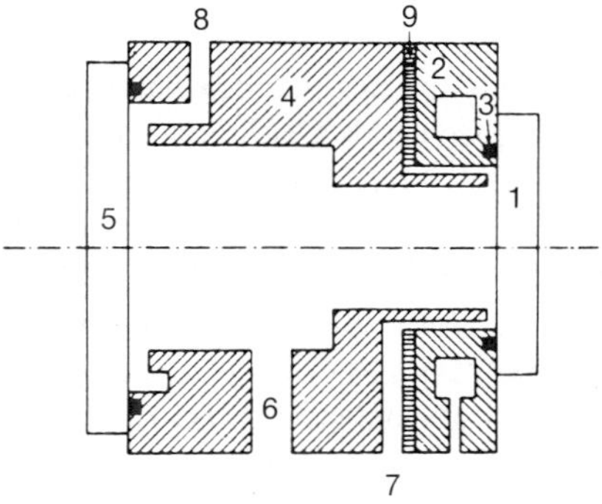

FIG. 15.7. The Grimm[8]-type glow discharge lamp used for GDOS by Ohashi *et al.*[9] (1) Sample. (2) Cathode block. (3) O-ring. (4) Anode. (5) Quartz window. (6), (7) Evacuation ports. (8) Gas inlet. (9) Teflon insulator. The sample is clamped against the cathode block via an O-ring, and forms the cathode. Argon is admitted to pressures between 1 and 10 torr, and a DC voltage of about 1000 V applied between anode and cathode (sample). A glow discharge is established, and positive ions so produced are attracted to the sample and bombard it continuously. Most of the sputtered material leaves as neutral atoms, which pass into the discharge region and are collisionally excited with high efficiency. The light emitted as a result of the decay of the excited state is analysed in a multichannel spectro-photometer, after passing out of the cell through the quartz window.

Application of a DC voltage of about 1000 V between the anode block (4) and the cathode, separated by an insulator (9), establishes a glow discharge in the argon; the mode of operation is in the 'abnormal' region of the current voltage characteristic. The essential feature of the Grimm lamp is the shaping of the anode in such a way that the discharge can occur only between the anode and the sample. As can be seen in the figure, the anode extends down to within a short distance of the sample surface, so that the sample is effectively at the end of a hollow cylindrical tube, which ensures that there is an almost constant discharge intensity adjacent to the sample. Positive argon ions produced in the discharge are attracted to the sample and erode it in a continuous fashion; the area eroded is delimited by the anode tube, and is normally about 8–9 mm diameter. As with any sputtering technique, most of the material leaves the sample as neutral atoms. The emitted neutrals pass into the discharge region, where they are collisionally excited with high efficiency, and then decay with the emission of light of wavelength characteristic of the elements in the sample, the light passing out through a quartz window (5), also sealed to the cell with an O ring.

By placing the cell with the quartz window facing the entrance slit of a multichannel spectrophotometer, with focusing lenses in between, the emitted light can be analysed. Of course the light from the glow discharge itself represents an interference, so it is dispersed by a grating, and the narrow emission lines from the sputtered atoms are then detected by photomultipliers. According to the choice of discharge conditions, i.e. gas pressure and applied voltage, the sputtering rate from the sample can be varied between about 30 nm min^{-1} and 1 μm min^{-1}.

The original design as typified by that shown in Fig. 15.7 suffered from some drawbacks. With the anode at earth potential and the cathode at high voltage, it was not easy to measure the temperature of the cathode during sputtering. In addition the pumping port ((7) in Fig. 15.7) acting on the gap between anode and cathode caused a non-uniform distribution of argon pressure across the sample surface, while the shape and proximity of the anode unavoidably caused non-uniform electric field distribution and therefore uneven erosion of the sample. Commonly, additional erosion is found at the edges of the erosion crater, as seen in Fig. 15.8 for some typical craters formed in various materials.[10]

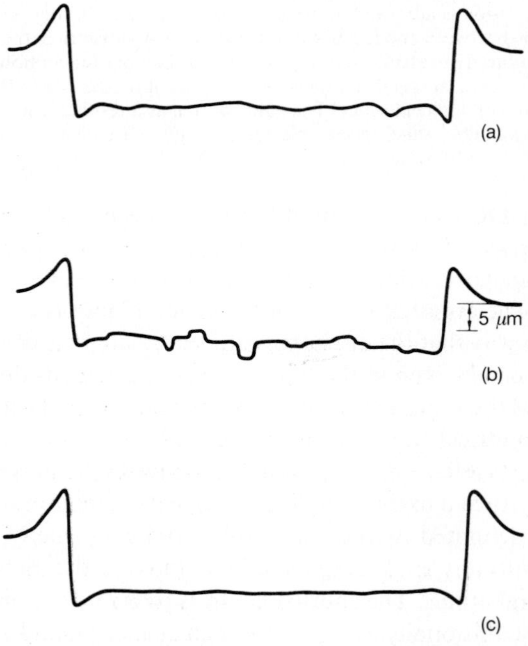

FIG. 15.8. Topographic profiles of craters eroded in various materials by bombardment in the Grimm type source shown in Fig. 15.7, recorded by Pons-Corbeau.[10] (a) Fe–15Ni; (b) Fe–34V; (c) FeO. The eroded crater is 4 mm in diameter. Additional erosion at the edges of the crater can be seen, due to the non-uniform shape of the electric field distribution from the proximity of the anode and the variation in argon pressure across the surface.

Modifications to the design to try and avoid these drawbacks have been proposed by Ko.[11] The most revolutionary is to replace the conducting hollow anode with a shaped tube of insulating material such as a ceramic, with the anode itself some distance from the cathode, and the cathode (sample) is now at earth potential. One of Ko's designs is shown in Fig. 15.9. The evacuation geometry has also been changed, so that there is no pumping adjacent to the sample surface. The surface of the ceramic acquires a floating potential during operation of the discharge, which, being self-regulating, ensures a more uniform electric field distribution across the sample. The potential acquired is in fact well above that of the cathode (at earth) but rather below that of the anode. The only apparent disadvantage of this design is that the applied voltage must be increased to maintain appropriate discharge conditions. However, the uniformity of the sputtered crater is certainly improved, as seen in Fig. 15.10, from Ko.

It will be obvious from the above description that an unavoidable disadvantage of all GDOS systems is the re-deposition of sputtered material around the interior of the lamp. Eventually this reaches the stage where electrical breakdown is caused, and the lamp must then be dismantled to be cleaned, but for the new design described above the time before cleaning is needed is extended since the material cannot re-deposit on the insulator surface as quickly as it would on an all-metal surface.

15.2.2 *Quantification*

Although the principle of operation of GDOS is well known and apparently straightforward, when one comes to attempt quantification there are several factors that must be taken into account. Since both the composition of the sample surface and the geometry (as a result of crater formation) may be changing during an analysis, the glow discharge operating conditions will also change, thus altering both the sputtering rate and the excitation processes in the plasma. Bengtson[12] has indicated how one can arrive at a formula for quantification by considering the effects of the excitation efficiency and of the sample sputtering rate separately. His formulation is based on the assumption of restriction to the analysis of low elemental concentrations, thus eliminating any interference from possible self-absorption of emitted light in the plasma.

Bengtson investigated first the dependence of the excitation processes by studying the effects of the discharge lamp voltage and current on the emission intensities of argon lines, since relative to metastable argon atoms the excited levels have excitation energies similar to those in metals. The intensities of all the argon emission lines fell to zero at the same threshold voltage V_0, and showed a non-linear increase with voltage. When the logarithm of the spectral intensity was plotted as a function of the logarithm of the voltage above threshold, $(V-V_0)$, straight lines resulted for each spectral line chosen, showing that the intensity–voltage relationship must be of the form:

$$I_\lambda = k_\lambda (V - V_0)^{x(\lambda)} \tag{15.5}$$

where k_λ is a constant, and $x(\lambda)$ is also a constant characteristic of the particular spectral line. Expression (15.5) is similar to that given by

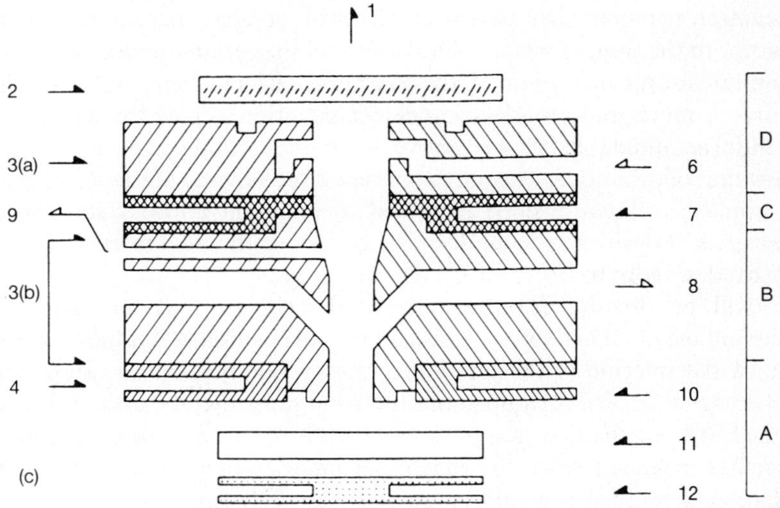

FIG. 15.9. Modification of the Grimm-type cell shown in Fig. 15.7 by Ko[11] in an attempt to overcome some of the drawbacks in the simple design. The construction has four sections as follows:

(A) The sample (11) is pressed against the cathode block (10), and the clamping device (12) and the cathode block are cooled (4).
(B) The main body of the lamp consists of an insulating block (3b) with pumping ducts (8, 9).
(C) A disc-shaped cooled anode (7).
(D) The window section consisting of an insulating block (3a) with gas inlet (6) and window (2) for observation of the radiation (1).

Items (10), (12), and (7) are made of copper, items (3a) and (3b) of ceramic material, and the window (2) of quartz.

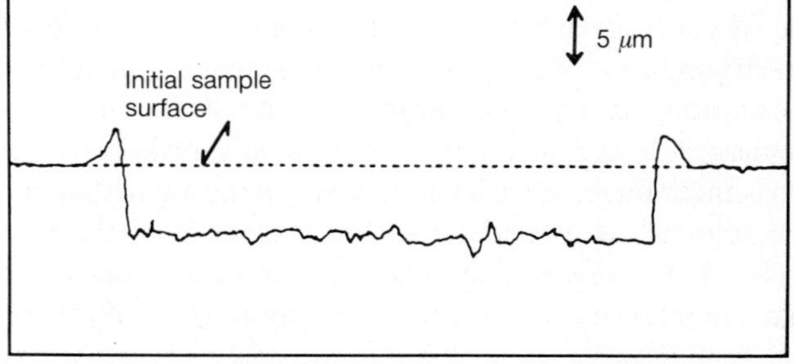

FIG. 15.10. Example of the improvement in uniformity of crater profile obtainable with the design of GDOS lamp by Ko[11] seen in Fig. 15.9. Topographic profile of the crater eroded in a Cu–Zn–Pb alloy after 60 s sputtering at an applied voltage of 940 V, a bombarding current of 45 mA, and a gas pressure of 240 Pa. The erosion rate was 103 nm s^{-1}.

Boumans[13] for the effective sputtering rate, where V_0 is the sputtering threshold. The spectral constant $x(\lambda)$ is in the range 0.2–0.5 for most lines.

The dependence on discharge current of the line intensity was found, however, to be strictly linear at least as far as 100 mA. Beyond that there was a deviation due to a consequent change in discharge voltage. If then the applied voltage is limited to 1000 V and the current to 100 mA, the combined dependence of the excitation on voltage and current can be written

$$I_\lambda = k'_\lambda i(V - V_0)^{x(\lambda)} \tag{15.6}$$

where i is the current and k'_λ is also a constant but of course different from k_λ in (15.5). Expression (15.6) is known as the excitation function.

The emission intensity of course depends on the sputtered atom concentration in the plasma, which in turn depends on the sputtering rate. The expression due to Boumans[13] for the effective sputtering rate in a glow discharge lamp is

$$q = Ci(V - V_0)\ \mu\text{g s}^{-1} \tag{15.7}$$

where C is a constant dependent on the sample and on the type of sputtering ion, and both C and V_0 depend on the geometry of the lamp. As well as the sputtering rate, the atom concentration will depend also on the rate of diffusion from the sample to the plasma. The diffusion process is complex and is found to have another non-linear dependence on the applied voltage. Bengtson therefore suggested the following simplified expression for the atom concentration n_A of atomic species A:

$$n_A = c_A q k_A (V - V_0)^{-y(A)} \tag{15.8}$$

where c_A is the concentration of element A in the sample, and k_A, $y(A)$ are constants characteristic of the atomic species A. Since the atom concentration should be equal to the observed emission intensity divided by the excitation function (15.6), one can combine these various expressions into:

$$I_\lambda = k_{\lambda A} c_A C i^2 (V - V_0)^{X(\lambda, A)} \tag{15.9}$$

where now $k_{\lambda A} = k'_\lambda k_A$ and $X(\lambda, A) = x(\lambda) - y(A) + 1$. Thus $k_{\lambda\alpha}$ contains instrumental factors and atomic factors related to both spectroscopic and diffusion processes, while $X(\lambda, A)$ contains the influence of applied voltage on excitation and sputtering as well as on diffusion.

If all the assumptions made in arriving at expression (15.9) are correct, then a plot of $\ln(I_\lambda / i^2)$ against $\ln(V - V_0)$ should be a straight line with slope $X(\lambda, A)$. Bengtson has shown that the predicted linear relationship is obeyed rather well, some examples for two Mn emission lines in two matrices being given in Fig. 15.11 from his paper. Furthermore, the slopes for the two lines are different, but independent of the matrix, which confirms one of the assumptions. For a variety of elements $X(\lambda, A)$ was found to vary in the range 0.3–0.5.

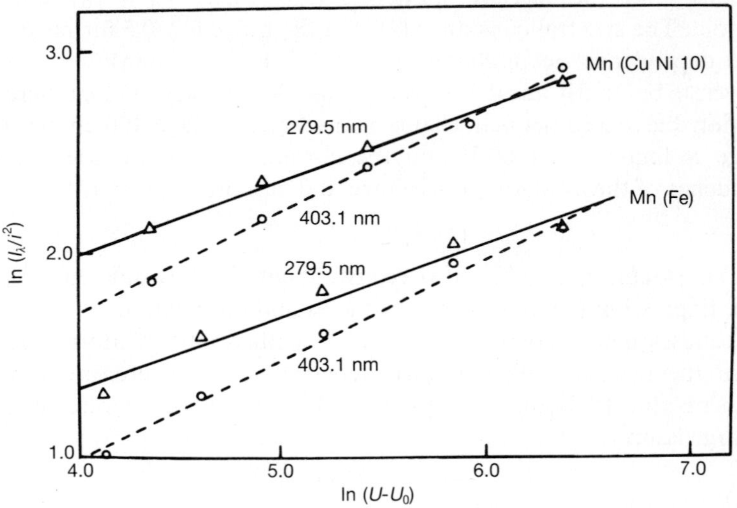

FIG. 15.11. Plot of $\ln(I_\lambda/i^2)$ against $\ln(V-V_0)$ for two Mn emission lines at 279.5 nm and 403.1 nm during the bombardment of two alloys containing Mn as a trace impurity. The alloys were Cu–10Ni (upper plots) and a steel standard (lower plots). Linearity is found over a couple of orders of magnitude, confirming the application of the Bengtson[12] approach to constituents in low concentration. Although the slopes for the plots for each emission line are different, they are independent of the matrix.

Expression (15.9) can be used to measure the concentration of an element in an unknown material if the parameters of emission from a standard reference material are known. Thus if the unknown matrix is M and the reference material is R, one can write for the relative emissions of atomic species A:

$$\frac{I_{\lambda M}}{I_{\lambda R}} = \frac{c_{AM} C_M i_M^2}{c_{AR} C_R i_R^2} \left(\frac{V_M - V_{0M}}{V_R - V_{0R}} \right)^{X(\lambda, A)}. \tag{15.10}$$

Normally the discharge current is kept constant when making relative analyses, so that the current dependence in (15.10) disappears, and the unknown concentration is then:

$$c_{AM} = c_{AR} \frac{C_R}{C_M} \frac{I_{\lambda M}}{I_{\lambda R}} \left(\frac{V_R - V_{0R}}{V_M - V_{0M}} \right)^{X(\lambda, A)}. \tag{15.11}$$

The applicability of expression (15.11) was tested by analysing for Mn, Ni, and Fe in some Cu-based alloys of already known composition, using a single reference material in the form of a steel standard. The results are given in Table 15.1.

The accuracy of the GDOS analysis is good, especially for the element, Mn, in low concentration. Where Ni is high, GDOS tends to give values that are too low by about 20%, probably due to self-absorption. Similarly all the Fe

Table 15.1

Results of GDOS analysis of Mn, Ni, and Fe in Cu-based alloys. Concentrations are in wt.% (from Bengtson[12]).

Material	Element and wavelength of line (nm)			
	Mn 403.1	Mn 279.5	Ni 341.5	Fe 372.0
CuNi10				
GDOS	0.86	0.72	8.5	1.70
Theor.	0.80	0.80	10.8	1.07
CuFe				
GDOS	—	—	0.009	3.5
Theor.	—	—	0.006	2.2
CuZn30				
GDOS	—	—	0.004	0.019
Theor.	—	—	0.003	0.016
CuZnNi				
GDOS	0.12	0.12	9.6	0.068
Theor.	0.12	0.12	11.7	0.056

determinations are too high, again probably due to self-absorption, but in this case in the standard, where Fe is about 90%.

As a further test of the validity of the approach, the discharge current for the steel standard was varied between 40 and 105 mA while that for the alloy CuZnNi was kept constant. The average Mn analysis using the 279.5 nm line was 0.128 ± 0.010, thus showing a very small scatter, and confirming that eqn (15.9) is substantially correct.

If no convenient standard is available, it is possible to use the emission from the substrate once the film has been penetrated, assuming that the bulk composition of the substrate is known.

15.2.3 *Applications*

As remarked earlier, because of the speed with which GDOS can provide a profile through a relatively thick (up to 15 μm) film, coupled with its high sensitivity, it has been used almost entirely in analysing steel and other metal surfaces after a variety of surface treatments. The technique probably lends itself more closely to quality control than any other surface analytical technique, with the possible exception of the sister technique GDMS (q.v.). Thus the GDOS literature is full of profile diagrams such as those shown in Figs 15.12 and 15.13, both from the paper by Koch *et al.*[14] The first example, in Fig. 15.12 shows the profile through the oxide film on a high-Cr steel after annealing at 900°C, primarily in order to determine the Si distribution. The

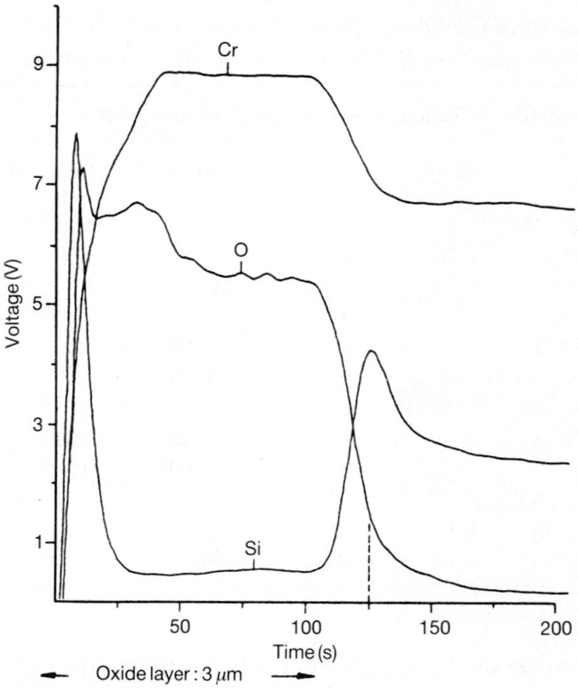

FIG. 15.12. Profile by GDOS through the surface of the oxide film on a steel containing a high chromium content, formed by annealing in a protective atmosphere at 900°C, by Koch *et al.*[14] The silicon distribution and the thickness of the film were of interest. Two peaks in Si concentration were found, one at the surface and the other at the oxide/metal interface, and the thickness was calculated to be 2.9 μm, in good agreement with a metallographically determined thickness of 3 μm. Only 200 s of sputtering were needed to record the profile.

film thickness was about 3 μm, and as can be seen the erosion boundary is well into the underlying steel after 200 s of sputtering. Two peaks in Si concentrations are found, one at the surface and another at the oxide/metal interface; other techniques have confirmed this distribution but have taken at least an order of magnitude more time to do so. The other example in Fig. 15.13 is that of the profiles through layers of two materials commonly applied either as protection or as hardening, to steels, namely TiN and AlN. Both layers were of thickness about 1.6 μm. It can be seen that the TiN layer is reasonably uniform in composition, but that the composition of the AlN layer is varying continuously, and what is more that Fe quickly appears after the start of sputtering. In the latter case the underlying steel was clearly not being protected adequately. The line marked U-GDOS is simply a record of the variation in applied voltage across the discharge lamp required to keep the discharge current constant. Clearly the conditions in the discharge are changing significantly during the AlN-layer analysis, but not so much during

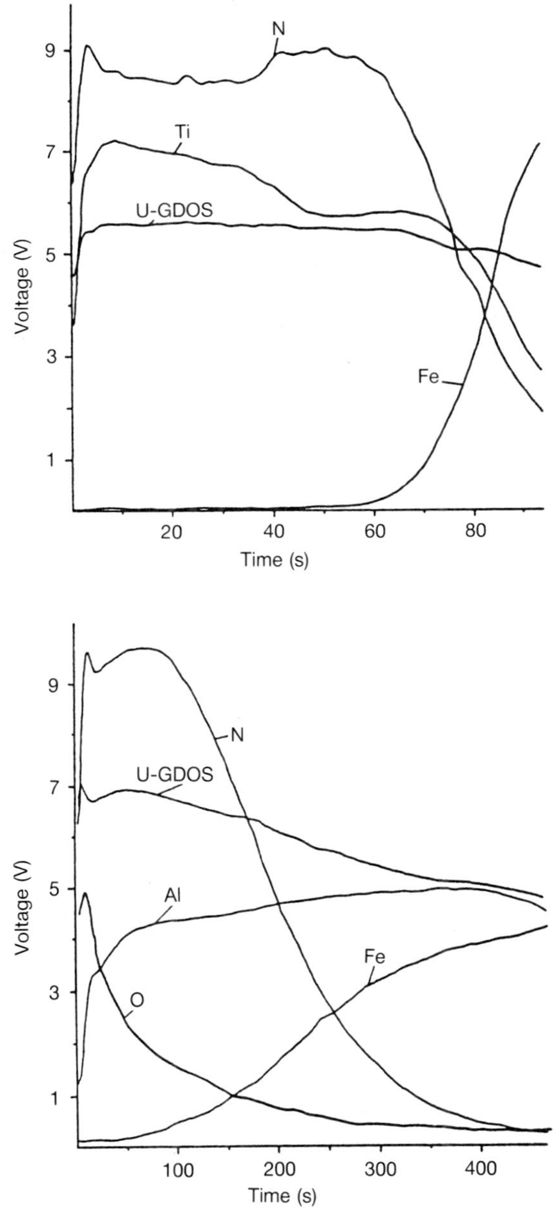

FIG. 15.13. Profiles by GDOS through (upper) a TiN layer on steel, and (lower) an AlN layer on steel, recorded by Koch *et al.*[14] The TiN layer appears to be a layered structure of two approximately equal thicknesses of different stoichiometries. The AlN layer, on the other hand, seems to have a continuously varying stoichiometry. The thicknesses were of the order of 1.6 μm. The lines marked U-GDOS indicate the variation in applied voltage necessary to maintain the sputtering current constant. For the TiN layer it is almost constant, confirming the constancy of the composition, but for the AlN layer it is varying continuously.

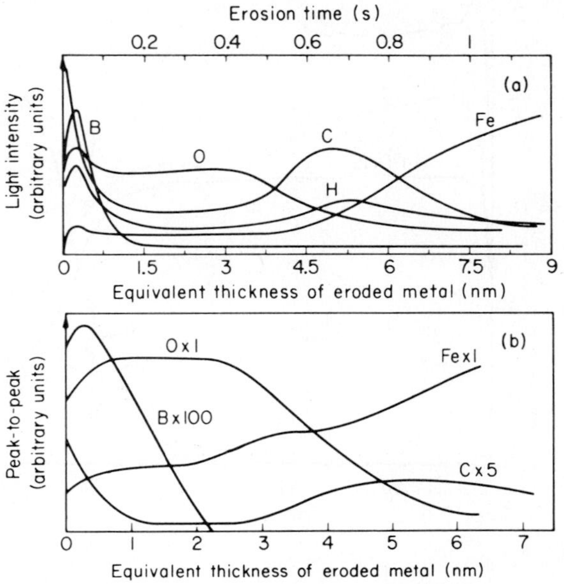

FIG. 15.14. Comparison of depth profiles obtained by GDOS (upper) and by AES (lower) through the passive film formed anodically on pure iron in boric acid buffered solutions, recorded by Berneron and Charbonnier.[15] The profiles are qualitatively similar. Note, however, the ability of GDOS to detect hydrogen, which AES cannot. The profile shows that there is a pile-up of hydrogen at the film/metal interface.

that of the TiN-layer. Note that in neither Fig. 15.12 or 15.13 is the profile given in terms of concentration, but merely as emission intensity in arbitrary units; this is the most common way of presenting the analysis.

Despite the normal application as a very fast profiling technique, it is in fact perfectly possible to operate GDOS at a much slower erosion rate, and with its great sensitivity it is certainly capable of recording signals from the first few atom layers. Thus in one of their examples, Berneron and Charbonnier[15] compared the results of GDOS when operated in a surface-specific manner with the result of a depth profile by AES as seen in Fig. 15.14. The profiles are rather similar, which is encouraging. Note that GDOS is apparently capable of detecting B beyond the point in the profile at which AES cannot, and, most importantly, that GDOS can detect H, which of course is impossible for AES. The profiles in this example were of passive films formed anodically on pure iron in solution buffered with boric acid, and as in all passivation studies, the location of H is important. According to GDOS there is a pile-up of H at the passive film/metal interface.

Another example of the usefulness of GDOS when used in a much slower mode concerns the recording of ion implantation profiles. The ranges of

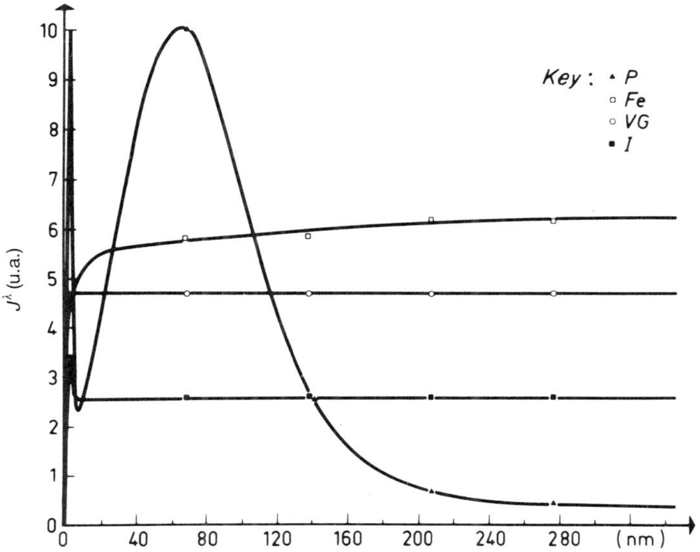

FIG. 15.15. Profile of phosphorus implantation into pure iron recorded by Pons-Corbeau.[10] The symbols VG and I stand for lamp voltage and current density, respectively. The observed range of the P at 125 keV implantation energy was 65 nm, which compares well with the theoretical value of 63 nm. The atomic concentration of P at the maximum in the profile was about 1.8%.

implanted ions are typically many tens of nm so a quasi-surface-specific technique is obviously needed to be able to record an accurate profile. Pons-Corbeau[10] has given the profiles observed for the implantation of P at 125 keV into Fe and FeO, and they are shown in Figs 15.15 and 15.16. The theoretical ranges of P at that energy in Fe and FeO are 63 and 74 nm, respectively, which compare well with the observed ranges of 65 and 88 nm, although the standard deviations (measure of the distribution widths) of the analysed profiles were some 45% greater than the theoretical. Note that the atomic concentration of P at the maximum of the profile is only about 2%, highlighting the sensitivity of GDOS. In another publication[16] dealing with ion implantation in more detail, it was found that, for implantation of B in Ni, there was a peak in the H profile at the same depth as that of the B. This was explained on the basis of the simultaneous implantation of $^{11}B^+$ and $^{10}BH^+$ ions, since the mass separator would not be able to discriminate between the two.

In summary, then, GDOS is widely applied as a very rapid, easy to use, profiling analytical technique, often in a purely qualitative way although it can easily be made quantitative. However, due to its sensitivity and flexibility it could equally often be used as a surface-specific technique, particularly if re-designed to the standards of cleanliness associated with UHV. Its information is purely elemental, with no direct chemical information available.

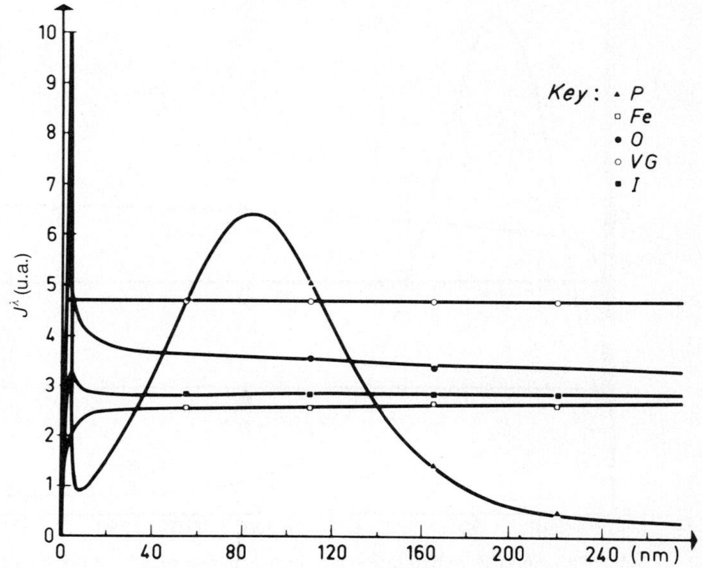

FIG. 15.16. Profile of phosphorus implantation into wustite, FeO, recorded by Pons-Corbeau[10] using GDOS. The symbols have the same meanings as in Fig. 15.15. The observed range of the P at 125 keV implantation energy was 88 nm, which compares reasonably well with the theoretical value of 74 nm. However, for implantation into both pure iron and wustite, the observed profile widths are wider than theoretical by about 45%. The atomic concentration of P at the maximum in the profile was about 1.5%.

References

1. Loxton, C. M. and Tsong, I. S. T. (1984). *Surf. Sci.* **139**, 453.
2. Wright, R. B. and Gruen, D. M. (1980). *Nucl. Instrum. Meth.* **170**, 577.
3. Hamilton, J. C. and Anderson, R. J. (1985). *Surf. Sci.* **149**, 81.
4. Martin, P. J. and MacDonald, R. J. (1977). *Surf. Sci.* **62**, 551.
5. Andersen, C. A. and Hinthorne, J. R. (1973). *Anal. Chem.* **45**, 1421.
6. MacDonald, R. J. (1982). *Applic. Surf. Sci.* **13**, 63.
7. Baucke, F. G. K. and Bach, H. (1982). *J. Am. Ceram. Soc.* **65**, 534.
8. Grimm, W. (1968). *Spectrochim. Acta* **23B**, 443; and (1967). *Naturwiss.* **54**, 588.
9. Ohashi, Y., Yamamoto, Y., Tsunoyama, K., and Kishidaka, H. (1979). *Surf. Interface Anal.* **1**, 53.
10. Pons-Corbeau, J. (1985). *Surf. Interface Anal.* **7**, 169.
11. Ko, J. B. (1984). *Spectrochim. Acta*, **39B**, 1405.
12. Bengtson, A. (1985). *Spectrochim. Acta* **40B**, 631.
13. Boumans, P. W. J. M. (1972). *Anal. Chem.* **44**, 1219.
14. Koch, K. H., Sommer, D., and Grunenberg, D. (1985). *Mikrochim. Acta* (Vienna), **Suppl. 11**, 137.
15. Berneron, R. and Charbonnier, J. C. (1981). *Surf. Interface Anal.* **3**, 134.
16. Takadoum, J., Pivin, J. C., Pons-Corbeau, J., Berneron, R., and Charbonnier, J. C. (1984). *Surf. Interface Anal.* **6**, 174.

16

Ion excitation: ion scattering spectroscopy (ISS)

The use of ion scattering from solids as an analytical technique has now a long history because in fact 'ion scattering' actually encompasses several techniques according to the chosen energy range of the ion. According to the classification of Armour,[1] these techniques can be labelled Low Energy Ion Scattering (LEIS), with primary ion energies in the range 100 eV to 10 keV, Medium Energy Ion Scattering (MEIS) from 100 to 200 keV, and High Energy Ion Scattering (HEIS), from 1 to 2 MeV. Of these LEIS is more usually called Ion Scattering Spectroscopy (ISS) and HEIS is always called Rutherford Back-Scattering (RBS). For the reasons given later ISS is the only surface-specific variation, although the higher energy forms are certainly surface sensitive, and this chapter is therefore devoted to ISS alone.

16.1 Operation

As with most of the other surface analytical techniques, ISS requires basically only three components, a source, a target, and a detector. For an ion source to be used in the low energy regime of ISS, either the thermal emitter type (Fig. 3.12) or the duoplasmatron type (Fig. 3.13), already described under Section 3.2.3, are suitable. An arrangement in which the former type has been used is shown in Fig. 16.1 from Martin and Netterfield;[2] notice here also the electron flood gun for neutralizing the surface charge on insulators with low energy electrons. For really detailed basic studies, particularly angle-resolved ones, it is desirable to work with as 'clean' an incident ion beam as possible, in which case the beam is passed through a mass filter before it strikes the target, as seen in Fig. 16.2 from Taglauer.[3] It is also possible when using an ion source with several gases simultaneously, in conjunction with a mass filter, to switch from one type of ion to another very quickly. Ion currents used are normally in the range 10^{-9}–10^{-8} A, into areas that depend very much on the type of gun being used, but are usually 1–10 mm^2, much the same as in SSIMS (q.v.). When using He^+ as the incident ion, the total fluence necessary to record a spectrum would be of the order of 10^{13} ions cm^{-2}. The typical energy range used is 0.5–4.0 keV, although energies up to 10 keV have sometimes been employed.

The scattered ion detector is normally one of the types of electron energy analyser described in Section 3.3.1, but of course with the polarity reversed to detect positive rather than negative particles. Because it is important in ISS to be able to carry out angularly dependent analysis, to obtain structural

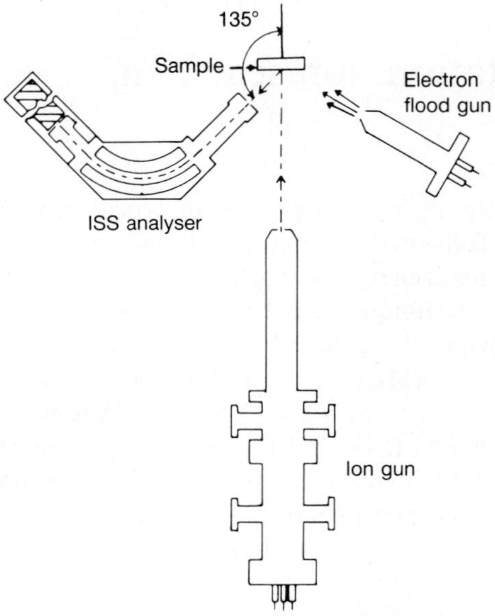

FIG. 16.1. Experimental arrangement of ISS system used by Martin and Netterfield.[2] The ion gun is of the thermal emission type, with the primary beam of current typically 60 nA and energy a few keV. Spot size on the sample is $\approx 300\mu m$. Energy analyser is of the 127° electrostatic hemispherical-sector type, and the scattering angle is 135°. An electron flood gun was found to be essential in the authors' study of zirconia surfaces.

information, analysers of the spherical sector type such as the CHA (Section 3.3.4) are much favoured since they have small angular acceptance angles. Analysers of the CMA type (also under Section 3.3.4) have also been used, particularly where angular studies are not required and only an elemental composition at maximum sensitivity is sought, and an interesting and compact ISS analysis system based on a CMA is shown in Fig. 16.3, from Brongersma *et al.*[4] The ion gun is co-axial with the CMA, and has a mass filter consisting of crossed electrical and magnetic fields, followed by a focusing lens that concentrates the ion beam on the sample. Ions scattered into the acceptance angle of the analyser, $42.3° \pm 6°$, and over a 2π solid angle, are energy selected and then detected on a ring detector rather than the normal electron multiplier. Because of the large solid angle the device has unusually high sensitivity. If a CMA is to be used for ISS, it is normal to use a single-pass analyser, as seen in Fig. 16.3, but Ellis and Taylor[5] have demonstrated how a double-pass CMA intended for high resolution UPS can be used for ISS.

Recent improvements in ion energy analysis have been made as a result of the need to reduce the amount of surface damage by sputtering that occurs during the acquisition of a spectrum. Even He^+ ions at 1 keV can cause

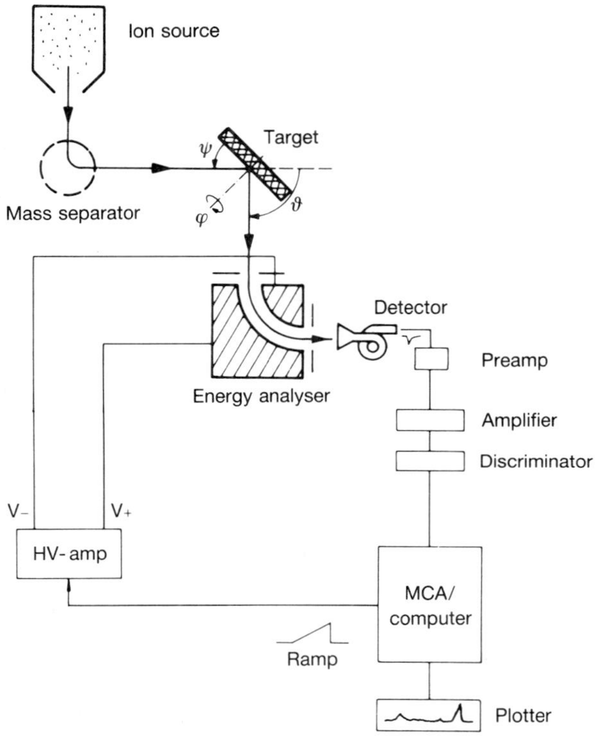

FIG. 16.2. Experiment arrangement for ISS constructed by Taglauer.[3] In this system the incident beam is passed through a mass separator before hitting the target, thus producing a 'clean' beam in terms of ion species. With such a separator it is possible to produce several ion species at the same time from the ion source, and then to be able to switch quickly from one ion species to another. The energy analyser is a 90° hemispherical-sector type. The scattering angle is fixed by the geometry as 90°, but the target can be tilted to vary the angle of incidence ψ, or rotated to vary the azimuthal angle ϕ.

damage, although obviously far less than would be caused by Ar^+ or Ne^+, and the collection efficiencies of the commonly used analysers are such that it is necessary to work with as small an incident current as possible spread over as large an area of the target as possible in order to minimize damage. A monolayer of adsorbed atoms contains about 10^{15} atoms cm^{-2}, and if, as in the worst case, the sputtering yield is 1 atom/incident ion, and it is required to maintain the extent of damage below 1%, then the total dose should be $\leq 10^{13}$ ions cm^{-2}. This is possible with a current of a few nA into a few mm^2, giving measuring times of up to 1 hour before the damage becomes unacceptable, but it would be desirable to be able to carry out analyses rather more rapidly.

An analysing system that is claimed to be able to perform both energy and angular resolution simultaneously has been described by Hellings *et al.*[6] The electrostatic analysing section consists of a number of toroidal deflection

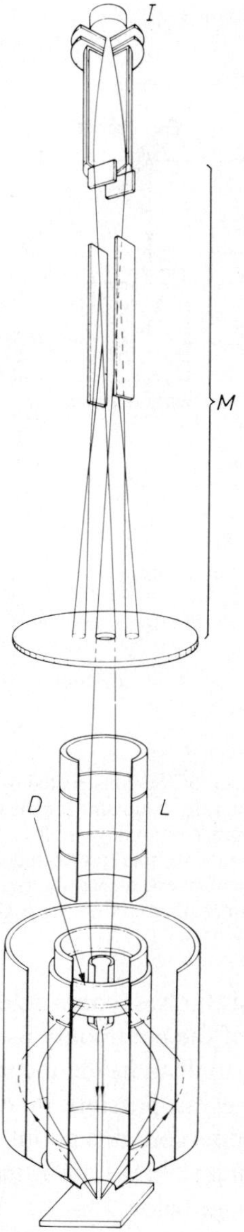

FIG. 16.3. A compact system for ISS analysis devised by Brongersma *et al.*[4] It is based on a cylindrical mirror analyser (CMA). The ion gun I is co-axial with the CMA, and the ion beam passes through a mass filter M consisting of crossed electrical and magnetic fields, before being focused by lens L onto the sample. Ions scattered back from the sample surface enter the analyser over a solid angle of 2π, and an acceptance angle of $42.3° \pm 6°$. They are energy selected by the operation of the CMA (see Chapter 3) and the selected ions are detected on a ring detector D. The device has high sensitivity because of the high solid angle of acceptance.

plates that constrain the ions into spiral paths; at the end of the analyser there is a 100 mm wide position-sensitive multi-detector that allows the angle of incidence into the analyser to be determined. A zoom lens at the front end can accelerate or decelerate ions, thus changing the pass energy and the resolution. The resolving power is quoted as being 1000, i.e. a relative resolution of 0.1%. Figure 16.4 shows a cross-section of the arrangement.

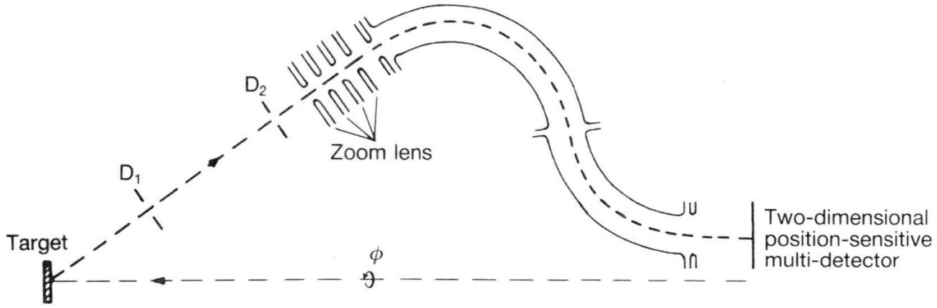

FIG. 16.4. Schematic of system designed by Hellings *et al.*[6] for simultaneous energy and angular resolution in ISS. The analyser is a double toroidal system that constrains the scattered ions into spiral paths. Apertures D_1 and D_2 define the scattering angle into the analyser, which is fixed at 145°. The azimuthal angle may be varied by rotating the target through angle ϕ. A zoom lens at the entrance to the analyser can accelerate or decelerate ions, thereby changing the pass energy and resolution. At the end of the analyser is a position-sensitive multi-detector about 10 mm wide consisting of 128 detectors arranged radially. The effective resolving power is about 1000.

For further reduction of the effects of beam damage time-of-flight (TOF) analysers have been developed. A recent example of such an analyser is shown in Fig. 16.5; it was designed and built by Rabelais and co-workers.[7] Ions from the source pass through Wien filter plates which have the combined task of velocity selection and pulse imposition. The pulsed beam then passes through an energy analyser to remove any unpulsed neutrals, and is directed onto the target surface via *x*–*y* deflection plates. At each pulse the output from the trigger circuit of the pulse generator is fed to the start of a time-to-pulse height converter, with a variable delay adjusted according to the time for the ion bunch to travel from the Wien plates to the target. On arrival at the detector after travelling down a flight tube the converter is stopped. Voltage pulses proportional in height to the elapsed time are sent to a multichannel pulse height analyser which then produces a histogram of the flight time distribution.

The TOF analyser operates typically with 3 keV Ar^+, current densities of 1.2 nA cm^{-2}, a pulse width of 500 ns, and a pulse rate of 20 kHz. Scattering angles of 90° or 135° can be chosen, with respective path lengths of 125 or 74 cm. The count rate for scattered particles is about 1.5 kHz. The current density is one or two orders of magnitude less than in the experimental arrangements previously described.

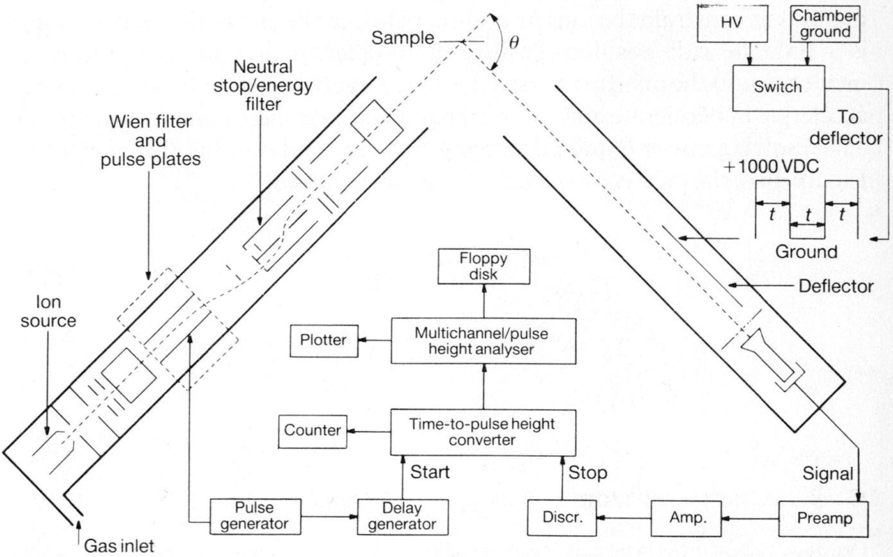

FIG. 16.5. Time-of-flight analyser and ion source used in the ISS system constructed by Kumar *et al.*[7] The ion source is a gas discharge type, from which ions are velocity selected in a Wien filter and energy analysed before being directed onto the sample via x–y deflection plates. In the Wien filter the ion beam is pulsed at 20 kHz with a pulse width of 500 ns, and the trigger output of the pulse generator goes to the start of a time-to-amplitude converter after a slight delay to allow the ion pulse to reach the sample. The converter is stopped when a particle arrives at the channeltron detector. A histogram of the distribution of particle flight times is produced by feeding into a multichannel pulse height analyser a voltage pulse proportional to the time between start and stop signals.
Two geometries are available, one with scattering angle 90° and flight path 125 cm, and the other with scattering angle 135° and flight path 74 cm. The incident argon ions had energy 3 keV and current density 1.2 nA cm^{-2}.

16.2 Theory and spectral interpretation

ISS is attractive as a surface-specific technique because of the consequences of the strong electron affinity of inert-gas ions. As a positive ion approaches a surface and starts to interact with the electron cloud extending from the surface, the probability of electron transfer by any one of several processes, some already described in Chapter 14, is very high even at the initial collision with a surface atom. After two or more collisions there will be virtually no more ions still existing as ions. The proportion of ions of the same mass scattered from the surface is only 10^{-4}–10^{-6} of the initial density. If a detector is set to detect ions of the same type as those in the incident beam, it will therefore detect almost entirely those ions that have had only one collision with a surface atom; hence the extreme surface specificity of the method.

A further very important factor in determining the usefulness of ISS is the large difference in the time of interaction of an ion and an atom, 10^{-15}–10^{-16} s, and the characteristic time of vibration of an atom in the solid,

10^{-13} s. The interaction time is so short in comparison that the scattering event is over long before the atom realizes that it is still attached to the surface. Under these circumstances ordinary binary collision ('billiard-ball') theory can be applied. Since in such collisions both energy and momentum must be conserved, the relationship between the energy E_1 of the ion of mass M_1 with incident energy E_0, after scattering from an atom of mass M_2 through an angle θ into the detector, is

$$E_1/E_0 = \frac{1}{\{1+(M_2/M_1)\}^2} [\cos\theta \pm \{(M_2/M_1)^2 - \sin^2\theta\}^{1/2}]^2. \quad (16.1)$$

The positive sign in (16.1) applies when $M_2 > M_1$, and both signs when $M_2 \leq M_1$. Note that if the scattering angle θ is chosen to be 90°, then eqn. (16.1) reduces simply to

$$E_1/E_0 = (M_2 - M_1)/(M_2 + M_1). \quad (16.2)$$

Of course the target atom in the surface receives energy in the scattering process as well, and if this energy transferred is E_2, with an angle α between the direction of the incident ion and the direction in which the atom recoils, then

$$E_2/E_0 = \frac{4(M_2/M_1)}{\{1+(M_2/M_1)\}^2} \cos^2\alpha. \quad (16.3)$$

It can be seen from eqn (16.1) that when the energy and mass of the incident particle, and the scattering angle, are all fixed, E_1 depends only on the mass M_2 of the atom from which the ion is scattered. In other words, an energy spectrum measured in an ISS experiment can be converted immediately into a mass spectrum via eqns (16.1) or (16.2), according to the scattering angle, which therefore provides an elemental analysis of the surface. ISS spectra are invariably recorded as scattered intensity vs. the ratio E_1/E_0, and some typical examples are shown in Fig. 16.6 of such spectra from a set of Al_2O_3, MoS_2, and Co_9S_8 standards used by Beuken and Bertrand[8] in their study of sulphide cobalt–molybdenum catalysts. These spectra were taken with He^+ at 2 keV over a scattering angle of 138°, at a primary current of 15 nA, and the acquisition time was about 5 min. Low energy flood electrons were used to neutralize surface charging as described above.

Notice from Fig. 16.6 what should be obvious from the form of eqns (16.1) and (16.2) that as M_2 increases, E_1/E_0 tends to unity, and that for a light ion such as He^+ the heavier elements become difficult to separate spectroscopically. This difficulty is a function also of the widths of the peaks in the scattered spectrum, broad peaks obviously being less easy to resolve; this will depend on the energy and angular resolution capabilities of the experimental arrangement. However, if the angular acceptance is small enough, then the mass resolution is governed only by the energy resolution available, and is given by

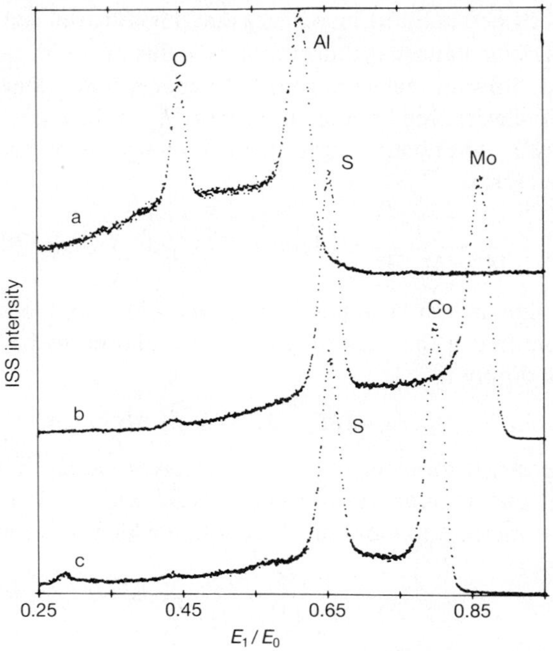

FIG. 16.6. ISS spectra using 2 keV He^+ ions at 138° scattering angle, from the standard materials (a) Al_2O_3, (b) MoS_2, and (c) Co_9S_8, measured by Beuken and Bertrand.[8] Primary current was 15 nA and acquisition time 5 min per spectrum. A flood gun was used to neutralize surface charging. The pronounced tails in the spectra at energies below the peaks are said to be due to inelastic scattering without neutralization, and the effect is most prominent when the metal atoms are not directly in the surface. For example, Al in Al_2O_3 is surrounded by 6 oxygen atoms, and ions cannot scatter directly from it. Notice from the figure that as M_2, the mass of the scattering atom, increases, E_1/E_0 tends to unity, so that for light primary ions the heavier elements are difficult to separate.

$$(M_2/\Delta M_2) = (E_1/\Delta E_1)\,\frac{2A}{A+1}\,\frac{A+\sin^2\theta - \cos\theta(A^2-\sin\theta)^{1/2}}{A^2-\sin^2\theta+\cos\theta(A^2-\sin\theta)^{1/2}} \qquad (16.4)$$

where M_2/M_1 has been put equal to A. In Fig. 16.7 the mass resolution from eqn (16.4) is plotted[9] in two ways, first in the upper half as a function of the target mass M_2 for a fixed scattering angle 90° and constant resolution 100, and in the lower half as a function of the scattering angle for different mass ratios A, again at fixed resolution. Clearly the mass resolution improves as one increases the incident mass M_1 (i.e. on going from He^+ to Ar^+ to Xe^+) but attendant on that is the problem of greatly increased sputtering and broader peaks. It can be seen also that the mass resolution starts to decrease sharply below a scattering angle of 90°, and indeed all ISS geometries are for scattering angles ≥90°. An example of the improved resolution found for the heavier surface elements by switching the primary ion is shown in Fig. 16.8 from

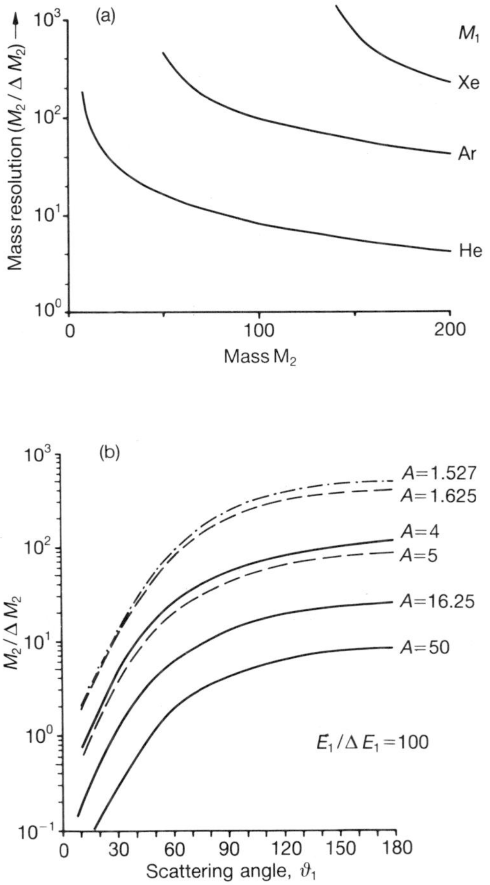

FIG. 16.7. (a) Mass resolution for different primary projectile ions M_1 as a function of the target mass M_2, for a scattering angle 90° and fixed energy resolution of 100.
(b) Mass resolution for different mass scattering ratios A ($=M_2/M_1$) as a function of the scattering angle, also at a fixed energy resolution of 100.
The mass resolution improves as the mass of the incident ion is increased, but accompanying that is increased sputtering and broader peaks. From the lower half of the figure it can be seen that the mass resolution decreases sharply below scattering angles of 90°. From Taglauer and Heiland.[9]

Smith.[10] On an Mo-contaminated Au–Ni alloy the use of Ne^+ allows the Mo and Au peaks to be separated.

Confirmation of the virtual monolayer specificity of ISS when operated under conditions of low energy primary ions of light mass comes from the classical experiment of Brongersma and Mul.[11] Bromine reacts readily with a clean silicon (111) surface, the reaction ceasing when a monolayer is formed; since bromine atoms are so large they form a dense monolayer in which all the bromine atoms touch. The ISS observation of the surface before and after

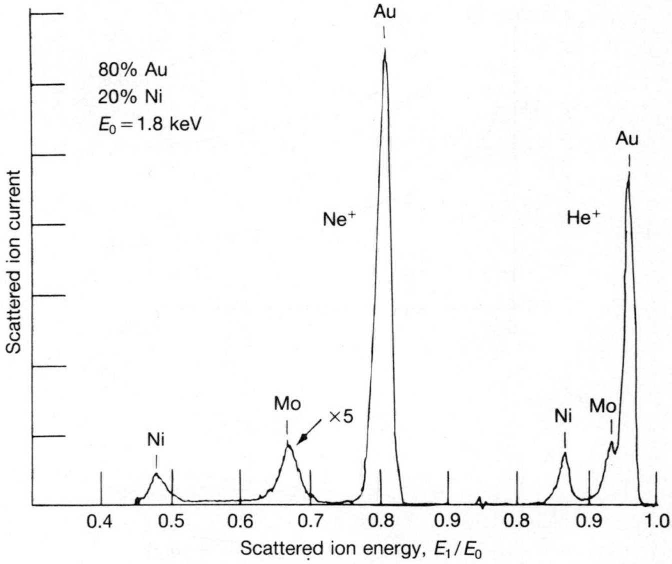

FIG. 16.8. Example from Smith[10] of the increased mass resolution for heavy elements obtainable by switching to a heavier primary ion, in this case from He^+ to Ne^+. In the He^+ ISS spectrum Mo and Au can barely be resolved, but in the Ne^+ spectrum are easily resolved.

bromination is given in Fig. 16.9. After the reaction the Si atoms were virtually invisible to the primary He^+ at 1 keV, the monolayer of Br having completely masked the Si.

16.3 Quantification

If a simple binary collision process from atoms in the first layer only is considered, then the intensity I_s of the scattered atom yield can be written down immediately as

$$I_s = I_0 N \sigma (1 - P)\, \Delta\Omega T \tag{16.5}$$

where I_0 = intensity of the primary beam
N = density of scattering atoms in the surface
σ = scattering cross-section of scattering atom
P = probability of neutralization of a primary ion
$\Delta\Omega$ = acceptance angle of analyser
T = transmission function of analyser.

In any one analysis the experimental geometry, the transmission function, and the primary ion current are kept constant, so that to obtain N from I_s involves knowledge of σ and P.

The differential cross-section for the scattering angle θ is defined as

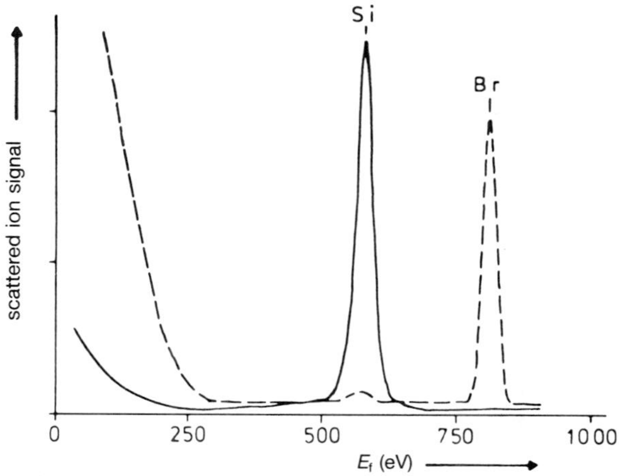

FIG. 16.9. Demonstration of the surface specificity of ISS by Brongersma and Mul.[11] The solid line is the spectrum from clean Si(111) using 1 keV He^+ ions scattered through 142°. The dashed line is the spectrum from the same surface after reaction with bromine, for which adsorption saturates at exactly one monolayer. Because of the size of the bromine atoms in the monolayer they are in fact in contact, and mask the underlying silicon atoms completely. Thus in the brominated spectrum virtually no silicon peak appears.

$$d\sigma(\theta) = 2\pi p \, dp \tag{16.6}$$

in which p is the so-called impact parameter. $d\sigma(\theta)$ has the form[12] of an annular area formed by the region between the impact parameters p and $p + dp$, bounded by the scattering of incident ions into a cone between the angles θ and $\theta - d\theta$. This is equivalent to a solid angle $d\Omega = 2\pi \sin\theta \, d\theta$, since the geometry is that of rotational symmmetry around the direction of the incident beam. The differential cross-section for scattering into unit solid angle is therefore

$$d\sigma(\theta)/d\Omega = (p/\sin\theta)(dp/d\theta). \tag{16.7}$$

To evaluate (16.7) requires a relationship between p and θ. Without going into a great deal of mathematical detail, the expression required for the 'scattering integral' can be derived from considerations of conservation of energy and momentum, and is usually expressed in terms of a scattering angle ψ, not the same as θ, defined in the centre-of-mass, or barycentric, frame of reference. The integral is

$$\psi = \pi - 2p \int_r^{\infty} \frac{dr}{r^2 \left\{ 1 - (p^2/r^2) - \dfrac{V(r)}{E_r} \right\}^{1/2}} \tag{16.8}$$

where $V(r)$ is the interaction potential and E_r is the reduced energy

$\{M_2/(M_1+M_2)\}E_0$. Here r_0 is the closest distance of approach of incident ion and scatterer, and is given by

$$1-p^2/r_0^2-V(r_0)/E_r=0. \qquad (16.9)$$

The system can be transformed from the barycentric back to the laboratory framework via the relation

$$\psi=\theta+\sin^{-1}\{(\sin\theta)/A\}. \qquad (16.10)$$

The form of the interaction potential $V(r)$ has been the subject of much discussion. When two atomic particles approach each other, the interaction potential arises from the mutual forces felt by each due to the atomic nuclei and the electron clouds around them. It is repulsive at very close distances and attractive at large distances. In ISS the energies are low enough for the distance of closest approach of the particles to be so small that the attractive part of the potential can be ignored, and it is usual to express the interaction in terms of a purely repulsive potential, or Coulomb repulsion. However, the low energies also mean that the interpenetration of electron clouds during scattering is not complete, that is, some electronic screening of the nucleus still occurs, and so the interaction potential must include a screening function as well. Most of the discussion has been concerned with the form of the latter function. The general form of the interaction potential is thus

$$V(r)=\frac{Z_1Z_2e^2}{r}\,\phi(r/a) \qquad (16.11)$$

where Z_1 and Z_2 are the atomic numbers of the ion and the scattering atom, ϕ is a screening function, and a is a characteristic screening length. Most screening functions are based on Thomas–Fermi statistics. The most widely used function is that of Molière,[13] which is approximated by the analytical expression

$$\phi(r/a)=0.35\exp^{-0.3x}+0.55\exp^{-1.2x}+0.1\exp^{-6x} \qquad (16.12)$$

where $x=r/a$, and the screening length is given by

$$a=0.468C(Z_1^{2/3}+Z_2^{2/3})^{-1/2} \qquad (16.13)$$

where C is an empirical adjustable constant for the individual scattering system, with values in the range 0.6 to 0.8. With the above approximations it is possible to compute ψ and hence the differential scattering cross-section in (16.7). It is also possible for a chosen ion–atom pair and primary energy to calculate the ion trajectories. Such a set of trajectories based[14] on the Molière screening function is shown in Fig. 16.10, in which S is the impact parameter appropriate to a particular scattering angle and R is the radius of the shadow cone cast by the atom. R is of the order of interatomic spacings, which implies that atoms lying in a row behind the scatterer are invisible to the primary ions.

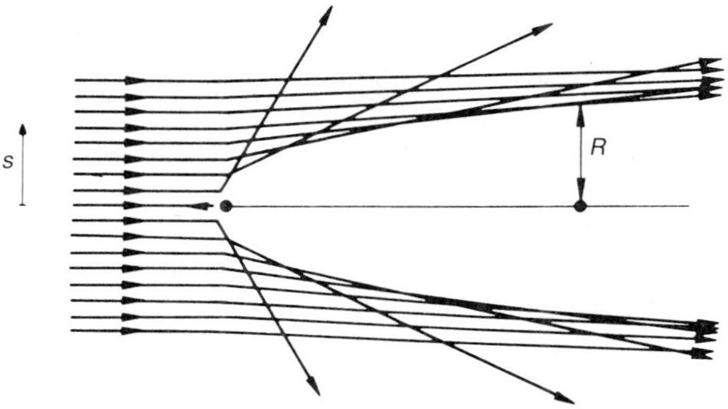

FIG. 16.10. Set of ion trajectories calculated by Van Der Veen[14] for a parallel beam of light ions hitting an atom. The interaction potential due to the atom's nuclear charge causes a shadow cone to arise, as seen in the figure; the radius of the cone is R. Since R is of the order of interatomic spacings, atoms in a row behind the scattering atom are invisible to the primary ions. S is the so-called impact parameter associated with a particular scattering angle (see Chapter 18 and Fig. 18.16 for a fuller description of the impact parameter).

The other parameter in eqn (16.5) that needs evaluation is the neutralization probability P. A noble gas ion approaching a surface represents a very deep potential well (e.g. -22.4 eV for He^+) compared to the typical value, ≈ 4.5 eV, of the work function of a solid. Thus neutralization is a highly probable event, and can take place by the same processes of Auger neutralization or resonance neutralization already described for much lower energy ions in some of the immediately preceding chapters. Based on the theory of neutralization by Auger electron ejection developed by Hagstrum,[15] the probability of an ion escaping neutralization, that is $(1-P)$, has the form

$$(1-P)=\exp\left\{\left(\frac{1}{V_{0\perp}}+\frac{1}{V_{1\perp}}\right)V_c\right\} \tag{16.14}$$

where $V_{0\perp}$ and $V_{1\perp}$ are the components of the velocities of the initial and scattered ions normal to the surface, and V_c is a characteristic velocity given by

$$V_c=\int_{r_0}^{\infty} S(r)\,\mathrm{d}r \tag{16.15}$$

where $S(r)$ is the rate of the transition giving rise to neutralization. The values of the parameters that appear in (16.14) and (16.15) are not well known but the Hagstrum model seems to be adequate since it predicts from eqn (16.14) that $\ln(1-P)$ should be linear with V^{-1}, and indeed Brongersma and Buck[16] have found this proportionality when plotting experimental parameters proportional to those in (16.14). Both Hagstrum's estimates and many empirical observations have indicated that $(1-P)$ is of the order of 10^{-3}. When instrumental factors such as angular acceptance and transmission are taken into account, this factor drops to 10^{-6}–10^{-7}.

Although it is possible from the above considerations to arrive at a quantification in ISS based on first-principle calculation, in practice, just as with most other surface analytical techniques, it is much more usual to base quantification on standards. Even then calibration is difficult, simply because ISS is so surface-specific; in the majority of multi-elemental materials such as alloys or compounds, the composition of the first atomic layer or two is different from the average composition in the bulk. Nevertheless, ISS has been used successfully in a quantitative manner either by using pure elemental standards or by calibration against another technique. A classical example of the former method is the use by Nelson[17] of Au and Ag elemental standards in the study of a range of Au/Ag alloys. The precise compositions of the alloys, that is the bulk compositions, were first established with electron microprobe analysis, and then the alloys were cleaned by sputtering until the surface composition had reached equilibrium. Figure 16.11 shows that a linear

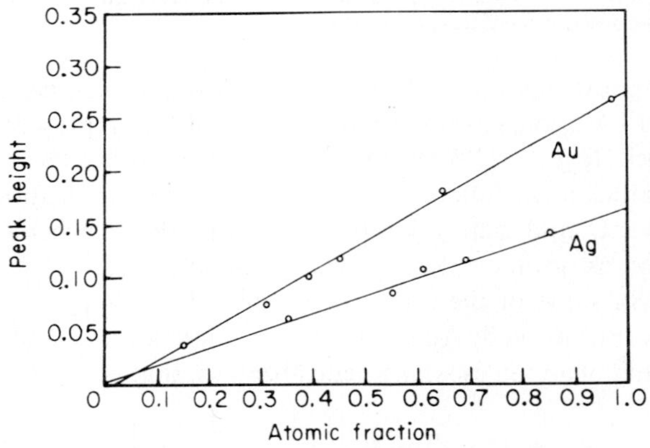

FIG. 16.11. Calibration by Nelson[17] of the peak heights of Au and Ag against the known bulk composition of a range of Au/Ag alloy standards. The bulk compositions were established by electron probe micro-analysis, and for ISS analysis the alloy surfaces were cleaned by sputtering until equilibrium was reached.

relationship between the peak heights in ISS and the bulk composition was found. Such a relationship could be used in reverse to work back from an observed peak height to a concentration in an unknown material.

The correlation between ISS intensities and those from another surface-specific technique is demonstrated in Fig. 16.12 from the work of Taglauer and Heiland.[18] During the exposure of a Ni(111) surface to oxygen both ISS and AES were used. Clearly the ISS scattered ion intensities and the Auger signals, from the increasing coverage of oxygen, are directly proportional to each other, at least as far as the experiment went. The same authors[18] have also shown the proportionality between the ISS and AES signals as a function of coverage with sulphur. Notice also in Fig. 16.12 that the scattered signal from

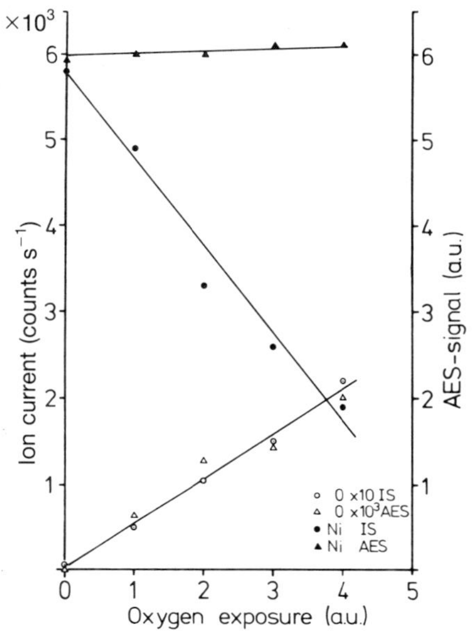

FIG. 16.12. Calibration by Taglauer and Heiland[18] of ISS peak intensities against AES intensities, for oxygen adsorbed on a Ni(111) surface, as a function of oxygen exposure. The oxygen signals from ISS and AES are seen to be directly proportional to each other. On the other hand, the nickel ISS intensity starts to decrease immediately on exposure to oxygen, whereas the AES intensity remains constant. This is a further indication of the extreme surface specificity of ISS, since the oxygen starts to 'shadow' the nickel atoms from the incident ions as soon as adsorption starts. Analysis conditions were: for ISS, 1 keV He^+ ions and a scattering angle of 90°; for AES, 2 keV primary energy and 50 μA beam current.

the Ni atoms in the surface starts to decrease immediately oxygen exposure starts, whereas the Auger signal from Ni remains constant within the range of exposures in the figure. This is a striking confirmation of the surface specificity of ISS; the first few oxygen atoms start to cause 'shadowing' of the top layer of Ni atoms immediately, whereas of course the Auger signal originates from several monolayers depth.

16.4 Applications

16.4.1 *Catalysis*

Catalytic reactions are known to occur principally at the outer surface of the catalyst; although most technologically interesting catalysts are in the form of porous pellets or compacts in the production of which the aim is to achieve a very high specific surface area, there is still sufficient exposed external surface available to be able to make use of surface-specific techniques. It has already been described in an earlier chapter how XPS has been able to make significant contributions to knowledge of catalytic mechanisms, through the provision of chemical information, but even XPS includes in its analysis several atomic layers. Often the ability to study the outer layer, without having to disentangle information from subsequent layers, is of great value. Beuken and Bertrand[8] have applied ISS to analysis of sulphided cobalt–molybdenum catalysts, used in the hydrodesulphurization process, and in the course of so doing have carried out some interesting data processing.

Before they can be used, catalysts have to go through an activation procedure, which may mean simply heating in air to an appropriate temperature, or heating in hydrogen or some other reducing atmosphere. The ISS spectra recorded by Beuken and Bertrand during the activation of a cobalt–molybdenum catalyst are shown in Fig. 16.13, at three different

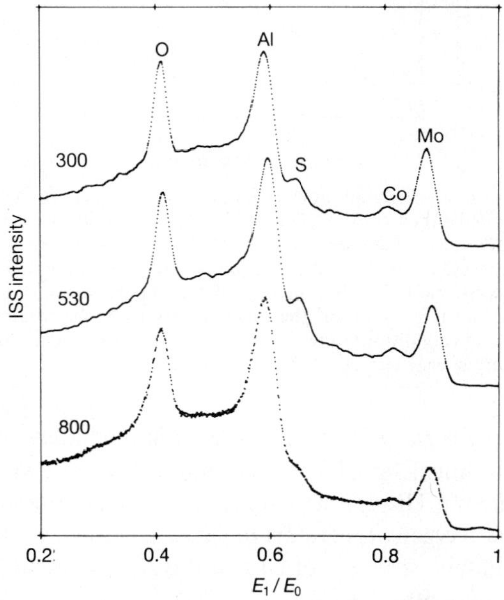

FIG. 16.13. ISS spectra after three different temperatures of activation of a $CoMoS/\gamma\text{-}Al_2O_3$ catalyst, recorded by Beuken and Bertrand.[8] He^+ ions at 2 keV, and a scattering angle of 138°, were used. The S, C, and Mo concentrations in the surface all decrease with increasing temperature. However, the large background tails associated with some of the peaks prevented accurate quantification, and the authors had to perform data processing as indicated in Fig. 16.14.

temperatures. It can be seen, of course, that the S, Co, and Mo contributions all decrease with increasing temperature, but an important aspect of the spectra that needs consideration before any interpretation can be made is the large background tail to the low energy sides of some of the metal atom peaks. According to the authors, this background arises when the scattering atoms are not located exactly in the top layer, but a little way below it, and in order for the He^+ to be able to scatter from such atoms it has to penetrate the region of influence of the lighter atoms, in this case oxygen and sulphur, and thus lose some of its energy. In Al_2O_3 the Al atoms are each surrounded with six O atoms, while in MoS_2 the Mo atoms are in the second basal plane.

To deal with the situation so that an analysis could be extracted from the data, Beuken and Bertrand first recorded spectra from the standards Al_2O_3, MoS_2, and Co_9S_8, relevant to the catalyst composition, and then fitted each peak either to a pure Gaussian for the non-metals or to a Gaussian with a tail of mathematical form for the metals. Fit parameters were evaluated by least-squares minimization, and their usefulness checked by recording ISS spectra from mixtures of known proportions of the standards. The deviation in determining the metal ratios was less than 7%. Figure 16.14 from their paper

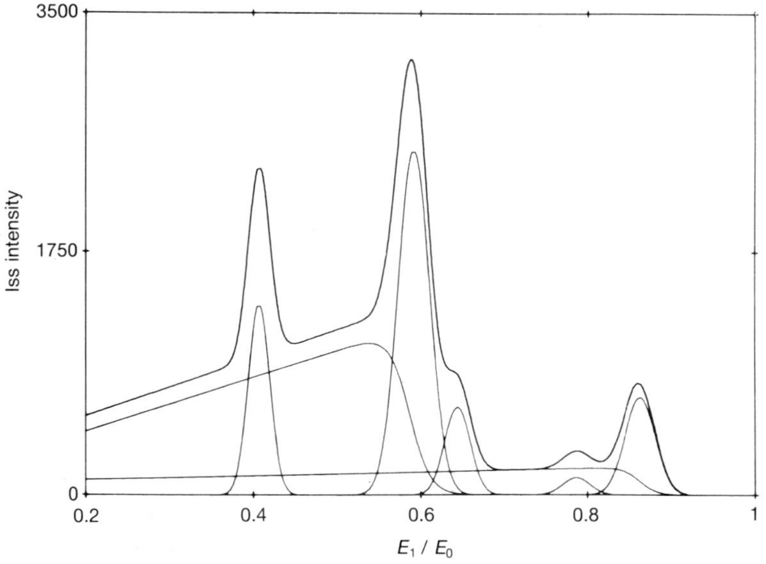

FIG. 16.14. Curve synthesis method applied by Beuken and Bertrand[8] to the ISS spectra of Fig. 16.13. For the catalyst $CoMoS/\gamma\text{-}Al_2O_3$ the standard materials Al_2O_3, MoS_2, and Co_9S_8 were first analysed individually, and their ISS spectra fitted either to a pure Gaussian for the non-metals O and S or to a pure Gaussian with a tail of the form $(ax+b)\{1+c\exp(x-x_0)\}^{-1}$ for the metals. Fitting parameters a, b, and c were evaluated for the standards, and then checked against known mixtures of the standards, from which the deviation was less than 7%. The peak areas under the Gaussians could then be used in analysis.

shows an example of the curve synthesis when applied to the cobalt–molybdenum catalyst itself. Having obtained the Gaussian peaks themselves by this method, it was then an easy matter to measure the areas under them and plot the areas or area ratios as a function of activation temperature, which has been done in Fig. 16.15. Note that the O/Al ratio remains virtually constant during

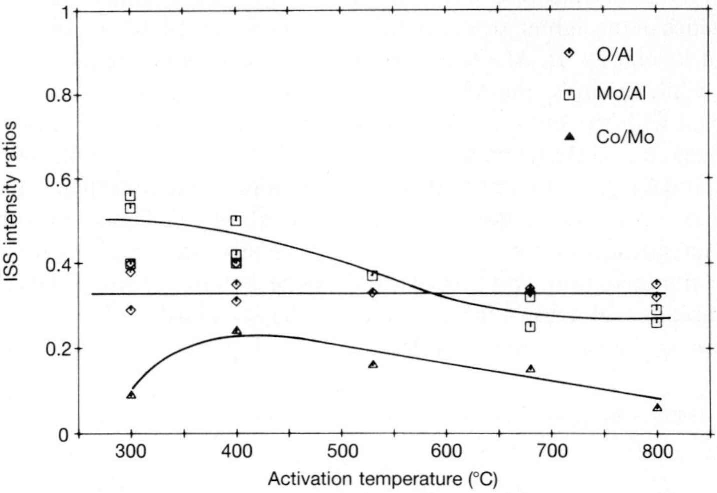

FIG. 16.15. ISS peak area ratio derived according to the method illustrated in Fig. 16.14, during activation of a CoMoS/γ-Al_2O_3 catalyst, by Beuken and Bertrand.[8] Analysis conditions He^+ ions at 2 keV and scattering angle 138°. The O/Al ratio remains virtually constant, indicating no change to the alumina support, and the Mo/Al falls progressively, suggesting a reduction in dispersion of MoS_2 as a result of sintering. The behaviour of the Co/Mo ratio is thought to be due to phase changes taking place in the Co–Mo–S system.

activation, but that the Mo/Al ratio falls progressively with temperature. The latter indicates that the dispersion of Mo over the Al_2O_3 surfaces is decreasing, probably as a result of sintering of MoS_2 crystallites. The behaviour of the Co/Mo ratio was not so easy to interpret; the authors suggest that phase changes in the Co–Mo–S system are likely to be responsible. Each phase would have a different surface contribution of Co atoms, and possibly a different screening factor as well, which would lead to different sensitivity factors. The important point is that all these changes are taking place in the outer surface layer or two layers, and could only have been established unambiguously by using a highly surface specific technique like ISS.

16.4.2 *Coatings*

Harnessing of solar energy requires concentration of the solar flux onto the heat exchanger or onto the industrial process (e.g. furnaces, melting pots, etc.); such concentration is performed with mirrors, and of course should occur with the absolute minimum of loss of flux by absorption. In addition for cost-effective application the life of a mirror should be measured in tens of years, rather in only a few years. Too rapid environmental degradation of the mirror means that the cost of using solar energy would be too high.

The metal with the highest solar reflectance is silver, but in practice mirrors made from the deposition of a silver coating on glass or silica do not at present have a sufficiently long life before the reflectance diminishes to an unacceptable level. The coating is normally produced by a wet chemical electroless process, and part of the reason for the lack of environmental stability may be due to the nature of the Ag–Si bonds formed during coating. Since such bonds must be a property of the surface on which the silver is to be deposited, the problem is clearly one that should be amenable to surface analytical techniques, and it has been investigated by Pitts *et al.*[19] using a combination of ISS and XPS.

Before silver will deposit on silica, the silica surface must of course be clean and must also be 'sensitized'. The latter procedure aims at producing a thin film of tin on the silica before deposition of the silver is carried out, and consists of treatment in an acidic solution of stannous chloride. Silvering then occurs when the sensitized surface is dipped in alkaline ammoniacal silver nitrate, with a reducing agent. In order to restrict the deposition of silver to a monolayer or less in thickness, Pitts *et al.* did not use a reducing agent.

ISS spectra from fused silica after cleaning with pumice and then sensitizing are shown in Fig. 16.16. Spectrum a was recorded after 50 s of bombardment

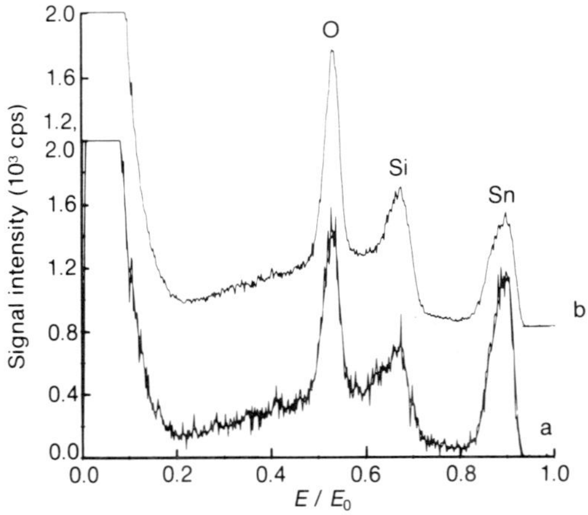

FIG. 16.16. ISS spectra from a fused silica surface to be used as the substrate for a high reflectance silver coating for solar mirror applications, recorded by Pitts *et al.*[19] The silica was first scrubbed with pumice and then 'sensitized' by treatment in an acidic solution of $SnCl_2$. Spectrum (a) was measured after 50 s of bombardment by 1 keV He^+ ions under the ISS analysis conditions, and (b) after 500 s such bombardment. Evidently the Sn is concentrated very near the surface since even under the weak bombardment conditions in ISS the Sn signal has dropped greatly relative to the O and Si signals. By comparing the ISS data with XPS analyses it was established that the Sn atoms are situated on the surface in sites where they 'shadow' Si atoms and not O atoms.

by 1 keV He^+, and b after 500 s. Firstly, the spectra show that Sn is concentrated at the outer surface, since after 50 s bombardment less than half the monolayer would have been removed, and after 500 s the Sn signal has dropped dramatically relative to O and Si. Secondly, in order to obtain agreement with the Sn/Si ratio obtained by XPS quantification, it was necessary to make assumptions about the positions of the Sn atoms on the surface. Thus, without any assumptions, the Sn/Si ratio by ISS was 0.21 as against 0.30 by XPS, but if sputtering during the 50 s analysis was taken into account then the ISS ratio was equal to the XPS ratio only if Sn shadowed Si sites and not O sites.

After silvering the ISS spectrum changed to that in Fig. 16.17. Again it is

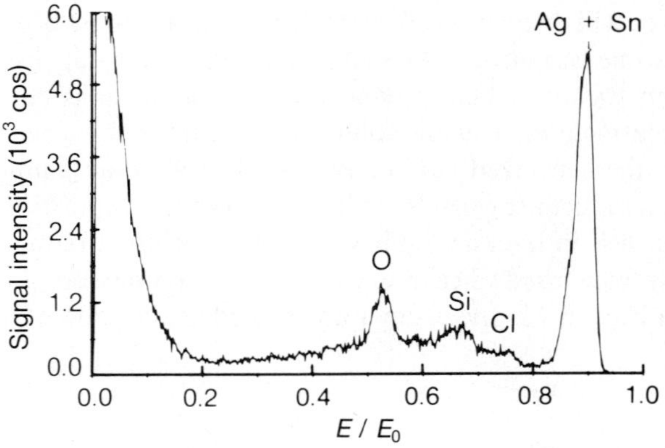

FIG. 16.17. ISS spectrum from a 'sensitized' fused silica surface after silvering from a commercial silvering solution, recorded by Pitts *et al.*[19] When using He^+ ions at 1 keV for the analysis, Ag and Sn peaks cannot be resolved, but reference to Fig. 16.16 shows that almost all the peak labelled Ag+Sn must be attributed to Ag. As with Sn, agreement with parallel XPS analyses could be obtained only if it were assumed that the Ag was bonded to Si sites on the silica surface, via an intermediate layer of Sn.

clear that the Ag is localized at the external surface; with He^+ the Ag and Sn scattered peaks cannot be resolved, but Fig. 16.16 shows that most of the combined peak must be attributed to Ag. As before, agreement with the XPS-derived coverages could be obtained only if sputtering was considered, *and* if the Ag was also bonded only to Si sites on the surface.

Thus ISS was able to provide the information additional to that from XPS that both the Sn and the Ag were strictly localized at the external surface, and that both were bonded to Si rather than to O, suggesting that these elements were sitting just above surface Si atoms. The quantification under these conditions showed that about 30% of the surface Si was covered with Sn after sensitization, and about 60% with Ag after silvering. These data were able to provide valuable information in elucidating the nature of the bonding, and therefore of the effect of environmental exposure.

16.4.3 *Adhesion*

Bonding of materials to each other by adhesives has now developed to such an extent that the adhesive joint is often stronger than the adjoining materials.

Joint failures, however, are still sufficiently common to justify an enormous amount of research into the mechanisms of failure, and many of the tools used in this research are surface analytical techniques.

Surfaces to be joined together with adhesives have always to be prepared, or 'primed', and this is particularly so for aluminium surfaces, on which the normal protective oxide film would make adhesion difficult if not impossible if it were not removed or modified. If care is not taken in the preparation, the situation may be made worse by the leaving behind of contaminating residues on the surface, preventing good adhesion. This situation has been investigated by Baun[20] using ISS. As shown in the upper part of Fig. 16.18, he found from

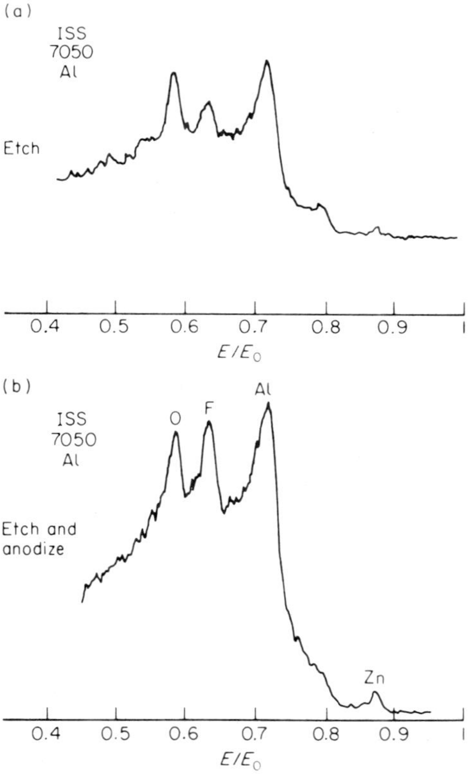

FIG. 16.18. ISS spectra from the surface of a commercial aluminium alloy, after (a) etching in HF/HNO_3 dilute solution, and (b) anodizing following etching, measured by Baun.[20] The etching treatment has left a residue of F on the alloy surface, but the anodizing treatment has not removed the F, in fact increased it if anything. Thus in this case adhesion bonding would not be to a pure Al_2O_3 surface as hoped after anodizing, but to an F-contaminated surface, with consequent effects on bondability and on long-term durability.

the scattering spectrum of He^+ from the surface of an aluminium alloy etched in dilute HF/HNO_3 that fluorine was left behind on the surface. Before adhesive bonding the alloy had then to be anodized, but instead of losing the fluorine in the process it was in fact concentrated even further at the external surface, as revealed in the lower spectrum of Fig. 16.18. In other words, subsequent application of the adhesive was not to an anodized aluminium surface (i.e. Al_2O_3) at all, but to a less well characterized surface containing impurities. This would certainly have affected the strength of the bond and its durability.

When an adhesive bond does fail by fracture, it is often instructive to examine the fracture faces to try and determine the exact path of failure. Baun shows an example of the application of ISS to this aspect of adhesion as well, as seen in Fig. 16.19. In this case the bond was to a titanium alloy with

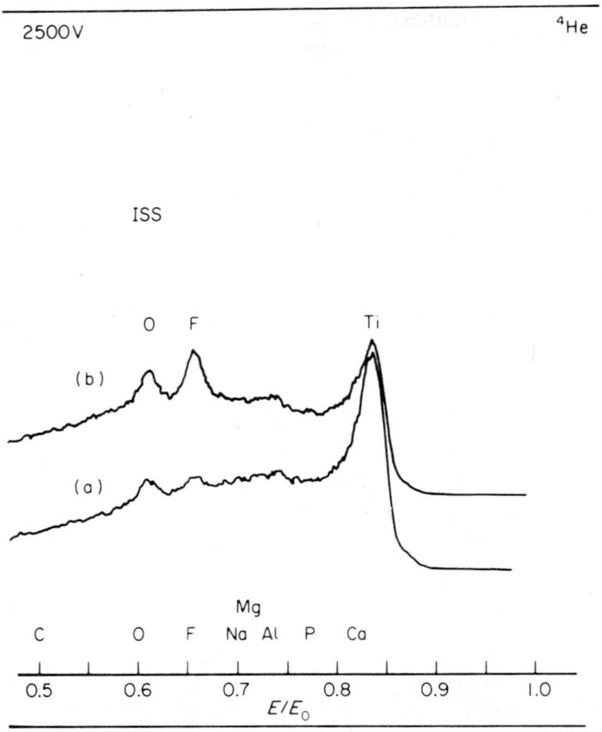

FIG. 16.19. ISS spectra from the fracture faces of a Ti6A14V alloy after adhesive bonded joint failure, from Baun.[20] (a) the Ti adherend face, (b) the commercial adhesive face. The peak due to F should not have been present, and its occurrence was traced to an F-containing contaminant inadvertently present in the anodizing solution. Analysis of these faces by AES proved to be impossible since on the adherend side the F was desorbed as soon as the electron beam hit the surface, while on the adhesive side charging effects were too severe. Analysis conditions for ISS were He^+ ions at 2.5 keV and a scattering angle of 90°.

commercial adhesive. The lower spectrum is that from the fracture face on the alloy side, and the upper spectrum is from the adhesive side. Once again an unwanted fluorine impurity is present, this time in a thin layer left behind by an inadvertent contaminant in an anodizing solution.

Adhesives are difficult to analyse by AES due to charging and beam damage, and XPS is not often useful since spatial resolution is usually required in studying fracture, so ISS is here particularly applicable. In addition the extreme surface specificity is essential in analysing fracture surfaces on which the impurity elements will normally be present in less than monolayer coverage.

16.5 Surface structural analysis

The effect of 'shadowing' by a surface atom of the atoms lying behind it along the direction of the primary ion beam has already been mentioned, and illustrated schematically in Fig. 16.10. This effect can be put to use in studying the surface arrangement, or two-dimensional structure, of atoms in a surface; the obvious complementary technique is low energy electron diffraction. The technique consists of varying the scattering geometry, either the angle of incidence or the target azimuthal angle, while keeping the scattering angle constant, and recording the scattered intensity variation. What one is doing in effect by this angular variation is sweeping the shadow cone (see Fig. 16.10) from a surface atom through the rows of substrate atoms behind it. If a light primary species such as He^+ is used, and a large scattering angle, it is possible, for example, to establish the positions of adsorbed atoms on the surface, since if the adsorbed species has an ordered surface arrangement in correspondence with the substrate structure, then in certain well defined directions the adsorbate will shadow the substrate atoms.

When a very low angle of incidence is used, it is also possible to investigate structural properties by using multiple scattering. Such a multiple scattering event is illustrated in Fig. 16.20 from Suurmeijer and Boers.[21] The scattering angle after the first event is sufficiently small that the scattered ion passes through the sphere of influence of a neighbouring atom on the surface, and suffers a second scattering event. The total scattering angle is thus the sum of the angles appropriate to each event. Figure 16.20 shows a double scattering event, but multiple events involving chains of atoms are possible if the conditions are chosen carefully. Detailed analysis of multiple scattering processes can usually be performed only by recourse to a model of the surface structure, which can again lead to the elucidation of the structure if the model predictions agree with the experimental observations.

This volume is not the appropriate place in which to cover the structural analytical aspects of ISS in any greater depth, since it was stated at the outset that discussion would be restricted to techniques of compositional analysis.

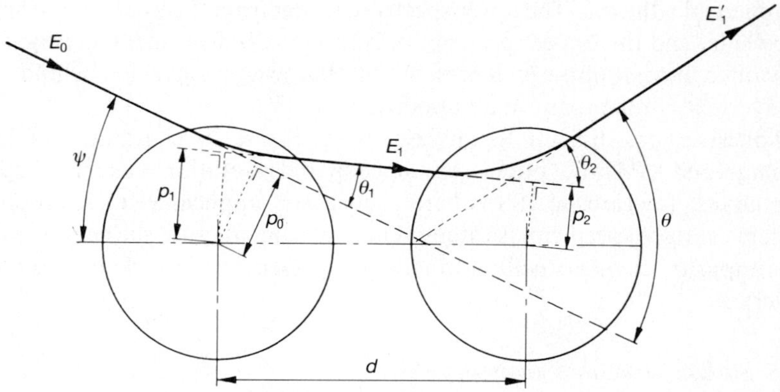

FIG. 16.20. Schematic of double scattering event sequence, as depicted by Suurmeijer and Boers.[21] In this case the scattering is assumed for simplicity to take place in the same plane, that of the page. E_0, E_1, and E_1' are the primary energy, and the energies after the first and second events, respectively, of the incident ion. θ_1, θ_2, are the laboratory scattering angles of the first and the second events, respectively, with θ the total angle, i.e. $\theta = \theta_1 + \theta_2$. d is the interatomic distance in the surface. ψ is the angle of incidence to the surface of the primary ion. p_0 and p_2 are the impact parameters for the first and second events, respectively, and p_1 is a notional parameter referred to the first scatterer by reverse extrapolation of the first scattered trajectory. For very small θ_1, $p_1 \approx p_0$.

For further reading on the structural applications the reader is referred to various excellent review articles.[22]

References

1. Armour, D. G. (1981). *Vacuum* **31**, 417.
2. Martin, P. J. and Netterfield, R. P. (1987). *Surf. Interface Anal.* **10**, 13.
3. Taglauer, E. (1985). *Appl. Phys.* **A38**, 161.
4. Brongersma, H. H., Meijer, F., and Werner, H. W. (1974). *Philips Tech. Rev.* **34**, 357.
5. Ellis, W. P. and Taylor, T. N. (1978). *J. Vac. Sci. Technol.* **15**, 679.
6. Hellings, G. J. A., Ottevanger, H., Boelens, S. W., Knibbeler, C. L. C. M., and Brongersma, H. H. (1985). *Surf. Sci.* **162**, 913.
7. Kumar, R., Mintz, M. H., and Rabelais, J. W. (1984). *Surf. Sci.* **147**, 15.
8. Beuken, J.-M. and Bertrand, P. (1985). *Surf. Sci.* **162**, 329.
9. Taglauer, E. and Heiland, W. (1976). *Appl. Phys.* **9**, 261.
10. Smith, D. P. (1971). *Surf. Sci.* **25**, 171.
11. Brongersma, H. H. and Mul, P. M. (1972). *Chem. Phys. Lett.* **14**, 389.
12. Van den Berg, J. A. and Armour, D. G. (1981). *Vacuum* **31**, 259.
13. Molière, G. (1947). *Z. Naturforsch.* **2a**, 133.
14. Van Der Veen, J. F. (1985). *Surf. Sci. Rep.* **5**, 199.
15. Hagstrum, H. D. (1954). *Phys. Rev.* **96**, 336; and (1961). *Phys. Rev.* **123**, 758.
16. Brongersma, H. H. and Buck, T. M. (1976). *Nucl. Instrum. Meth.* **132**, 559.
17. Nelson, G. C. (1976). *J. Vac. Sci. Technol.* **13**, 974.
18. Taglauer, E. and Heiland, W. (1975). *Surf. Sci.* **47**, 234.

19. Pitts, J. R., Thomas, T. M., Czanderna, A. W., and Passler, M. (1986). *Appl. Surf. Sci.* **26**, 107.
20. Baun, W. L. (1981). *Surf. Interface Anal.* **3**, 243.
21. Suurmeijer, E. P. Th. M. and Boers, A. L. (1974). *Surf. Sci.* **43**, 309.
22. e.g. Taglauer, E. (1985). *Appl. Phys.* **A38**, 161; Armour, D. G. (1981). *Vacuum* **31**, 417; Van den Berg, J. A. and Armour, D. G. (1981). *Vacuum* **31**, 259.

17

Ion excitation: static secondary ion mass spectrometry (SSIMS)

SSIMS ranks with XPS and AES as one of the three principal techniques currently used in surface analysis, although judging from the published applications its usage is still probably less widespread than that of the other two. Since its origination by Benninghoven[1] in 1969 coincided roughly with that of AES, it must be regarded as a mature technique, although as will be seen below there are even now some considerable problems in its interpretation. As remarked earlier (Section 2.3.5), 'static' SIMS is at one end of the spectrum of SIMS-type techniques, while 'dynamic' SIMS is at the other, and the two should not be confused. Unfortunately, they often are, since the published literature frequently uses the general name SIMS for both. However, it is usually sufficient to look at the stated experimental conditions and mode of operation in any one paper in order to decide which variation is being used. SSIMS operates at incident ion current densities of the order of 1 nA cm^{-2}, while dynamic SIMS uses much higher current densities, as high as 1 mA cm^{-2}. Also the latter usually operates in either an imaging or a scanning mode, providing secondary ion images, whereas scanning SSIMS is only just appearing.

17.1 Operation

Positive ions from a noble gas, usually argon, at energies between 0.5 and 4.0 keV and currents of tenths of a nA, irradiate a surface over an area of between 1 and 10 mm^2. Secondary ions emitted from the surface, either positive or negative, are accepted into, and mass analysed, by a quadruple mass spectrometer or, more recently, a time-of-flight mass spectrometer. Typical ion sources for SSIMS would be the low energy types already described under Section 3.2.3, e.g. either the hot filament type of Fig. 3.12 or the duoplasmatron of Fig. 3.13. However, many workers, particularly in the early days of SSIMS, designed and constructed their own sources. An outline of the principles of operation of a quadrupole mass spectrometer has been given in Section 3.3.2.

The above description of the way in which SSIMS is operated is very basic and does not mention two important additional considerations. One of these is the desirability of mass filtering the incident ion beam, and the other is the necessity of energy selecting the secondary ion emission.

17.1.1 *Primary beam mass filtering*

Some ion sources, e.g. the discharge types such as the duoplasmatron, when operated at high currents, produce a proportion of ions other than the singly charged species. For example, Ar^{2+} as well as Ar^{+}. In addition, during adsorption experiments in which reactive gases are admitted to the system, it is possible for foreign ions to be produced in the source and find their way down the accelerating system to the sample surface. If the pressure of an admitted gas becomes too high, it is also possible for the performance of the source to be affected owing to changes in the space charge conditions inside it. All these possibilities will change the characteristics of the primary ion beam and therefore affect the SSIMS analysis, either by altering the physical processes leading to secondary ion ejection or by implanting unwanted impurities in the surface, or both.

In general this is not a critical problem in most SSIMS applications, since the concentrations of impurity species likely to be implanted will usually be below the detection limits, but there are situations, such as in depth profiling, where it is desirable to operate with a pure incident ion beam. This is achieved by mass filtering the primary ions. For this purpose the universally employed device is a Wien filter, which is a straight-through velocity filter consisting of crossed electrostatic and magnetic fields. Since its sole function in this case is to remove all masses except $^{40}Ar^{+}$ from the beam, the Wien filter can be made relatively crude, that is, with a low mass resolution but with a high transmission.

A Wien filter, often called a 'velocity analyser', or an '$E \times B$ filter', is shown schematically in Fig. 17.1 from Wilson and Brewer.[2] It consists of long parallel magnetic pole pieces, often shaped, along whose axis the magnetic field is uniform; this is the z axis in the figure, and the y axis is normal to the faces of the pole pieces. In the x direction, an electrostatic potential is set up between two parallel plates, arranged so that the equipotential line is along the filter axis, i.e. along the z axis. The principle of the analyser is that the electrostatic field normal to both the magnetic field and the filter axis is adjusted to counterbalance the Lorentz force on a charged particle so that particles of a particular velocity pass straight through the analyser. Particles with other velocities, higher or lower, will be deflected and lost from the beam. If E and B are the electrostatic and magnetic fields, respectively, and v the general velocity of particles entering the filter, with energy eV and mass m, then the velocity v_0 for which the above force balance occurs will be equal to:

$$v_0 = E_0/B_0 \tag{17.1}$$

where E_0 and B_0 are the particular values of E and B. The velocity of the transmitted particle is given in terms of its energy eV_0 and its mass m_0 by

$$v_0 = (2eV_0/m_0)^{1/2}. \tag{17.2}$$

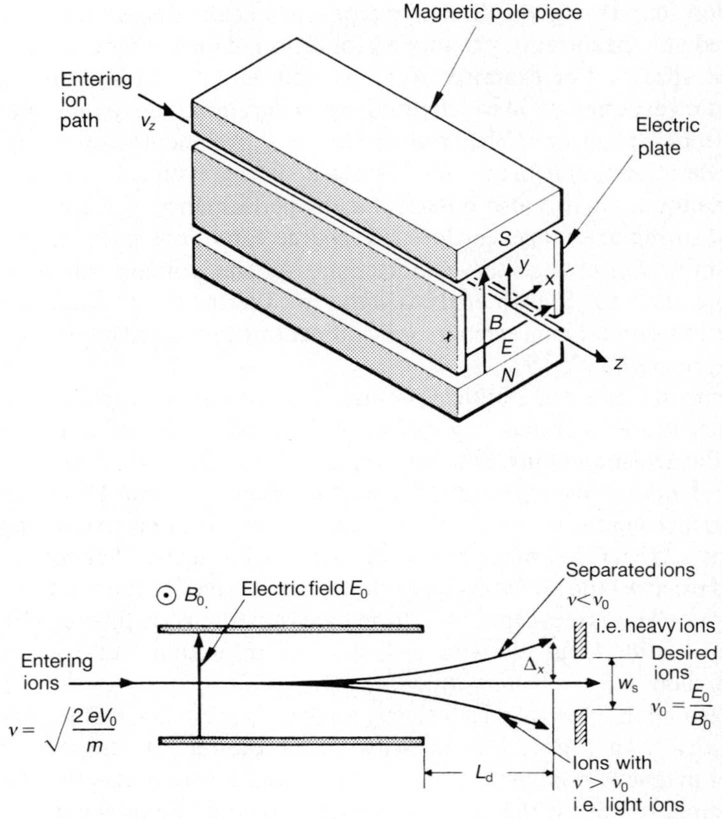

FIG. 17.1. Schematic of a Wien filter or velocity analyser, as described by Wilson and Brewer.[2] Parallel to the z axis, the ion path, are two long magnetic pole pieces with faces normal to the y axis, and two electrostatic plates with faces normal to the x axis. The electrostatic field normal to both the axis and to the magnetic field direction is adjusted to balance the Lorentz force from the magnetic field acting on the charged particles, in order that particles of a particular velocity pass straight through while those with other velocities are deflected and lost. The lower diagram illustrates the principle, and the governing expressions (1) to (4) are given in the text. The Wien filter is also called an '$E \times B$ filter'.

The lower part of Fig. 17.1 shows schematically the particle paths for transmission or deflection. If the length of the filter is L_m, then at a distance L_d from the exit end, the deflection Δx of ions of mass $m_0 \pm \Delta m$ is given by

$$\Delta x = (\Delta m \beta L_m / 4m_0)(L_m + 2L_d) \tag{17.3}$$

where $\beta = \omega / v_0$, ω being the cyclotron radian frequency of the ion. Relation (17.3) holds for the approximation that $\Delta m / m_0 \ll 1$, i.e. the separation of ions of closely similar masses. If this is not so, then $\Delta m / m_0$ should be substituted by

$$\Delta m/m_0 = 2\{1-(m_0/m_d)^{1/2}\} \tag{17.4}$$

where m_d is the mass of the deflected ion, i.e. $m_d = m_0 \pm \Delta m$.

Because of the necessity to correct for field distortions due to the finite dimensions of the filter, Wien filters are in practice not as simple in design as that shown in Fig. 17.1, and the reader is referred to papers dealing with SSIMS and dynamic SIMS instrumentation for details of the correction procedures.

An example of the use of a Wien filter to purify the incident beam is given in Fig. 17.2 in which the arrangement of the quadrupole-based SSIMS instrument of Magee *et al.*[3] is shown schematically. With this arrangement not

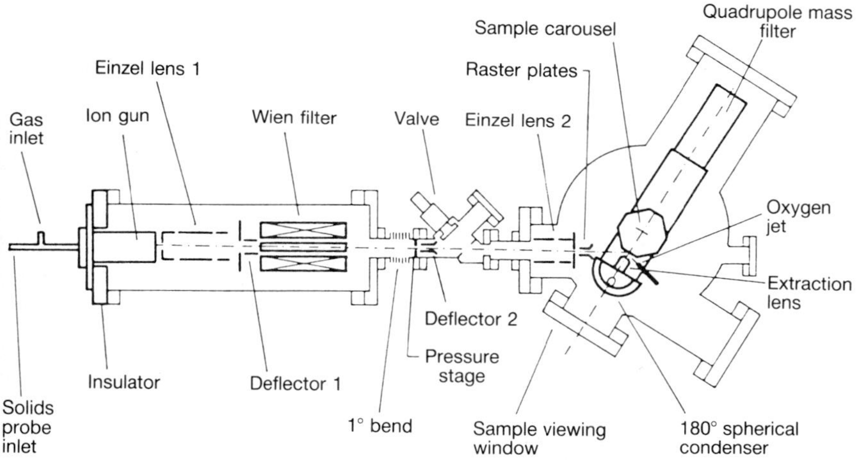

FIG. 17.2. Experimental arrangement used by Magee *et al.*[3] for SSIMS. The Wien filter (see Fig. 17.1) reduces the content of the primary beam to $^{40}Ar^+$ only, and the emergent beam is then deflected through 1° before entering the chamber, so that energetic neutrals are removed from it. The mass analysed beam is then focused onto the sample by an einzel lens, and rastered through a maximum speed of 80 frames s^{-1} over an area greater than 1 × 1 mm.

only is the content of the ion beam reduced to $^{40}Ar^+$ only, but a deflection of the purified beam through 1° immediately after leaving the source region ensures that energetic neutrals do not strike the specimen as well.

17.1.2 *Secondary ion energy selection*

In the earliest days of SSIMS, the secondary ion emission was accepted directly into the quadrupole mass spectrometer with virtually no energy selection, as seen in the arrangement of Benninghoven *et al.*[4] (in 1974) of Fig. 17.3. At the time that did not matter very much since the early instruments used quadrupoles of somewhat low resolving power (e.g. 60 in ref. 4). As the technique developed and higher mass resolution became necessary, it was found that the wide spread of energies of the secondary ions limited the resolution achievable. Examples of some typical secondary ion

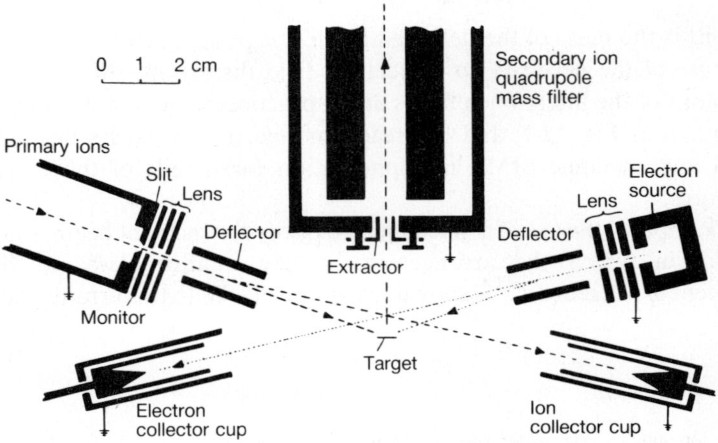

FIG. 17.3. An early experimental arrangement for SSIMS analysis, by Benninghoven *et al.*[4] Secondary ions were extracted directly into the quadrupole mass spectrometer without any energy selection. Dashed lines: ion beam paths. Dotted lines: electron beam paths.

energy distributions are given in Figs. 17.4 and 17.5 from the work of Dawson,[5,6] for titanium and molybdenum surfaces, respectively, under various surface conditions. Notice several features about these distributions. Firstly, for some masses, usually of type M^+ or M_n^+, there is a long tail in the distribution extending to many tens of eV. Secondly, for other masses, often those containing one or more oxygen atoms, the distributions are much sharper, with little tailing. Thirdly, there are shifts in the distribution peak energies between various masses. Now for fixed operating conditions of frequency, quadrupole rod length, and field radius between the rods, the mass resolution in terms of the separable mass difference, is proportional to the maximum axial and radial ion injection energies into the quadrupole (Paul *et al.*[7]). Thus to obtain acceptable mass resolution the ion energy should be as low as possible; high energy ions cannot be analysed with sufficient resolution. On the other hand neither the median ion energy nor the energy spread (called the band width) accepted into the quadrupole should be so low that the overall transmission, and hence sensitivity, of the detecting system deteriorates significantly. In other words, in view of the situation shown in Figs 17.4 and 17.5 where different masses peak at different energies in their distributions, a compromise must be chosen between mass resolution and a bandwidth that includes as much of the areas under the distributions as possible. Wittmaack[8] has shown that ions with energies in the range of 4–10 eV can be analysed with a reasonable resolving power $M/\Delta M$ of 200, based on the peak width at 10% of peak height, and with a reduction in transmission to an acceptable 10% of total. Since most operators are concerned with sensitivity rather more than with mass resolution, a typical band width used is about 20 eV, centred on

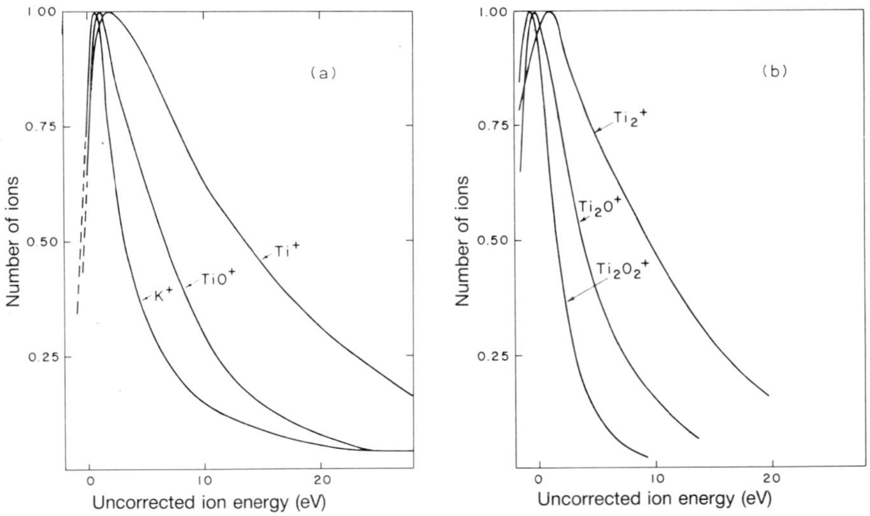

FIG. 17.4. Energy distributions of secondary ions emitted from a titanium surface, recorded by Dawson.[5] (a) 'Clean' titanium surface. (b) Titanium exposed to CO_2. Note that the Ti^+ and $Ti_2{}^+$ distributions have long tails extending over many tens of eV, while the oxygen-containing ion distributions are much sharper.

about 10 eV. Such conditions do not impose any great restrictions on the design of an energy selector, and most of those used are indeed quite crude in their operation compared, say, to the operation of the energy analysers described in chapter 3.

The majority of energy selectors are in fact miniature variations on the various well known energy analysers. Thus dispersive analysers in various geometries have been popular, particularly the 90° spherical analyser, an example of which is given in Fig. 17.6, from the paper by Gnaser *et al.*[9] The latter has a mean radius of 2.75 cm and an energy resolution of about 3%, and is accompanied on the input and output sides by multi-element lenses, which act to maintain focusing of the secondary ion beam on the entrance apertures of the analyser and quadrupole as the pass energy of the energy selector is varied. The ioniser between the sample and the first lens is there to ionize neutral atoms sputtered from the sample in the technique of Secondary Neutral Mass Spectrometry (SNMS) to be described in the next chapter. With the arrangement in the figure, ions from molecules in the residual gas can be discriminated against with respect to postionized neutrals by off-setting the energy selector against the purely thermal energies of the gas molecules.

Although retarding-field analysers have not been used in the electron spectroscopies for many years, because of their poor signal-to-noise characteristics compared to those of dispersive analysers, their very high transmission makes them attractive for energy selection in secondary ion

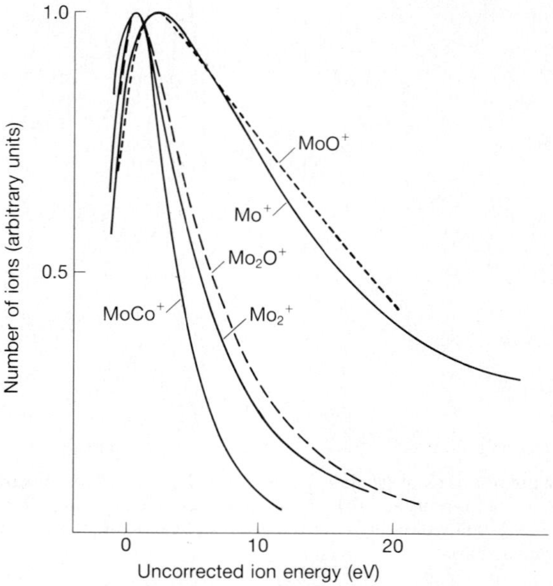

FIG. 17.5. Energy distributions of secondary ions from a molybdenum surface after exposure to CO to saturation coverage, recorded by Dawson.[6] Again some masses have long tails in their distributions, others short tails, but the distinction between Mn^+ ions and oxygen-containing ions is not as obvious as for titanium in Fig. 17.4. Notice in both figures that the maximum in the energy distributions show considerable shifts.

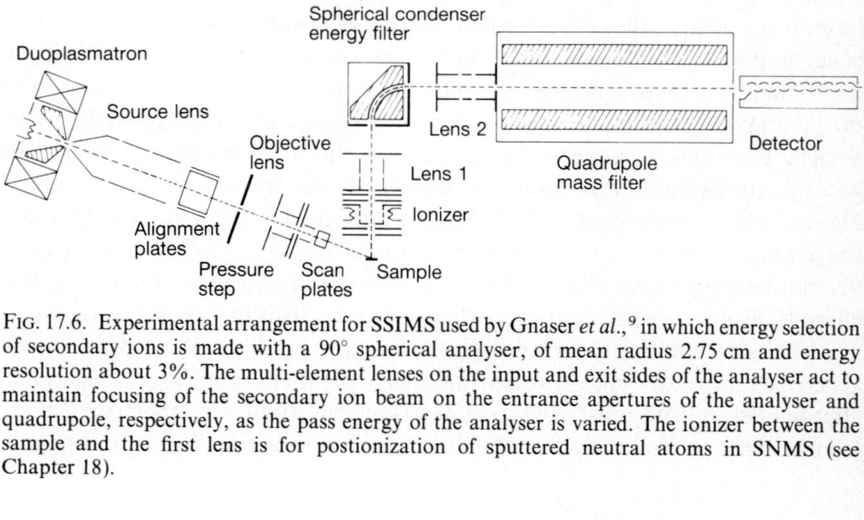

FIG. 17.6. Experimental arrangement for SSIMS used by Gnaser *et al.*,[9] in which energy selection of secondary ions is made with a 90° spherical analyser, of mean radius 2.75 cm and energy resolution about 3%. The multi-element lenses on the input and exit sides of the analyser act to maintain focusing of the secondary ion beam on the entrance apertures of the analyser and quadrupole, respectively, as the pass energy of the analyser is varied. The ionizer between the sample and the first lens is for postionization of sputtered neutral atoms in SNMS (see Chapter 18).

analysis. In particular the retardation/acceleration design originally intended for electron spectroscopy by its originator Staib[10] has been found[11,12] to be useful. The version employed by Krauss and Gruen[12] is shown in Fig. 17.7. Like the Staib analyser it consists of two pairs of spherical grids, the first pair centred on the point of emission of ions from the sample and the second on the exit aperture. The first pair impose a retarding field on the emitted ions, the first grid being at ground potential to maintain a field-free region around the sample, and the second biased positive with respect to it. The second pair of grids provides a radial accelerating field that picks up the ions passed by the first pair with nearly zero kinetic energy and injects them into the quadrupole. Neutrals and high energy ions are little affected and do not reach the aperture. A disc-shaped central stop blocks straight-through transmission. The analyser can accept ions from a larger area, $\approx 1\ cm^2$, while maintaining adequate energy resolution ($<1\%$), and is thus ideal for SSIMS. The grid pairs are symmetric, with radii of 6.25 and 8.9 cm, and are constructed from mesh of 90% transparency. With an acceptance half-angle of 15°, the energy selector transmission is 1.5% when all the ions entering it are collected.

17.1.3 *Time-of-flight mass spectrometers*

Although quadrupole mass spectrometers are almost universally used in SSIMS, having characteristics that make them ideal for most applications, they do have some drawbacks that prevent a ready extension of their applicability to the detection of high mass numbers. The principal drawback is the low transmission, of the order of 10^{-4}, which in fact drops off with increasing mass number. This factor, combined with the requirement of using very low primary ion current densities, has meant that attempts to carry out analyses of large organic molecules on surfaces with quadrupole-based systems have been unsuccessful. For similar reasons the maintenance of adequate mass resolution in the high mass (e.g. $> \sim 200$) range is also impossible in instruments of conventional design and convenient size. In principle an increase in transmission can be gained by making the quadrupole much larger, but practical limitations set in very quickly.

To improve the situation for organic surface analysis, Benninghoven and co-workers[13] have developed a time-of-flight (TOF) mass spectrometer for use in SSIMS. Their instrument is shown schematically in Fig. 17.8. The sample (6) sits in the middle of a spherical chamber within which it can be manipulated and transferred in the usual way. To the left-hand-side of the diagram (items numbered 1–5) is the ion optical arrangement for producing the primary ion beam. The source (1) is of electron impact type with magnetic focusing, and provides a continuous beam of argon ions. The ions are accelerated through 8 keV into a parallel plate deflection system (2), with focusing via slit electrodes at the entrance. By applying a short ($\sim 1\ \mu s$) 5 keV pulse to the deflection system, a packet of ions is removed from the continuous

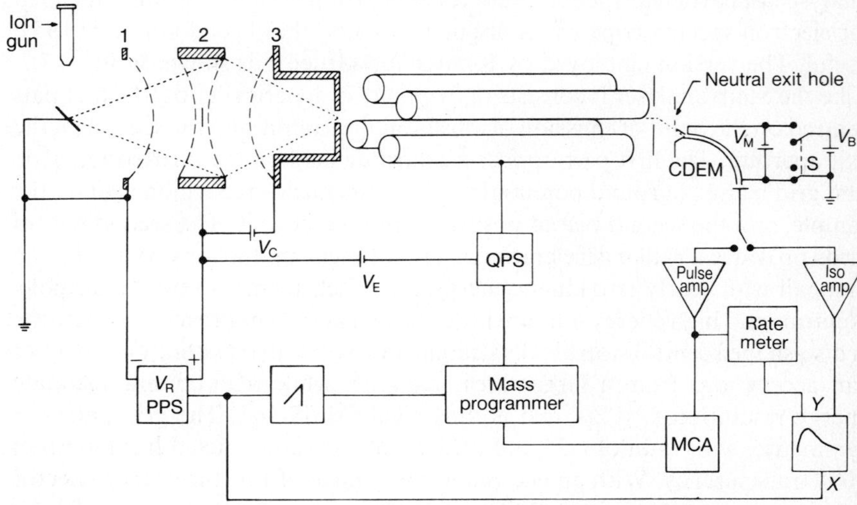

FIG. 17.7. Experimental arrangement for SSIMS used by Krauss and Gruen,[12] in which energy selection of secondary ions is performed with a retarding-field analyser of the Staib[10] type. Two pairs of spherical grids are arranged back-to-back, the first pair (1, and left half of 2) centred on the point of emission of ions from the specimen, and the second pair (right-half of 2, and 3) centred on the exit aperture. The first grid is at earth potential, and the second biased positive with respect to the sample, thus providing a radial retarding field. The second pair of grids provides a radial accelerating field, injecting ions into the quadrupole. A stop in the middle blocks any straight-through transmission of neutrals. The analyser accepts ions from a large area, ≈ 1 cm^2, and has a resolution of $\approx 1\%$, so is eminently suitable for SSIMS. Transmission of the grid system is 1.5%.

beam, and accelerated to ≈ 12 keV through a hole in the right-hand grounded plate. As it leaves the hole the ion packet is elongated across its path, and this undesirable elongation is reduced by a bunching magnet (3) into a much smaller dimension, although with an increased angular spread. The magnet also removes any ions of different masses since they would not be focused at the exit slit. The bunched packet is then focused onto the sample by lens (5) over an area of about 0.5×0.7 mm. Owing to the bunching, the pulse width is also reduced at the sample to about 7 ns measured normal to the beam. A monitoring probe (4) checks the pulse shape. In normal operation the sample is not normal to the ion beam, and the pulse length is about 10 ns; although the average current can be as high as 1 nA, detection is in the single-ion counting mode, and normally the current would be of the order of 1 pA.

Mass analysis is performed in the TOF arrangement above and to the right of the central chamber, comprising components 7 to 13. The ionic time-of-flight is measured with a combination of electrostatic toroidal deflector (8) and of two linear drift tubes (7) and (9). The deflector has a mean radius of 20 cm and a total deflection angle of 164°, while the total drift length is 157 cm. After passing through an aperture at the end of the second drift tube (9), the ions are

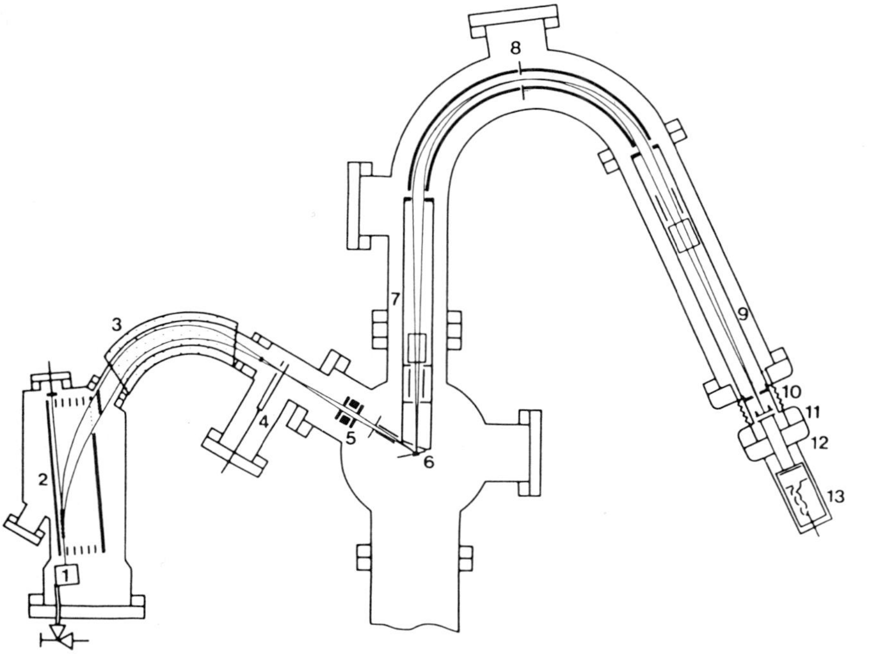

FIG. 17.8. SSIMS system using a time-of-flight (TOF) mass spectrometer, described by Steffens *et al.*[13] Primary ions from the source (1) are accelerated to 8 keV in the parallel plate deflection system (2). When a short (~1 μs) pulse of 5 keV is applied to the deflection system, a packet of ions leaves the continuous beam and is accelerated to ≈12 keV through an aperture in the right-hand grounded plate. A bunching magnet (3) imposes the right shape on the packet as it passes round the focusing system towards the specimen; ions of different masses are also removed. Before it reaches the specimen (6), the ions pass through a slit (4), which monitors the shape of the packet. A lens (5) carries out final focusing onto the specimen, into an area approximately 0.5 × 0.7 mm.
The TOF mass spectrometer consists of elements (7) to (13), and consists of an electrostatic toroidal deflector (8) and two linear drift tubes (7) and (9). The total length is 157 cm, and the total deflection angle 164°. Ions are accelerated across the gap (10) to about 20 keV, and detected by a channel-plate (11)-scintillator (12)-photomultiplier (13) combination. Application of the high voltage pulse to the deflection system (2) triggers a 100 Mc clock, which is switched off at the end of a pulse. The flight time is determined by a fast discriminator coupled to the photomultiplier output, the pulses being normalized and synchronized with the clock.

accelerated to 20 keV across the gap (10), thereby increasing the detection efficiency and thus the mass range. The first stage of detection is a single channel plate (11), which is followed by scintillator (12), the latter then being coupled to a photomultiplier (13) via a light pipe.

When the high voltage pulse is applied to the plate deflection system in the source, a 100 Mc clock is triggered. The flight time is determined by a fast discriminator coupled directly to the photomultiplier output, the normalized discriminator pulses being counted and synchronized with the clock. The clock is switched off at the end of the pulse. Up to two counts can be processed in each 10 ns pulse, and counts are accumulated in a 16k × 24 bit memory array. For low intensity high mass peaks accumulation times of up to 5 min might be necessary.

Several advantages of the TOF spectrometer, some of them self-evident, must be pointed out. They are the following:

1. Provided the flight path is long enough (in practice >100 cm) and the timing circuitry can cope, the mass range can be very large, probably $>10\,000$ amu.
2. There are no defining slits needed, and so the transmission is high. A typical overall figure of 10% is many orders of magnitude greater than that of a quadrupole mass spectrometer.
3. In any one pulse the *whole* mass spectrum from the molecular species on the surface is available. Detection of all masses is simultaneous, again unlike the modes of operation of other types, in which the spectrum has to be scanned. This capability coupled with the high transmission listed under (2) means that in practice the increase in sensitivity compared to that of a quadrupole is some five orders of magnitude!
4. In SSIMS applications, where the ions are being produced on a planar equipotential surface, the mass resolution can be good. Resolving powers as high as 3000 have been used.

Examples of the extended mass range available and the excellent resolution at high masses, are given in Figs 17.9 and 17.10 respectively, from the work of Bletsos *et al.*[14] The spectrum of polydimethoxysilane shows masses resolved beyond 9000 amu, while in Fig. 17.10 for polystyrene, the 104 mass repeat unit is very clearly observable to beyond 7000 amu.

Since the initiation of TOF SSIMS by Benninghoven and collaborators various other instruments based on the same principle have appeared, some[15,16] using a high spatial resolution liquid metal ion source as the source of primary ions, and commercial instruments are available.[15]

17.2 Spectral identification

In SSIMS spectral interpretation is probably the least difficult aspect of the technique, since it is identical to that of any other mass spectrometric technique. The spectrum consists of peaks of varying intensity appearing at various mass numbers. The masses can be identified as arising either from the substrate material itself, or from adventitious contamination and impurities on the surface, or from deliberately introduced molecular or other species adsorbed on the surface. Identification is made either on the basis of atomic weight and charge, in the simplest case of ions of the type M_i^{n+}, where n and i can each equal 1, 2, 3, . . . ; or of molecular weight and charge, in the much more complex case of ions of the type $M_iX_j^{n+}$, where again n, i, and j can each equal 1, 2, 3, ... Fortunately under the conditions of SSIMS, the most

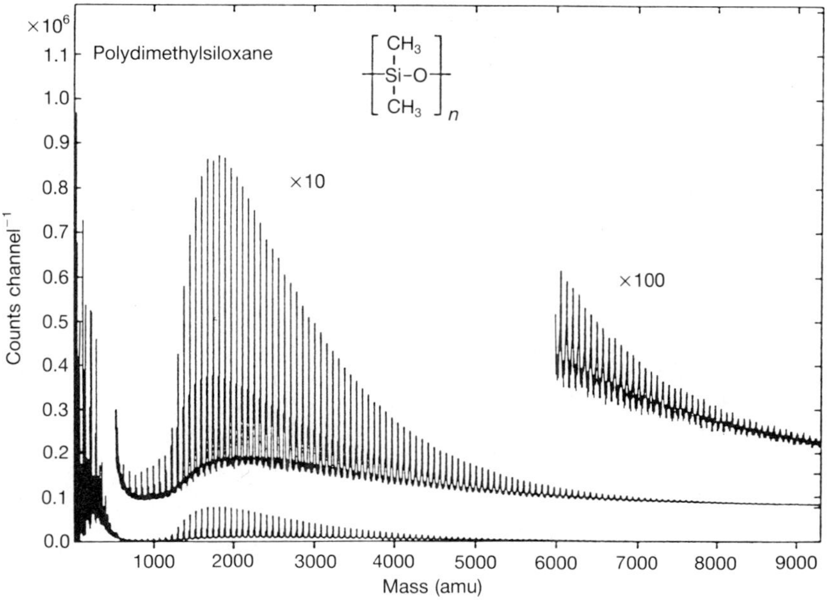

FIG. 17.9. Example of the extension of the mass range available with a TOF mass spectrometer, compared with that of a quadrupole mass spectrometer, whose transmission at high masses is poor, from Bletsos *et al.*[14] Masses from polydimethoxysilane extend to beyond 9000 amu with clear visibility.

prominent secondary ions have $n=i=1$. Since SSIMS is a mass sensitive technique there are of course also isotopic effects to be considered; many elements have only one principal isotope, but there are also many in which the natural isotopic abundance implies potential complications in the mass spectrum. In general the isotopic contributions to a spectrum have been ignored, since much research using SSIMS has tended to concentrate on the behaviour of selected peaks as a function of surface treatment, e.g. oxidation, but as SSIMS becomes an increasingly quantitative analytical technique then peaks arising from minor isotopes must be taken into account. A good example of spectra showing the potential problems that might occur is that in Fig. 17.11, from Benninghoven *et al.*[17] The positive and negative SSIMS spectra show, respectively, peaks corresponding to the isotopes of K, Ni, and of Cl, in their correct natural abundances. In the positive ion spectrum can also be seen peaks arising from M_2^+ ions, where the isotopic effect introduces additional complications. These spectra are on logarithmic scales, which exaggerate the minor contributions, but it is clear that an unravelling problem exists.

Other problems of overlapping spectra have also always existed in mass spectrometry, and are normally resolved by using big mass spectrometers of

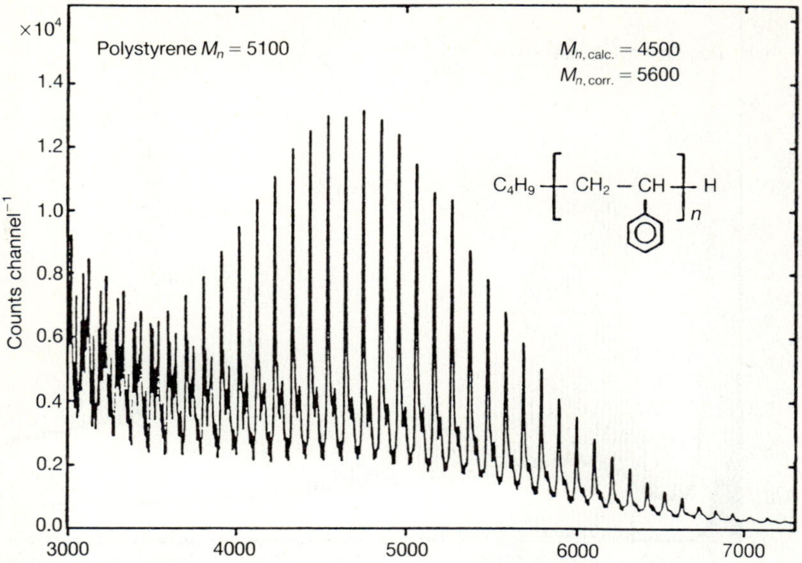

FIG. 17.10. Example of the mass resolution routinely available from a TOF mass spectrometer such as that in Fig. 17.8, from Bletsos *et al.*[14] The 104 mass repeat unit from polystyrene can be observed to beyond 7000 amu.

large resolving power. With the quadrupoles used conventionally in SSIMS, however, that is not possible, and the identification of a particular peak with one or other of the possible attributions must be made on grounds of other information about the surface (e.g. from another technique or from the probable chemical condition). Alternatively one can make use of some of the spectral stripping procedures available in the data processing packages provided by some manufacturers. These packages depend for their operation on the existence of a large data base of stored standard spectra, and on an unambiguous starting point, i.e. a major peak that can be positively identified. If the data base does not already include a standard spectrum from a component on the surface, or if there is no unambiguously identifiable major peak, then the SSIMS analysis must itself be ambiguous, but fortunately that situation is not common.

17.3 Theory

The interaction of a relatively heavy particle, such as an argon ion or atom, possessing a kinetic energy of some hundreds or thousands of eV, with a solid surface, leads to the most complex sequence of events occurring in any surface analytical technique. Following such an interaction it is possible to detect the emission of a greater variety of secondary particles than in any other

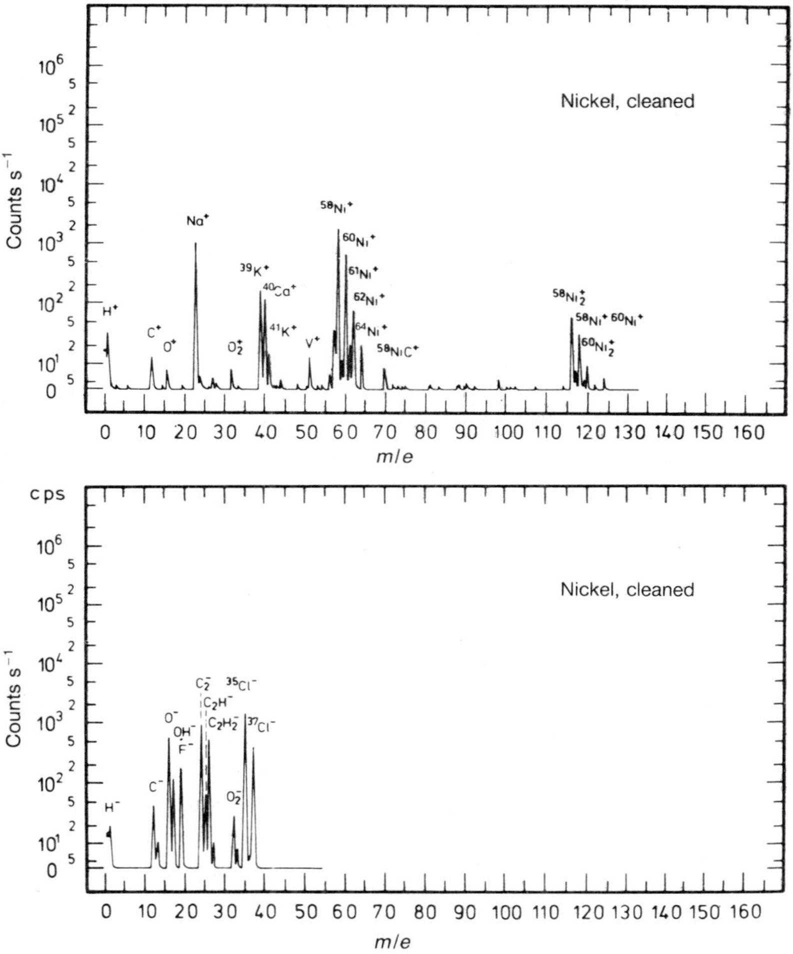

FIG. 17.11. Positive and negative secondary ion spectra from a 'cleaned' nickel target; cleaning was carried out by the normal procedure of bombardment/heating cycles. Recording conditions were those of 3 keV primary argon ion energy, 5×10^{-9} A primary current, into an area of 0.1 cm^2. Note the potential complication in interpretation of SSIMS spectra due to the appearance of isotopic masses, in both the M^+ and in the $M_2{}^+$ spectra. From Benninghoven *et al.*[17]

technique. The variety includes positive and negative secondary ions, backscattered primary ions, neutral atoms and clusters in various stages of excitation, low energy secondary electrons and Auger electrons, visible light and (if the primary energy is high enough) X-rays. Each combination of incident and secondary particle gives rise to an individual surface analytical technique, most of which have already been described in this volume.

With so many different types of particle emitted during the interaction or bombardment process, it is self-evident that the process must be causing considerable damage in the surface regions of the solid. This is hardly surprising. A heavy energetic particle cannot be stopped immediately by the first atoms it encounters at the surface, but continues on into the surface and eventually comes to a halt as a result of losing energy via elastic scattering off atomic cores and inelastic scattering off electronic shells as well as off the sea of free electrons. The depth at which the ion stops depends on its mass and that of the material of the solid, on its energy, and on its angle of incidence. Along its path to its final resting-place the ion, as a result of the scattering interactions, displaces some host atoms from their normal positions in the lattice. As they recoil these atoms in turn displace other atoms, which also in turn displace others, and so on, resulting in a complex sequence of collisions, which will result in some atoms becoming permanently displaced from their equilibrium lattice positions, while others will be knocked temporarily out of place but will then return elastically afterwards. This sequence is called a collison cascade. As this cascade spreads out from the path of the intruding ion, its effect eventually reaches the atoms actually in the surface layer; if there is still sufficient energy left in the collisional interactions at that stage then the bonds binding atoms, or groups of atoms, to the surface, may be broken, and material will leave the surface. The material is then said to be sputtered from the surface. Some sputtering may also occur by direct interaction of an incoming particle with a surface atom, but this contribution is thought to be minor. A schematic picture of a collision cascade and other processes is shown in Fig. 17.12, from Pivin.[18]

The collision cascade model of the sputtering process, described above, is due to Sigmund.[19] It has been able to explain total yields and energy distributions of sputtered material, and their dependence on primary mass and energy, angle of incidence, and target material characteristics. Strictly speaking it applies only to elemental structureless targets, but has found applicability over a wide range of target compositions and structures, and is now supported by a substantial body of experimental evidence. There therefore appears to be a general understanding of the collision processes within the target and of the mechanism of the ejection of material. However, there is nothing in Sigmund's theory that makes any statement about the charge state of sputtered material; he was concerned only with the process in the solid leading to removal of material and not with the processes that might lead to ionization of some of it. Since the theory dealing with total sputtered yield is on a good, fairly quantitative, basis it will not be dealt with any further here but instead the much more parlous state of the theory of secondary ion formation will be examined. An excellent description of the various factors involved in sputtering can in any case be found in the volume by Benninghoven *et al.*[20]

It was remarked by Vasile[21] in 1984 that 'The difficulties encountered in

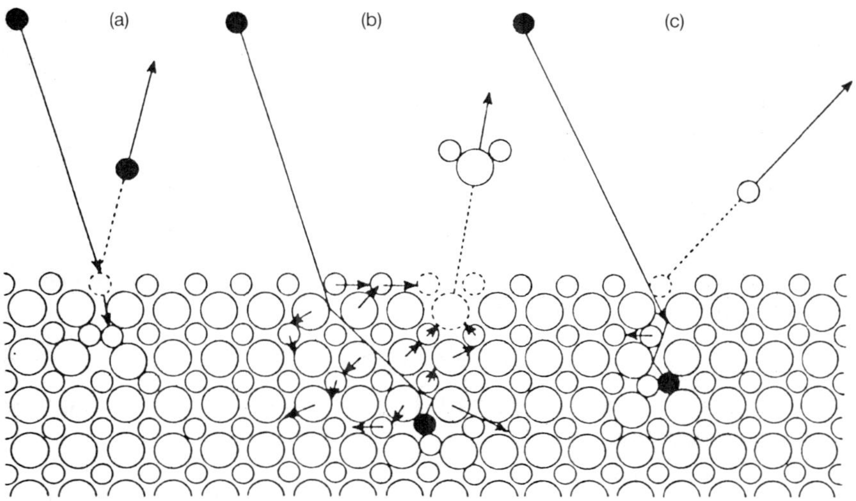

FIG. 17.12. Schematic of the various possible heavy particle emission processes that can take place in and at a solid surface on ion bombardment, as depicted by Pivin.[18] (a) Ionic backscattering, as described in Chapter 16 on ISS. (b) A possible collision cascade sequence leading to the ejection of material from the surface, in the process called sputtering. (c) The process known as recoil sputtering.

obtaining even rudimentary agreement between theory and experiment in secondary-ion emission are formidable'. The situation does not seem to have changed significantly since then. It is precisely because of these difficulties that so many theories and models of secondary ion formation have been propounded and still exist; no one theory or model has yet stood out from all the others as being universally acceptable simply because the experimental tests are inadequate. Part of the problem stems from the complexities of the theories themselves, which means that approximations have to be made in order to achieve any experimental comparison at all. Thus any agreement found between theory and experiment may be fortuitous because of the degree or nature of the approximation. Under these circumstances, which still exist, it would be tempting fate to single out any one proposed mechanism and describe it in a volume of this nature as representing the accepted theory. Instead brief resumés of the ideas that have been put forward to explain secondary ion formation will be given, with some indication of their apparent success and current acceptance.

17.3.1 *Local thermodynamic equilibrium (LTE) model*

Surprisingly, the model that has had the widest application to date in terms of attempts to use it quantitatively is also that with the flimsiest physical basis. This is the LTE model proposed by Andersen and Hinthorne.[22] It takes no account at all of any detailed description of electronic processes leading to ionization, but regards the surface region involved in the sputtering process as equivalent to a dense plasma in local thermodynamic equilibrium. The plasma consists of a cloud of sputtered atoms and ions, molecules and electrons whose spatial extent is not well defined. Under the assumption of LTE, one can write for the reversible dissociation reaction between neutral atoms M^0, positive ions M^+ and electrons.

$$M^0 \rightleftarrows M^+ + e^-. \tag{17.5}$$

If the respective concentrations are n^0, n^+ and n_e, then the dissociation constant associated with (17.5) is

$$K^+ = n^+ n_e / n^0. \tag{17.6}$$

The constant K^+ can be derived from statistical mechanics via the Saha–Eggert equation

$$K^+ = \frac{2P^+(T)}{P^0(T)} (2\pi m_e kT/h^2)^{3/2} \exp\{-(I_p - \Delta E)/kT\} \tag{17.7}$$

where $P^+(T)$ and $P^0(T)$ are the temperature-dependent internal partition functions for ions and neutral atoms, respectively, the factor 2 being the corresponding function for electrons, m_e is the electronic mass, T the absolute plasma temperature, I_p the first ionization potential of the neutral atom and ΔE the reduction in the ionization potential due to Coulomb interaction in the plasma. h and k have their usual meanings.

The ionization coefficient, n^+/n^0, for ionization during sputtering is thus

$$n^+/n^0 \propto n_e^{-1} \frac{P^+(T)}{P^0(T)} T^{3/2} \exp\{-(I_p - \Delta E)/kT\}. \tag{17.8}$$

The ionization potential reduction ΔE is itself given by

$$\Delta E = e^3 (8\pi n_e / kT)^{1/2}. \tag{17.9}$$

Similar equations can be set up for ionization leading to negative ions. The ionization coefficient can be calculated from eqn (17.8), using the electron density and plasma temperature as variable parameters. The conceptual problem that arises when effective plasma temperatures are extracted in this way by using experimentally observed quantities in eqn (17.8) is the magnitude of the temperatures. To obtain adequate fits values of T in the range 5×10^3–10^5 K have to be used; the precise value of T is also very sensitive to the chemical nature of the matrix. These and a variety of other experimental observations indicate that the basic assumption in the model of local thermal equilibrium is untenable. Intuitively, the picture of the sputtered region as a dense plasma is difficult to accept. Each termination of a collision cascade at a surface resulting in the emission of a great variety of particles must constitute a highly individual event, in which the various mechanisms should be examined individually. Nevertheless, as will be described below, the LTE model has had some success in quantification. Wittmaack[23] attributes this to the presence of the exponential Boltzmann factor in eqns (17.7) and (17.8), which in fact appears in other models as well.

17.3.2 *The kinetic model*

One of the early steps taken to try and include a description of an atomic process in the ionization mechanism was that of Joyes,[24] who put forward the kinetic model. It applies only to collision processes occurring within the target after penetration of the primary beam, and depends on the well known mechanism of electron promotion. An example of such a mechanism has already been given in Chapter 13, for the production of Auger electrons as a result of impact by ions of low to medium energy.

The model can be understood by reference to Fig. 17.13, which shows

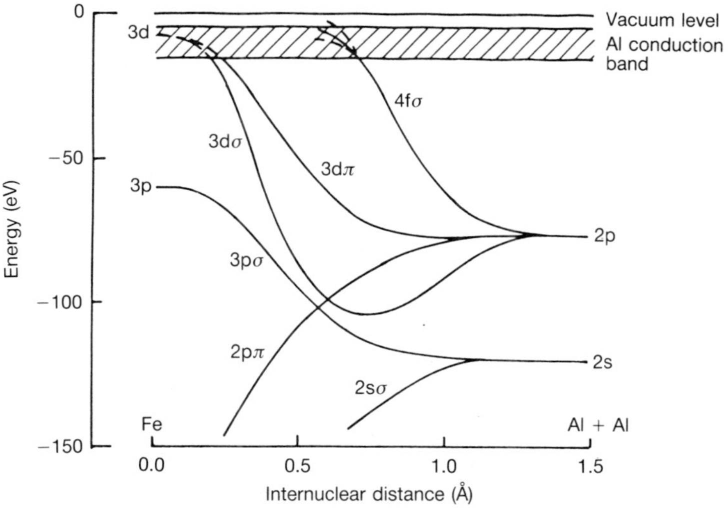

FIG. 17.13. The ionization mechanism of Joyes,[24] adapted by Wittmaack,[25] to explain ionization occurring within a target after penetration of the primary beam. It is based on the electron promotion mechanism already described in another context in Chapter 13 on IAES. On the right is the energy level diagram of aluminium, and on the left that of iron, and internuclear distance decreases from right to left. As two Al atoms are forced together, the atomic levels split, and at hypothetical zero separation the atoms coalesce to form an Fe atom. Before that point is reached, one of the split-off 2p states becomes a molecular orbital state 4fσ, and crosses the conduction band of metallic Al. The electron in that orbital is then 'promoted', and transfers into the free electron sea, leading to ionization.

schematically what happens when two aluminium atoms are forced together during recoil from an energetic penetrating particle. The diagram is due to Wittmaack,[25] who adapted it from an original one of Joyes. As the Al atoms approach each other, increasing overlap of their outer atomic orbitals occurs, and at some critical separation the atomic levels are split into molecular-type levels. Further forcing together of the two atoms increases the complexity of the splitting, with the outermost fully occupied atomic level, the 2p, splitting into four separate states. In the hypothetical case of zero separation, i.e. coalescence, the Al orbitals merge to give an Fe atom, whose outer levels are shown on the left. The important factor in the model is that long (in interatomic terms) before minimum separation is reached, one of the split-off 2p states crosses the conduction band of metallic Al; that is, the electron occupying that orbital is 'promoted'. This electron may then simply transfer into the free electron sea, leading to ionization of the 2p level.

Once the 2p core level has become ionized, de-excitation by an Auger process can occur at any of various stages in the passage of the ionized atom to the outside, i.e. while still within the solid, actually at the surface, or in vacuo after ejection. Since the Auger lifetime is very short, the ionization probability is correspondingly low.

The analysis in the kinetic model applies only to ions ejected with relatively high energies from pure metals under noble gas excitation, and therefore has less relevance to SSIMS than some other models. In that restricted regime, however, the agreement between experimental and theoretical secondary ion yields is good, and clearly the mechanism is operative in parallel with many others.

17.3.3 *The auto-ionizing model*

The ideas behind this model arose from the observation that light emission during ion bombardment of transition metals was spectrally very similar to that seen from arc sources of those metals. In particular, spectral lines originating from excited states in which normally unoccupied energy levels were occupied as a result of promotion of electrons from deeper lying levels have been found to be both numerous and relatively intense. For example, copper, whose ground state is $3d^{10}4s(^2S_{1/2})$, has been observed[26] in the excited states $3d^94s^2$ and $3d^94s4p$; it seems energetically easy to create a d-hole in the 3d transition metals. Thus it seems reasonable to assume that during ion bombardment there will be a certain number of copper atoms leaving the metal surface in excited states involving holes in the 3d level. However, it should be noted that the energies of states such as $3d^94s^2$ and $3d^94s4p$ in copper are significantly less than the ionization energy, and so they themselves cannot act as the starting-point for the creation of an ion. Such states can only be the decay products from other excited states of much higher energy.

The proposal forming the basis of the model, which is due to Blaise and Slodzian,[27] is that ions can be formed by autoionization following formation of a so-called 'superexcited' autoionizing state. With the example of copper again, this state is regarded as being the $3d^94s5s(^4D)$, whose energy of 7.8 eV is fractionally greater than the ionization energy 7.72 eV. The energy of the state is gained by an effective double promotion, that of a $4s(^2S_{1/2})$ electron into a $5s(^4D)$ state (5.2 eV) and of a 3d electron into a $4s(^3D_3)$ state (2.6 eV). The populating of the excited state does not in fact occur by direct promotion within the atom, but by simultaneous quasi-resonant transitions of two electrons from the copper conduction band into the $4s(^3D_3)$ and $5s(^4D)$ states. The schematic energy diagram showing the various relationships is set out in Fig. 17.14. Note that the quasi-resonant transitions from the conduction band

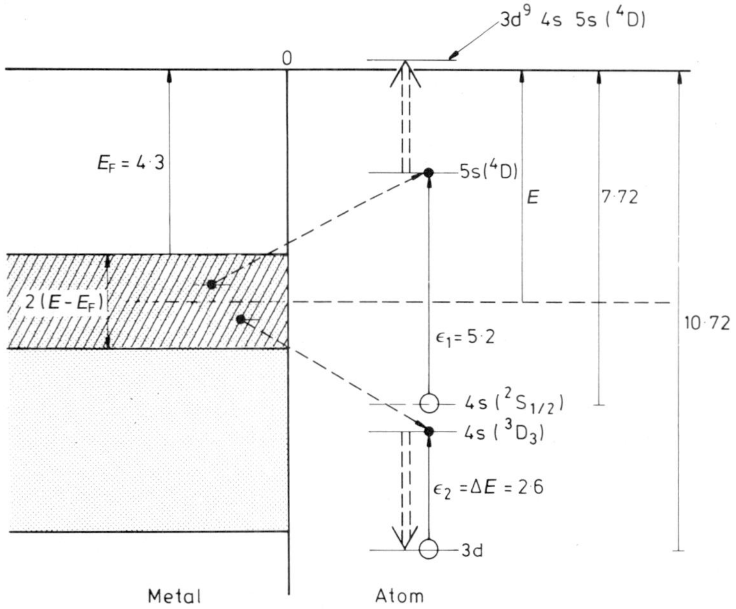

FIG. 17.14. Energetic diagram of Blaise and Slodzian[27] illustrating their autoionization model, in this case with copper as an example. They postulate the existence of a 'superexcited' state, the $3d^94s5s(^4D)$ for copper, reached by simultaneous quasi-resonant transitions of two electrons from the copper d-band into the $5s(^4D)$ and $4s(^3D_3)$ states. Since the energy of the superexcited state, 7.8 eV, is marginally greater than the ionization energy, 7.72 eV, autoionization from the former can occur. Such quasi-resonant transitions can take place only if the weighted mean E of the levels falls into the occupied part of the conduction band; in the case of the copper 4s and 5s levels it does.
The figure also shows two routes of de-excitation. The dotted arrows show an Auger de-excitation process; the other is by direct recombination of the 3d hole from the conduction band.

can take place only if the weighted mean E of the levels 4s and 5s falls into an occupied part of the conduction band.

Figure 17.14 also indicates two routes for the de-excitation of the superexcited autoionizing state. One is by Auger de-excitation, in which a $4s(^3D_3)$ electron drops into the 3d hole and the 5s electron is ejected; in this case the resultant entities are an Auger electron and a positive ion. The Auger electron will of course have very low energy, only a fraction of an eV, and would be undetectable. Another de-excitation process is by direct extinction of the 3d hole via a transition from the copper conduction band.

The probability that a positive ion will actually be produced near the surface in this model thus depends firstly on the probability of creation of a deep (e.g. 3d) core hole in an atom near the surface, and secondly on the probability of an electronic transition into an autoionizing state. Note that it is still assumed in this model that the excitation leading to the creation of the core hole occurs within the metal, although near the surface. If the yield of positive ions is S^+, one can then write

$$S^+ = P_H P_A S_T \tag{17.10}$$

where P_H and P_A are the two above-mentioned probabilities, respectively, and S_T is the total sputtering yield. The probability P_A depends very much on the relationships between the energetic positions of atomic levels in the free atoms, and the occupied densities of states in the conduction band, for each transition metal. Blaise and Slodzian[27,28] have provided very detailed accounts of how P_A can be calculated, not only for pure metals (i.e. atom A emitted from A) but for dilute alloys as well (i.e. atom B emitted from $\underline{A}$B). Naturally, simplifying assumptions had to be made, as in all SSIMS theory. Nevertheless, for the rather restricted range of comparison represented by the pure transition metals and their dilute alloys, under argon ion bombardment at a few keV energy, and for secondary ions of high energy (>30 eV), agreement between calculated and experimental relative secondary ion yields is quite good. Figure 17.15[27] shows the extent of this agreement for the pure metals; not shown in the interests of clarity are the error bars on each point, which amount to $\sim 25\%$ on experimental points, and to 25–40% on theoretical points. The theoretical treatment predicts the observed trends particularly well, even if quantitative agreement is not perfect. Similarly for dilute alloys of one transition metal in another, the predicted trends are mostly correct, although again there are considerable quantitative gaps between the theoretical and experimental yields. Figure 17.16 is adapted from Blaise and Slodzian,[28] and illustrates not only the agreement in trends, but also the pronounced matrix effects to be expected.

17.3.4 *The 'surface effect' or tunnelling models*

The kinetic and autoionization models described above assume implicitly or explicitly that the initial excitation process occurs below the surface in the bulk of the material, and that formation of an ion also takes place either in the bulk or while the excited particle is still effectively part of the bulk electronic structure. Yet another series of theoretical models considers the processes leading to ionization as occurring at the point at which the excited particle is leaving the surface, i.e. in the vacuum just outside the surface. Rüdenauer[29] has classified these models by the term 'surface effect'. They are all also

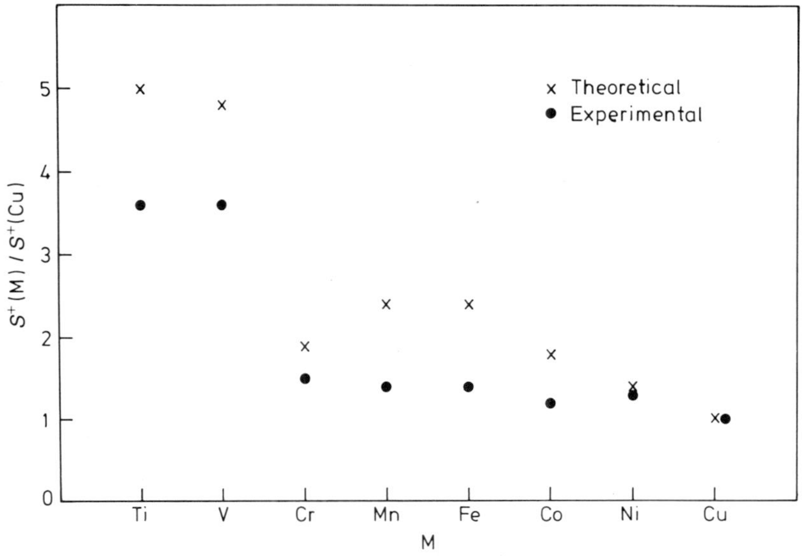

FIG. 17.15. Comparison of theory with experiment, of pure metal ion yields relative to Cu, based on the Blaise and Slodzian[27] model of Fig. 17.14. The error bars on experimental points are about 25%, and those on theoretical points 25–40%, thus agreement can be said to be reasonable.

quantum-mechanical models, in that the emitted particle is supposed to change its electronic structure in a continuous and progressive manner as it moves further away from the surface, from that characteristic of an atom in the bulk material to that characteristic of a free atom or ion. On the way out, electronic transitions may take place in either direction to produce either ionization or neutralization, as the case may be, by tunnelling to or from the atom. Since interactions with both free conduction electrons and the electronic structure of adjacent atoms should be considered, both of which will vary with distance from the surface, a complete formulation of the problem would involve time-and-space-dependent theory. Each model therefore incorporates one or more approximations; most of the models date from the middle to late 1970s, which implies a somewhat static theoretical situation in the last few years. One of the problems may be the provision of experimental data of adequate quality against which to test the theories.

The progressive change in the electronic structure of the emitted atom is shown very schematically in Fig. 17.17, from Schroeer *et al.*,[30] who have proposed one of the above models. While the atom is still part of the bulk metal (on the left) its wave function ψ_f forms part of the valence band of the metal. At a sufficiently large distance into vacuum the wave function ψ_i is now that of the valence electron of an atom in its ground state. In the surface region ψ_i will undergo some continual change as sketched in the figure. While still under the influence of the surface, electronic transitions may take place leading

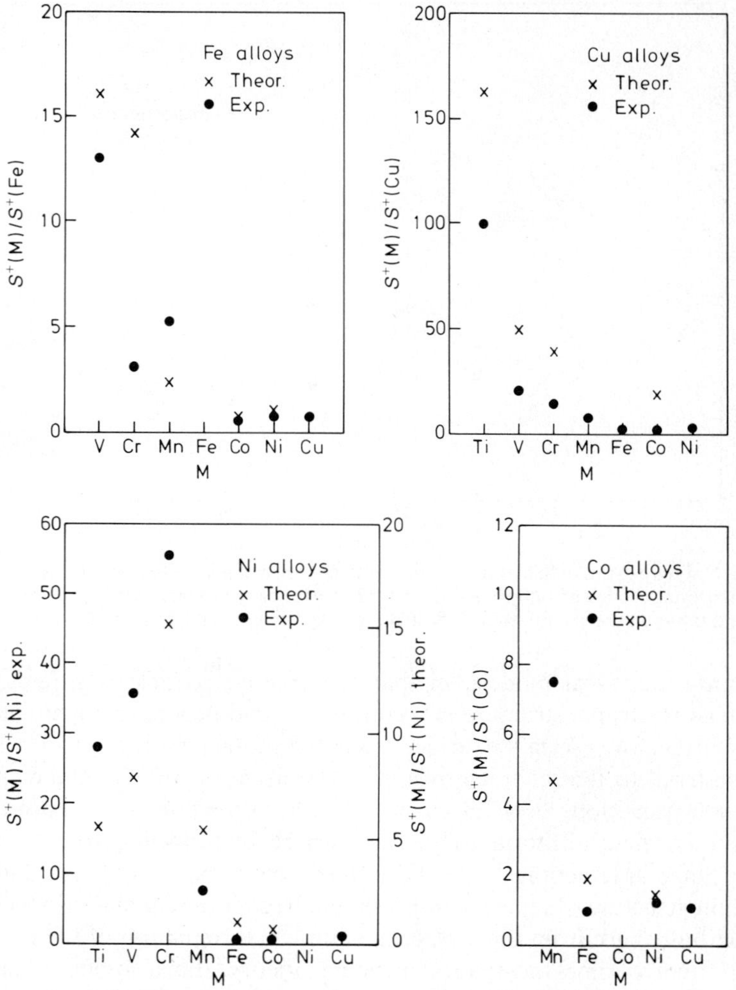

FIG. 17.16. Comparison of theoretical with experimental ion yields from a series of dilute alloys, relative in each case to the host metal, based on the Blaise and Slodzian[27] model, adapted from data given by Blaise and Slodzian.[28] The trends are predicted well, but there are also some pronounced matrix effects (e.g. Cr in Fe, Ti in Cu, and Mn in Co).

to ionization either by direct transitions to the vacuum level (energy $\Delta E = I$) or by back transition to the conduction band (energy $\Delta E = I - \phi$), or to an excited state of the free atom (energy ΔE).

The basic assumption made by Schroeer *et al.* is that the emitted atom is moving sufficiently slowly through the surface region into vacuum that its dwell time in that region is large compared with the time of an electronic transition. Under these conditions the probability of any one transition can be calculated with the co-called adiabatic approximation, which gives the probability as

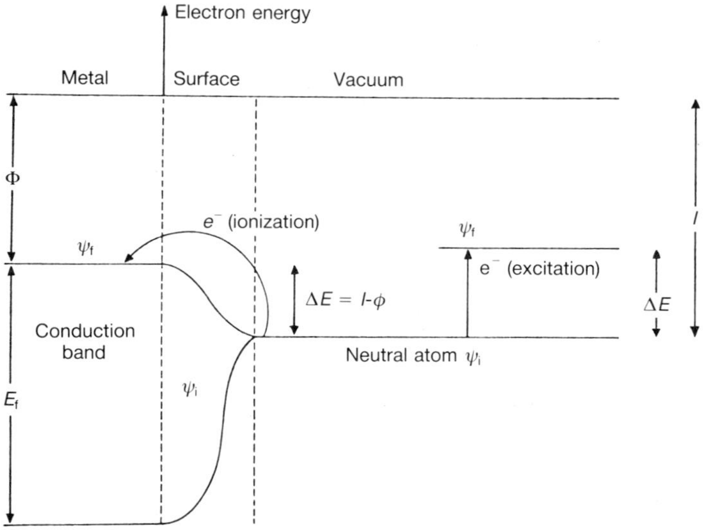

FIG. 17.17. Schematic energy diagram illustrating the progressive change in the electronic structure of an emitted atom according to the quantum-mechanical ionization model of Schroeer *et al.*[30] The bulk metal band structure is on the left, and a valence energy level of a neutral atom with corresponding wave function ψ_i on the right. While still a part of the bulk metal the atomic wave function ψ_f forms part of the valence band of the metal, but as the atom moves away from the metal ψ_i undergoes a continuous change as indicated in the figure. In the surface region while still under the influence of the metal, electronic transitions can take place either by direct ionization to the vacuum level ($\Delta E = I$) or by back transition to the Fermi level ($\Delta E = I - \phi$). Another possibility is the formation of an excited state. The basic assumption in the model is that the atom is sufficiently slow moving that its time spent in the surface region is long compared with the time of an electronic transition. Transition probabilities can then be calculated according to expression (17.11), leading to expression (17.12).

$$R = \left\{ \int_{-\infty}^{\infty} \frac{\langle \psi_f^* | \partial H/\partial t | \psi_i \rangle}{\Delta E(t)} \exp(\mathrm{i}\Delta E(t)t/h)\, \mathrm{d}t \right\}^2 . \qquad (17.11)$$

Here $\partial H/\partial t$ is the rate of change of the potential operating on the valence electron as the atom moves away from the surface, and $\Delta E(t)$ is the change in the excitation energy with time. The authors were able to show that for an assumed surface region thickness of ≈ 0.3 nm, and an excitation energy ΔE of ≈ 3 eV, the adiabatic approximation held reasonably well for incident ion energies < 100 keV.

In order to evaluate expression (17.11), Schroeer *et al.* had to make several approximations. Firstly $\Delta E(t)$ was assumed to be independent of time, i.e. constant and equal to the free atom value, $(I - \phi)$. Secondly, since the matrix elements $\langle\ \rangle$ are very difficult to evaluate, they assumed various analytical expressions for them that enabled (17.11) to be integrated. In the most general case, the expression for the probability of ionization derived from the integration of (17.11) was found to be

$$R^+ = \{E_b/(I-\phi)\}^2\{hv/a(I-\phi)\}^n \tag{17.12}$$

in which E_b is the binding energy of a surface atom and v the velocity of the sputtered atom. The thickness a of the surface region, and the exponent n are now adjustable parameters.

Using some earlier measurements by Beske,[31] published values for I and ϕ, the bulk sublimation energy for E_b, and calculated average values $\bar{v}$ for v, Schroeer *et al.* plotted $R^+(I-\phi)^2/E_b^2$ vs. $(I-\phi)\bar{v}$ on logarithmic scales, and found the dependence shown in Fig. 17.18, for 13 pure metals under 12 keV

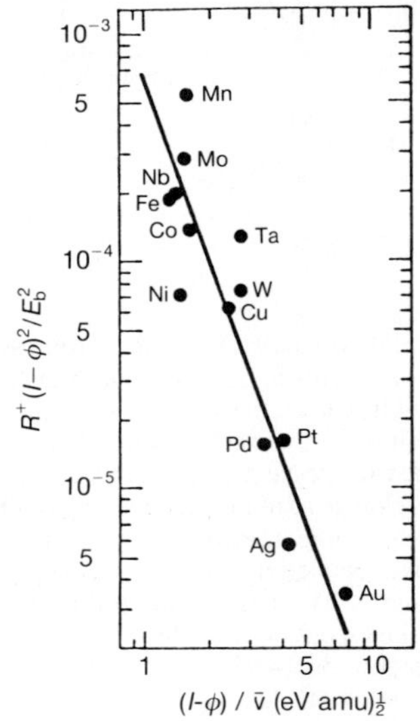

FIG. 17.18. Plot by Schroeer *et al.*[30] of the quantities $R^+(I-\phi)^2/E_b^2$ and $(I-\phi)\bar{v}$ from expression (17.12), for 13 metals under bombardment by 12 keV argon ion bombardment, based on the results of Beske.[31] A fair linear agreement is found, from which a least-squares fit produced values of 0.142 nm for a in expression (17.12), the thickness of the surface region, and of 2.43 for the exponent n.

argon ion bombardment. A reasonable linear dependence was found. A least-squares fit yielded values of 0.142 nm for a and of 2.43 for n, and for these constants the calculated and experimental R^+ values agreed within a factor of 3.

However, there are difficulties with this model. When Benninghoven *et al.*[20] fitted other (equally early) experimental data to it, values of n of about unity were found while values of a of many tens to hundreds of nm were needed! Clearly a at least is very ill-defined. The problems may stem not only from the concepts of the model itself, but on the use of such global quantities as work function and sublimation energy in what is really a highly localized surface event. Such problems almost certainly beset the other models to be described below as well.

In his first model Šroubek[32] adopted a slightly different approach from that of Schroeer *et al.* He regarded the combined emitted atom–metal surface system as a giant molecule and applied molecular orbital (MO) theory to the calculation of the probability of the mixing of atomic and valence wave functions, leading to an electronic transition and hence ionization. The simplest form of the model was that of a non-interacting metal-emitted atom system, i.e. with no perturbation of the atomic levels as a function of time, and in this form the result for the probability of ionization was formally the same as found by Schroeer *et al.*[30] If, however, the system was allowed to be interactive with the energy level of the emitted atom varying according to the assumed expression in Fig. 17.19, and with interaction occurring with a continuum of

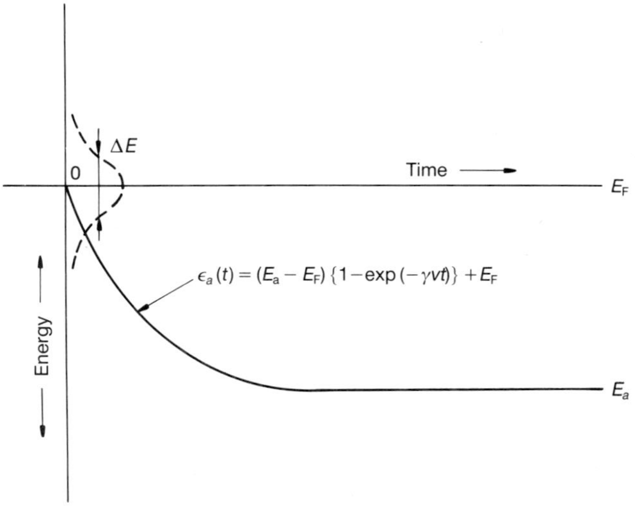

FIG. 17.19. Schematic energy diagram of the atom-surface quantum-mechanical interaction model of Šroubek[32] to account for ionization in SSIMS. The emitted atom-metal surface is regarded as a giant molecule in his model, to which molecular orbital theory can be applied. The figure shows the form of the interaction as the atom leaves the surface at velocity v, as a function of time t. At the Fermi surface the energy levels of emitted atom and surface atom are assumed the same, and the broadening of the ground state of the atom in the surface region is ΔE. The model then leads to expression (17.14) for the ionization probability.

states below the Fermi level E_F, then the ionization probability was deduced to be

$$R^+ = \left\{\frac{(E_a - E_F)\gamma vh}{n\,\Delta E^2}\right\}^{1/2} \tag{17.13}$$

in terms of the quantities in Fig. 17.19, or, more usually,

$$R^+ = \left\{\frac{(\phi - I)hv}{na\,\Delta E^2}\right\}^{1/2} \tag{17.14}$$

where all symbols now have the same meanings as in expression (17.12), but where n is an integer and ΔE the width of the interaction-broadened ground state of the emitted atom. The latter quantity does not seem directly calculable, but must vary considerably from one atom–metal pair to another, which might account for the observed variations in ion yields. Šroubek does not give any comparison with experimental measurements, possibly for that reason.

Expressions (17.13) and (17.14) were derived from a model in which the nuclear motions in the substrate resulting from the collision cascade initiated by heavy particle impact (i.e. the primary ion) were ignored; as Šroubek puts it, they describe ionization processes from a 'quiescent' substrate. Subsequently, Šroubek[33] extended the model to include electronic excitations resulting from substrate collisions. He found it possible to calculate an ionization probability based on the collision of two adjacent substrate atoms by describing the effect of such a collision as a small perturbation of the Hamiltonian operator appropriate to the quiescent state. The collision was assumed to take place close to the site of desorption, and it was also assumed to simplify the calculation that the emitted atom states were those near the ionization energy, i.e. deep below the Fermi level of the substrate. For a two-level emitted atom system (the earlier model was a one-level system) the calculated ionization probabilities without and with collisional electronic excitation were

$$R^{+} = \frac{2}{\pi}\exp\left\{-\frac{\pi(I-\phi)}{2c\gamma_{a}v_{a}}\right\} \tag{17.15}$$

and

$$R^{+} = \frac{(2\pi V)^{2}}{(\delta E)}\frac{\gamma_{a}v_{a}}{\gamma v}\exp\left\{-\frac{\pi(I-\phi)}{\gamma v}\right\} \tag{17.16}$$

respectively. In these expressions γ_a and γ are the spatial extents of the emitted atom–substrate and the substrate–substrate atoms interaction, respectively, v_a and v are the velocities of the emitted atom perpendicular to the substrate surface and of the colliding atoms, respectively, V is a hopping integral, and δE is the width of the substrate valence band. c is an adjustable constant that should equal unity if the ionization energy of the atom is time-independent, i.e. constant, but has been found to be equal to ≈ 10 to fit some experimental data.

Note that both expressions (17.15) and (17.16) contain the exponential dependence of R^+ on $(I-\phi)$ that seems to appear in several models. However, the factor $\pi/\gamma v$ in (17.16) is significantly smaller than $\pi/2\gamma_a v_a$ in (17.15), so that (17.16) predicts much larger R^+ values, in closer agreement with experiment. On the other hand, the dependence of R^+ on v_a, the velocity of the emitted atoms, in (17.16) is small, which is contrary to some experimental evidence, e.g. that of Vasile.[21]

In yet another extension of his model Šroubek[33] referred back somewhat to the Andersen–Hinthorne[22] concept by generalizing the effects of multiple substrate collisions leading to multiple electronic excitations, in terms of an effective temperature T_e. This temperature was assumed to be constant during the emitted atom–substrate interaction, in other words, the interaction time was regarded as very short compared to the time that the atom spent in the vicinity of the surface. With inclusion of the time variation of the valence level of the emitted atom, in other words the time-dependent ionization energy as pictured in Fig. 17.19, the expression for the ionization probability then became

$$R^+ = \frac{P^+(T_c)}{P^0(T_e)} \exp\left\{ -\frac{(I^* - \phi)}{kT_e} \right\} \tag{17.17}$$

where as before (eqn (17.8)), P^+ and P^0 are the partition functions for the ionic and neutral states. Note that (17.17) has a form similar to that of (17.8). I^* is the value of the ionization energy at the time after emission corresponding to the maximum atom–substrate interaction; the contribution to the total interaction at that time is assumed to be dominant.

The term kT_e in (17.17) now replaces the terms $2c\gamma_a v_a$ and γv in (17.15) and (17.16), that is the velocity of the emitted atom does not appear explicitly, but it is implicit in the time-dependence of I. Šroubek shows that consideration of the form of this time-dependence leads to

$$R^+ \propto v_a^n. \tag{17.18}$$

Such a velocity dependence with n in the range of 0.5 to 2 has been established in some experiments, and n can also be negative when $\phi < I$, which has also been found in some cases. Even so, the increased velocity dependence derived from this modified model is still lower than observed in other experiments.

The approaches adopted by Schroeer *et al.* and by Šroubek were criticized by Blaise and Nourtier[34] as being physically unrealistic. The latter authors claimed that the simple models developed by the former which assumed two-level interactions between atom and substrate, were quite inadequate since the complete treatment requires many-body time-dependent perturbation theory. This is because the time-dependent variation of the emitted atom–substrate interaction, i.e. the time-dependent probability of electron tunnelling, can lead to de-excitation as well as excitation, and since the processes compete correlations between the two must be included. To be fair, Blaise and Nourtier also criticized the earlier model of Blaise and Slodzian as being questionable. They attempted to improve the situation by extending the formalism of the quantum mechanical treatment to the time-dependent coupling between a localized state (i.e. on the emitted atom) and a continuum (i.e. the conduction states in the metallic substrate). However, they still used a one-electron Hamiltonian, and had to assume no interaction between the electrons, and a constant density of conduction states, so the extended model still included some drastic approximations.

If, as before (eqns (17.13) and (17.14)), ΔE is the interaction-broadened ground state of the atomic localized level, then ΔE must be a function of distance from the surface, that is, a function of velocity and time. Blaise and Nourtier[34] took at first the simplest conceptual form of the variation of ΔE, the exponential dependence, namely

$$\Delta E(t) = (\Delta E)_0 \exp(-\gamma vt) \tag{17.19}$$

where $(\Delta E)_0$ is the width of the state at $t=0$, and where γ has the same meaning as before. γ^{-1} has dimensions about 0.2 nm. The effective ionization energy, which is some analytical mean of ΔE, will of course also vary, so for convenience it was allowed to vary in the same way as ΔE. To simulate the effect of a collisional expulsion of the atom some 'sudden' δ-like terms were added to the expression similar to (17.19) for the ΔE and I variations. For the situation where $h\gamma v$ was much smaller than either $(\Delta E)_0$ or $(I-\phi)$, which is true physically, the authors found two regimes for the ionization probability R^+, as follows:

$$R^+ = \frac{\sin\alpha}{\alpha} \exp\{-2\alpha(I-\phi)/h\gamma v\} \quad \text{for } \alpha(I-\phi) < (\Delta E)_0 \tag{17.20}$$

$$R^+ = p \exp\{-2(\Delta E)_0/h\gamma v\} \quad \text{for } \alpha(I-\phi) > (\Delta E)_0 \tag{17.21}$$

where $0<\alpha<\pi$, α being a parameter linking the time-dependent variations of ΔE and I, and where p is a factor that includes the δ-like terms mentioned above. The two regimes stem basically from the mathematical discontinuity at $t=0$ (i.e. a sudden change of slope) plus the asymptotic time variation of the atom–metal interaction, which allows some additional tunnelling of an electron from the atom to the metal as a result of a basic uncertainty in the energy. The latter effect is of course operative only if $(I-\phi)$ is small enough.

The appearance of the two regimes led the authors to look more closely at the precise form of the ΔE time variation. Replacing the exponential decay of expression (17.19) by a linear decay led to a linear dependence of R^+ on v, which is then similar to the Šroubek result of expression (17.18) for $n=1$, but for a physically unrealistic form of ΔE. If the form of (17.19) were adjusted to try and eliminate the discontinuity problem, by letting $\Delta E \rightarrow (\Delta E)_0$ at $t=-\alpha$ rather than $t=0$, then the following expression for $\Delta E(t)$ could be used

$$\Delta E(t) = (\Delta E)_0/2\{1-\tanh(\gamma vt/2)\}. \tag{17.22}$$

For an assumed constant ionization energy (a departure from the time-dependent I used above) the result obtained was then

$$R^+ \approx p \exp\{-c(I-\phi)/h\gamma v\} \tag{17.23}$$

where $p \approx 1$, and c ranges from π to 2π. (17.23) is very similar in form to the Šroubek expression (17.15).

It will have been noted how frequently the exponential factor enters into the derived expression for the positive ion yield. A similar exponential dependence is derived for the negative ion yield, with I replaced by A, the electron affinity of the neutral atom, and with the sign in brackets reversed. Tests of this dependence have been made by several workers, with very mixed results; if it is true then of course a plot of log R^+ against ϕ should yield a straight line, for ions of a particular velocity. Such a linear dependence has indeed been observed in certain situations, as seen in Fig. 17.20 from the work of Yu.[35] In

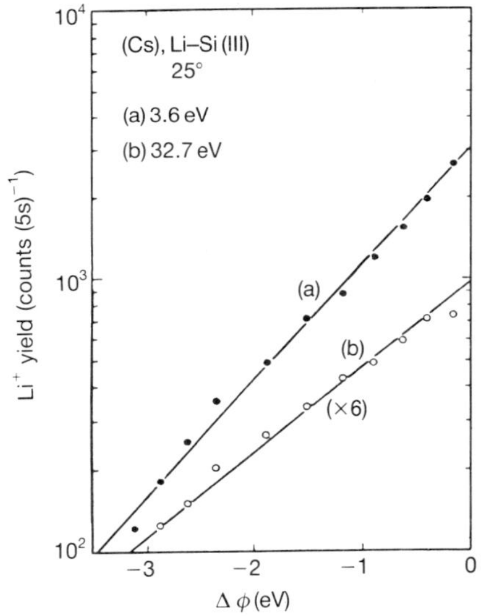

FIG. 17.20. Linear dependence found by Yu[35] of the sputtered Li^+ yield from Li, deposited on a Si surface whose work function was modified by subsequent deposition of caesium, on the work function of the surface. The two plots (a) and (b) correspond to the two secondary ion energies 3.6 eV and 32.7 eV, respectively. Ions emitted at the higher energies (i.e. higher velocities) have a weaker dependence on ϕ, as expected from the form of expression (17.23).

that experiment a small amount of lithium was deposited first on a silicon surface, followed by increments of caesium to lower the work function progressively, during which the Li^+ yield was monitored as the surface was bombarded with Ne^+ ions. The figure shows the logarithmic yield as a function of the change in work function, for two secondary ion energies, and in both cases a close linear relation is obtained, over 1–1.5 orders of magnitude. For the same elements on the same substrate, a similar closely linear dependence has been observed[36] for the Si^- negative ion yield over more than three orders of magnitude. Thus there are undoubtedly situations in which the exponential dependence on work function is the dominant factor, although it has to be said that the concept of the work function does itself include various contributions, any one of which might in fact be the controlling factor.

In the experiments reported above in which deposition of alkali metal was used, there would have been little or no perturbation to the surface chemistry. The situation changes when there is a strong chemical interaction as, typically, in the case of oxygen adsorption or oxide formation. Figure 17.21, also from

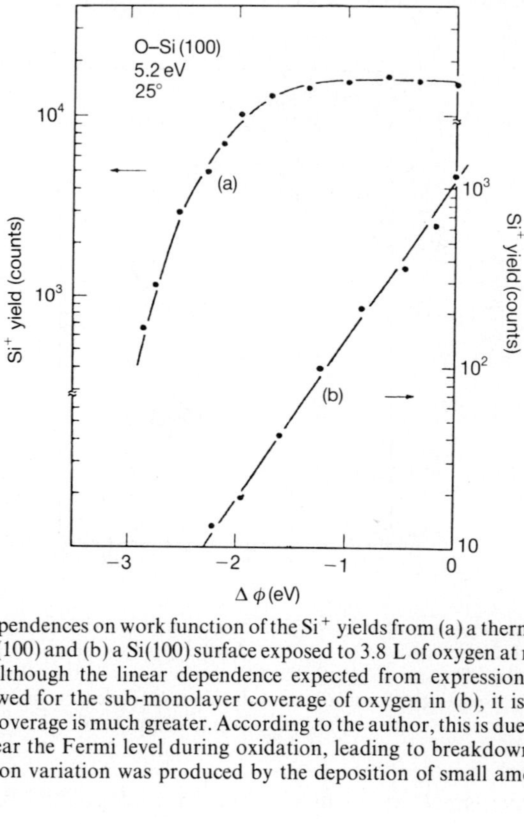

FIG. 17.21. The dependences on work function of the Si^+ yields from (a) a thermally grown silicon dioxide layer on Si(100) and (b) a Si(100) surface exposed to 3.8 L of oxygen at room temperature, found by Yu.[35] Although the linear dependence expected from expression (23), and seen in Fig. 17.20, is followed for the sub-monolayer coverage of oxygen in (b), it is not obeyed in (a) where the oxygen coverage is much greater. According to the author, this is due to the depletion of electronic states near the Fermi level during oxidation, leading to breakdown of the model.

The work function variation was produced by the deposition of small amounts of Li on the surface.

Yu,[35] compares the logarithmic dependence of the Si^+ yield on change in work function for two conditions of the Si surface, (a) for a thin film of thermally grown silicon oxide, and (b) for the silicon surface exposed to only 3.8 L (1 L $=10^{-6}$ torr s) of oxygen at room temperature in the presence of a hot filament. The nascent oxygen so formed in the latter case would have adsorbed to a fraction of a monolayer. For the fractional monolayer condition the strict linear dependence was again observed but for the oxide film there was independence of the work function up to a 1.5 eV decrease, followed by a change to a quasi-linear dependence beyond a decrease of ≈2.5 eV. The reason for the work function-independent portion of the yield curve has been postulated by Yu and Lang[37] on the basis of further experiments, and is illustrated by the schematic energy diagram in Fig. 17.22.

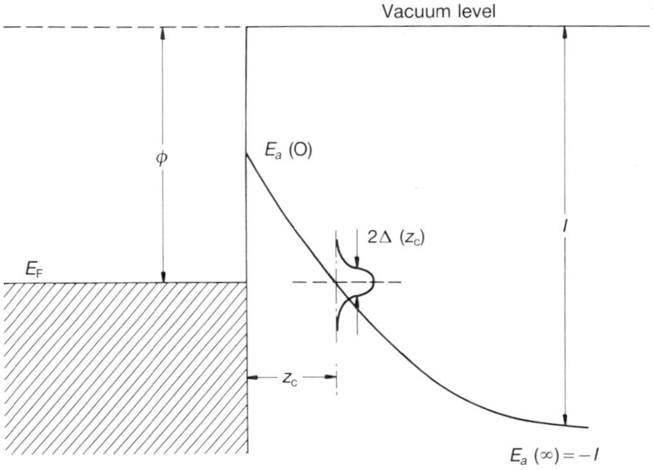

FIG. 17.22. Schematic energy diagram of the changing conditions as an atom leaves a metal surface, in the model of Yu and Lang.[37] The notional metal band structure is to the left, the free atom to the right. Initially the atomic level is broadened, and lies above E_F due to screening by conduction electrons. As the atom moves away the atomic level both narrows, and shifts down in energy. Once $E_a \leq E_F$, at a critical distance z_c, electron tunnelling leading to ionization, can occur. Before that, while $I < \phi$, no tunnelling can occur, and therefore no ionization or indeed neutralization either.

As an atom leaves a metal surface its ionization energy, which is I at an infinite distance, is shifted upwards owing to screening by the conduction electrons in the metal, and the atomic level is broadened as discussed earlier. On moving away from the surface the atomic level shifts down and narrows as a result of an image potential of a form similar to that in Fig. 17.19. Now electrons can tunnel resonantly between atomic and conduction levels within the same energy bands, so that if the atomic level starts off *above* the Fermi level as shown in Fig. 17.22, no tunnelling is possible and therefore no ionization or neutralization, as the case may be. Once the atomic level has shifted down to the position of the Fermi level, at some crossing point z_c, tunnelling can occur and therefore ionization or neutralization. Clearly, then, for $I < \phi$ ionization or neutralization could never occur, which means that a neutral atom leaving the surface would never become ionized and an ion leaving the surface would never become neutralized. For $I > \phi$, ionization or neutralization can start near the distance z_c, and again the extent of tunnelling will be governed by the relative magnitudes of I and ϕ since if they are similar the crossing point z_c might be too far away from the surface for effective tunnelling.

Yu and Lang[37] have verified the tunnelling model for the case of ions rather than atoms leaving a surface. It is known that Cs, which has a low ionization potential (3.9 eV) is adsorbed as ions on most metallic surfaces. In addition there is no strong chemical interaction between adsorbed Cs and a metal surface. Yu and Lang therefore adsorbed Cs on three different metallic substrates, Al, Si, and Au, all having average work functions greater than the ionization potential of Cs. On the bare metals, therefore, one would expect on the above model to find no neutralization of sputtered Cs^+, and a high yield. Yu and Lang then varied the work functions downwards by adsorbing small amounts of Li as well, and found as predicted that near the value 3.9 eV, the Cs^+ yield suddenly started to drop as neutralization became possible. The results of their experiment are shown in Fig. 17.23. For Au and Al substrates the constant yield for $\phi > I$ can be seen, followed by an approximately linear drop of the logarithm of the yield with work function to be expected from the exponential term in expressions such as (17.23). The Si substrate behaved rather differently since Si is of course not a metal but a semiconductor with a band gap of ≈ 1.1 eV and states in the band gap. Additional tunnelling possibilities thus appear. For neutral atoms, the reverse effect would be expected i.e. negligible yield for $\phi > I$, with the positive ion yield increasing as ϕ decreases. However, there are many measurements that conflict with those conclusions, particularly during oxidation during which there seems to be no correlation with the work function.

Variations on the tunnelling model have been described by other authors, notably Nørskov and Lundqvist[38] and Brako and Newns,[39] using different quantum-mechanical approaches. The former authors used a linear interpolation scheme to approximate a smooth transition from the region outside the metal surface where the image potential is dominant to the region very close to the surface where the atomic level has reached its interaction-broadened limit, i.e. $E_a(0)$ in Fig. 17.22. The result that they obtain for the probability of the emission of a positive ion is

$$R^+ = \frac{2}{\pi} \exp\{-C_1\pi(I-\phi)/h\gamma v\}\exp\{-C_2\pi/h\gamma v\}. \qquad (17.24)$$

Here C_1 and C_2 are the 'constants' of interpolation. C_1 is determined by the position of the change-over point, i.e. the distance at which the asymptotic image potential takes over, and is weakly dependent on γ, v, and ΔE (the interaction broadening as before). C_2 involves the value of the image potential at that position as well as the energy difference between $E_a(0)$ and the Fermi level. An expression similar to (17.24) was derived for negative ion emission.

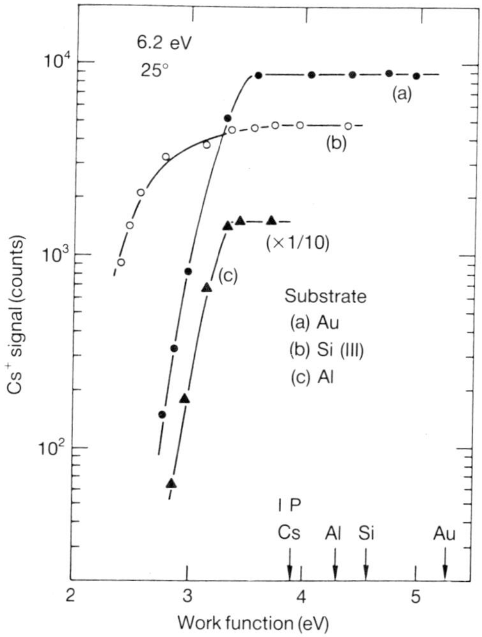

FIG. 17.23. Verification by Yu and Lang[37] of their tunnelling model for the case of *ions* rather than atoms leaving a surface. Caesium is adsorbed as ions on most surfaces, because of its low ionization potential, and there is no strong chemical interaction. According to the Yu and Lang model, adsorption of Cs on surfaces having work functions higher than the ionization potential of Cs should lead to *no* neutralization of Cs^+, and therefore a high yield of that ion. This can be seen in the three curves in the figure where the work function is still high. If the work function of the surface is lowered by the deposition of small amounts of Li, the point is reached where $I = \phi$, when neutralization becomes possible, and the Cs^+ yield drops suddenly. The situation is more complicated on the silicon surface since there is a band gap of ≈ 1.1 eV and states in the gap.

Compared to some other results, expression (17.24) involves a strong dependence on the ion velocity v. For emission from several pure clean metal surfaces Vasile[21] has shown that this strong velocity dependence is obeyed.

Brako and Newns[39] were not concerned so much with the ionization probability following ion impact as in SSIMS, but with charge exchange in atom scattering as a neutral atom approached a surface. The model was basically that of Blaise and Nourtier,[34] and the conclusions reached were much the same.

17.3.5 *Summary of the state of the theory*

As the reader may have gathered from the foregoing, the theory of the mechanisms of secondary ion formation during positive ion (or indeed any other particle) bombardment is in a far worse state than the theories of the other two principal surface analytical techniques, AES and XPS. Indeed it seems that few new ideas have appeared since the late 1970s/early 1980s. One of the problems, perhaps the main problem, is that of having sufficiently good, in the sense of unambiguous, experimental data against which to test theoretical ideas. There are so many different factors that could conceivably play a role in the mechanism that few experiments seem yet to have been able to control all of them at once. For example, almost no mention has been made in this Section of the very large enhancing effect on the yields of the presence of oxygen either as the bombarding ion itself or as an auxiliary jet allowed to play on the surface, simultaneously with the bombardment. In SSIMS oxygen is hardly ever used in this way, hence the lack of mention, but in dynamic SIMS it is commonplace, and its effects need to be taken into account in any theoretical model. There are indeed models that attempt to do this, but it is not appropriate to discuss them as well in this volume. The point to be made here is that the effect of oxygen can be so large that its concentration in, say, a material for use as a standard in SSIMS could lead to quite erroneous conclusions if not very strictly controlled. That is just one parameter among the many that one suspects have not been under adequate control in otherwise carefully designed experiments.

Clearly some of the theoretical models have been successful, but only in very limited sets of conditions, and they have failed when applied outside those conditions. Nevertheless, any degree of success must imply that there is some aspect of a model that is physically correct. What is needed, but does not yet exist, is a unified theory that will be able to include all these parts of earlier models that have been successful, but with the ability to cover the entire range of material variability and instrument operating conditions.

17.4 Quantification

Just as with the approaches to quantification made in the other major surface analytical techniques, so with SSIMS there are two distinct paths that can be followed in an attempt to extract quantitative information. These are the empirical and the theoretical. As before, quantification means the conversion of an experimentally measured signal, in this case an ion current, either positive or negative, into the atomic concentration in the surface of the species corresponding to that ion. The ion current at a particular mass number from element A in a matrix B can be written, after correction for isotopic abundance, as

$$I^{\pm}_{\mathrm{A(B)}} = I_{\mathrm{p}} S_{\mathrm{tot}} A c_{\mathrm{A}} \tau_{\mathrm{A}} \epsilon_{\mathrm{A}} \alpha^{\pm}_{\mathrm{A(B)}} \int_{E}^{E+\Delta E} \int_{\Omega} n_{\mathrm{A}} \, \mathrm{d}E \, \mathrm{d}\Omega. \qquad (17.25)$$

Expression (17.25) contains all the factors at present known to affect the measured ion current. They are:

I_p	= primary current
S_{tot}	= total sputtering yield
A	= effective field of view on specimen by spectrometer
c_A	= concentration of element A in B
τ_A	= spectrometer transmission of the particular mass from A
ϵ_A	= efficiency of detector of the particular mass from A
$\alpha^{\pm}_{A(B)}$	= probability of positive or negative ionization of element A in matrix B
$E \ldots E+\Delta E$	= selected energy window of the secondary ions
Ω	= solid angle of acceptance of ions
n_A	= energy and angular distribution of sputtered material.

In any one experiment or analysis, the quantities in the above list that can be regarded as constant are the primary ion current I_p, the effective field of view A, the energy window ΔE at a selected pass energy $E+(\Delta E/2)$, and the solid angle of acceptance Ω into the analyser. Very often the transmission τ_A and the detection efficiency ϵ_A are also assumed to be constant, but since both τ and ϵ refer not just to element A but to the mass number chosen as representative of A, they could well alter if the mass number chosen as representative of another element is in a mass range considerably removed from the first. The sputtered distribution n_A is a function primarily of angles of incidence and emission as well as of secondary ion energy, and since these quantities are fixed, n_A will be virtually constant although it may also vary slightly with widely separated masses. However, if to a first approximation the latter quantities can be assumed constant then (17.25) reduces to

$$I^{\pm}_{A(B)} = F_A S_{tot} c_A \alpha^{\pm}_{A(B)} \tag{17.26}$$

in which F_A includes now all the instrumental factors appropriate to A. Expression (17.26) is in the form used by many writers.

Virtually all the problems of quantification of SIMS (both static and dynamic) stem from the behaviour of $\alpha^{\pm}_{A(B)}$ with the nature of the matrix B and with the concentration of A in B. The ionization probability α is often very matrix- and concentration-dependent, i.e. is a function of c_A, sometimes a rapidly varying function. An example of matrix effects on the ionization probability is shown in Fig. 17.24, from the work of Loxton and Tsong.[40] A series of CuNi alloys was bombarded with 7 keV Ar^+ ions until an equilibrium surface composition was reached in each case, that is, the steady state condition in which the sputtering yields reflected the bulk concentrations of the constituents. Ion yields from Cu and Ni were normalized to the yields of the pure elements, so that the probabilities α appearing in the figure are normalized ionization and excitation probabilities. The latter were measured by recording the photon yields from characteristic light emission lines of Cu and Ni, as in the technique of IBSCA described in Chapter 15. It can be seen from Fig. 17.24 that neither the Cu nor the Ni ionization probabilities are

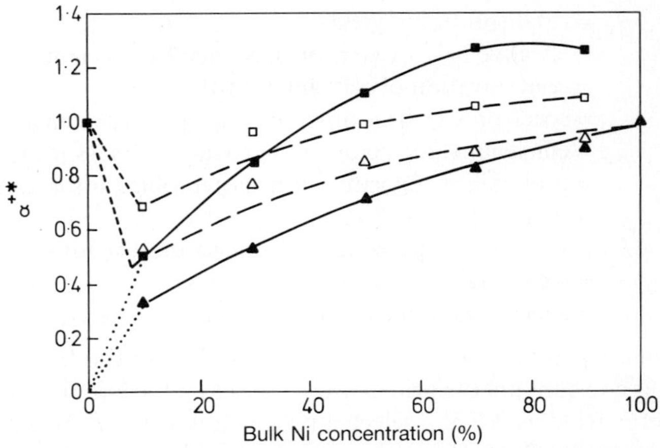

FIG. 17.24. Illustration of the intrusion of matrix effects into the quantification of SSIMS. Ionization, $\alpha^{+}(\phi)$, and excitation, $\alpha^{*}(\phi)$, probabilities, as a function of the bulk concentration of Ni in a series of CuNi alloys under bombardment by 7 keV argon ions until an equilibrium surface concentration had been reached, measured by Loxton and Tsong.[40] The meanings of the symbols are ▲: $\alpha^{+}_{Ni}(\phi)$. ■: $\alpha^{+}_{Cu}(\phi)$. △: $\alpha^{*}_{Ni}(\phi)$. □: $\alpha^{*}_{Cu}(\phi)$. Neither the Cu nor Ni ionization probabilities are linear with concentration, that of Cu in particular showing some erratic variations.

linear with the concentration of the constituent in the alloy, although at least α^{+}_{Ni} and α^{*}_{Ni} show monotonic variations with c_{Ni}. Both α^{+}_{Cu} and α^{*}_{Cu} drop by a factor of about 2 on addition of a small amount of Ni to Cu, however, and α^{+}_{Cu} eventually becomes some 30% *greater* than in pure Cu in alloys containing less than 30 at.% Cu! It is the existence of erratic variations of this type that makes quantification such a problem.

The two approaches to quantification attempt to deal with the matrix effects in different ways.

17.4.1 *Empirical methods*

These methods make no assumptions about any physical processes involved in secondary ion creation and emission, and seek to avoid any problems connected with the ionization probability variations by restricting themselves to concentration ranges in which the ionization probability seems to be well behaved. All the methods are based on expression (17.26), that is, with the assumption of constant instrumental parameters.

The first empirical method is the so-called absolute sensitivity factor, or calibration curve, method. Expression (17.26) is written[41] as

$$I^{\pm}_{A(B)} = \rho^{\pm}_{A(B)} c_A \tag{17.27}$$

where $\rho^{\pm}_{A(B)}$ is now the absolute sensitivity factor of element A in matrix B, linking directly the ion current and the concentration. The method consists in generating a calibration curve for element A in matrices of type B, based on a set of carefully prepared standards in which only the concentration of A is varying, the other components, and hence the nature of the matrix, being kept as constant as possible. Obviously, if the concentration of A starts to become appreciable, then the other concentrations must also be varying appreciably, and the matrix will be changing its physical character. Thus the method is to first order applicable only to dilute concentrations of A in B.

Because of the various factors that have been rolled into ρ, it is unlikely that ρ will be constant, and it is usual to find that it depends on the concentration of A, because of the aforementioned matrix effects. Usually the calibration curves are monotonic and continuous, but occasionally discontinuities or sudden changes of slope are found. The method is useful in analysing unknowns only when the matrix of the unknown matches that of the nearest standard very closely, and is therefore best suited to the analysis of a particular element or set of elements in dilute concentration in a matrix of otherwise fixed composition. A good example would be the analysis of impurity elements in a steel of a particular composition. Under these circumstances the precision can be good. Naturally, the instrumental parameters must be maintained accurately constant.

The second method is that of relative sensitivity factors. The nomenclature will be familiar from Chapters 4 and 10 on AES and XPS, respectively, but in this case there is a difference. Whereas in AES and XPS all elemental sensitivities are quoted relative to that of a single element, usually fluorine, in SIMS the relativity is to the signal from an internal, selected, reference element. Normally the latter would be the principal matrix element. If the signal from the chosen reference is $I^{\pm}_{\mathrm{R(B)}}$, then, using expression (17.26) again, one can write

$$I^{\pm}_{\mathrm{A(B)}}/I^{\pm}_{\mathrm{R(B)}} = \frac{F_{\mathrm{A}} S_{\mathrm{tot}} c_{\mathrm{A}} \alpha^{\pm}_{\mathrm{A(B)}}}{F_{\mathrm{R}} S_{\mathrm{tot}} c_{\mathrm{R}} \alpha^{\pm}_{\mathrm{R(B)}}} \tag{17.28}$$

or

$$I^{\pm}_{\mathrm{A(B)}}/I^{\pm}_{\mathrm{R(B)}} = \rho_{\mathrm{AR}} c_{\mathrm{A}}/c_{\mathrm{R}} \tag{17.29}$$

where ρ_{AR} is now the relative sensitivity factor of A with respect to R. As before ρ_{AR} contains matrix-sensitive parameters, but the effects of the variations in them are damped by virtue of taking ratios, and ρ_{AR} is normally assumed constant if the atomic concentration of A is low, e.g. <1 at.%. In dynamic SIMS the matrix-dependent variations, particularly that of the ionization probability, are also suppressed by using incident oxygen ions rather than argon ions, but since SSIMS has been concerned principally with the study of surface reactions the use of oxygen is out of the question.

It is possible to obtain a value of ρ_{AR} for a particular matrix from a single measurement on a sample with accurately known concentrations of A and R, but for better accuracy clearly a range of standards should be used just as in the absolute sensitivity method. Even so, the accuracy can leave a lot to be desired. For example, Smith and Christie[42] used the relative sensitivity method in analysing two sets of standard materials, one of glasses and the other of iron, provided by the National Bureau of Standards, basing the relativity on silicon in the former and on iron in the latter. The variations in the values of the relative sensitivity factors for several minor constituents in the iron standards, as a function of concentration, are shown in Fig. 17.25, (a) referring to the

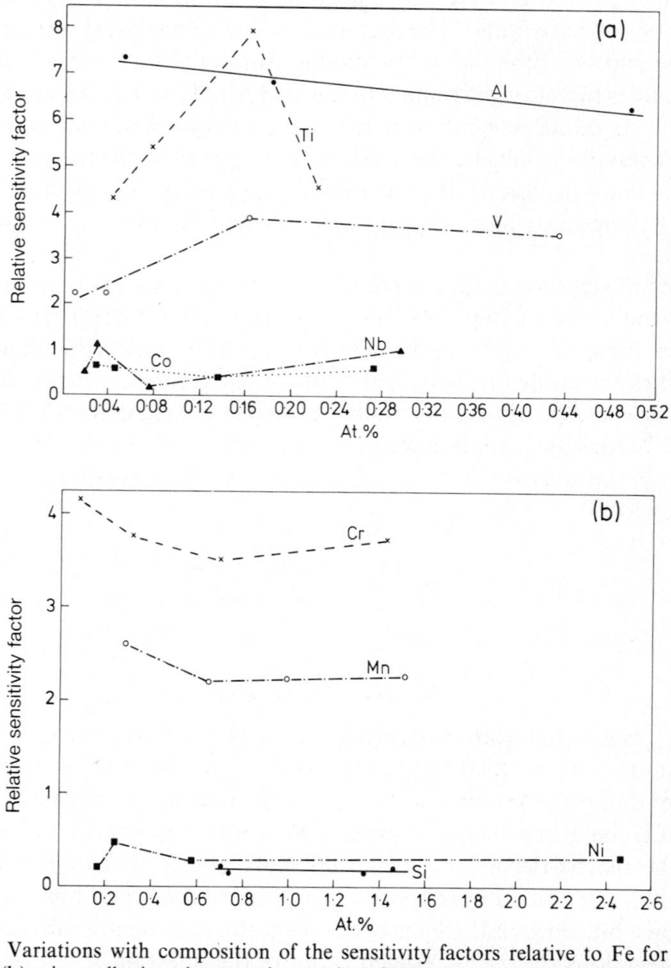

FIG. 17.25. Variations with composition of the sensitivity factors relative to Fe for (a) trace impurities, (b) minor alloying elements, in a series of NBS iron standards, based on the data provided by Smith and Christie.[42] Although some factors are almost independent of concentration, those of a few elements, e.g. Ti, show significant excursions.

constituents at very low concentrations, and (b) to those at slightly higher concentrations. Although the factors for some constituents (e.g. Al, Co, and Si) do indeed seem to be almost independent of concentration, for most of them there are significant variations from constancy, Ti being a particularly bad example. For the glasses the situation was, if anything, worse. Using sensitivity factors averaged over all the standards (except that one or two particularly severe excursions from constancy were excluded), only about 85% of the analyses were within a factor of 2 of the certified compositions. Such an accuracy can be regarded at best as only semiquantitative. Nevertheless the relative sensitivity method, as in the other major techniques, is the most widely used approach.

The third empirical method is really only a variation on the relative sensitivity method. It is clear from the example given above and from many others that ρ_{AR} is matrix-dependent, even if less so than ρ_A, and in a purely empirical approach ρ_{AR} should strictly speaking be measured for a very large number of different compositions and types of sample. However, if for any one set of standards ρ_{AR} is found to vary continuously but not wildly, the likelihood is that there is a matching variation in one of the parameters contained in ρ_{AR}. If this variation can be linked to some measurable index of the dependence on the matrix properties, then there would not be the need to carry out such a multidimensional array of measurements on standards. The index usually consists of a ratio, or set of ratios, of mass spectral peak heights from some major element in the material that provides sufficiently prominent peaks. Such ratios might well be of the form M^{n+}/M^+, M_n^+/M^+, MO^+/M^+, M_nO^+/M^+, MO_n^+/M^+, etc., or based on M_n^+ in the denominator. It is known, for example, that different oxides of various metals can be characterized quite accurately by the correct choice of ratio of molecular to atomic or to cluster ions.

Christie *et al.*[43] have used the matrix-indexed relative sensitivity factor method to analyse thermocouple wires and their sheaths after decalibration as a result of operation at temperatures above 1100°C, i.e. in excess of their normally accepted range of operation. Since alumel and chromel contain nickel as a major element ($\approx$88%) they used the NiO^+/Ni_3^+ peak height ratio as the index for the thermocouple material, and for the analysis of the stainless steel sheath the ratio FeO^+/Fe^+, iron in that case being the major matrix element. Figure 17.26 shows the relationship between the sensitivity factors for chromium and iron relative to nickel, and the chosen matrix indexing ratio. Note that the ρ_{AR} values for Cr fall on the same curve for three different matrices, a chromel standard wire and two NBS nickel-based standards, which supports the use of this particular method. The calibration curves based on the ratio FeO^+/Fe^+ for the steel are given in Fig. 17.27, also from Christie *et al.*[43] A comparison of the compositions of wire and sheath with their nominal compositions, found by the authors, is given in Table 17.1. It can be

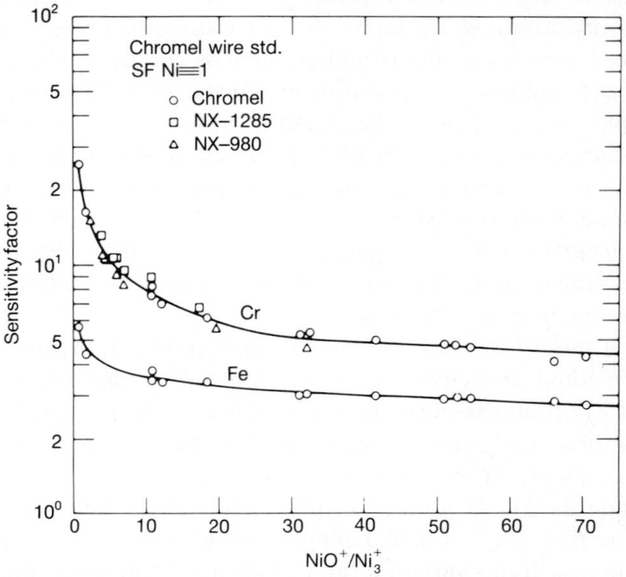

FIG. 17.26. Calibration plots derived by Christie *et al.*[43] for the sensitivity factors of Cr and Fe in chromel and in two standard NiCr alloys, using the matrix-indexed relative sensitivity factor method described in the text. The matrix indexing ion ratio chosen was $NiO^+/Ni_3{}^+$. The solid lines are best-fit curves to the data points. The variation in $NiO^+/Ni_3{}^+$ was produced by increasing the partial pressure of oxygen around the sample.

seen that the accuracy with this method can be good for certain ranges of composition within certain materials.

17.4.2 *Theoretical methods*

As stated earlier, the theoretical approach to quantification seeks to establish the values and ranges of validity of $\alpha^{\pm}_{A(B)}$ in eqn (17.26) from first principles. In so doing, the various expressions for probability of secondary ion formation or for ion yield, already described in Section 17.3, should be used, and some indication has already been given there of their applicability in some restricted material regimes. However, it should be noted that where quantification in SIMS has been attempted using physical models as a basis, it is almost without exception the Andersen–Hinthorne LTE model (see Section 17.3.1) that has been used, despite the fact that the model itself has now been discredited. The reason for this seems to be that a computer program based on the LTE model was made available to users of a particular make of ion microprobe, in the early 1970s, and some success was found with its use before doubt was thrown on the physical reality. In other words the mathematical formalism seemed to work even when there was no obvious reason why it should. In a sense, therefore, the estimation of $\alpha^{\pm}_{A(B)}$ via the LTE model is not strictly a theoretical

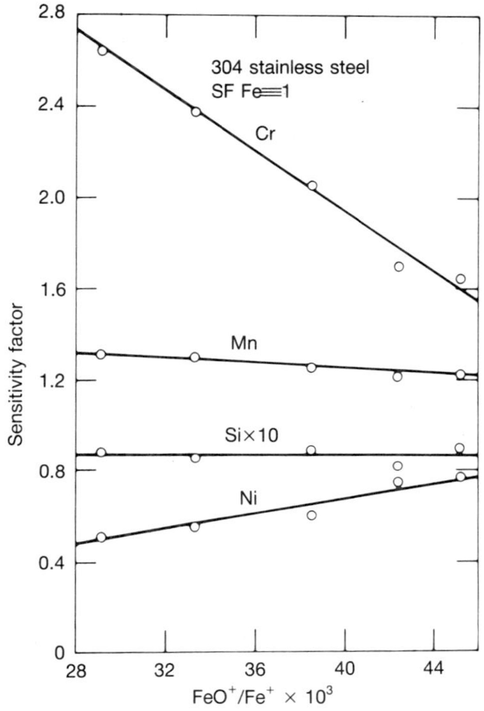

FIG. 17.27. Calibration plots similar to those in Fig. 17.26 for the sensitivity factors of several alloying elements in 304 stainless steel, derived by Christie *et al.*[43] In this case the matrix indexing ion ratio was FeO^+/Fe^+, and again the ratio was varied by changing the oxygen partial pressure.

approach since the algorithms based originally on theory are now used in an empirical manner, with adjustable parameters (merely 'fudge factors', according to some authors[44]).

Equation (17.8) can be re-written as

$$n^+n_e/n^0=\{2P^+(T)/P^0(T)\}\,\{2\pi m_e kT/h^2\}^{3/2}\exp\{-(I_p-\Delta E)/kT\} \tag{17.30}$$

where all symbols have the same meanings as defined in Section 17.3.1. With temperature in K, energies in eV, and particle concentrations expressed per cm^3, (17.30) reduces to

$$n^+n_e/n^0=4.81\times10^{15}\,T^{3/2}\{P^+(T)/P^0(T)\}\exp\{-11605(I_p-\Delta E)/T\} \tag{17.31}$$

where

$$\Delta E=2.95\times10^{-8}(n_e/T)^{1/2}. \tag{17.32}$$

Usually $n^+\ll n^0$, so that n_X^0 may be taken as proportional to the concentration c_X of element X in the surface; n_X^+ will also of course be proportional to the positive secondary ion current I_X^+. If expressions (17.30)

Table 17.1

Analysis of thermocouple wires and sheath by Christie *et al.*[43] using the matrix-indexed relative sensitivity factor method, at.%.

Material	Al	Si	Cr	Mn	Fe	Ni	Co
Alumel wire							
Ref. Section	4.38	3.99	8×10^{-4}	2.80	0.07	88.26	0.50
Nominal	4.19	3.53	8×10^{-4}	3.01	0.031	88.76	0.48
Chromel wire							
Ref. Section	0.023	1.06	10.80	0.012	0.29	87.78	0.035
Nominal	0.021	0.92	10.37	0.010	0.35	88.33	0.040
304 S/S							
Ref. Section	1.4×10^{-3}	1.87	20.60	2.08	66.59	8.85	0.014
Nominal	—	1.95	19.96	1.99	66.42	9.32	—

and (17.31) are valid, then a plot of $\ln\{I_X^+ P^0(T)/c_X P^+(T)\}$ will be proportional to the ionization energy I_p of the element X, with slope $-11605/T$. Since both the pre-exponential factor and ΔE contain n_e and T these two quantities are the adjustable parameters in the procedure. Such a proportionality has been observed in some instances. For example, in Fig. 17.28 $(I_X^+/c_X)\{P^0(T)/P^+(T)\}$ has been plotted on a logarithmic scale against the ionization energy I_p for several constituents in two special iron alloys, by Schroeer.[45] One set of data points refers to the clean alloy surfaces, the other to the surfaces in the presence of oxygen at 10^{-6} torr. The data were normalized to the secondary ion signal from clean iron, the 'oxidized' data being multiplied by 130 because of the difference in yields from clean and oxidized iron. The two sets fall on approximate straight lines, and the figure also provides a good impression of the changes in secondary ion yield (in those particular matrices) on going from clean to oxidized surfaces. The plasma temperature T was established iteratively after an initial guess. It is interesting to note that for those particular alloys Schroeer attempted to apply his own[30] quantum-mechanical model (see Section 17.3.4) to the quantification but obtained a much worse fit.

Quantification proceeds by the use of the general reduced eqn (17.26). If in the matrix B there are several component elements $X, Y, \ldots$, etc., of concentrations $c_X, c_Y, \ldots$, then one can write

$$I^+_{X(B)}/I^+_{Y(B)} = S_{tot}c_X\alpha^+_{X(B)}/S_{tot}c_Y\alpha^+_{Y(B)} \tag{17.33}$$

or

$$c_X/c_Y = (I^+_{X(B)}/I^{+Y}_{(B)})\,(\alpha^+_{Y(B)}/\alpha^+_{X(B)}). \tag{17.34}$$

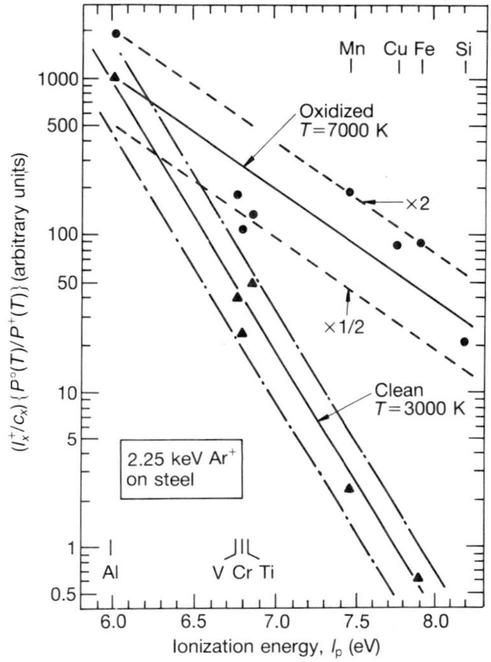

FIG. 17.28. Application by Schroeer[45] of the Anderson–Hinthorne[22] LTE model to the yields, weighted by the atom to ion partition function ratios (see expressions (17.30) and (17.31)), of several alloying constituents in two special ferrous alloys. The results marked 'clean' were from surfaces that had been cleaned by alternate ion bombardment and exposure to oxygen. Those marked 'oxidized' were measured in the presence of 10^{-6} torr oxygen. The alloying element concentrations were all about 1 wt.%. The solid lines are visual best fits to the data points. The indicated plasma temperatures were evaluated iteratively.

This assumes that the instrumental factor F in (17.26) is the same for X and Y, which will not be strictly true if the mass numbers m_X and m_Y being used differ widely. However, since for a quadrupole mass spectrometer the transmission is proportional to m^{-1}, a correction factor could be introduced if necessary.

Assuming that the concentrations c_X, c_Y, etc., are sufficiently dilute that the sputtering yields of X, Y, . . . , in B are the same, and introducing the Saha–Eggert eqn (17.31), one can write for the ratio of concentrations of X and Y.

$$c_X/c_Y = (I^+_{X(\mathrm{B})}/I^+_{Y(\mathrm{B})})\{P^0_X(T)P^+_Y(T)/P^0_Y(T)P^+_X(T)\}\exp\{11605(I_{\mathrm{p}(X)} - I_{\mathrm{p}(Y)}]/T\} \tag{17.35}$$

in which T is apparently the only adjustable parameter. One then has a coupled set of equations for the ratios $c_X/c_Y, \ldots, c_M/c_Y$, together with $c_X + c_Y + \cdots c_M = 100\%$, so that in principle once T is evaluated the concentrations can be found. This was the method used, for example, by MacDonald and Garrett,[46] in analysing some NBS standard alloys, and their results are shown in Table 17.2.

Notice from Table 17.2 that the accuracy of quantification is good even though the constituents are in concentrations far from dilute. The agreement in this particular case must be regarded as a special case, since the sputtering yields of Fe, Cr, and Ni are not dissimilar, while the transmission correction mentioned above would also be very small. Nevertheless, steels are a common subject of analysis, so it seems that in some circumstances the empirical application of the Saha–Eggert equation arising out of the Andersen-Hinthorne theory, does work.

Most quantification schemes using the same equation have in fact consisted of two-parameter fitting procedures, that is, the effective plasma temperature T as above, plus the effective electron density n_e. To do this requires so-called internal standards, that is, there must be at least two constituents in the matrix of accurately known concentrations. Measurement of their ion currents under fixed operating conditions then gives two Saha–Eggert expressions in which T and n_e are the only unknowns, from which values of the parameters may be deduced. However, it should be noted that on going to a matrix of significantly different composition, the values of T and n_e needed for fitting purposes can alter dramatically, n_e in particular showing variations by orders of magnitude in some cases. Clearly the actual physical significance of T and n_e is obscure.

Several simplified models based on the overall assumption of LTE and on the Saha–Eggert expression have been developed by SIMS practitioners usually to enable iterative computer programs to be applied more easily. However, they are all concerned with quantification in dynamic SIMS, in which the use either of a primary oxygen ion beam, or of argon ion bombardment in an oxygen atmosphere is standard practice in order to increase the secondary ion yields and to reduce the range of elemental sensitivities. Since SSIMS does not employ oxygen ion probes, the description of such models is inappropriate here, and the reader is referred to more general treatises on the subject, e.g. Benninghoven *et al.*[20]

17.5 Applications

17.5.1 *Polymers*

Although the analysis of polymer films and of polymer-coated surfaces has always been the prerogative of XPS, in the last few years it has been shown by several workers[47–49] that SSIMS is capable of providing useful and indeed sometimes quantifiable, information about polymer surfaces. At first glance this is surprising. The two obstacles that spring to mind at once are of course the irrecoverable beam damage and the problem of charging. The two are not independent, since the extent of surface charging is liable to change with the

Table 17.2

Analysis of some NBS standard alloys, using a single T parameter in the Saha-Eggert equation, by MacDonald and Garrett.[46]

	Fe	Cr	Ni	Effective plasma temperature T (K)
Certified Conc.	87	13	0.3	
Exp. Conc.	88 ± 3	12.3 ± 4	—	4325
Certified Conc.	62.1	23.7	13.3	
Exp. Conc.	60 ± 2	21 ± 4	19 ± 5	4450
Certified Conc.	85	5.48	6.62	
Exp. Conc.	87.6 ± 2	5.7 ± 1.4	6.7 ± 1.5	4130
Certified Conc.	72	3	24.8	
Exp. Conc.	68.8 ± 3	1.9 ± 1	29.4 ± 7	4230
Certified Conc.	64	—	36	
Exp. Conc.	68 ± 3	—	32 ± 4	4190

extent of damage. The problems have been studied exhaustively by Briggs and co-workers in a series of papers, particularly in that by Briggs and Hearn.[48] By using an electron flood gun to stabilize the surface potential during ion bombardment, and by varying the primary electron and ion currents independently, they were able to establish the conditions necessary to analyse the surfaces of several typical polymeric materials. Damage, arising from either electron or ion impact, or both, was assessed using both the secondary ion spectra and XPS. The general conclusions reached in that and other papers were that the total ion dose at an argon ion energy of 4 keV should not exceed 10^{13} ion cm^{-2}, and that the neutralizing electron beam current density should be greater than the incident ion current density. It is also preferable to use an incident ion energy of less than 4 keV, but then the signal strength is less, and the acquisition times correspondingly longer. The extent of ion beam damage in the material poly-methyl-methacrylate, as monitored by the C/O ratio from XPS analysis, is shown in Fig. 17.29, for several ion beam energies and for two types of ion. The need to limit total ion dose to $\approx 10^{13}$ ions cm^{-2} is clear from the figure.

Polymer films, coatings, and fibres are used so extensively that the nature of their surfaces in terms of composition and chemistry is vitally important. For example, unless the surface condition is correct, there may be problems of too little or too great adhesion, of inability to accept printing ink or dyes, and of undesirable transfer of material to other surfaces in contact.

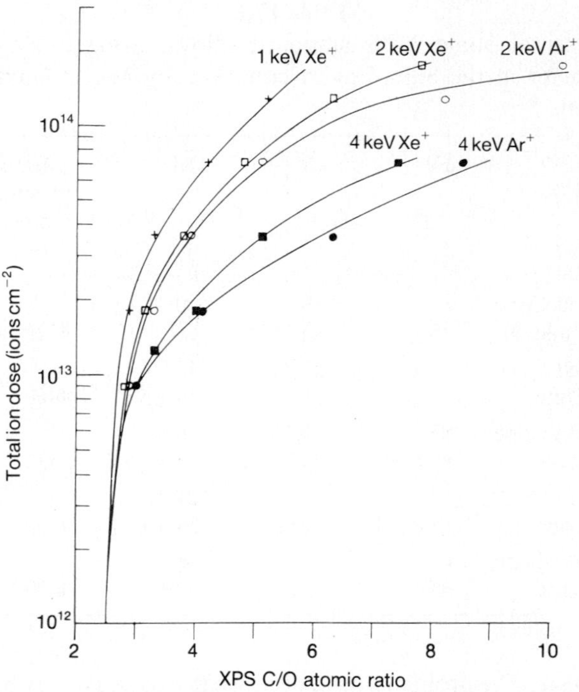

FIG. 17.29. Ion beam damage in polymethyl-methacrylate monitored by the C/O ratio from XPS analysis, for two types of bombarding ion at several kinetic energies, recorded by Briggs and Hearn.[48] It is clear from these plots that to minimize damage, at least in this material, ion doses should be restricted to about 10^{13} ions cm^{-2}.

An example of the way in which SSIMS has been used to analyse the change in the surface condition of a polymer film is that of Briggs,[50] who was concerned with the segregation of an unwanted species to the surface of polyethyleneterephthalate (PET) film on heating to $\approx$170C. The unwanted species is in fact an oligomer, the cyclic trimer of the polymer repeat unit. Since the oligomer has carbon and oxygen in exactly the same chemical situation as in the polymer, the segregation could not have been studied by XPS. At high surface concentrations of the trimer, the segregant crystallizes and can then be seen in an SEM, but at low concentrations there is no crystallization and it is therefore not visible.

The clue to the presence of the trimer came in the observation of much higher intensities at certain high mass numbers from the pure trimer than from PET itself after solvent extraction to remove the oligomer. The mass of the polymeric repeat unit is 192, and in the positive SSIMS spectra peaks at 385 and 577 daltons were found, corresponding to twice and three times the basic unit, plus one hydrogen, respectively. When normalized in each case to the intensity of the principal peak at 104 daltons, it was found that the relative difference in the intensity of the 577 dalton peak between the extracted PET and the pure trimer was about 25. This difference is demonstrated in Fig. 17.30. The reason for the higher relative intensities at 385 and 577 daltons is of course that the cyclic trimer is effectively a quasi-molecular ion, even when situated on the surface of the polymer containing the same repeat unit. Using this relative difference Briggs was able to establish the presence of the oligomer on the surface of the PET after only a short treatment at temperature, as shown in Table 17.3.

The results in the table suggest that even after 1 h at 170°C the non-crystallized trimer is contributing substantially to the 577 dalton signal, since at that stage the surface coverage by the oligomer is only $\approx 2\%$.

17.5.2 *Surface reactions*

This is the area of major application of SSIMS to date; in the early days of the technique especially nearly all publications were devoted to following the variations in intensity of several selected ion signals, both positive and negative, as a function of exposure to a reactive gas. Many of these experiments were purely phenomenological in nature, with little effort being devoted to any explanation as to why certain ion intensities increased or decreased at various stages in the reaction, nor indeed to the question as to why certain ions appeared at all. Latterly such measurements have been more systematic in nature, with attempts to correlate features in the secondary ion spectra with features in the results from the application of other techniques in parallel.

A good example of such an approach has been that of Vickerman and co-workers in a series of papers dealing with CO adsorption on a variety of metal surfaces. When CO is admitted to an iron surface at a pressure of 10^{-8} torr and a temperature of 195 K, and allowed to equilibrate, the positive secondary ion spectrum shown in Fig. 17.31 results.[51] In addition to the usual elemental ions Fe_n^+ prominent peaks of the multiple ions Fe_nCO^+ can be seen, as well as much smaller ones associated with the dissociation products of CO. Similar fragments, but with different relative intensities, are found for adsorption of CO on nickel, but when adsorption takes place on Pd no dissociation products are found. The problem addressed by the authors was to find a relationship between the nature and intensities of secondary ions emitted from a surface, and the chemical and structural properties of that surface. Their approach was initially entirely empirical in character, but was found to have some basis in probable desorption mechanisms.

When CO adsorbs on a metal surface, it may either dissociate into its constituent C and O or remain as a molecule. If it adsorbs molecularly, there are several possibilities for its geometric arrangement on the surface. It may be singly bonded, i.e. to one metal atom, in the so-called 'linear' or 'on-top' configuration, in which bonding is through the oxygen and the molecule sticks

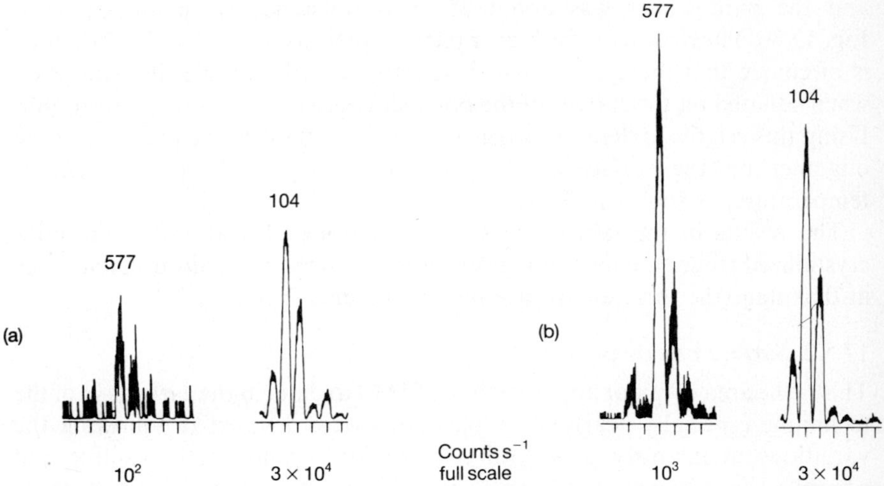

FIG. 17.30. Narrow scans over the 104 dalton and 577 dalton regions from (a) the surface of polyethyleneterephthalate (PET) film that had been solvent-extracted to remove the unwanted cyclic trimer, or oligomer, and (b) the pure cyclic trimer. The differences in the relative intensities of the peaks at the two mass numbers could be used as a fingerprint in establishing the presence of the oligomer on the PET surface, down to a level of ≈2% coverage. From Briggs.[50]

Table 17.3

Intensity of the 577 dalton peak relative to that of the 104 dalton peak in the SSIMS spectra from polyethyleneterephthalate film, compared to the relative intensity from the pure cyclic trimer. The relative intensities have been normalized to that of PET from which the trimer has been extracted.

Sample	Relative intensity 577 D/104 D
Extracted PET	1.0
Unextracted PET	1.7
PET after 1 h at 170°C	6.3
PET after 93 h at 170°C	20
Pure cyclic trimer	25

up from the surface; it may be doubly bonded to two metal atoms, in which case the molecule lies down roughly parallel to the surface in the so-called 'bridged' configuration; or it may even be triply bonded to three metal atoms, in which the oxygen is doubly bonded and the carbon singly, again roughly

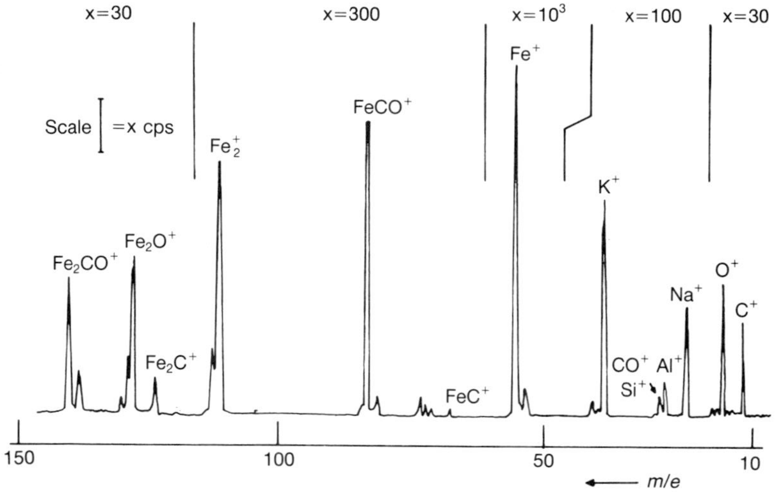

FIG. 17.31. Positive ion spectrum recorded by Barber *et al.*[51] from a polycrystalline iron surface after exposure to CO at 195 K and at a pressure of 10^{-8} torr, to saturation. The incident argon ion beam was at an energy of 3 keV, and the primary current was 5×10^{-10} A cm^{-2}. Prominent Fe_nCO^+ peaks appear, as well as others due to the dissociation of CO.

parallel to the surface. Barber *et al.*[51] established that secondary ions of the types M_nC^+ and M_nO^+ were characteristic of dissociative adsorption of CO, while species such as M_nCO^+ appeared following molecular adsorption. The problem in using the intensities of individual species for interpretation of surface reactions and for quantification is that the work function changes will affect the ionization probabilities and hence the ion yields, of all species. The effect of the work function changes can be partly cancelled by using *ratios* of selected peak intensities rather that the absolute intensities themselves.

In their study of CO adsorption Brown and Vickerman[52–54] have used intensity ratios either of the type $\sum(M_nCO^+/M_n^+)$ or of the type $M_nCO^+/\sum M_nCO^+$ as monitors of structural changes occurring at the surfaces of Ru, Ni and Pd single-crystal faces. They measured relative coverage of CO by using the areas under thermal desorption peaks after each exposure. When they plotted $\sum(M_nCO^+/M_n^+)$ as a function of relative coverage for each of the three surfaces, they found the result shown in Fig. 17.32. For all three there was an accurately linear dependence up to $\theta/\theta_{max} = \sim 0.6$, at which coverage there was a break in slope followed by a second linear dependence. On the Ru and Ni surfaces the second slope was slightly lower than the first, but in Pd there was a sharp increase in slope. On all three surfaces CO adsorbs at room temperature in an ordered structure, the $(\sqrt{3} \times \sqrt{3})R30°$, as established by LEED, which reaches its optimum ordering near $\theta/\theta_{max} = 0.6$. Beyond that coverage further adsorption of CO leads to compression in the adsorbed layer,

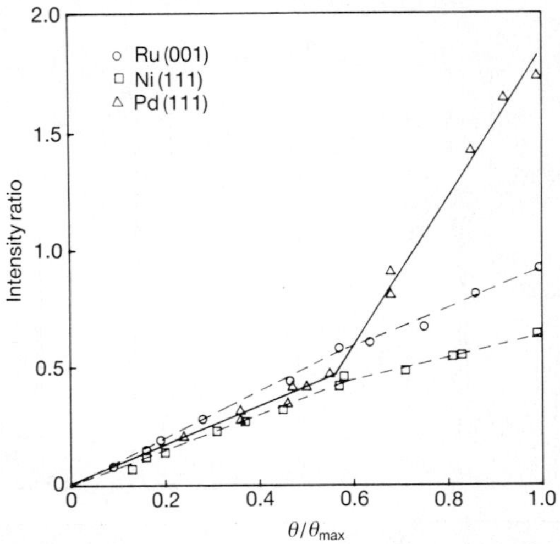

FIG. 17.32. Plot by Brown and Vickerman[53] of the ion ratio $\sum_n(M_nCO^n/M_n^+)$, against coverage determined by thermal desorption spectroscopy, for CO adsorption at 300 K on the metals Ru(001), Ni(111), and Pd(111). At the coverage $\theta=0.6$ the CO adsorption reaches its optimum surface ordering in one particular structure. Beyond that coverage the adsorbed layer becomes compressed, leading to the formation of another ordered structure. The ion ratio is thus an indicator of structural changes, and, because of the accurate linear dependences, also a good measure of CO concentration.

with the appearance of another structure, the $c(4\times2)$. Thus the ratio $\sum_n(M_nCO^+/M_n^+)$ is both a good measure of CO concentration on those surfaces, and an indicator of structural changes.

The interest lies in whether or not the SSIMS features can be used to estimate just what structural changes are taking place. According to other types of measurement, principally LEED, HREELS and RAIRS, CO adsorbs exclusively in 'on-top' sites on Ru (001) up to $\theta/\theta_{max}=0.6$, and at higher coverages still in 'linear' but not necessarily 'on-top' sites. On Ni (111) the CO adsorption is mainly of the doubly or 'bridge'-bonded type. The third type of bonding, the triple bridging, is found on Pd(111) at low coverages, giving way to the doubly bonded adsorption as the coverage increases. It is tempting to allocate MCO^+ to linear sites, and M_2CO^+ and M_3CO^+ to doubly and triply bonded sites, respectively, but that would be too naïve. Figure 17.33, in which the ratio $MCO^+/\sum_n M_nCO^+$ for Ru(001) and Ni(111) is compared[53] with the intensity of the linear stretching frequency of the CO molecule from HREELS, shows that for Ru (001) both ratio and intensity are approximately constant at ≈0.9, while for Ni(111) the two follow one another closely, as a function of CO exposure. Thus the SSIMS spectrum from linearly bonded CO is mostly but not exclusively MCO^+, while that from doubly- or bridge-bonded CO has a variable proportion of MCO^+ and M_2CO^+ according to the coverage. For CO on Pd the situation is even more complex, with contributions from

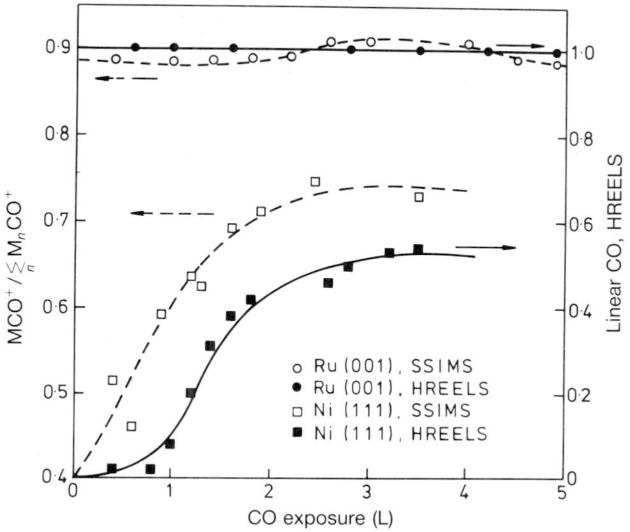

FIG. 17.33. Comparison of the ion ratio $MCO^+/\sum_n M_nCO^+$ for the adsorption of CO on Ru(001) at 300 K (circles) and on Ni(111) at 195 K (squares), with the proportion of linearly adsorbed CO calculated from relative band intensities in HREELS, by Brown and Vickerman.[53] For Ru(001) the ratio and intensity are constant, indicating that linearly bonded CO gives rise mostly to MCO^+ in the SSIMS spectrum, while for Ni(111), where it is known that CO adsorption is mainly of the 'bridge-bonded' type, it can be deduced that bridge-bonded CO gives rise to a variable proportion of MCO^+ and M_2CO^+ according to the coverage.

$PdCO^+$, Pd_2CO^+, and Pd_3CO^+, as seen in Fig. 17.34, from Brown and Vickerman.[53] However, because M_3CO^+ was not observed from linear or bridged sites, the conclusion was that Pd_3CO^+ was indeed characteristic of the triply bonded adsorption on Pd. The authors were able on the basis of these results to go further and calculate as a function of exposure the proportion of linear bonded CO, which was then compared[52] with the HREELS estimate of the same quantity. Figure 17.35 shows the result. On Ru(001) there is linear bonded CO at all exposures, but for Ni(111) the bonding is almost entirely bridged at low coverages, changing to approximately equal properties of linear and bridged as the CO adsorption reaches saturation. On the other hand, the analysis for Pd(111) indicated that at low coverages adsorption into all three types of site occurred, but once the coverage reached the point where structural ordering could take place, the doubly and triply bonded configurations predominated.

17.5.3 *Passive oxide films*

One of the major advantages of SIMS over most of the other surface analytical techniques is its ability to detect hydrogen and hydrogen-containing species. In dynamic SIMS this advantage is put to good use in depth profiling, e.g. in

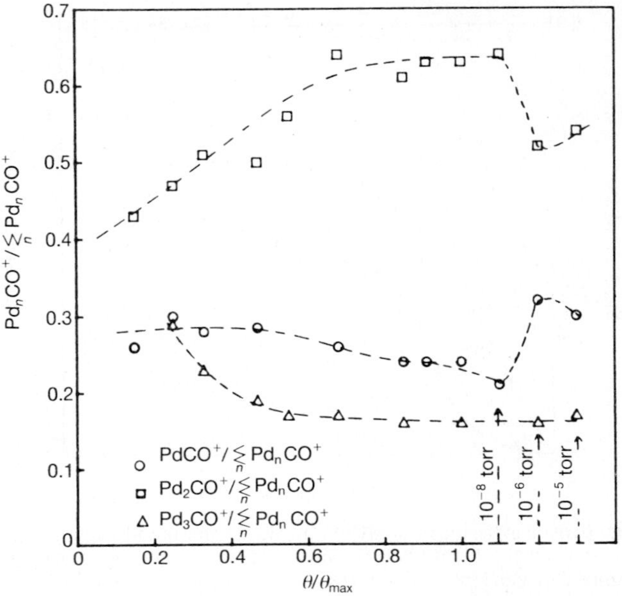

FIG. 17.34. Variations of the various ion ratios $Pd_nCO^+/\sum_n Pd_nCO^+$ as a function of coverage of CO adsorbed on Pd(111) at 300 K, recorded by Brown and Vickerman.[53] The situation is more complex than for Ru and Ni in Fig. 17.33, in that there are contributions from $PdCO^+$, Pd_2CO^+, and Pd_3CO^+ at all coverages. However, infra-red studies indicate that the initial adsorption is principally into a triply-bonded or threefold site, in which case the species Pd_3CO^+ can be identified as characteristic of CO in that site since the Pd_3CO^+ drops with coverage as adsorption goes partially into twofold sites near $\theta = 0.3$. At saturation infra-red data suggest a mixture of two- and threefold sites, which would account for the persistence of Pd_3CO^+.

amorphous silicon–hydrogen alloys[55] but in static SIMS there are problems to be overcome. Hydrogen is one of the major residual gases in an ultra-high vacuum environment, and water vapour is generally present as well although at a much lower relative level. Because the rate of erosion in dynamic SIMS is so high, the contribution to the secondary ion signal from the background gases is proportionately insignificant, but the reverse is true in SSIMS, and if hydrogen, hydroxyl or any other secondary ion that has a counterpart in the residual gas spectrum is to be analysed unambiguously, special precautions must be taken.

Such precautions had to be taken very seriously by Mitchell *et al.*[56] who used SSIMS to attempt to establish whether or not protective passive films on metals did indeed contain hydroxyl ions. The first problem that arose was due to the fact that the electrochemically formed passive films, which were introduced on sample holders from outside the vacuum system, could not tolerate any heating and thus the system could not be baked after sample introduction. In order to achieve a sufficiently good vacuum in the region of

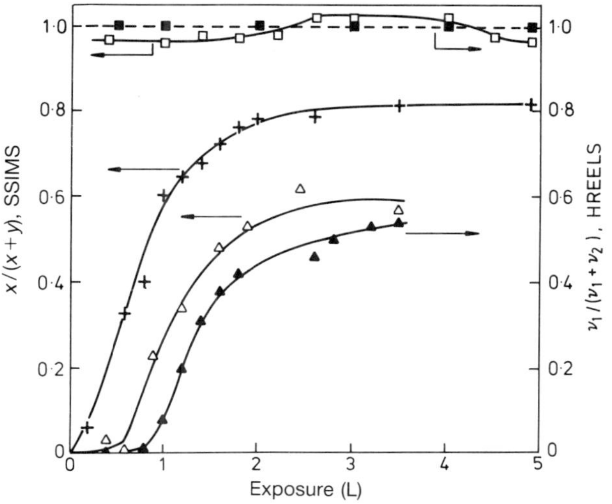

FIG. 17.35. Comparison of the proportion of linearly adsorbed CO from SSIMS, in terms of the ratio $x/(x+y)$, and from HREELS, in terms of the ratio $\nu_1/(\nu_1+\nu_2)$. Here $x/(x+y)$ is derived from the ion ratio $MCO^+/(MCO^+ + M_2CO^+)$, and will be 0.9 for entirely linear adsorption and 0.5 for entirely bridged adsorption. ν_1 and ν_2 are the vibrational intensities in HREELS in bands attributed to linear and bridged CO respectively. The other symbols have the meanings:

■: Ru(001) at 300 K. □: Ru(001) at 195 K.
△: Ni(111) at 195 K. ▲: Ni(111) at 140 K.
+: Ni(111) at 300 K.

The trends observed by SSIMS follow closely those by HREELS, indicating the correctness of the approach by Brown and Vickerman.[52] For Ru(001) the proportion of linear CO remains unity throughout the exposure. For Ni(111) at low temperature bridged CO predominates, but at room temperature there is a mixture of types of site.

the sample it was therefore necessary to place a cryogenic panel just above it, cooled to 20 K. On pumping for half a day after insertion the base pressure near the sample had dropped to 3×10^{-10} torr. At that stage very intense hydroxyl signals were obtained during SSIMS analysis, and their true origin had to be established before any conclusions could be drawn about the sample itself. By varying the sputter rate and the sample-to-quadrupole distance, the authors were able to show that most of the hydroxyl signal was in fact originating within the analyser itself! A simple calculation based on analyser volume and pumping conductance to the internal volume showed that the analyser could never be adequately pumped. Modifications that included increasing the conductance by a factor of ≈ 20, insertion of a cryogenically cooled surface in the lens region, and additional auxiliary heating of the lens by electron bombardment in order to outgas it properly were all necessary in order to eliminate the very large hydroxyl contribution previously arising from inadequately baked and pumped regions of the analyser. Further investigation of the origin of the remaining hydroxyl signal showed that once

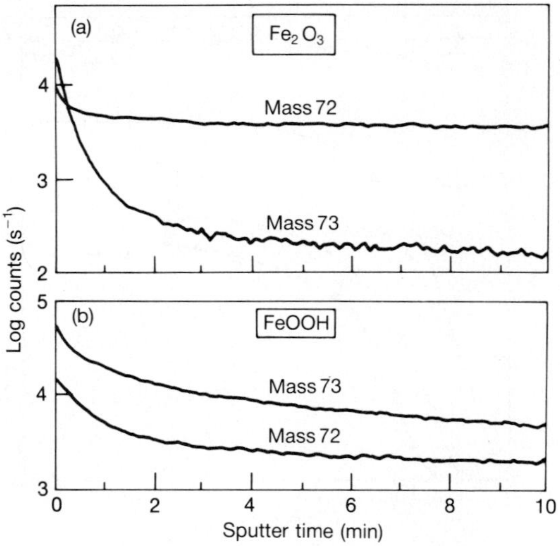

FIG. 17.36. Profiles using SSIMS performed by Mitchell *et al.*[56] to establish the ions to be used as indicators of the presence or absence of hydroxyl. The standards employed were 'dry' sputter-deposited Fe_2O_3 that should contain no hydroxyl, and an anodically formed 'wet' FeOOH film on platinum, that should contain a great deal of hydroxyl. Mass 72 is $^{56}FeO^+$, and mass 73 is ($^{56}FeOH^+ + {}^{57}FeO^+$), but ^{57}Fe is known to be 2.5% of ^{56}Fe, so the contribution of $^{57}FeO^+$ to mass 73 could be accurately evaluated. At equilibrium, after about 7 min sputtering, the ratio $FeOH^+/FeO^+$ was 0.05 for Fe_2O_3, but 3.5 for FeOOH (note the log intensity scale). Thus these ions can be used as good 'fingerprints' for the presence or absence of hydroxyl.

again much of it was coming from a source other than the sample, in this case from the outgassing of components of the ion gun. Reduction was achieved by local baking of the source followed by packing it with solid CO_2. With these modifications the background hydroxyl was brought down to a level at which the analysis could be made unambiguously. The above steps that were found to be necessary are described in detail in order to emphasize the lengths to which one might have to go in SSIMS in order to achieve unambiguous interpretation.

In analysing for hydroxyl in SSIMS, there are ions other than OH^- that can be used as indicators, for example MO^+, MOH^+, $MOOH^-$, etc. OH^- itself will probably have the greatest count rate, but can as described above suffer severely from interferences from the instrumental background. Mitchell *et al.* decided to use the metal-containing ions MO^+ and MOH^+, since they can be recorded in the same (positive) spectrum, and being separated by only one dalton will avoid any quantification problems associated with mass-dependent analyser transmission. For calibration they used two standards, a sputter-deposited Fe_2O_3 film that should contain no hydroxyl, and an FeOOH film anodically (i.e. wet) deposited on to a platinum substrate, that by definition should contain a large amount of hydroxyl. The SSIMS profiles

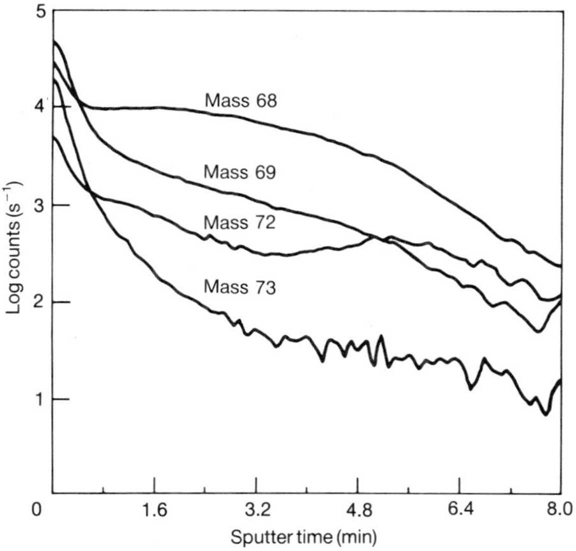

FIG. 17.37. Profile recorded by Mitchell *et al.*[56] through a passive film on Fe–25Cr alloy formed by anodization in H_2SO_4 in the passive potential range. SSIMS analytical conditions were the same as for the standardization experiments of Fig. 17.36, viz. Xe^+ at 1 keV and a sputtering rate of $\simeq 4$ Å min^{-1}. The masses used correspond to: 68, $^{52}CrO^+$; 69, ($^{52}CrOH^+ + ^{53}CrO^+$); 72, $^{56}FeO^+$; 73, ($^{56}FeOH^+ + ^{57}FeO^+$). ^{53}Cr is known to be 11.3% of ^{52}Cr. On the basis of the calibration shown in Fig. 17.36, and of similar calibration performed for Cr-containing standards, it could be estimated that the hydroxyl content in the outer iron-rich part of the film was about 1.1% and in the inner chromium-rich part, about 0.6%. Thus passive films on Fe–25Cr are almost hydroxyl free, in contrast to films formed by electropolishing, which were found to contain up to 15% hydroxyl at the surface.

obtained from these two standards, using masses 72 ($^{56}FeO^+$) and 73 ($^{56}FeOH^+ + ^{57}FeO^+$), are shown in Fig. 17.36. Since ^{57}Fe is 2.5% of ^{56}Fe, the contribution to mass 73 from FeO^+ could be accurately evaluated and subtracted. For both standards there was an initial drop in signal intensities of the two mass numbers, as the surface layers were removed, but at equilibrium (after ≈ 7 min sputtering) the differences between the standards are quite marked. The ratio $FeOH^+/FeO^+$ changed from 0.05 for Fe_2O_3 to ≈ 3.5 for FeOOH; clearly the chosen polyatomic species represent good 'fingerprints' of the presence or absence of hydroxyl. Similar calibratory profiles were obtained for $^{52}CrO^+$ and $^{52}CrOH^+ + ^{53}CrO^+$, and for $^{58}NiO_2^-$ and $^{58}NiOOH^-$.

A profile through a film on an Fe–25Cr alloy anodized in H_2SO_4 in the passive potential range, under the same SSIMS conditions as for Fig. 17.36, is shown in Fig. 17.37, using the mass numbers appropriate to Fe and Cr. In order to extract from the data in Fig. 17.37 an estimate of the hydroxyl content in different regions of the passive film, it is assumed that the system factors affecting the film are the same as those affecting the standards. These are, the ^{57}Fe and ^{53}Cr contributions, the degree of hydroxyl adsorption on the surface, and the continuous reaction of the sputtered area with residual water vapour. When these are taken into account the percent hydroxyl in the film as a function of sputtering time can be estimated from expressions containing the time-dependent 68, 69, 72, and 73 signals from the film and the standards. The result is that the outer part of the passive film contains about 1.1% OH^-, and the inner part about 0.6% OH^-. On pure Fe and Ni no OH^- was found in a passive film. Thus it is clear that on ferrous materials the films formed under passivating conditions contain essentially no hydroxyl. By contrast analysis of films formed by electropolishing showed hydroxyl contents ranging from 2 to 15% on Fe–25Cr and from 20% to 50% in Ni. On this basis one of the principal differences between passivating and non-passivating films lies in the presence or absence of hydroxyl.

References

1. Benninghoven, A. (1969). *Z. Phys.* **220**, 159.
2. Wilson, R. G. and Brewer, G. R. (1973). *Ion beams*, Ch. 3, p. 214. John Wiley and Sons, New York.
3. Magee, C. W., Harrington, W. L., and Honig, R. E. (1978). *Rev. Sci. Instrum.* **49**, 477.
4. Benninghoven, A., Plog, C., and Treitz, N. (1974). *Int. J. Mass. Spectrom.* **13**, 415.
5. Dawson, P. H. (1977). *Surf. Sci.* **65**, 41.
6. Dawson, P. H. (1978). *Surf. Sci.* **71**, 247.
7. Paul, W., Reinhard, H. P., and von Zahn, U. (1958). *Z. Phys.* **152**, 152.
8. Wittmaack, K. (1982). *Vacuum* **32**, 65.
9. Gnaser, H., Fleischhauer, J., and Hofer, W. O. (1985). *Appl. Phys.* **A37**, 211.
10. Staib, P. (1972). *J. Phys. E.* **5**, 484.
11. Dawson, P. H. and Redhead, P. A. (1977). *Rev. Sci. Instrum.* **48**, 159.
12. Krauss, A. R. and Gruen, D. M. (1977). *Appl. Phys.* **14**, 89.
13. Steffens, P., Niehuis, E., Friese, T., Greifendorf, D., and Benninghoven, A. (1985). *J. Vac. Sci. Technol.* **A3**, 1322.
14. Bletsos, I. V., Hercules, D. M., Benninghoven, A., and Greifendorf, D. (1986). *SIMS V* (eds. A. Benninghoven, R. J. Colton, D. S. Simons, and H. W. Werner). Springer-Verlag, Berlin.
15. Waugh, A. R., Kingham, D. R., Hearn, M. J., and Briggs, D. A. (1988). *SIMS VI* (eds. A. Benninghoven, A. M. Huber, and H. W. Werner), p. 231. John Wiley and Sons, Chichester.
16. Eccles, A. J. and Vickerman, J. C. (1988). *SIMS VI* (eds. A. Benninghoven, A. M. Huber, and H. W. Werner), p. 239. John Wiley and Sons, Chichester.
17. Benninghoven, A., Müller, K-H., Schemmer, M., and Beckman, P. (1977). Proc. 3rd Int. Conf. Solid Surfaces, Vienna, p. 1063.
18. Pivin, J.-C. (1983). *J. Mater. Sci.* **18**, 1267.
19. Sigmund, P. (1977). *Inelastic ion-surface collisions* (eds. N. H. Tolk, J. C. Tully, W. Heiland, and C. W. White), p. 121. Academic Press Inc., New York.
20. Benninghoven, A., Rüdenauer, F. G., and Werner, H. W. (1987). *Secondary ion mass spectrometry*, ch. 2. John Wiley and Sons, New York.
21. Vasile, M. J. (1984). *Phys. Rev. B.* **29**, 3785.

22. Andersen, C. A. and Hinthorne, J. R. (1973). *Anal. Chem.* **45**, 1421.
23. Wittmaack, K. (1980). *Nucl. Instrum. Meth.* **168**, 343.
24. Joyes, P. (1973). *Rad. Effects* **19**, 235.
25. Wittmaack, K. (1977). *Inelastic ion-surface collisions* (eds. N. H. Tolk, J. C. Tully, W. Heiland, and C. W. White), p. 153. Academic Press, London.
26. Fluit, J. M., Friedman, L., Van Eck, J., Snoek, G., and Kistemaker, J. (1961). Proc. Int. Conf. Ionization Phenomena in Gases, Munich, p. 131.
27. Blaise, G. and Slodzian, G. (1970). *J. Phys.* **31**, 93.
28. Blaise, G. and Slodzian, G. (1974). *J. Phys.* **35**, 243.
29. Rüdenauer, F. G. (1977). *Mikrochim. Acta* (Vienna), **Suppl. 7**, 85.
30. Schroeer, J. M., Rhodin, T. N., and Bradley, R. C. (1973). *Surf. Sci.* **34**, 571.
31. Beske, H. (1967). *Z. Naturforsch.* **22a**, 459.
32. Šroubek, Z. (1974). *Surf. Sci.* **44**, 47.
33. Šroubek, Z. (1983). *Int. J. Mass Spectrom.* **53**, 289.
34. Blaise, G. and Nourtier, A. (1979). *Surf. Sci.* **90**, 495.
35. Yu, Ming L. (1983). *J. Vac. Sci. Technol.* **A1**, 500.
36. Yu, Ming L. (1982). *Phys. Rad.* **B26**, 4731.
37. Yu, Ming L. and Lang, N. D. (1986). *Nucl. Instrum. Meth. Phys. Res.* **B14**, 403.
38. Nørskov, J. K. and Lundqvist, B. I. (1979). *Phys. Rad.* **B19**, 5661.
39. Brako, R. and Newns, D. M. (1981). *Surf. Sci.* **108**, 253.
40. Loxton, C. M. and Tsong, I. S. T. (1984). *Surf. Sci.* **139**, 453.
41. Werner, H. W. (1980). *Surf. Interface Anal.* **2**, 56.
42. Smith, D. H. and Christie, W. H. (1978). *Int. J. Mass Spectrom.* **26**, 61.
43. Christie, W. H., Eby, R. E., Anderson, R. L., and Kollie, T. G. (1979). *Appl. Surf. Sci.* **3**, 329.
44. Werner, H. W. and Boudewijn, P. R. (1984). *Vacuum* **34**, 83.
45. Schroeer, J. M. (1977). *J. Vac. Sci. Technol.* **14**, 343.
46. MacDonald, R. J. and Garrett, R. F. (1978). *Surf. Sci.* **78**, 371.
47. Gardella, Jr., J. A. and Hercules, D. M. (1980). *Anal. Chem.* **52**, 226.
48. Briggs, D. and Hearn, M. J. (1985). *Spectrochim. Acta*, **40B**, 707.
49. van Ooij, W. J. and Brinkhuis, R. H. G. (1988). *Surf. Interface Anal.* **11**, 430.
50. Briggs, D. (1986). *Surf. Interface Anal.* **8**, 133.
51. Barber, M., Vickerman, J. C., and Wolstenholme, J. (1977). *Surf. Sci.* **68**, 130.
52. Brown, A. and Vickerman, J. C. (1982). *Surf. Sci.* **117**, 154.
53. Brown, A. and Vickerman, J. C. (1983). *Surf. Sci.* **124**, 267.
54. Brown, A. and Vickerman, J. C. (1985). *Surf. Sci.* **151**, 319.
55. Lundquist, T. R., Burgner, R., Swann, P. R., and Tsong, I. S. T. (1981). *Applications Surf. Sci.* **7**, 2.
56. Mitchell, D. F., Sproule, G. I., and Graham, M. J. (1985). *Applic. Surf. Sci.* **21**, 199.

18

Ion excitation: secondary neutral mass spectrometry (SNMS) and glow discharge mass spectrometry (GDMS)

It has already been stated earlier that most of the particles ejected as a result of ion bombardment are neutrals, either atomic or molecular species. Since the neutral flux does not depend on any ionization mechanism but on the sputtering yield, which is only weakly dependent on matrix composition, it is obvious that if the neutral component could be measured independently, an analysis technique would be available that would be both more sensitive and more quantifiable than SSIMS. To measure the neutral signal requires conversion of the neutral flux to ions after ejection from the surface, followed by mass analysis in the usual way, taking care that interference from true secondary ions is avoided. Both SNMS and GDMS depend on postionization of neutrals, and indeed the two techniques are very similar, differing mainly in the precise conditions used for the ionization.

18.1 Secondary neutral mass spectrometry

Three approaches to postionization have been made in the last few years. These are (a) electron impact ionization, (b) gas discharge ionization, and (c) multiphoton ionization. Each has its own advantages and disadvantages.

18.1.1 *Operation*

Electron impact postionization As the name suggests, this approach employs direct ionization of neutrals by an electron beam from a hot filament. Two designs of postionization optics have recently been developed. The first, due to Lipinsky *et al.*,[1] is shown diagrammatically in Fig. 18.1. The target (1) is positioned opposite a 3 mm hole in a stainless steel cap (2) that encloses the ionizer. Just inside the cap is a deflection capacitor (3). The cap, which is insulated via a Macor insulator (7), and the deflection capacitor, can be employed together or separately to suppress the passage of secondary ions in the SNMS mode, or to allow them to pass through to the mass filter in the SSIMS mode. The ionizer itself (4) has dual rhenium filaments on either side of the axis, the electron beam being confined by a small permanent magnetic field so that the potential, and its span, of ion formation are readily controlled. Before and after the ionizer are placed shielding apertures, whose function is to maintain field-free conditions in the ionization volume when the deflection

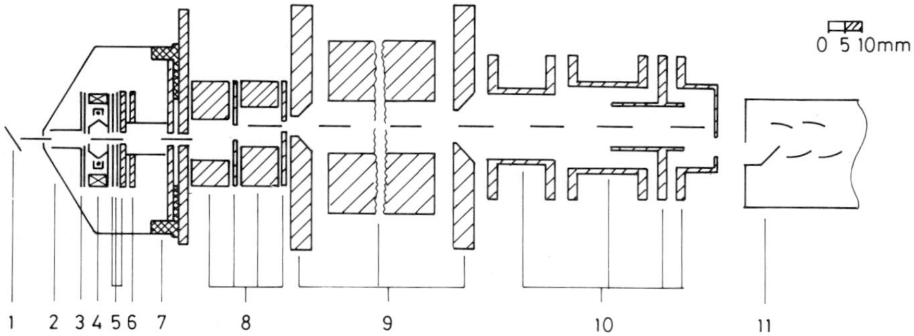

FIG. 18.1 Experimental arrangement for electron impact postionization of neutrals constructed by Lipinsky *et al.*[1] (1) Target (i.e. specimen). (2) Stainless steel cap enclosing ionizer, with a 3 mm diameter hole opposite the target. (3) Deflection capacitor, used to suppress secondary ions in the SNMS mode. (4) Ionizer with twin rhenium filaments and a small permanent magnet to confine the electron beam. (5) Retardation and extraction electrodes with apertures. (6) Cylindrical lens that focuses ions onto an aperture. (7) Macor insulator on which ionizer assembly is mounted. (8) Double lens system with offset halves to provide a deflection field. (9) Quadrupole mass spectrometer. (10) Immersion lens that focuses ions onto the input to the multiplier. (11) Electron multiplier with offset entrance. The device can be switched between SNMS and SSIMS operation within a few ms.

capacitor, and the retardation and extraction electrodes (5), are in use. Since the sputtered neutrals have energies in the range 2–5 eV, whereas the residual gas molecules have thermal energies of a fraction of an eV, as shown in Fig. 18.2, a weak retarding voltage of ≈ 1 eV on the retardation electrode is

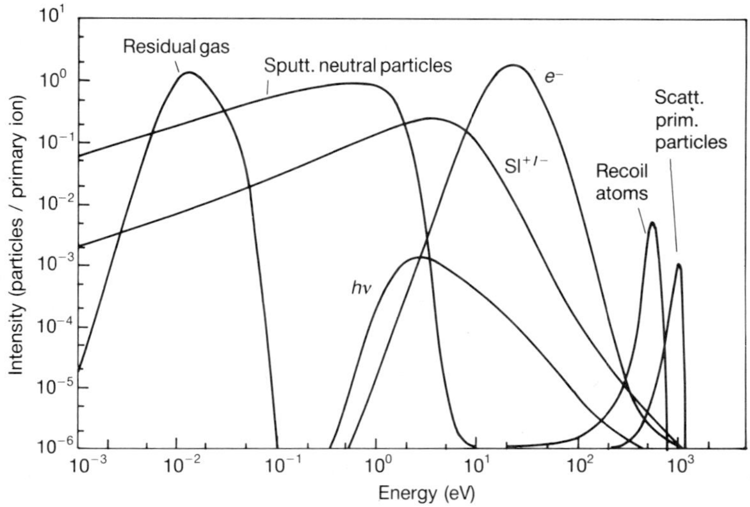

FIG. 18.2. Schematic indication of the energy distributions of the variety of particles emitted from a surface under ion bombardment, compared with the thermal energy distribution of ions created by ionization of residual gas molecules. From Lipinsky *et al.*[1]

sufficient to suppress further any ions formed in the ionizer from the residual gas. After acceleration by the extraction electrodes ions are focused onto an aperture by a cylindrical lens (6), and then pass into a double lens system (8) whose two identical halves are offset in order to provide a deflection field. This field acts as a low-pass filter that prevents secondary ions of high energy (see Fig. 18.2) from entering the quadrupole mass spectrometer (9), while the off-axis arrangement serves as a stop for high energy neutrals and light. Finally, after mass separation in the mass spectrometer, the ions are focused by an immersion lens (10) onto the (again) offset entrance of the multiplier (11).

With the above instrument, discrimination factors of $>10^5$ and $>10^6$ were obtained against unwanted signals from residual gases and from secondary ions, respectively. Maximum count rates in the SNMS mode were in the range $1\text{–}2 \times 10^4$ per μA ion current and per mA postionizing electron current, and in the SSIMS mode were between 10^5 and 10^6 per μA ion current. Switching between the two modes could be performed electrostatically within a few ms. The respective SSIMS and SNMS spectra obtained under identical bombardment conditions from GaAs are shown in Fig. 18.3, from Lipinsky *et al.*[2] Note the difference by many orders of magnitude between the Ga^+ and As^+ signals in SSIMS, but the virtual equality of the Ga and As signals in SNMS, emphasizing the severe intrusion of matrix effects into the ionization probability in SSIMS.

The second recent version of an SNMS instrument using electron impact postionization, and also capable of being switched rapidly to SSIMS operation, is that of Wilson *et al.*[3] Their aim was to optimize secondary neutral collection and detection without compromising the SSIMS capability, but as with others concerned with SNMS the principal design difficulty encountered was that of the removal of interferences from secondary ions and residual gas ions.

The design of Wilson *et al.* is shown in Fig. 18.4. The first ion optical element is in fact similar to that of Lipinsky *et al.*, being a hemicylindrical grid attached to a plate, in which there is a 7 mm aperture. The function of the first element is to discriminate against secondary positive ions, a potential of +150 V being applied to it, which the authors state leads to a secondary ion suppression of some six orders of magnitude. Following this 'discriminator' is a screen plate with a larger aperture, whose function is principally to create an attractive field for positive ions when operating in the SSIMS mode, but according to the authors it has an unexpectedly beneficial effect on secondary ion suppression in SNMS when a bias of −180 V is applied. A further order of magnitude suppression is obtained, thought by them to be due to the prevention of the passage of secondary ions through the ionizer, that arise by bombardment of the front of the ionizer by high energy particles. The ionizer itself consists of a hot filament, an ionization 'cage' or volume, and a surrounding filament box. The filament is wound from tantalum wire and can provide 2 mA emission.

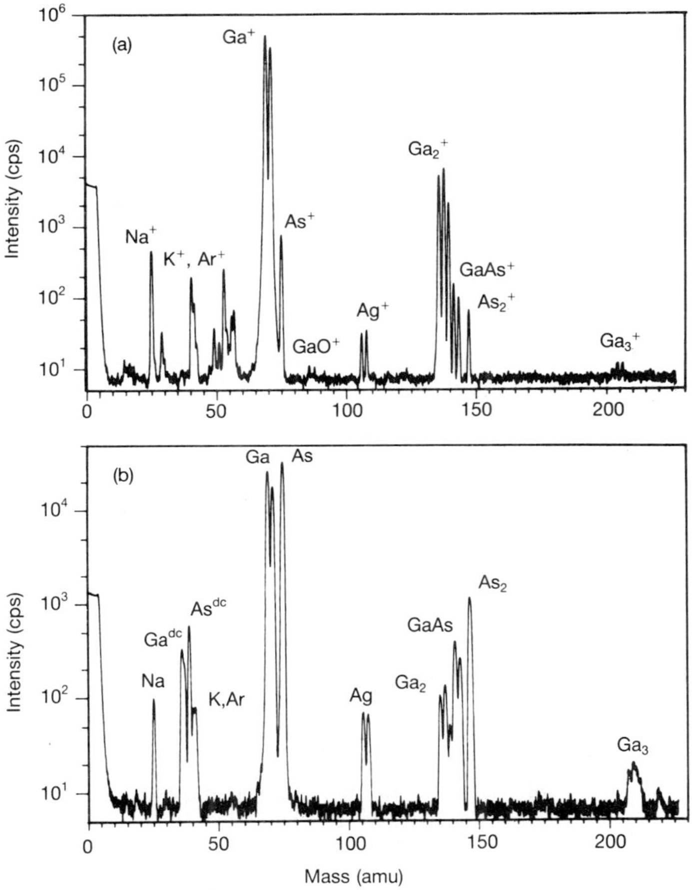

FIG. 18.3. The positive secondary ion spectra from GaAs obtained by switching the instrument described in Fig. 18.1 from (a) SSIMS to (b) SNMS, measured by Lipinsky *et al.*[2] Note the logarithmic intensity scale. The relative intensities of the Ga^+ and As^+ signals are virtually equal in SNMS, but differ by orders of magnitude in SSIMS, owing to severe matrix effects.

The cage is probably the most critical element, since the ionization probability of neutrals passing through the cage must be maximized if the design performance is to be achieved. It is constructed from open high transmission grid material, and is of diameter 3 mm. Thus it acts as a collector for electrons emitted from the filament and accelerated through 60 eV, but the high transmission of the grid allows easy access of electrons to its internal volume, which ensures long oscillatory electron paths within the ionization volume. The lifetime of electrons within this volume is increased significantly by the application of a permanent magnetic field of 1300 gauss at right angles to the ionizer axis. The grid potential defines the energy of the postionized particles

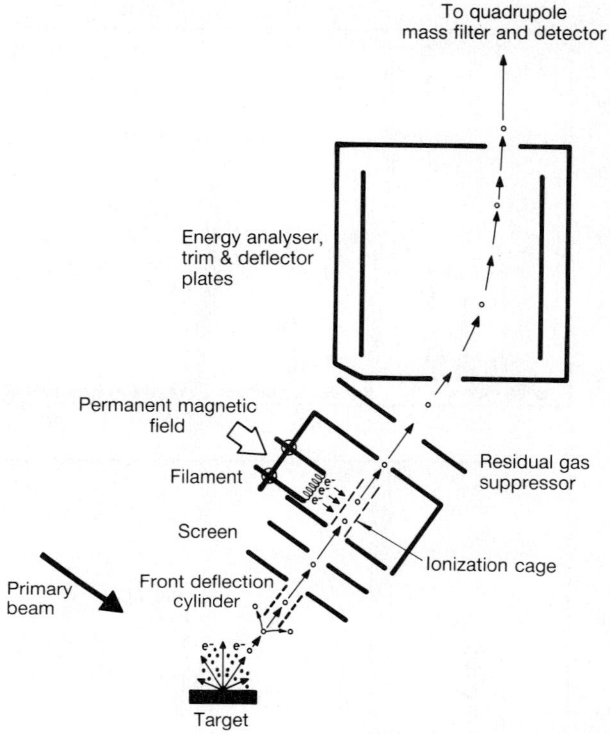

FIG. 18.4. Experimental arrangement for electron impact postionization of neutrals constructed by Wilson *et al.*[3] Similarly to the device shown in Fig. 18.1, it is also capable of switching rapidly between SSIMS and SNMS. The first element after the target is a hemicylindrical grid attached to a plate, in which there is a 7 mm aperture, acting as a deflection cylinder to suppress secondary ions in SNMS. After that there is a screen plate with a bias that attracts positive ions in the SSIMS mode, but that also suppresses secondary ions in the SNMS mode. The ionizer consists of a hot filament, an ionization cage, and a box surrounding both. The cage is of diameter 3 mm and is constructed from high transmission grid mesh. Electron lifetime is increased by a permanent magnetic field at right angles to the axis. The relative potentials on the filament box and on the cage impose an extraction field on the ions created in the ionizer. On leaving the ionizer, the positive ions pass through an apertured plate acting as a residual gas ion suppressor, and thence into an energy analyser of the Wittmaack[4] type. The analyser deflects ions, those within a particular energy interval passing through the exit aperture into the quadrupole mass spectrometer.

as they pass into the subsequent energy analyser. The filament box is at the same potential as the filament, which, when combined with the grid potential, acts as an extraction field for the newly created positive ions.

After leaving the ionizer, the positive ions pass through another element in the form of an apertured plate, termed the 'residual gas suppressor'. As can be seen from Fig. 18.2, residual gas ions have only thermal energies, whereas the ionized neutrals have energies, relatively, of a few eV, thus the potential of the residual gas suppressor can be adjusted with respect to both the ionization cage and the energy analyser, to discriminate against the residual gas ions. The purified ionized neutral beam then passes into the energy analyser, which is of the so-called 'Wittmaack-box' type, first described by Wittmaack *et al.*[4] The entrance and exit apertures are offset, and asymmetric deflection voltages of ≈ -20 V and ≈ -50 V are applied to the plates inside the box; all ions are deflected, those in a particular energy interval passing through the exit aperture, while the remainder are rejected by striking the walls. Application of a negative potential to the box further suppresses high energy secondary ions. After energy selection, the ionized neutral beam then moves on into a standard quadrupole mass spectrometer for mass analysis, with detection in the normal way by a channel electron multiplier.

During evaluation of the above instrument Wilson *et al.* achieved a residual gas and secondary ion suppression of $> 10^6$, but found significant signals at masses corresponding to O, F and Cl. The latter species are all those with high electron desorption cross-sections, which suggested that they were originating from somewhere within the optics themselves. The most likely source was from the ionization cage as a result of electron bombardment during operation of the ionizer. Confirmation of this came by adjusting the discrimination potential between ionization cage, filament potential, and residual gas suppressor, to discriminate against ions of higher energies, when the parasitic signals disappeared.

Gas discharge postionization This approach has been developed by Oechsner and co-workers.[5] His innovation was to use a plasma produced by high frequency discharge at a relatively low pressure as the postionizing medium, with surface bombardment also principally from ions in the plasma, although latterly the possibility of focused ion beam bombardment in the usual way has been added.

The arrangement developed by Oechsner and co-workers[6] is shown diagrammatically in Fig. 18.5. The sample, or target, is clamped in a holder and positioned opposite an aperture in a vacuum vessel of dimensions about 15 cm diameter and 10 cm length. Argon gas is admitted to the vessel at a typical pressure of 5×10^{-4} mbar, and a plasma is generated in the argon by inductive excitation at a frequency of 27.12 MHz. The plasma is maintained under electron cyclotron resonance conditions by superimposing a small DC magnetic field of about 15 Gauss, following the method described by Oechsner.[7] With the sample surface held negative with respect to the plasma, positive ions created in the plasma are accelerated towards the sample and act as the bombardment source. Neutrals sputtered from the sample surface are emitted into the plasma volume, where a fraction of them is ionized by the electron component in the plasma, and then mass analysed in a quadrupole mass spectrometer. A lens and deflection system at the input to the quadrupole ensures that any negative ion component in the plasma volume is rejected.

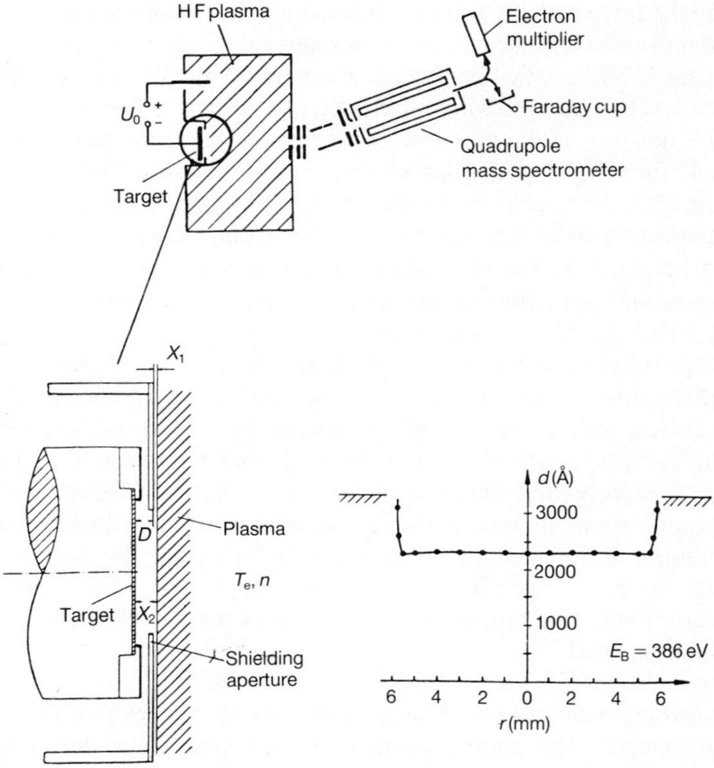

FIG. 18.5. Experimental arrangement for plasma postionization of neutrals developed by Oechsner and co-workers, the diagram from Beckmann *et al.*[6] The sample is clamped in a holder opposite an aperture in a vacuum vessel. Argon is admitted to the vessel to a pressure of about 5×10^{-4} mbar, and a plasma is generated by inductive excitation at a high frequency. Electron cyclotron resonance conditions are maintained by a small magnetic field. The sample is held negative with respect to the plasma, whereupon positive ions created in the plasma are accelerated to the sample and bombard it with energies variable between 20 eV and several keV. Neutrals sputtered from the sample pass into the plasma volume, where some are ionized by the electron gas in the plasma, and are then mass analysed in a quadrupole mass spectrometer, via an extraction and deflection system.

The overriding requirement in the method of plasma postionization is to ensure adequate electron density for ionization while at the same time minimizing the frequency of collisional interactions between the neutrals and argon species in the gas phase. At a pressure of 5×10^{-4} mbar, the mean free path for such interaction in argon is of the order of 15 cm; the length of travel from the target to the first element of the input lens to the quadrupole is therefore kept down to ≈ 5 cm, so that the influence of collisions can be neglected. The electron density and temperature under conditions of electron cyclotron resonance, as a function of argon pressure, are shown in Fig. 18.6 from Oechsner and Stumpe.[5] Although the electron density is starting to drop

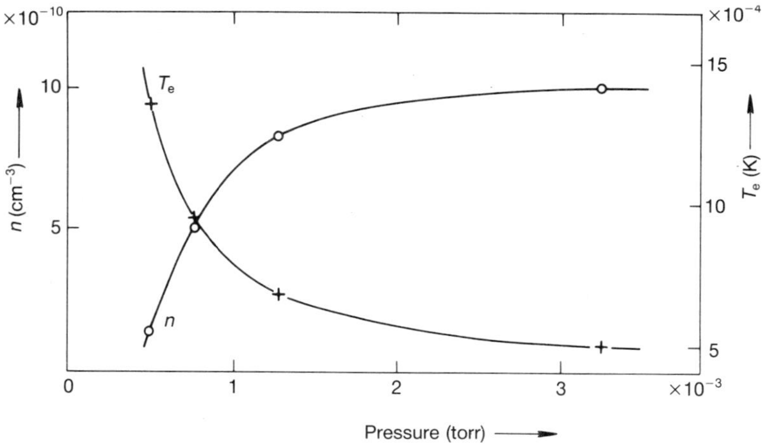

FIG. 18.6. Plasma density n and electron temperature T_e of the postionizing plasma (see Fig. 18.5) as a function of argon pressure, under electron cyclotron wave resonance conditions, plotted by Oechsner and Stumpe.[5] At the typical operating conditions for plasma postionization SNMS, the plasma density is 1.5×10^{10} cm^{-3} and the electron temperature 1.3×10^5 K.

rapidly at a pressure of a few $\times 10^{-4}$ mbar, nevertheless it is still in the region of 10^{10} cm^{-3}, which is enough to ensure a postionization probability of 10^{-2}. This figure compares well with the probability of $\approx 10^{-4}$–10^{-5} obtained in postionization by electron beam interaction.[1,3]

The primary ion energy is determined by the potential U_0 (see Fig. 18.5) applied between the target and an electrode inserted into the centre of the plasma discharge. This energy E_B can be varied between the limit imposed by the self-biasing of the floating electrode, about 20 eV, and several keV. The current density of bombardment is in the range 2–3 mA cm^{-2}.

For uniform erosion of the target surface, very important in depth profiling, it is necessary to pay careful attention to the electrostatic conditions close to the target surface. When a negative bias is applied to a target with respect to the plasma, a space charge layer forms just above the target surface at a distance X_2 (Fig. 18.5). The uniformity or otherwise of the bombarding current density depends not only on the lateral distribution of the plasma density just above the space charge layer, but also on ion optical (i.e. electrostatic) conditions at the edges of the layer. Stumpe *et al.*[8] have shown that these effects can be controlled by inserting a circular shielding aperture a distance D in front of the target, and of diameter about 11 mm. The target is electrically isolated from the aperture, which is at the same potential as the walls of the plasma volume. If the thickness of the space charge layer above the shielding aperture is X_1, then for uniform lateral intensity, i.e. a planar boundary opposite the target, clearly $X_2 = X_1 + D$. For an electron temperature of about 1.2×10^5 K, as indicated in Fig. 18.6, the authors calculated a potential difference of about 50 V between the plasma and the shielding aperture, which value was confirmed with a Langmuir probe inserted in the plasma. With this potential difference and a separation D of 1.3 mm, an optimum target bias of 378 V was calculated. The flattest crater formed by ion erosion from the plasma was that shown on the right-hand side of Fig. 18.5 for an applied bias of 386 V, agreeing excellently with the calculated value, and thereby confirming the correctness of the ion optical analysis of the target–aperture–plasma system. For other applied biases the crater was either concave or convex. Such ion energies are low compared to those used in SSIMS, and the damage caused in the near-surface region in SNMS is therefore reduced compared to that caused in SSIMS.

A variation of the SNMS arrangement, that allows the ion bombardment and the postionization to be decoupled, has also been used by Oechsner and co-workers.[9,10] In this arrangement bombardment is performed with the same type of ion source as in SSIMS, outside and separate from the plasma volume, as shown schematically[10] in Fig. 18.7. An essential additional component is an

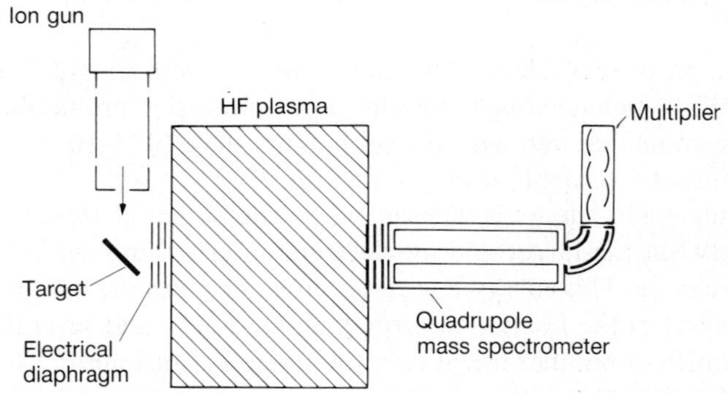

FIG. 18.7. Variation of the SNMS arrangement of Oechsner, to allow bombardment either by an external ion gun or by the positive ion component of the plasma, from the paper by Müller and Oechsner.[10] In the so-called 'direct' bombardment mode the bombardment and postionization processes are decoupled. An electrical diaphragm at the entrance to the plasma volume can be operated either in a 'shut' mode to prevent secondary ions passing into the plasma during SNMS, or in an 'open' mode to admit ions to the mass spectrometer in SSIMS, when there is no plasma present.

electrical diaphragm between the target and the plasma; this diaphragm can be used either in the 'shut' mode to prevent secondary ions passing into the plasma, or in the 'open' mode, when there is no plasma present, to allow ions to pass through to the quadrupole mass spectrometer for SIMS measurements. However, although the primary ion energy used, 4 keV, is similar to that used in SSIMS, the current densities quoted in the publications put the SIMS side of the arrangement more in the region of dynamic SIMS; for example, 4.7×10^{-5} A cm^{-2} into an area of 4.5 mm^2 implies a primary ion current of $\approx 2\ \mu A$, which is some three orders of magnitude greater than the primary currents used in SSIMS.

SNMS with plasma postionization and plasma ion bombardment also uses high current densities, in the dynamic SIMS regime, but if the ion optical parameters are adjusted very carefully as described above, then the actual depth resolution achieved during profiling can be made comparable to that in SSIMS. Across the bottom of the crater shown in Fig. 18.5 the flatness is estimated to be within ± 1 nm, and that is after removing about 100 nm of a film of thickness 330 nm. Provided the plasma bombardment conditions can be maintained at optimum at all depths, there is no reason why the depth resolution should not be maintained right through the film.

The plasma postionization method of Oechsner has been taken up commercially, and the commercial version is shown schematically in Fig. 18.8,

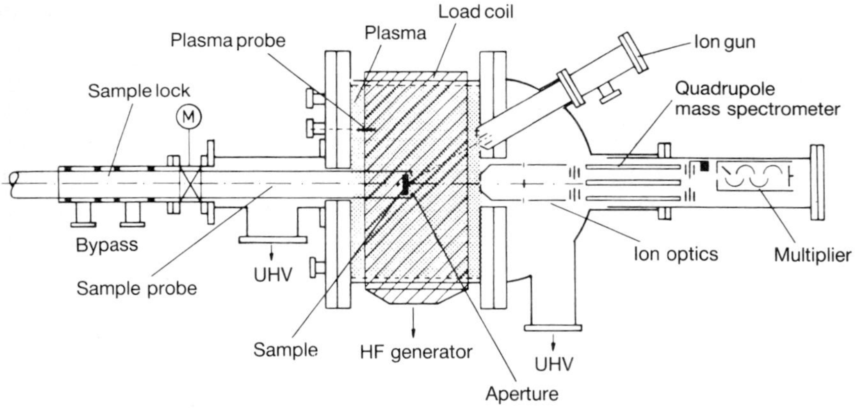

FIG. 18.8. Commercially realized version of the plasma postionization method of Oechsner, from Müller *et al.*[11] It is a combined SNMS/SSIMS instrument, and bombardment can be performed either with the positive ions from the plasma or with a separate, focused, ion gun. It is designed to operate with a base pressure in the UHV range, so that it is basically clean, and the plasma volume is made of glass, allowing good visibility.

from the paper by Müller *et al.*[11] It is also a combined SNMS/SSIMS instrument, and bombardment can be performed either with the positive ion component of the plasma or with a separate, focused and rastered, ion beam. Unlike most commercial instruments using ion bombardment and ion analysis, it is constructed to operate with a base pressure in the UHV range, so that it is basically a very clean device. The central portion of the plasma volume is constructed from glass, giving excellent visibility of the specimen during loading and analysis. The arrangement is very similar to that of Oechsner, with some additional sophistication in the form of automatic specimen selection and loading.

Multiphoton postionization Neutrals ejected from a surface by ion bombardment can also be ionized by high intensity laser beams, involving both resonant and non-resonant multiphoton ionization. In both cases adequate sensitivity can be achieved only by ensuring high ionization probabilities (i.e. by saturating the various allowed multiphoton transitions) which requires very high power lasers, in the range 10^6–10^{10} W cm^{-2}. It is the development of the latter that has in turn allowed development of these variations of SNMS.

The application of multiphoton postionization to SNMS has spawned yet another crop of technique acronyms, largely because the initial publications all appeared at about the same time, and there was therefore no possibility of any consensus with respect to a title. Thus there exist at the moment, for resonant excitation, Surface Analysis by Resonance Ionization of Sputtered Atoms (SARISA), Sputter Initiated Resonance Ionization Spectroscopy (SIRIS), and Multiphoton Resonance Ionization (MPRI), and for non-resonant excitation, Surface Analysis by Laser Ionization (SALI). The last one is somewhat ambiguous since it obviously includes the others, and in addition might be mistaken for techniques using direct irradiation of the surface with a laser beam.

In resonant multiphoton ionization a tunable dye laser is tuned to exactly the right wavelength to excite a particular atom from its ground state to an excited state by absorption of a photon (resonance). While in the excited state, a second photon then also interacts with the atom, leading to ejection of an electron (ionization). The processes are depicted diagrammatically in Fig. 18.9, from Pellin *et al.*,[12] in an Fe atom. Resonance radiation is produced by synchronously mixing the output of a Nd:YAG laser with that from a dye laser, giving, after doubling, photons at about 303 nm, which matches exactly the a^5D_4 to $y^5D_4^0$ transition energy in Fe (marked LASER in the figure). A second photon from an XeCl excimer laser then couples the excited $y^5D_4^0$ state to the continuum, resulting in release of an electron and hence ionization. Note that if the second photon is absent, or does not arrive at the right moment, the excited atom can decay by fluorescence, to the state a^5F_5 in Fe, and the detection of the emitted light constitutes yet another technique called[13] 'static mode' Laser Fluorescence Spectroscopy (LFS). The term 'static' is employed in the same way as in 'static' SIMS.

It is important in the resonant form of the technique to choose excited states that can be easily excited resonantly and that have large ionization cross-sections, otherwise the resultant intensities will be too low. With medium high power lasers it is in fact possible to saturate all the processes in selected atoms, so that every laser pulse produces an ionized atom. Under these conditions the sensitivity can be high; detectabilities in the ppb range have been reported.

The experimental arrangement used by Pellin *et al.*[12] for resonant multiphoton ionization is shown in Fig. 18.10. Note that the detector in this case is not a mass spectrometer, but a simple ion extraction system into an electron multiplier. The reason for this is that resonant ionization is so

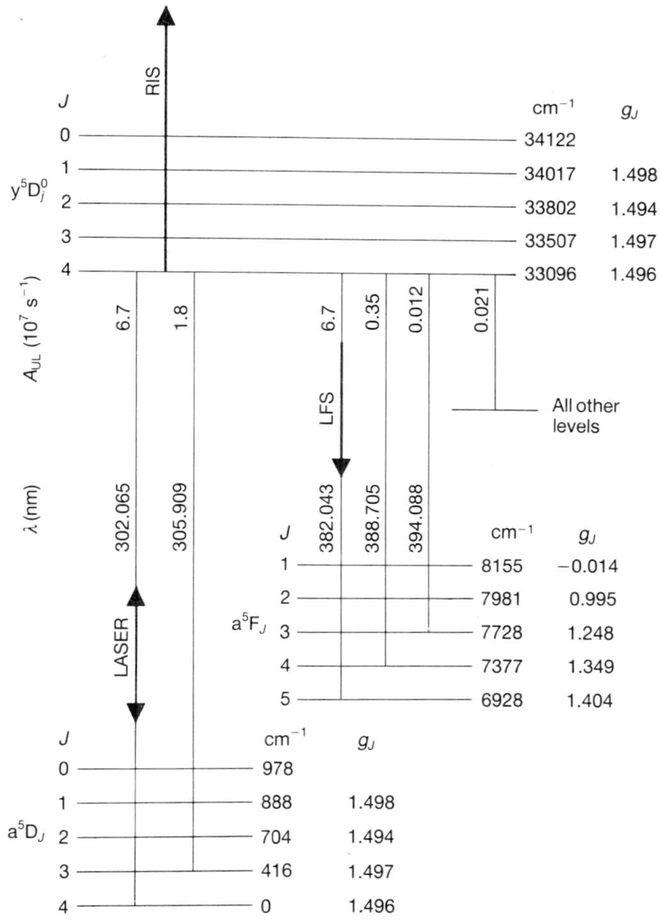

FIG. 18.9. Partial energy level diagram illustrating the processes taking place in resonant multiphoton ionization of gaseous Fe atoms, from Pellin *et al.*[12] Resonance radiation produced by synchronous mixing of the outputs from a Nd:YAG laser and a dye laser, and doubling, gives photons with a wavelength of about 303 nm, which matches exactly the a^5D_4 to $y^5D^0_4$ transition energy in Fe (marked LASER). Irradiation from a second XeCl excimer laser then couples the excited $y^5D^0_4$ state to the continuum, resulting in release of an electron and hence ionization (marked RIS). If a second photon is not used, the excited atom decays by fluorescence to the a^5F_5 state, the emitted light constituting another technique called laser fluorescence spectroscopy. The resonant multiphoton ionization technique is usually called Surface Analysis by Resonance Ionization of Sputtered Atoms (SARISA).

selective that there is little point in using mass discrimination since only one charged species is involved, and it is then possible to avoid the relatively poor transmission of the quadrupole mass spectrometer. Clearly, however, in such a method of ion extraction there is no discrimination between the various positively charged species emitted from the sample, including backscattered primary ions. The method used to separate the species is that of very careful

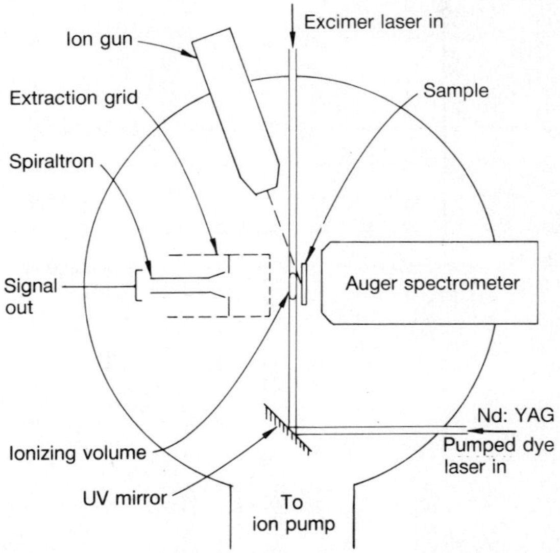

FIG. 18.10. Experimental arrangement used by Pellin *et al.*[12] for resonant multiphoton ionization (SARISA). The ion beam is deflected away from the sample at the beginning of an analysis, and a pulse of length 2 μs is applied to the deflection plates to produce a cloud of ions immediately above the surface. About 1–5 μs after the end of the ion pulse backscattered ion intensity has dropped but most of the sputtered material, which has much lower velocity, is still near the surface. At that moment the laser pulses arrive simultaneously, the laser timing sequence having been triggered some time before. The sputtered atoms are resonant ionized in an interval of about 8 ns, and their arrival at the detector is spread over some 300 ns. Their arrival time is used to discriminate further by acceptance at the detector only within a 55 ns window, via a gating pulse.

timing sequence of operation, and depends on the fact that the backscattered ions will still have most of their initial energy of several keV and will therefore reach the detector long before the ionized sputtered atoms, which have energies of only a few eV. The ion gun is pulsed and produces ion pulses at the surface of 2 μs duration and peak current 25 nA, with a 10 Hz repetition rate. About 1.5 μs after the ion pulse the backscattered ion intensity has dropped to an acceptable level, while there still remain the bulk of the sputtered atoms in the vicinity ($\sim$2 mm) of the surface. At that point the two laser pulses from the previously triggered Nd:YAG pumped dye and XeCl excimer lasers arrive in the ionizing volume, from opposite directions. The sputtered atoms are ionized in an interval of about 8 ns, but their variations in energy and trajectory cause a spread in arrival time at the detector of about 300 ns, located about 8.5 μs after the start of the ion pulse. Their arrival is used for further discrimination by accepting them only within a 500 ns gating pulse to the charge integrator or pulse counter in the detecting circuit. A subsequent 500 ns gating pulse provides a measure of the background count level for quantification purposes.

Observations made by Pellin *et al.* on sputtering from pure Fe and Fe implanted in Si showed that the ratio atoms detected/atoms sputtered for Fe in both cases was $3\text{–}5 \times 10^{-4}$, which compares favourably with the typical SSIMS figure of $\approx 2 \times 10^{-5}$. They indicated various ways in which the technique could be improved which would substantially increase the ratio; one obvious way would be to impose a retarding pulse on the extraction grid immediately after the ion pulse, in order to reject the backscattered ions from the ionization region before the laser pulses arrive.

The same group has used[13] the so-called 'static mode' LFS in a similar experimental arrangement to the above, but with the addition of optics and filters for light collection and focusing on to the entrance slit of a spectrometer. Figure 18.11 shows the arrangement. The timing sequence followed was also

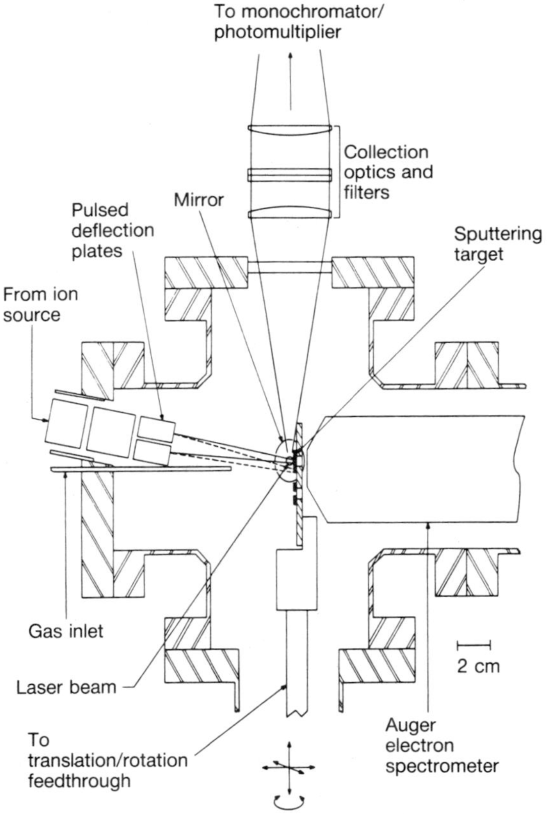

FIG. 18.11. Experimental arrangement used by Pellin *et al.*[13] for LFS operated in ultra-high vacuum. Instead of a spiraltron collector of ions as seen in Fig. 18.10, the detector of the fluorescent light is a monochromator/photomultiplier onto which the light is focused via optics and filters. The timing sequence of ion beam pulse and laser firing was much the same as in the system in Fig. 18.10, except that photon counting was delayed for 50 ns to allow scattered laser light to disperse.

similar to the multiphoton ionization version, except that photon counting was delayed for 50 ns after the laser pulse in order to allow scattered light to dissipate. Gating was again used, one for the signal over an interval of 3 μs, and another for the background some 100 μs after the laser pulse. The latter had a duration of 18 ns, and a peak intensity greater than 10^6 W.

The authors investigated the oxidation of titanium with the technique, and for the excitation of titanium had to tune the laser in the 520 nm region, as indicated from the spectroscopic energy diagram of Fig. 18.12. Oxygen

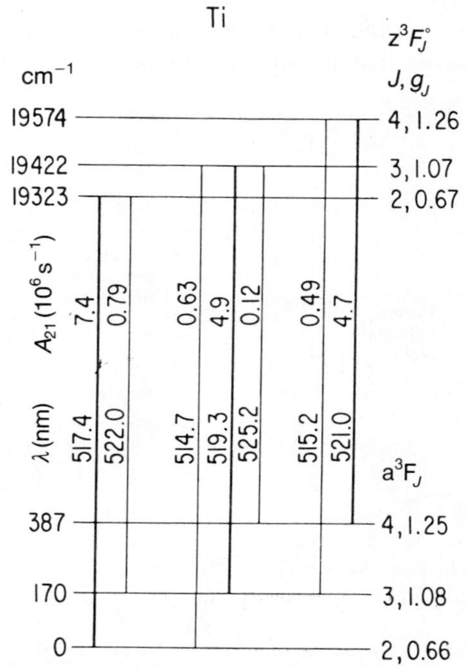

FIG. 18.12. Spectroscopic data based on the energy level diagram of titanium used to decide on the exact laser wavelength needed to excite a particular ground state multiplet, according to Pellin *et al.*[13] All excitation and fluorescent wavelengths for the $a^3F \rightarrow z^3F^0$ system are in the region of 520 nm, so very precise laser tuning is needed.

coverage was estimated from Auger measurements performed between each LFS measurement, the object of the experiment being to measure the sputtering yield of titanium as a function of oxygen coverage. By exact tuning of the pumped dye laser, the titanium yield as a function of the particular ground state multiplet of Fig. 18.12, and as a function of the oxygen/titanium ratio from AES, could be determined, as seen in Fig. 18.13. Close examination of the data in the latter figure (and also those from bombardment with He^+

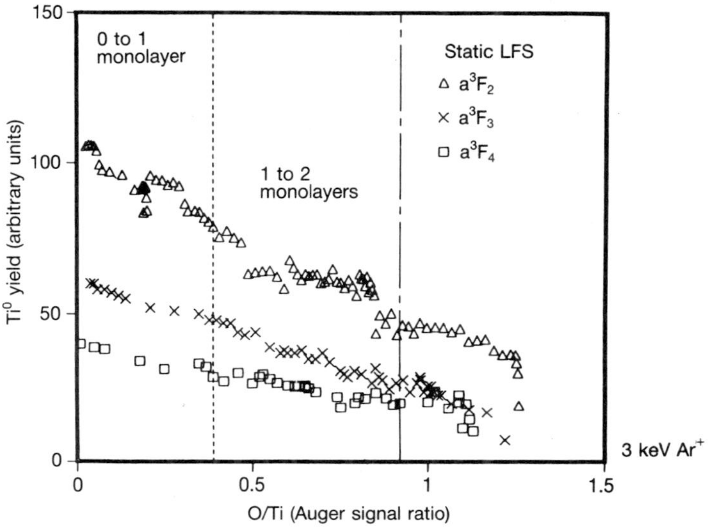

FIG. 18.13. The titanium sputtered neutral yield in each of three ground state multiplet channels a^3F_2, a^3F_3, and a^3F_4 (see Fig. 18.12) measured as a function of the oxygen coverage by Pellin *et al.*[13] Primary argon ion energy was 3 keV. The yield data show steps at each integral oxygen monolayer coverage, followed by plateaux. The results, taken in conjunction with others, indicate that room temperature oxidation of titanium occurs by the formation of oxygen monolayers *beneath* the external surface of the titanium.

rather than Ar^+) shows steps in the yield at each integral oxygen monolayer coverage, followed by plateaux. Similar observations on the oxygen sputtering yield, when taken in conjunction with those on titanium, lead to a model in which oxidation of titanium at room temperature occurs by *under*layer formation of oxygen monolayers for the first two monolayers formed. The oxygen is suggested as occupying octahedral holes between titanium layers near the surface, since TiO has a density similar to that of the metal, and diffusion could occur without necessitating any significant atomic rearrangement.

It will have been appreciated from the foregoing description of resonant multiphoton ionization that, despite the apparent high sensitivities available, it is an extremely selective technique. Prior knowledge of the particular atomic species to be analysed is essential in order that the exciting laser radiation can be tuned to exactly the right resonant frequency. Thus, not only might it not be possible to analyse for some elements at all, but the technique is obviously unsuitable for the overall type of analysis available from, say, XPS and AES. Resonance ionization is successful at moderate laser intensities because at resonance the ionization probabilities are enormously enhanced; with very high intensities, however, it should be possible to achieve adequate ionization even if the frequency of the laser photons is a long way from resonance for most atoms in the material to be analysed. Such non-resonant multiphoton ionization has been demonstrated by Becker and Gillen.[14]

The arrangement used by Becker and Gillen is shown schematically in Fig. 18.14. As with all these variations on multiphoton ionization the technique

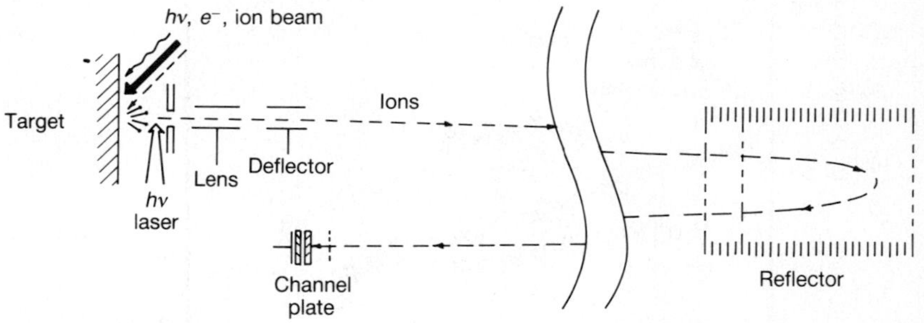

FIG. 18.14. Schematic of experimental arrangement used by Becker and Gillen,[14] for non-resonant multiphoton ionization (also called SALI, Surface Analysis by Laser Ionization). Operation is in ultra-high vacuum. The incident ion beam is pulsed at 10 Hz with pulse widths of 1–5 μs at an ion energy of 2 keV and a continuous current of 1 μA. About 100 ns after each ion pulse a KrF excimer laser fires for a pulse of about 10–15 ns, the firing also triggering the detection circuitry. Ionized neutrals and secondary ions are accelerated and focused into a reflecting time-of-flight mass spectrometer. The potential of the reflecting grid is adjusted to allow high energy secondary ions to pass through, but to reflect low energy ionized neutrals back towards a channel plate detector. The mass resolution is about 500 at mass 200. The system can also be used for SSIMS by increasing the reflecting potential to reflect all ions, and by not using the laser pulse.

operates in UHV. Although in principle material from the target surface can be removed or desorbed by any one of a number of different particle beams, in fact the probe species used was Ar^+, as in the other techniques. The argon ion probe beam is pulsed at 10 Hz as above, with pulse widths 1–5 μs, and with an ion energy of 2 keV and a current of 1 μA when measured continuously. About 100 ns after the end of the ion pulse the KrF excimer laser is fired for a 10–15 ns pulse, which also triggers the mass spectrum acquisition circuitry. The focal diameter of the laser beam is 0.2 mm, and its distance in front of the target surface is ≈ 1 mm. Ionized neutrals, as well as secondary ions, are accelerated, focused, and then deflected slightly into a reflecting time-of-flight (TOF) mass spectrometer. Since the sputtered secondary ions have higher initial energies than the postionized neutrals, the potential of the reflecting grid is adjusted to act as a high pass filter, that is, to allow the high energy ions to pass through and the low energy ions to be reflected back towards the channel plate detector. The mass resolution of the TOF spectrometer is 500 at mass 200. By increasing the reflection voltage to reflect secondary ions, and without the laser pulse, the arrangement can also be used for SSIMS.

To demonstrate that non-resonant multiphoton ionization, or SALI as they named it, is non-selective and can cover the whole mass range, Becker and Gillen[15] analysed an NBS copper standard containing certified amounts of impurities. Part of the mass spectrum from the standard is shown in Fig. 18.15. Note first the simplicity of the spectrum compared to spectra produced by SSIMS. For most elements there is only one peak, corresponding to the principal isotope of that element, and where that isotope is close to 100% abundance. Dimers between atoms of the same major element, or between two major elements, are found, but they are almost the only complicating factor. The minor impurities are easily identified; the certified concentrations were, for ^{107}Ag, 51 p.p.m., for ^{123}Sb, 9 p.p.m., for 197Au, 11 p.p.m., for ^{209}Bi, 6 p.p.m., and for ^{206}Pb, 4 p.p.m. With the signal-to-noise levels shown in the figure, and with improvements in the ionization and detection efficiencies, analysis at the p.p.b. level should be perfectly possible.

18.1.2 *Theory*

Since SNMS depends for its application on the flux of neutral species emitted from a surface by ion bombardment, in other words the sputtered species, the theory of sputtering is appropriate here rather than any considerations of ionization mechanisms. The currently accepted model of the sputtering process is that of Sigmund,[16] already described in qualitative terms in the preceding chapter on SSIMS. Some of the more quantitative predictions of his model need to be examined here.

It will be appreciated from the description of Sigmund's cascade model that only atoms in the outermost few layers of a solid will have enough energy to be scattered in the reverse direction, that is away from the solid, and to overcome any surface potential barrier. The number n_s of such sputtered atoms will depend on the characteristics of the cascade, that is, on the number n_d and average imparted energy of the displaced atoms, and on their spatial distribution. These characteristics depend in turn on the energy E_p of the incoming primary ions and on the energy E_d needed to cause displacement of an atom in the solid. The displacement energy E_d is a function of both the composition and the structure of the solid. There will be a cross-section σ_d for displacement that will be a function of the energy (greater than E_d) of the displacing entity, either the primary projectile as it loses energy in the solid or a secondary projectile already displaced from its equilibrium position. Both E_d and σ_d are connected with what is termed the 'stopping power' of a particular solid for an incident ion at a particular energy, i.e. the ability of that solid to slow down and stop the ion at some characteristic depth.

The stopping power is defined as the increment dE_p of energy loss per incremental depth dx into the target, that is (dE_p/dx). The energy losses must include the effects of both elastic and inelastic scattering, and in the theory of Lindhard, Scharff, and Schiøtt[14] the two effects are considered separable and therefore can be evaluated independently. Elastic scattering is that involved in

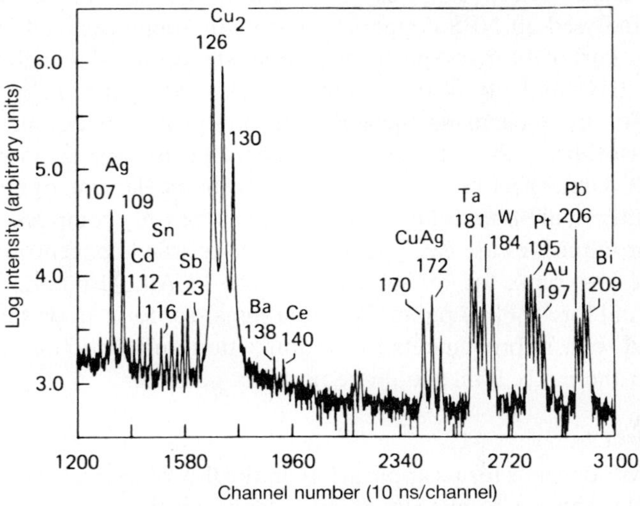

FIG. 18.15. Demonstration by Becker and Gillen[15] of the non-selectivity of SALI compared to SARISA, by analysing an NBS copper standard (C1252) containing certified quantities of impurities. The spectrum is simple compared to that produced by, e.g. SSIMS, with only one peak for most elements. The minor impurities Ag, Sb, Au, Bi, and Pb are easily identified and quantified on the logarithmic intensity scale.

collisions or near-collisions between incident heavy particles such as noble gas ions, and the atoms of the target. If the distance of approach of two particles is so close that the influence of surrounding atoms in the target can be neglected, then binary collision theory can be used, and this has already been discussed in Chapter 16 on ion scattering spectroscopy. Where the closest distance of approach is not small, then interactions with several atoms have to be taken into account, and perturbation theory has to be used. In elastic collisions both energy and momentum are conserved. When a collision results in the transfer of energy to electrons bound to an atom, energy and momentum are not conserved, and the scattering event is then regarded as inelastic. The total stopping power $(\mathrm{d}E_p/\mathrm{d}x)$ can thus be expressed to a first approximation as a sum of nuclear and electronic stopping powers, i.e.

$$(\mathrm{d}E_p/\mathrm{d}x) = (\mathrm{d}E_p/\mathrm{d}x)_n + (\mathrm{d}E_p/\mathrm{d}x)_e \tag{18.1}$$

which can be expressed in terms of the nuclear and electronic stopping cross-sections $\sigma_n(E_p)$ and $\sigma_e(E_p)$ as

$$(\mathrm{d}E_p/\mathrm{d}x) = N\{\sigma_n(E_p) + \sigma_e(E_p)\} \tag{18.2}$$

where N is the number of scattering centres in the incremental depth $\mathrm{d}x$.

It will be recalled from Chapter 16 that the collisional interaction between two particles of masses M_1 and M_2, M_1 being the projectile and M_2 a target atom at rest, can be described in terms of an impact parameter p. The latter is defined schematically in Fig. 18.16, and it can be seen there that it is the minimum distance between the extrapolated tangent to the trajectory of M_1 and the original position of M_2. The actual closest distance of approach is r_0, while the energy transferred to M_2 is E_t, and the deflection and recoil angles are ϕ_1 and ϕ_2, measured in the laboratory frame of measurement. Note from the smaller diagram in Fig. 18.16 that the differential cross-section $d\sigma_n$ is an annular area bounded by $p+dp$ and $\theta_1+d\theta_1$, where θ_1 is now the deflection angle measured in the centre of mass, or barycentric, frame of measurement. ϕ_1 and θ_1 are of course related, as described in eqn (16.10). The impact parameter, the closest distance of approach, and the primary energy are related by

$$1-p^2/r_0^2-V(r_0)/E_r=0 \tag{18.3}$$

where E_r is the reduced energy $E_p\{M_2/(M_1+M_2)\}$, and $V(r)$ is the interaction potential, also already discussed in Chapter 16. $V(r)$ has the general form

$$V(r)=\frac{Z_1Z_2e^2}{r}\,\phi(r/a) \tag{18.4}$$

where Z_1 and Z_2 are the atomic numbers of the projectile and of the target atom, respectively, ϕ is a screening function, and a is the characteristic Lindhard screening radius. The general form of $V(r)$ proposed by Lindhard is a power law approximation of the form

$$V(r)=\frac{Z_1Z_2e^2}{r}\,km(r/a)^{m/(1-m)} \tag{18.5}$$

where k is a constant and m is a fitting parameter. In the range of primary energies used in SNMS m is the range 0.3 to 0.5, corresponding to a Thomas–Fermi type potential.

In the barycentric frame the energy E_t transferred to the recoil atom is given by

$$E_t=E_p\left[\frac{4M_1M_2}{(M_1+M_2)^2}\right]\sin^2(\theta/2) \tag{18.6}$$

from which

$$E_t\,(\max)=E_p\left[\frac{4M_1M_2}{(M_1+M_2)^2}\right]. \tag{18.7}$$

If the collisions can be regarded as 'weak', i.e. small θ and large p, then integration over r of the scattering integral given by expression (16.8), using the power law potential of expression (18.5) above, gives the relation between θ and p:

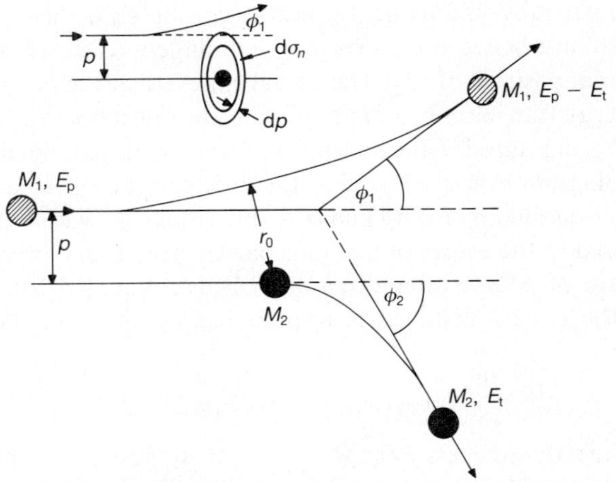

FIG. 18.16. Diagrammatic definition of the impact parameter p describing the collisional interaction between a projectile of mass M_1 and energy E_p, and a stationary target atom of mass M_2. The energy transferred to M_2 in the collisions is E_t, and the deflection and recoil angles of M_1 and M_2, respectively, are ϕ_1 and ϕ_2. The closest distance of approach is r_0. p is thus the minimum distance between the tangent extrapolated to the trajectory of M_1, and the original position of M_2. The differential cross-section $d\sigma_n$ is an annular area surrounding M_2. p, r_0, and E_p are related through expression (18.3).

$$\theta = (\gamma k/\epsilon)\,(p/a)^m \tag{18.8}$$

where $\gamma = m\{(3-m)/2m\}^{1/2}$ and where ϵ is a dimensionless parameter given by

$$\epsilon = \frac{aM_2E_\mathrm{p}}{Z_1Z_2e^2(M_1+M_2)}. \tag{18.9}$$

From (18.8) one can derive an expression for p as a function of θ, and also $\mathrm{d}p/\mathrm{d}\theta$, which when inserted in the expression for $\mathrm{d}\sigma$ similar to that of expression (16.6), namely

$$\mathrm{d}\sigma_\mathrm{n}(\theta) = -2\pi p(\theta)\,\frac{\mathrm{d}p}{\mathrm{d}\theta}\,\mathrm{d}\theta \tag{18.10}$$

gives

$$\mathrm{d}\sigma_\mathrm{n}(\theta) = Ka^2\epsilon^{-2m}\theta^{-(1+2m)}\,\mathrm{d}\theta. \tag{18.11}$$

For small θ (i.e. for $\sin\theta \sim \theta$), insertion of (18.6) and (18.9) then leads to the differential energy cross-section for elastic or nuclear scattering,

$$\mathrm{d}\sigma_\mathrm{n}(E_\mathrm{t}) = CE_\mathrm{p}^{-m}\,E_\mathrm{t}^{-(1+m)}\,\mathrm{d}E_\mathrm{t} \tag{18.12}$$

where the constant C is given by

$$C = \frac{\pi}{2}\,\lambda a^2(M_1/M_2)^m(2Z_1Z_2e^2/a)^{2m} \tag{18.13}$$

in which

$$\lambda = 2^{(1-2m)}(\gamma k)^{2m}.$$

The electronic or inelastic stopping cross-section $\sigma_e(E_p)$ is much more difficult to calculate, but Lindhard and Scharff[18] have shown to a first approximation that in the energy regime appropriate to SSIMS and SNMS

$$\sigma_e(E_p) = kE_p^{1/2}. \quad (18.14)$$

Here k is a complicated material constant. Thus the total stopping cross-section can be written, from expression (18.2),

$$(\mathrm{d}E_p/\mathrm{d}x) = N\left\{kE_p^{1/2} + C\int_{E_t} E_p^{-m}\, E_t^{-(1+m)}\, \mathrm{d}E_t\right\} \quad (18.15)$$

or, using expression (18.6) linking E_t and E_p, and assuming small θ,

$$(\mathrm{d}E_p/\mathrm{d}x) = N\left\{kE_p^{1/2} + C'\int_{E_t,\theta} E_t^{-(1+2m)}\, \theta^{-2m}\, \mathrm{d}E_t\, \mathrm{d}\theta\right\}. \quad (18.16)$$

Because of the relative magnitude of the constants k and C' in expression (18.16), the stopping power in the energy range used in SSIMS and SNMS is dominated by the elastic or nuclear stopping cross-section. This is shown in Fig. 18.17, from Pivin,[19] in which the nuclear and electronic differential

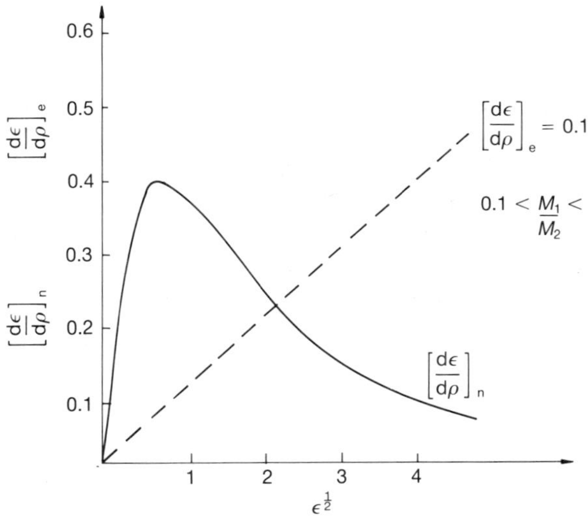

FIG. 18.17. The nuclear ($[\mathrm{d}\epsilon/\mathrm{d}\rho]_n$) and the electronic ($[\mathrm{d}\epsilon/\mathrm{d}\rho]_e$) stopping powers plotted as a function of the reduced energy ϵ (see expression (18.9)) and the reduced path length ρ given by expression (18.17), from Pivin.[19] For low energy Ar^+ bombardment of lighter elements, $\epsilon^{1/2}$ varies between 0.08 and 0.13 and k (see expression (18.16)) between 0.1 and 0.3, which means that both SNMS and SSIMS operate very close to the vertical axis in the figure. Thus most of the energy loss is in nuclear scattering and there is negligible ionization within the target.

cross-sections are plotted as a function of the reduced energy ϵ (see expression (18.9)), and the reduced path length ρ given by

$$\rho = 4\pi n x\{M_1 M_2/(M_1 + M_2)^2\} \tag{17}$$

n being the target atom density. Note that for low-energy Ar^+ bombardment of elements in the lower half of the periodic table, $\epsilon^{1/2}$ varies between ≈ 0.08 and 0.13 and k between 0.1 and 0.3, so that the SSIMS and SNMS conditions are very close to the vertical axis in the figure. The implication from the fact that most of the energy loss is involved in nuclear scattering is that there will be little ionization within the target.

Having established the derivation and form of the stopping cross-sections appropriate to the energy regime of SSIMS and SNMS, it is now necessary to relate them to the factors governing the sputtering yield. That there must be a direct relation between the stopping power and the sputtering yield is intuitive, since virtually all the stopping power is due to the atomic nuclei, as seen above, and therefore the number of displaced atoms per incident projectile must depend on the stopping power. The sputtering yield in turn depends directly on this number, and their transferred energy. Such a direct relationship has been established experimentally, as demonstrated in Fig. 18.18, from Sigmund,[16] in which the calculated nuclear stopping power is compared with the sputtering yield over a wide range of primary energies. Clearly, to a reasonable approximation

$$S(E_p) \propto (dE_p/dx)_n. \tag{18.18}$$

The precise relationship between the sputtering yield and the nuclear stopping power can be evaluated by considering the depth distribution of the energy deposited by the primary particles, the density of atoms displaced by the primary particles and in the cascade, the number of such displaced atoms in the target surface, and the probability that an atom displaced in the surface will have enough energy to leave the surface.

Since the energy lost by the incoming ions is primarily in elastic collisions, as has been shown above, their depth distribution $f_1(x)$ can be approximated by a Gaussian, as seen[19] in Fig. 18.19a. The mean implantation depth is then $\bar{x}_1$, and the width of the distribution Δx_1. Similarly the depth distribution $f_2(x)$ of the energy deposited during a cascade can also be approximated by a Gaussian, where the mean depth is $\bar{x}_2$ and the width Δx_2. Integration of $f_1(x)$ and $f_2(x)$ for successive increments of x (successive atom layers) then gives the total energy deposited at each depth by the primary particles and by the cascade, respectively, shown as the functions $F_1(x)$ and $F_2(x)$ in Fig. 18.19b. For $F_2(x)$ the tail projecting backwards from the surface represents the energy lost by sputtered atoms. Obviously, the mean value of $F_1(x)$ is the nuclear stopping power of the target material for ions of energy E_p. F_1 and F_2 differ by

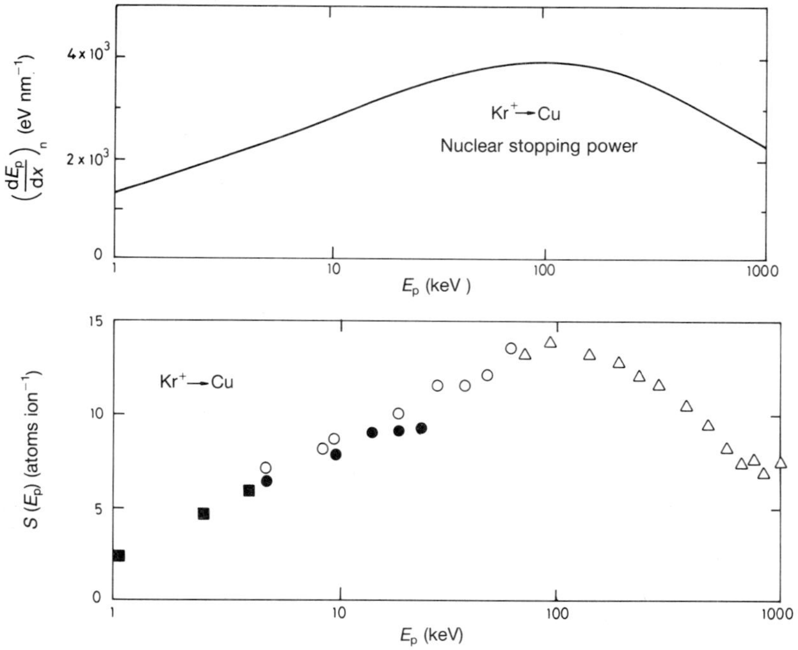

FIG. 18.18. Comparison by Sigmund[16] of the calculated nuclear stopping power of krypton ions in copper with the experimentally observed sputtering yield, over a wide range of primary energies. The symbols in the lower half refer to various published experimental values. Clearly to a first approximation the sputtering yield is proportional to the nuclear stopping power.

a factor α, which contains dependence on M_1 and M_2, the angle of incidence of M_1, and the crystal nature of the target. One can therefore write

$$S(E)_p) = \Lambda\alpha(\mathrm{d}E_p/\mathrm{d}x)_n = \Lambda\alpha\sigma_n(E_p). \tag{18.19}$$

The factor Λ in (18.19) is a measure of the probability that a displaced surface atom will leave the target. It takes into account the velocity of an atom after displacement, the angular distribution of the velocity, the nuclear stopping cross-section at that velocity, and the resultant energy in comparison with the height U_s of the surface potential barrier. The total sputtering yield is then an integration over all angles and energies ($\geq U_s$) during the duration of the cascade. Such an integration has been performed by some workers by computer cascade simulation, using Monte Carlo methods, which have the advantage of allowing the dependence of sputter yield on the crystallographic nature of the target to be included. Integration can also be carried out based on the assumptions, such as that of a planar surface potential barrier, made in the Sigmund model. The expression for Λ is then

$$\Lambda = (3/4\pi^2)\,(CU_s)^{-1} \tag{18.20}$$

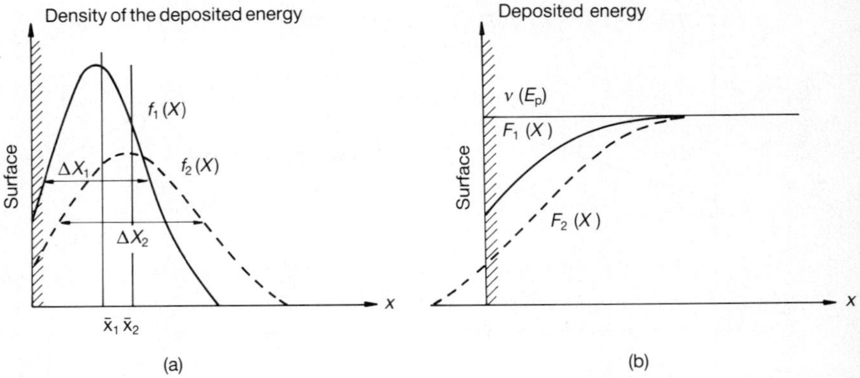

FIG. 18.19. (a) Solid line: depth distribution $f_1(x)$ of energy lost by incoming ions. Mean implantation depth is $\bar{x}_1$ and distribution width is Δx_1. Dashed line: distribution $f_2(x)$ of energy deposited near surface during a cascade process. Mean depth of distribution $\bar{x}_2$ and width Δx_2. (b) Solid line: result of integration of $f_1(x)$ for successive atom layers, giving $F_1(x)$, the total energy deposited at each depth by incoming particles. Dashed line: result of integration of $f_2(x)$, for successive atom layers, giving $F_2(x)$, the total energy at each depth deposited during a cascade. The tail in $F_2(x)$ projecting outwards from the surface is the energy lost from the cascade by sputtered atoms. From Pivin.[19]

where C is given by expression (18.13), and the final expression for the sputtering yield

$$S(E)_p) = (3/4\pi^2)(CU_s)^{-1}\alpha_N(M_2/M_1)(\cos\theta)^{-5/3}N^{-1}\sigma_n(E_p), \quad (18.21)$$

θ being the angle of incidence with respect to the surface normal. It should be noted that expression (18.21) contains no reference to the crystallinity of the target, since the Sigmund model applies to a random or isotropic target, nor to multiple processes. The linear cascade is considered in isolation, and the effect of the arrival of a subsequent projectile or projectiles while the initial cascade is still in motion is ignored. Nevertheless, the model has had considerable success with the situation in which a low energy non-chemically-interactive ion strikes a target of relatively low Z, i.e. a typical SSIMS or SNMS situation. Figures 18.20 and 18.21 show two examples of comparison of experiment with calculation. In Fig. 18.20 the sputter yield from elemental silicon as a function of primary energy for four noble gas ions is compared with the predictions of the Sigmund model, while in Fig. 18.21 the variation in sputter yield from a large number of atomic species bombarded by 1 keV Ar^+ ions is also compared with the theoretical values. Precise agreement is not always found, but the trends in both cases are certainly predicted accurately.

18.1.3 *Spectral interpretation*

Since the output from analysis by SNMS is a mass spectrum just as in SSIMS, the allocation of peaks at various mass-to-charge ratios to specific fragments

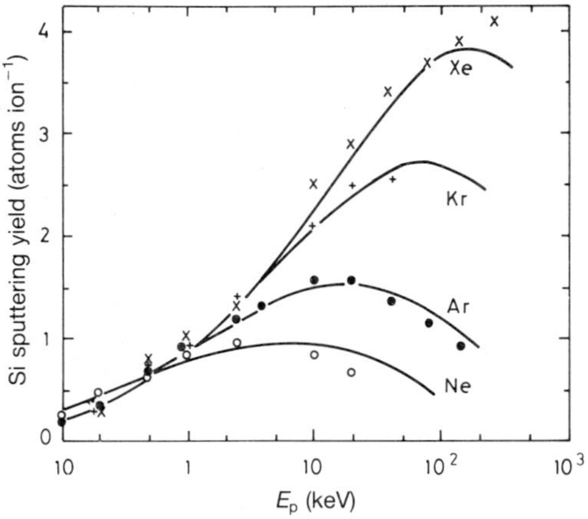

FIG. 18.20. Comparison of the measured sputter yield from elemental silicon as a function of the primary energy of each of four noble gas ions, with the predictions (solid lines) of the Sigmund[16] model.

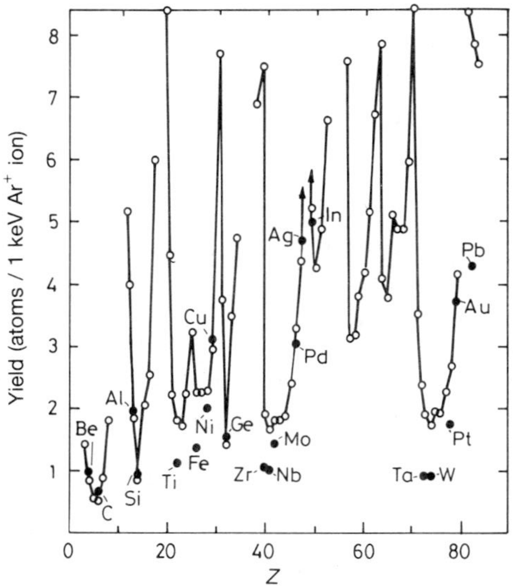

FIG. 18.21. Comparison of the measured sputter yields using 1 keV argon ions from 20 elements (full circles) with the predictions of the Sigmund[16] model (open circles).

or molecular ions is made in the same way. However, the spectra produced by the two techniques from any one surface, particularly a multi-component surface, will look quite different, and often the SNMS spectrum will be simpler. Common to both techniques are isotope effects, and of course postionized monomers will be found at the same place in the SNMS spectrum as singly charged monomers in the SSIMS spectrum, but there the similarities on the whole end. As has been seen in Chapter 17, SSIMS spectra from complex surfaces will contain a vast array of peaks at mass numbers corresponding to all the molecular and atomic combinations that can be produced in the secondary ion production process. On the other hand, SNMS spectra will contain mostly singly charged monomers, but with a significant contribution from both elemental and mixed dimers, and from doubly charged species. A comparison of SNMS and SSIMS spectra recorded from the same surface, an NBS standard alloy, in the commercial version of Oechsner's gas discharge post-ionization method, is given in Fig. 18.22, from Jede *et al.*[20] In addition to the differences listed above, there are other obvious ones, deriving from the different conditions of ion formation; that is, the SNMS spectrum will reflect the relative sputtering yields of the elements in the surface, SSIMS their ionization probabilities. Thus in the SNMS spectrum the peaks corresponding to the transition metals Cr to Ni are of approximately the same magnitude, but in the SSIMS spectrum Cr and Mn are much more prominent. Similarly, Nb, Mo, Ta, and W features are of relatively greater intensity in SNMS because of their low ionization probabilities in SSIMS, whereas on the other hand Na and K are much more intense in SSIMS, for the opposite reason.

Note from Fig. 18.22 that the dynamic range in SNMS can be similar to that in SSIMS, although in this case the comparison is slightly misleading because the bombarding sample current for SNMS was 200 μA as against 3 μA for SSIMS, over similar areas.

18.1.4 *Quantification*

An equation very similar to (17.26), relating measured intensity of postionized neutrals to the atomic concentration, can be written as

$$I^0_{\mathrm{A(B)}} = F_{\mathrm{A}} \cdot S_{\mathrm{tot}} \cdot C_{\mathrm{A}} \cdot \alpha^0_{\mathrm{A}} \tag{18.22}$$

where $I^0_{\mathrm{A(B)}}$ = current of postionized neutrals of element A in matrix B
F_{A} = instrumental factor, including primary ion current, transmission of spectrometer, detector efficiency, angular acceptance, and angular sputter yield distribution, appropriate to A
S_{tot} = total sputtering yield
C_{A} = concentration of element A in B
α^0_{A} = postionization probability of sputtered neutrals from A.

Note that (18.22) is correct only if the proportion of sputtered neutrals is very much greater than that of secondary ions, in other words, using the nomenclature of Chapter 17, if $\alpha^{\pm}_{\mathrm{A(B)}} \ll 1$. Since $\alpha^{\pm}$ is likely to be comparable to unity only for species such as alkali metals that are easily ionized, it will generally be true that $\alpha^0 \approx 1$. However, where $\alpha^{\pm}$ is significant then a term of the form $(1-\alpha^+ - \alpha^-)$ must be included in (18.22).

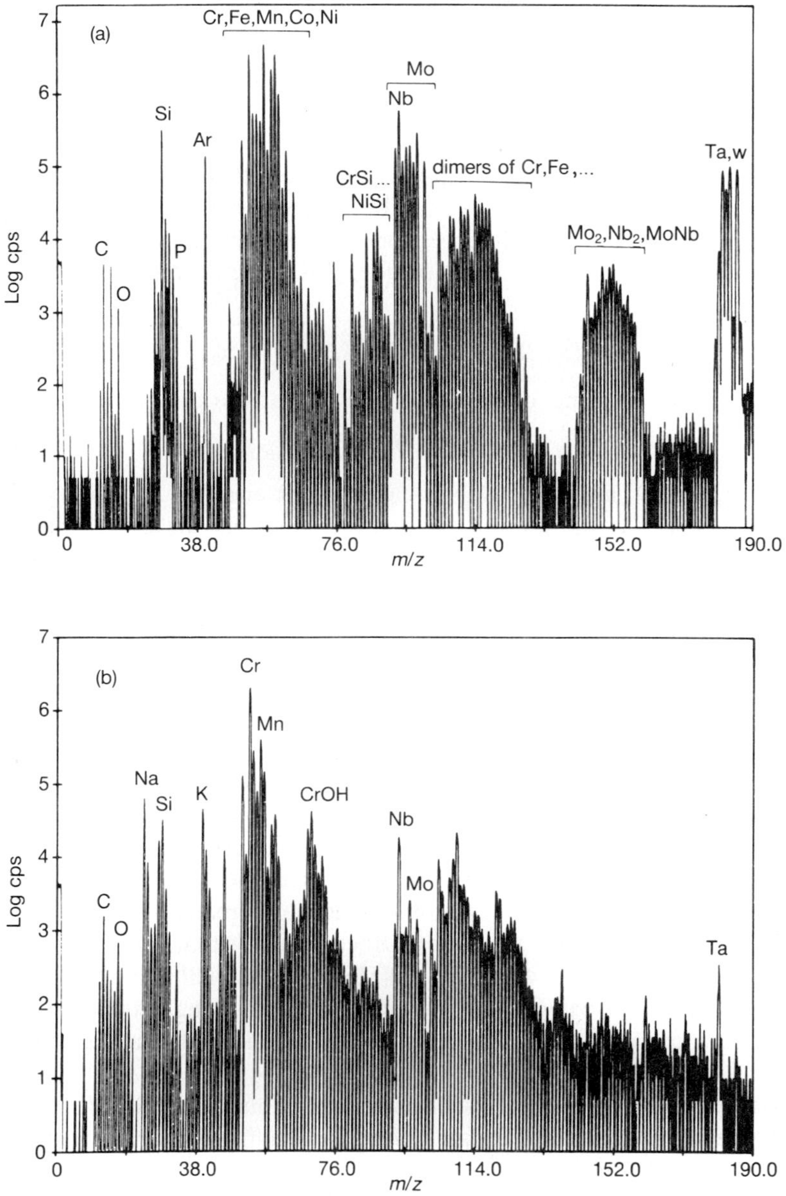

FIG. 18.22. Comparison of SNMS (a) and SSIMS (b) spectra recorded by Jede *et al.*[20] from the same surface, that of an NBS standard alloy. The instrument used was the commercial version of Oechsner's plasma postionization method, seen in Fig. 18.8. Note that in SNMS peaks from the transition metals Cr, Fe, Mn, Co, Ni are of approximately equal magnitude, reflecting similar sputtering yields, but in SSIMS are quite dissimilar in magnitude due to the differences in ionization probabilities.

The method of employing eqn (18.22) for quantification has up to now been that exclusively of relative sensitivity factors. No attempt appears to have been made to use values of S, the sputtering yield, from, say, Sigmund's theory as outlined in Section 18.1.2. There is thus no equivalent at the moment of the theoretical methods of quantification described for several of the other techniques.

For the extraction of relative sensitivity factors either reference elements or standards are required, and these can be external or internal. Most workers seem to have used internal reference elements, invariably in any particular case that element at the highest concentration. If R is the reference element, then eqn (18.22) gives

$$I^0_{A(B)}/I^0_{R(B)} = (F_A/F_R)(C_A/C_R)(\alpha^0_A/\alpha^0_R) \tag{18.23}$$

since S will be the same for both and R. This leads to the expression

$$c_A = c_R(D_R/D_A)(I^0_{A(B)}/I^0_{R(B)}) \tag{18.24}$$

where D_R and D_A are the sensitivity factors of R and A. The validity of this approach depends on several assumptions, such as the establishment of sputter equilibrium, the neglect of all secondary ion signals, the invariance of postionization efficiency and transmission with the neutral species, and the neglect of molecular sputtering, i.e. dimers and trimers. The relative sensitivity factor is thus $D_{A,R} = (D_R/D_A)$.

The commercial version of the gas discharge postionization method has been on the market now for long enough for a reasonably consistent set of relative sensitivity factors to have been derived. Figure 18.23, from Jede *et al.*,[21] shows a compilation of the sensitivity factors relative to R = Fe for 30 elements based on measurements from some 200 materials over a couple of years. It is important to note that most of these materials were homogeneous ferrous alloys, and that sputter equilibrium was reached before a measurement was made. Problems arise in inhomogeneous materials with matrix-dependent sputtering characteristics. Other materials that can give problems are oxides, nitrides, etc., where the SNMS spectrum will contain significant contributions from oxygen, or nitrogen, etc., containing fragments. In the latter cases corrections have to be made, and universal relative sensitivity factors are not so applicable.

For 'well-behaved' materials such as homogeneous alloys, SNMS using gas discharge postionization is capable of good analytical accuracy. Table 18.1 shows a comparison of the analyses of an FeCr alloy by SNMS and by wet chemical methods, respectively, made by Müller and Oechsner,[10] using the relative sensitivity factor method.

The maximum and mean percentage deviations from the wet chemical analysis in Table 18.1 are 23, and 9.5, respectively, which for a surface analytical technique is reasonable.

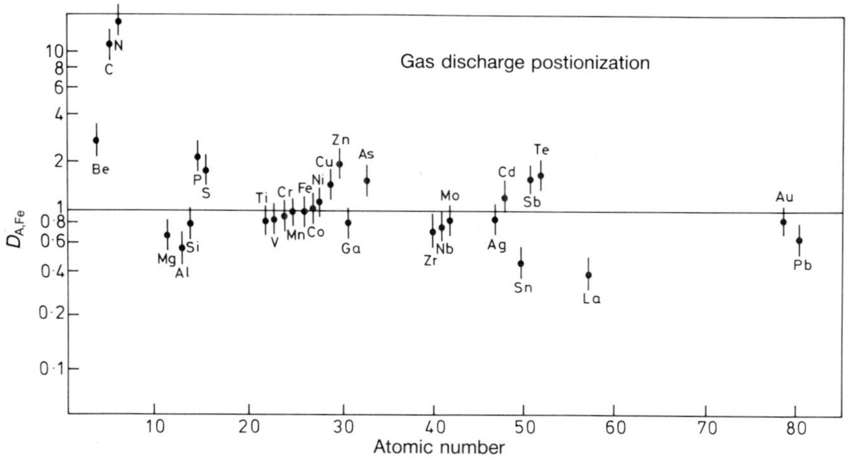

FIG. 18.23. Compilation by Jede *et al.*[21] of the sensitivity factors relative to Fe of 30 elements, from measurements using the plasma postionization method of Oechsner in the commercial instrument shown in Fig. 18.8. Most of the measurements were made on homogeneous ferrous alloys, and sputter equilibrium was allowed to be attained before a measurement was made.

Table 18.1

Compositional analysis of an FeCr alloy by SNMS and by wet chemistry, at.%. From Müller and Oechsner[10].

Element	SNMS	Wet chemistry
C	0.156	0.156
Si	0.76	0.765
Mn	1.37	1.54
P	0.051	0.043
S	0.0069	0.0086
Cr	18.2	18.9
Ni	8.73	8.91
Mo	0.13	0.16
Cu	0.128	0.104
Co	0.35	0.31
Ti	0.036	0.035
V	0.13	0.12
Fe	69.9	69.0
As	0.03	—

Some progress towards quantification has also been achieved using electron impact postionization methods. A compilation of sensitivity factors relative to Fe, similar to that in Fig. 18.23, for electron impact postionization, can be constructed from the tabulation of Lipinsky *et al.*[1] and is shown in Fig. 18.24.

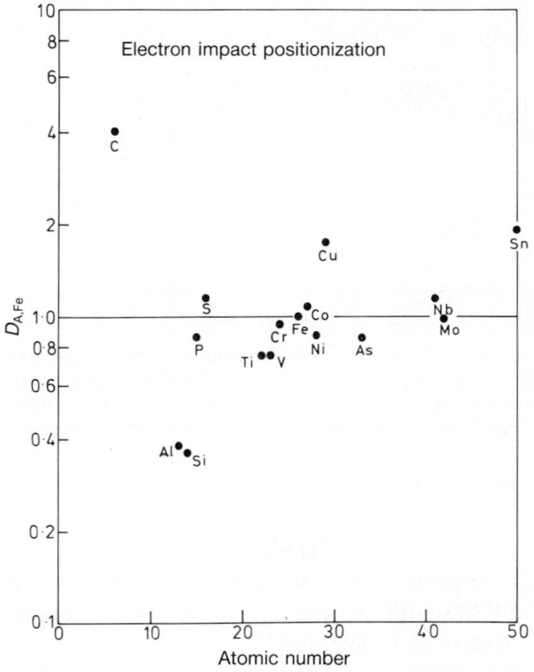

FIG. 18.24. Compilation of sensitivity factors relative to Fe from the tabulation by Lipinsky *et al.*[1] from measurements using electron impact postionization in the instrument shown in Fig. 18.1. The qualitative variation of relative sensitivity with atomic number is similar to that in Fig. 18.23, particularly for the transition metals.

Again the bulk concentrations of the elements were measured first by wet chemical analysis. Although there are only half as many elements in Fig. 18.24 compared with 18.23, the qualitative variation with atomic number is similar, particularly amongst the important elements in the first transition series. Quantitative agreement is not so good, which is perhaps hardly surprising in view of the major differences in the two methods of postionization.

Tümpner *et al.*[22] used the relative sensitivity method in electron impact postionization to analyse a series of low alloy steel targets, in which all elements other than Fe were at concentrations lower than 3.2%, mostly below 1%. They found excellent correlation between the intensities relative to Fe, and the bulk concentrations of the impurity and alloying elements as determined by wet chemistry. A sample of their correlation is shown in Fig. 18.25 for carbon and sulphur. Quantification limits in the range 10–30 p.p.m. were said to be possible.

One of the important features of Figs 18.23 and 18.24, when considered in comparison with the situation in SSIMS, is that the range of relative sensitivities covers only one to one-and-a-half orders of magnitude, rather similar to that in XPS and AES. Contrast this with the range of 4 to 5 orders of magnitude of relative sensitivity in SSIMS.

As far as can be ascertained the resonant and non-resonant multiphoton postionization methods have not yet been applied widely in a quantitative way, although measurements by Becker and Gillen[23] have demonstrated the same narrow range of relative elemental sensitivity factors for non-resonant MPI as found for the other two methods of postionization. In addition, using SIRIS, or resonant MPI, Parks *et al.*[24] have shown excellent correlation over two orders of magnitude of concentration, between the *B* levels detected in a series of NBS standard steels, and the certified levels. The lowest *B* concentration studied was 1.3 p.p.m.

18.1.4 *Applications*

Since, as described earlier, the geometry and experimental conditions in the gas discharge postionization method can be adjusted to provide extremely uniform erosion over a large area on a sample, the method is ideally suited to accurate and quantitative depth profiling. It has been used in this way for a variety of types of material, from metals to semiconductors to insulators.

The high depth resolution attainable in SNMS is demonstrated in Figs 18.26 and 18.27. In the former the interface width between oxide and metal of a Ta_2O_5 film on Ta metal is found[10] to be 3.3 nm after sputtering through the thickness of the film, which was 336 nm. This compares more than favourably with the width established in the same film by a combination of AES and sputter profiling. The energy of the bombarding Ar^+ ions from the discharge was only 202 eV, thus minimizing atomic mixing effects, which can degrade the depth resolution when the more normal Ar^+ energies of several keV are used in depth profiling. Another example of accurate profiling is that of Fig. 18.27, from the paper by Chu *et al.*[25] The profile across a layer structure consisting of 24 alternate layers of W and Si is shown, using the neutral species ^{184}W; the substrate is silicon. Each layer pair (W + Si) was 10.5 nm thick. It can be seen that there was no loss of depth resolution down to the total structure thickness of ≈ 250 nm. Again a low Ar^+ energy of 280 eV was used.

SNMS profiling has been used extensively as a rapid means of checking the quality of films grown by various epitaxial methods. An example is that of the (Al, Ga)As films grown on GaAs for use in optoelectronics and microwave devices. The design characteristics of the latter depend on knowledge of the variation of composition with depth, and the sharpness or otherwise of

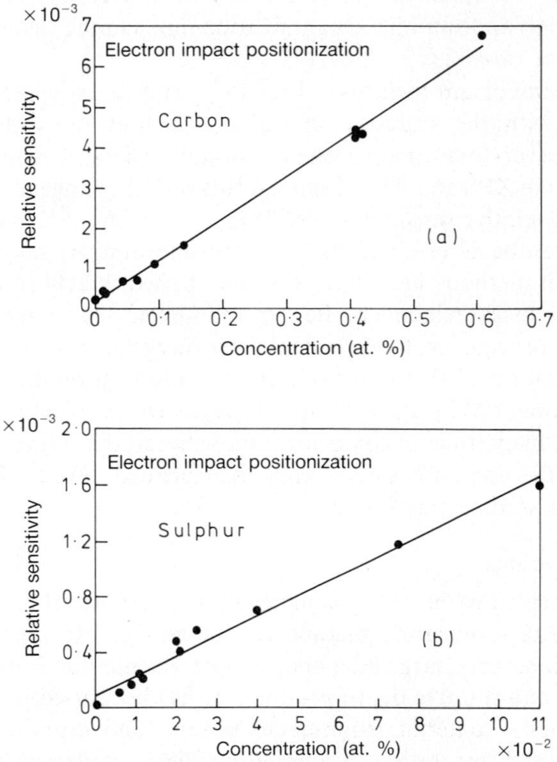

FIG. 18.25. Correlation found by Tümpner *et al.*[22] between the intensities relative to Fe and the bulk concentrations determined by wet chemical methods, in a series of low alloy steels, for the impurity elements carbon and sulphur. Electron impact postionization was used. On the basis of such correlations for many elements it was stated that quantification in SNMS down to 10–30 p.p.m. was possible.

interfaces. Kaiser *et al.*[26] have applied SNMS to such films. To achieve quantification they derived relative sensitivity factors by plotting I_{Ga}/I_{As} and I_{Al}/I_{As} for a series of samples with different Ga/Al ratios, and found a linear dependence which confirmed the absence of matrix effects. Using these factors they determined the Al fractional concentrations in the series, and compared them with the known concentrations from other techniques such as RBS, with the result seen in Fig. 18.28. The accuracy is obviously excellent. A profile through a solar cell deposited by metal–organic vapour phase epitaxy, using the Al neutrals, is compared with the ideal profile in Fig. 18.29. Within experimental error the ideal structure is confirmed; even the peak at 0.75 μm, which was expected from the details of the growth process, is found.

The ability of SNMS to analyse insulators, since the sputtering and ionization processes in the technique are decoupled and the charging

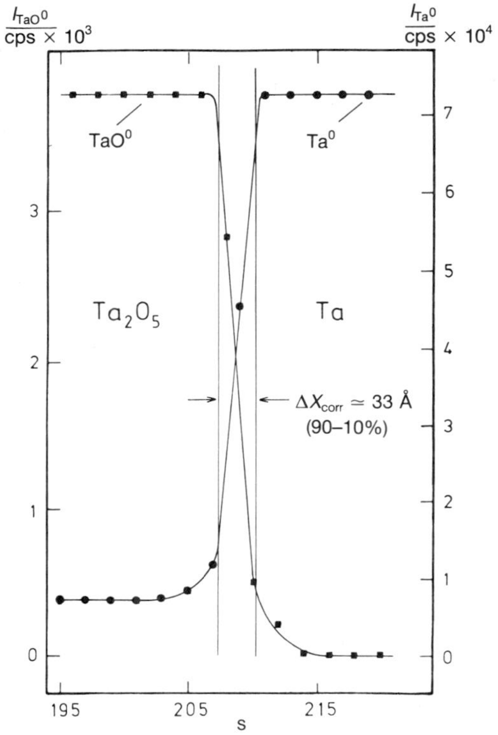

FIG. 18.26. Demonstration of the depth resolution obtainable in SNMS, by Müller and Oechsner,[10] during profiling through a standard Ta_2O_5 film on Ta metal, of thickness 336 nm. The interface width between oxide and metal was found to be 3.3 nm, based on the 10%–90% definitions of width. This is in fact better than the widths found by AES during sputter depth profiling.

problems found in other techniques are thereby minimized, is demonstrated[25] in the spectrum of Fig. 18.30. This is from an NBS glass standard. The procedure used here to avoid sample charging altogether is to bias the sample slightly positive in order to reject Ar^+ ions from the plasma discharge, and then to use a separate beam of exciting ions for the sputtering. With the positive sample bias, electrons are attracted to the sample surface and neutralize any charge build-up. The composition of the glass is given in the figure, and it can be seen that most elemental peak intensities are approximately proportional to the corresponding elemental concentrations, the major exception being that of oxygen. The sputtering yield of oxygen is found to vary substantially with chemical state, and it remains difficult to quantify.

Using electron impact postionization, Wilson *et al.*[3] have analysed a series of samples of $InAs_xP_{1-x}$ grown epitaxially on InP. It was of interest to check whether the surface compositions of the series as a function of x correlated

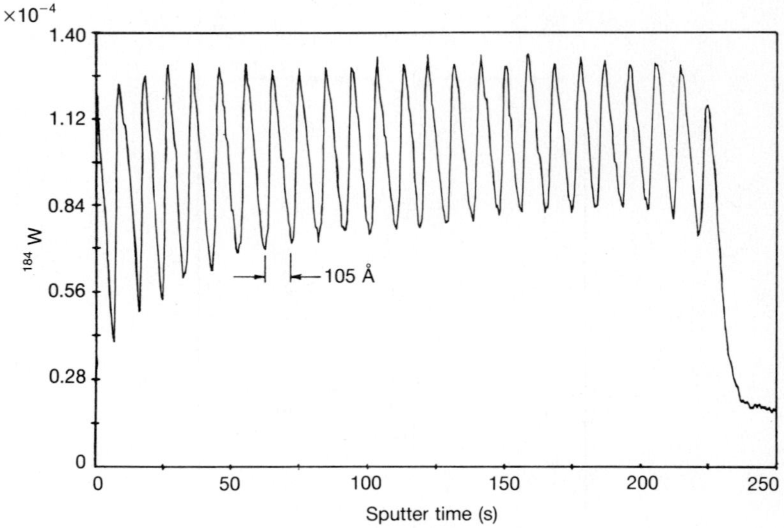

FIG. 18.27. Example of accurate profiling by SNMS across a layer structure consisting of alternate W and Si layers (each layer pair was 10.5 nm thick), recorded by Chu *et al.*[25] The neutral species ^{184}W was used, and the argon ion energy was 280 eV. The interface resolution even after the total structure thickness of 250 nm has been penetrated is still excellent.

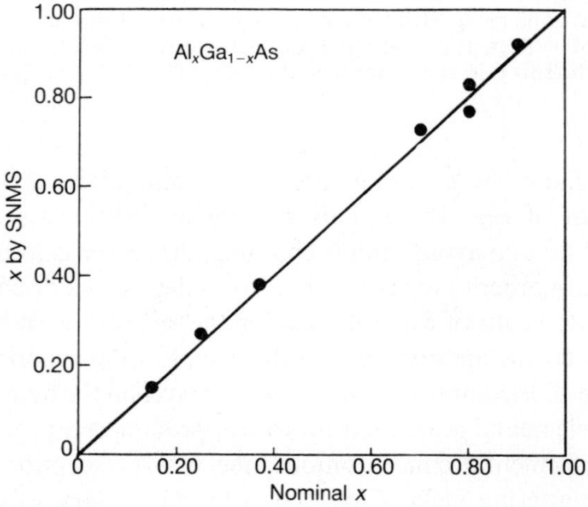

FIG. 18.28. Linear correlation found by Kaiser *et al.*[26] between the ion ratios Al/As and Ga/As in a series of semiconductor samples of (Al, Ga)As with different Ga/Al ratios. Matrix effects were therefore absent, and from the plot relative sensitivity factors could be derived that could then be used in quantification of other compositions. The bulk compositions of the samples were measured initially by Rutherford Backscattering Spectroscopy (RBS).

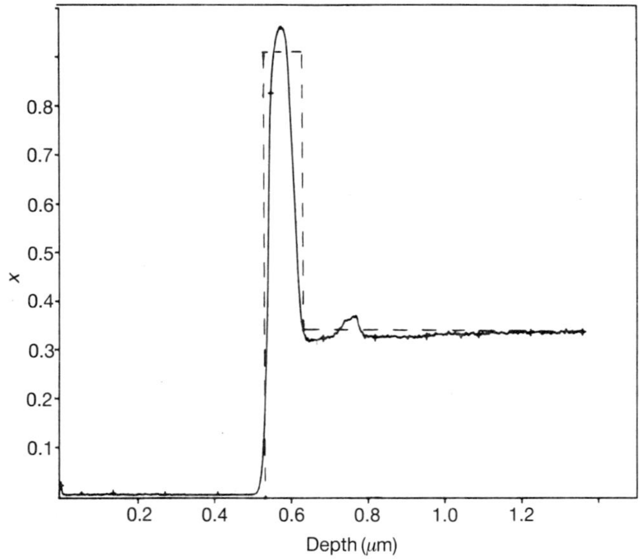

FIG. 18.29. Profile by SNMS using plasma postionization of an $Al_xGa_{1-x}As$ solar cell, using Al neutrals as the indicator, from Kaiser *et al.*[26] The ideal profile is shown for comparison, and is seen to be virtually confirmed. The additional peak at 0.75 μm depth was expected from knowledge of the growth process of the material, and is also confirmed.

with the bulk composition, or departed from it. Figure 18.31 from their paper compares the SNMS integrated peak intensity ratios of As to In with the composition in terms of x derived from X-ray diffraction data. Clearly the correlation is excellent over the whole range of x, confirming the uniformity of the $InAs_xP_{1-x}$ films and homogeneous growth.

18.2 Glow discharge mass spectrometry

Although in its normal mode of operation GDMS is not a surface-specific technique, it can be and has been operated in a surface-specific mode, and in addition it is a natural partner to SNMS in the classification of techniques. It is therefore included in this volume, and its description follows.

18.2.1 *Operation*

The basic principles of the technique were laid down by Coburn,[27,28] and current designs vary little from the original, although of course there have been the usual improvements in resolution, detectability, etc.

Operation is shown schematically in Fig. 18.32 from Harrison *et al.*[29] In a small chamber a glow discharge is set up between an anode and a cathode. The gas is normally argon, at a pressure of 0.1 to 1.0 torr, and the cathode is the

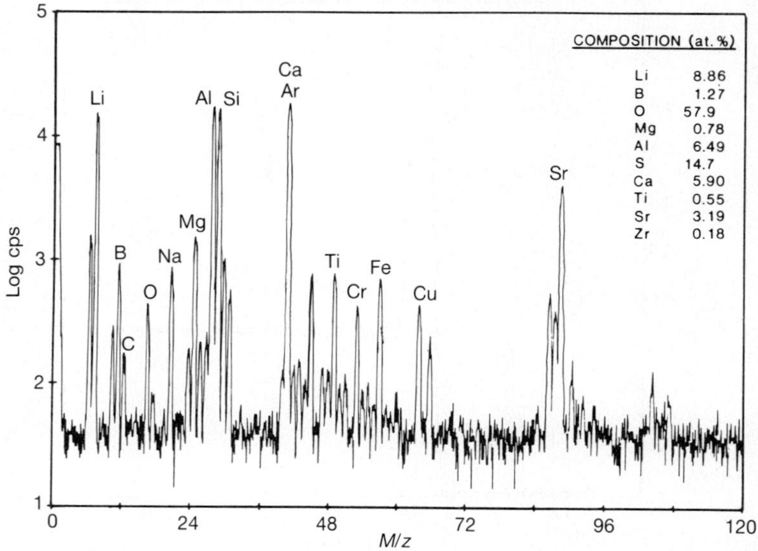

FIG. 18.30. Demonstration of the ability of SNMS to analyse insulators, by Chu *et al.*[25] The spectrum is from an NBS glass standard (K1080), and the composition given at the right of the figure is seen to correspond quite well with the relative peak heights, with the major exception of oxygen, which remains difficult to quantify. A separate ion beam was used for sputtering, and the sample surface was biased slightly positive to attract electrons to the surface for charge neutralization.

sample to be analysed. The applied voltage, which may be DC or pulsed DC, is of the order of 500 V. Under these conditions the glow discharge separates into several zones, of which only two, the negative glow region and the so-called cathode dark space, are of interest here. The former is a region of diffuse bright light emission extending away from the cathode and filling most of the chamber. The latter, on the other hand, shows little or no emission, and consists of a thin layer separating the cathode surface from the edge of the negative glow region. Virtually all the applied voltage is dropped across the cathode dark space, and is called the cathode fall.

Positive Ar^+ ions produced in the discharge space are accelerated through the cathode fall, and both they and fast neutral atoms also created in the discharge strike the cathode (sample) surface, causing sputtering. As in other techniques employing discharges, positive ions produced by sputtering return to the surface by the cathode fall field, while neutral atoms move away into the discharge region. Because in the glow discharge region at a pressure of $\approx$0.5 torr mean free paths are very short, all species are subject to very frequent collisions. This ensures not only that the energies of the bombarding ions have been reduced from the initial 500 eV to 100–200 eV, but that the sputtered neutrals are very efficiently ionized. Since, as remarked several times previously, the great majority of the sputtered species are in the form of neutral

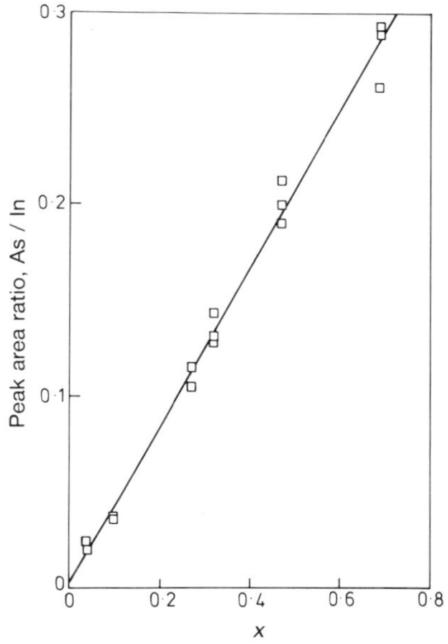

FIG. 18.31. Correlation found by Wilson *et al.*,[3] using electron impact postionization, between the ion peak ratio As/In and the value of x in the series of semiconductor materials In As_xP_{1-x} grown epitaxially on InP. The parameter x was derived from X-ray diffraction data. The observation of almost exact linear correlation confirms that the compositions of the $InAs_xP_{1-x}$ films are maintained right up to their surfaces.

atoms, detection of the postionized neutrals can lead to very high sensitivity in elemental analysis. Also, of course, the severe matrix effects attendant on the detection of secondary ions are avoided.

The two principal processes of ionization thought to account for postionization are those of electron impact

$$M + e^- \rightarrow M^+ + 2e^- \tag{18.25}$$

and of Penning exchange

$$M + Ar^* \rightarrow M^+ + Ar + e^-. \tag{18.26}$$

Because metastable noble gas atoms have relatively long lifetimes, the Penning mechanism is believed to be the dominant one, even though the population of electrons in the discharge is much greater than that of metastables. As a result the postionization probability is almost uniform, since the energies of the metastable states of, say, argon, are much higher than most first ionization potentials. This leads in turn to a much narrower range of relative elemental sensitivities than in SIMS.

A variety of sample and ionization chamber configurations have been used in GDMS, and these have been summarized by Harrison *et al.*[30] into the

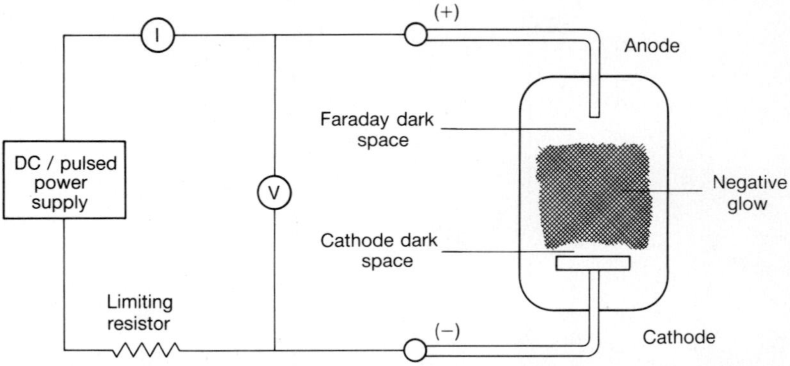

FIG. 18.32. Schematic diagram of the principle of operation of Glow Discharge Mass Spectrometry (GDMS), from Harrison *et al.*[29] The sample to be analysed is made the cathode, and a glow discharge is set up between it and an anode in a small chamber filled with argon at 0.1 to 1.0 torr pressure. The applied voltage is about 500 V. The glow discharge separates into various regions, including the negative glow which extends through most of the chamber, and the cathode dark space, a thin layer with almost no emission of light next to the cathode. Most of the applied voltage drop is across the cathode dark space, and is called the cathode fall. Ar^+ ions from the discharge are accelerated through the cathode fall to the cathode, causing sputtering. Positive ions produced by sputtering return to the cathode, while neutrals pass into the discharge region and become ionized. The postionized neutrals are then detected and measured in a mass spectrometer as in SNMS.

categories of pin, disc, hollow-cathode, hollow-cathode plume, and Grimm. The last of these has already been described in the chapter on GDOS. An example of a commercial version that uses pin geometry is shown in Fig. 18.33 (by courtesy of VG Isotopes Ltd.). As the name implies, the sample is in the form of a pin or rod or wire gripped in a pin chuck, and projecting a few mm beyond the end of an insulator. It is pushed on a rod into the centre of a small ionization chamber. On the other side of the chamber opposite the pin is the exit slit for the extraction of the postionized atoms, which is performed with an electrostatic field, following which the ions are focused onto the entrance slit of a mass analyser. The analyser used in this particular equipment is double focusing (i.e. both angle and energy) consisting of a magnetic mass analyser followed by an electrostatic analyser for energy selection. One instrumental design aspect that has to be considered is that of the high gas pressure in the ionization chamber. The flow of gas into the mass analyser from the chamber at 0.1–1.0 torr would impose far too high a pressure for operation of the mass analyser, so successive stages of differential pumping have to be arranged to bring the analyser pressure into the 10^{-6}–10^{-7} torr region.

In other systems quadrupole mass spectrometers have been used rather than magnetic sector spectrometers. The trade-off is well known—lack of mass resolution in the quadrupole, much greater expense in the magnetic sector, and the choice obviously depends on the type of analysis required and on the financial circumstances.

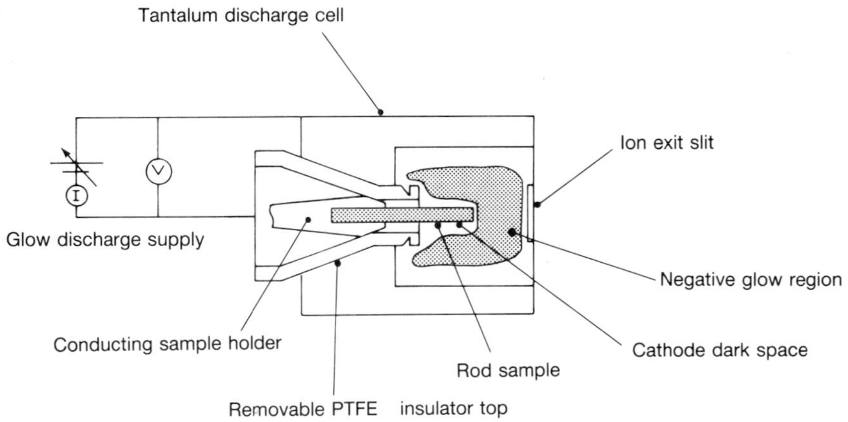

FIG. 18.33. Example of a GDMS configuration that uses pin geometry (by courtesy of VG Isotopes Ltd.). The sample is in the form of a pin or rod gripped in a chuck, that projects a few mm beyond the end of an insulator into a small ionization chamber. Opposite the end of the pin is the exit slit for extraction of postionized neutrals, through which the ions are focused into a double focusing magnetic mass analyser. Because of the high pressure of gas in the ionization chamber, the mass spectrometer has to be differentially pumped.

18.2.2 *Theory and spectral identification*

Since GDMS depends on the same sputtering processes as in SNMS, and the nature of the mass spectra produced in the two techniques is identical, the two topics under this heading have already been covered in sections 18.1.2 and 18.1.3.

18.2.3 *Quantification*

The path to quantification is also much the same as in SNMS but there are some slight differences in usage. The basic equation, which includes the same assumptions, is again expression (18.24), repeated here

$$c_A = c_R(D_R/D_A)\,(I^0_{A(B)}/I^0_{R(B)}) \qquad (18.27)$$

where c_A = concentration of element A in matrix B
c_R = concentration of reference element in B
D_R = sensitivity factor of R
D_A = sensitivity factor of A
$I^0_{A(B)}$ = postionized neutral yield from A in B
$I^0_{R(B)}$ = postionized neutral yield from reference element in B.

The different approaches to quantification based on expression (18.27) have been summarized by Sanderson *et al.*,[31] who used the results contained in Fig. 18.34. There the ion yields relative to iron are plotted for 18 elements analysed in a variety of different matrices; the measurements were all made in the

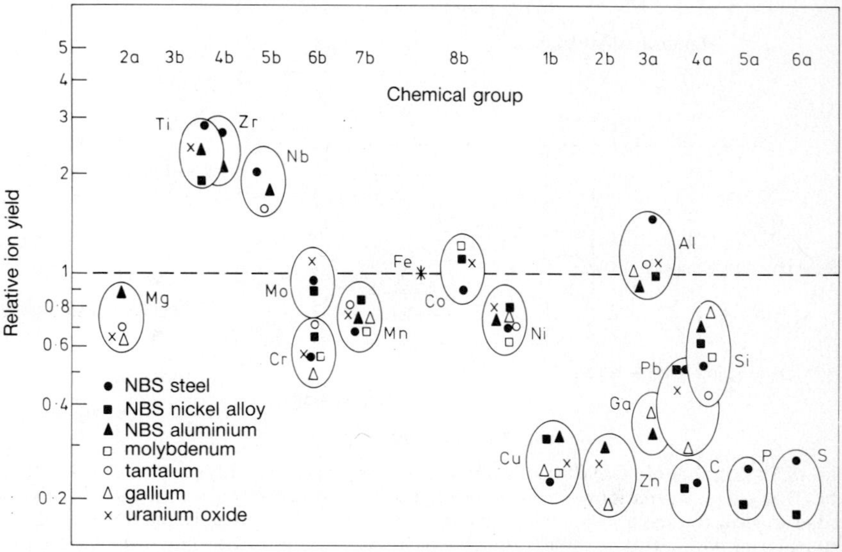

FIG. 18.34. Compilation by Sanderson *et al.*[31] of GDMS ion yields relative to iron analysed in a variety of different matrices (see list in figure) and in several different laboratories. The measurements were all made on the instrument using a magnetic mass spectrometer (by VG Isotopes Ltd.) mentioned in Fig. 18.33. For any one element the relative ion yields in different matrices overlapped by $\pm 30\%$, while the spread in yield from greatest to least was a factor of about an order of magnitude.

GDMS equipment using the magnetic mass spectrometer (VG Isotopes Ltd.) mentioned above, at four different international laboratories. The relative ion yield is defined in this case as $(I^0_{A(B)}/I^0_{R(B)})(c_R/c_A)$, and is therefore equal to (D_A/D_R). The matrices were all those of certified standard materials in which both c_R and c_A were known. It can be seen that for any one element the overlap of the relative ion yields in different matrices is about $\pm 30\%$, while the spread of yields from top to bottom is about an order of magnitude, i.e. within a factor of three of unity. In other words, the relative sensitivity factors $D_{A,Fe}$ also clustered for each element in different matrices, and were also within a factor of three of unity.

The first approach could be termed the order-of-magnitude approach, since it uses no standards or relative sensitivity factors, but simply takes the ion beam ratio (referred to a principal constituent) as equal to the concentration ratio. Since, as observed above for the case of the reference element Fe, all relative sensitive factors are within a factor of three of unity, this approach will be inaccurate to that extent. However, compared to the uncertainties in SIMS, which can easily be much more than an order of magnitude, even this very approximate approach is useful; it is also, of course, very fast, since the only quantification output needed is a set of ion ratios. A typical result of this approach is shown in Table 18.2, which compares the ratios of impurity ions to

Table 18.2

Comparison of ion ratio ($I^0_{A(Ga)}/I^0_{Ga(Ga)}$) with the known concentrations of impurities in a gallium sample. (VG Isotopes Ltd.)

Element	Concentration (chemical analysis) (p.p.m.)	Ion ratio (GDMS) (p.p.m.)
Si	4.16	3.54
Ti	0.42	1.39
Cr	0.58	0.24
Mn	0.86	0.61
Fe	8.40	11.10
Ni	0.41	0.43
Cu	0.91	0.44
Zn	1.75	0.67
Sn	0.43	0.40
Sb	0.42	0.17
Pb	0.42	0.53

Ga^+ ions in a sample of high purity gallium to which known amounts of contaminants had been added deliberately, with the known concentrations (by courtesy of VG Isotopes Ltd.).

The ratios of GDMS ion ratio to the known concentration vary from 3.3 (Ti) to 0.4 (Zn), which confirms what has been said above about the narrow range of relative elemental sensitivities, and therefore about the order-of-magnitude accuracy of the approach.

While the first approach assumes that all relative sensitivity factors are unity, the second uses relative sensitivity factors averaged over all the standard matrices used for the derivation of the factors. Referring again to the results of Fig. 18.34, such an approach would then have an accuracy of about $\pm 30\%$, that is, an improvement by an order of magnitude over the first, no-standard, approach. Sanderson *et al.*[31] call this approach the semi-quantitative approach, but in many cases it is perfectly adequate for trace element analysis. They have provided an example of its application, in the analysis of trace elements in a molybdenum sample in which the trace concentrations had been previously established by neutron activation analysis. Table 18.3 shows the comparison, along with the sensitivity factors relative to Mo, the matrix reference. Note that the relative sensitivity factor used as a multiplier for conversion to concentration will be approximately the inverse of the corresponding ion yield in Fig. 18.34, since $D_{Mo} \approx D_{Fe}$.

If the analysis for V is ignored, since the actual numbers are very small, the largest deviation between the results from the two analyses is $\approx 18\%$. This and

Table 18.3

Comparison of GDMS analysis using averaged relative sensitivity factors, with the concentration measured by neutron activation analysis, of impurities in a molybdenum sample. From Sanderson *et al.*[31]

Element	Ion ratio (p.p.m.)	RSF (average)	Conc. (GDMS) (p.p.m.)	Conc. (NAA) (p.p.m.)
Co	31	0.77	24	21
Cr	30.3	1.7	51	45
Cu	8.7	3.8	33	28
Fe	2285	0.99	2125	1880
Mn	15	1.3	19	18
Ni	22.6	1.3	29	28
Pb	(1–4)	2.5	(2.5–10)	7
Si	96	1.6	151	133
V	3.4	0.4	1.4	1
W	4614	1.1	5075	4700
Zr	180	0.55	99	120

the single order of magnitude range of relative sensitivity factors confirms again what has been said about the ease of quantification of GDMS.

The third approach is called 'full' quantification, although that is to some extent a misnomer. For complete quantification one would need to use standard materials whose composition matched the sample very accurately, whereas in this approach, although individual standards are used, their matching is only approximate. The philosophy is again that the insensitivity to matrix is such that accurate matching is not needed. Robinson and Hall[32] have demonstrated the application of the use of standards in their analysis of Zr alloy (zircalloy) materials used as fuel cladding in nuclear power production. Using the same GDMS equipment (VG Isotopes Ltd.) as the authors quoted above, they first derived relative sensitivity factors for a large number of minor constituents in a Zr alloy of certified composition. They then used the factors to quantify the compositions of other Zr alloys in which the concentrations of the minor constituents varied only slightly. The results for some of the alloys whose compositions were also certified are given in Table 18.4.

It will not have escaped notice that the elements appearing in the quantification results of Tables 18.2 to 18.4 are metallic only. Analysis of non-metals is feasible, in the sense that features arising from the presence of C, N, S, O, etc., can be measured in the mass spectra, but the problem is that of the relative sensitivity factors to be used. Although relative ion yields for C, P, and

Table 18.4

Comparison of GDMS analysis using relative sensitivity factors from a matched standard, with the certified compositions, of alloy and trace elements in Zr alloys. From Robinson and Hall.[32]

Element	Alloy 1		Alloy 2		Alloy 3	
	GDMS	Cert.	GDMS	Cert.	GDMS	Cert.
			%			
Sn	0.73	0.92	1.22	1.48	0.45	0.47
Fe	0.122	0.129	0.126	0.136	0.086	0.093
Cr	0.012	0.013	0.087	0.098	0.155	0.150
Ni	0.097	0.094	0.057	0.058	0.0005	0.001
			p.p.m.			
Hf	121	128	69	72	219	220
B	2.6	3.7	0.03	0.2	0.14	0.2
U	0.92	1.1	0.22	0.5	0.18	0.5
Cu	89	98	7	8	11	11
Co	22	20	0.8	3	55	49
Mn	28	28	4	4	6	7
Pb	12	11	4	5	3.5	3
Al	111	86	63	53	139	130
Si	82	95	13	21	113	124
Ti	98	93	2	4	2	3
Mo	44	46	0.3	2	0.2	2
W	33	32	5	7	48	43

S are shown in Fig. 18.34, they were able to be derived because the bulk analyses for the particular certified standards used happened to include those elements. In most cases certification might include C, but not usually many other non-metals, and rarely O. Thus quantification of the non-metals is at present not available in general at the same level of accuracy as that of the metals, probably only order-of-magnitude at best. Even so, that represents an advantage over SIMS.

One aspect of trace element quantification that needs to be emphasized is in fact an instrumental one. This is the need for high mass resolution. The mass spectrum from a sample containing a large number of trace impurities is certain to contain many spectral interferences. A couple of examples will suffice. When looking for Cu and Fe impurities in the semiconducting material CdHgTe, an instrument with insufficient resolution would be unable to separate $^{63}Cu^{+}$ from $^{126}Te^{++}$, or $^{56}Fe^{+}$ from $^{112}Cd^{++}$, since in both cases these would merely be features at masses ≈ 63 and ≈ 56, respectively. However, the two respective coincidences can be separated with good mass resolution since their mass-to-charge ratios are not exactly the same. This separation is demonstrated in Fig. 18.35, recorded[33] with the double focusing

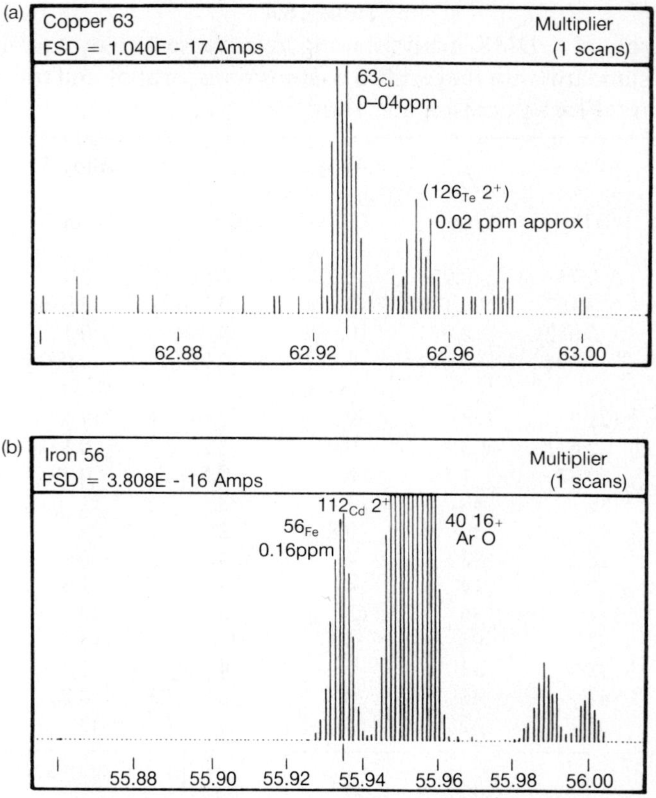

FIG. 18.35. Example given by Hall and Robinson[33] of the very high mass resolution often needed in GDMS to identify and measure impurities. In CdHgTe copper and iron are impurities that would impair the semiconducting properties and that need to be analysed. However, $^{63}Cu^{+}$ would be indistinguishable from $^{126}Te^{++}$, and $^{56}Fe^{+}$ from $^{112}Cd^{++}$ without adequate resolution. With the resolution $M/\Delta M$ of 4000–6000 available in the double focusing analyser (VG Isotopes Ltd.), the figure shows that $^{63}Cu^{+}$ at 62.9296 daltons can be separated from $^{126}Te^{++}$ at 62.9516 daltons, and $^{56}Fe^{+}$ at 55.9349 daltons from $^{112}Cd^{++}$ at 55.9514 daltons. The levels of concentration of the Cu and Fe impurities are in the range of tens of p.p.b.

analyser (VG Isotopes Ltd.) mentioned earlier, in which the mass resolution $M/\Delta M$ is in the range 4000–6000. With this resolution the $^{63}Cu^{+}$ peak, at 62.9296 daltons, and the $^{126}Te^{++}$ peak, at 62.9516 daltons, are readily separated, as are the $^{56}Fe^{+}$ at 55.9349 and the $^{112}Cd^{++}$ at 55.9514. Note in the latter case that the complex $^{40}Ar^{16}O^{+}$ ion is also separated. The levels of detection of the impurity Cu and Fe constituents are of the order of some tens of p.p.b.

18.2.4 *Applications*

There are two principal areas of application in which GDMS scores heavily—analysis of trace metallic elements and depth profiling. In both areas the analysis is both rapid and quantitative. Examples of the former application have already been given in the previous section on quantification, so the following examples concern depth profiling.

Since, as noted several times in the previous section, the range of relative elemental sensitivities in GDMS is narrow, of the order of one order of magnitude, merely recording the ion currents as a function of time (i.e. depth) of erosion will provide a depth profile that is reasonably informative. Such a profile is shown in Fig. 18.36, through a multilayer sample of Cr and Cu on an

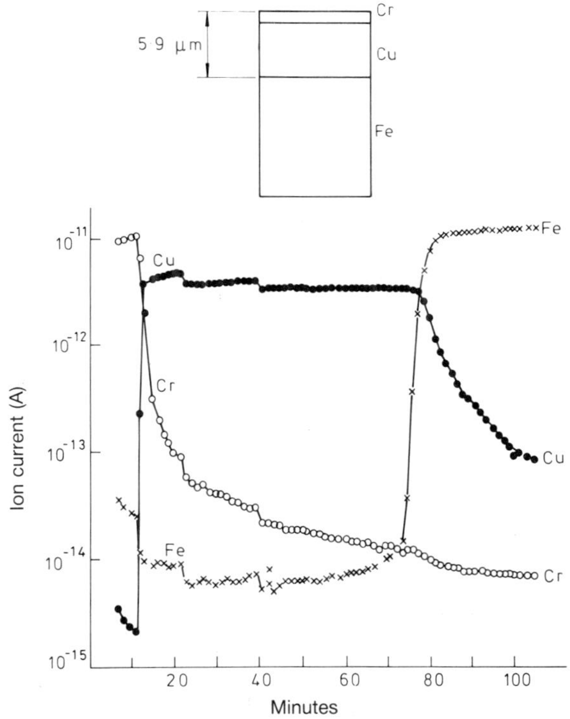

FIG. 18.36. Profile by GDMS through a multilayer sample of Cr and Cu on an Fe substrate, the total thickness of the Cr + Cu layer being 5.9 μm (by courtesy of VG Isotopes Ltd.). The thickness is too great for normal surface analytical techniques, with which the profile would take days or weeks to be obtained. With GDMS, however, it took less than 2 h. The depth resolution in terms of interface sharpness is nevertheless good. Note also the excellent sensitivity for Cu in GDMS; in SIMS it would have been very difficult to record such a profile for Cu.

Fe substrate (by courtesy of VG Isotopes Ltd.). Although the thickness of the layer, 5.9 μm, puts it out of the realm of surface analysis, since any of the surface analytical techniques such as XPS or AES combined with ion profiling would take many days to duplicate the result, it can be seen that the depth resolution in terms of interface sharpness is nevertheless similar to what would be obtained by them. From the known thickness and the time taken to reach the Fe substrate, the removal rate in this case can be estimated to be 80 nm min^{-1}. The good sensitivity for Cu should be noted; in SIMS it would have been almost impossible to achieve such a profile for Cu.

The other example, from Sanderson *et al.*,[31] is that of the silicon levels in two epitaxially grown layers on GaAs, itself doped with Si. Each epilayer had been doped with Si by a different amount, the notional concentrations as well as the thicknesses being shown in the inset. In this case the Si quantification was of the semi-quantitative variety, that is, averaged relative sensitivity factors were used, unmodified, at all points in the profile. For flat surfaces such as semiconductor wafers, a slightly different discharge chamber geometry is used, very similar in fact to that of the Grimm lamp used for GDOS, and shown in Fig. 15.6. Figure 18.37 shows the result of the profile. Each epilayer

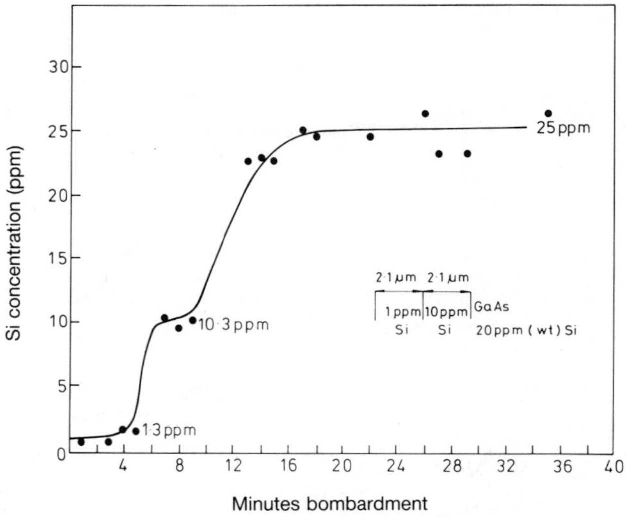

FIG. 18.37. Profile through two epitaxially grown layers of GaAs grown on a GaAs wafer, with different levels of Si doping (1 p.p.m. and 10 p.p.m., respectively) in the two layers, recorded by Sanderson *et al.*[31] Each epilayer can be distinguished clearly in terms of the amount of Si in it, and the concentrations derived by semi-quantitative methods agree very well with the notional amounts incorporated during growth. Again the thicknesses of the epilayers were too great for conventional surface analytical methods, and the Si levels too low, but GDMS provided an analysis in about 35 minutes.

can be clearly distinguished, and the semi-quantitatively derived Si concentrations in the layers and in the bulk accord well with the notional amounts incorporated during growth. The rapidity and accuracy of such an analysis means that GDMS can easily be used for quality control in device production.

It will be appreciated from the above examples and from the description of

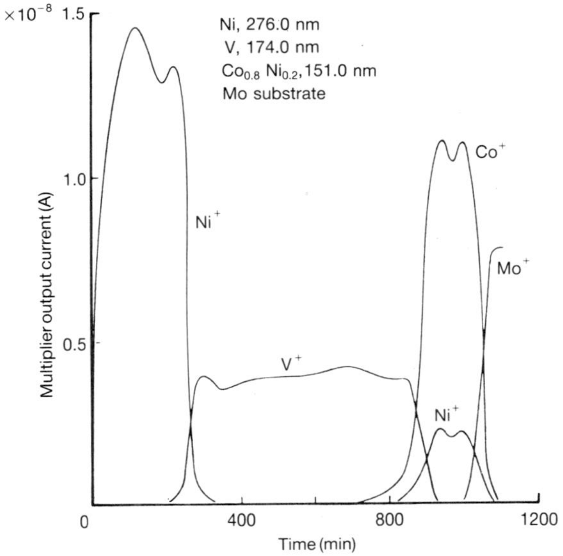

FIG. 18.38. Profile of a multilayer film on a molybdenum substrate obtained by Coburn *et al.*[34] by GDMS operated in an ultra-high vacuum system. A radio-frequency rather than a glow discharge was used, but the principle was exactly the same. Note that the detection sensitivity for Ni in pure Ni was almost the same as in the Co–Ni layer, and that the interfaces appeared sharp even after some hundreds of nm had been removed by sputtering.

the technique that the normal mode of operation is not surface-specific, although the depth resolution can be made very good. Indeed the commercially available equipment is not intended for surface-specific analysis, even though in principle the rate of erosion could be reduced to levels similar to those employed in, say, normal ion profiling in XPS and AES. However, GDMS has been used in a manner close to that of the other techniques described here, but in the earlier work by Coburn and co-workers.[27,28] The discharge conditions used were slightly different from those in the commercial GDMS, in that a radio-frequency discharge was established, but the principle was exactly the same. It is also worth noting that the technique was set up in ultra-high vacuum, with base pressures better than 10^{-8} torr achieved without bakeout after running the discharge to remove adsorbed gases. During analysis the pressures of noble gases were in the range 60–120 mtorr. Sputtering rates were normally much lower than in the commercial GDMS described above, and could be of the order of 1 nm min^{-1}, as seen in the profile of a multilayer film in Fig. 18.38, from Coburn *et al.*[34] This particular profile took almost 20 h to record at the low sputtering rate, and for such a long analysis the rf discharge conditions could not be maintained accurately constant, which accounts for some of the variations in the profiles.

Nevertheless, it should be noted that the detection sensitivity for nickel in pure nickel was the same, within experimental error, for nickel in the cobalt–nickel alloy layer. Notice also the sharpness of the interfaces even after sputtering away many 100's of nm. The profiling rate could be increased controllably by increasing either the argon pressure or the rf power.

References

1. Lipinsky, D., Jede, R., Ganschow, O., and Benninghoven, A. (1985). *J. Vac. Sci. Technol.* **A3**, 2007.
2. Lipinsky, D., Jede, R., Tümpner, J., Ganschow, O., and Benninghoven, A. (1985). *J. Vac. Sci. Technol.* **A3**, 2035.
3. Wilson, R., Van den Berg, J. A., and Vickerman, J. C. (1989). *Surf. Interface Anal.* **14**, 393.
4. Wittmaack, K., Maul, J., and Schulz, F. (1973). *Int. J. Mass Spectrom.* **11**, 23.
5. Oechsner, H. and Stumpe, E. (1977). *Appl. Phys.* **14**, 43.
6. Beckmann, P., Kopnarski, M., and Oechsner, H. (1985). *Mikrochim. Acta (Vienna)* **Suppl. 11**, 79.
7. Oechsner, H. (1974). *Plasma Physics* **16**, 835.
8. Stumpe, E., Oechsner, H., and Schoof, H. (1979). *Appl. Phys.* **20**, 55.
9. Oechsner, H., Rühe, W., and Stumpe, E. (1979). *Surf. Sci.* **85**, 289.
10. Müller, K-H. and Oechsner, H. (1983). *Mikrochim. Acta (Vienna)*, **Suppl. 10**, 51.
11. Müller, K-H., Seifert, K., and Wilmers, M. (1985). *J. Vac. Sci. Technol.* **A3**, 1367.
12. Pellin, M. J., Young, C. E., Calaway, W. F., and Gruen, D. M. (1984). *Surf. Sci.* **144**, 619.
13. Pellin, M. J., Young, C. E., Gruen, D. M., Aratono, Y., and DeWald, A. B. (1985). *Surf. Sci.* **151**, 477.
14. Becker, C. H. and Gillen, K. T. (1984). *Anal. Chem.* **56**, 1671.
15. Becker, C. H. and Gillen, K. T. (1984). *Appl. Phys. Lett.* **45**, 1063.
16. Sigmund, P. (1977). *Inelastic ion-surface collisions* (eds. N. H. Tolk, J. C. Tully, W. Heiland and C. W. White), p. 121. Academic Press Inc., New York.
17. Lindhard, J., Scharff, M., and Schiøtt, H. E. (1963). *Kgl. Danska Vid. Selskob. Mat. Phys. Medd.* **33**, 1, 10, and 14.
18. Lindhard, J. and Scharff, M. (1961). *Phys. Rev.* **124**, 128.
19. Pivin, J-C. (1983). *J. Mater. Sci.* **18**, 1267.
20. Jede, R., Seifert, K., and Dünnebier, G. (1987). *Fresenius Z. Anal. Chem.* **329**, 116.
21. Jede, R., Kaiser, U., Peters, H., and Ganschow, O. (1988). *Proc. 6th Int. Conf. SIMS (SIMS VI)* (eds. A. Benninghoven, A. M. Huber, and H. W. Werner), p. 343. John Wiley & Sons, Chichester.
22. Tümpner, J., Wilsch, R., and Benninghoven, A. (1987). In *Secondary ion mass spectroscopy* (eds. A. Benninghoven, F. G. Rüdenauer and H. W. Werner), pp. 946–949. John Wiley & Sons, New York.
23. Becker, C. H. and Gillen, K. T. (1986). *Proc. 5th Int. Conf. SIMS (SIMS V)* (eds. A. Benninghoven, R. J. Colton, D. S. Simons, and H. W. Werner), p. 85. Springer-Verlag, Berlin.
24. Parks, J. E., Beekman, D. W., Schmitt, H. W., and Taylor, E. H. (1985). *Nucl. Instrum. Meth.* **B10/11**, 280.
25. Chu, P. K., Huneke, J. C., and Blattner, R. J. (1987). *J. Vac. Sci. Technol.* **A5**, 295.

26. Kaiser, U., Meeker, G. P., and Huneke, J. C. (1988). *J. Vac. Sci. Technol.* **A6**, 1082.
27. Coburn, J. W. (1970). *Rev. Sci. Instrum.* **41**, 1219.
28. Coburn, J. W. and Kay, E. (1971). *Appl. Phys. Lett.* **19**, 350.
29. Harrison, W. W., Hess, K. R., Marcus, R. K., and King, F. L. (1986). *Proc. 5th Int. Conf. SIMS (SIMS V)* (eds. A. Benninghoven, R. J. Colton, D. S. Simons and H. W. Werner), p. 75. Springer-Verlag, Berlin.
30. Harrison, W. W., Hess, K. R., Marcus, R. K., and King, F. L. (1986). *Anal. Chem.* **58**, 341A.
31. Sanderson, N. E., Hall, E. F. H., Clark, J., Charalambous, P., and Hall, D. (1987). *Microchim. Acta (Vienna)*, **I**, 275.
32. Robinson, P. K. and Hall, E. F. H. (1987). *J. Metals*, **39**, 14.
33. Hall, D. J. and Robinson, P. K. (August 1987). *American Laboratory*.
34. Coburn, J. W., Taglauer, E., and Kay, E. (1974). *J. Appl. Phys.* **45**, 1797.

19

Neutral excitation: fast atom bombardment mass spectrometry (FABMS)

In its origination this technique was really only an offshoot of SSIMS, arising from efforts to minimize or avoid the charging problems encountered with SSIMS when attempting to analyse non-conducting materials, and the damage problems when analysing large organic molecules. Indeed the very first papers by Barber *et al.*[1,2] dealt specifically with the mass spectrometric analysis of biochemicals. The same authors found subsequently[3] that a very useful variant of the technique was to disperse the high molecular weight organic in a matrix of low vapour pressure and good dispersing properties, and for this glycerol has been found to have excellent characteristics. Fragments from the glycerol itself are of course found in the mass spectrum, but their interference can be subtracted by computer. However, this chapter is not concerned with the way in which fast atom bombardment has been used routinely for organic analysis in conventional mass spectrometers, but with its development into a surface-specific analytical technique with very close affinities with SSIMS. In this respect it is interesting that the authors of a recent major work[4] on SIMS dismiss FABMS as merely 'one of the many modes of SIMS', and accuse the originators of introducing confusion through the acronym FAB. On the other hand, one could argue that they themselves were in fact introducing confusion, since strictly speaking one can have *secondary* ions only if there are *primary* ions present as well. The development of FABMS into a technique with general surface analytical application is due principally to Vickerman and co-workers.

19.1 Operation

The design of a combined SSIMS/FABMS system is shown in Fig. 19.1, from Eccles *et al.*[5] The intention in both techniques was to produce a rastered, microfocused, particle beam; the philosophy in design was to concentrate on the ion beam production in the initial stages of the source, with conversion of ions to neutrals taking place at the end of the beamline.

After production in a hot-filament, magnetically confined, source (1), argon ions are extracted through a few hundred volts by (2), and then accelerated by an immersion lens (3) to about 10 keV. The first beam alignment is imposed by the set of plates (4), after which the ions are mass analysed in a Wien filter (5, and see Section 17.1) and corrected for astigmatism by the octupole stigmator (6). The valve (7) is used to isolate the column, which has its own miniature

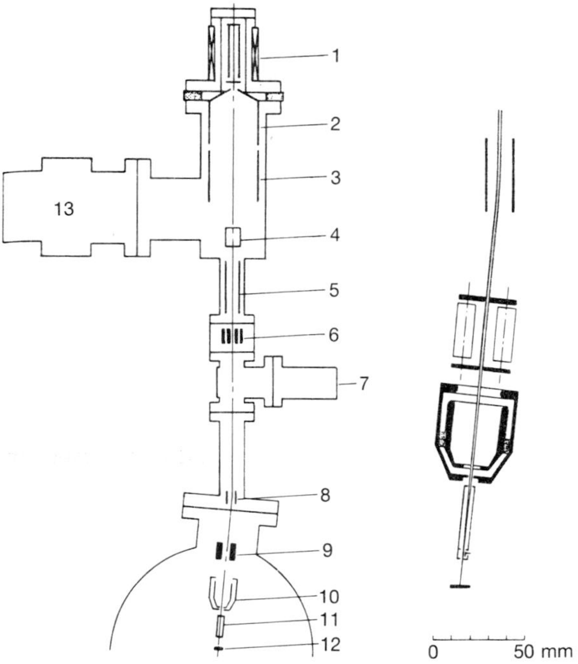

FIG. 19.1. Diagram of experimental arrangement designed by Eccles *et al.*[5] for combined SSIMS/FABMS. (1) Hot filament, magnetically confined, ion source. (2) Extractor. (3) Immersion lens for acceleration to about 10 keV. (4) Beam alignment plates. (5) Wien filter. (6) Octupole stigmator. (7) Isolation valve. (8) Deflection plates to remove neutrals from beam. (9) Scan rod system. (10) Asymmetric unipotential lens. (11) Charge-transfer cell. (12) Specimen. (13) Miniature turbo-molecular pump. Components 8–12 are shown on a magnified scale to the right.

turbomolecular pump (13), for maintenance and filament replacement. As the beam enters the analysis chamber it is bent through an angle of 5°, which removes the neutral component for SSIMS operation, and passed through a 1.5 mm diameter skimming aperture, so that only the centre of the beam, with uniform current density, is used as the final probe. Since the charge-transfer cell for ion neutralization was to be placed at the end of the beamline, the focal length of the final focusing lens had to be twice as long as strictly desirable, and therefore the beam rastering took place before the lens, in the scan rod system (9), shown on an enlarged scale in Fig. 19.1. The rastered beam then passes through an asymmetric unipotential lens (10), chosen because most of the focusing takes place near the lens exit. After leaving the lens the rastered, focused, ion beam then enters the charge-transfer cell (11) which consists of a tube of length 40 mm filled with argon gas to a pressure of $\approx 10^{-3}$ mbar. The principle of gaseous charge-transfer is that of the resonance between fast argon ions and argon atoms with only thermal energy, according to the process

$$\mathrm{Ar}^{+}_{\mathrm{fast}} + \mathrm{Ar}^{0}_{\mathrm{kT}} \rightarrow \mathrm{Ar}^{0}_{\mathrm{fast}} + \mathrm{Ar}^{+}_{\mathrm{kT}}. \tag{19.1}$$

Ion neutralization efficiencies of 15–25% can be achieved if the dimensions of the charge-transfer cell are matched to the collisional mean free path in the gas appropriate to the pressure in the cell. At the end of the cell a deflection electrode removes any remaining ions from the beam, which then passes on as a rastered, focused, neutral beam to strike the specimen (12). With a relatively high argon pressure in the cell the pressure in the analysis chamber can be maintained in the 10^{-6} mbar region.

It will be clear from the above description that three possible modes of operation are possible. Without gas in the cell or final deflection, the incident beam consists of ions only; with gas in the cell and with deflection, of atoms only; with gas in the cell but without deflection, both ions and atoms. However, the latter mode is not particularly useful for analytical purposes.

With the above equipment beam energies can be varied from 0.5 to 3.0 keV, and specimen currents from 0.05 to 100 nA cm^{-2}. At the higher energies and lower currents the spot sizes on the specimen are of area 0.3 cm^2, i.e. about 6 mm diameter.

Beams of fast atoms are also produced by some ion sources as a natural result of their mode of operation. This is particularly true of the saddle field source described in Section 3.2.4, and in Fig. 3.19. When run at high voltages (e.g. ≈ 8 kV) at least 50% of the emission from the aperture consists of fast neutrals. The ion component can be removed either by deflection or by passage through a charge transfer cell as above. There is little possibility of rastering or focusing with such a source and it must be regarded as useful principally for fast and uniform erosion over a large area.

19.2 Theory

Insofar as a mechanistic theory exists at all for FABMS, it is identical to that for SSIMS. This lack of a separate theory is based on the assumption implicit in SSIMS theory that the incoming energetic ion is actually neutralized by an Auger process *before* it hits the surface and is, in effect, a bombarding neutral. It must be implicit because nowhere in SSIMS theory is the charge state of the incident particle taken into account; the particle is regarded merely as a projectile of a certain mass and energy, and any considerations of electronic transfer mechanisms are restricted entirely to how secondary ions can be formed, i.e. on the way out rather than on the way in. However, there can be little doubt that the charge state of an incident ion is significant, because otherwise there would not be the difference in surface charging behaviour between SSIMS and FABMS. Neutralization of surface charge by low energy electron irradiation is often used in SSIMS when attempting to analyse insulators, but does not always work and in any case is undesirable because of possible electron stimulated desorption from the sample, or even decomposition of the sample. It has been shown conclusively now by several workers[6,7] that the use of neutral beams when studying insulators virtually eliminates the charging problem.

The other significant aspect of the use of a charged particle in SSIMS is that of beam damage. Of course, by its very nature bombardment of a surface by a relatively heavy particle will cause damage to an extent dependent on both the mass and energy of the particle, as described in the chapter on SNMS, but if it carries a charge as well there is evidence for additional, or more rapid, damage. This has been shown by Brown *et al.*[8] in experiments using the combined SSIMS/FABMS equipment shown in Fig. 19.1. For direct comparison of the effects of ion and of atom bombardment respectively, on a material, they went to considerable lengths to ensure that the fluxes (i.e. particles cm^{-2} s^{-1}) in the two beams were identical, and they also checked the beam profile in the two cases and found them to be identical.

Brown *et al.*[8] compared the relative effects of ion and atom bombardment on two materials, a 'typical' polymer, polystyrene, and a 'typical' compound, Nb_2O_5. For the polystyrene the comparison between the two spectra obtained in the earliest stages is given in Fig. 19.2. The spectra are qualitatively similar, with the same set of characteristic fragment ions, but significant differences are nevertheless apparent, in that the overall intensity of the ion-induced spectrum is a factor of 2–3 less than that of the atom-induced spectrum, and the clusters at low mass numbers are relatively more intense under ion than under atom bombardment. When bombardment was continued for many minutes, with continuous mass spectral recording, the variations in the intensities of the principal mass peaks as a function of time are shown in Fig. 19.3. As indicated, the respective bombardment conditions were identical. Quite clearly, the rate of damage, based on the rate of decay of the $C_3H_3^+$ and $C_7H_7^+$ secondary ions, is much greater during ion than during atom bombardment, in fact by a factor of about four. In other words, since all bombardment conditions except the charge state of the bombarding particle, are the same, there must be a charge-induced damage mechanism operating in addition to the heavy energetic particle-induced damage.

Somewhat similar observations and deductions were made for bombardment of Nb_2O_5. In Fig. 19.4 are shown the profiles for the fragment ions Nb^+, NbO^+, and NbO_2^+, again as a function of time of bombardment under identical conditions. For both ion and atom bombardment all mass intensities show initial increases, due to removal of surface contamination, and thus leading to an overall enhancement. However, whereas the atom-induced profile shows a general levelling off of intensities to plateaux, within the bombardment period employed, the ion-induced profiles of NbO^+ and NbO_2^+ go through maxima and then start to decrease rapidly. The Nb^+ profile is much the same for each. These results demonstrate once again, for a

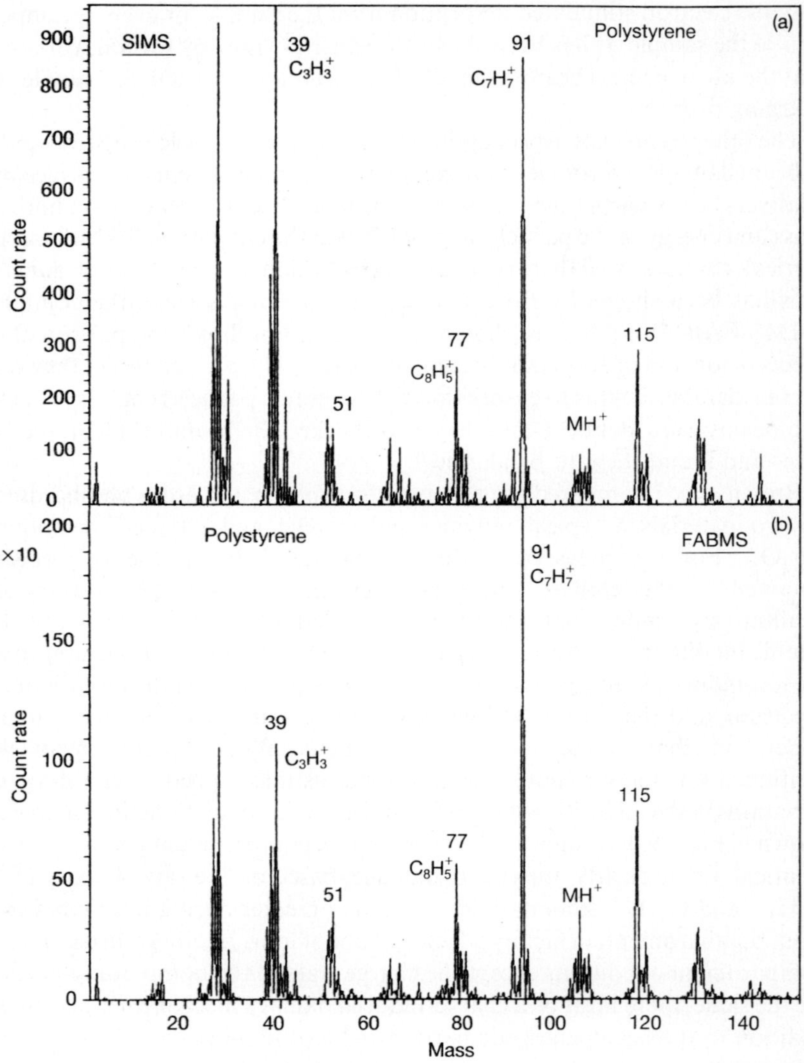

FIG. 19.2. Comparison of secondary ion spectra excited by (a) primary ions Ar^+ 2 keV in SSIMS, 3×10^9 ps cm^{-2} s^{-1} and (b) fast neutrals Ar^0 in FABMS, from polystyrene, under the same irradiation fluxes, by Brown *et al.*[6] Note that the overall intensity of the ion-induced spectrum is a factor of 2–3 less than that of the atom-induced spectrum, and that low mass number clusters are relatively more intense under ion compared to atom bombardment.

totally different type of material, the additional rate of damage attributable to the presence of the positive charge on the bombarding particle.

Although the observations described above really have more bearing on the theory of primary particle interaction in SSIMS than in FABMS, they have

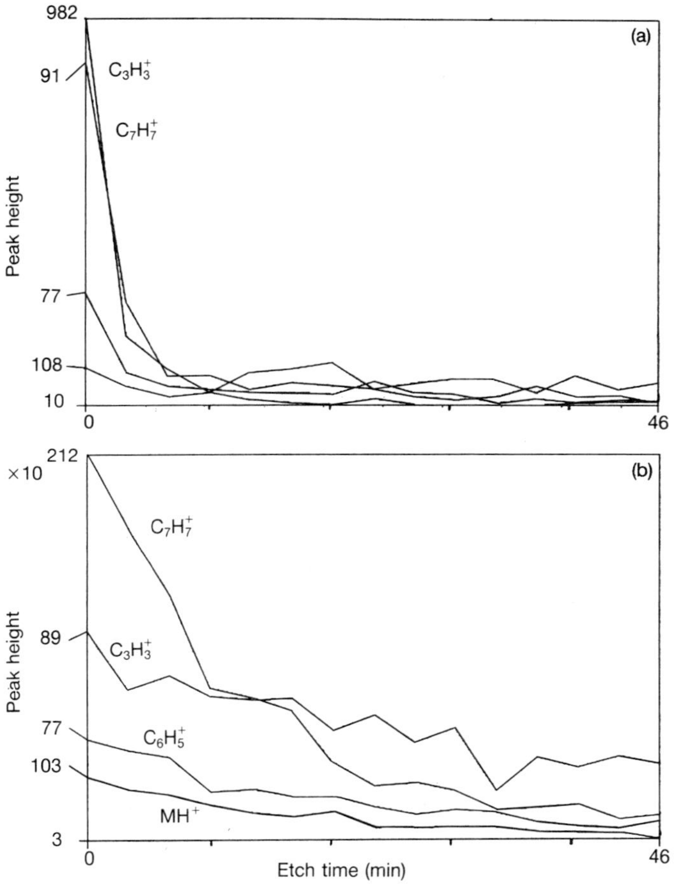

FIG. 19.3. The variation of secondary ion intensity with bombardment time for (a) ions, and (b) atoms, under the same irradiation fluxes, from polystyrene, observed by Brown *et al.*[6] The rate of damage of the polymer surface is much greater using ions compared to atoms, by a factor of ≈ 4.

been brought in at this point because without the possibility of performing FABMS the charge-induced effects would not have been realized. These effects will be particularly serious in an insulating medium. Any excess charge appearing in a conducting medium as a result of implantation of positively or negatively charged species can be screened sufficiently quickly by movement of conduction electrons, but in an insulator any redistribution of charge is very local and cannot be screened efficiently. The result is local electrostatic stress leading to bond breaking and hence to fragmentation over and above the purely collisional fragmentation. Thus Brown *et al.* conjecture that in the case above of the polymer, an electron can be donated from a molecular orbital, thereby destabilizing the polymeric bonding and leading to easier release of

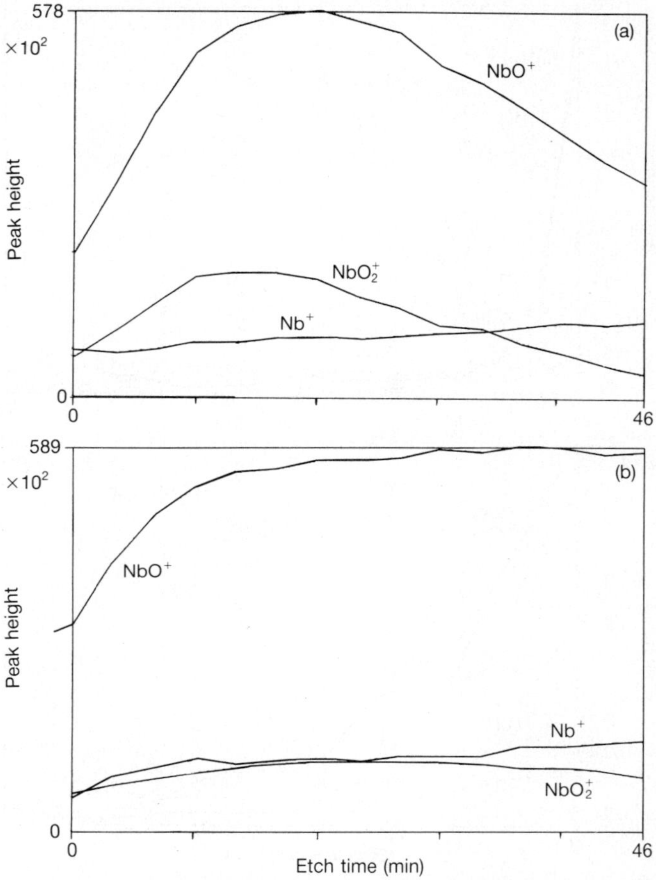

FIG. 19.4. The variation of secondary ion intensity of the fragment ions Nb^+, NbO^+, and $NbO_2{}^+$ with bombardment time for (a) ions, and (b) atoms, under the same irradiation fluxes, from a thin layer of Nb_2O_5 on steel, observed by Brown *et al.*[6] Whereas the ion-induced profiles of NbO^+ and $NbO_2{}^+$ go through maxima with time and then decrease, the corresponding atom-induced profiles level off at plateaux.

fragments; and in the case of the oxide, an electron could be lost from a metal–oxygen bond, again leading to weakening of the bond and thus to more rapid loss of oxygen.

The present state of the theory thus appears to be as follows. For processes occurring during bombardment of conducting materials, both the theories of particle–solid interaction, and of secondary ion formation, previously described, apply equally to FABMS as to SSIMS. During the bombardment of insulating and semi-insulating materials, on the other hand, the ideas concerning particle–solid interaction need amendment in the case of SSIMS, where the incident particle is charged. For all one knows, since the theory of secondary ion formation is in such a parlous state, that may need amendment as well. On this basis, what has been said in Chapter 18 about the sputtering process is probably rather more applicable to FABMS than to SSIMS when considering non-conductors. Clearly many more experiments need to be carried out to elucidate in more detail the precise nature and extent of the effect of charge.

19.3 Spectral interpretation and quantification

As with SSIMS, any problems in interpretation of mass spectra are the same as those in any mass spectrometric technique, and since the nature of the spectra obtained in FABMS is the same as that in SSIMS, the reader should refer to Section 17.2. Similarly, as far as is known the mechanisms of secondary ion formation are the same in the two techniques, and therefore the possibilities or otherwise of quantification are also the same. In this case the reader should therefore consult Section 17.4.

19.4 Applications

So-called liquid SIMS, that is the version mentioned earlier in which organic or biochemical compounds are dispersed in, typically, glycerol, and bombarded with fast atoms, has been applied to a great variety of such compounds with considerable success. However, that technique is considered to be more an extension of general mass spectrometric analysis than a technique that can be applied widely to surface analysis of all types of involatile surface, and to be outside the scope of this volume. The reader is referred to review articles such as that of Rinehart[9] for details of the applications of liquid SIMS or FABMS.

The principal advantage that FABMS has over SSIMS, that of being able to analyse the surface of non-conducting materials without having to worry about charging problems, has meant that most of the applications that have so far been described have been concerned with polymeric materials, glasses, and other insulators. For all such materials the normally employed method of charge neutralization by electron beam irradiation has given rise to continual doubts about interpretation of spectra, owing to the possibility of additional and perhaps sometimes synergistic, effects of electron beam damage.

Depth profiling can be achieved in SSIMS if the rate of erosion is increased, without going into the regime of dynamic SIMS, and has been used extensively on conducting materials. With the growing interest in the properties of polymers in thin film applications, the surface characterization and profiling of polymer films becomes progressively more important, but in SSIMS there will always be severe problems for the reasons mentioned above. The potential use of FABMS in depth profiling polymers has been demonstrated by Michael and Stulik.[10] They created a polymer layer structure by depositing a film of polymethylmethacrylate (PMMA) of thickness 1.25 μm on to a substrate of polytetrafluorethylene (PTFE) sheet, and then recorded positive secondary ion spectra as the structure was profiled by FAB. The removal rate was of the order of 12 nm min^{-1}, or 2 Å s^{-1}, which is rather faster than the normal rate in SSIMS, but not as fast as in dynamic SIMS (cf. 100–500 nm min^{-1}).

Selected mass spectra from different stages in the profile are shown in Fig. 19.5, a to d. The most prominent peaks in the series can be attributed as follows in Table 19.1.

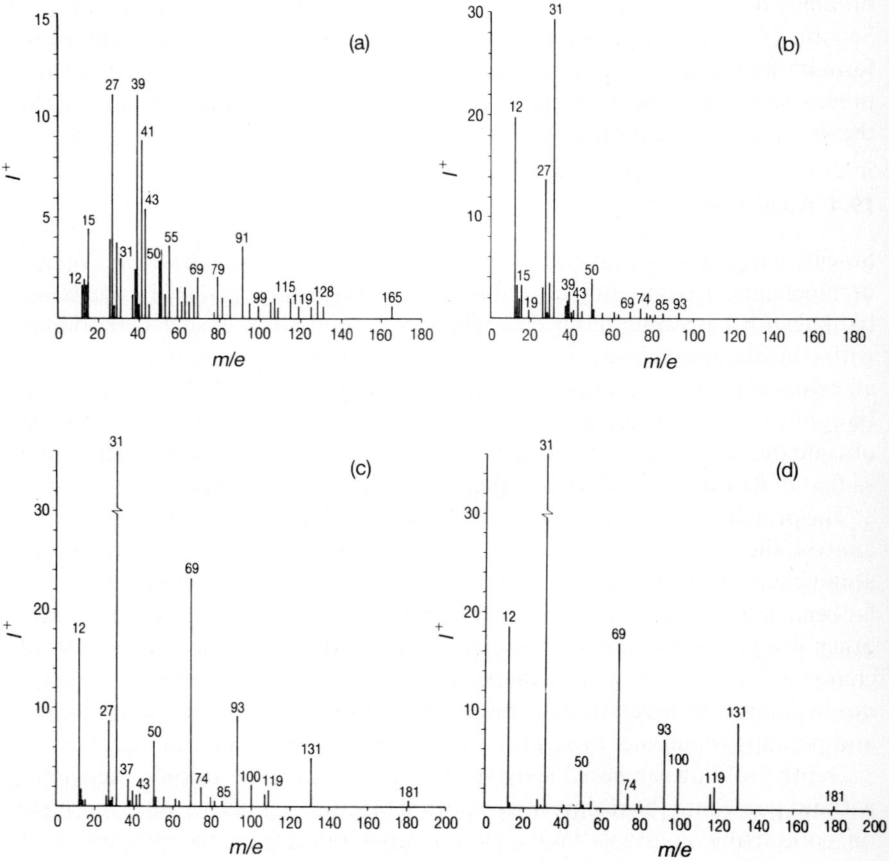

FIG. 19.5. Atom-induced secondary ion spectra from a PMMA–PTFE sandwich structure at different times of bombardment, recorded by Michael and Stulik.[10] The removal rate was about 12 nm min^{-1}, i.e. about 2 Å s^{-1}. (a) 0.08 min; (b) 10 min; (c) 105 min; (d) 125 min. The spectra in (a) and (d) are characteristic of pure PMMA and pure PTFE respectively, the others being mixtures of the two.

Table 19.1
Attribution of mass numbers to fragments in Fig. 19.5.

m/e	Fragment	*m/e*	Fragment
12	C^+	50	$C_4H_2^+$, CF_2^+
15	CH_3^+	55	$C_4H_7^+$, $C_3H_3O^+$
27	$C_2H_3^+$	69	CF_3^+
31	$C_2H_7^+$, CH_3O^+, CF^+	91	$C_7H_7^+$
39	$C_3H_3^+$	93	$C_3F_3^+$, $C_7H_9^+$
41	$C_3H_5^+$	100	$C_2F_4^+$
43	$C_3H_7^+$, $C_2H_3O^+$,	131	$C_3F_5^+$

Spectra 19.5a and 19.5d are characteristic of those from pure PMMA and pure PTFE, respectively, while the others are mixtures of the two. Inspection of the spectra show that for PMMA only masses 15 and 27 are not found in the PTFE spectrum and conversely, for PTFE only mass 131 is unique. If the intensities, as percentages of the total current, of masses 27 and 131 are then plotted as a function of erosion time (i.e. depth) then the profiles in Fig. 19.6 are found. The anomalous ion yields for the two masses on either side of the interface are said to be due to adsorbed water or oxygen present on the original PTFE surface, which can diffuse into the PMMA film either as it is deposited or during the subsequent heat treatment. Improved vacuum conditions and surface pre-treatment eliminated the anomalies.

Analysis of glass surfaces has also proved much easier with FABMS, as demonstrated by Brown and Vickerman[11] and by Surman *et al.*[12] For example, in Fig. 19.7 is the positive ion spectrum[11] under FAB from a glass made up of calcium, silicon, and aluminium oxides. Several aspects of the surface composition can be established from such a spectrum simply by inspection. Firstly, it is known that the spectrum of pure amorphous silica contains the ions Si^+, $SiOH^+$, SiO_2H^+, $SiO_3H_2^+$, etc., so that if a glass contained bonding configurations similar to those in silica one would expect the complex ions $MO_nH_m^+$ and $M_1M_2O_nH_m^+$, etc., M_1, M_2, . . . , being the different cations present, and $n, m = 0, 1, 2, 3, \ldots$. It can be seen from the spectrum that most of the simpler ($n = 1, 2$; $m = 0, 1$) ions are indeed present, which at once gives an indication of the state of vitrification of the surface. Secondly, the ions $SiOH^+$ and $CaOH^+$ are relatively intense, giving the information that the surface is hydrolysed to a certain extent. On heating in vacuo to 750 K nearly all the above features are lost from the spectrum, only CaO^+, SiO^+, AlO^+, and Si_2O^+ remaining as the molecular ion contributions, showing that the glass had then separated into its component oxides.

Depth profiling of glasses was achieved with FABMS by Surman *et al.*[12] The usefulness of glasses, principally optical, depends critically on the

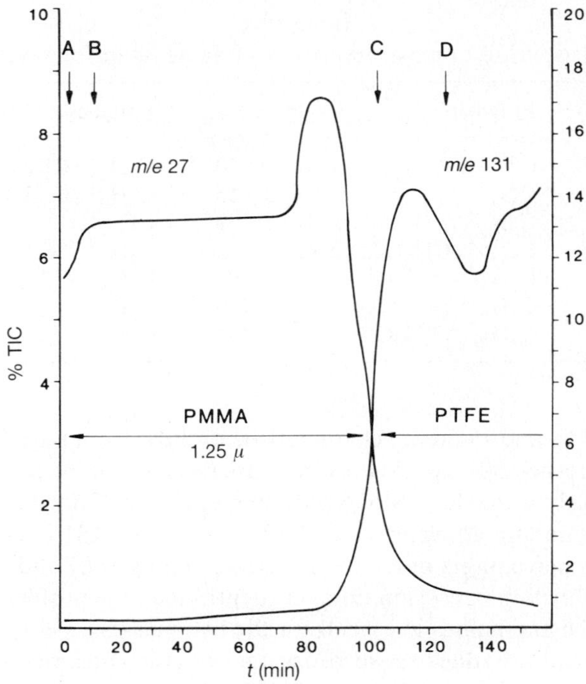

FIG. 19.6. Intensities of the ions at $m/e=27$ (characteristic of PMMA) and at $m/e=131$ (characteristic of PTFE) plotted as a function of bombardment time, i.e. depth, by Michael and Stulik.[10] The letters A to D correspond to the respective spectra in Fig. 19.5. The anomalous peaks in the profiles near the interface are attributed to adsorbed water or oxygen present on the PTFE surface before deposition of the PMMA.

composition and durability of their surfaces, two interdependent properties. It is of interest to know, for instance, whether there is an excess of any particular component in the surface, and, if possible, what the surface chemical structure is. The latter is still difficult to decide merely from a secondary ion spectrum, but the former can be derived from a depth profile. Such profiles for two glasses, both containing silicon and aluminium oxides, inter alia, are given in Fig. 19.8, in terms of the Al^+/Si^+ ratio. The very high initial values of the ratio are probably spurious, owing to interference from the ion $C_2H_3^+$ originating in surface contamination, but quickly the ratio became constant in each case, indicating either no or virtually no surface enrichment of aluminium. Such information is vital in evaluating the weathering characteristics of any particular glass, but could not have been obtained easily, or even at all, with techniques using charged particle beams.

Another class of non-conducting materials of considerable technological interest are the zeolites, which are alumino-silicate compounds active in catalysis. In their normal form they are highly porous powders, which makes

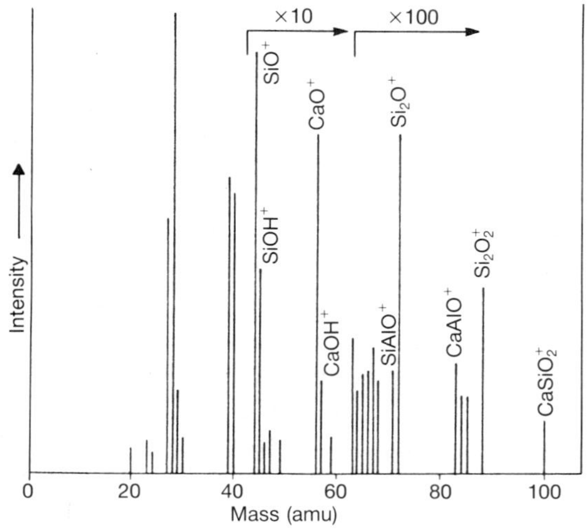

FIG. 19.7. Atom-induced positive secondary ion spectrum from the surface of a glass composed of calcium, silicon, and aluminium oxides, from Brown and Vickermann.[11] The presence and intensities of the ions $SiOH^+$ and $CaOH^+$ indicate that the surface is slightly hydrolyzed. The spectrum contains most of the simpler ions of type $M_1M_2O_nH_m^+$, where $n=1$ or 2, $m=0$ or 1, indicating a state of vitrification similar to that of amorphous silica.

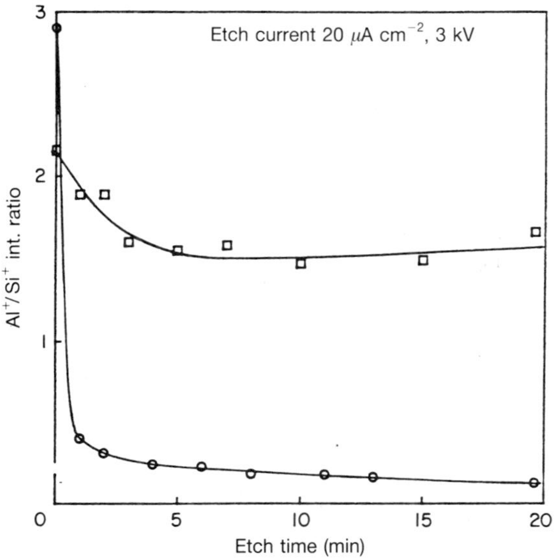

FIG. 19.8. Profiling of two glass surfaces by FABMS to establish the degree of enrichment, if any, of one or other component in the surface, carried out by Surman *et al.*[12] Both glasses contained silicon and aluminium oxides, in different relative concentration. Apart from the very high initial values of the Al^+/Si^+ ratio, probably spurious due to interference from surface contamination, the ratio quickly became virtually constant with depth, indicating little or no surface enrichment.

application of techniques such as AES and SSIMS even more difficult. Surman *et al.*[12] and Dwyer *et al.*[13] have shown how FABMS can provide surface characterization without the problems attendant on the other techniques.

Zeolites, like other catalytic materials, are generally subjected to some form of pre-treatment before use. Such treatments include acid leaching, heating in steam, and dealumination in $SiCl_4$, and it is essential to know what effect any treatment is having on the surface composition and chemistry so that treatment conditions can be adjusted if necessary. In their study of these effects the above authors were able to achieve a certain degree of quantification by comparing the silicon/aluminium ratio, either from Si^+/Al^+ or from SiO_2^-/AlO_2^-, with the known ratio in a wide range of sodium zeolites. In each case the ratio was measured first in the unbombarded condition, then after ion etching for 75 μA min, and again after etching for 150 μA min, to allow for any possible surface segregation. The results of this calibration procedure are given in Fig. 19.9, where it can be seen that the two ratios correlate very closely, the slope of the linear best fit being about 0.2. (Interestingly, the same correlation and slope were also found[12] for the same ratios in glasses.) Thus direct intercomparison of FABMS analyses on the surfaces of widely differing zeolites is permissible.

A selection of depth profiles of zeolites treated in various ways is shown in Fig. 19.10. In the left hand panel are the profiles of two zeolites with a range of Si/Al ratios from 6 (Na mordenite) to 18 (ZSM-5), treated in 4M HCl for long periods at or just above room temperature. It is obvious that the treatment has resulted in severe de-alumination of the surface layers, even though other techniques such as XRD and IR had apparently indicated no change compared with the starting material. This ready de-alumination of zeolites is clearly an effect that needs strict control. The right-hand panel compares the depth profiles of hydrogenated ZSM-5 in the untreated condition, and after treatment in steam at 600°C. The reverse effect occurs on steam treatment, i.e. the surface becomes rich in aluminium. This implies, according to the authors,[13] that aluminium is dislodged from the alumino-silicate framework, and accumulates at the external surface, either by segregation at the high temperature, or by redeposition. In another example, that of a zeolite with an initially very high external silicaceous content, the Si^+/Al^+ ratio became constant with depth after steam treatment.Again, aluminium had migrated to the surface, balancing the excess silicon in this case.

The above analyses are of course qualitative or at best semi-quantitative, but it must be emphasized again that such analyses would be difficult or impossible to obtain by use of other methods, and that where charging problems are likely to occur FABMS is often the only recourse. Some progress has been made in the avoidance of charging interference in SIMS by the use of liquid metal ion sources (LMIS, see under Section 3.2.3 and Fig. 3.16), but experience to date is variable, and the applicability to any one material

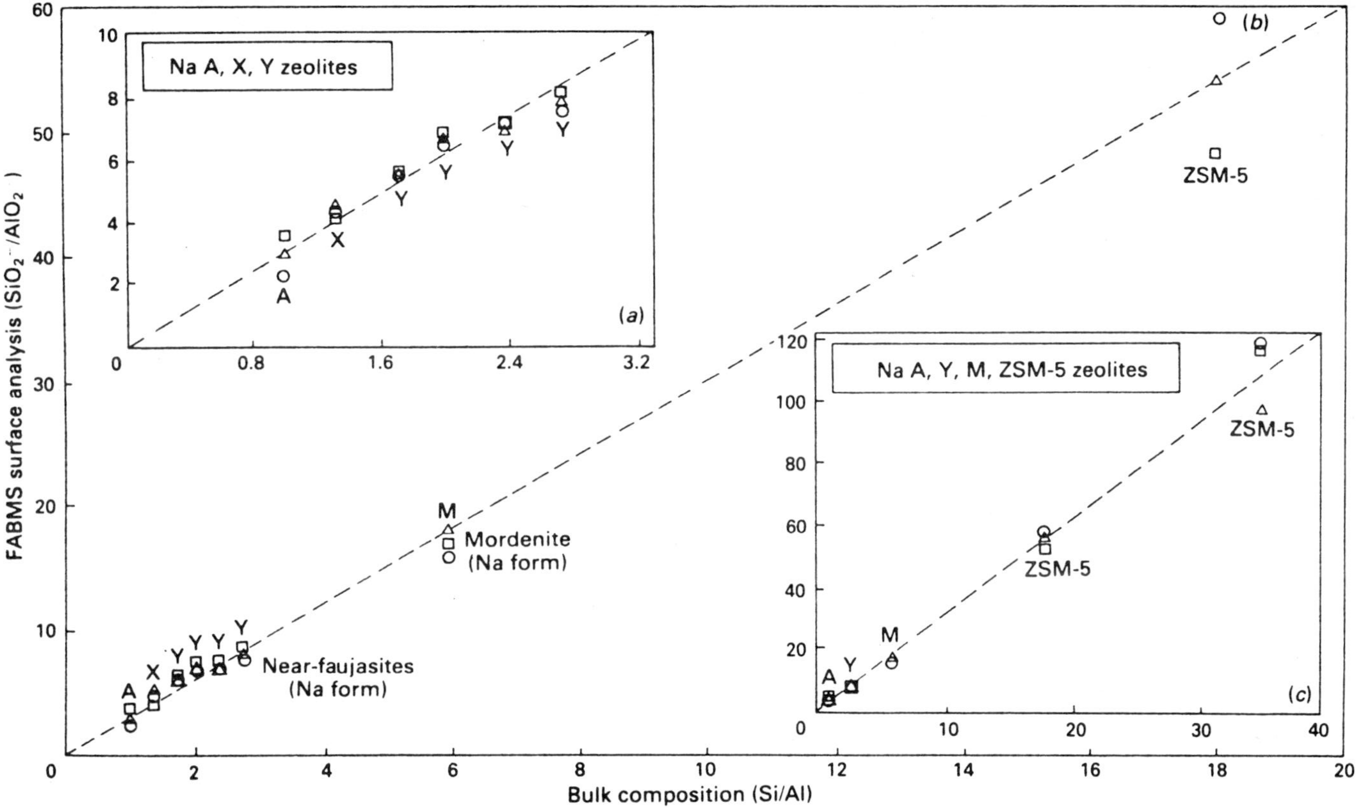

FIG. 19.9. Calibration of the atom-induced secondary ion ratio SiO_2^-/AlO_2^- against bulk composition in terms of the Si/Al ratio, for a wide range of insulating sodium zeolites, from Brown and Vickerman.[11] The ion ratios were measured first before bombardment, then after 75 μA min ion etching, and again after 150 μA min ion etching, to check for possible surface segregation effects. The calibration allows intercomparison of FABMS analyses on zeolite surfaces of widely differing compositions.

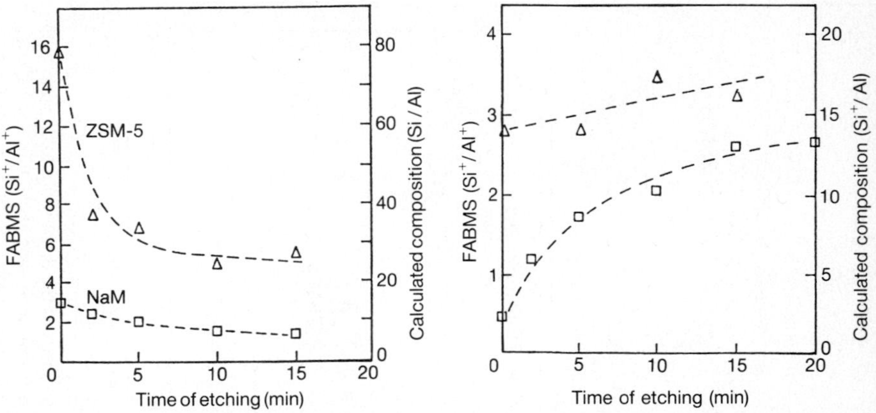

FIG. 19.10. Depth profiles by FABMS of zeolites that have had a variety of surface treatments, in terms of the Si^+/Al^+ ratio, measured by Surman *et al.*[12] Left-hand side. △ = ZSM-5 heated in 4M HCl for 38 h at 50°C. □ = sodium mordenite heated in 4M HCl for 24 h at 25°C. Right-hand side. △ = H-ZSM-5 untreated. □ = H-ZSM-5 heated in steam for 14 h at 600°C. The acid treatment has resulted in severe de-alumination of the zeolites to a depth of many tens of nm. On the other hand, the steam treatment of hydrogenated ZSM-5 results in the reverse effect, the enrichment of the surface in aluminium.

unpredictable. Extension of the use of FABMS as an essential adjunct to SSIMS seems certain.

References

1. Barber, M., Bordoli, R. S., Sedgwick, R. D., and Tyler, A. N. (1981). *J. Chem. Soc. Chem. Commun.*, 325.
2. Barber, M., Bordoli, R. S., Sedgwick, R. D., and Tyler, A. N. (1981). *Nature* **293**, 270.
3. Barber, M., Bordoli, R. S., Elliott, G. J., Sedgwick, R. D., and Tyler, A. N. (1982). *Anal. Chem.* **54**, 645A.
4. Benninghoven, A., Rüdenauer, F. G., and Werner, H. W. (1987). *Secondary ion mass Spectrometry*, p. 894. John Wiley, New York.
5. Eccles, A. J., van den Berg, J. A., Brown, A., and Vickerman, J. C. (1986). *J. Vac. Sci. Technol.* **A4**, 1888.
6. Brown, A. and Vickerman, J. C. (1984). *Surf. Interface Anal.* **6**, 1.
7. Surman, D. J. and Vickerman, J. C. (1981). *Appl. Surf. Sci.* **9**, 108.
8. Brown, A., van den Berg, J. A., and Vickerman, J. C. (1985). *Spectrochim. Acta.* **40B**, 871.
9. Rinehart, K. L., Jr. (1982). *Science* **218**, 254.
10. Michael, R. and Stulik, D. (1987). *Appl. Surf. Sci.* **28**, 53.
11. Brown, A. and Vickerman, J. C. (1984). *Analyst* **109**, 851.
12. Surman, D. J., van den Berg, J. A., and Vickerman, J. C. (1982). *Surf. Interface Anal.* **4**, 160.
13. Dwyer, J., Fitch, F. R., Quin, G., and Vickerman, J. C. (1982). *J. Phys. Chem.* **86**, 4574.

20

High field excitation: inelastic electron tunnelling spectroscopy (IETS)

Although certainly a surface-specific technique, IETS is unique amongst the techniques described in this volume in that it does not require an ultra-high vacuum environment for its operation, and has hardly ever been used in vacuum. The description that follows below will make it clear why vacuum is not necessary, and also why it should be regarded as a high field technique. IETS was originated in 1966 by Jaklevic and Lambe,[1] and measures the vibrational spectra of molecules adsorbed on the insulator surface of a metal/insulator/metal sandwich, usually called a junction. The technique therefore has very close affinities with those other techniques such as HREELS and RAIRS (q.v.) that also measure vibrational spectra. As will be seen later, there are advantages and disadvantages in each of the vibrational spectroscopic techniques when inter-compared, and the choice of which to use depends on the particular system to be analysed.

20.1 Operation

The basic principle of IETS is shown diagrammatically in Fig. 20.1(a), from Khanna and Lambe.[2] When a voltage V is applied between two metal electrodes separated by an insulating layer, the two Fermi levels become shifted apart by an energy eV. Energy conservation then says that electrons from metal I can make horizontal (i.e. no loss) transitions into empty states above the Fermi level in metal II by tunnelling through the insulator. If the thickness l of the insulating layer is small enough (e.g. ≈ 3 nm) this tunnelling process has a high probability. Since such transitions involve no energy loss the process is known as elastic tunnelling. Provided the applied voltage V is low enough (usually 0–0.5 V) the elastic tunnelling current increases linearly with V, as shown in Fig. 20.1(b).

Now suppose that molecules of some material are adsorbed on the insulator surface, between the insulator and one of the metal electrodes, and that there exists a vibrational state of energy $h\nu$ in the molecule. As long as $eV > h\nu$ it is possible for the tunnelling electron to give up an amount of energy $h\nu$ and still find an empty state in metal II to which to go. Since such a process involves energy loss it is called *in*elastic electron tunnelling. In effect, at the threshold voltage $V_{th} = h\nu/e$ a new tunnelling channel becomes available for passage through the insulating layer, and the tunnelling current recorded as a function of V will show an increase at V_{th}, as indicated in Fig. 20.1(b). Obviously, for

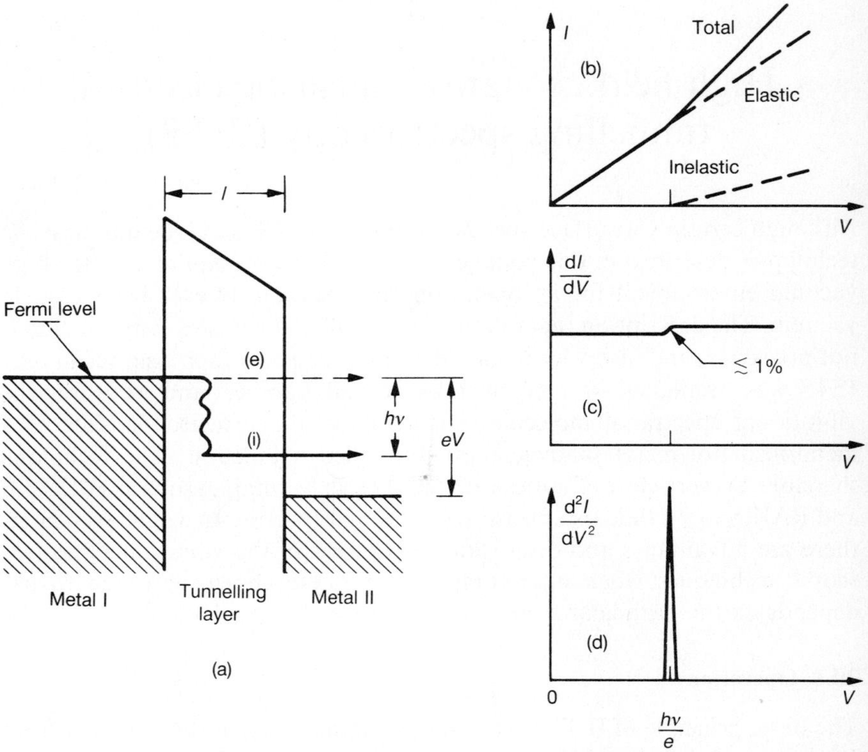

FIG. 20.1. Basic principles of IETS, as given schematically by Khanna and Lambe.[2] Two metal electrodes I and II are separated by an insulating layer of thickness l. Application of a voltage V between them shifts the Fermi levels apart by an energy eV. Electrons from I may than make horizontal transitions into empty states in II above the Fermi level in the latter, by tunnelling. If l is small enough (in practice ≤ 3 nm) the process has a high probability, and is known as elastic tunnelling. Provided V is low enough (e.g. 0–0.5 V), the elastic tunnelling current increases linearly with V as seen in Fig. 20.1(b).

A molecule adsorbed at the interface between one of the metals and the insulator, with a vibrational state of energy $h\nu$, will allow the tunnelling electron to give up an amount of energy $h\nu$ and still find an empty state as long as $eV > h\nu$. This involves energy loss and is called inelastic tunnelling. At the threshold $eV_{th} = h\nu$ the tunnelling current will show an increase, as indicated in 20.1(b). Figures 20.1(c) and 20.1(d) show the first and second derivatives of the tunnelling current with respect to voltage. The step in (c) is of the order of only 0.1 to 1%, thus the second derivative is normally used. For several vibrational states of a molecule there will therefore be a vibrational spectrum.

each vibrational state of the molecule there will be a different V_{th}, and hence for several vibrational states the current–voltage characteristic of the junction will show several increases in slope, each one adding to the previous ones.

The change in slope drawn in Fig. 20.1(b) is of course exaggerated. The actual change is very small, and difficult to distinguish. Even if the I–V characteristic is differentiated with respect to V, as shown in Fig. 20.1(c), the step is of the order of only 0.1 to 1% in height, and again is not easy to see. It is normal, therefore, to use the second derivative, d^2I/dV^2, in which the step in 20.1(c) then appears as a sharp peak, as shown in Fig. 20.1(d). The observational difficulties and the method used to overcome them are in fact identical to those found in AES (q.v.). To record the second derivative a small modulating voltage, typically between 1 and 4 mV, and of frequency typically 50 kHz, is superimposed on the applied voltage as it is ramped upwards from 0 to about 0.5 V. A lock-in amplifier then amplifies the second harmonic voltage, i.e. at 100 kHz, and either applies the output to the y axis of an x–y recorder, or feeds it into a D-to-A converter before storage in a computer. A simplified schematic of the detection circuitry is shown in Fig. 20.2, from Comyn *et al.*[3] This synchronous detection method actually measures d^2V/dI^2

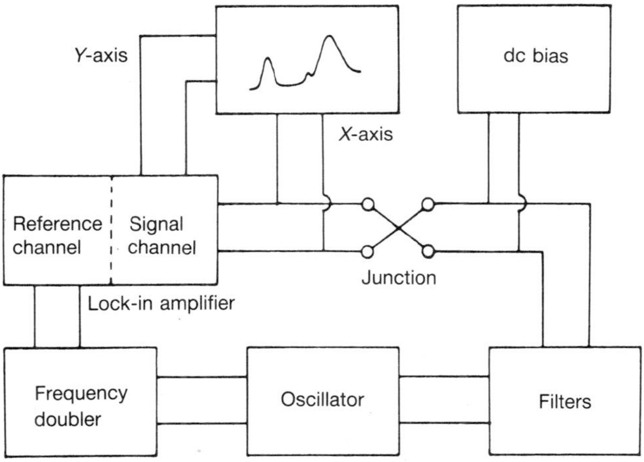

FIG. 20.2. Schematic of the synchronous detection circuitry used in IETS by Comyn *et al.*[3] A small modulation of 1–4 mV is superimposed on the applied voltage as the latter is ramped, and a lock-in amplifier amplifies the second harmonic of the tunnelling current. The output is fed either to the Y axis of an X–Y recorder or to a D-to-A converter before computer storage.

rather than d^2I/dV^2, but in fact the two are equivalent from the spectral observational point of view since

$$d^2V/dI^2 = -(dV/dI)^3 d^2I/dV^2, \tag{10.1}$$

and $(dV/dI)^3$ is a slowly varying function without any sharp structure in the voltage range of interest.

Preparation of the tunnel junction, including the doping with adsorbed molecules, appears to have as many variations as there are practitioners of the technique. Indeed, Hansma[4] points out that the closest analogy is with haute cuisine, in which every chef has his own recipe. It is sufficient here to mention the general approach. One aspect of the technique on which all operators are in agreement is that sample preparation is at once the most important and the most difficult part.

The junction base for fabrication is a glazed glass or ceramic such as a microscope slide. Before starting any fabrication the base must be scrupulously clean; the slightest organic contamination will ruin the junction, and the methods of cleaning that have been adopted are varied and individual. The base is placed in a vacuum evaporator capable of achieving an *oil-free* ultimate vacuum of better than 10^{-7} mbar, and coated through a mask with a strip of evaporated metal to act as the first electrode (metal I in Fig. 20.1). Aluminium is used almost universally for two reasons—it is easy to evaporate, and easy to oxidize with the formation of a very uniform, thin, oxide layer. The thickness of the deposited aluminium layer is not critical, but it is normally 60–80 nm thick, while the width of the strip is 0.5–1.0 mm. Formation of the first electrode is followed by oxidation, which can be carried out either by thermal oxidation or by glow discharge oxidation. In the former it is sufficient to expose the film to air or oxygen at room temperature for a few minutes to form a uniform oxide layer of 2–3 nm thickness, but again it is essential to exclude contaminating organic vapours, which means admitting filtered air or pure oxygen to the evaporator. In glow discharge oxidation a DC glow discharge is operated at about 500 V and in an oxygen pressure of about 5×10^{-2} mbar, for about 20 min; the advantage of using the latter method is that the aluminium does not have to be exposed to air.

Once the oxide film has been formed on the first electrode, the surface of the oxide must be 'doped' with the molecular species to be studied. Doping can again be done in several ways, according to the preference and ingenuity of the operator. The object is to coat the oxide surface with approximately a monolayer of chemisorbed molecules. Before the second electrode is deposited (see below) there are two principal methods. Firstly, for liquid doping, either the film can be immersed in a solution containing the desired species, followed by removal and evaporation of solvent, or a drop of solution is placed on the film and the excess removed by spinning. In either case the concentration in solution must be adjusted to produce the required final thickness. Secondly, for gas phase doping, the film is coated with the molecules of interest either by admitting the molecular vapour directly into the evaporator vacuum chamber, or by suspending the film over a solution of the molecular species, if the partial vapour pressure is high enough. A third doping method is available after formation of the second electrode, and will be described later.

After doping by liquid or vapour phase methods, the final stage in junction preparation is the deposition of the second, or top, electrode. The doped film is returned to an evaporator (preferably not the same as that used for the aluminium deposition) and a strip of lead is deposited at right angles to the strip of aluminium. The thickness of lead is 200–400 nm, and the width of the strip again 0.5–1.0 mm. Lead is used for several reasons. The most important of these is connected with the requirement that the IETS measurements be made at liquid helium temperature, 4.2 K, in order that the energy resolution be adequate; at room temperature the line width is 0.15 eV, which is useless, at 77 K the lines can be observed but are not well resolved, while at 4.2 K the line width is of the order of 2 meV. The significance of lead is that it is superconducting at liquid helium temperature, which has the effect of improving the resolution. The other reasons for using lead are that it is chemically inert, so that it does not oxidize rapidly after deposition, and that its presence on top of the adsorbed layer very rarely affects the vibrational characteristics of the molecules in the layer, unlike other metals that have been tried.

It is normal practice, for economy of effort, to fabricate several junctions simultaneously; there is always a finite probability of a junction not having the correct *I–V* characteristics, so that if a number are produced at the same time there is a high chance of success. Figure 20.3 shows the five-junction configuration adopted by Comyn *et al.*[3] Once the junction is complete the electrical connections are made and the device is inserted at once into a liquid helium Dewar vessel, and measurements are commenced once it has reached a temperature below 4.2 K.

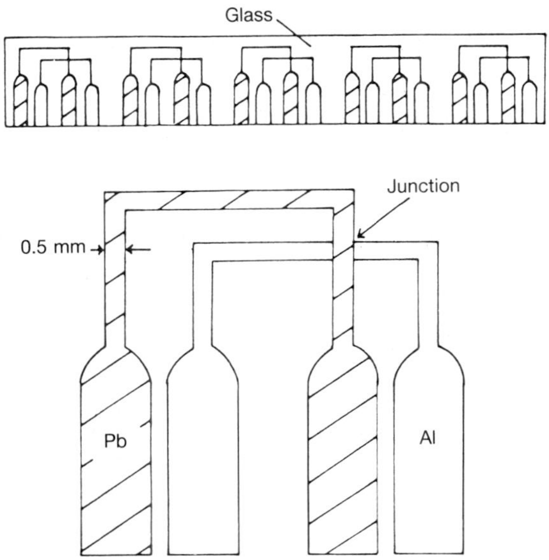

FIG. 20.3. The five-junction configuration used by Comyn *et al.*[3] The five individual tunnelling junctions are prepared simultaneously under the same conditions, thus ensuring that at least one junction will have the correct *I–V* characteristics for use. Underneath is shown the arrangement of a single junction in detail.

The third method of doping of the alumina/lead interface is by so-called infusion doping, and is carried out *after* the lead electrode has been deposited. The completed junction is exposed to an atmosphere with a relative humidity of 50 to 95% containing the molecules to be doped, for times between 1 min and 1 hr. Another variation is to apply the molecular species as a liquid to the outer surface of the lead electrode, followed by exposure to humidity. Evidently under the influence of the water vapour the lead film develops microporosity, which allows the dopant molecules to be carried through with the water, acting as a carrier. The method allows doping of molecules that could not easily be adsorbed by the other methods, and also, of course, sequential doping, if a dopant layer already exists at the interface.

20.2 Theory

The theory associated with any electron spectroscopy must be able to predict line positions, widths, and intensities to good approximation if it is to be acceptable. In IETS the loss peak positions are based only on energy conservation and are independent of any details of interaction between the electron and the adsorbed molecule. Note, however, that the large electric fields (of the order of 10^5–10^7 V cm^{-1}) present under applied voltage conditions have a significant perturbing effect on selection rules. Thus the usual spectroscopic selection rules are relaxed, and both Raman and infra-red active modes can appear in the IETS spectra with similar intensities. Figure 20.4, from Kirtley,[5] demonstrates the appearance of Raman spectral features in the IETS spectrum from an Al–AlO_x–4-pyridine–carboxylic acid–Ag junction. The prominent Raman features at ≈845, 985, 1150, 1380, and 1610 cm^{-1} are all reproduced in the IETS spectrum although with rather different relative intensities. Another effect of the high electric field existing across the junction is that of peak shift on reversal of the polarity of the applied voltage. The effect is not large, but can be important when subtle spectral differences between adsorbed molecules of closely related structure are being studied; it has been demonstrated, and explained on the basis of a simple model, by Reynolds *et al.*[6] Figure 20.5 from their paper indicates the magnitude of the effect, in this case 2.2 mV, on the spectral region corresponding to the O–H stretching vibrational mode in an undoped Al–Al-oxide–Pb junction. The shift is not necessarily uniform for all peaks in a spectrum. When the authors examined a similar junction doped with benzoate anion they found that the reverse polarity shifts varied from 0.02 mV to 0.45 mV, always in the same direction. Their simple model, which will not be described in detail here, explains the shifts through the interaction of an anharmonic molecular oscillator, with a non-zero dipole moment, with the large electric field. They were able to obtain order-of-magnitude agreement with the observed O–H stretching shift, while for the more complex case of

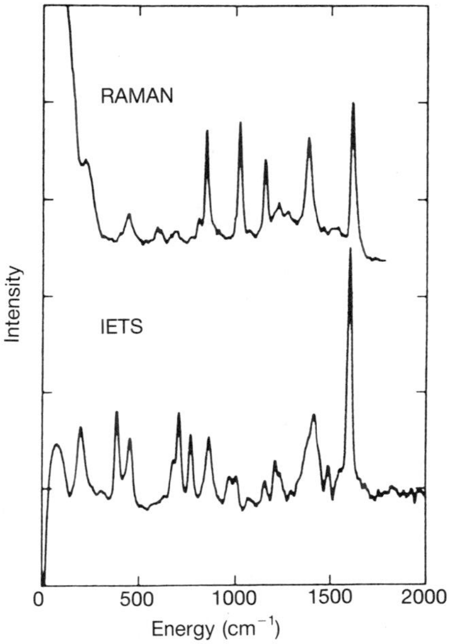

FIG. 20.4. Comparison by Kirtley[5] of the Raman and IETS spectra from an Al–AlO_x–Ag junction doped with 4-pyridine-carboxylic acid. The Raman features at ≈845, 985, 1150, 1380, and 1610 cm^{-1} are reproduced in the IETS spectrum, although with different intensities.

benzoate ion postulated that the electric field was causing anisotropic redistribution of charge density in the molecule, which would alter the vibrational frequencies in a non-uniform manner.

Every peak in an IETS spectrum has an associated intensity and width, and any theory should be able to predict these characteristics from a detailed consideration of the physical processes occurring during interaction between the tunnelling electron and the adsorbed molecule. Since the basic physical process is the tunnelling of an electron through a potential barrier, the starting point is the calculation of elastic tunnelling currents from the complete time-independent Schrödinger equation for given barrier dimensions. Incoming and reflected waves on one side of the barrier are matched to transmitted waves on the other side. Electrons passing through the barrier then go from an initial state i with energy ϵ_i to a final state f with energy ϵ_f. Solutions are assumed as usual to take the form of linear combinations of initial and final state wave functions. If the electron is tunnelling in the z-direction, with momentum parallel to the surface (i.e. in the x and y directions) conserved, and if superconductivity is neglected, then the total elastic current under the influence of the applied voltage V is given by the sum

$$j_{el} = \frac{4\pi e}{h} \sum_{if}\sum |M_{if}|^2 \{f(\epsilon_i) - f(\epsilon_f + eV)\} \, \delta(\epsilon_i - \epsilon_f) \qquad (20.2)$$

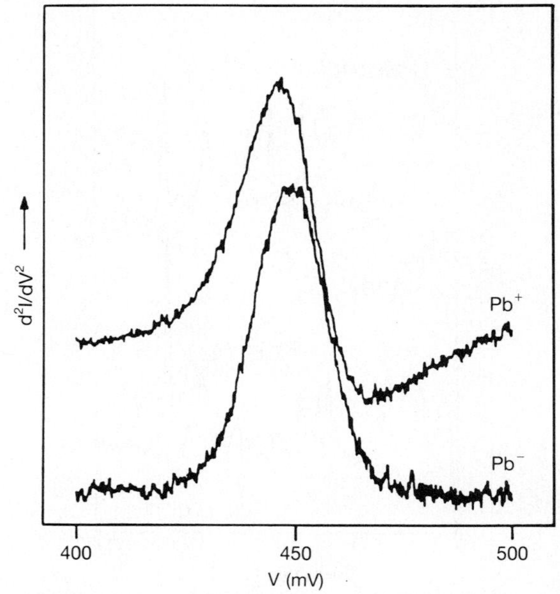

FIG. 20.5. Shift in the energy of a vibrational peak in the O–H stretching region from an undoped Al–AlO_x–Pb junction, recorded by Reynolds *et al.*[6] when the applied bias voltage polarity was reversed. In this instance the polarity shift is 2.2 mV, but it is not always the same for all peaks in a spectrum. See the text for a description of the authors' explanation of the effect.

where M_{if} is the matrix element for the initial to final state transition. This is called the transfer Hamiltonian formalism, and has been used as the basis for all calculations of inelastic tunnelling current; it uses the so-called WKB approximation to the form of the wave functions.

The extension of the theory to cover inelastic tunnelling has by now become fairly sophisticated, but in fact the original approach by Scalapino and Marcus,[7] although limited in scope, demonstrates enough of the physical concepts. In this approach a vibrational mode in a molecule adsorbed at the oxide–top metal interface is represented by a simple isolated dipole; because of the proximity of the metal the image dipole induced in the metal opposite the molecule must be taken into account. The situation is depicted schematically in Fig. 20.6, from Cederberg,[8] in which a coordinate system is also defined. The direction of tunnelling, taken to be the z-direction, is from metal 1 to metal 2, and the thickness of the oxide layer is l; for mathematical convenience the zero of z is taken to be at the oxide/metal 2 interface, so that at the other interface $z = -l$. The position of the dipole, assumed to be a point dipole at the interface, and of its image in metal 2, are shown. The co-ordinates of the tunnelling electron are $z, r_\perp$, where $r_\perp$ is the radial distance from the point dipole in the $z=0$ plane, that is $r_\perp = (x^2 + y^2)^{1/2}$. The principal assumptions

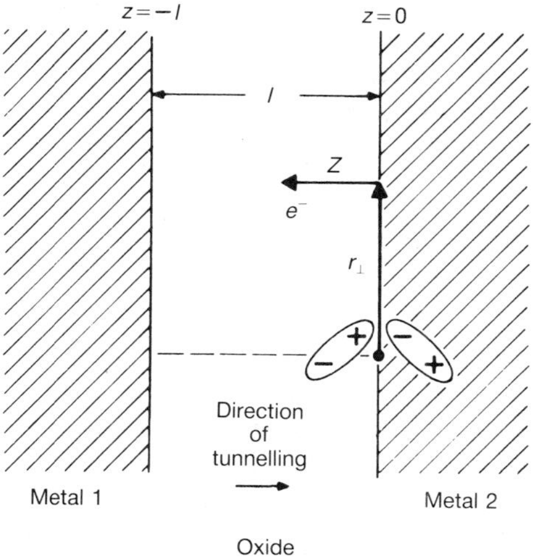

FIG. 20.6. The model of Scalapino and Marcus[7] to explain inelastic tunnelling, as depicted schematically by Cederberg.[8] Tunnelling is from metal 1 to metal 2, which is taken to be the z-direction; the zero of z is taken for mathematical convenience to be at the oxide/metal 2 interface, so that the other interface is at $z=-l$. The molecule adsorbed at the oxide/metal 2 interface is assumed to be a point dipole at $z=0$, and to have a dipole image in metal 2. The co-ordinates of the tunnelling electron are $z, r_\perp$, where $r_\perp$ is the radial distance from the point dipole in the $z=0$ plane. Scalapino and Marcus assumed that the electron-dipole interaction had a z-component only, and that the interaction could be treated as a small perturbation on the elastic tunnelling mechanism.

made by Scalapino and Marcus were that the interaction potential between the electron and the dipole had a z-component only, of magnitude p_z, and that the interaction could be treated as a small perturbation to the elastic tunnelling formalism given above. The electron/dipole interaction potential is then

$$U_{\text{int}}(z) = \frac{2ep_z z}{(z^2 + r_\perp^2)^{3/2}}. \tag{20.3}$$

Assuming a rectangular tunnelling barrier of height U, the total potential acting on the electron is therefore $U + U_{\text{int}}$, as long as the electron is inside the barrier, and for an electron energy ϵ_k the matrix element in the transfer Hamiltonian formalism then becomes

$$M_{\text{if}} \propto \left\{1 + \left(\frac{2m}{\phi}\right)^{1/2}\left(\frac{ep_z}{l}\right) g(r_\perp/l)\right\} \exp\left\{-\left(\frac{2m\phi}{h^2}\right)^{1/2} l\right\} \tag{20.4}$$

in which $\phi = U - \epsilon_k$, and the function g is given by

$$g(a) = \frac{1}{a} - \frac{1}{(a^2+1)^{1/2}}. \tag{20.5}$$

Expression (20.4) is derived in a simple way from the fact that in the interaction region ϕ will be greater than U_{int}, so that the quantity $(\phi + U_{\text{int}})^{1/2}$ that appears in the matrix element can be expanded ignoring powers of U_{int}/ϕ higher than two, followed by integration between the limits $z=-l$ to $z=0$. An expression analogous to (20.2) for the inelastic current can then be written

$$j_{\text{in}}^{k} = \frac{4\pi e}{h} \sum_{\text{i}}\sum_{\text{f}} |M_{\text{if}}|^2 f(\epsilon_{\text{i}}) \{1 - f(\epsilon_{\text{f}} + eV)\}\, \delta(\epsilon_{\text{i}} - \epsilon_{\text{f}} - \hbar\omega_k) \tag{20.6}$$

where k refers to the kth vibrational mode of the adsorbed molecule, with frequency ω_k.

Since the inelastic conductance is given by the second term in curly brackets in (20.4), and the elastic conductance by the first term in curly brackets, the ratio of the two eliminates the exponential term, and is then simply the square of the change in the elastic matrix element due to the vibrational potential, divided by the total elastic matrix element. The terms are squared because the tunnelling probability is given by $|M_{\text{if}}|^2$. The ratio is then

$$\frac{\Delta\sigma_{\text{in}}^{k}(r_\perp)}{\sigma_{\text{el}}} = \frac{2me}{\phi} \left(\frac{e}{hl}\right)^2 p_z^2 g^2 \left(\frac{r_\perp}{l}\right) \delta(eV - \hbar\omega_k). \tag{20.7}$$

Here the δ function allows conductance through the inelastic vibrational channel only when the applied voltage V provides the electrons with enough energy to excite the vibration ω_k. The total inelastic conductance is obtained by integrating (20.7) over unit area and multiplying by the number n of adsorbed molecules per unit area. The integration of g^2 has to be approximated by introduction of a cut-off distance r_0 since the model starts to fail very close to the point dipole. However, as the authors point out, r_0 enters logarithmically, so its value is not critical. Integration of (20.7), with these approximations, then gives

$$\frac{\Delta\sigma_{\text{in}}}{\sigma_{\text{el}}} = \frac{4\pi nme^2}{\phi h^2} \ln\left(\frac{l}{r_0}\right) \sum_k |\langle \text{f}|p_z|\text{i}\rangle|^2\, \delta(eV - \hbar\omega_k). \tag{20.8}$$

The squared term in the summation is of course the oscillator strength of the kth vibrational mode. If (20.8) is then integrated over a restricted applied voltage range covering a particular vibrational mode, then the change in conductance is

$$\frac{4\pi nme^2}{\phi h^2} \ln\left(\frac{l}{r_0}\right) \frac{e^2 f_k}{2m\hbar\omega_k}. \tag{20.9}$$

When Scalapino and Marcus inserted some reasonable values of the quantities in (20.9), for the change in conductance imposed by an OH dipole at the interface, they found a change of the order of 1%, in close agreement with experiment.

The Scalapino and Marcus model, for all its various approximations, represented an important step forward. It established, for example, that the ratio $\Delta\sigma/\sigma$ was the obvious quantity to take to define IETS intensities, because of its convenience of theoretical calculation, and indeed intensities are now defined by that ratio. Also, it is clear from the reasonable success of their model that many of the concepts must be at least qualitatively correct, in particular that inelastic interactions will occur only with dipoles whose net moment is normal to the surface, with the implication that there will be a strong dependence of the orientation of the adsorbed molecule at the interface. Furthermore, the model predicts that only vibrational modes involving changes in the molecular dipole moment, with intensities proportional to p_z^2, will be observed. Since this is the same factor as appears in infrared spectroscopy, the vibrational modes that are strong in the latter should also be strong in IETS, and this is often found, but without precise correlation. Extensions of the theory, by Lambe and Jaklevic,[9] to include interaction with the induced as well as with the intrinsic dipole, via bond polarizability, and by Kirtley *et al.*[10] who used an empirical potential made up of a set of localized atomic charges, have improved agreement but not contributed much further to the physical insight.

The simple theory thus appears to explain relative peak intensities in an acceptable way, but says nothing directly about peak widths. In IETS this omission does not really matter. Intrinsic widths of vibrational modes are known from high resolution vibrational spectroscopies such as FTIR to be of the order of 1 meV. What does matter in IETS, when considering peak shapes or profiles, are the effects of the imposed voltage modulation and the temperature on peak broadening. These are really experimental, rather than theoretical, considerations, so suffice it to say here that the two above contributions need to be deconvoluted from a spectrum if the original high-resolution peak shape is required. However, this does not appear to be done very often, because at low temperature and small modulating voltage the broadenings are in fact also small. Hansma[4] has given some typical figures. At a junction temperature of 1 K, the effective energy resolutions obtainable for modulations of 2, 1, and 0.7 meV were stated to be 3.4, 1.8, and 1.3 meV respectively. There is, of course, a price to pay for increased energy resolution in terms of spectrum acquisition time, which at optimum resolution could be a couple of orders of magnitude greater than at the resolution, ≈ 3.5 meV, usually used for analytical applications.

Some semi-theoretical calculations of peak shape have been carried out by various authors. Thus Kirtley and Hansma[11] used Gaussian, Lorentzian and δ-functions to synthesize shapes, while Magno and Adler[12] used Lorentzians only. The degree of agreement between experiment and theory in the latter case is shown in Fig. 20.7, for the IETS lines characteristic of lead (the top electrode) in both normal and superconducting states. In the experimental spectra the contribution from the aluminium oxide spectrum alone (i.e. the Al–O stretching vibrational mode) has been subtracted, so that they are differential spectra (labelled D by convention). The choice of background to be subtracted is also important, not only in the determination of shape but, as in all the other spectroscopies, in arriving at the correct integrated peak intensity for quantification. This is particularly clear in Fig. 20.7, where for lead in the superconducting state there is what is termed an 'undershoot' at energies just above that of the peak. Hansma[4] has shown how a physically realistic background can be constructed for subtraction by taking account of the superconducting density of states, almost always of lead, which would normally contain phonon structure as well. The latter structure is evident in Fig. 20.7 in the form of the broad features near 139 and 145 meV. Clearly subtraction of the correct background in the superconducting case requires spectrum acquisition extending at least 25 meV beyond the peak position.

20.3 Spectral identification

As in any of the vibrational spectroscopies, identification of spectral features in IETS proceeds by a combination of comparisons with the results of the other techniques and careful consideration of the selection rules pertaining to the particular molecular system. The point has already been made briefly in the previous section that both Raman and infra-red active vibrational modes can appear in IETS spectra, but additional features often appear and there can also be shifts in position from those found in the photon spectroscopies. A good example is that of the ferrocyanide ion, frequently studied, and whose spectra have been discussed thoroughly by Hipps and Mazur.[13]

Some comparative spectra from selected regions, measured by IRS, IETS, and Raman spectroscopy, respectively, are given in Fig. 20.8, from the work of Hipps and Mazur. Note, as before, that in general the IETS spectrum reflects the features in both the IR and Raman spectra. There are, however, some significant differences, which can be better appreciated by a direct comparison of line positions, as has been made in Fig. 20.9, from Hipps and Mazur.[14] At the top of the figure are the assigned positions of lines from ferrocyanide ions in solution, based on a mixture of IR, Raman, and theoretical considerations. The nomenclature in the top line is that adopted conventionally from group theory to describe the nature of the possible vibrational modes in atomic structures of the various available symmetries. The nomenclature in the second line has already been encountered in the chapter containing an account of the technique HREELS. In the present context it is essential to note that symmetry considerations impose sharply defined restrictions on which of the vibrational modes can actually be observed in either of the photon techniques

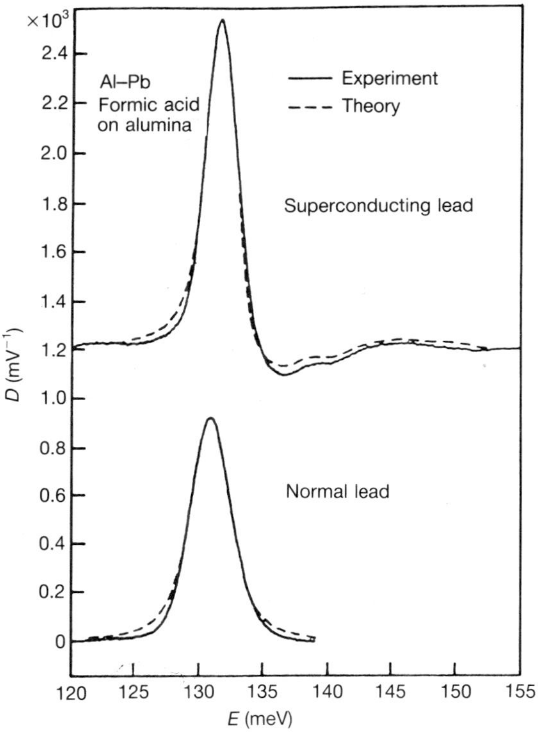

FIG. 20.7. Comparison of theoretical and experimental IETS line shapes for normal and superconducting lead electrodes, by Magno and Adler.[12] The AlO_x spectrum contribution has been subtracted, as has a background based on the superconducting density of states for the upper spectrum.

IRS and Raman spectroscopy. For a species of high symmetry such as the ferrocyanide ion the only infra-red active mode would be the t_{1u}, the only Raman active modes would be the a_{1g}, e_g, and t_{2g}, while the a_{2g}, t_{1g}, a_{1u}, a_{2u}, e_u, and t_{2u} would be inactive. Figure 20.9 shows that these restrictions are obeyed for the IR spectra and for the Raman spectra as well with the exceptions of the lines at 414 cm^{-1}, 480 cm^{-1}, and 584 cm^{-1}. The first and last of these can be identified from the IR spectrum as the normally forbidden t_{1u} modes, arising from the fact that the ferrocyanide material used for the Raman measurements was the half-hydrate, rather than the tri-hydrate which is similar to the aqueous ion. The half-hydrate structure is severely distorted and the appearance of these modes is attributed to the absence of a centre of symmetry of the ferrocyanide ion in that structure. More will be said below about the line at 480 cm^{-1}.

The surprising finding in the IETS spectrum in Fig. 20.8 is the appearance of the peaks at 358 cm^{-1} and 480 cm^{-1}, labelled a and b, respectively, in

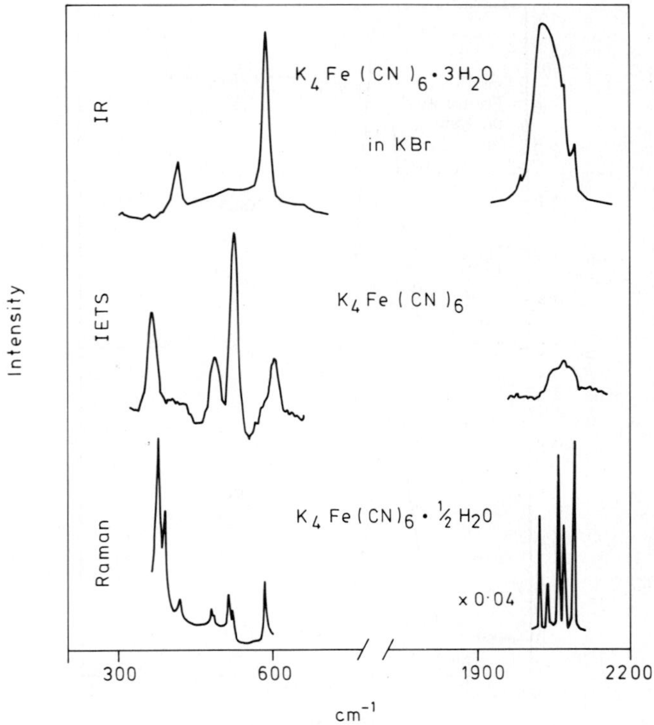

FIG. 20.8. Comparison of vibrational spectra of the ferrocyanide ion measured by infra-red spectroscopy (upper), IETS (middle), and Raman spectroscopy (lower) by Hipps and Mazur.[14] In general the IETS spectrum reflects features in both IRS and Raman spectra, although there are some significant differences as emphasized in Fig. 20.9.

Fig. 20.9. The former does not appear in either the IR or Raman spectra, and would not be expected to do so, since it is close to the theoretical position of the t_{1g} mode, which is doubly forbidden in photon excitation. It must appear, therefore, purely as a result of electron interaction with the adsorbed ferrocyanide, rather than from photon interaction, i.e. through a dynamic interaction in which selection rules are relaxed to the extent that normally completely forbidden modes become allowed. The line at 480 cm^{-1} is found in the Raman spectrum only from the distorted half-hydrate structure, and since it has no counterpart in the IR spectrum, must also arise from a normally doubly forbidden mode, probably one of the t_{2u} modes. However, since it can become Raman active under special circumstances, Hipps and Mazur identify it with a static rather than a dynamic interaction.

The above example demonstrates, firstly, that both IR and Raman active vibrational modes can give rise to features in an IETS spectrum, secondly, that additional normally forbidden modes can sometimes produce additional

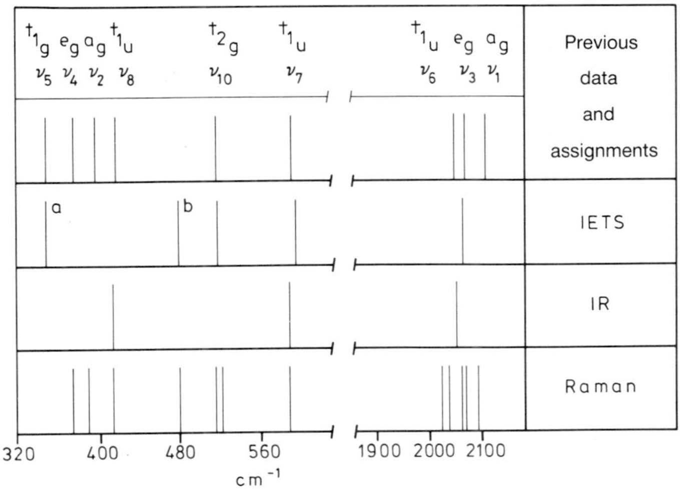

FIG. 20.9. Comparison of the IETS, IRS, and Raman line positions from Fig. 20.8, with the assignments in terms of the possible vibrational modes, for ferrocyanide ions in solution, made by Hipps and Mazur.[14] Refer to the text for a discussion of the active and inactive vibrational modes in IRS and Raman spectroscopy and for the reasons for the appearance of the lines labelled a and b in the IETS spectrum.

features, and thirdly, that identification of all features in an IETS spectrum can be performed only by reference to the results of IRS and Raman spectroscopy coupled with a careful consideration of the selection rules in operation. The extent of modification or breakdown of selection rules required to explain a spectrum can in turn give valuable information about the orientation of an adsorbed molecule, but discussion of that aspect is beyond the scope of this volume.

20.4 Quantification

As with the other vibrational spectroscopies, IETS appears to have been used hardly at all in any quantitative way. Peak energy positions and their shifts on isotopic substitution are used to identify an adsorbed species and the nature of its interaction, if any, with the aluminium oxide substrate or with other adsorbed species, but no information ever seems to be requested as to adsorbed molecular concentration. The principal reason for this is probably the difficulty of establishing exactly the quantity of adsorbed material, in terms of coverage, against which to compare the measured intensities of the IETS peaks. Normally doping is assumed to have reached monolayer coverage, but if the precise coverage is needed then methods of calibration must be devised.

One obvious method of calibration is by the use of radioisotopes, enabling the concentration to be measured directly simply by counting for a fixed period of time. This has been done by Langan and Hansma,[15] using tritiated benzoic acid in aqueous solution as the doping medium. The coverage of benzoic acid on the aluminium oxide was varied by varying the concentration of benzoic acid in the solution. Figure 20.10 contains a selection of IETS spectra from their experiments, showing how the peak intensities associated with the dopant increased with concentration of the latter. The undoped

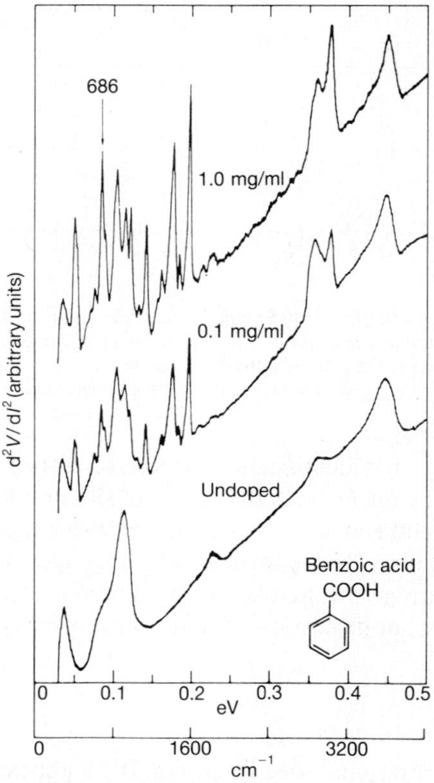

FIG. 20.10. IETS spectra from junctions doped with liquid benzoic acid, as a function of concentrations of the acid in solution, recorded by Langan and Hansma.[15] The lower spectrum of the undoped junction contains only the phonon structure due to the aluminium oxide and the superconducting lead.

spectrum simply contains the phonon structure due to the aluminium oxide and the superconducting lead. The dependences of the IETS 686 cm^{-1} signal, and of surface concentration, on the concentration of benzoic acid in solution, are shown in Fig. 20.11, on log–log scales. On these scales both dependences are linear, and the surface concentration saturates at one monolayer even though the concentration of the solution is progressively increased. The IETS signal saturates at approximately the same coverage. Notice the dynamic range of IETS. Tunnelling spectra could be acquired down to 3% of a monolayer.

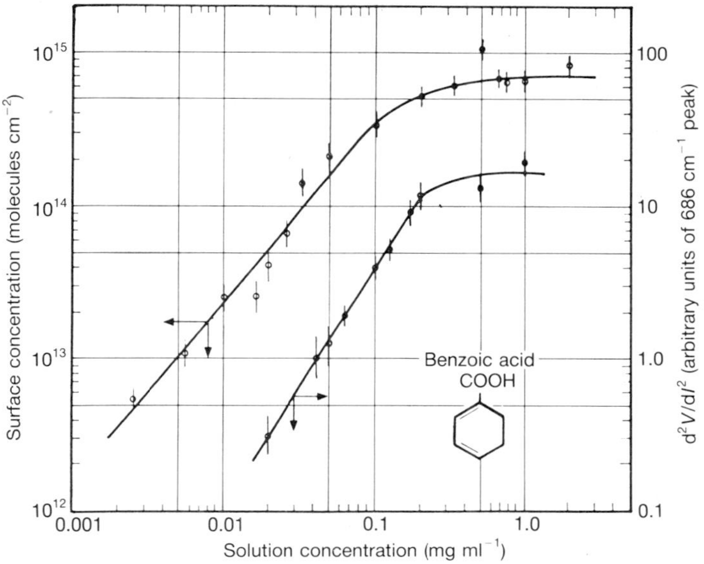

FIG. 20.11. The dependence[15] of the intensity of the IETS peak at 686 cm^{-1} in Fig. 20.10, and of the surface concentration, on the concentration of benzoic acid in solution. Notice that the plots are on log–log scales. Both dependences are linear, and the surface concentration saturates at one monolayer. The dynamic range of the signal intensity in IETS is about two orders of magnitude, which means that spectra could be acquired at coverages down to 2–3% of a monolayer.

The authors did not regard such a level as the lower limit of sensitivity, and claimed that with special precautions lower coverages could be studied.

The linear dependences recorded in Fig. 20.11 on log–log scales are of course somewhat misleading, since what is needed is the IETS intensity as a function of coverage. When the data in that figure are re-plotted in this way, one finds the dependence given in Fig. 20.12 by the full circles. The IETS signal intensity is clearly not a linear function of surface coverage. In fact, even for the surface concentration vs. solution concentration data in Fig. 20.11, Cederberg and Kirtley[16] have found a good fit to a statistical mechanical model for the isotherm for absorption from a liquid, that gives a non-linear dependence. In view of these non-linear relationships, Cederberg[8] set out to develop a model that would provide at least a qualitative explanation.

In his model he proposed that a junction could be viewed as having three independent tunnelling channels, in parallel. These were the channels for (a) elastic tunnelling through an undoped region, (b) for elastic tunnelling through a doped region, and (c) for inelastic tunnelling through a doped region. Associated with each of these channels are conductances σ_e (empty), σ_f (filled), and σ_i (inelastic), respectively. If the elastic and inelastic conductances of a completely doped junction are represented by Σ_f and Σ_i, respectively, and

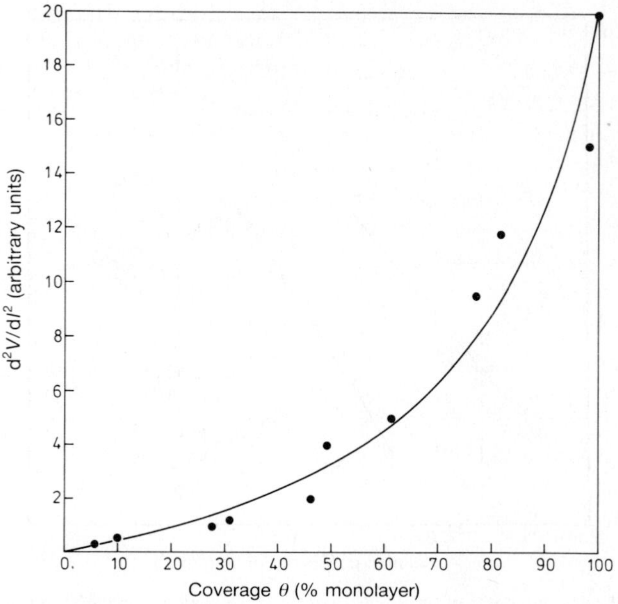

FIG. 20.12. Plot of the IETS signal intensity in Fig. 20.11 against the surface coverage. The dependence in this case is clearly not linear. The solid line is the result of applying Cederberg's[8] analysis to the data (see text). A value of Σ_e/Σ_f of 5.12 was required.

the elastic conductance of an undoped junction by Σ_e, then for a fractional coverage θ of dopant the following relationships hold

$$\sigma_f = \theta\Sigma_f \tag{20.10}$$

$$\sigma_e = (1-\theta)\Sigma_e \tag{20.11}$$

and

$$\sigma_i = \theta\Sigma_i. \tag{20.12}$$

Then, with the definition of Section 20.2,

$$I = \Delta\sigma/\sigma = \sigma_i/(\sigma_f + \sigma_e) \tag{20.13}$$

that is,

$$I = \frac{\theta\Sigma_i}{\theta\Sigma_f + (1-\theta)\Sigma_e} \tag{20.14}$$

or

$$I = \left(\frac{\Sigma_i}{\Sigma_f}\right)\left\{\frac{\theta}{\theta + (1-\theta)(\Sigma_e/\Sigma_f)}\right\}. \tag{20.15}$$

The first factor in (20.15), (Σ_i/Σ_f), is the quantity that was the subject of earlier discussion, since it is the intensity for a completely doped junction (i.e. when $\theta = 1$). The second factor gives the coverage dependence in this model. If $\Sigma_e = \Sigma_f$, i.e. if the elastic conductance through undoped and doped regions were the same, then I would simply be proportional to θ. In general $\Sigma_e \neq \Sigma_f$, so expression (20.15) predicts non-linear behaviour.

Cederberg[8] tested the model by carrying out measurements on junctions with varying amounts of dopant, again chosen as benzoic acid since there was already so much information about it. The relationship between surface coverage and the solution concentration was in this case established on the basis of a Langmuir adsorption isotherm. Chemisorption leading to monolayer coverage is generally found to obey the Langmuir behaviour well, provided the molecular binding to the surface is strong enough. In IETS the latter proviso is met more or less automatically, since after doping the junction is placed in a vacuum system for deposition of the top electrode, and of course evacuation will remove any loosely bound material. Thus adsorption from solution, followed by evacuation, leads to a chemical binding situation similar to that of gaseous chemisorption, and the statistical mechanical treatment that leads to the Langmuir isotherm is valid.

The expression analogous to the gaseous isotherm has been derived by Cederberg as

$$\theta = y/\{y + \exp(\gamma_{sol}/kT)\} \tag{20.16}$$

where y = number of dopant molecules/number of solvent molecules in the solution, i.e. a dimensionless concentration, and $\gamma_{sol} = \epsilon - \psi_{sol}$.

Again, ϵ is the binding energy of a molecule at a site on the aluminium oxide surface, and ψ_{sol} is the Helmholtz free energy per solute molecule in solution. As remarked above, Cederberg and Kirtley[16] were able to fit the isotherm (20.16) to the tritiated benzoic acid adsorption data in Fig. 20.11, with the two adjustable parameters N, the number of binding sites per unit area of oxide, and γ_{sol}/kT. A binding energy of benzoate ions to aluminium oxide of -1.0 ± 0.2 eV was derived.

Using the results of his own experimental measurements, Cederberg[8] first derived the fractional surface coverage θ from the above isotherm, and then fitted expression (20.15) to the I–θ values so obtained, with Σ_e/Σ_f as an adjustable parameter. The result is shown in Fig. 20.13, for the vibrational mode with peak energy at 985 cm^{-1} (122.1 meV). If the same analysis is applied to the data in Fig. 20.12 from Langan and Hansma,[15] which refer to the intensity of the 686 cm^{-1} (85 meV) peak, then the solid line in that figure is obtained. The value of Σ_e/Σ_f required for optimum fit in the former case was 3.09 and in the latter 5.12. Cederberg analysed the intensity vs. coverage dependences of nine peaks in the benzoic acid spectrum, and found a variation of Σ_e/Σ_f from 1.04 to 4.27. The possible reasons for this wide variation in the

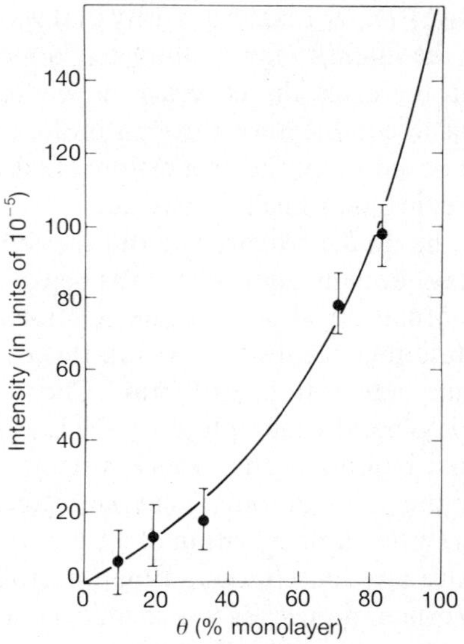

FIG. 20.13. The fit by Cederberg[8] of eqn (20.15) to the results of his own measurements of the adsorption of benzoic acid on a tunnel junction. In this case the value of Σ_e/Σ_f required for best fit was 3.09. Note that there are only five data points in this figure compared with 10 in Fig. 20.12.

ratio values were discussed by Cederberg, but no firm conclusion was reached; however, it should be noted that his fit of expressions (20.15) and (20.16) was made with only five different concentration values, whereas in Langan and Hansma's measurements at least 10 values can be extracted, covering the whole range of θ.

Despite the approximations involved, and the uncertainties surrounding the physical meaning of the considerable variations in the ratio Σ_e/Σ_f, it is clear that the non-linear θ dependence can be accounted for in the model. In particular, it has emphasized the importance of the different tunnelling behaviour that should occur when a junction is only partially doped, and the tunnelling electrons have to pass through both doped and undoped interfaces. However, it is also clear that the number of experimental measurements against which the model can be tested are extremely limited, and many more such are needed.

20.5 Applications

Although on the face of it IETS might seem severely restricted in its potential range of application, since the most successful junctions are still those constructed from aluminium/aluminium oxide/lead, nevertheless the technique has in fact been widely applied. This is mainly because of the happy coincidence that a great many systems of technological interest involve alumina in one way or another, either as a catalyst support material, or as a natural oxide on aluminium, or as the dielectric in thin film devices, etc. In addition, as pointed out earlier, the relaxation of infra-red and Raman selection rules in IETS enables vibrational modes of molecules to be studied that could not otherwise be recorded.

20.5.1 *Adhesively bonded aluminium joints*

Aluminium is an important structural material, and is usually joined to itself with adhesive, as it cannot be welded. Since oxide forms almost immediately on aluminium on exposure to air, or as a resultant of the various pre-treatments prescribed before bonding, the chemical conditions at the adhesive/oxide interface are of enormous importance to the strength of the joint and its durability. In particular the penetration of moisture into the joint invariably leads to progressive deterioration and eventual failure. One of the methods of preventing or at least slowing down this penetration is to treat the oxidized aluminium surface with so-called 'hydration inhibitors', and one of the class of materials proposed for this purpose is the phosphonic acids. Their interaction with oxidized aluminium has been studied by Ramsier *et al.*,[17] whose paper is also a good demonstration of how IRS and IETS should be used together.

Ramsier *et al.* prepared their junctions by the route of glow discharge oxidation of the evaporated aluminium, followed by doping with a drop of solution and removal of excess by spinning. Lead top electrodes were then deposited as usual, and the temperature of measurement was 4.2 K. IR spectra for comparison with the results of IETS were recorded from the solution by FT-IRS.

Figures 20.14 and 20.15 from their paper show the IETS spectra in comparison with the IR spectra on the same energy scale, for phosphonic acid, $H_2PO(OH)$ and phosphonic acid, $HPO(OH)_2$, respectively. The strong absorption band near 1650 cm^{-1} in the IR spectra is due to the solvent, which was water, and will therefore not appear in the IETS spectra, but there are other important, and diagnostic, differences as well. In particular, in both figures there is an absence in the IETS spectra of the peaks due to the P=O stretching vibration that appears prominently in the IR spectra at 1185 cm^{-1} in Fig. 20.14 and at 1179 cm^{-1} in Fig. 20.15. Additionally, for phosphinic acid in Fig. 20.14 the P–OH stretching vibration at 979 cm^{-1} in the IR spectrum is missing in the IETS, while for phosphonic acid in Fig. 20.15 various bands arising from vibrational modes of $P(OH)_2$ are present in the IR spectrum at 536, 946, and 1026 cm^{-1}, but not in the IETS spectrum. On the other hand, vibrations due to PO_2 are seen at 474 cm^{-1} and 1142 cm^{-1} in the IETS spectrum of Fig. 20.14 that are not in the IR spectrum, and in the IETS spectrum of Fig. 20.15 the P–H bending and stretching modes at 1034 cm^{-1} and 2428 cm^{-1} are far more intense than in the IR spectrum.

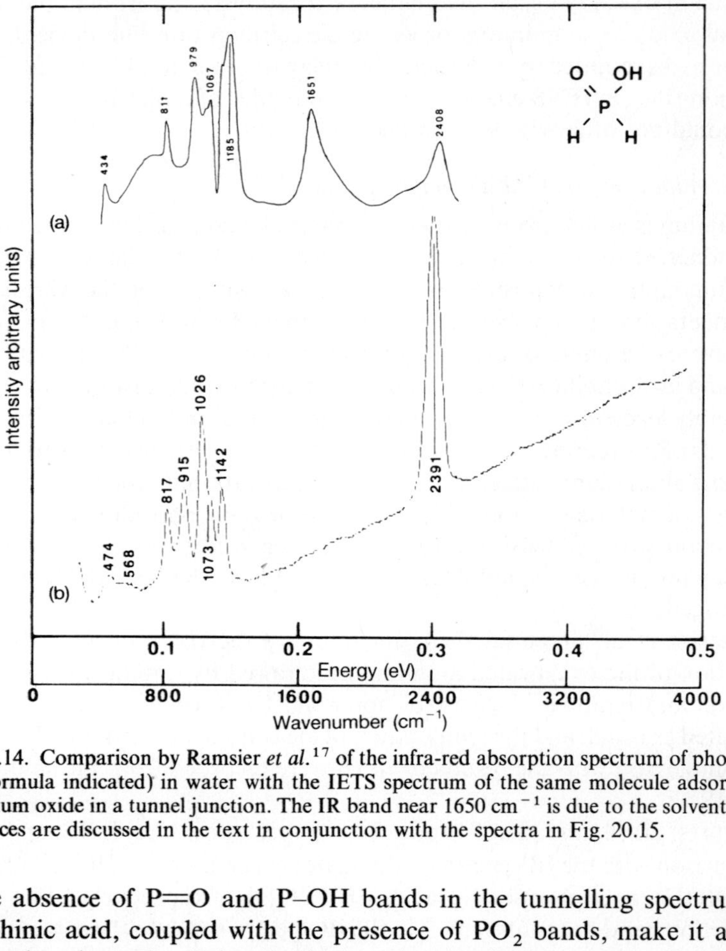

FIG. 20.14. Comparison by Ramsier *et al.*[17] of the infra-red absorption spectrum of phosphinic acid (formula indicated) in water with the IETS spectrum of the same molecule adsorbed on aluminium oxide in a tunnel junction. The IR band near 1650 cm^{-1} is due to the solvent. Other differences are discussed in the text in conjunction with the spectra in Fig. 20.15.

The absence of P=O and P–OH bands in the tunnelling spectrum for phosphinic acid, coupled with the presence of PO_2 bands, make it almost certain that the molecule is adsorbed on the aluminium oxide in a resonance stabilized form as shown schematically in Fig. 20.16(a). This form is probably reached via a condensation mechanism in which an acidic proton (hydrogen ion) in the molecule combines with a surface hydroxyl to form water, which then disperses. The very low intensity of the OH stretching band near 3610 cm^{-1}, which is obvious on the undoped aluminium oxide surface, supports the mechanism. The bending form proposed is also supported by the enhancement of the peak near 2400 cm^{-1}, due to PH_2 stretching vibrations.

For phosphonic acid the mechanism of adsorption is the same, that of condensation to eliminate water, since the P=O and $P(OH)_2$ bands are both absent in the IETS spectrum. However, for the form in which the molecule is adsorbed the very intense PH modes at 1034 and 2428 cm^{-1} in the IETS spectrum are diagnostic, since such an intensity could only be achieved if the

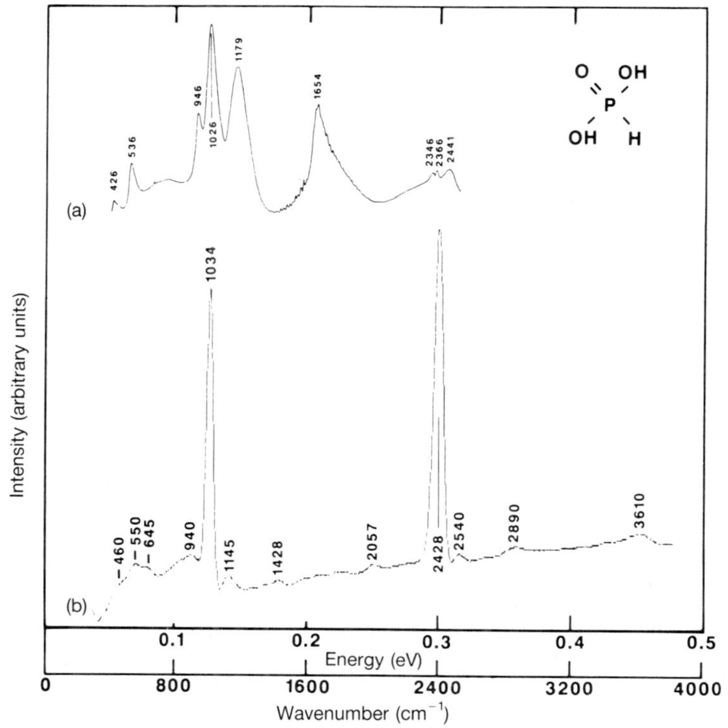

FIG. 20.15. Comparison by Ramsier *et al.*[17] of the infra-red absorption spectrum of phosphonic acid (formula indicated) in water with the IETS spectrum of the same molecule adsorbed on aluminium oxide in a tunnel junction. The IR band near 1650 cm^{-1} is due to the solvent. Note in both Figs 20.14 and 20.15 the absence of the P=O stretching vibration that appears in the IR spectra at 1185 cm^{-1} and 1179 cm^{-1}, respectively. Other differences are discussed in the text.

particular vibrational mode was oriented perpendicular to the surface, according to the selection rules. In addition the minor peaks at 460 and 550 cm^{-1} are associated with deformation of a symmetrical phosphite anion $(HPO_3)^{2-}$. Thus the adsorption form is likely to be that shown in Fig. 20.16(b). Again, the very low intensity of the hydroxyl peak at 3610 cm^{-1} supports the mechanism of dehydroxylation. The alternative mode of adsorption, that in Fig. 20.16(c), is ruled out by virtue of the absence of P–OH bands in the IETS spectrum.

When several methylphosphonic acids of more complicated structure, and of larger dimensions, were adsorbed, the results were very similar, apart from a slightly increased 3610 cm^{-1} peak, indicating more residual OH left on the aluminium oxide, due to steric effects, since the larger molecules could not react with all the surface hydroxyls. Thus the phosphonic acids act as hydration inhibitors by removing the residual OH group on the aluminium

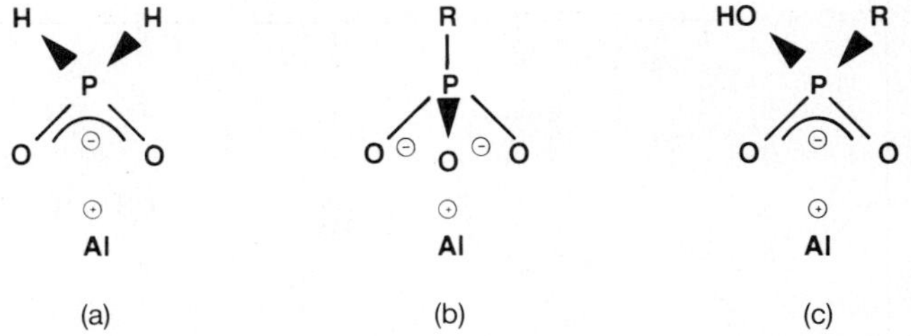

FIG. 20.16. Possible bonding models of phosphorus acids to alumina, put forward by Ramsier *et al.*[17]

(a) Structure resonance stabilized with the phosphonyl oxygen (phosphinic acid).
(b) Structure resonance stabilized via the deprotonated hydroxilic oxygens (phosphonic acid), following a condensation reaction.
(c) Bidentate structure similar to (a) that might exist as an intermediate species.

oxide, and by forming a strong P–O–Al bond with the oxide surface that cannot be attacked by water molecules.

20.5.2 *Amorphous vs. microcrystalline aluminium oxide for the adsorption of formic acid in UHV*

In the preparation of catalysts by the dispersion of active metals such as Pt and Rh on alumina supporting material, the crystalline nature of the oxide support may be of equal importance to the physical nature of the active metal clusters. That is, there may be significant differences in behaviour towards adsorbates between crystalline and amorphous surfaces of aluminium oxide. These differences have been studied by Liehr and Ewert[18] for the adsorption of formic acid on to aluminium oxide surfaces that could be prepared in either amorphous or microcrystalline form according to the method of formation. Their study is again a good example of the interpretation of IETS spectra by comparisons with IRS and Raman spectroscopy, and, what is almost unique in the way in which IETS is normally used, was carried out entirely in UHV.

Liehr and Ewert found that if the base pressure in their vacuum system was above 10^{-6} Pa amorphous aluminium oxide was formed, but below that pressure microcrystalline oxide resulted, during the evaporation of aluminium. To form amorphous oxide tunnel barriers, they therefore evaporated aluminium in an oxygen pressure of 5×10^{-6} Pa, while microcrystalline oxide barriers were formed by evaporation of aluminium at pressures below 10^{-7} Pa, followed by an oxygen glow discharge. After formation of the oxide, the surface was doped with formic acid by exposure to the vapour in vacuo, the amorphous requiring about 10^9 L to saturate, and the microcrystalline 10^7 L. The top electrode of lead was then evaporated on top of the formic acid layer without removing the junction from the vacuum system. The IETS measurements themselves were *also* performed in the UHV system without the necessity of removal of the junction. Both normal, HCOOH, and deuterated, DCOOD, forms of formic acid were used for doping.

Their IETS spectra are shown in Fig. 20.17, for the amorphous oxide on the left and for the microcrystalline on the right. The broad peaks below 500 cm^{-1} in each spectrum are due to phonons in aluminium oxide. It is to be expected

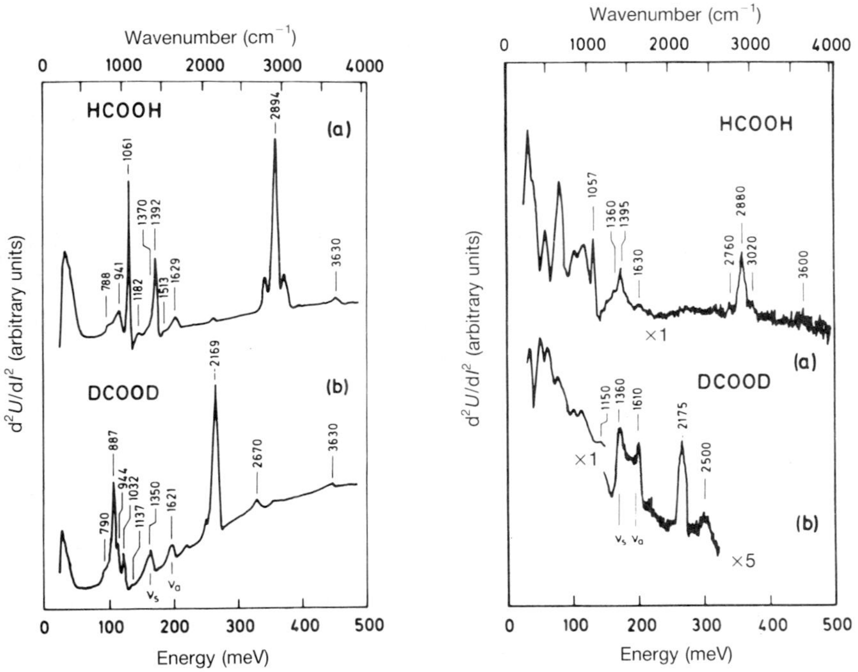

FIG. 20.17. IETS spectra recorded by Liehr and Ewert[18] for the adsorption of normal and deuterated formic acid at 300 K on amorphous aluminium oxide (left-hand side) and on crystalline aluminium oxide (right-hand side). The measurements were performed *in ultra-high vacuum*. A comparison of the peak positions observed with results from IRS and Raman spectroscopy, together with the assignments, is given in Table 20.1.

that formic acid will adsorb as a formate ion, being a fairly strong acid, in which case six molecular vibrations should be seen. The nomenclature and the schematic representation of the six are given in Table 20.1, in which both infra-red and Raman data for the two formate ions are compared with the data from IETS.

The amorphous and microcrystalline spectra appear different because of the varying backgrounds from the contributions from the phonon structure from the underlying aluminium oxide, and also because the signal-to-noise level is worse in the case of the microcrystalline oxide. However, the peak positions for the adsorption of HCOOH and of DCOOD are in fact much the same on the two substrates, even if some of the relative peak intensities are not.

Table 20.1

Comparison by Liehr and Ewert[18] of the peak positions measured in IETS for the room temperature adsorption of normal and deuterated formic acid on amorphous and crystalline aluminium oxide, with IRS measurements from formic acid adsorbed on γ-alumina and with Raman measurements from crystalline formate.

Type	Form	Wavenumbers (cm^{-1})					
		$HCOO^-$			$DCOO^-$		
		IR	Ra	IETS	IR	Ra	IETS
ν_s[OCO)		1390	1352	1370[a]	1385	1329	1350
ν(CH)		2915	2825	2894	2220	2122	2169
δ(OCO)	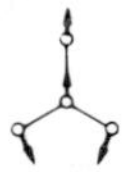	769	773	788	762	757	790[a]
ν(OCO)		1625	1584	1629	1625	1579	1621
δ(CH)		1407	1386	1392	1029	1029	1032
π(CH)	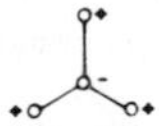	1062	1069	1061	912	918	887

[a] Approximated value.

The initial peak assignment comes from the observation of large shifts on deuteration from 2894 cm^{-1} to 2169 cm^{-1} and from 1061 cm^{-1} to 887 cm^{-1} in some of the most prominent and sharpest peaks. Because the only change has been from H to D, these peaks must therefore be associated with CH (CD) vibrational modes and comparison with the IR and Raman data confirms this allocation. The high relative intensities of the peaks also suggests strongly that the CH bond is oriented nearly perpendicular to the surface. The minor peaks at 1392 cm^{-1} for HCOOH and at 1032 cm^{-1} for DCOOD, can also be allocated to CH and CD modes, respectively.

There remain the modes associated with the O–C–O group. Two of these, ν_a and ν_s, have dipole moments perpendicular to each other, so that if the group were adsorbed with the plane through O–C–O normal to the surface, then ν_a

should have low, and ν_s high, observed intensities. The positions assigned to these modes, on the basis of the observation that there is little change in position on deuteration, and by comparison with the IR and Raman data, are 1629 cm^{-1} and 1370 cm^{-1}, respectively, for HCOOH, and 1621 cm^{-1} and 1350 cm^{-1} for DCOOD. Because of interference from a CH mode in the HCOOH spectra, the positions are easier to see in the DCOOD spectra. There it is obvious that, for the amorphous oxide, the relative intensities are about the same, while for the microcrystalline oxide the ν_s peak (at 1360 cm^{-1}) is more intense than the ν_a peak (at 1610 cm^{-1}).

Liehr and Ewert suggest that the correct interpretation of these observations is that on the amorphous oxide much more formic acid can be adsorbed but that the molecule adsorbs in a nearly random orientation, whereas on the microcrystalline oxide less is adsorbed but the molecules have a more preferred orientation. Thus in an applied situation it would in principle be possible to adjust the nature of the alumina support either so that as much reacting material as possible was accommodated, or so that the reactant had an orientation to the surface that allowed greater overall efficiency of reaction due to configurational matching of the relative molecular orientations.

References

1. Jaklevic, R. C. and Lambe, J. (1966). *Phys. Rev. Lett.* **17**, 1139.
2. Khanna, S. K. and Lambe, J. (1983). *Science* **220**, 1345.
3. Comyn, J., Horley, C. C., Oxley, D. P., Pritchard, R. G., and Tegg, J. L. (1983). *Brit. Polymer J.* **15**, 50.
4. Hansma, P. K. (1982). *Tunneling spectroscopy: capabilities, applications, and new techniques* (ed. P. K. Hansma), p. 26. Plenum Press, New York.
5. Kirtley, J. R. (1980). *Vibrational spectroscopies of adsorbed species*, American Chemical Society Symposium Series, Washington D.C., (eds A. T. Bell and M. L. Hair), vol. 137, p. 217.
6. Reynolds, S., Peasgood, A., Oxley, D. P., Pritchard, R. G., Walmsley, D. G., Tomlin, J. L., and Turner, R. J. (1987). *J. Phys. C: Solid State Phys.* **20**, 4297.
7. Scalapino, D. J. and Marcus, S. M. (1967). *Phys. Rev. Lett.* **18**, 459.
8. Cederberg, A. A. (1981). *Surf. Sci.* **103**, 148.
9. Lambe, J. and Jaklevic, R. C. (1968). *Phys. Rev.* **165**, 821.
10. Kirtley, J., Scalapino, D. J., and Hansma, P. K. (1976). *Phys. Rev. B* **14**, 3177.
11. Kirtley, J. and Hansma, P. K. (1976). *Phys. Rev. B* **13**, 2910.
12. Magno, R. and Adler, J. G. (1978). *J. Appl. Phys.* **49**, 5571.
13. Hipps, K. W. and Mazur, U. (1982). *Tunneling spectroscopy: capabilities, applications and new techniques* (ed.. P. K. Hansma), p. 252 ff. Plenum Press, New York.
14. Hipps, K. W. and Mazur, U. (1980). *J. Phys. Chem.* **84**, 3162.
15. Langan, J. D. and Hansma, P. K. (1975). *Surf. Sci.* **52**, 211.
16. Cederberg, A. A. and Kirtley, J. R. (1979). *Solid State Commun.* **30**, 381.
17. Ramsier, R. D., Henriksen, P. N., and Gent, A. N. (1988). *Surf. Sci.* **203**, 72.
18. Liehr, M. and Ewert, S. (1983). *Surf. Sci.* **126**, 208.

21

High field excitation: atom probe field ion microscopy (APFIM)

The technique of atom probe field ion microscopy has grown out of the older technique of Field Ion Microscopy (FIM), originated by Müller,[1] which grew in turn out of field emission microscopy, also originated by Müller.[2] Of all the surface-specific analytical techniques, APFIM is the only one with the ultimate in compositional sensitivity, i.e. a single atom. On the other hand, as will be seen below, analysis using APFIM is performed under somewhat extreme conditions of local electrostatic field, which has meant that nearly all its applications have been metallurgical. APFIM was also the brainchild of Müller and co-workers,[3] and it is probably true to say that there is no other area of surface analytical technique that has been so dominated in its early and developmental stages by one man, E. W. Müller, as that of ultra-high resolution imaging and analysis.

21.1 Operation

In order to give an adequate description of APFIM it will be necessary first to say something about the operation of FIM, even though the latter is an imaging and not an analytical technique.

The principle of FIM is shown schematically in Fig. 21.1, from the review article by Miller.[4] The material to be studied is formed into a very sharp needle whose apical radius is between 10 and 100 nm. Very specialized electro-chemical or chemical polishing techniques are used for the needle fabrication, and for each material an individual technique or sequence of techniques has to be developed. The needle is positioned in a vacuum chamber about 5 cm from either a fluorescent screen or a micro-channel plate backed by a phosphor screen, with the axis of the needle approximately perpendicular to the screen. The shank of the needle is in thermal contact with a cryogenic reservoir, usually at liquid nitrogen temperature although liquid helium is sometimes necessary. A noble gas, called the imaging gas, is admitted to the chamber at a pressure in the region of 10^{-5}–10^{-4} torr; helium is used most often, but other noble gases, especially neon as indicated in Fig. 21.1, are used for special purposes where helium is unsuitable. An increasing positive potential of some kilovolts (typical range 5–20 kV) is then applied to the needle, and at a particular potential, the 'imaging' potential, a field-ion image of the needle tip appears on the screen.

The generally accepted process of imaging formation is shown in Fig. 21.2,

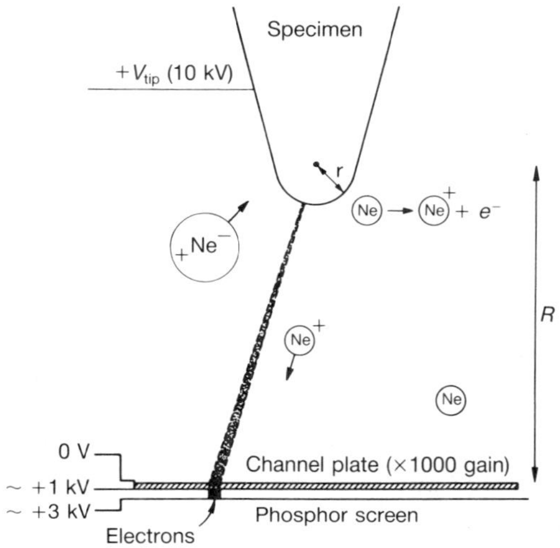

FIG. 21.1. Schematic of the principle of operation of an FIM, according to Miller.[4] The specimen is formed into a very sharp needle with apical radius 10 to 100 nm, and positioned in an ultra-high vacuum chamber about 5 cm from a micro-channel plate backed by a phosphor screen. The shank of the needle is immersed in liquid nitrogen or helium. A noble gas, often helium, but neon in this figure, is admitted to a pressure in the region 10^{-5}–10^{-4} torr, and an increasing positive potential is applied to the needle, in the range 5–20 kV. At some particular potential, called the 'imaging' potential, an image of the needle tip appears on the screen. The process of image formation is shown in Fig. 21.2.

from Smith.[5] Gas atoms near the tip of the needle become polarized by the very high electric field (e.g. 15–45 V nm^{-1}), and are then attracted to the surface. On reaching the surface, the polarized atoms lose some of their energy by thermal accommodation and are trapped within the high field region. However, they are mobile over the surface and can move across it by a series of hops until they happen to reach a position where the field is locally accentuated, when they are ionized by electron tunnelling. This is known as field ionization, and there is a critical distance from the surface for field ionization given by

$$x_c = (I - \phi)/eF \tag{21.1}$$

where I = ionization potential of gas atom
ϕ = work function of the material of the needle
and F = electric field gradient.

The result of field ionization is of course to leave a positively charged gas ion at the surface, which is immediately repelled and travels in a straight line along the potential gradient to the screen or micro-channel plate, where it creates an image of the point in the surface at which ionization took place. The whole strength of the technique is that the points of local accentuation of the field will

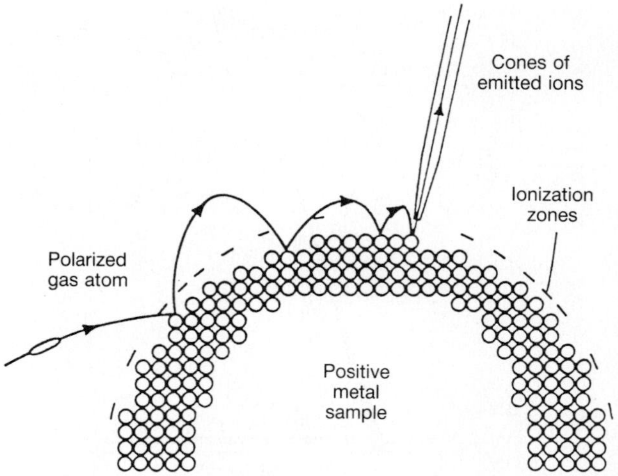

FIG. 21.2. Schematic of the process of field-ion image formation, as depicted by Smith.[5] In the very high electric field (15–45 V nm^{-1}) near the tip gas atoms become polarized and attracted to the surface. When they reach a point on the surface where the field is locally accentuated, they become ionized by electron tunnelling, and are then repelled in straight line paths to the channel plate, where an image is created of the point of ionization on the surface. The overall magnification is about 10.[6] Each point of ionization will give rise to a bright spot on the phosphor screen.

be just above the most prominent atoms in the surface, so that from each of these atoms there will be a narrow cone of repelled ions giving rise to an image in the form of a bright spot on the screen. In a highly curved surface such as that of a tip of radius about 50 nm, many crystallographic planes will intersect the surface at different angles, giving rise to sets of protruding atoms in the form of rings. The way in which this happens is shown schematically in Fig. 21.3, from Miller,[4] in which the lower half shows a cross-section through the curved surface, and the upper half shows a projection of that surface, with the protruding atoms indicated. Thus a field-ion image from a crystalline specimen will consist of intersecting rings, representing a stereographic projection of the particular orientation that is perpendicular to the axis of the needle. Such a field-ion image, and its corresponding stereographic projection, for a tungsten sample oriented [011], is shown in Fig. 21.4, from Smith.[5] Successive atom layers parallel to any one crystallographic plane will image as sets of concentric rings, while the ring systems associated with the various exposed crystallographic planes will intersect in a more or less complex way. Note that the overall magnification of the tip is $\approx 10^6$.

Of course, a needle when first inserted in the field-ion miscroscope will not have a tip that has a perfect crystallographic orientation. Indeed, after the tip formation by electrochemical and/or chemical methods, the surface will be far too rough and too contaminated to show a field-ion image with any recognizable order in it. The production of crystallographic perfection at the

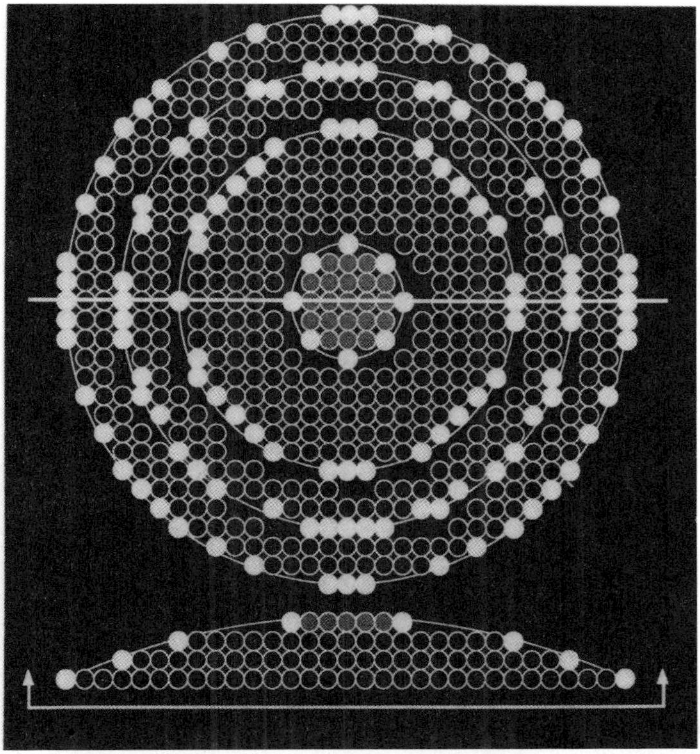

FIG. 21.3. Diagram of the image to be expected from a crystallographically ordered needle tip, from Miller.[4] The lower half of the figure shows a cross-section through a tip of small radius, showing that crystallographic planes will intersect the surface at different angles, and at each intersection point there will be protruding atoms. The upper half shows a two-dimensional projection of the curved surface, with the protruding atoms indicated. The latter atoms form a pattern of intersecting rings, which is exactly what a field-ion image looks like, as seen in Fig. 21.4.

surface takes place *after* the needle is inserted in the system, by the process called field evaporation. This process is also the basis of APFIM. If the applied voltage is increased slightly over the imaging voltage, as in FIM, or if a high voltage pulse is superimposed on the imaging voltage, as in APFIM, then surface atoms can be removed. The evaporation field required is a strong function of the particular material, and can be estimated in the following way. In the absence of a field, the amount of energy required to remove a positive ion of charge ne from a surface is

$$Q_0 = \Delta H + \sum I_n - n\phi \qquad (21.2)$$

where

ΔH = sublimation energy

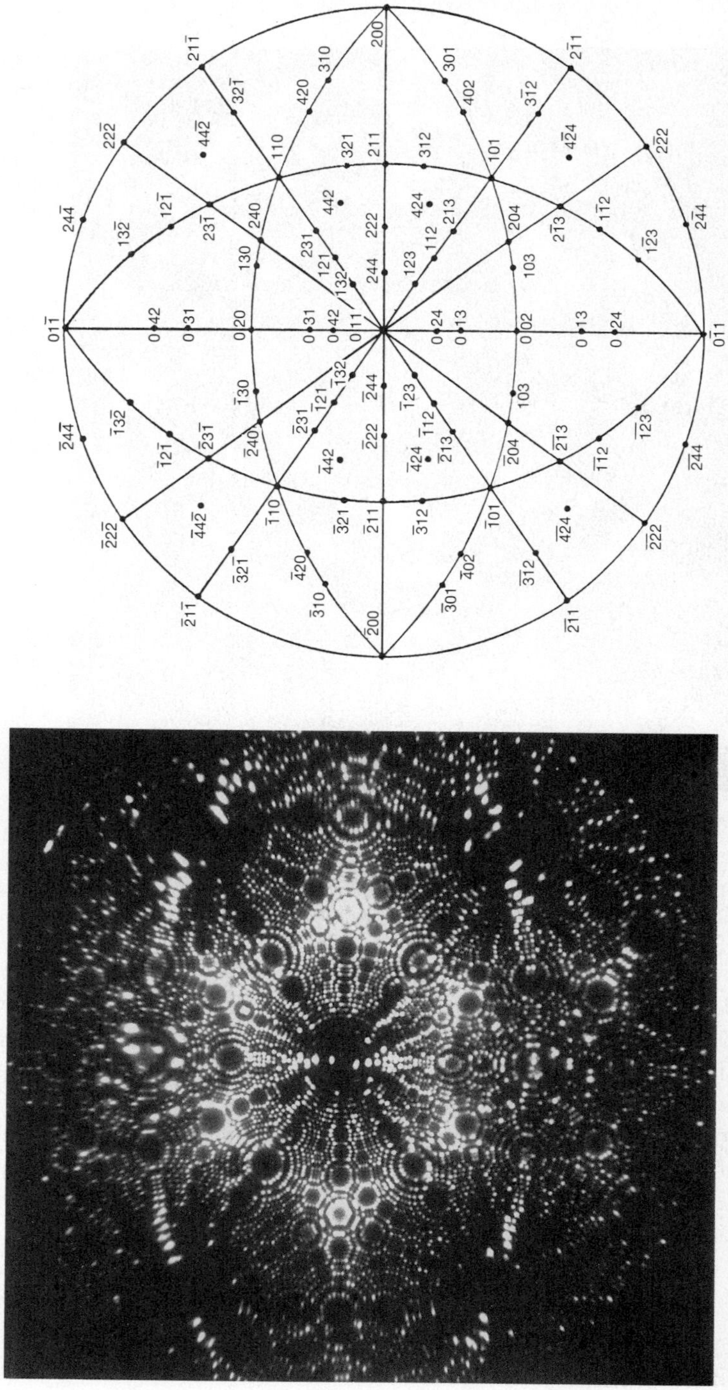

FIG. 21.4. Field-ion image of a tungsten tip oriented [011] along its axis, with the corresponding stereographic projection alongside. The intersecting ring systems can be seen clearly, as can also the concentric rings on the central (011) plane. By courtesy of T. J. Godfrey, University of Oxford.

$\sum I_n$ = sum of ionization potentials up to charge ne
ϕ = work function.

In the presence of a field F, the required energy is lowered, since the potential of the ion reduces progressively with distance from the surface. The energy barrier to be overcome is then

$$Q = Q_0 - (n^3 e^3 F)^{1/2}. \tag{21.3}$$

For sufficiently high F, Q can become zero, which means that field evaporation of atoms will occur even at cryogenic temperatures. The critical evaporation field is then given by

$$F_{\mathrm{ev}} = \frac{1}{n^3 e^3} (\Delta H + \sum I_n - n\phi)^2. \tag{21.4}$$

Since ΔH is normally the largest contribution to (21.4), it can be concluded that the more refractory materials need greater evaporation fields, in agreement with observation.

In FIM, the successive removal of atomic planes, or portions of planes, by repetitive field evaporation, is a powerful method of studying the structure of a sample in three dimensions, and has been used very widely in the study of alloys, precipitates, grain boundaries, slip planes, etc. However, such structural studies are not the subject of this volume, which set out with the stated intention of describing methods of compositional analysis. It is at this point that the technique of APFIM takes over.

An atom probe microscope consists of two parts, an FIM and a time-of-flight mass spectrometer. The principle of APFIM, and its main components, are shown in Fig. 21.5, from Brenner and Miller.[6] The first step is to produce an acceptable field-ion image of a specimen in the way described above; this will probably involve not just cleaning and ordering by field evaporation, but the removal of a certain number of atom layers as well to expose a feature of interest to be analysed. Then the sample, which in this case is mounted on a turntable, or some other form of manipulator, is rotated and tilted until the image of the area to be analysed falls over an aperture, of diameter about 2 mm, in the screen. This aperture is in fact the entrance to the time-of-flight spectrometer. Atoms from the surface are then field evaporated by applying a pulse of high voltage on top of the imaging voltage. Most of the ionized atoms will be stopped by the screen, but those from the selected area will pass through the aperture into the spectrometer. Compositional identification of individual ions is achieved by measuring their flight time to the particle detector. For a probe hole size of 2 mm, the area analysed on the tip is about 2 nm.

During acceleration from the tip as a result of field evaporation, an ion loses potential energy, which is converted into kinetic energy. For an ion of mass m and charge ne accelerated through a potential V, the final velocity v is given by

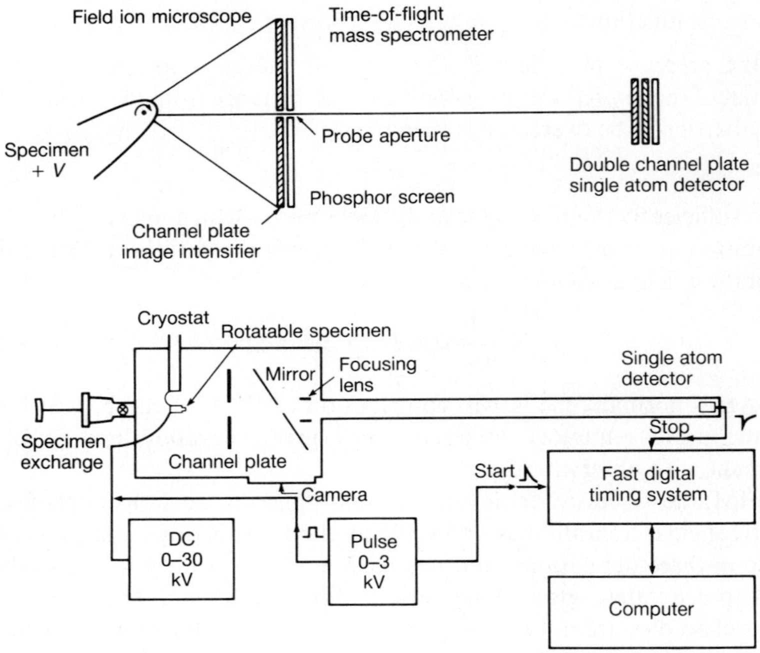

FIG. 21.5. Schematic diagram of the principle of operation of the atom probe field ion microscope, from Brenner and Miller.[6] The first step is the production of an acceptable field-ion image in the normal way by field evaporation either to clean the tip or to expose a desired feature such as a grain boundary or a precipitate. The tip, on a suitable manipulator, is then rotated and/or tilted until the image of the area on the feature of interest to be analysed falls over a small aperture (2 mm diameter) in the phosphor screen. A high voltage pulse applied to the tip over and above the imaging voltage then field evaporates atoms, those from the selected area passing through the aperture into the drift tube of a time-of-flight mass spectrometer. Measurement of the flight time to the detector by very fast counting and timing circuitry then gives the mass-to-charge ratio of an ion, and hence the composition.

$$\tfrac{1}{2}mv^2 = neV. \tag{21.5}$$

Here V is the total applied voltage, i.e. the sum of the constant imaging voltage and of the voltage pulse (typically 2 kV), i.e.

$$V = V_{\mathrm{DC}} + \beta V_{\mathrm{pulse}} \tag{21.6}$$

where β is a factor accounting for the attenuation of the pulse due to distortion on transmission to the sample, and must be measured experimentally. If the ion is assumed to reach its final velocity v close to the sample surface, then one can write

$$v = L/t \tag{21.7}$$

where L is the length of the flight path (typically 1 m), and t is the time between application of the pulse and arrival of the ion at the detector. For a flight path of ≈ 1 m, t is of the order of 1–10 μs. Thus the mass-to-charge ratio of the ion is given by

$$m/n = (2et^2/L^2)(V_{DC} + \beta V_{pulse}). \tag{21.8}$$

If t is in μs, L in metres, and V in kilovolts, then

$$m/n = (0.193\, t^2/L^2)(V_{DC} + \beta V_{pulse}). \tag{21.9}$$

Once L and β have been established by experimental calibration, the mass-to-charge ratio of the analysed ion is therefore given simply by the flight time and the applied voltages.

For adequate mass-to-charge resolution, it is essential that the rise time of the applied pulse and its duration be kept as short as possible, and that the pulse when it arrives at the tip be as uniform as possible, i.e. flat-topped. Modern pulse generators can deliver pulses with rise times of 1 ns, and durations of 10 ns, which would imply a mass-to-charge resolution of about 1000, but in fact this is reduced in practice to 100–150 because of the distortion of the pulse shape. Such a resolution can probably not be improved further in the 'simple' APFIM of Fig. 21.5, and any improvement can be achieved only by using an energy-compensated time-of-flight mass spectrometer such as that of Poschenrieder.[7] A modern instrument using a Poschenrieder energy compensator is shown diagrammatically in Fig. 21.6, from Miller.[4] The compensator uses a combination of electrostatic toroidal sector analyser and linear drift paths to allow focusing free of first-order time dispersion. The deflection system is symmetric, in the sense that the flight distances on the input and output sides of the compensator are equal; the total deflection angle is 163°. With correct adjustment, the toroidal sector analyser becomes time-focusing, which means that ions of the same mass-to-charge ratio but with different energies have the same flight times. Improvements in resolution to 1000–2000 have been achieved, allowing, for example, complete separation of isotopes, as seen in Fig. 21.7, from Cerezo and Smith.[8]

Also shown in Fig. 21.6 is a pulsed laser focused on the sample tip. Since the evaporation field of a material is a function of temperature, evaporation can be achieved equally effectively by holding the applied voltage constant and instantaneously raising the temperature. Such a momentary increase can conveniently be induced by a nanosecond laser pulse. It is sufficient to increase the local temperature from cryogenic temperatures to about room temperature to produce adequate evaporation, in which case bulk thermal damage to the sample is negligible. This version of APFIM is called the Pulsed Laser Atom Probe or PLAP.

It will be clear from the above description that, since the operation of APFIM is destructive in the sense that material must be removed for analysis just as in, e.g. SSIMS, continued analyses of the same area produce a depth

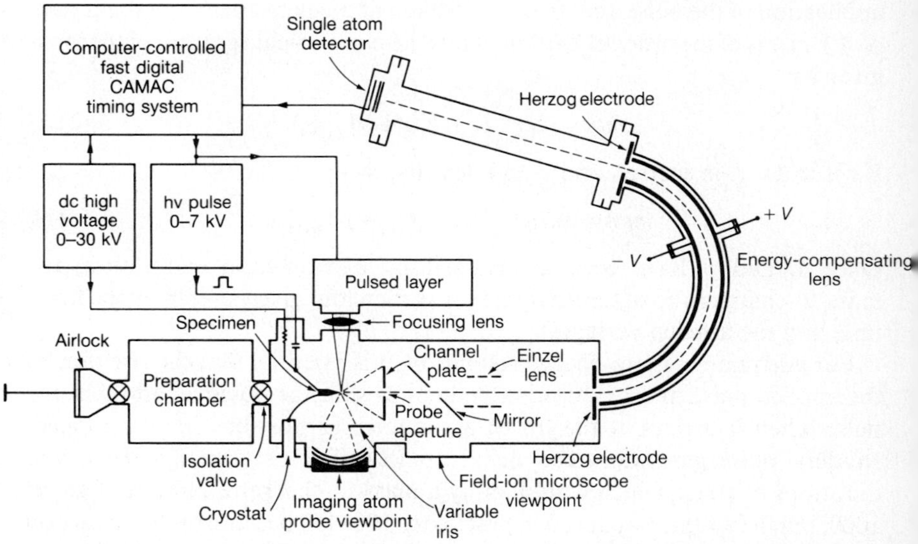

FIG. 21.6. An energy-compensating time-of-flight mass spectrometer using a Poschenrieder[7] energy compensator as described by Miller.[4] A toroidal sector analyser with equal linear drift paths at either end has a total deflection angle of 163°. The analyser is time-focusing, that is, ions with the same mass-to-charge ratio but different energies have the same flight times. Also shown is a pulsed laser focused on the sample tip, used in the variation of APFIM called Pulsed Laser Atom Probe, or PLAP.

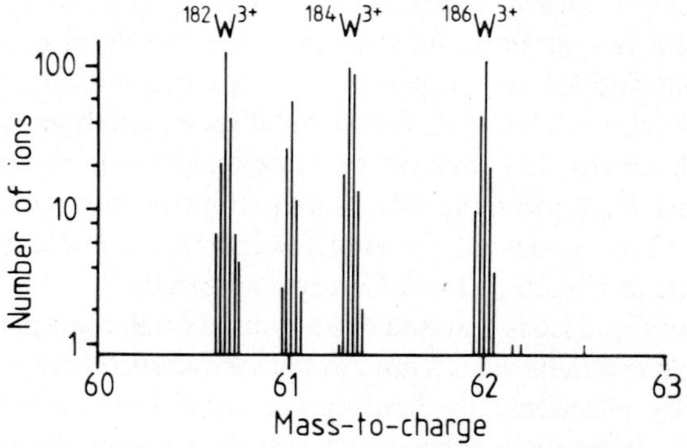

FIG. 21.7. Example from Cerezo and Smith[8] of the high mass resolution available in the energy-compensated time-of-flight atom probe. Complete separation of the four main isotopes of tungsten has been achieved.

profile. Removal of one atom layer for analysis uncovers a second layer, analysis of which uncovers a third layer, etc. The volume analysed is thus effectively a cylinder of diameter about 2 nm, rather akin to the cylindrical core of material removed from a drill during geological analysis of the earth's crust. Such a cylinder is depicted diagrammatically in Fig. 21.8, from Cerezo

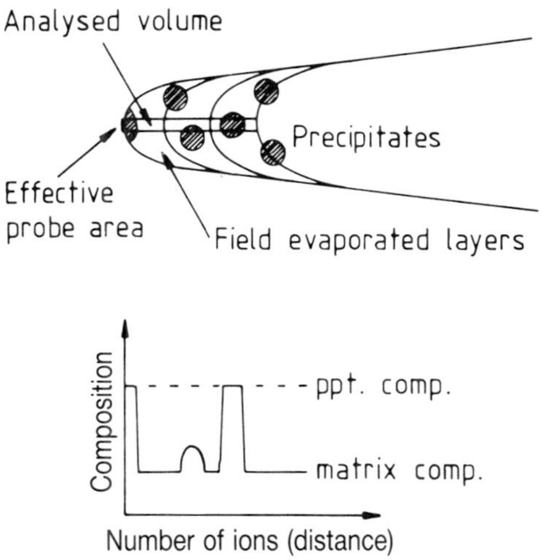

FIG. 21.8. Illustration by Cerezo and Smith[8] of the analysed volume in APFIM. Field evaporation pulses strip the surface layer by layer with atomic layer resolution, and since only those atoms from an area of about 2 nm diameter are analysed, the volume is that of a narrow pipe, or core, into the surface. In the sketch below are shown the signals that might be expected as the analysis passes through a precipitate in the matrix.

and Smith,[8] in which the signals to be expected from the material in a precipitate, and from the matrix, respectively, are indicated. It is this manner of operation that has found so many valuable metallurgical applications. Unlike other types of metallurgical analysis, however, APFIM has the capability of single atom detection, and of single atomic spacing resolution.

21.2 Theory

Expressions (21.1) and (21.4) contain the dependences on field strength in the models for field ionization and field evaporation. Figure 21.9, adapted from Smith,[5] shows the notional potential energy diagrams on which these models are based. In Fig. 21.9(a) a neutral atom is approaching the surface of the tip, and as it does so its potential well is distorted by the high electric field and is also affected by the image potential in the sample surface owing to the field polarization. At distances of approach sufficiently close, the system of polarized atom–surface barrier merges to form a potential barrier of height that decreases progressively with separation. At a critical distance x_c from the surface the filled atomic level is raised sufficiently, and the barrier is low enough, that electron tunnelling from atom to vacant levels in the solid above the Fermi level can occur. This distance is given by expression (21.1). For significant field ionization due to tunnelling, x_c must be of the order of 0.5 nm,

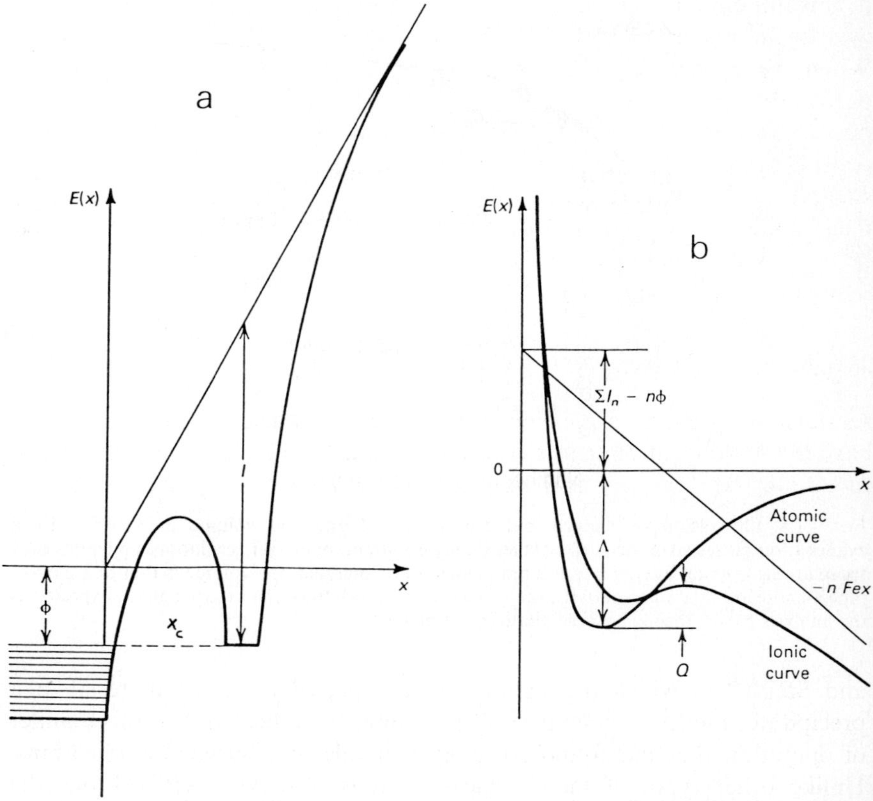

FIG. 21.9. Notional potential energy diagrams showing the effect of the high electric field on neutral atoms approaching a surface, adapted from Smith.[5] In (a), the potential well of a neutral atom has become distorted by the field to such an extent that at some critical distance x_c from the surface the potential barrier has been lowered sufficiently for electron tunnelling to occur, leading to ionization. In (b) are shown potential energy curves of an atom and of an ion as a function of distance from the surface, with a field *Fex* applied. For an atom to leave the surface as an ion requires the energy barrier of Q to be overcome, given by expression (21.3). I_n is the nth ionization potential, ϕ is the work function, and Λ the heat of sublimation.

which implies from (21.1) fields of the order of tens of volts per nm. The field strength is related to the end radius r (Fig. 21.1) and the applied voltage by

$$F = V/kr \tag{21.10}$$

where k is a constant related to the included angle of the needle, and is of order 5.

In Fig. 21.9(b) are shown the potential energy curves of an atom and of an ion as a function of distance from a surface, in the presence of a high field. Removal of an atom as an ion from the surface in the absence of the field is given by Q_0 in expression (21.2), which reduces to Q (expression (21.3)) as the field is imposed, being the distance from the bottom of the atomic potential well to the maximum in the ionic potential beyond the atomic minimum. When the imposed field is high enough for Q to be reduced to zero (no potential barrier to ionization) then expression (21.4) holds. It should be noted, however, that expression (21.4) is only approximate. In a full consideration of the effects of a very high applied field, other consequences such as the atomic energy level shift and broadening, field penetration into the metal, and the different polarizabilities of surface atom and ion, should be taken into account. The article by Tsong[9] goes into these effects more fully than is justifiable here.

21.3 Spectral interpretation

As stated earlier, once the calibration constants L and β in expression (21.9) have been established, then the output from an APFIM is simply a mass spectrum, and the usual problems of interpretation of mass spectra will apply. Many of these have already been discussed in the chapters on SSIMS, SNMS, FABMS, etc., but it should be noted that the physical processes governing field evaporation lead to differences in the character of the spectra arising in secondary ion emission and in field evaporation, respectively. In particular, for those materials that need especially high evaporation fields, e.g. the refractory metals, a higher proportion of multiply-charged states can be found in the APFIM spectrum compared to the SSIMS spectrum. Thus molybdenum, for example, can appear in the APFIM spectrum in the Mo^{2+}, Mo^{3+}, and Mo^{4+} states, whereas in SSIMS only Mo^{+} and Mo^{2+} are normally seen. The reason for the appearance of the higher charge states can be deduced rather simplistically from Fig. 21.9(b). For each charge state there will be an individual potential energy curve, the 'ionic curve' in the Figure, for which under the influence of an applied field there will be a particular barrier Q associated with the cross-over with the 'atomic curve'. The magnitude of Q will increase with the charge multiplicity, but if the field is high enough, then Q for even the multiply charged ions becomes small enough for the atom-to-ion transition to take place. Evaporation fields needed for W, Mo, Ir, Pt, etc., are of the order of 50–55 V nm^{-1}, whereas those for, say, the transition metals Fe, Co, Ni, Cu, are all in the range 30–35 V nm^{-1}.

The other difference to be noted between the species appearing in APFIM and in SSIMS spectra, respectively, is that far fewer cluster ions are found in the former. This difference might be expected intuitively in view of the quite different mechanisms of ion formation and emission, in that field evaporation in APFIM is a process taking place at an individual exposed atomic site, whereas in SSIMS the residual energy from the primary ion impact that finds its way back to the surface by a cascade process is more diffusely distributed and just as likely to eject small clusters of two or three atoms as it is a single atom. Some exceptions occur, of course. Figure 21.10, from Smith,[5] shows the PLAP spectrum from a ternary III–V semiconductor $Ga_{(1-x)}Al_xAs$, in which the cluster ions As_3^{2+}, As_2^+ and As_3^+ appear. The reasons for this are not well established, but are thought to be associated with the preferential removal of Ga, which has a lower evaporation field than As, followed by a reconstruction of the GaAs structure around the Ga vacancy leading to association of two or three As atoms in quasi-molecules.

21.4 Quantification

Unlike the situation in many of the techniques described in this volume, no complicated calibration or correction procedures are required in APFIM when quantifying the composition of a material. Each pulse arriving at the detector corresponds to an ion, therefore if all the pulses are counted and the associated mass-to-charge ratios identified with elements, quantification follows automatically. Conditions of analysis must be adjusted so that preferential field evaporation and field-induced segregation are minimized. There are no atomic number related differences in sensitivity as with other techniques, since the ion detector is sensitive to charged particles of all mass-to-charge ratios; even the lightest elements can thus be detected along with the rest. In addition, all elements at the surface within the selected area will be evaporated and analysed during each voltage or laser pulse, and to achieve increased precision of measurement for any one element it is sufficient to continue the analysis for a large enough number of pulses, although an infinitely high level of precision cannot be reached due to a statistical limitation. Since atoms have to be removed for analysis, there is a statistical error in any measurement of composition based on the number of ions that have been removed.[8] If the total number of ions removed is N, of which n are those of a particular species, the standard deviation on the determination of n is given by

$$\sigma = \{n(N-n)/N\}^{1/2} \tag{21.11}$$

from which the error in measurement of the atomic fraction n/N is

$$\pm\{n(N-n)/N^3\}^{1/2}. \tag{21.12}$$

The implication of this is, to give an example, that to analyse an impurity at the 0.1 at.% level of concentration to an accuracy of 0.1 at.% would require removal of $\approx 10^5$ ions, which could take several hours. The minimum detection limit, as opposed to the analytical precision limit, is much lower and is governed by factors such as detector dark current, electronic noise in the timing circuitry, and random gas atoms hitting the detector. The energy compensation in the time-of-flight APFIM reduces the latter contribution, but cannot eliminate it completely. Miller[4] quotes the current effective detection limit as 10 p.p.m.

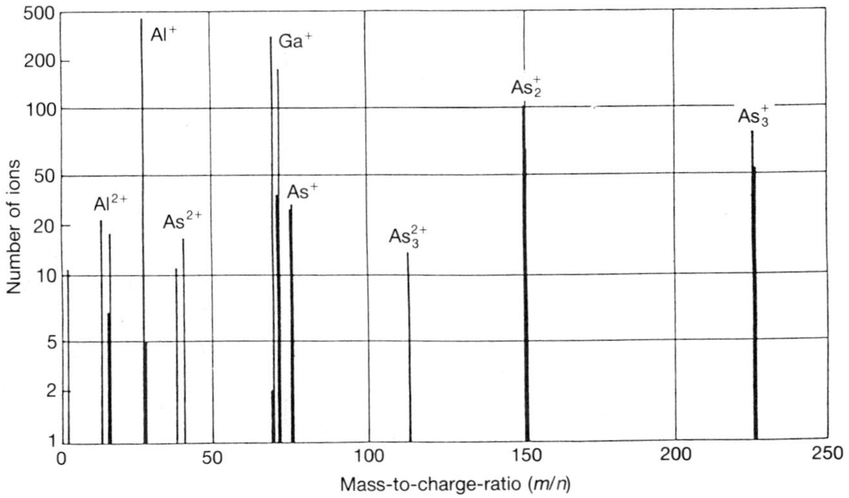

FIG. 21.10. PLAP spectrum from $Ga_{1-x}Al_xAs$, recorded by Smith,[5] showing the appearance of the cluster ions $As_3{}^{2+}$, $As_2{}^{+}$, and $As_3{}^{+}$.

21.5 Applications

In the early days of FIM, there was a general feeling that the technique would never have any wide application because it was restricted to the study of refractory metals only, and should be regarded as an interesting but exotic plaything of a few dedicated experimenters in pure surface science. These doubts have long since been dispelled, partly as a result of the developments in imaging techniques that have enabled materials of ever lower cohesive strengths to be imaged, and partly as a result of the origination and development of APFIM. Although the necessary high fields still impose certain restrictions on which materials can be studied, the opening up of the imaging capability to nearly all metallic materials has led the way to the study of a vast number of metallurgical problems. As will be evident from the description of APFIM above, the overwhelming strength of the technique in materials analysis is the depth resolution of a single atom layer and the lateral resolution of only ≈ 2 nm, which makes APFIM the only technique that can, for example, image and analyse material at a grain boundary. The applications described below have been selected to demonstrate the power of the technique.

21.5.1 *Grain boundary segregation*

The precise composition of material segregated to a grain boundary in a metal, either as a thin film or as a precipitate, has been the source of argument ever since it was established that such segregation could affect the mechanical strength of the metal drastically. Surface analytical techniques have had a major part to play in this discussion, as has already been described in the chapter on AES (q.v.). However, because of limitations on the spatial resolution of most scanning or imaging techniques, in comparison with a required resolution of 1–2 nm to analyse a grain boundary directly, it has normally been the practice to attempt to expose the boundary by fracture in order to produce a larger area for analysis. Considerable success has been achieved by such methods, but there are always questions remaining as to the validity of the analysis when, as in most cases, only one side of a fracture face can be analysed. In addition, there are often precipitates in the fracture faces that might either be too small to be analysed by the other techniques, or be lost from the face during fracture. The *direct* analysis of a grain boundary by APFIM avoids these problems.

An example of the imaging and then analysis of a grain boundary segregant is that given by Waugh and Southon[10] for carbon and carbides in heavily drawn steel wire, as used in the manufacture of springs. Their paper also demonstrates one of the materials problems to be taken into account in quantitative APFIM analysis. The atom-probe equipment that they used was very similar to that of Fig. 21.6, i.e. with energy compensation that allowed removal from the time-of-flight spectra of pulses due to ionized imaging gas. Figure 21.11 from their paper is a composite of FIM images obtained with neon as the imaging species ((a) and (b)) and then with Fe^{2+} and C^{2+} as the imaging species selected in the time-of-flight mass spectrometer, (c) and (d), respectively. In Fig. 21.11(a) can be seen the probe hole into the spectrometer showing up in the micrograph as a central dark circle. Images in selected ions from a sample can be obtained by a gating technique in which the channel plate detector in the FIM section is switched to high gain only during the duration of a high voltage pulse, with a subsequent variable delay that selects only those ions of appropriate flight time. The mass resolution in this variation is much worse than in normal APFIM or PLAP, but is very useful in rapid imaging of the whole tip in one elemental ion or perhaps combination of two ions closely spaced in mass.

The two neon FIM images in Fig. 21.11(a) and (b) reveal various grain boundaries running across the specimens in the form of dark lines. The scale in (b) shows that these lines are of average width 2 nm, typical of a grain boundary. Analysis by APFIM of the two grain boundaries in (b) is shown in Fig. 21.12, in which it can be seen that the boundaries contain appreciable quantities of carbon, the right-hand boundary containing about five times as much as the one on the left. Point analyses from areas in the right-hand boundary where the width was greatest, i.e. with minimum interference from adjoining steel regions, indicated that the precipitate composition was approaching that of Fe_3C, cementite. The C^{2+} image in Fig. 21.11(d) shows

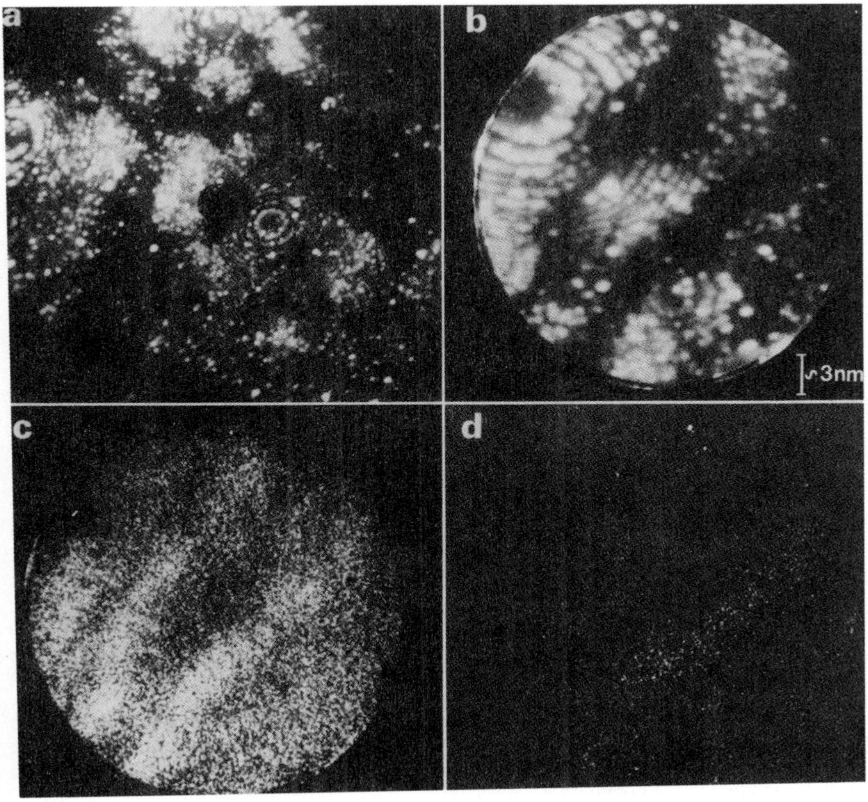

FIG. 21.11. Imaging and analysis of a grain boundary in a steel wire used for spring manufacture, by Waugh and Southon.[10] (a) FIM image using neon as imaging species. (b) Same as (a) but at higher magnification. (c) APFIM image of (b) using the Fe^{2+} ion as imaging species. (d) APFIM image of (b) using the C^{2+} ion as imaging species. The probe hole into the TOF mass spectrometer can be seen as a dark circle in (a). The scale on (b) shows that the grain boundaries, imaging as dark strips, are of average width 2 nm.

the qualitative differences in carbon concentration between the two boundaries, but it is interesting and important to note that in the Fe^{2+} image, (c), the carbon-rich areas actually appear *brighter* than the surrounding steel, in which iron should have a higher concentration. The explanation put forward by the authors is a topographical one, in that where the boundaries intersect the surface there are sharp changes in surface profile, leading to a pronounced decrease in local magnification and hence a smearing of the magnification along the lines of the boundaries. In other words, Fe^{2+} ions in those regions will be arriving at the detector from points outside the area of the probe hole, as well as from within it. Such an effect can clearly lead to quantification problems if not realized, and in such a situation it is essential to have point analysis capability with the ultimate spatial resolution.

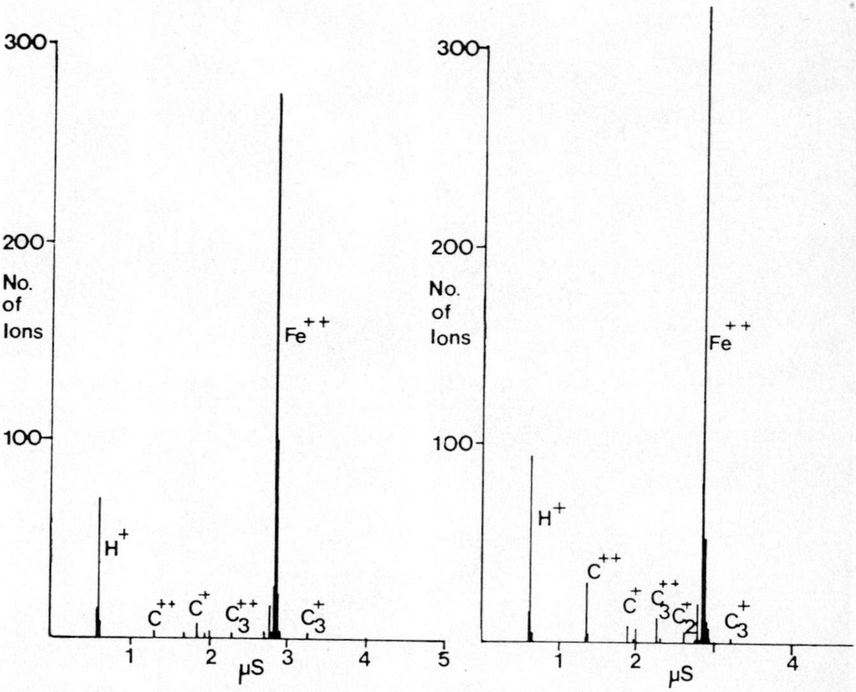

FIG. 21.12. Analysis by APFIM of the two grain boundaries seen imaged in Fig. 21.11(b), from Waugh and Southon.[10] The right-hand boundary contains about five times as much carbon as the left-hand. Point analyses where the boundary width was greatest indicated that the composition of the precipitate in the boundary was probably Fe_3C, i.e. cementite.

21.5.2 *Particles*

Just as with grain boundary segregation direct analysis can be made for many of the particulate precipitates occurring in metallurgical transformation reactions only with near-atomic spatial resolution. In the search for steels that combine ever higher strength and ever greater toughness after tempering, the steel composition Fe 14Co 10Ni 2Cr 1Mo 0.15C (AF1410) has been found to have outstanding qualities. During quenching to the martensite structure, the high Co content is thought to hold back the rate of recovery of the dislocation network, the effect of which on tempering is to produce a very fine dispersion of secondary carbides. These carbides are in the form of particles too small to be seen except in section by TEM, and are therefore suitable for study by the APFIM, as carried out by Chang *et al.*[11] After tempering for 5 h at 510°C, and then over-ageing for 24 h at 593°C, the steel was formed into a fine needle by the usual polishing treatments.

The FIM images from two of the coarser carbides (c. 5–10 nm in diameter) are shown in Fig. 21.13, and the APFIM mass spectra from them in Fig. 21.14.

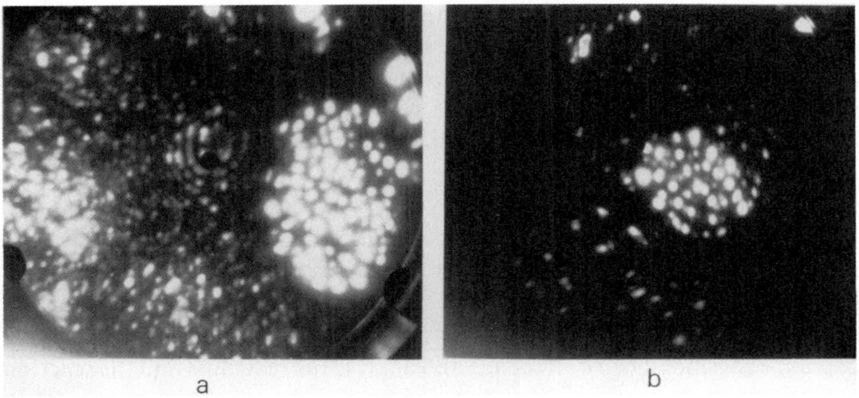

FIG. 21.13. FIM images of two coarse carbide particles in over-aged AF1410 steel, recorded by Chang *et al.*[11] The steel was tempered for 5 h at 510°C, and then over-aged for 24 h at 593°C before forming into a fine needle for FIM. The carbides are too small to be studied by any other technique except TEM in section. The imaging gas was neon, and voltages were (a) 9 kV, (b) 6 kV.

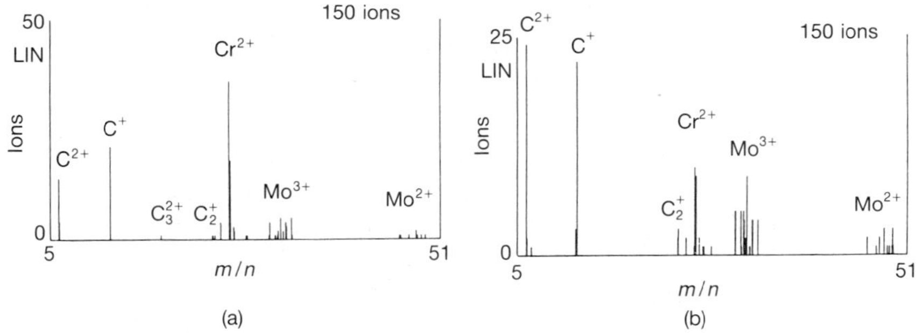

FIG. 21.14. Analysis by APFIM of the two carbide particles imaged in Fig. 21.13, from Chang *et al.*[11] In each case a total of 150 ions was used for analysis. The particle in 21.13(a) approximates in composition to the carbide M_2C, and that in 21.13(b) to MC. Compared with similar carbides present before over-ageing, those in Fig. 21.13 contained much less Fe and far more Cr and Mo.

After correction to eliminate any contribution from the matrix steel, the composition of the larger particle, in Fig. 21.13(a) was found to be almost exactly (Cr, Mo, Fe)$_2$C, and the smaller one in (b) to be an Mo-rich but slightly C-deficient MC. Interestingly, it was found experimentally that the best imaging voltage for the M_2C carbides was greater (9 kV) than for the MC carbides (6 kV), allowing them to be distinguished easily when occurring together. Compared to similar carbides observed after the standard heat treatment, i.e. no over-ageing, the carbides shown in Fig. 21.13 contained far less Fe and much more Cr and Mo. At that stage all the Mo and about half the Cr has apparently migrated to the carbides, rejecting Fe. Thus, dependent on the Mo:C ratio in the matrix, Cr can act as a carbide-former. Ni and Co carbides would of course be unstable at the temperatures of heat treatment, and their function may well be to control either the chemical activity in solution, or the diffusivity of the other alloying constituents, thus controlling the reinstatement of the dislocation structure on tempering.

21.5.3 *Alloys*

The solid state reactions taking place in an alloy during the various heat treatments needed to produce the desired properties, are often complex and lead to phase changes, precipitation, interface reactions, etc., whose scale is near-atomic. The APFIM is an ideal tool for the study of such metallurgical phenomena, since it can both image and analyse the reactants and the reaction products.

One example[8] of such a solid state reaction is that of the material Alnico 5, of composition Fe–24Co–14Ni–8Al–3Cu, which is used in permanent magnets. On ageing, the alloy decomposes to an interconnected microstructure containing so-called α and α' phases. In Fig. 21.15 is seen the FIM

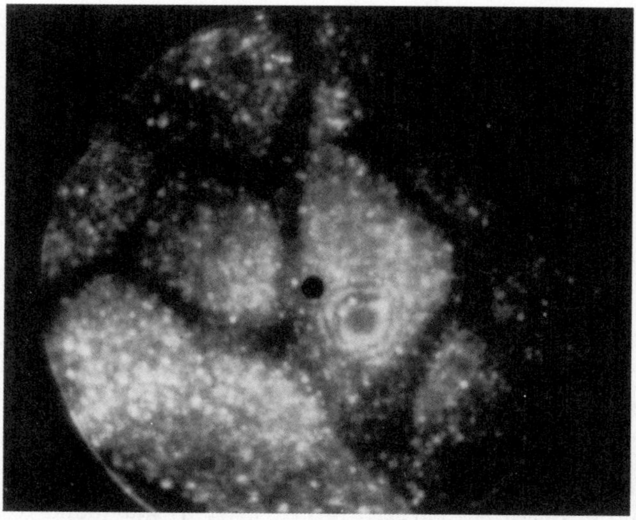

FIG. 21.15. FIM image of the alloy Alnico 5, used in permanent magnets, after ageing to produce a solid state reaction causing decomposition into two phases, from Cerezo and Smith.[8] The phases are the α (brightly imaging) and α' (darkly imaging). The imaging gas was neon at an imaging potential of 15 kV.

image of this material after ageing, revealing clearly the two phases, one brightly imaging (α) and the other darkly imaging (α'). The size of the probe hole in the centre of the image shows that it is perfectly possible with the APFIM to analyse each phase without interference from the other, and the result of such an analysis is given in Fig. 21.16. Clearly the α (bright) phase is Fe–Co rich and the α' (dark) phase is Ni–Al rich, in fact the analysis showed that in the α phase the atomic percentages of Fe and Co were approximately 60 and 23, while in the α' phase those of Ni and Al were approximately 30.5 and 40, both respectively. Thus the suspicion that the α' phase was non-ferromagnetic was confirmed.

In the study of ordering in alloys one of the alternative forms of data presentation that has been found to be particularly informative is the so-called 'ladder diagram'. The diagram is based on two axes, each of which represents an element or a group of elements present in the alloy being studied. During successive removal of atomic layers in the field evaporation mechanism of the APFIM, a step of unit length is recorded on the diagram each time an ion of one of the selected elements is detected. The individual steps will obviously be either horizontal or vertical. If an alloy is perfectly ordered, then the change in gradient of the ladder as one atomic layer is removed completely, and atoms from the next layer start to be removed, should also be either perfectly horizontal or perfectly vertical. In addition, the average slope of the ladder should reflect the proportion of the selected atoms in the alloy composition. Such a ladder diagram is shown in Fig. 21.17, from Brenner and Miller,[6] for an ordered equiatomic alloy Fe–Co containing 2% V. For 100% ordering, the atom planes should be alternately pure Fe and pure Co. The diagram shows that this is almost, but not quite, true; the sections are not exactly horizontal or vertical, and it could be calculated from this ladder that the degree of order was in fact 83% and not 100%. It can be seen that the vanadium was not associated particularly with either the Fe or the Co layers, and it can also be seen that the average slope of the ladder is indeed unity, as it should be for an equiatomic alloy.

Another example of the usefulness of ladder diagrams in examining both the local degree of order and the location of substituting elements is that of the A_3B $L1_2$-ordered Ni_3Al alloy. In the ladder diagram in Fig. 21.18, from Miller,[4] numbers of nickel atoms are plotted against numbers of aluminium atoms, with the occurrence of hafnium atoms indicated separately, since this particular alloy contained 3% Hf. The diagram indicates at once that the alloy is ordered in such a way that pure nickel layers (horizontal sections) alternate with NiAl layers (sections with slope 45°), but it is also interesting to note that the Hf occurs only in the NiAl layers. Since the nickel sites are all equivalent, the Hf must therefore be substituting for the Al and not the Ni. Note again that the average slope is 1/3, corresponding to the bulk compositional ratio.

Ladder diagrams can also be used to check the presence and composition of interfaces, and the extent of any preferred segregation of alloying or impurity elements at an interface. For example, in the superalloy IN939, which is a nickel-based alloy containing high Cr(24.5 at.%) and Co(18 at.%), the final heat treatment at 700°C is needed to produce the desired properties, although

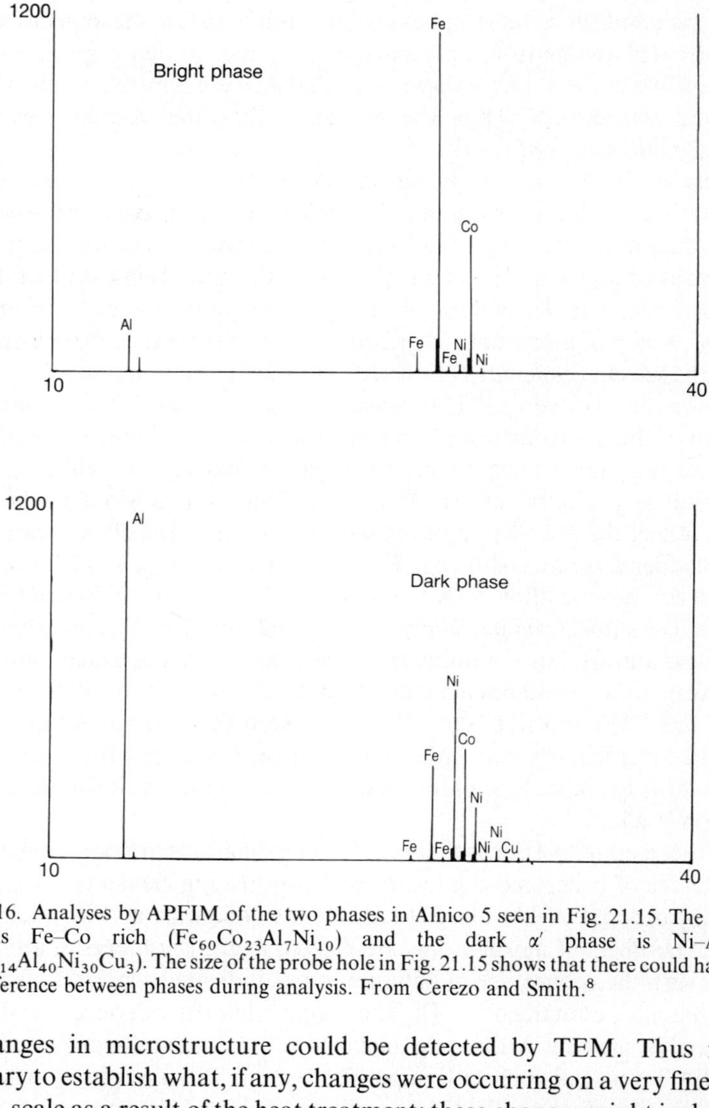

FIG. 21.16. Analyses by APFIM of the two phases in Alnico 5 seen in Fig. 21.15. The bright α phase is Fe–Co rich ($Fe_{60}Co_{23}Al_7Ni_{10}$) and the dark α' phase is Ni–Al rich ($Fe_{13}Co_{14}Al_{40}Ni_{30}Cu_3$). The size of the probe hole in Fig. 21.15 shows that there could have been no interference between phases during analysis. From Cerezo and Smith.[8]

no changes in microstructure could be detected by TEM. Thus it was necessary to establish what, if any, changes were occurring on a very fine, near-atomic, scale as a result of the heat treatment; these changes might include the formation of very small precipitates not easily resolvable in the TEM, the alteration of the compositions of existing phases, and the segregation of elements (in this case refractory elements) to interphase interfaces.

FIM images[5] of the superalloy revealed primary γ' precipitates, also visible by TEM, of diameter approximately 100 nm, but after the final heat treatment an additional distribution of much finer precipitates of average diameter 3 to 5 nm. Analysis by APFIM showed[5] that the ultra-fine precipitates were also of

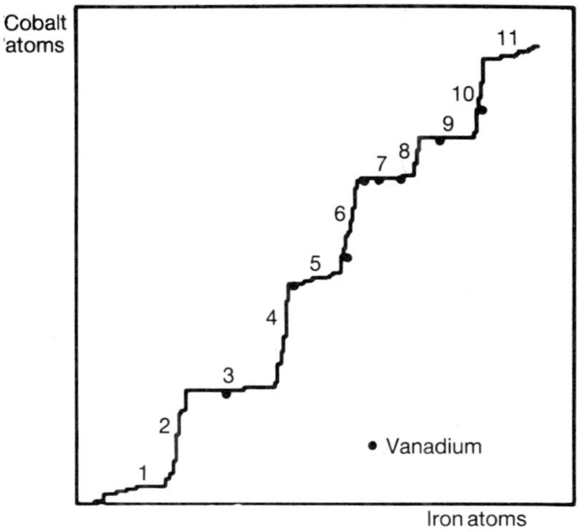

FIG. 21.17. A 'ladder diagram' constructed by Brenner and Miller[6] during APFIM analysis of an ordered equiatomic Fe–Co alloy containing 2% V. Each axis represents an element present in the alloy. As atomic layers are removed successively, a step of unit length is recorded in the diagram each time an ion of one of the elements is recorded. Individual unit steps will thus be either horizontal or vertical. For an equiatomic alloy, the average slope should be 45°, as seen, and for a perfectly ordered alloy in which the atom planes are alternately one element and then the other, the step should be perfectly horizontal or perfectly vertical. The diagram shows that the latter was not quite true for this alloy, and the degree of ordering was calculated to be 83% and not 100%. Note that the vanadium atoms are not associated particularly with either Fe or Co, but appear randomly.

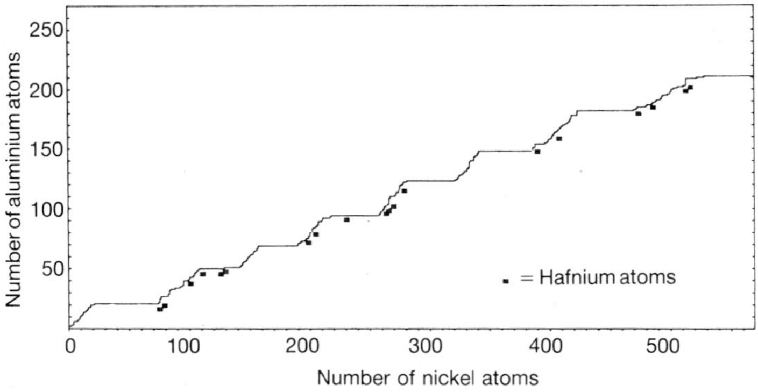

FIG. 21.18. Ladder diagram constructed by Miller[4] during APFIM analysis of the A_3B $L1_2$-ordered Ni_3Al alloy containing some Hf. The ladder shows conclusively that pure Ni layers (horizontal sections) alternate with NiAl layers (sections at 45°), but the Hf occurs only in the NiAl layers. The conclusion was that, since the Ni sites in the alloy are all equivalent, the Hf must be substituting for the Al. The average slope of the ladder is $\frac{1}{3}$, as required by the bulk Ni/Al ratio.

the γ' type, but with a lower Co content than the larger ones. Table 21.1 shows the average analyses of the various phases in comparison with the bulk composition of the alloy.

As can be seen from Table 21.1, there is strong partitioning of certain elements between the γ matrix phase and the γ' precipitates. In particular, high Cr is characteristic of the γ phase, and high Al+Ti is characteristic of the precipitates. If now a ladder diagram is constructed that has combined numbers of ions of Al and Ti along one axis, and numbers of Cr ions along the other axis, and if the probe hole is traversed across a matrix/precipitate interface, it should be possible to reveal conditions at the interface. The resultant diagram is presented in Fig. 21.19. It shows, as expected, that there are very few Cr atoms in the γ' precipitates, and equally very few Al and Ti atoms in the γ matrix. The interface region contains contributions from both axes, in the approximate ratio $(\mathrm{Al}+\mathrm{Ti})/\mathrm{Cr} \approx 2$, indicating the existence of an intermediate composition only a few nanometres wide. Particularly interesting however, is the observation that no refractory metal ions were found in the interface; this contradicts previous speculation, and shows that segregation is not a significant factor in the mechanism of creep strengthening of this alloy.

21.5.4 *Semiconductors*

Although the voltage-pulsed form of the atom-probe has been applied to the study of semiconducting materials, there can be problems with specimens of low electrical and thermal conductivity, and as a consequence the number of successful applications using this version of APFIM is relatively small. With the advent of the pulsed laser form (PLAP), however, the conducting nature of the material is not so important, and the range of application has been widened as a result to include most compound and elemental semiconductors. The information sought is that both of stoichiometry of a compound semiconductor at and near the surface, and, more importantly, of the composition and width of interface regions formed between thin metallic films and semiconductor surfaces. The latter topic is of course also being investigated by various other surface analytical techniques, but APFIM and PLAP have been able to make their own unique contributions.

Just as for metallurgical investigations, so for semiconductors has the ladder diagram form of information presentation been instructive. For InAs in the [111] direction the ladder diagram produced[12] by PLAP is shown in Fig. 21.20. The slope to be expected based on the nominal composition of the binary compound is indicated, and it can be seen that the average slope of the ladder conforms closely to expectation. Underneath is drawn an ideal representation of the ladder, assuming adequate resolution in the atom probe to resolve alternate planes of In and As. Clearly the alternate planes are not completely resolved, and this is hardly surprising, since the [111] planes are very closely spaced, being only 0.09 nm apart, and cannot therefore be

Table 21.1
Analyses of γ and γ' phases in IN939 by APFIM, at.%

Element	γ	γ', primary	γ', secondary	Bulk*
Nickel	40.9	61.4	68.1	46.5
Cobalt	22.4	9.8	3.2	18.2
Chromium	32.9	1.5	1.4	24.5
Aluminium	1.3	11.7	13.8	4.0
Titanium	0.8	13.8	11.4	4.4
Tungsten + tantalum	1.6	1.7	2.1	1.1

* Also Nb, C, B and Zr.

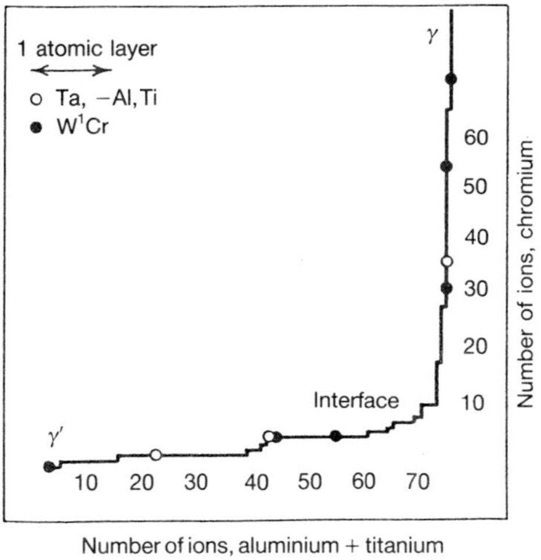

FIG. 21.19. Ladder diagram from Smith[5] constructed from APFIM analysis of the superalloy IN939. In this case the probe hole was not maintained stationary over one area, but traversed across the interface between the γ matrix and the γ' precipitates formed after final heat treatment at 700°C. Individual analyses of these phases, as in Table 21.1, showed that the γ' phases were characterized by enriched Al and Ti, whereas high Cr was characteristic of the γ phase. The ladder was therefore plotted with number of (Al + Ti) ions as one axis, and number of Cr ions as the other. As expected, the γ' precipitates contained almost no Cr, and the γ phase almost no Al or Ti, but a very narrow interface region was observed in which (Al + Ti)/Cr $\approx$ 2, only a few nm in width. Some refractory metal ions were found in the γ and γ', but none at all at the interface, precluding their segregation as a mechanism of creep strengthening.

resolved with the standard aperture. A much smaller aperture would improve the resolution, but acquisition times would then increase drastically, and in any case a limit would eventually be imposed by aberrations. Nevertheless, it is clear that PLAP can certainly provide an accurate compositional analysis of

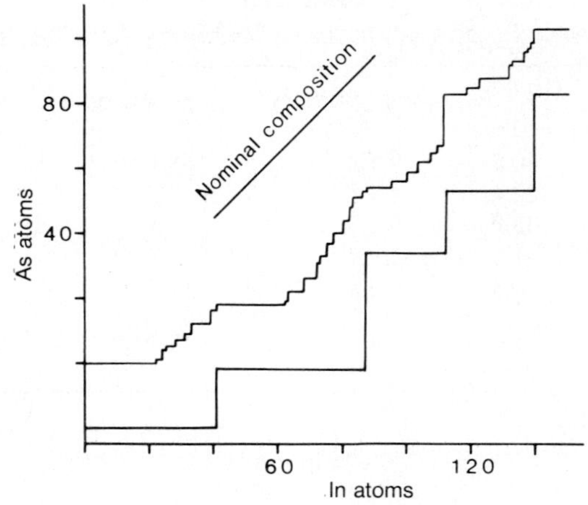

FIG. 21.20. A ladder diagram constructed using PLAP analysis, of the semiconductor InAs in the [111] direction, by Cerezo *et al.*[12] The slope expected based on the nominal composition is indicated, and the average slope is seen to conform to it. At the bottom is the idealized expectation of what would be observed if the probe hole provided enough resolution to resolve alternate planes of In and As. However, since the latter are only 0.09 nm apart, they cannot be resolved by PLAP. Nevertheless an accurate compositional analysis could be provided with an indication of the alternate layers.

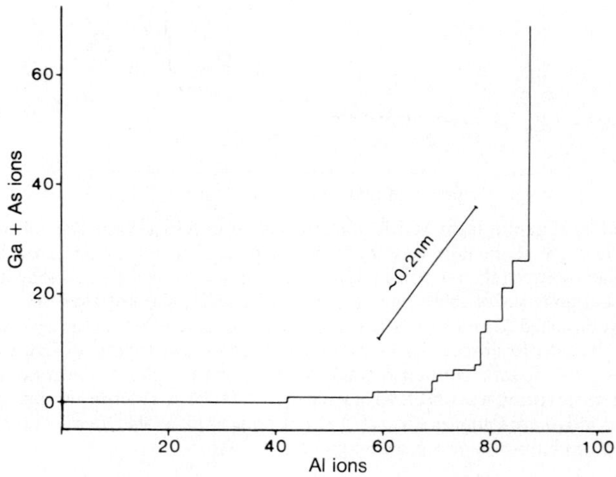

FIG. 21.21. Ladder diagram recorded by Cerezo and Smith[8] by APFIM analysis of an unannealed Al film on GaAs, deposited at 100 K. Here the ladder axes are numbers of (Ga + As) ions and numbers of Al ions, respectively. Rather unexpectedly, a transition region is found at the Al/GaAs interface, about 0.2 nm in width, and with a slope of about $\frac{5}{8}$, corresponding approximately to a composition $GaAl_3As$. Thus, even at 100 K some solid state reaction has occurred between the Al and the GaAs.

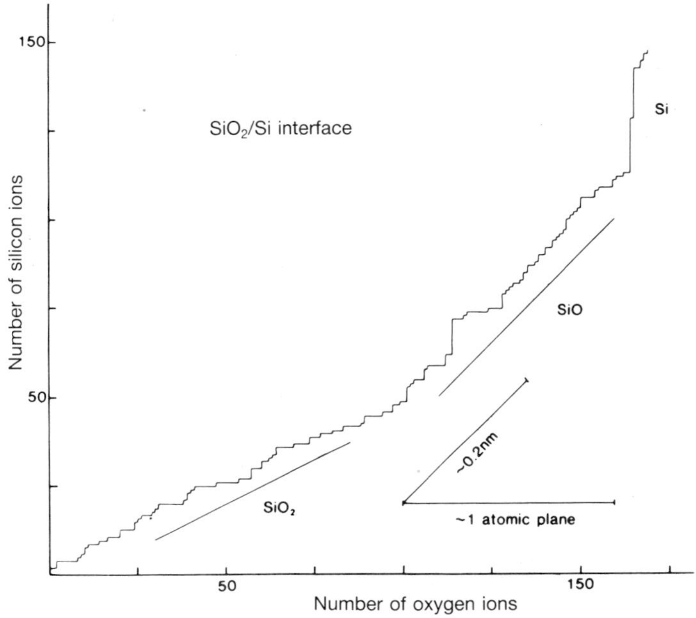

FIG. 21.22. The transitional region between silicon dioxide and silicon investigated by Cerezo and Smith[8] using PLAP analysis. The oxide had been formed by wet oxidation at 800°C for 10 min. The axes of the ladder are numbers of Si and of O ions, respectively. The analysis shows clearly a transitional region of width about 0.3 nm in which the ladder slope agreed very accurately with the nominal composition SiO. The slope was constant over the region, and the changes at the outer and inner edges were quite abrupt, allowing excellent statistics to be achieved within the region.

compound semiconductor surfaces without any apparent artefacts being introduced due to excessive heating of the tip by the laser pulse. Similar analysis of a GaAs specimen gave a ratio of Ga/As of 0.99 ± 0.03.

Interface analysis on semiconductors has also provided some interesting results. A ladder diagram recorded by Cerezo and Smith[8] from an unannealed film of aluminium on GaAs, with evaporation taken down through the interface, is reproduced in Fig. 21.21. One axis is the number of Al ions, the other the number of Ga + As ions. In the transition region of the interface between the film and the semiconductor there appears to be a layer formed with a constant (Ga + As)/Al slope of about 5/8, which corresponds to an approximate composition $GaAl_3As$. Note that although the width of the interfacial region is only 0.2 nm, the time-of-flight counting statistics are such that the value of the slope of (Ga + As)/Al, and hence the composition, can be determined with reasonable accuracy.

Another interfacial example concerns the much-debated question as to

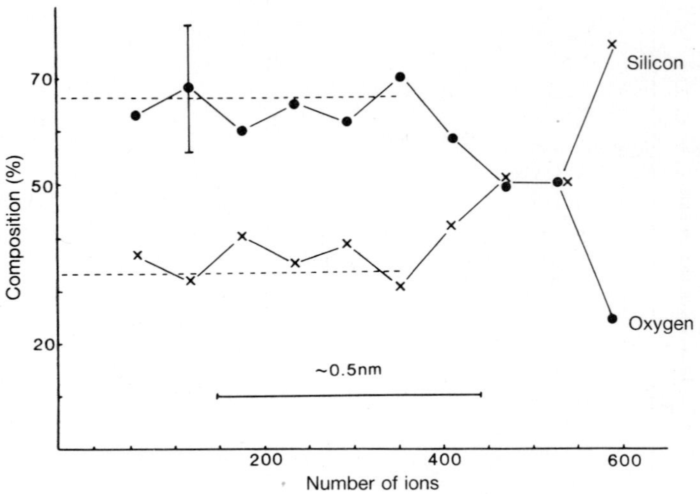

FIG. 21.23. An alternative method of analysis of the SiO_2/Si interface region which gives the same result as that of the ladder diagram approach of Fig. 21.22. Here the actual composition in terms of percentage of Si and O ions is evaluated, and again the interface region is seen to be of width (statistically) about 0.3 nm and of composition SiO. From Cerezo *et al.*[12].

whether there really is an intermediate transitional silicon oxide in the interface between the thermal oxide grown on a silicon surface, and the substrate silicon. Again, various other surface analytical techniques have been brought to bear on the question. The importance lies in the existence and population of localized electronic states at the interface, which may affect the performance of the Si–SiO_2 interface in MOS devices. The ladder diagram for evaporation by PLAP of a film of SiO_2 formed on silicon by wet oxidation at 800°C for 10 min, from the work of Cerezo and Smith,[8] is shown in Fig. 21.22.The slopes to be expected for the compositions SiO_2 and SiO are shown alongside the experimental ladder. Also shown below is the spatial resolution. Once more, the statistics are perfectly adequate to reveal a marked change of slope in the interface region, the slope remaining remarkably constant over the region, and conforming very closely to that of the composition SiO. The width of the interface was $\approx$0.3 nm. Other techniques have suggested an interface region consisting of SiO_x, without being specific about the value of x, but PLAP indicates firmly that it is unity, i.e. corresponding to SiO. A more conventional method of representation of the analysis is that of the profile shown in Fig. 21.23, from Cerezo *et al.*,[12] which again is unequivocal in giving the interface composition as SiO.

The recent studies of semiconductors of which a few have been selected above, suggest that eventually APFIM and its variation PLAP may have as many useful applications in the semiconductor as in the metallurgical field.

References

1. Müller, E. W. (1951). *Z. Phys.* **131**, 136.
2. Müller, E. W. (1936). *Z. Phys.* **37**, 838.
3. Müller, E. W., Panitz, J. A., and McLane, S. B. (1968). *Rev. Sci. Instrum.* **39**, 83.
4. Miller, M. K. (1987). *Int. Mater. Rev.* **32**, 221.
5. Smith, G. D. W. (1986). *Metals Handbook, 9th edn.* **10**, 583.
6. Brenner, S. S. and Miller, M. K. (1983). *J. Metals* **35**, 54.
7. Poschenrieder, W. P. (1972). *Int. J. Mass Spectrom.* **6**, 413; **9**, 357.
8. Cerezo, A. and Smith, G. D. W. (1987). *Mater. Forum* **10**, 104.
9. Tsong, T. T. (1980). *Prog. Surf. Sci.* **10**, 165.
10. Waugh, A. R. and Southon, M. J. (1979). *Surf. Sci.* **89**, 718.
11. Chang, Li, Smith, G. D. W., and Olson, G. B. (1986). *J. Phys. Colloq. C2*, Suppl. 3, **47**, 265.
12. Cerezo, A., Grovenor, C. R. M., and Smith, G. D. W. (1986). *J. Microsc.* **141**, 155.

22

High field excitation: scanning tunnelling microscopy and spectroscopy (STM and STS)

Just as APFIM (see previous chapter) arose out of its parent technique FIM, so Scanning Tunnelling Spectroscopy (STS) has arisen out of Scanning Tunnelling Microscopy (STM). The similarities in origin go further, because while both FIM and STM are surface topographical imaging techniques, i.e. are concerned principally with effects due to surface structure, APFIM and STS provide information about the chemical nature of the surface, either by atomic compositional analysis or by surface state spectroscopic analysis. Again, in the same way that it was necessary in Chapter 21 to describe FIM in order to describe APFIM, so here it will be necessary to go into the details of the STM technique in order to be clear about the associated spectroscopic technique. There the similarities end, however, because although APFIM is now a well established technique, both STM and STS, the latter in particular, are relatively new. The inventors of STM were Binnig and Rohrer,[1] who received the Nobel Prize for their contribution to physics.

22.1 Operation

In principle the operation of a scanning tunnelling microscope is simple. A metal tip is brought close enough (0.5–1.0 nm) to a surface that an electron tunnelling current becomes measurable at applied voltages in the range 2 mV–2 V, and is then scanned across the surface to be studied. The tunnelling current is monitored, and used in either of two ways, shown schematically in Fig. 22.1 from Hansma and Tersoff.[2] In the constant current mode a feedback circuit changes the height of the tip in order to keep the current constant, while in the constant height mode the tip is scanned at constant height above the surface while the current variations are recorded. Since in the one-dimensional formalism the tunnelling current is a very sensitive function of the tip/surface separation through the experimental dependence

$$I \propto V_T \exp(-A\phi^{1/2}s) \tag{22.1}$$

where V_T is the applied voltage, ϕ is the average work function, s is the separation, and A a constant, variations in I reflect variations in s to very high resolution. With work function values normally $\approx$4.5 eV, the current I changes by about an order of magnitude for each 0.1 nm (1 Å) change in the separation s. Thus in the constant current mode the tip follows the atomic contours of the surface by rising or falling, as seen in the left-hand side of

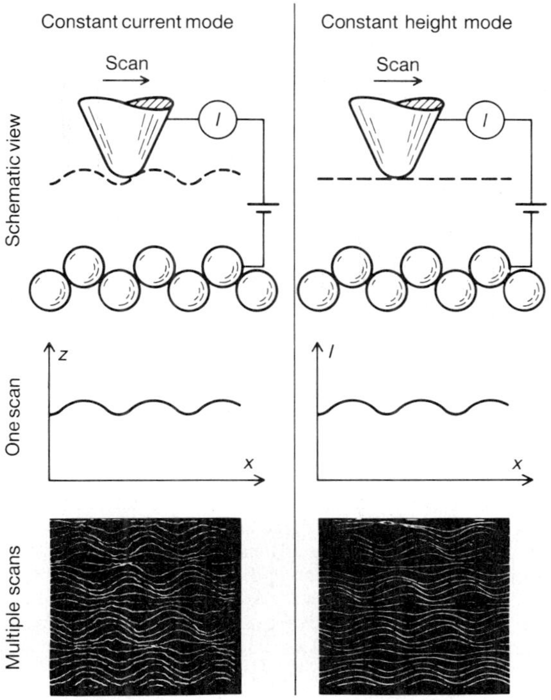

FIG. 22.1. Schematic illustration of the modes of operation of a scanning tunnelling microscope, from Hansma and Tersoff.[2] On the left, the height of the tip above the surface is altered via a feedback circuit to keep the tunnelling current constant at fixed applied voltage. On the right, the height is kept fixed while the tunnelling current variations are monitored. The two images are of a graphite surface, recorded in each of the two modes. In either case the image is related to the surface topography to a first approximation (but see text).

Fig. 22.1, while in the constant height mode it is the current that rises or falls in the same way, as seen in the right-hand side of that figure. An image related to the surface topography can thus be obtained, to a first approximation, by constructing a map of linear scans displaced successively in the y direction by equal amounts. Examples of such images are given at the bottom of Fig. 22.1. In the constant height mode an alternative mode of operation is to keep the current constant and monitor the variations in the applied voltage, again via feedback circuitry.

It will not have escaped notice that the inter-dependence of I, V_T, and s in eqn (22.1), that is the basis of STM, holds true only if ϕ, the average work function, is constant. This will not necessarily be true, and represents a complication in the interpretation of images. Thus, as shown schematically in Fig. 22.2 from Baratoff *et al.*,[3] the presence of a patch C of different work function (lower than average in the example) will give rise to a feature B in either the z vs. x or the I vs. x scans. Feature A, on the other hand,

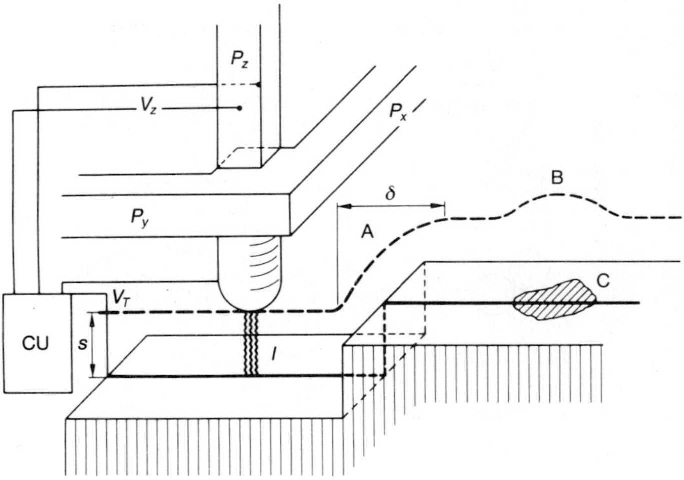

FIG. 22.2. Diagram of the principle of operation of a scanning tunnelling microscope, as used by Baratoff *et al.*[3] The tip is mounted on an orthogonal piezo-electric framework that allows it to be driven vertically via P_z, or scanned horizontally via P_x and P_v. In the constant current mode the control unit CU supplies a voltage V_z to the P_z drive to keep the current I constant for a fixed applied voltage V_T. Variations in height such as the step at A, when line scanning along y, combined with sequential shifts along x, then provide a three-dimensional map. Note that a patch of lower work function such as that at B will also cause an apparent change in height, according to expression (22.1).

corresponds to a genuine topographical feature, a step of width δ. Such work function induced features can, however, be separated experimentally from topographical features by modulation of the separation s through a very small distance Δs, at a frequency well beyond that of the feedback loop, but below that of the lowest characteristic vibration of the device. The current I_s at the frequency of modulation is then a measure of the work function, since

$$I_s = \Delta(\ln I)/\Delta s \propto \phi^{1/2}. \tag{22.2}$$

Figure 22.2 also shows schematically the early experimental arrangement used by Binnig and co-workers. The probe tip is mounted on a piezo-electric framework that allows drive in an orthogonal P_x, P_y, P_z system. In the constant current mode the control CU supplies a voltage V_z to the P_z drive to keep I constant for a fixed applied voltage V_T when line scanning along the x axis. Steps along the y axis combined with line scanning then provide a three-dimensional V_z, V_x, V_y plot that, as stated above, is proportional to the surface topography to a first approximation. In the constant height mode V_z is fixed, and either the variation in I for constant V_T, or the variation in V_T required to keep I constant, are recorded. Although the constant current mode was the first to be used, in fact the constant height mode is more convenient since scanning can be performed much faster; in the latter case it is only the feedback electronics and not the mechanical system that must have a sufficiently rapid response. Typical scan rates in the constant height mode are of the order of 1 kHz.

The above description of the technique raises some obvious questions about experimental realization. The first concerns vibration. With tip-to-surface separations of 0.5–1.0 nm, and the tip movement measurable to a vertical resolution of 10^{-3} nm, vibration must clearly be eliminated. To be more specific, the possibility of exciting any resonant frequency in any of the components supporting the tip must be eliminated. This necessity has led to enormous ingenuity in the devising of anti-vibration mountings for STM devices, and the variety is such that a full description here is not appropriate. Suffice it to say that the variety has included superconducting levitation, damped springs, rubber alone, rubber interleaved with steel plates, magnetic damping, air tables, and so on. It is now accepted, however, that it is more important to increase the general rigidity of the mounting system than it is to improve vibration damping, in order that the characteristic resonant frequencies are very high. Local resonances associated with buildings are usually of very low frequencies that are easily attenuated at the high resonant frequencies of sufficiently rigid mounts.

The second question is of course that of how the tip is brought up to the surface, without actually coming into physical contact with it. (In one of the off-shoots of STM, that of nano-hardness testing, the tip is deliberately pressed into the surface.) Again there has been scope here for some remarkable mechanical engineering. The early STM devices used a coarse positioning drive called a 'louse', pictured in Fig. 22.3 from Binnig and Rohrer.[1] It consists of a piezoplate (PP), on top of which is the sample (not shown), which sits on three metal feet (MF) separated by insulators (I) of high dielectric constant from the groundplates (GP). Application of a voltage V_F between the feet and the groundplate clamps the feet to the latter electrostatically. By an appropriate sequence of unclamping and clamping of the feet, together with piezo-electric induced elongation or contraction of the plate, the louse can be moved in any direction in steps of 10 nm to 1 μm, at a rate of up to 30 steps per second, until the sample is brought within range of the tip piezo-drives, which can cover only a few μm in each direction. The range of movement of the louse is such that the sample can be moved right away from the tip for cleaning or surface treatment.

A variation in method of coarse positioning that is purely mechanical is that devised by Demuth *et al.*,[4] and shown in Fig. 22.4. The sample G is mounted on a pair of arms F that pivot about a pin. The pin is in turn fixed to two rods C that can slide backwards and forwards through two ball races E, attached to the base plate B. Upward movement of the rods tilts the arms and the sample away from the tip J by virtue of restriction by the rotatable stop H, as seen in 22.4(c). In the other direction of movement, towards the tip, a stop D prevents

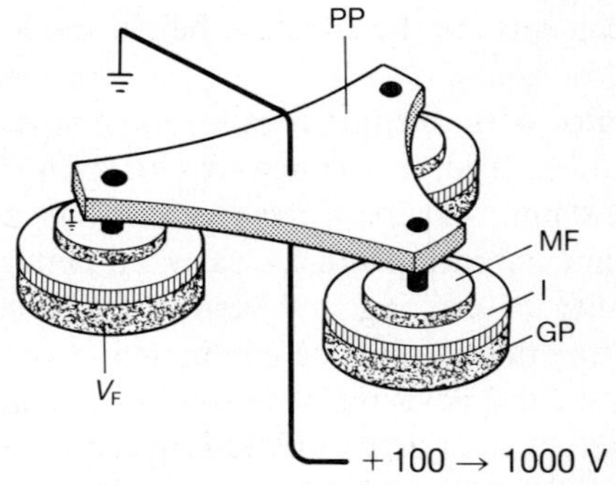

FIG. 22.3. One of the earlier methods of coarse positioning of the sample in front of the tip, devised by Binnig and Rohrer.[1] A piezoplate PP holds the sample (not shown) and sits on three metal feet MF separated by insulators I of high dielectric constant from the groundplate GP. When a voltage V_F is applied between feet and groundplate, they are clamped together electrostatically. Elongation or contraction of the piezoplate by appropriately applied voltages, followed by clamping and unclamping of the feet in the right sequence, then causes the positioner to move in the required direction in steps of 10 nm to 1 μm, at a rate of 30 step s^{-1}. Once within a few μm of the tip, the tip piezo-drives can then perform the final fine positioning. The device was nicknamed a 'louse'.

further coarse travel, and by increasing the force through the rods, that are driven by a lead screw not shown, the sample is pushed up against a 'foot' labelled I. Foot and tip are separated by about 0.5 mm, and the end of the tip is about 0.1–0.02 mm behind the end of the foot. Further pressure then rotates the sample about the foot towards the tip. The dimensions of the device and the pitch of the lead screw provide about 25 nm of forward motion per degree rotation of the screw. Actuation from outside the vacuum system of the screw is performed via a rotary motion feedthrough. The tip is driven by three orthogonal piezo-electric elements as before, and the drive is mounted on a Macor block A.

It might be thought from the description of STM that the most difficult aspect of operation would be the preparation of a tip of sufficiently minute radius. After all, the lateral resolution is of the order of 0.5 nm. In fact the tip preparation is not so critical as appears at first sight. The reason for this is that the tip itself selects the precise point at which tunnelling occurs, since any surface produced by chemical or mechanical sharpening of a metal to produce a point (as in FIM) will possess many irregularities on the atomic scale, any of which can act as 'nano-tips'. On approach to the sample surface whichever of these nano-tips happens to be nearest will carry the tunnelling current. Thus it is not necessary to go to extreme lengths to attempt to create a tip of near-atomic radius, and indeed preparation often consists simply of rough grinding the end of a tungsten rod! However, if no images can be obtained at first after insertion in vacuum it can be found necessary to treat the tip by application of a much higher voltage than normally used, when retracted a substantial

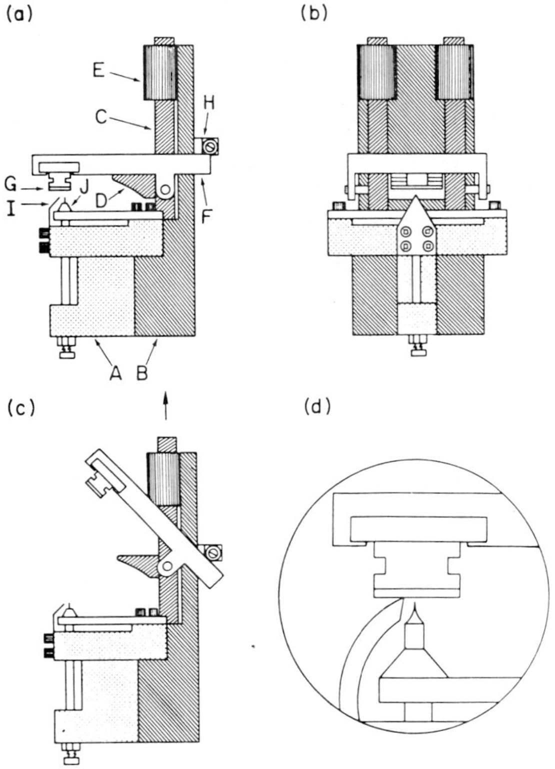

FIG. 22.4. A mechanical method of coarse positioning designed by Demuth *et al.*[4] A, macor block carrying the piezo-electric drives. B, base plate. C, rods sliding backwards and forwards through the ball races E. D, mechanical stop. F, arms pivoting about a pin. G, sample. H, rotatable stop. I, foot. J, tip mounted in a collet.

In (a) the sample has been brought into the vicinity of the tip by downward movement of the rods C, actuated by lead screws. In (c) the sample has been tilted away from the tip, e.g. for surface treatment, by upward movement of the rods against the stop H. The magnified diagram (d) shows the sample in contact with the foot, preparatory to the final positioning.

distance, in order both to remove nano-tips of the wrong size and position and to create others by whisker growth. One of the disadvantages of STM arises as a consequence of the likely presence of several nano-tips at the end of a tip, and that is that when scanning across a particularly rough surface a nano-tip some distance from the one that first carried the tunnelling current may find itself closer to the surface, in which case the tunnelling position will switch. This has caused confusion in image interpretation in some instances.

Once methods had been invented of bringing a tip reproducibly to within a nm of a surface, it was inevitable that STM would spawn several associated techniques. One of these, that of nano-hardness measurement, in which the force necessary to push a tip into a surface is measured, has already been mentioned. Surface structure with a lateral resolution better than 0.25 nm has been achieved[5] with the extension of STM called the atomic force microscope (AFM). By moving a tip away slightly from a surface, and increasing both tunnelling voltage and current, so that very localized thermal stresses and perhaps melting are induced, Staufer *et al.*[6] have produced reproducible surface features and called the technique nanometer fabrication. Gimzewski *et al.*[7] have observed photon emission from the tunnelling region and shown that the spectrum as a function of tunnelling voltage has very close affinities with an inverse photoemission (q.v.) spectrum from the same surface. Finally, of particular relevance to this volume, is the technique of scanning tunnelling spectroscopy in which the spectrum of both occupied and unoccupied electronic states below and above the Fermi level can be probed, as in photoemission and inverse photoemission, but in this case with near-atomic resolution. STS was also originated by Binnig and co-workers.[8]

STM images are obtained as a function of position (x, y) by keeping either I, or V_T, or s constant, and monitoring one of the other parameters. If, however, at a particular position (i.e. chosen x and y) the tunnelling current I is recorded as a function of the applied voltage V_T, then the I–V dependence will reveal local surface electronic properties. This follows from the general first-order expression for tunnelling current from the Bardeen[9] formalism, used by Tersoff and Hamann[10] in the form

$$I \propto \sum_{s,t} f(E_t)\{1-f(E_s+eV_T)\}\,|M_{st}|^2\,\delta(E_t-E_s) \tag{22.3}$$

where $f(E)$ is the Fermi function, E_t and E_s are the energies of states ψ_t and ψ_s in the tip and surface, respectively, and M_{st} is the tunnelling matrix element between states ψ_t and ψ_s. If, as did Tersoff and Hamann[10] and Selloni *et al.*,[11] it is assumed that the wave functions ψ_t of the tip are localized, i.e. a constant density of states, and if only s-waves are considered, then (22.3) reduces to

$$I \propto \sum_{s} |\psi_s(\mathbf{r})|^2\,\delta(E_s-E_f). \tag{22.4}$$

The quantity on the right-hand side of (22.4) is simply the surface local density of states at the Fermi energy E_F, and of course the LDOS exhibits steps at the edges of bands, thus the variation of the tunnelling current I with the applied voltage V_T will also show such structure. To enhance the structure it is normal to record the derivative $\delta I/\delta V_T$, as a function of V_T. Because of the extremely localized nature of the interrogation of energy states, it should be possible to find peaks in $\delta I/\delta V$ due to saddle points in surface and bulk bands, and to localized states, e.g. dangling bonds. According to Selloni *et al.*,[11] there should be a further term in expression (22.4) to account for the modification of the surface potential barrier by the applied electrostatic field, that is, a transmission factor, and they showed how to evaluate it for a barrier of simple geometric form.

There is, of course, a practical problem in recording the variation of I with V_T in that the maintenance of constancy of the gap tip-to-surface is controlled by the current itself. In other words, as V_T is varied, and I with it, so the gap will alter, which is undesirable. The ways in which that problem has been avoided are either by keeping the ratio V_T/I fixed, i.e. in a constant-gap-resistance mode, or by superimposing on V_T a small modulation of frequency well beyond the bandwidth of the central feedback system. The consequent current modulation can then be amplified at that frequency by phase-sensitive detection.

Once a prominent feature in a tunnelling spectrum has been identified, then the value of V_T corresponding to that feature can be fixed, and the tip scanned over the surface at constant separation s to provide an image of the distribution of the energy state corresponding to the feature. Hence *scanning* tunnelling spectroscopy. It will be appreciated immediately that fixing V_T and s, and recording variations of I as the tip is scanned is indeed exactly what is done in STM in one of the constant-height modes. In other words all STM images in that mode are not just topographical images, but images of variations in surface electronic structure that of course have a direct relationship to the topography. For that reason the relative contrast in STM images of a particular area of surface will vary with the chosen applied voltage.

22.2 Theory and image interpretation

Some mention of the theoretical background has already been made in the last section, but needs to be expanded here. Expression (22.1) relating the tunnelling current to the applied voltage and the separation is a simplification of the results of the model of Simmons[12] based on one-dimensional tunnelling through a planar barrier. For tunnelling between two conductors separated by a gap s and of the same work function ϕ, the tunnelling current I is related to the applied voltage V_T by

$$I = \frac{e}{2\pi h}\{\bar{\phi}\exp(-A\bar{\phi}^{1/2}s) - (\bar{\phi}+eV_T)\exp\{-A(\bar{\phi}+eV_T)^{1/2}s\}\} \qquad (22.5)$$

where $\bar{\phi} = (\phi - eV_T/2)$, appropriate for the normal STM regime in which $eV_T < \phi$, and where $A = (4\pi/h)(2m)^{1/2}$.

If eV_T is sufficiently small, then (22.5) reduces to

$$I = -\frac{e^2 V_T}{2\pi h}\exp(-A\phi^{1/2}s) \qquad (22.6)$$

which is the same as expression (22.1). The linear I–V_T behaviour predicted by (22.6) at very low applied voltages is obeyed, but at higher voltages the behaviour becomes more nearly exponential, i.e. decidedly non-linear. Such behaviour is characteristic of field emission.

The one-dimensional model breaks down because in STM the tunnelling current is a function of the three-dimensional structure of the surface, i.e. the local band structure. As indicated earlier, the most complete theoretical exposition of the origins of the tunnelling current in STM has been that of Tersoff and Hamann,[10] who used first-order perturbation theory based on the Bardeen[9] formalism. From this expression (22.3) resulted, in which the summation is performed over all states ψ_s and ψ_t in the surface and tip, respectively. Reduction of (22.3) to expression (22.4) was valid in the limit where the tip radius vanishes, i.e. the tip is replaced by a point. In general, however, the matrix element M_{st} appearing in (22.3) must be evaluated, and Bardeen[9] has shown that it can be written as

$$M_{st}=\frac{h^2}{2m}\int \mathrm{d}\boldsymbol{S}\cdot(\psi_t^*\nabla\psi_s-\psi_t\nabla\psi_s^*). \tag{22.7}$$

The quantity in brackets is simply the current operator, and integration can be performed across any surface, most conveniently a plane, within the vacuum region between the two sides. For the surface wave function Tersoff and Hamann took a completely general expression of the form

$$\psi_s=\Omega_s^{-1/2}\sum_L a_L \exp\{(\kappa^2+|\boldsymbol{\kappa}_L|^2)^{1/2}z\}\exp(\mathrm{i}\boldsymbol{\kappa}_L\cdot\mathbf{X}). \tag{22.8}$$

Here Ω_s is the sample volume, $\kappa=(2m\phi)^{1/2}/h$, which is the minimum inverse decay length for ψ_s in vacuum, ϕ is the work function, a_L are coefficients, and $\boldsymbol{\kappa}_L=\boldsymbol{\kappa}_\parallel+\boldsymbol{L}$. $\boldsymbol{\kappa}_\parallel$ is the surface Bloch wave vector of state s, and $\boldsymbol{L}$ is a surface reciprocal lattice vector. On the other hand, the detailed atomic structure of the tip is ill-defined, so Tersoff and Hamann approximated it by a spherical potential well at its closest distance of approach, as shown in Fig. 22.5. The sphere of radius R has centre located at $\boldsymbol{r}_0$, and the separation is s as before. The tip wave functions are then assumed to have the standard asymptotic spherical form

$$\psi_t=\Omega_t^{-1/2}c_t\kappa\,\mathrm{Re}^{\kappa R}(\kappa|\boldsymbol{r}-\boldsymbol{r}_0|)^{-1}\,\mathrm{e}^{-\kappa|\boldsymbol{r}-\boldsymbol{r}_0|}. \tag{22.9}$$

Ω is the tip volume, and c_t is also a normalizing coefficient.

Substitution of (22.8) and (22.9) into (22.7), and expansion term by term in L gives

$$M_{st}=\frac{h^2}{2m}4\pi\kappa^{-1}\Omega_t^{-1/2}\kappa\,\mathrm{Re}^{\kappa R}\psi_s(\boldsymbol{r}_0). \tag{22.10}$$

Finally, substitution into expression (22.3) provides the result

$$I=32\pi^3h^{-1}e^2V_T\phi^2D_t(E_F)R^2\kappa^{-4}\mathrm{e}^{2\kappa R}\sum_s|\psi_s(\boldsymbol{r}_0)|^2\,\delta(E_s-E_F). \tag{22.11}$$

D_t is the density of states per unit volume of the tip. Note from the form of

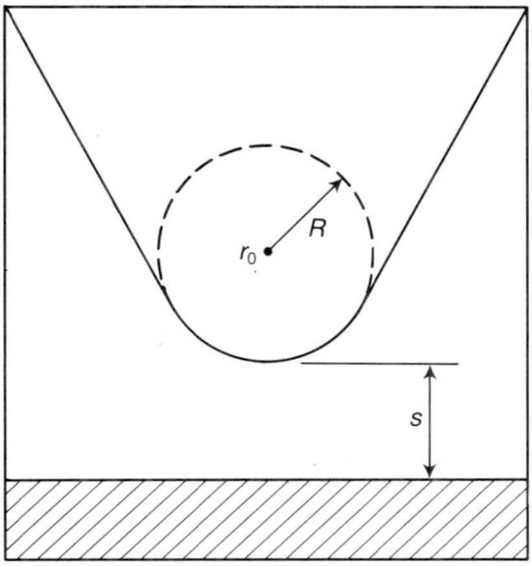

FIG. 22.5. Schematic diagram of the model of the tip-surface geometry used by Tersoff and Hamann.[10] The tip is approximated by a spherical potential well of radius R, whose centre is located at the system coordinate position r_0. The tip–surface separation is s as before.

(22.10) that evaluation of the surface wave function at a distance $r_0 = R + s$ describes the lateral averaging of the current due to finite tip size.

The critical approximation in the above treatment is not so much the assumption of a finite tip size as that of dealing only with s-wave functions for the tip. Tip wave functions with angular dependence can be included through the use of appropriate weighting factors in the Fourier expansion of ψ_s.

When Tersoff and Hamann inserted some typical values of the constant terms in (22.11) for metals, they found for the tunnelling conductance that

$$\sigma \sim 0.1 R^2 \, e^{2\kappa R} \, \rho(\boldsymbol{r}_0, E_F)$$

where

$$\rho(\boldsymbol{r}_0, E_F) = \sum_s |\psi_s(\boldsymbol{r}_0)|^2 \, \delta(E_s - E_F). \tag{22.12}$$

Notice that expression (22.12) is identical to (22.4). $\rho(\boldsymbol{r}_0, E_F)$ is the local density of states at the surface for E_F at the point $\boldsymbol{r}$. Thus for constant current the moving tip follows *contours of constant LDOS* at the Fermi level of the surface. This result has been obtained on the basis that the tip wave functions can be approximated adequately by s-waves, and that $\boldsymbol{r}_0$ is the centre of curvature of the tip wave function, and is valid for tips of arbitrary size as long as these approximations hold. It may well be valid for an effective tunnelling tip consisting of a single atom, and will certainly be good for a small cluster of atoms, assuming a correct $\boldsymbol{r}_0$ can be chosen.

Since the above procedure removes the properties of the tip from considerations of interpretation of STM images, and the image can thus be regarded as a property of the surface alone, steps can be taken to attempt to calculate images. Tersoff and Hamann[10] have shown that this can be done in what they term a crude scheme, by calculating the total charge density of the surface through a superposition of atomic charge densities. They noted that the $\rho(\boldsymbol{r}, E_F)$ contours calculated for two gold surfaces bore strong resemblances to the total charge density contours used in interpretation of helium atom scattering. Although the superposition approximation had never been tested at the distances from the surface normally used in STM, they felt that the consequent ease of calculation made an attempt at application to STM worthwhile.

The approximation to the total charge takes the form

$$\rho(r) = \sum_{R} \phi(\boldsymbol{r} - \boldsymbol{R}) \tag{22.13}$$

where $\phi(R)$ is the charge density of a free atom, and the R are atom positions, not necessarily periodic. This is in fact a rather drastic assumption to make, since it assumes that the atomic charge densities can simply be summed at distances from the surface so great that only states near E_F make any contribution. The next assumption is to put

$$\rho(\boldsymbol{r}, E_F) \approx \rho(\boldsymbol{r})/E_0 \tag{22.14}$$

where $\rho(\boldsymbol{r})$ is the total charge. Expressing $\rho(\boldsymbol{r}, E)$ as

$$\rho(\boldsymbol{r}, E) \approx A \exp\{-h^{-1}(2mE)^{1/2}z\}, \tag{22.15}$$

and using

$$\rho(\boldsymbol{r}) = \int_{-\infty}^{E_F} \rho(\boldsymbol{r}, E)\, \mathrm{d}E, \tag{22.16}$$

one finds that $E_0 \approx E_F/\kappa z$ if A is assumed invariant with E over the energy range considered. Thus $\rho(\boldsymbol{r}, E_F)$ can be calculated in this approximation.

It might be expected that such a calculation would produce theoretical images bearing only qualitative resemblance to experimental STM images, since surface-induced electronic states are not included, and expression (22.13) reflects only the atomic structure of the surface. On the other hand, Tromp *et al.*[12] have found remarkable quantitative agreement between calculated and experimental images that has enabled them to establish once and for all the true nature of the reconstructed Si(111)$-(7 \times 7)$ surface. For each of five recent geometric models[13–17] of this surface, they calculated the STM image to be expected at an applied voltage of $+2$ V, and compared it with the observed image. The result is seen in Fig. 22.6. The model of Takayanagi *et al.*[13] is seen to fit experiment exceptionally well. Interestingly, in the context of the information derivable from STS, it is only at a positive tunnelling voltage that the experimental image reflects only the geometric structure. At other

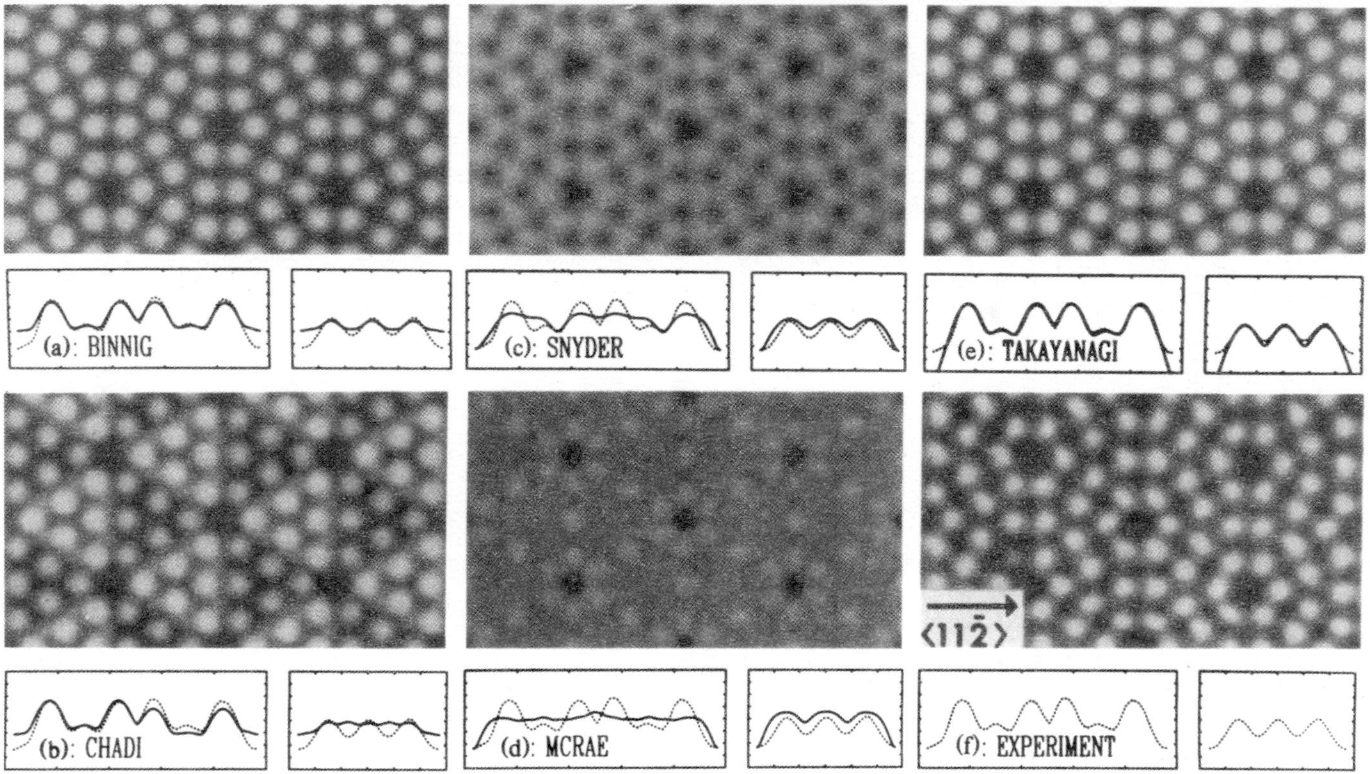

FIG. 22.6. Comparison of the STM images calculated by Tromp *et al.*[12] for several proposed models of the Si(111) surface (7 × 7) reconstruction, with the experimentally observed image. The model by Takayanagi *et al.*[13] shows almost exact agreement with experiment.

voltages, e.g. a negative bias of −2 V, the image looks different, as seen in Fig. 22.7, from Tromp *et al.*,[12] and in this case arises from asymmetries in the *electronic* nature of the unit cell. Such a comparison enables separation of geometric and electronic contributions to the image.

In summary, then, an STM image in either the standard constant-current or constant-height modes of operation will provide a direct topographical image of the surface *if* the variations in local electronic structure are themselves reflecting faithfully the variations in local topography. Many of the beautiful images do indeed appear to be of this nature. However, under certain conditions of applied voltage, surface reactivity, surface reconstruction, etc., the STM image will be a complex convolution of electronic and topographical features, and there will therefore almost certainly not be a one-to-one correspondence between the image and the topography. It is in just that situation that STS provides valuable additional chemical information.

22.3 Quantification

Just as in, say, UPS and IPES, the question as to whether an STS signal is quantifiable in any way with respect to surface concentration is largely irrelevant. The chemical information that has been sought in STS as in the above techniques has not been compositional in nature but electronic, i.e. in terms of local densities of states and such surface features as dangling bonds. It is difficult to see at the moment how any information from STS could be sufficiently element-specific to produce a surface composition, but that situation may change.

22.4 Applications

When attempting to study variations in electronic structure on a scale as localized as that of a single atom, it is self-obvious that the interpretation of images and spectra must not be complicated by the presence of major structural features such as steps, pits, etc., on the surface being studied. For this reason all applications to date of STS have been confined to materials that can provide virtually atomically flat surfaces, either by cleavage or by careful growth procedures. Even if the technique were always to be confined in its application to such materials the scope is extensive since most semiconductors used technologically are of that type.

22.4.1 *Comparison with other techniques*

Since, according to the voltage range, STS can give information on the densities of both filled and unfilled states near a surface, it is pertinent to ask as to how the ST spectra compare with those from associated techniques, in particular UPS and IPES (q.v.). The comparison may well not be exact. After

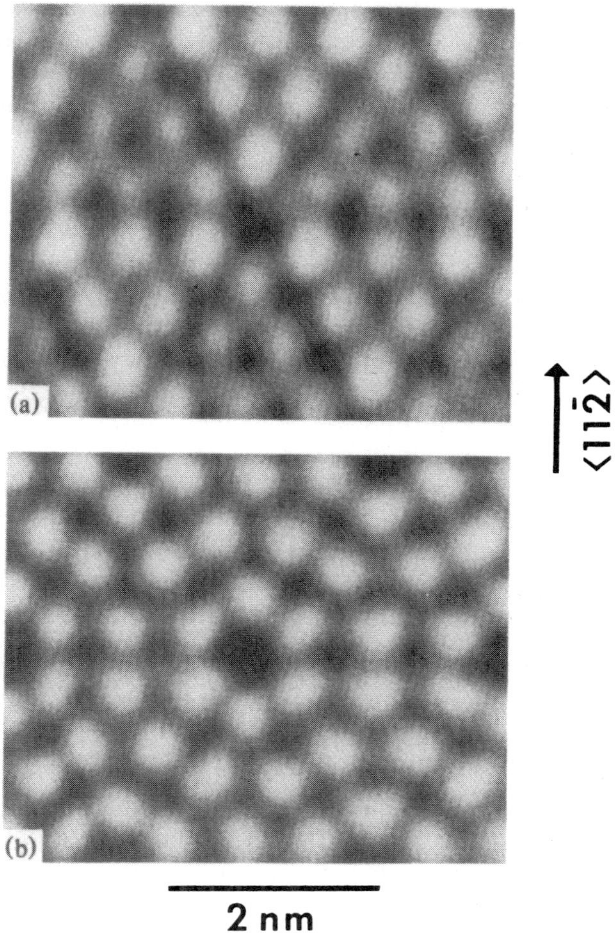

FIG. 22.7. Demonstration by Tromp *et al.*[12] that an STM image will look different according to the voltage applied, due to the different contributions of topographical and electronic variations at various voltages. Thus for the Si(111)$-(7\times7)$ surface at -2 V (a) the image is dominated by electronic asymmetries in the unit cell, whereas at $+2$ V (b) the image is largely topographical. Thus the comparisons in Fig. 22.6 were made for a positive applied voltage.

all, the coupling that takes place in STS is between the tunnelling electrons and surface states whose wave functions extend sufficiently far into the gap, and this will certainly favour states with significant density amplitude out to a few Å from the surface. In UPS and IPES, on the other hand, the inelastic mean free paths of the electrons are of the order of 30 Å. Nevertheless, if the bulk band structure is reflected strongly in the surface densities of states, there may well be agreement between the spectra from the different techniques.

Such an agreement between STS and UPS spectra was found by Kaiser and

Jaklevic[18] for the Pd(111) surface, as seen in Fig. 22.8. Panel (a) on the left shows the constant-resistance conductance vs. voltage spectrum, derived as described earlier, and panels (b) and (c) the experimentally determined[19] band structure, and the UPS spectrum,[20] respectively. The conductance peak at -1.4 V in (a) corresponds to the position of the d-band edge at about -1.2 eV in (b), and the conductance peak at 1.0 V in (a) to the saddle point at 0.94 eV in (b). In the UPS spectrum surface-state peaks are found at ≈ -0.5 eV and ≈ -2.1 eV in (c), with corresponding conductance peaks at -0.6 V and -2.1 V in (a). The identification of the latter with surface-states is supported by the observation that their intensities vary considerably from one position to another over the surface, while the band structure-derived features do not. The minor feature in (a) at ≈ 0.54 V is also variable and may be identifiable with a surface state as well.

Comparison of STS and IPES spectra has been performed for graphite by Reihl *et al.*[21] To achieve the requisite flatness on an atomic scale they used highly oriented pyrolytic graphite (HOPG), from which layers could be peeled off in vacuo by a knife operated through a bellows. Figure 22.9 shows their comparison. On the right are several IPES spectra for intercomparison, including one from polycrystalline graphite. The two central spectra are from different surface regions of HOPG, while the spectrum at the bottom is that of the surface in (b) after exposure to activated O_2 or H_2. Spectrum (b) is believed to be that most characteristic of the clean crystalline surface, since treatment with the activated gases caused loss of intensity around 2.5 eV, which was already missing in (c). The features at ≈ 1.7 eV and ≈ 3.5 eV are labelled $\pi^*(Q)$ and G, respectively, since the former can be identified with π^* bands split at the Q symmetry point in the bulk band structure, while the latter is allocated to non-direct transitions to the high density of states at the bottom of the interlayer band in the graphite band structure. The loss of intensity around 2.5 eV was postulated as due to a surface-sensitive feature in that region, probably a surface state.

The STS spectra provide confirmation of the above allocations. In this voltage range only bulk states or intrinsic surface states contribute to the tunnelling current, and image states of the type described later do not appear. The two spectra in the left hand panel of the figure are from different areas of HOPG. In spectrum (a) in particular, three peaks are seen at 1.7, 2.5, and 3.3 V, marked with π^*, an arrow, and G, respectively. With the increased sensitivity of STS to surface states, the feature at 2.5 V is more prominent than in the IPES spectra, confirming its identification as a surface state. The peak at 1.7 V is again a sharp feature as in IPES, and since it appears at exactly the same energy with respect to E_F, must once more be allocated to the π^* bands. If the feature at G had been due to surface states, not only would it have been extinguished by reaction with active gases, but it should also have increased in intensity relative to the bulk feature at 1.7 V on increasing the tip/surface

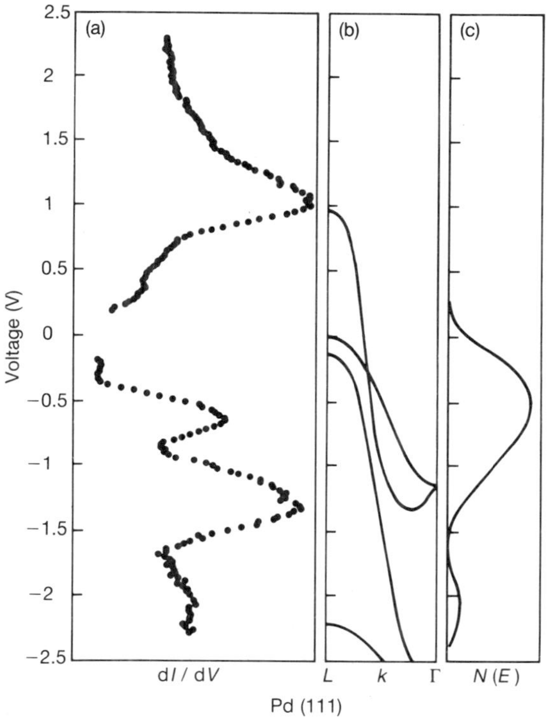

FIG. 22.8. Comparison by Kaiser and Jaklevic[18] of a scanning tunnelling spectroscopy spectrum from Pd(111), in panel (a), with the experimentally determined band structure in panel (b), and the UPS spectrum in panel (c). The STS spectrum was obtained as constant-resistance conductance vs. voltage, as described in the text. The conductance peak at −1.4 V correlates with the d-band edge at about −1.2 eV, and that at 1.0 V with the saddle point at 0.94 eV, in the band structure. Similarly, conductance peaks at −0.6 V and −2.1 V correlate with surface-state peaks at ≈ −0.5 V and ≈2.1 eV in the UPS spectrum.

separation; neither of these effects was observed. The allocation to another set of transitions within the bulk band structure is thus confirmed.

If the applied voltage range in STS is extended, it is, as remarked earlier, possible to enter the regime of field emission. When that happens, the STS spectrum at voltages beyond that corresponding to the work function has a more or less oscillatory appearance, as seen in Fig. 22.10 for several different surfaces, from Binnig *et al.*[22] This also has a counterpart in the IPES spectrum. For example, in Fig. 7.12 in Chapter 7, there is a feature labelled IS, for image states. As described there, these are hydrogen-like states due to the image of the interacting electron in the surface, which are bound to the surface but kept just outside it by the reflective properties of the surface. Binnig *et al.* have shown how their energies can be calculated assuming that electronic motion parallel and perpendicular to the surface is independent. In IPES the

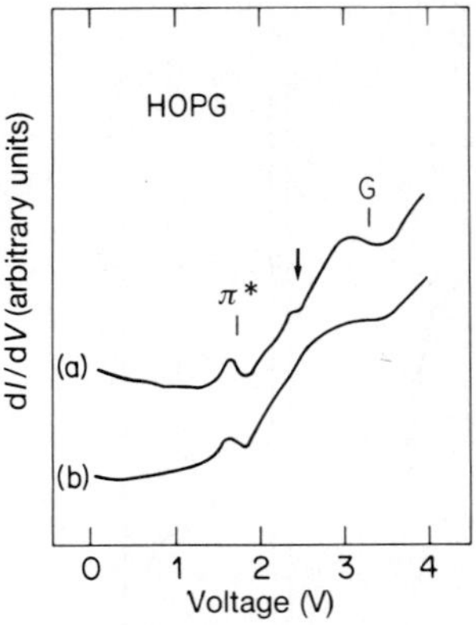

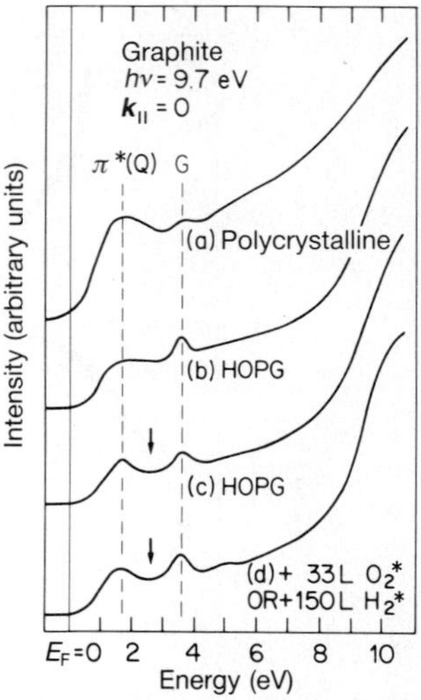

FIG. 22.9. Comparison of scanning tunnelling spectra and inverse photoemission spectra from graphite surfaces, by Reihl *et al.*[21] On the left, STS spectra from two different graphite areas, on the right, normal-incidence IPES spectra from several different graphite surfaces. HOPG refers to **h**ighly **o**riented **p**yrolytic **g**raphite. The features at ≈ 1.7 eV and ≈ 3.4 eV in both sets of spectra are identified with π^* bands and with non-direct transitions to the bottom of the graphite interlayer band, respectively. The feature arrowed in the IPES spectra cannot be resolved separately, but the intensity in that region is dependent on surface preparation and treatment, and can be reduced by reaction with activated gases (lowest spectrum). In the STS spectrum, however, with the greatly increased surface sensitivity, the feature is rather better resolved, and its allocation to a surface state is confirmed.

image states cannot be individually resolved since they are too close together in energy, but in STS the high electric field between tip and surface has a powerful effect on the dispersion of the hydrogen-like spectrum of states.

This effect is illustrated schematically in Fig. 22.11, from Binnig *et al.* On the left is the notional projected band structure of the Ni(100) surface, showing a band gap of 7.9 eV across the vacuum level. Next to it, (a), are shown, again notionally, a couple of hydrogenlike image states with principal quantum numbers $n=1$ and 2, in energetic positions corresponding to an undistorted surface structure and in the absence of any applied field. In (b) the surface structure is imposed, which has the effect of pulling the states down slightly. IPES records their positions in this condition, but without adequate resolution to distinguish them. When a field is applied between tip and surface, of 3 V nm^{-2} in the example (c), there is a Stark shift in the hydrogenic states, resulting in a continuous upward shift and an expansion along the energy axis. The reason that these shifted states are so readily picked out by STS is that there is a very pronounced dependence of the spatial extent of the image state wave function on its energy. Binnig *et al.* give the example that for the hydrogenic level with $n=5$ a change in binding energy of only 0.05 eV changes

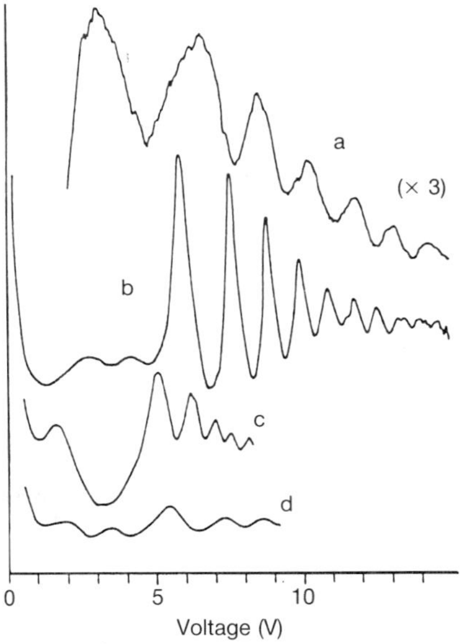

FIG. 22.10. Demonstration by Binnig *et al.*[22] of the oscillatory appearance of STS spectra when the applied voltage is extended into the field emission regime. The spectra relate to the following surfaces:

(a) low oxygen coverage on Ni(100)
(b) clean Ni(100)
(c) gold on Si(111) to coverage giving $(\sqrt{3} \times \sqrt{3})$ structure
(d) Si(111)$-(7 \times 7)$.

The oscillations arise from bound hydrogen-like states due to the image of the tunnelling electron in the surface, as explained in Fig. 22.11.

the energy position at which the state may be detected in a field of 3 V nm^{-1} by about 1 eV. What this amounts to, they say, is that the large ratio of spatial extent to energy range characteristic of a hydrogenic spectrum is converted into a large ratio of energy range to spatial extent by the high applied field. The hydrogenic spectrum is effectively spread out by a large factor along the energy axis, allowing its resolution by STS.

22.4.2 *Highly localized chemical reactions*

Techniques such as XPS, UPS, IPES, etc., can provide a great deal of valuable information about the nature of the interaction between a gas and a solid surface, but the information is averaged over a very large number of reaction sites. In many cases this does not matter since there may well be little variation in reactivity between sites, but in other cases, such as for example the surfaces of some semiconductors, there is likely to be such a variation. With the advent of STS there now seems to be the possibility available of studying the reactivity of individual sites, and some initial experiments along these lines have been

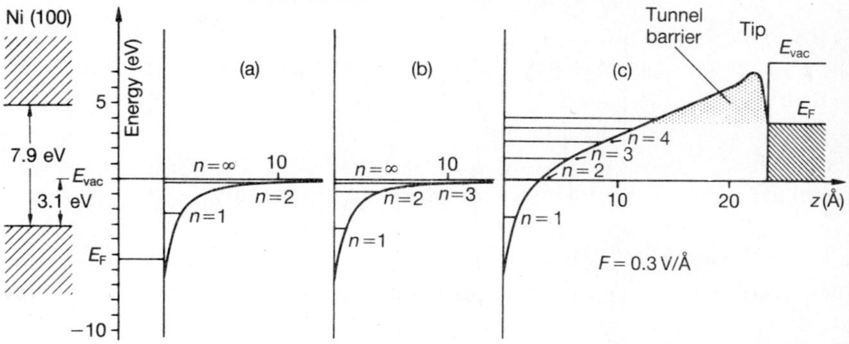

FIG. 22.11. Schematic energy diagram given by Binnig *et al.*[22] to explain the oscillatory behaviour of the STS spectrum often observed at high applied voltages, as seen in Fig. 22.10. On the left, the notional projected band structure of Ni(100). In (a), two notional hydrogen-like image states with principal quantum numbers $n = 1$ and 2, in the absence of an applied field and with an undistorted surface structure. In (b), the surface structure is established, which has the effect of pulling the states down in energy slightly. In (c), the field of about 3 V nm^{-1} is applied, which results in a strong Stark shift in the hydrogenic states upwards in energy. They also expand considerably along the energy axis, that is, the spatial extent of the image state wave function is expanded, which allows resolution by STS.

carried out by Wolkow and Avouris,[23] who studied the reaction of NH_3 with the Si(111)−(7 × 7) surface.

The currently accepted model for the well known (7 × 7) reconstruction of the Si(111) surface is that of Takayanagi *et al.*,[13] already mentioned earlier. In this model, Fig. 22.12, the surface unit mesh is based on regular missing atoms

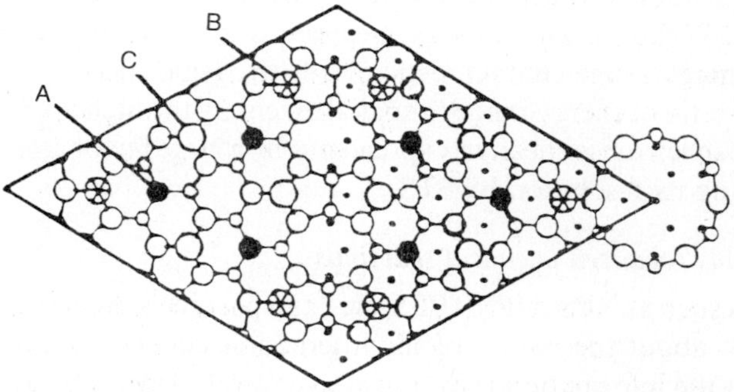

FIG. 22.12. Model of the (7 × 7) reconstruction of the Si(111) surface according to Takayanagi *et al.*[13] The surface unit mesh is based on silicon atoms missing in a regular array, with other silicon atoms, also in a regular array, lying on top of atoms in the underlying unreconstructed substrate. In the figure, atoms A are so-called 'rest atoms', i.e. part of the underlying layer visible in the surface, while atoms B and C are adatoms, with B atoms at 'corner' sites and C atoms at 'centre' sites.

at the corners, plus silicon adatoms, also in a regular array, on top of atoms in the underlying unreconstructed layer. The precise details are not of concern here, but the important feature is that a clean silicon surface has surface states associated with dangling bonds, and that it is quite likely that these states may have different occupancies (and hence reactivities) for atoms in different positions. In the Takayanagi model there are 19 dangling bonds in the (7×7) unit cell, 6 on the so-called 'rest atoms' which are part of the first layer but visible in the surface, 12 on the adatoms, and one on the atom at the bottom of the corner hole. In Fig. 22.12, the rest atoms are labelled A, while the adatoms are divided into two types, 'corner' adatoms labelled B, which are next to the corner holes, and 'centre' adatoms labelled C, comprising all the others.

A series of STM images of the Si(111)$-(7 \times 7)$ surface is shown in Fig. 22.13, from Wolkow and Avouris.[23] At the left, (a), is an image of the

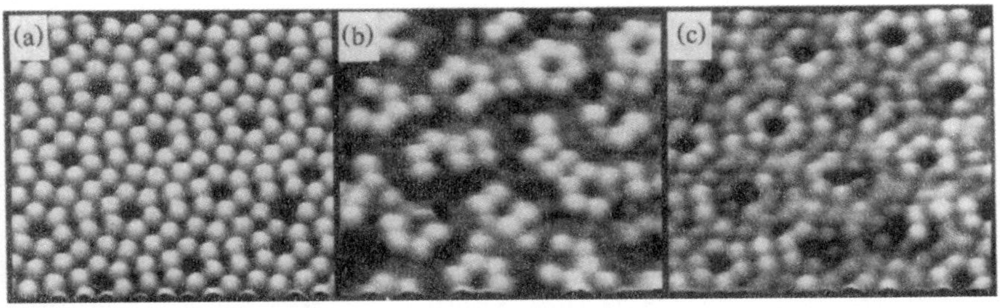

FIG. 22.13. A series of STM images of the Si(111)$-(7 \times 7)$ surface taken by Wolkow and Avouris[23] under different acquisition and surface conditions. (a) Clean surface image at an applied voltage of +0.8 V; (b) image, also at +0.8 V, taken after exposure to 1 L of NH_3; (c) same condition as (b), but taken at an applied voltage of +3.0 V. In (b) it seems that about half the adatoms are missing, but they reappear, although at different contrast, in (c). Thus the basic structure of the reconstructed surface has been preserved, but some *electronic* distortion has been introduced by the reaction with ammonia.

clean surface taken at an applied voltage of +0.8 V, which condition maps the local density of states up to 0.8 eV above the Fermi level. In the middle, the image (b) is from the same surface after exposure to 1 L of NH_3, also recorded at +0.8 V. Apparently about half the adatoms are missing if the image were to be interpreted purely on topography. On the other hand, it is known from other experiments that the reaction of NH_3 with silicon does not etch the silicon surface, but rather that the NH_3 dissociates with the formation of both Si–H and Si–NH_2 species on the surface. The apparent disappearance of some adatoms must therefore be an electronic effect, and indeed if the image is taken again at an increased voltage of +3.0 V, as in (c), all the adatoms become visible once more. Thus the basic structure of the reconstructed (7×7) surface has been preserved during the reaction, but some distortion has been introduced. When many comparisons of 'before-and-after' image sequences at different energies are made, it is found that most of the unreacted adatoms are the 'corner' adatoms. The rest atoms always react first with NH_3, followed by the 'centre' adatoms. Clearly such image comparisons can in themselves provide much fascinating additional information about variability in site reactivity, but the analysis can be taken further. Tunnelling spectra can be obtained from each type of atomic site, before and after reaction, a unique capability.

The left-hand side of Fig. 22.14, from Wolkow and Avouris, shows the tunnelling spectra obtained over each type of atom marked in Fig. 22.12, for the silicon surface in the clean state. There are marked differences between

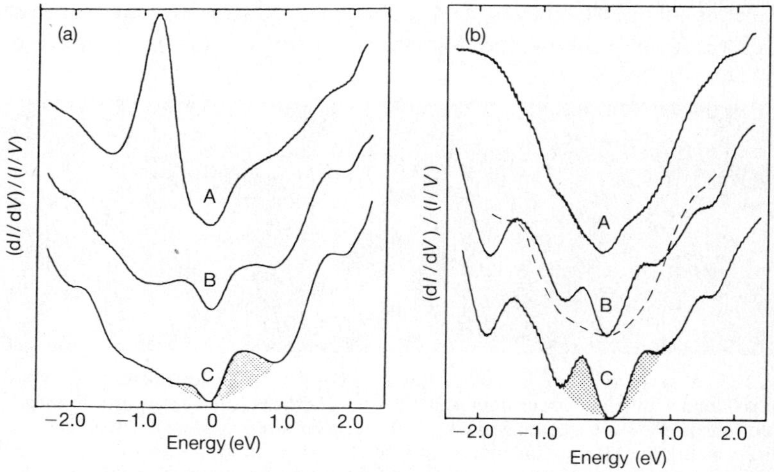

FIG. 22.14. (a) STS spectra recorded by Wolkow and Avouris[23] over each of the types of atom A, B, and C in the reconstructed Si(111) surface model of Takayanagi *et al.*[13] of Fig. 22.12. The strong band at ≈ -0.8 eV in the spectrum from A atoms suggests a fully occupied dangling bond state. On the other hand, the spectra from adatoms B and C show almost negligible occupied dangling bond intensity. The implication is that the dangling bond state in 'rest' atoms A is filled by charge transfer from adatoms B and C, particularly from C according to the spectra.
(b) STS spectra recorded after reaction with NH_3, over the same types of atom site as in the left-hand spectra. The spectra after reaction are those for A and for B (dashed line). The solid lines for B and C are from adatoms that were not *themselves* reacted, but were adjacent to reacted atoms. The intense dangling bond state at A has disappeared, and the same state for reacted B atoms has also disappeared. Note that in spectrum from C atoms adjacent to reacted atoms there is an indication of a transfer of charge *into* the dangling bond state, presumably from A, allowing the latter to react with NH_3.

them. At a rest atom site, A, there is a strong band at ≈ -0.8 eV, suggesting that the dangling bond state is fully occupied. Over a 'corner' adatom B, there are weaker occupied and unoccupied bands at ≈ -0.4 eV and $\approx +0.5$ eV, respectively, corresponding to the occupied and unoccupied portions of the dangling bond state at that site. For the 'centre' adatom C, the spectrum shows an almost negligible occupied intensity and a slightly increased unoccupied intensity, compared with B. The indications from these results are then that the rest atom dangling bond state is filled by charge transfer from the adatom states, particularly from those of the 'centre' adatom.

Following reaction with NH_3, tunnelling spectra from the same sites are shown in the right-hand side of Fig. 22.14. In the spectrum from the rest atom A it can be seen that the intense occupied band present before reaction has been eliminated, and the dashed line for reacted 'corner' adatoms B shows that from that site the occupied band has also been lost. This elimination of dangling bond surface states accounts for the apparent disappearance of the reacted atoms from the image at an applied voltage of only +0.8 V. More information can be gleaned from the spectra with solid lines in Fig. 22.14, for adatoms B and C. These spectra were recorded from 'corner' and 'centre' adatoms that were themselves unreacted, but were adjacent to reacted rest atoms. The change in the spectrum from C, compared with the 'clean' spectrum, is particularly significant, since there has been a transfer of intensity from the unoccupied part of the dangling bond state into the occupied part, suggesting that during reaction there is a charge transfer from a rest atom to an adatom, allowing the rest atom to react with the NH_3. Other new bands appear in the spectra from B and C sites, in Fig. 22.14, that can be interpreted by similar considerations of localized charge transfer.

These preliminary experiments are very important, because they demonstrate conclusively that the spatial distribution of reactivity can be followed on an atomic scale, and that the degree of localization or otherwise of such surface-specific features as dangling bond states, with associated charge transfer, can now be measured.

References

1. Binnig, G. and Rohrer, H. (1983). *Surf. Sci.* **126**, 236.
2. Hansma, P. K. and Tersoff, J. (1987). *J. Appl. Phys.* **61**, R1.
3. Baratoff, A., Binnig, G., Fuchs, H., Salvan, F., and Stoll, E. (1986). *Surf. Sci.* **168**, 734.
4. Demuth, J. E., Hamers, R. J., Tromp, R. M., and Welland, M. E. (1986). *J. Vac. Sci. Technol.* **A4**, 1320.
5. Binnig, G. Gerber, Ch., Stoll, E., Albrecht, T. R., and Quate, C. F. (1987). *Surf. Sci.* **189/190**, 1.
6. Staufer, V., Wiesendanger, R., Eng, L., Rosenthaler, L., Hidber, H. R., Güntherodt, H-J., and Garcia, N. (1987). *Appl. Phys. Lett.* **51**, 244.
7. Gimzewski, J. K., Reihl, B., Coombs, J. H., and Schlittler, R. R. (1988). *Z. Phys. B.* **72**, 497.
8. Baratoff, A., Binnig, G., and Rohrer, H. (1983). *J. Vac. Sci. Technol.* **B1**, 703.
9. Bardeen, J. (1961). *Phys. Rev. Lett.* **6**, 57.
10. Tersoff, J. and Hamann, D. R. (1985). *Phys. Rev. B* **31**, 805.
11. Selloni, A., Carnevali, P., Tosatti, E., and Chen, C. D. (1985). *Phys. Rev. B* **31**, 2602.

12. Tromp, R. M., Hamers, R. J., and Demuth, J. E. (1986). *Phys. Rev. B* **34**, 1388.
13. Takayanagi, K., Tanishiro, Y., Takahashi, M., and Takahashi, S. (1985). *J. Vac. Sci. Technol.* **A3**, 1502.
14. Binnig, G., Rohrer, H., Gerber, C., and Weibel, E. (1983). *Phys. Rev. Lett.* **50**, 120.
15. Chadi, D. J. (1984). *Phys. Rev. B* **30**, 4470.
16. Snyder, L. C. (1984). *Surf. Sci.* **140**, 101.
17. McRae, E. C. and Petroff, P. M. (1984). *Surf. Sci.* **147**, 385.
18. Kaiser, W. J. and Jaklevic, R. C. (1986). *IBM J. Res. Dev.* **30**, 411.
19. Himpsel, F. J. and Eastman, D. E. (1978). *Phys. Rev. B* **18**, 5236.
20. Demuth, J. E. (1977). *Surf. Sci.* **65**, 369.
21. Reihl, B., Gimzewski, J. K., Nicholls, J. M., and Tosatti, E. (1986). *Phys. Rev. B* **33**, 5770.
22. Binnig, G., Frank, K. H., Fuchs, H., Garcia, N., Reihl, B., Rohrer, H., Salvan, F., and Williams, A. R. (1985). *Phys. Rev. Lett.* **55**, 991.
23. Wolkow, R. and Avouris, Ph. (1988). *Phys. Rev. Lett.* **60**, 1049.

23

Thermal excitation: thermal desorption spectroscopy (TDS)

TDS has been used very extensively, but not as a general analytical tool in the same sense as have XPS or AES. In basic experiments concerned with the study of the reactions between pure gases and clean solid surfaces, TDS is now regarded as an indispensable adjunct to the other techniques used principally in fundamental surface science, such as UPS, LEED, and work function measurement. Thus TDS is to be found in use in virtually every academic laboratory carrying out research into the reactivity of surfaces, but not at all in analytical laboratories associated with technological or trouble-shooting activities. Note that TDS is also known as Temperature Programmed Desorption (TPD) and Temperature Programmed Mass Spectrometry (TPMS).

23.1 Operation

Of all the techniques described in this volume, TDS is in principle one of the simplest, in the sense that the number of necessary experimental components is reduced from three, namely probe–specimen–detector, to two, namely specimen–detector. No external probe is needed because the particles, in this case ions, to be detected are emitted from the specimen surface simply by raising the temperature. In its basic form an experiment consists of adsorbing gas on a surface at some chosen temperature, and then recording the current of a particular ionic species via a mass spectrometer tuned to that mass-to-charge ratio, as a function of temperature as the latter is increased at a strictly linear rate. An even more basic approach, and one used in the early days of TDS as well as in its predecessor, flash desorption, is to use an ionization gauge to measure changes in total rather than partial pressure. In that case, however, one would have to be sure that only one species was being desorbed.

Since the TDS experimental arrangement is so simple, almost trivial, no one ever bothers to provide a diagram of his particular arrangement, which would in any case be only a small part of a system containing several other techniques. However, there are possible artefacts that must be avoided. The most important of these is the desorption of gas from surfaces other than the desired one. Specimen heating is normally resistive, i.e. by passage of current through the specimen itself, which implies that the specimen resistance must be sufficiently high, and therefore that it must be in the form of a thin foil or a filament. The first obvious step in ensuring that gas is desorbed only from the specimen surface is therefore so to design the mounting of the specimen that the specimen itself is by far the hottest component during heating. Not only that, but the temperature distribution across the hot area should be as uniform as possible, otherwise there will be a continual overlapping of events occurring at different temperatures that will make interpretation difficult.

The two latter requirements are largely incompatible. If a specimen is mounted in such a way that it is the hottest part, then it is likely that the components in contact with the specimen are acting as heat sinks, and there will therefore be an inevitable temperature gradient between them. Various ingenious methods of mounting have been devised to minimize the problem, but in general the most successful approach seems to be to use a long thin ribbon of material so that the temperature gradient along a substantial length is sensibly zero, and then to accept ions into the mass spectrometer only from that section. An alternative approach is not to heat resistively but by electron bombardment from behind, in which case the specimen is mounted on, or suspended from, leads of low thermal conductivity such as thin wires.

Many workers emphasize the need to restrict the angle of acceptance into the mass spectrometer, irrespective of the method of specimen mounting, and such restriction is usually accomplished by placing the centre of the specimen a short distance (3–5 mm) from an aperture (1–2 mm) at the end of a drift tube. Obviously the specimen must not be too near the aperture otherwise desorption will take place from the aperture plate as well. The drift tube is attached to the end of the mass spectrometer and is about 25 mm long, and the only communication between the specimen and the mass spectrometer is via the aperture, the mass spectrometer being differentially pumped. Where differential pumping is not possible or too costly, a compromise on angular acceptance restriction is occasionally reached by interposing a variable aperture between specimen and mass spectrometer, on a retractable drive. However, for optimum background suppression it is generally advisable to shield the spectrometer in some fashion.

Temperature measurement is normally performed by means of a thermocouple on the back of the specimen, although infra-red pyrometry has been used on occasion. Since for proper thermal contact it is necessary to spot-weld the end of the thermocouple to the specimen, additional problems arise in positioning of the thermocouple. If the specimen is a single crystal, or in any case if contamination of the front face is to be avoided, the thermocouple cannot be positioned in the centre of the specimen but over to one side near the mounting or clamping arrangement. In such a position it will clearly not be recording the correct centre temperature, and its response will need to be calibrated by means of, for example, an optical pyrometer. To that extent the quoted temperatures of desorption have a built-in systematic error. The control of temperature in order to impose a strictly linear rate of rise throughout the range is of course accomplished by standard feedback circuitry controlling the power input either to the specimen if resistively heated or to the electron bombardment source. In most cases this linear temperature ramp is computer controlled, and the computer also allows multiplexing of the mass spectrometer, so that at least 10 mass-to-charge ratios can be recorded quasi-simultaneously. Heating rates employed have varied enormously from 0.7 to 70 K s^{-1}, but the usual range is between 1 and 15 K s^{-1}. The actual temperature range covered depends on the particular system being studied, i.e. on the binding energies involved, and would typically be between 300–600 K and 300–1000 K. Often adsorption is carried out at temperatures below room temperature in cases where there is no or insufficient adsorption at 300 K.

One or two other obvious experimental artefacts have to be avoided. The heating rate must be fast enough that the proportion of desorbed species accepted by the mass spectrometer is indeed characteristic of the desorption mechanism, that is, there must not be time for a significant adsorption and re-emission from other surrounding surfaces to interfere with the signal. Also the heating rate must not be so fast that the pressure rise around the specimen is so great that re-adsorption occurs. The effective pumping speed of the system near the specimen is important in this respect, and the heating rate has to be selected to match that pumping speed. However, these artefacts are easily avoided by good design.

23.2 Theory and interpretation

The dependence of the rate of evolution of adsorbed material from a surface, on the temperature, is given by the general Arrhenius equation

$$-\frac{dN}{dt} = k_m N^m \exp(-E_d/RT) \qquad (23.1)$$

where N = surface concentration of adsorbed particles per unit area
k_m = frequency factor
m = order of the desorption reaction
E_d = activation energy of the desorption process
T = absolute temperature.

Here $m = 1$ for a first order reaction in which the particles desorb independently of each other, and $m = 2$ for a second-order reaction in which the particles must combine before desorption and in which their recombination is rate-determining. It is possible to have fractional orders for complex desorption processes. Thus

$$\left(-\frac{dN}{dt}\right)_1 = k_1 N \exp(-E_{d_1}/RT) \qquad (23.2)$$

and

$$\left(-\frac{\mathrm{d}N}{\mathrm{d}t}\right)_2 = k_2 N^2 \exp(-E_{\mathrm{d}_2}/RT). \tag{23.3}$$

Now in TDS, there is always a sufficiently high pumping speed around the specimen compared with the desorption rate that the increase in pressure at any instant is proportional to the desorption rate. Thus a peak at some temperature, say $T_{\max}$, corresponds to a maximum desorption rate at that temperature. Maxima in the desorption rate can be derived by equating the second derivative $\mathrm{d}^2N/\mathrm{d}T^2$ to zero, giving

$$E_{\mathrm{d}_1}/RT_{\max}^2 = k_1(\mathrm{d}t/\mathrm{d}T)\exp(-E_{\mathrm{d}_1}/RT_{\max}) \tag{23.4}$$

for a first-order reaction, and

$$E_{\mathrm{d}_2}/RT_{\max}^2 = 2k_2N_{\max}(\mathrm{d}t/\mathrm{d}T)\exp(-E_{\mathrm{d}_2}/RT_{\max}) \tag{23.5}$$

for a second-order reaction. Note that the surface concentration N does not enter eqn (23.4), and that the desorption peak at $T_{\max}$ is not therefore coverage-dependent, a feature that is diagnostic of a first-order reaction. An example of such a spectrum is given in Fig. 23.1 from Kirstein *et al.*[1] for the

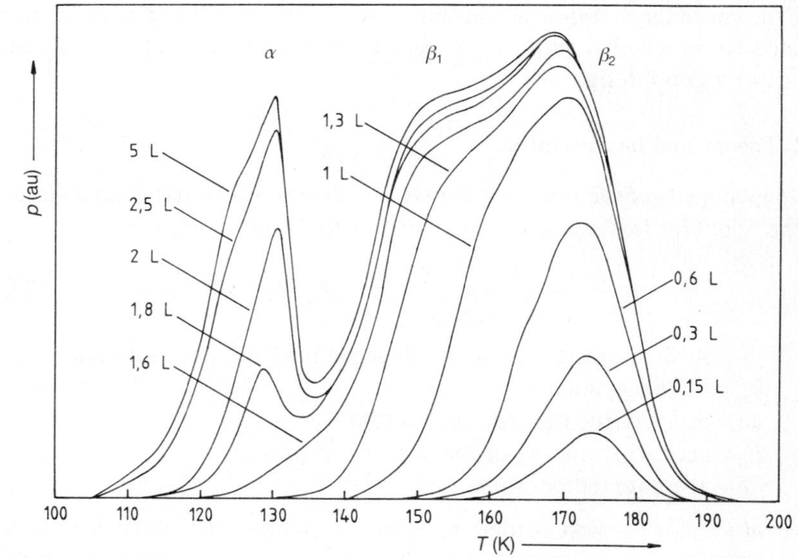

FIG. 23.1. The thermal desorption spectra of CO from Cu(111) after adsorption at 105 K, from Kirstein *et al.*[1] The heating rate was 0.7 K s^{-1}. The spectra show first-order desorption behaviour, i.e. the desorption peaks α, β_1, and β_2 are virtually independent of CO coverage.

desorption of CO from Cu(111) after adsorption at 105 K. Three desorption peaks $T_{\max}$ are found, labelled α, β_1, and β_2, developing in reverse order with increasing coverage of CO; the positions of $T_{\max}$ are almost coverage-independent, with β_2 showing just a slight reduction in $T_{\max}$ with coverage.

Since the heating rate (dT/dt) is known, and a reasonable guess can be made for k_1, values of the activation energies E_{d_1} can be extracted from the measurements. Alternatively the pre-exponential factor k_1 can be determined by a series of experiments at different heating rates, but such experiments do not seem to have been performed very often.

For second-order reactions, the form of eqn (23.5) shows that the position of T_{max} can be expected to change with surface concentration. Figure 23.2, from

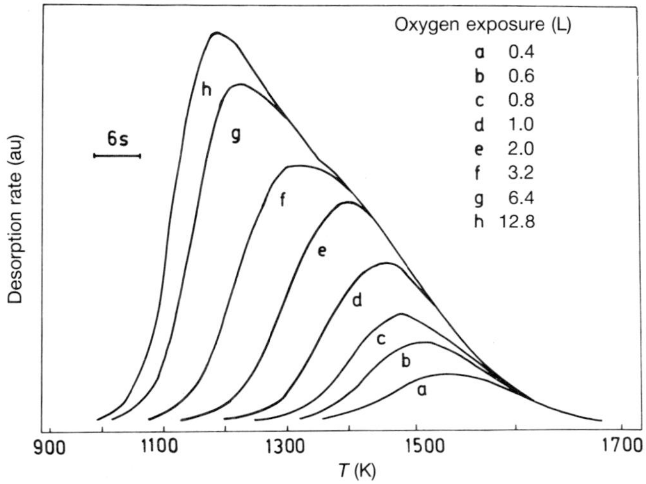

FIG. 23.2. Thermal desorption of oxygen from Ru(001) after adsorption at 880 K, from Surnev *et al.*[2] Heating rate unknown. The progressive downward shift in desorption peak maximum temperature from exposures a to e is characteristic of second-order desorption kinetics. At higher exposures the desorption peak is more complex, but still has second-order characteristics.

Surnev *et al.*,[2] shows such a variation with coverage, at least up to a coverage corresponding to an oxygen exposure of ≈ 2 L (i.e. spectra a to e). Using parallel AES measurements to calibrate the areas under the desorption curves in terms of oxygen concentration, the activation energy for the initial state was found to be approximately 400 kJ mol^{-1}. As can be seen in Fig. 23.2, at higher oxygen exposures the desorption curves changed shape, indicating a more complex, multistate desorption, but still with second-order kinetic characteristics. Another approximate activation energy for the second state appearing at higher exposures was estimated as 315 kJ mol^{-1}.

Interpretation of thermal desorption spectra is rarely as straightforward as for those in Figs 23.1 and 23.2. Use of eqns (23.4) and (23.5) without modification implies that the rate constant k is independent of the coverage. There are by now many systems studied for which it is realized that this constancy is not correct; for example k can vary with coverage according to any lateral interactions between adsorbed species as the coverage increases. If this variation were significant, the erroneous conclusion might be reached that the desorption was following second-order kinetics rather than first-order. In addition, it is possible for additional structure to appear in a desorption peak, apparently deriving from desorption from another site, when in fact the desorbing atoms are still all coming from the same site.

There is another parameter that can assist in interpretation, and that is the desorption peak shape, or symmetry. Based on an original analysis by Redhead,[3] Madix[4] has given a schematic indication, seen in Fig. 23.3, of the

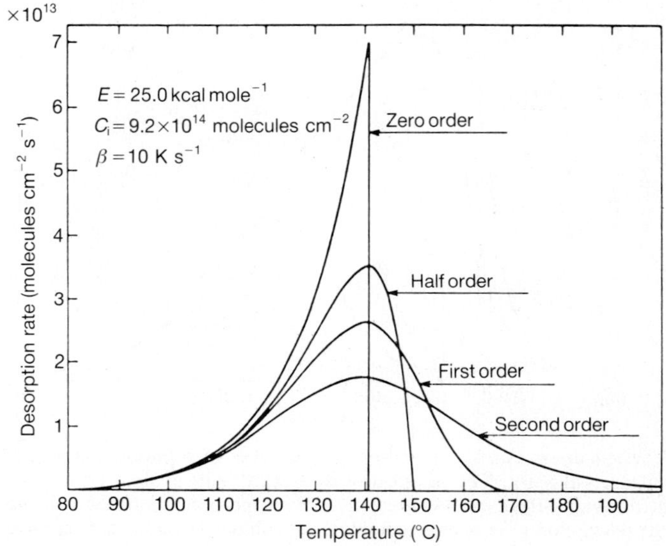

FIG. 23.3. Schematic indication by Madix[4] of the dependence of desorption peak shape on the order of the desorption reaction. Compare the predicted shapes for first and second-order reactions with the experimental shapes in Figs 23.1 and 23.2, respectively.

peak symmetries to be expected for desorption reactions of various orders. The peaks are of course shown on a zero background, whereas in practice they would be superimposed on some sort of background. To be of any use, such an analysis can be applied only to well resolved peaks thought to be free of overlaps and experimental artefacts. Examples of such shapes can be found in the literature in sufficiently well resolved spectra to be recognizable. For instance, Fig. 23.4, from Zhou *et al.*,[5] for the desorption of iron from tungsten (notice the unusually high temperature range for this system) shows how the desorption from several layers of iron is consistent with zero-order kinetics (peak β_1). Peak β_2 corresponds to desorption from a monolayer of iron. The activation energy, 80 kcal mol^{-1}, calculated from the spectra, is close to the known heat of sublimation of iron, 84.6 kcal mol^{-1}. Pure first-order reaction

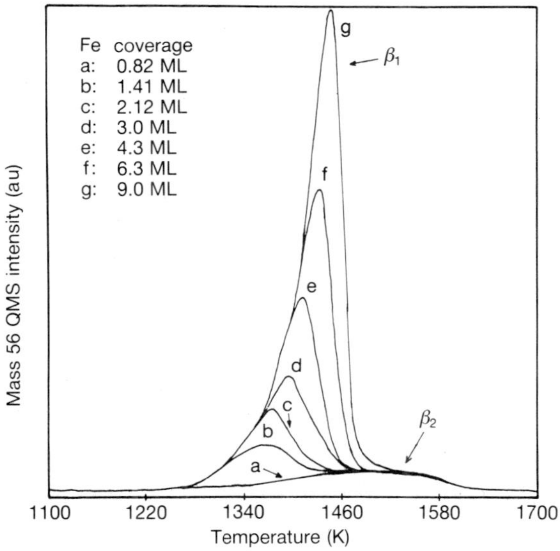

FIG. 23.4. Thermal desorption spectra of Fe from W(100) after deposition of Fe at 200 K, from Zhou *et al.*[5] The heating rate was 28 K s^{-1}. The shape of the desorption peak (cf. Fig. 23.3) is characteristic of a zero-order desorption reaction.

kinetics are shown by the low-coverage regime of the adsorption of CO on Ni(110), demonstrated in Fig. 23.5, from the work of Falconer and Madix;[6] compare the peak shapes with the predicted shape in Fig. 23.3. The activation energy for the state giving rise to that peak was calculated to be 32.7 kcal mol^{-1}. Finally, compare the predicted peak shape for a second-order reaction in Fig. 23.3 with those already shown in Fig. 23.2. It can be seen that the symmetric shapes of spectra a–e in the latter are indeed what would be expected of a second-order process.

Other interpretational difficulties can sometimes intrude. Although the aim of a TDS experiment is to study the nature of the binding states existing *at the temperature of adsorption*, it has to be borne in mind constantly that the very act of increasing the temperature, at whatever rate, may itself change the nature of the binding. Thus at the higher temperatures reached, it may be that the desorption seen is characteristic of the states existing at some intermediate temperature. In addition, for some materials an increase in temperature beyond some critical temperature may cause an irreversible change in surface structure, either by a transition from one crystallographic phase to another, or by changing the defect concentration and distribution. Thus, as with most surface-specific techniques, it is not sufficient in any one study to rely on TDS alone, but to use it as part of a team of techniques providing complementary information.

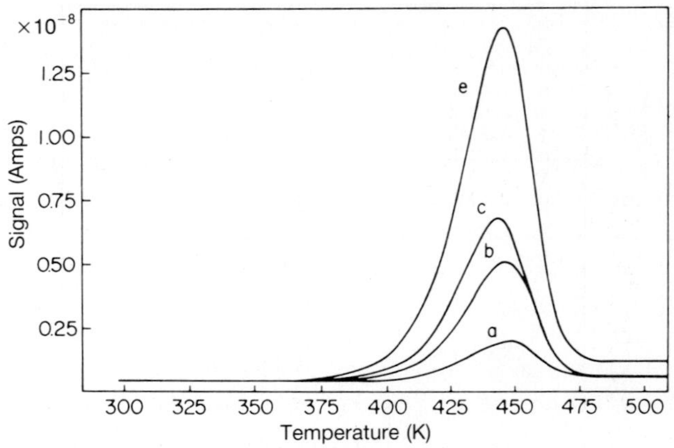

FIG. 23.5. Thermal desorption spectra of CO from Ni(110) after adsorption to low coverages (below 0.04 monolayers) at room temperature, measured by Falconer and Madix.[6] The peaks show pure first-order behaviour in both shape and coverage dependence.

23.3 Quantification

In addition to providing information about the activation energies of desorption and about the order of a desorption reaction, TDS can also give quantitative information about the amount of material adsorbed. Indeed, in amenable situations the information can be accurate, but is accurate in a relative rather than absolute sense, as will be seen below.

Since, if the temperature is increased far enough, all the adsorbed species of a particular type will be desorbed, and since the mass spectrometer is recording directly the number of ions of that species leaving the surface at each instant, it follows that the area under the desorption peak for that species will be a measure of its surface concentration. This area is simply $\int p(t)\,dt$, where $p(t)$ is the partial pressure of the species as a function of the time t after the start of desorption; naturally, this assumes that the absolute temperature T is a strictly linear function of t. Then the relative coverage of the species at a time t_x after the start of desorption is given by

$$\theta_{\text{rel}}=\theta_x/\theta_{\text{sat}}=\int_{t_x}^{\infty} p(t)\,\mathrm{d}t \bigg/ \int_0^{\infty} p(t)\,\mathrm{d}t. \tag{23.6}$$

Infinite temperature is equivalent, of course, to the temperature at which the desorption rate has become zero.

Just as with consideration of peak shape, so in the integration of desorption peaks must it be necessary that the peak being integrated is well resolved and clear of interference from neighbouring peaks. An example of a set of spectra in which integration is straightforward, except at very low coverages, is that of

Fig. 23.6 from Paul,[7] for the adsorption of hydrogen and deuterium on Al(100) at 90 K. Only a simple peak appears at all coverages, superimposed on a flat background. At low coverages the signal-to-noise ratio is poor, which makes precise integration difficult. The raw peak areas are plotted as a function of hydrogen dose in Fig. 23.7, which shows that the desorption of hydrogen or deuterium follows the same course, i.e. there is no isotope effect.

Also indicated in Fig. 23.7 on the right-hand scale is the coverage of hydrogen or deuterium established from the TDS results. Now, examination of eqn (23.6) shows that the coverage can be determined only if θ_{sat} is known, i.e. the saturation coverage at the temperature of adsorption. There is no route by which a value for θ_{sat} can be reached directly by TDS, because the mass spectrometer accepts only a small fraction of the desorbed gas, and in addition the transmission characteristics of the instrument would have to be known very accurately. θ_{sat} has to be derived from measurements using other techniques that are able to provide a calibration, and those normally used are two or more of AES, XPS, and LEED. Most TDS experiments are carried out with single crystals, and observations made with LEED are therefore particularly useful. However, in the case of the results in Fig. 23.7, the hydrogen calibration was achieved by Paul[7] from the decomposition of monolayers of sodium or potassium hydrides or deuterides. Paul and Hoffmann[8] have shown that exposure of a monolayer of sodium or potassium on Al(100) to atomic hydrogen (or deuterium) produces the stoichiometric hydrides or deuterides. The total hydrogen or deuterium desorbed as a result of the decomposition of the NaH or KH layers could then be used as the calibration for θ_{sat}.

Another method of calibration of a TDS signal was used by Cameron and Dwyer.[9] They were studying the effect of the presence of small quantities of potassium on an Fe(100) surface, on the adsorption of CO. In this case LEED could not be used directly since potassium does not form ordered structures on the Fe(100) surface. Instead, they used one of their own[10] measurements on the saturation coverage of CO on pure Fe(100) at 120 K, determined self-consistently by combined XPS, TDS, and LEED, from which the C 1s XPS signal intensity corresponding to exactly one monolayer of CO could be extracted. Since the C 1s and the K 2p XPS energies are within 10 eV of each other, analyser transmission differences could be neglected, and the ratio of cross sections K 2p/C 1s was obtained directly from a K_2CO_3 standard. Thus the K 2p signal corresponding to a monolayer of potassium was established.

The calibration of potassium TDS signal against the K 2p XPS signal was carried out by first adsorbing potassium on the clean Fe(100) at low temperature so that a multilayer deposit was formed. From this deposit the TDS spectrum at a linear desorption rate of 10 K s^{-1} is shown in Fig. 23.8. The multilayers of potassium beyond a monolayer all desorb in the same temperature range, giving rise to a sharp peak at 330 K, leaving a monolayer

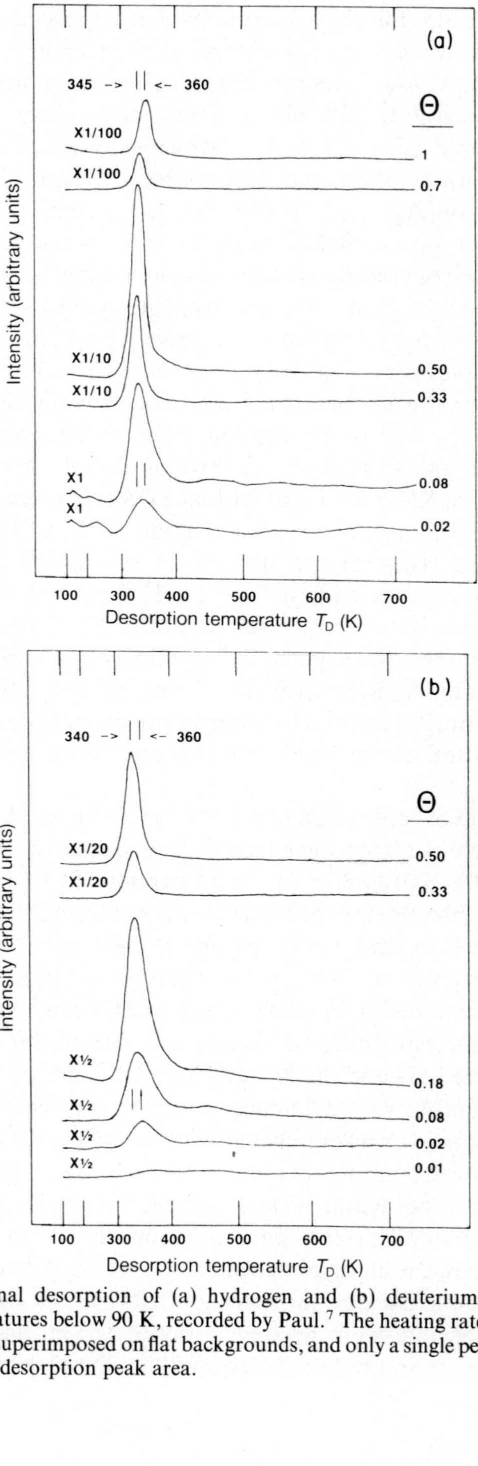

FIG. 23.6. The thermal desorption of (a) hydrogen and (b) deuterium from Al(100) after adsorption at temperatures below 90 K, recorded by Paul.[7] The heating rate was 1.3 K s^{-1}. The desorption peaks are superimposed on flat backgrounds, and only a single peak appears, allowing easy measurement of desorption peak area.

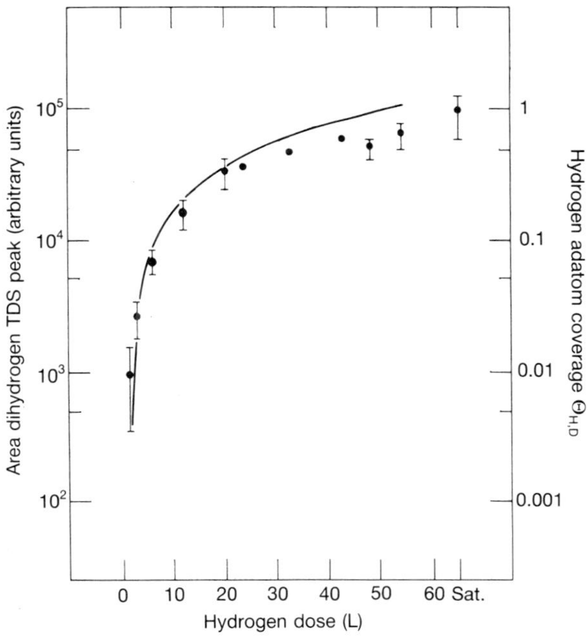

FIG. 23.7. Variation of peak area of the desorption peaks in Fig. 23.6 with dose of hydrogen or deuterium. The coverage scale on the right-hand side was calibrated from the decomposition of sodium and potassium hydrides, as described in the text. The solid line is a least-squares fit in the low-dose range. From Paul.[7]

which desorbs slowly up to 900 K. The monolayer desorption region is shown on an expanded scale, with a continuous decaying background from the multilayer peak, and is repeated in Fig. 23.9. The solid line in that figure is the result of applying eqn (23.6) to the K desorption data, that is, by integrating the area under the desorption curve above each desorption temperature, and then by dividing the result by the value of θ_{sat} derived in the calibration described above. This was performed for each of a series of annealing temperatures, and the fractional coverage of potassium so determined at each temperature is compared with the XPS-derived values. The agreement can be seen to be very good, implying that the calibration procedure was substantially correct.

23.4 Applications

A glance at the recent literature in which TDS has been used in conjunction with other surface-specific techniques shows that at least 75% of publications have been concerned more or less directly with catalysis-related topics. The fact that most of this work has been carried out on single-crystal surfaces

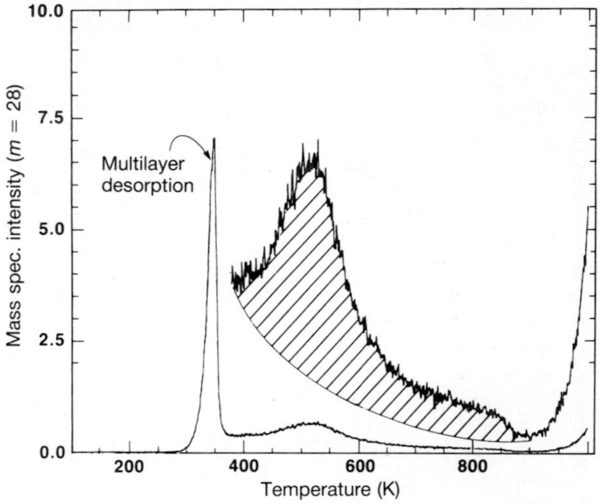

FIG. 23.8. Thermal desorption of K from Fe(100) after adsorption at ≈ 100 K, observed by Cameron and Dwyer.[9] Heating rate was 10 K s^{-1}. The spectrum shows a sharp peak at 330 K associated with desorption from multiple layers of K, followed by a slowly varying monolayer desorption continuing to 900 K.

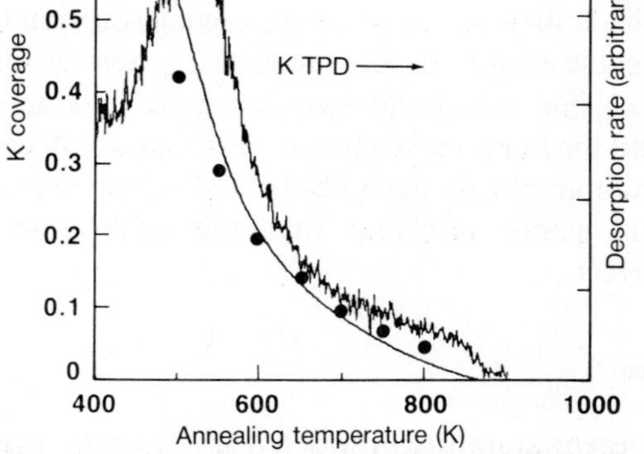

FIG. 23.9. Plot by Cameron and Dwyer[9] of the integrated area under the monolayer component of Fig. 23.8, as a function of annealing temperature of the deposited K film. The solid circles are results derived from XPS measurements. The calibration in terms of K coverage was carried out by the method described in the text.

might seem to cast doubt on its relevance, until one remembers that most active catalytic species are in a finely dispersed form consisting almost certainly of microcrystallites. It is hardly surprising that TDS should prove so useful in this area of study since it is able to give information not only about the nature of the products of reaction between a catalysing surface and one or more adsorbing gases, but also about the strength of their binding to the surface and about their concentration. In this respect TDS provides an essential contribution to the interconnected network of information from several techniques, that enables a self-consistent picture of the mechanism of a reaction to be constructed.

The two examples chosen here for illustration, from the many, are both concerned in different ways, with interactions with catalysing surfaces.

23.4.1 *Trace impurity effects on reactions*

Platinum is very widely used in heterogeneous catalysis, as well as in electrochemical and corrosion studies, and at some stage in the reactions involved in these processes water is bonded to the platinum surface. The nature of this bonding is thus of great interest. The reaction of water vapour with clean platinum has been studied extensively, with general agreement about the non-dissociative character of the water adsorption, but occasionally showing effects not fitting into the accepted pattern. Some of these effects were found and then studied by Griffiths and Bonnett.[11] In particular, according to the state of the Pt(111) surface, *dissociative* desorption of water was found to occur.

The authors showed that there were two distinct conditions of the Pt(111) surface as revealed by TDS. In one, termed by them the 'clean' condition, no impurities were present as judged by AES following the standard cleaning procedures of cyclic argon sputtering and heating in low pressures of oxygen. In the other, termed the 'oxidized' condition, the crystal was flashed (i.e. very rapidly heated) several times to 1273 K in 10^{-5} Pa (7.5×10^{-8} torr) O_2, followed by a flash to 1273 K in vacuo. In the latter a small peak due to oxygen could be seen in the AES spectrum, corresponding to 9% of a monolayer, and that oxygen could be removed only by treatment much more severe than that (1000 K) normally required to remove chemisorbed oxygen from platinum.

The TDS spectra from these two surface conditions are shown in Fig. 23.10, as a function of exposure to rigorously dissolved-oxygen-free D_2O at 131 K. The heating rate was quite low, 2 K s^{-1}. In both spectra the peak due to multilayers of D_2O, i.e. ice, appears at exposures beyond 60 s, and then grows continually. However, in the early, fractional monolayer, stages of adsorption, there are marked differences between the spectra. On the 'clean' surface a desorption peak appears first of all at 195 K, but quickly saturates, followed by the appearance of another at 184 K that grows without appreciable shift and also saturates after 2 min. On the 'oxidized' surface, on the other hand, the initial peak appears at 168 K and shifts progressively to 175 K at saturation,

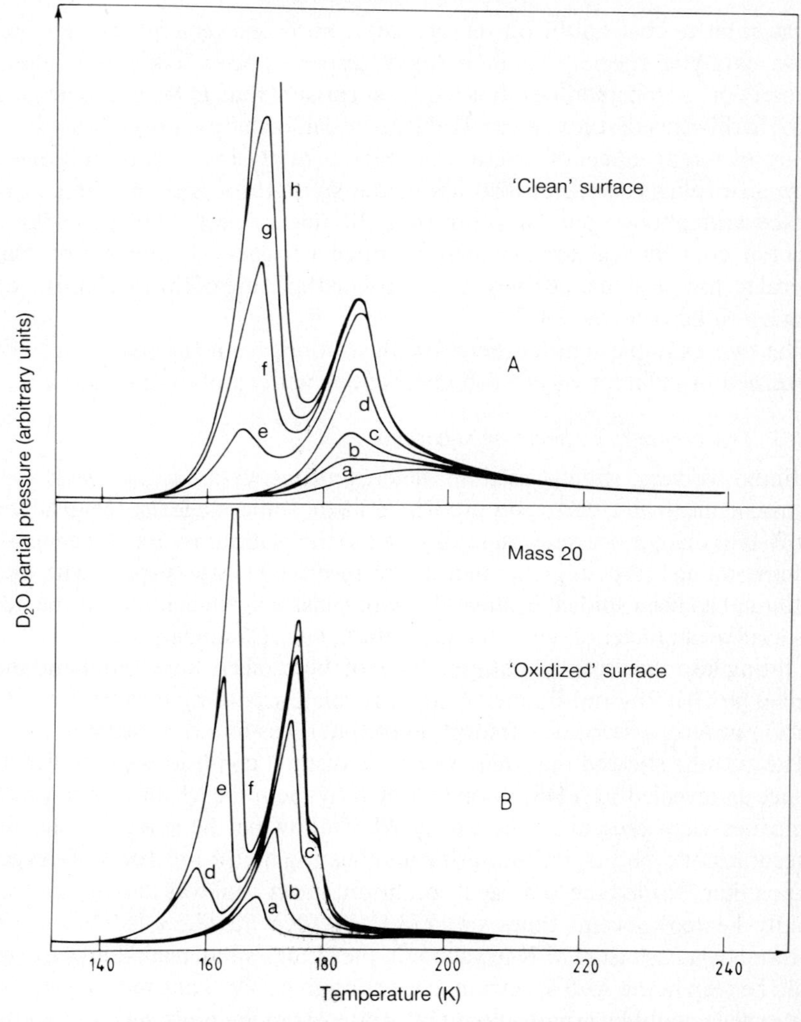

FIG. 23.10. Thermal desorption spectra of D_2O desorbed from Pt(111) surfaces in two different conditions (see text) after adsorption at 131 K, measured by Griffiths and Bonnett.[11] The heating rate was 2 K s^{-1}. The exposures to D_2O are:
A (a) 1 L, (b) 1.3 L, (c) 1.5 L, (d) 2 L, (e) 3 L, (f) 4 L, (g) 5 L and (g) 6 L.
B (a) 0.5 L, (b) 1 L, (c) 2 L, (d) 3 L, (e) 4 L and (f) 5 L.
Both sets of spectra show a desorption peak due to ice multilayers which grows continually with exposure. On the 'clean' surface, the first peak appears at 195 K, but saturates quickly, followed by another at 184 K, that grows without shifting and also saturates. On the 'oxidized' surface, on the other hand, the initial peak is at 168 K and shifts to 175 K before saturation. A shoulder appears on the latter at ≈ 180 K. These features are discussed in the text.

again after about 2 min; a shoulder also appears at ≈ 180 K. The shape of the latter peak is also different from that on the 'clean' surface, being narrower and having a sharp cut-off on the high temperature side. This shape and behaviour is characteristic of a fractional order desorption process, whereas the behaviour on the 'clean' surface is more characteristic of a first-order reaction.

Further evidence from TDS included the observation that if, after adsorption of water on the 'clean' surface, the crystal was not flashed to 1373 K but only to 1273 K, a shoulder appeared at ≈ 180 K on the monolayer desorption peak, and grew at each adsorption cycle, until eventually the desorption spectrum of the 'oxidized' surface was obtained. In addition, desorption of D_2 as well as D_2O was observed. Clearly the D_2O was dissociating to a significant extent and the residual oxygen was converting the surface into the 'oxidized' state. The dissociative nature of the adsorption on the 'clean' surface is confirmed by the observation of first-order desorption kinetics, since to desorb D_2O would require the recombination of D_2 and oxygen. On the 'oxidized' surface the fractional-order behaviour is that commonly observed for D_2O desorption from genuinely clean platinum surfaces.

Since neither AES nor XPS analysis of the 'clean' surface revealed any feature due to elements other than platinum, Griffiths and Bonnett had recourse to SSIMS analysis, which has much higher sensitivity for many elements. In addition to showing traces of the alkali metals and calcium, common contaminants for which SSIMS is exceptionally sensitive, the analysis also revealed traces of silicon. Sputtering the crystal removed the silicon, but annealing at 1200 K caused it to reappear in the SSIMS spectrum.

The presence of minute amounts of silicon on or near the surface would explain the anomalous TDS results. Although a genuinely clean platinum surface does not dissociate water, it is known that a silicon surface does dissociate it to form Si–H and Si–OH bands. When a water-covered silicon surface is heated silicon oxide is formed, which is stable to much higher temperatures than platinum oxide. Thus on the 'clean' surface the silicon traces are largely controlling the nature of the interaction of water with the platinum, whereas on the 'oxidized' surface, the silicon has been rendered inert with respect to interaction with water by virtue of complete oxidation. Since the silicon reappears at the surface on heating after sputtering, it is obviously segregating to the surface from a reservoir in the bulk.

23.4.2 *Catalytic oxidation of ethylene*

The oxidation of ethylene to ethylene oxide is an important industrial process for which silver has unique catalytic properties. In order to understand more fully the mechanism of the oxidation reaction it is necessary to have information on the nature of the adsorption of both ethylene and ethylene oxide on silver surfaces. Such information was sought by Krüger and Benndorf[12] in their study of the interaction of ethylene and ethylene oxide with Ag(110), using TDS, LEED, UPS, and work function measurements.

Initial adsorption was performed at 110 K, and the heating rate in TDS was 4.5 K s^{-1}.

Desorption spectra for ethylene oxide on Ag(110) for exposures up to 13.3 L are shown in Fig. 23.11, from their paper. At first, a single peak, labelled A, appears at 170 K, and increases to saturation at 3 L, at which stage it has shifted downwards slightly to 160 K. Beyond the saturation of A two other peaks appear, at 140 K (labelled B) and 133 K (labelled C). Peak C increases continuously with exposure without saturation, and is clearly due to multilayer condensed ethylene oxide. Since the saturation of the A state coincides with the appearance of a sharp $c(2\times2)$ pattern in LEED, the coverage at which A saturates is $\theta=0.5$. Compared with A and C, the intensity of the B peak (i.e. the population in the B state) is low. In the inset to Fig. 23.11b is shown the integrated desorption peak areas of peaks A to C as a function of ethylene oxide exposure. The coverage changes linearly with exposure, and there is no change in slope during the appearance of peaks B and C following saturation of A, indicating that the sticking probabilities in sites A, B, and C are identical.

The nature of the desorption of ethylene itself from Ag(100) is quite different, as can be seen from Fig. 23.12, also from Krüger and Benndorf. Only one desorption peak, or state, is observed, which starts at 148 K at low coverages and shifts progressively downwards to 123 K, broadening considerably as it does so. No multilayer peak is observed for ethylene, showing that it cannot be condensed at 100 K, the temperature of adsorption. The saturation coverage corresponds to $\theta=0.5$, since at saturation the LEED Pattern is again $c(2\times2)$. Up to $\theta\approx0.4$, the coverage increases linearly with exposure, as seen in the inset to Fig. 23.12, indicating once more a constant sticking probability in that range.

Since masses 29 and 28 (see Fig. 23.11 and 23.12) are characteristic of molecular ethylene oxide and ethylene, respectively, it is clear that molecular adsorption occurs on Ag(110) in both cases. (Although mass 29 is not the parent mass of C_2H_4O, it is the major fragment after cracking in the mass spectrometer ionization source.) On the other hand, the nature of the adsorption is different, since the initial ethylene oxide desorption peak is at 170 K, while for ethylene it is at 148 K, i.e. the binding of ethylene oxide to the silver is stronger. The greater strength of the ethylene oxide–silver binding is also confirmed by the appearance of peak B in the desorption spectra of Fig. 23.11, showing that it is possible for the oxide to take up additional adsorption sites that ethylene cannot.

Further information of a confirmatory nature came from the measurements of change in work function during adsorption, shown in Fig. 23.13. For both molecules, a *decrease* in work function was recorded as a function to exposure. Such an effect might be expected for ethylene adsorption, involving donation of electrons from the π-bond in the molecule to the metal, whose d-band is

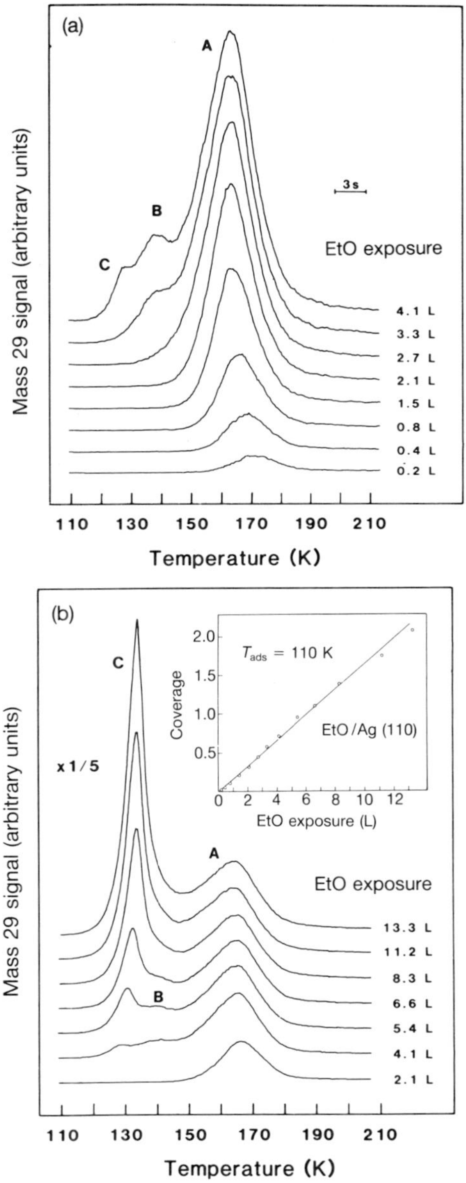

FIG. 23.11. Thermal desorption spectra of ethylene oxide from Ag(110) after adsorption at 110 K, from Krüger and Benndorf.[12] The heating rate was 4.5 K s^{-1}, and the maximum exposure was 13.3 L. A single desorption peak A appears at first at 170 K, which saturates at ≈ 3 L after shifting downwards to 160 K. Beyond that point two other peaks B and C appear. C increases continuously with exposure and is due to multilayer condensed ethylene oxide. The intensity of B, at ≈ 140 K, is low. The spectra are discussed in the text. Inset: Integrated peak areas converted to coverages of ethylene oxide.

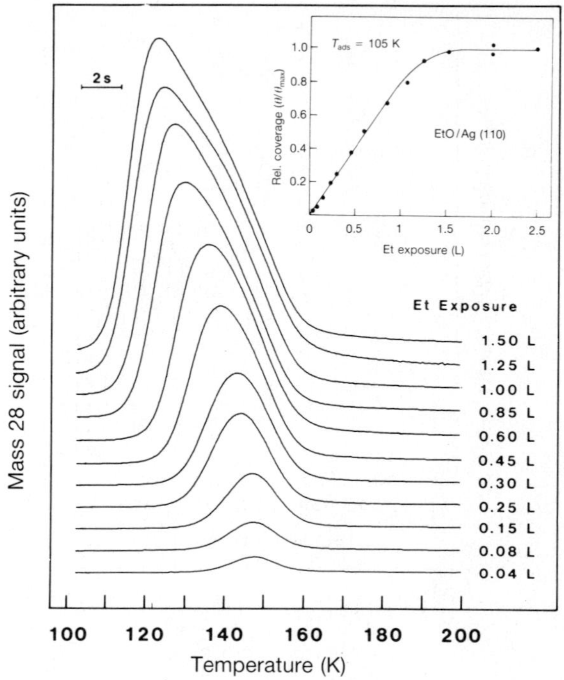

FIG. 23.12. Thermal desorption of ethylene from Ag(110) after adsorption at 110 K, from Krüger and Benndorf[12] up to an exposure of 1.5 L. Heating rate was 4.5 K s^{-1}. Only one desorption peak is observed, that shifts downwards progressively from 148 K to 123 K, and broadens. No multilayer peak is observed, and the coverage saturates at $\approx$1.5 L, as shown in the inset, unlike the behaviour of ethylene oxide.

empty near the Fermi level, leading to a dipole polarized positively outward. For ethylene oxide, however, observation of a decrease in work function leads inescapably to the conclusion that the molecule is bonded to the surface via the oxygen atom, with the axis of the molecule perpendicular to the surface. Only with such an orientation, i.e. with the negative end of the molecular dipole attached to the surface, could a decrease in work function ensue. Furthermore, such a bonding, through the oxygen atom, would indeed be stronger than that of ethylene by π-electron sharing, and would also allow multilayer formation. The broadening and shifting of the ethylene desorption peak in Fig. 23.12 has the characteristics of a second-order desorption reaction, but is more likely, according to the authors, to be a result of increasing lateral repulsion between the adsorbed ethylene molecules.

Thus the significant differences between the desorption behaviours of ethylene and ethylene oxide on silver can be attributed both to the nature of the bonding and to the orientation of the molecule with respect to the surface.

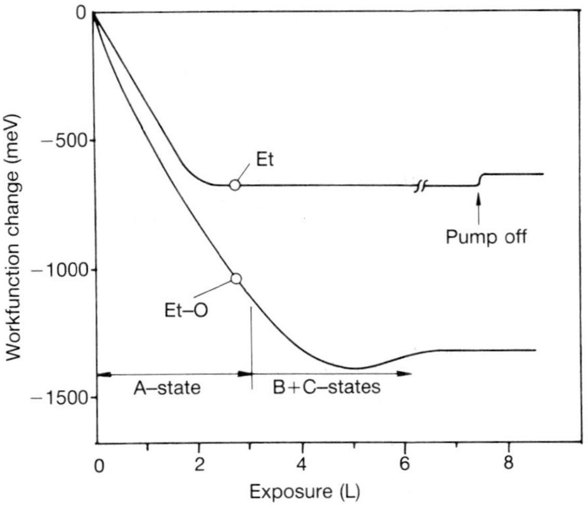

FIG. 23.13. Change in work function of Ag(110) during the adsorption of ethylene and of ethylene oxide at 95 K, as a function of exposure. The decrease in work function observed in both cases indicates the existence of molecular dipoles with the positive pole oriented outwards. For ethylene oxide this must mean bonding to the Ag(110) surface via the oxygen atom. From Krüger and Benndorf.[12]

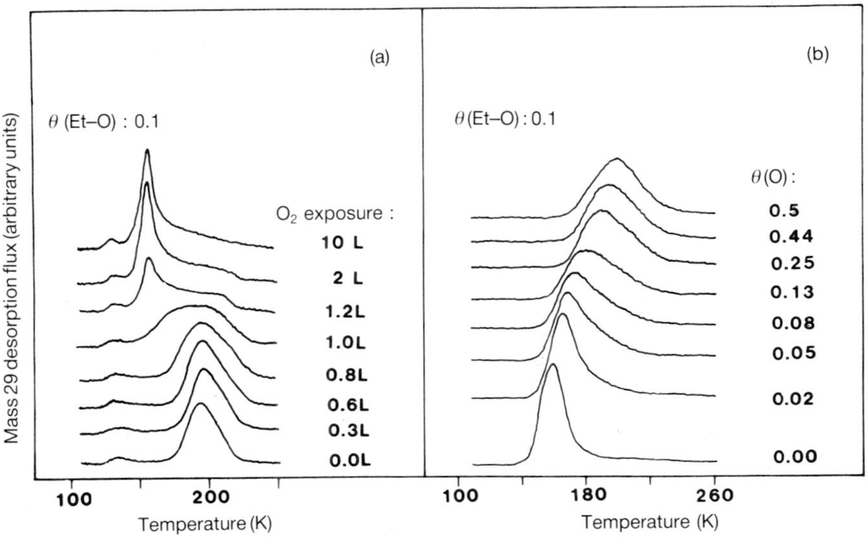

FIG. 23.14. Thermal desorption spectra of ethylene oxide from oxygen-covered surfaces of (a) Cu(110) and (b) Ag(110), as a function of oxygen coverage, after adsorption at 110 K, from Benndorf *et al.*[13] The heating rate was 5 K s^{-1}. Although on the clean Cu and Ag surfaces the desorption behaviour was similar, on the oxygen-covered surfaces it is strikingly dissimilar.

Such information is vital when considering the various possibilities for the mechanism of reaction between ethylene and oxygen to produce ethylene oxide. Subsequent experiments by the same authors[13] on the adsorption and desorption behaviour of oxygen-covered copper and silver revealed striking differences between the two metals; the binding energy of the A state (Fig. 23.11) increased continuously with oxygen coverage on silver, but on copper the A state disappeared beyond 1 L oxygen coverage, to be replaced by a new state at lower binding energy. Figure 23.14 shows the effect. Although much further work needs to be done to establish the reasons for the differences, TDS is obviously an essential component in the armoury of techniques needed for their elucidation.

References

1. Kirstein, W., Krüger, B., and Thieme, F. (1986). *Surf. Sci.* **176**, 505.
2. Surnev, L., Rangelov, G., and Bliznakov, G. (1985). *Surf. Sci.* **159**, 299.
3. Redhead, P. A. (1962). *Vacuum* **12**, 203.
4. Madix, R. J. (1977). *Cat. Rev.—Sci. Eng.* **15**, 293.
5. Zhou, X-L., Yoon, C., and White, J. M. (1988). *Surf. Sci.* **203**, 53.
6. Falconer, J. L. and Madix, R. J. (1975). *Surf. Sci.* **48**, 398.
7. Paul, J. (1988). *Phys. Rev. B* **37**, 6164.
8. Paul, J. and Hoffmann, F. M. (1988). *Surf. Sci.* **194**, 419.
9. Cameron, S. D. and Dwyer, D. J. (1988). *Surf. Sci.* **198**, 315.
10. Cameron, S. D. and Dwyer, D. J. (1988). *Langmuir* **4**, 282.
11. Griffiths, K. and Bonnett, D. (1986). *Surf. Sci.* **177**, 169.
12. Krüger, B. and Benndorf, C. (1986). *Surf. Sci.* **178**, 704.
13. Benndorf, C., Krüger, B., and Nieber, B. (1986). *Appl. Catalysis* **25**, 165.

Index

absorption cross-section, infra-red 358
absorption isotherm, IETS 609–12
activation energy for desorption 673–8
adhesion
  IETS 613–16
  ISS 464–7
  XPS 301–4
alkali metals, promoters in catalysis 369
alloys
  analysis by APFIM 638–42
  passive films 288–94
alumina
  catalyst support 616–19
  formic acid adsorption, IETS 616–19
  ISS spectrum 452
  phosphonic acid adsorption, IETS 613–16
aluminium
  adhesion pretreatment
    IETS 613–16
    ISS 465–6
  electron energy losses 130, 259
  film on GaAs
    PLAP 644–5
  IAES and PAES spectra 396
  oxidation, HREELS 164–6
  plasmon losses 96, 259
  pyridine adsorption, SERS 379–384
  SERS spectra 379–83
  SRPS spectra 333
  water reaction, ESD 230–4
aluminium nitride on steel, GDOS 440–1
ammonia
  molecular orbitals, UPS 346
  reaction with iron, HREELS 158–64
analyser transmission 60–1, 63, 82, 275
angle-resolved UPS (ARUPS) 199, 315, 317, 321, **324–31**
  band structure determination 327–31
angular distribution
  Auger electrons 387, 389
  ions 11, 447, 505
anodic film on iron, GDOS 442
antimony, Cooper minimum 335–6
anti-Stokes radiation 13
appearance potential spectroscopy (APS), *see* SXAPS
argon, gas-phase Auger spectra 392, 394–5
Arrhenius equation 673
arsenic, band structure, ARUPS 327–9
asymmetry parameter 274–5, 277
atom probe field ion microscopy (APFIM) 2, **620–47**, 648
  analysed area 625, 633–4, 636
  applications
    alloys 638–42
    grain boundary segregation 633–6
    particles 636–8
    semiconductors 642–6
  mass resolution 627, 634
  operation 620–9
  physical principles 19
  quantification 632, 635, 637, 643
  sensitivity 620, 629, 632
  spectral interpretation 631–2
  surface specificity 25–6, 620, 633
  theory 629–31
Auger electron appearance potential spectroscopy (AEAPS) 2, **167–92**
  applications 185–92
  energy resolution 56
  operation 168–70
  physical principles 5
  surface specificity 23
  theory and spectral identification 179–81
Auger electron spectroscopy (AES) 2, **81–124**
  applications
    ceramics 119–23
    interfaces 102–5
    materials characterization 116–23
    metallurgical 105–11, 634
    metglasses 117–19
    oxidation 112–16, 214–15, 230–4, 542–3
    semi-conductors 102–4
  characteristic spectra
    KLL 88
    LMM 89, 258
    MNN 90
  chemical effects 94
  diffraction features 97, 98
  electron sources 31–6, 81, 249
  energy resolution 56
  nomenclature 85–7
  operation 81–5, 595
  physical principles 5
  quantification 97–102
  spectral features and interpretation 85–97

Auger electron spectroscopy—*contd.*
surface specificity 22
use of CHA 83–4
use of CMA 61, 82–4
use of RFA 58, 82
Auger energies, theory 91
Auger features, X-ray excited 247–9, 258, 308
Auger fine structure 92–7, 250, 280–3
Auger intensities, theory 92
Auger maps 85, 107–11
Auger neutralization 14–15, 419–21, 457, 488, 489, 580
Auger parameter 12, 247–8, 278–80, 311
Auger process 6, 85, 223–4, 263, 279, 420
hole localization 263
auto-ionization 488–90

background
AES 93, 250
IETS 604, 617
ISS 460
TDS 676
XPS 244–5, 246, 248
background subtraction, XPS 244–6, 278, 604
backscattering factor (coefficient)
elastic 168
inelastic 97–100
ball cratering 78–9
depth resolution 79
band-edge luminescence, CdS 216–8
band structure
ARUPS 327–31
SRPS 332–3
STS 662–3, 664
band width 474
barium, ELS quantification during oxygen exposure 142–4
base width 54
benzoic acid
IETS spectra 608
coverage dependence 611–12
beryllium, Auger spectra in metglasses 117–9
binary collision theory 17, 451–2, 454, 546
binding energies
Al 308, 310
As 350
Au 350
C 298, 300
Fe 286, 289, 306
Ga 350
Mo 294–7
O 289, 298, 305
boron, Auger spectra in metglasses 117–9
bremsstrahlung isochromat spectroscopy (BIS) 193–5, 198
bremsstrahlung radiation 9, 40, 42, 172, 173, 193–209, 244, 247–8, 264–5, 280
Brillouin zone 327–31, 335, 411, 413

cadmium
quasi-atomic Auger spectra 93, 250
SRPS spectra 353
cadmium sulphide, laser annealed, CLS 215–18
caesium
adsorption on oxygen-covered tungsten, ELS 149–50
ion yield dependence on work function 502–3
calcium compounds
ion beam reduction 75
oxide, loss spectrum 127
vanadate, ionization losses 130
carbide particles, APFIM 636–8
carbon
analysis in steel, SNMS 558, 560
binding energies 298, 300
calculated and experiment chemical shifts 271
chemical effects in AES 95
carbon monoxide
adsorption modes 369–71, 517–521, 522, 523
desorption
from Cu 674
from Ni 677–8
molecular orbitals, UPS 339, 342, 343
catalysis
HREELS 157–64
ISS 459–62
TDS 681–90
XPS 294–300
cathode dark space 564, 566, 567
cathodoluminescence spectroscopy (CLS) 2 **210–8**
applications 212–8
buried interfaces 215–8
electron sources 210
operation 210
physical principles 9, 210–11
surface specificity 24, 211–13, 216–18
theory and spectral identification 210–12
cerium
oxidation, valence change 152–4
SXAPS and AEAPS spectra 181
charge-potential model, XPS 270–1
charge-transfer cell, FABMS 578–80
charging problems
electron excitation 119
ion excitation 445–6, 451–2, 514–15, 578 580, 585

neutral excitation 580, 583, 585
chemical effects, Auger spectra 94, 95
chemical shifts 259–61
theory 269–73
chemisorption 185, 203
chromium
enrichment in passive films 290–4
photon yield dependence on oxygen pressure 428–9
cluster ions 631–2
clusters, UPS 374–50
cobalt, catalyst spectrum, XPS 297
cobalt–molybdenum catalyst, activation 460–2
cobalt sulphide, ISS spectrum 452
collision cascade 25, 484–6, 496, 545, 550–2
concentric hemispherical analyser(CHA) 61–4, 125, 238–9, 256
AES 83–4
energy resolution 63
ISS 446
slit width 63
transmission 63
UPS 314
with lens 63, 239
XPS 238–9
constant analyser transmission (CAT) mode 55
constant retard ratio (CRR) mode 55
Cooper minimum 334–8
copper
adsorption of CH_3CN, RAIRS and HREELS 372–3
isolated atoms and clusters, UPS 347–50
reaction with CO, IPES 203–6
theoretical and experimental valence band structure 320
unoccupied band structure, IPES 206–9
XAES fine structure 280–1
XPS spectrum 244, 258
copper-nickel alloys, matrix effects, SSIMS 505–6
copper oxides, shake-up satellites 261, 262
core-electron energy loss spectroscopy (CEELS) 2, 96, **125–66**
applications 150–6
electron sources 125
energy analysis 125
energy resolution 56
physical principles 5
spectral features and interpretation 129–32
surface specificity 23
correlation diagram 397–8
corrosion and passivation, XPS 285–94
Coulomb interaction, energy 282–3, PAES 391
crater profiles, GDOS 434, 436
cross-section
elastic ion scattering 548
electron ionization 81, 83, 87, 97–100
photoionization 272–6, 334–8
cylindrical mirror analyser (CMA) 59–61, 82–4, 125, 386, 389, 392, 418, 446, 448
advantages and disadvantages 61
AES 82–4
double-pass 61–2, 446
energy resolution 61
IAES 386–7, 389
ion energy analysis 219, 446, 448
transmission 60–1, 82

dangling bond states 155–6, 654, 660, 667–9
density of states 319, 408, 411–13, 654–7, 660–1, 662, 664, 667
depth distribution, primary ion energy 550–2
depth profiling **70–9**, 628–9
angular variation 71, 290
applications 102–5, 108, 183, 290–3, 431, 432, 440–4, 524–5, 559–63, 573–6, 587–92
ball-cratering 78–9
ion bombardment 71–8, 535
quantification 72–5
depth resolution 537, 559, 561, 573–6
desorption
peak area 678, 686–8
peak shape, dependence on reaction order 676
rate 673–8
techniques, *see* ESD, ESDIAD, PSD
thresholds 223–5
yields 225
detector efficiency 275
deuteration
HREELS 135–40, 158
IETS 617–18
deuterium, desorption from Al 679–81
dielectric constant 126–9, 145, 326
loss function 126–8, 376–7
difference spectra
IETS 604, 605
UPS 315, 316, 322–3, 340, 346
differential ion scattering cross-section 455–6
differential pumping 388, 390, 566, 672
diffraction features
AEAPS and DAPS 172, 180–2, 189–92
AES 97, 98
dipole scattering
HREELS 133, 137, 141
loss intensities 133–4, 600–3

dipole selection rules
HREELS 133, 158
IETS 603
dis-appearance potential spectroscopy (DAPS) 2, **167–92**
operation 171–2
physical principles 8
surface specificity 23
Doniach–Šunjić line shape, XPS 267
doping of tunnel junctions 596–8, 611
double scattering, ISS 467–8
dye lasers, multiphoton ionization 538–42
dynamics SIMS 17, 470

elastic conductance, IETS 602, 609–10
electron affinity 17
electron beam effects 185, 186, 210, 219, 515, 585
electron energy loss spectroscopy (ELS) 2, **125–66**
angle-resolved 202
applications 144–50
electron sources 125
energy analysis 125
energy resolution 56
operation 125
physical principles 6
quantification 141–4
spectral features and interpretation 126–9
surface specificity 23
electron-induced luminescence (EIL), *see* CLS
electron promotion 397–8, 487–8
electron sources
field emission 32
monochromatic for HREELS 29–31, 125
thermionic 32–6, 171
electron spectroscopy for chemical analysis (ESCA), *see* XPS
electron stimulated desorption (ESD) 2, 138, **219–37**
applications 229–37
electron sources 219
operation 219–23
physical principles 9–11
surface specificity 24, 234
theory and spectral interpretation 223–9
electron stimulated desorption ion angular distribution (ESDIAD) 2, **219–37**
applications 229–37
electron sources 219
mapping 220–3, 228
patterns 226–8, 234–6
directionality of surface bonds 226–8
physical principles 9–11
surface specificity 24
theory and spectral interpretation 223–9
electron temperature, postionizing plasma 535
electron tunnelling 14–15, 19–20, 21, 32, 593–4, 599–604, 621, 629, 648, 652, 654–69
electrostatic imaging, XPS 254–6
energy analysers
electron 53–64, 238–9, 253, 387, 389, 424
ion 445–50, 475–7, 478, 532, 566
energy distributions, sputtered particle 529
energy resolution 31, 39, 54–6, 193, 194, 362, 475, 477, 597, 603
absolute 54
relative 55
enhancement mechanisms, SERS
chemical 377–8
electromagnetic 376–7
epitaxial films
GDMS 574
SNMS 559–60, 563, 564
equivalent cores method, XPS 271–3
erosion profile, plasma postionization 535–6
escape depth 22
ethylene
adsorbed on Ag, HREELS 136–41
desorption from Ag, TDS 685–6, 688–90
ethylene oxide, desorption from Ag, TDS 686–7, 689
europium, oxidation, valence change, CEELS 152–3
evaporation field 625, 627, 631
excitation probability, ions, SSIMS 506
extended energy loss fine structure (EELFS) 156–7
extended X-ray absorption fine structure (EXAFS) 156, 164

fast atom bombardment mass spectrometry (FABMS) 2, **578–92**, 631
applications 585–92
comparison with SSIMS
from Nb_2O_5 581, 583
from polystyrene 581, 582
operation 578–80
physical principles 18
surface specificity 25
theory 580–5
fast atom source, saddle field 51–4
ferrocyanide ion, vibrational spectra, IETS 604–7
field desorption and evaporation 19, 623–9, 631
field ion image, comparison with stereographic projection 624
field ionization 19, 621, 629

field ion microscopy (FIM) 620–3, 625, 648
field polarization 621, 629
first-order desorption 673–4, 678, 684–5
flash desorption 671
fluorescence 5, 180
fluorine, adhesive joint contaminant, ISS 465–7
formaldehyde, vibrational modes 135
formic acid, IETS spectra 617
Fourier transform methods 358–62
FT–RAIRS, *see* RAIRS
Fowler–Nordheim field emission expression 32
fractional order desorption 673, 676, 685
fracture
  adhesive joint 301–3, 466–7
  glass-ceramic 121–3
  grain boundary 68, 634
  silicon carbide 119–20
fracture stage 68–9
Fresnel reflection coefficient 364

gadolinium, loss spectra 128, 152–3
gallium aluminium arsenide, PLAP spectrum 633
gallium arsenide, SNMS and SSIMS comparison 530–1
gallium, Auger line shape during depth profiling 102–3
gas handling system 68–70
gas-phase spectra 322–4, 346, 392, 394–5, 399
Gausso–Lorentzian function 246
Geiger–Müller detector, IPES 194, 196
germanium, reaction with Ag 335–8
giant resonances 151–4
glasses
  analysis by FABMS 587–8, 589
    by SNMS 561, 564
    leaching and corrosion, IBSCA 430–1
glow discharge mass spectrometry (GDMS) 2, **563–76**
  analysis, non-metals 570–1
  applications 572–6
  depth profiling 573–6
  mass resolution 571–2
  operation 563–7
  physical principles 17–18
  quantification 567–72
  sensitivity 565, 569–71
  surface specificity 25, 563, 575
glow discharge optical spectroscopy (GDOS) 2, **431–44**, 566
  applications 439–44
  depth profiling 439–44
  operation 432–5
  physical principles 16
  quantification 435–9
  surface specificity 25, 431, 443
gold
  band structure, ARUPS 330–2
  reaction with GaAs 350–3
grain boundary
  analysis by APFIM 634–6
  attack by lithium, SAM 108–11
graphite, band structure, STS and IPES 662–3, 664
grating spectrograph, IPES 194
Grimm glow discharge lamp 432–5, 436

helium discharge 36–9
  line energies and widths 37, 314
helium, metastable atoms 417–18
high brightness electron sources 32
high pressure analysis, RAIRS 371–3
high resolution electron energy loss spectroscopy (HREELS) 2, **125–66**, 357, 372–3, 378, 520, 523, 593, 604
  applications 157–66, 372–3
  dipole selection rules 133
  dipole scattering loss intensities 133–4
  electron sources 125
  energy analysis 126
  energy resolution 56, 372
  instrumentation 29–31
  physical principles 7
  scattering mechanisms 132–3
  spectral features and interpretation 132–41
  surface specificity 24
hole localization, Auger process 263, 280–3
hopping integral 496
hydrogen
  desorption from Al 679–81
  detection by GDOS 442
  detection by SSIMS 521–2
  storage materials 181–2
hydrogen-like states 666
hydroxyl ion
  detection by SSIMS 522–6
  surface 290, 522–6

image dipole, IETS 600–3
image formation
  APFIM 634–5
  FIM 620–3, 624
image states 206–9, 662–5
imaging gas 620–3, 634–5, 638
impact parameter 455–7, 547–8
impact scattering, HREELS 133, 137, 141
indium arsenide, PLAP 642–4

inelastic background, IAES 391–2
inelastic conductance, IETS 602, 609–12
inelastic electron tunnelling spectroscopy (IETS) 2, **593–619**
  applications 612–9
  energy resolution 603
  line shapes 603–4, 605
  operation 593–8
  physical principles 19–20
  quantification 607–12
  selection rules 603–7
  sensitivity 608–9
  spectral identification 604–7
  surface specificity 26
  theory 598–604
inelastic mean free path 22–3, 97–101, 275–6, 325, 350, 354, 409, 661
infra-red reflection-absorption spectroscopy (IRAS), *see* RAIRS
infra-red spectroscopy (IRS) 135–40, 158–61, 604–7, 613–15, 616–19
instrumental de-broadening 243, 267–8
instrumentation **27–80**
insulator surfaces 18, 429–31, 445, 560–1, 564, 583, 585–92
interaction potential, ISS and SNMS 456, 547
interband losses 148
interfaces
  AES 102–5, 112–16
  APFIM and PLAP 639, 642–6
  CLS 215–18
  FABMS 585–7 588
  IETS 613–16
  SRPS 335–8, 350–5
  XPS 301–3
interferogram 360–2
intermediate coupling 87
intra-band losses 6
inverse photo-emission spectroscopy (IPES) 2, 150, 154, 173, **193–209**, 654
  angular resolution 193, 198–202, 205
  applications 202–9
  detectors 193–6
  electron sources 193
  operation 193–6
  physical principles 9, 197
  STS, comparison 662–3, 664
  spectra, graphite 664
  surface specificity 24
  theory and spectral identification 197–202
ion beam damage 446–7, 514–17, 578, 581–4
ion beam spectro-chemical analysis (IBSCA) 2, **423–31**
  applications 429–31
  ion source 49, 423–5
  operation 423–5
  oxygen effect on photon yield 428–9
  physical principles 16
  surface specificity 25, 429
  theory and quantification 425–9
ion bombardment
  chemical reduction 75–6
  matrix effects 74, 505–6
  statistical effects 78
  topographical effects 76–8
ion (excited) Auger electron spectroscopy (IAES) 2, **386–402**
  energy resolution 56
  ion source 49, 387–8
  operation 386–91
  physical principles 14
  surface specificity 22
  theory and spectral interpretation 391–401
ionic transfer, glasses, IBSCA 430–1
ion implantation, GDOS profiles 443–4
ionization
  coefficient, SSIMS 486–511
  electrons 5, 7–8, 96, 97
    cross-section 81, 83
  ions and protons 14
  loss spectroscopy(ILS), *see* CEELS;
  photons 11–12, 538–45
    cross-section 272–4, 276
  potential and energy 426, 486, 493–8, 502, 511–13, 621, 625, 630
  probability 490, 493–8, 501, 505–6, 512, 531, 543, 565
  thresholds 7, 125, 129, 156–7, 168–72, 223–4
ion neutralization probability 457, 501, 580
ion neutralization spectroscopy (INS) 2, **403–17**
  applications 415–17
  energy analyser 405
  energy resolution 56, 411
  ion source 404–5
  operation 403–6
  physical principles 14–16
  spectral interpretation 410–15
  surface specificity 22, 410
  theory and data processing 406–10
ion scattering spectroscopy (ISS) 2, 424, **445–69**, 485, 546
  applications 459–67
    adhesion 464–7
    catalysis 459–62
    coatings 462–4
  data processing 461
  energy resolution 451–3
  ion source 46–9, 445–50
  mass resolution 451–3, 454

ion scattering spectroscopy—*contd.*
operation 445–50
physical principles 17
quantification 454–9, 464
structural analysis 467–8
surface specificity 25, 450, 453, 455, 458–9
theory and spectral interpretation 450–4
ion scattering trajectories 456–7
ion sources 44–51
duoplasmatron 46–8, 386, 470–1, 423–5, 445, 476
electron impact 45–7, 445, 470, 477, 479, 578–9
glow discharge 50–1, 53
high frequency 49, 533–4
liquid metal 50–1, 480
Penning discharge 44–6
iron
adsorption of CO, SSIMS 517, 519
analysis, GDOS 439
binding energies 286, 289
oxidation, UPS 315, 316
XPS 303–8
valence band structure, UPS 318–19, 321–2
variation of photon yield with oxygen pressure 428
iron–chromium alloy, SNMS 556–7
isolated atoms, UPS 347–50
isotopic mass separation, APFIM 627–8

j–j coupling 85–6, 267–8, 280
joint density of states (JDOS) 126, 321–2, 326

kinetic model, SSIMS 487–8
Knotek–Feibelman model, electron impact ion desorption 223–5, 232
k-resolved inverse photoemission spectroscopy (KRIPES) 199–202, 206–9
krypton, UPS spectra 348

ladder diagrams, APFIM and PLAP 639, 641–5
lanthanides
ionization losses 132, 150–4
SXAPS spectra 175–7
laser fluorescence spectroscopy (LFS) 538, 541
laser irradiation 13, 215–18, 538–44
lead, tunnel junctions, IETS 596–7
light emission following ionization 9, 210
linear background, XPS 245
liquid metal, unoccupied states 186–9
liquid nitrogen traps 28
lithium
depth profile, protonated glass 432
grain boundary analysis, AES and SAM 108–11
ion yield, work function dependence 499
localization, holes, after Auger transition 92
localized states 131–2, 150–4, 175–7, 654
local thermodynamic equilibium (LTE) model, SSIMS 426, 485–8, 510–14
lock-in amplifier, *see* phase-sensitive detection
low energy electron diffraction (LEED) 36, 138, 164, 203–5, 219, 226–9, 368, 369–70, 415, 679
optics 57, 81–2, 171, 220–1, 403–4
low energy electron energy loss spectroscopy (LEELS), *see* HREELS
low energy ion scattering (LEIS), *see* ISS
low temperature adsorption 341–50, 368, 369–70, 373, 379–84, 674, 677, 680, 684, 687–8, 686–9
L–S coupling 86

magnesium, IAES spectra 391, 393, 400–1
manganese, GDOS analysis 437–8, 439
manipulator 66–7, 403–4, 625
mass analysers 64–5, 219–21, 445–6, 566
mass filters 471–3, 476, 478
mass separation 220, 473–4, 477, 480, 482, 571–2, 627
materials in UHV 28
matrix isolation technique 347–50
Menzel–Gomer–Redhead model, electron impact ion desorption 223
mercury, SRPS spectra 353
metastable atom source 417–18
metastable quenching spectroscopy (MQS) 2, **417–22**
mechanisms 419–20
operation 417–18
physical principles 16
surface specificity 22, 420
metglasses, AES 117–19
Michelson interferometer 358–62
modes of vibration 134–40, 160
nomenclature 136–40, 160
modulation techniques 58, 60, 83, 650, 655
molecular orbitals
NH_3 346
CO 339, 342, 343, 420–21
H_2S 415–17
H_2Se 415–17
NO 420–21
O_2 341, 344 415–17
pyridine 323–4

molybdenum
  binding energies 294–7
  catalyst, XPS 294–8
  secondary ion distribution 476
momentum conservation, photoemission 319–20, 326
monochromators
  light 424–5, 433, 541
  X-ray 42–4, 193, 195, 255–7
multiphoton ionization, non-resonant 538, 543–5, 546

negative glow region 546, 566, 567
neon discharge 36–9
  line energies 37, 314
    widths 37, 314
neutral beam damage 581–4
neutral sources, *see* fast atom sources
nickel
  GDOS analysis 439
  INS spectra 410–11
    carbon-contaminated 413–14
    mercury-contaminated 413–14
  MQS spectrum 421
  reaction with CO
    MQS 420–22
    RAIRS 369–72
    SSIMS 519–21, 523
    UPS 341–3
    with N_2, UPS 341–3, 344
    with NO, MQS 420–2
    with O_2
      INS 415–17
      ISS and AES 458–9
      UPS 341–3, 344
    with S, INS 415–17
    with Se, INS 415–17
  solid and liquid, unoccupied states 188–9
  unoccupied band structure, IPES 201–3
nuclear fuel cladding, oxidation behaviour 285–7, 288, 289

optical data, comparison with loss spectra 127–8
optical monochromator, CLS 210–11
order, desorption reaction 673–8
organic materials
  FABMS 585–7, 588
  SSIMS 477, 480–2, 514–17, 518, 585
  XPS 301–3, 515–16
oxide film on steel, GDOS 440
oxygen
  adsorption
    caesium-covered tungsten, ELS 147–50
    iron, XPS and UPS 303–8
    silver, HREELS 136–40
    semiconductors, XPS 308–11
    thorium, CLS 212–15
    transition metals, SXAPS 185–7
  binding energies 289, 298, 305
  desorption from Ru, TDS 675
  ions, SSIMS 504, 507, 514
  XPS spectra, smoothing 242

palladium
  adsorption
    CO, SSIMS 517–21, 522
    pyridine, UPS 323–4
  band structure, STS and UPS 662–3
  deposition on SiC, AES 104–5, 106
  film on tantalum, adsorption of CO, UPS 338–9, 342–3
pass energy 55, 61
passivation
  SSIMS 521–6
  XPS 287–94
peak, fitting, ISS 461
peak separation, XPS 245–7, 296
Penning ionization 420, 565
phase changes, reflected polarized light 362–7
phase-sensitive detection 58, 60, 83–4, 168, 169, 171, 595, 655
phonon scattering, HREELS 138–42
phonon structure, IETS 617
phosphonic acids, IETS 613–15
photoelectrons, energy distribution 326
photoelectron spectro-microscopy (PESM) 252–5, 257
photoemission spectroscopy (PES), *see* UPS
photoemission, three-step model 325
photoionization, cross-section 273–6, 334–8
photoluminescence 218
photon attenuation length 325
photon emission
  CLS 210–18
  GDOS 16, 435, 437–8
  IBSCA 16, 426, 428, 488
  STM 654
photon sources 36–44
  synchrotron 43–4, 315–18
  UV 36–9, 314–15, 404
  X-ray 39–43, 238, 248
photon stimulated desorption (PSD) 2, 219
  physical principles 13–14
  surface specificity 24
plasma density 535
plasma, gas discharge, SNMS 533–7
plasmon loss energies, theory 129
plasmon losses
  bulk 6, 94–6, 127–30, 145, 258–9

surface 94, 127–30, 145–6, 258–9
platinum
adsorption
methanol, UPS 322–3
NH_3, UPS 343–7
NO, ESD 224–5
surface condition, effect on water adsorption, TDS 683–5
poisoning, catalysts by sulphur, RAIRS 369–72
polarity shift, IETS 598–9, 600
polarized light 13, 358, 362–7, 374–5
polymers: FABMS 585–7, 588
SSIMS 514–17, 518
Poschenreider energy compensator 627
position-sensitive detector 449
post-ionization
electron impact 528–33, 558, 561, 563, 565
gas discharge 533–7, 556, 559–61, 562, 563
glow discharge 563–7
multiphoton 538–45, 559
potassium
catalysis promoter, XPS 298–300
desorption from Fe 679, 681–2
protonation of glasses, IBSCA 430–1
proton (or particle) (excited) Auger electron spectroscopy (PAES) 2, **386–402**
energy resolution 56
ion source 50, 52, 387–8, 390–1
operation 387–91
physical principles 14
surface specificity 22
theory and spectral interpretation 391–7
pulsed laser atom probe (PLAP) 627–8, 634, 642–6
pumps for UHV 28
pyridine, adsorbed on Ag, SERS 380–1, 383

quadrupole mass spectrometer 17, 18, 22, 64–5, 219, 386, 418, 424, 470, 473–4, 476, 477, 482, 529–30, 532, 533, 534, 536, 537, 539, 566, 671–3, 678, 679
quasi-atomic Auger spectra 92–3, 250, 280–3, 399

radial distribution function 156–7
radioisotope calibration, IETS 607
Raman scattering 373, 375–7, 598–9, 604–7, 616–19
rare earths, multiplet splitting 262–264
rare gas discharge, UV source 38
reciprocal lattice vector 327
reduction, molybdenum catalyst 294–7
reflection–absorption infra-red spectroscopy (RAIRS) 2, **357–73**, 593
applications 369–73
energy resolution 362, 372, 603
Fourier transform methods 358–62
operation 357–62
physical principles 13
quantification 367–9
selection rules 363, 367, 604–7
surface specificity 24–5
theory and spectral interpretation 362–7
reflectivity 366
refractive index 364–6
relative ion yields, GDMS 567–8, 569
relative sensitivity factors
AES 101
GDMS 568–71
SNMS 556–9, 560, 562
SSIMS 507–9
matrix-indexed 509–10
XPS 276–7, 285
relaxation energies 91, 269–70, 279
resolving power 55, 473–4, 482, 572
resonance
ionization 16, 419–21, 489, 538–43
neutralization 457
resonant enhancement, SERS 376–7
retardation, energy analysis 55, 61
retarding field analyser (RFA) 56–8, 404–5, 475, 477, 478
disadvantages 58, 82
energy resolution 58
Richardson equation, thermionic emission 32
Rowland sphere 42–4, 255–7
Russell–Saunders coupling, *see* L–S coupling
ruthenium, CO adsorption, SSIMS 519–21, 523
Rutherford backscattering spectroscopy (RBS) 445

Saha–Eggert equation 426, 486, 513, 514, 515
sample
cleaning 66–8
insertion and handling 66–70
manipulation, *see* manipulator
transfer 66
satellite removal, XPS 243
scanning Auger microscopy (SAM) 2, **81–124**
applications
metallurgical 105–11
ceramics 119–23
electron sources 31–5, 84
energy resolution 56

Scanning Auger microscopy—*contd.*
applications—*contd.*
physical principles 5
surface specificity 22
scanning tunnelling microscopy (STM) and spectroscopy (STS) 2, **648–70**
applications 660–9
operation 648–55
physical principles 21
spatial resolution 648, 652, 654
surface specificity 26
theory and image interpretation 649, 653, 655–60
Schottky barrier 350
screening length, ion scattering 456
secondary electron
coefficient 168
energy distribution 81–2, 170, 322, 399–401
microscopy (SEM) 85, 109, 113–16
secondary ion
energy
distribution 474–5, 505, 540, 544
selection 473–7
formation, SSIMS 484–504
suppression, SNMS 528–30, 532, 533
secondary neutral emission 18
secondary neutral mass spectrometry (SNMS) 2, 475, **528–63**, 631
applications 559–63, 564, 565
depth resolution 559, 561, 562
operation 528–45
physical principles 18
quantification 554–9, 567
sensitivity 541, 545
spectral interpretation 552–4, 631
SSIMS comparisons
GaAs 530–1
standard alloy 554–5
surface specificity 25
theory 545–52
second-order desorption 673–7, 688
selected area analysis, XPS 249, 251
self-convolution, empty state densities 173–5
self-deconvolution: INS 408–10, 411
SXAPS 177–9, 188–9
semiconductors: interfaces, PLAP 644–6
oxidation, SRPS 353–5
XPS 308–11
shadowing, ISS 459, 467
shake-up satellites, XPS 261, 262, 277–8, 286, 289
Shirley background, XPS 245
silica, ISS spectra 463
silicon
Auger excitation, X-rays 247–8, 265–6
Auger peak shape 106
carbide, AES and SAM 119–20
dangling bond states, CEELS 155–6
epitaxial layer, GDMS 574
interface with oxide, PLAP 646
IAES 391, 393–4, 396, 401
ion yield, work function dependence 500
metal-oxide interfaces 112–16
phonon scattering 142
platinum reactivity, TDS 685
reaction, NH_3, STS 666–9
reconstructed surface 666
STM images 667
silver
Cooper minimum 335–8
deposited Al, SERS 380–3
pyridine adsorption, SERS 378–84
SERS spectra 379–83
smoothing 240–2
soft X-ray appearance potential spectroscopy (SXAPS) 2, **167–92**
applications 181–5
instrumentation 37
operation 167–8
physical principles 7–8
surface specificity 23
theory and spectral interpretation 172–9
solar reflectance, ISS and XPS 462–4
spalling, steel, AES and SAM 112–16
spatial resolution 29, 31–5, 45, 49, 50, 81, 84, 125
spectroscopic notation 86–7
spinels 286–7, 292
spin polarization 193
sputtered particle, energy distributions 529
sputtering 484–5, 545, 564, 585, 683
rate 72, 437
yield 72–4, 484, 490, 505, 512, 550–3, 554
stainless steel, SXAPS spectrum 178
Stark shift 664
static secondary ion mass spectrometry (SSIMS) 2, 424, 428–9, **470–527**, 536–7, 544, 554–5, 578, 631
applications 514–26
data processing 482
FABMS, comparison
Nb_2O_5 581, 583
polystyrene 581, 582
ion sources 45–9, 536–7
isotopic effects 481
mass resolution 473
operation 470–80
physical principles 17
quantification 504–14, 515
spectral identification 480–2, 631
surface effect models 490–503
surface specificity 25, 429

theory 482–504, 580, 584–5
sticking probability 686
Stokes radiation 13, 376–8
stopping cross-sections 546–51
stopping power 545–6, 549–51
sublimation energy 623, 630
sulphur
analysis, steel, SNMS 558, 560
catalyst poisoning 369–72
surface composition by analysis of neutral and ion impact radiation (SCANIIR), *see* IBSCA
surface-enhanced Raman scattering (SERS) 2, **373–85**
applications 378–84
operation 374–5
physical principles 13
silver, Raman enhancement 375
surface specificity 25
theory and spectral interpretation 375–8
surface plasmons, *see* plasmon losses, surface
surface reactions
SRPS 350–5
SSIMS 517–21, 522, 523, 685
UPS 340–7
XPS 303–11
surface roughness 25, 275, 653
surface specificity 22–6
surface states 332–3, 334, 335, 654, 658, 662–4
surface topography 649–50, 655, 660
synchrotron radiation 43, 45, 316, 318
synchrotron radiation photoelectron spectroscopy (SRPS) 2, **314–56**
applications 350–5
band structure determination 332–3
energy resolution 55
operation 315–18
physical principles 11
theory and spectral interpretation 331–8

tantalum oxide
film, depth resolution 559, 561
ion beam reduction 76
tellurium, SRPS spectra 353
temperature measurement, TDS 672–3
theoretical methods, SSIMS quantification 510–4
thermal desorption spectroscopy (TDS) 2, 138, 343–5, 368, 519, **671–90**
applications 681–90
operation 671–3
physical principles 21–2
quantification 678–82
surface specificity 24
theory and interpretation 672, 673–8
thermal energy spread 29, 31
thorium, photon emission during electron iradiation 213–15
three-step model, photoemission 200–2
time-of-flight ion energy analyser 449–50
time-of-flight mass spectrometer 19, 220–3, 470, 477–80, 544, 625–9
tin
line shape, XPS 267–8
oxidation, ELS 145–7
titanium
adsorption, O_2, AEAPS 189–92
SNMS 542–3
AEAPS and DAPS spectra 180, 190
EELFS 156–7
secondary electron distribution 170
secondary ion distribution 475
XAES spectra 264–5
titanium–iron alloy, hydrogenation, SXAPS 182–4
titanium nitride, coating on steel, GDOS 440–1
trace element analysis, GDMS 569–71
transition density function, INS 408–9, 411–14, 416
transition intensity, light 427
transition metals, SXAPS spectra 176, 183, 184
theory 173–5
transition probability 492–3
triangular apodization, FT–RAIRS 361
triangulation method, ARUPS 330–1
true secondary electron background 322
tungsten
adsorption, CO, ESD 224–5
chemical shift 259–60
FIM image 624
peak fitting, XPS 248
tunnel junction 595–8, 613, 616

ultra-high vacuum(UHV) 27–8
ultra-violet photoelectron spectroscopy (UPS) 2, 9, 202, **314–56**
angle-resolved (ARUPS) 202, 315, 317, 321, 324–31
band structure determination 327–31
applications 340–50
comparison, with INS 414–15
with STS 662–3
CHA 64, 314–15
data processing 315
energy resolution 55
gas-phase 322–4, 346
operation 314–15, 317, 328
physical principles 11
quantification 338–40, 341, 660
surface specificity 22
theory and spectral interpretation 318–31

unoccupied states 7–9, 172–4, 185, 197–202, 206–9
uranium, SXAPS spectrum 169

vacuum conditions 27–8
valence band
  SRPS 335–7
  SSIMS theory 493–501
  UPS 318–22
  XPS 261, 263
valence-potential model, XPS 271–2
Van de Graaf accelerator 388–91
velocity filter 471
vibrational losses 7, 13, 19–20, 29, 133–41, 357, 373, 593, 598, 600, 603–7
vibrational loss spectroscopy (VLS), *see* HREELS

water
  desorption from Pt
    TDS 683–5
  semiconductor reaction, XPS 309–11
Wien filter 423–4, 450, 471–3, 578–9
work function 21, 32, 147, 164, 168, 172, 190–2, 303–8, 318, 322, 325, 338, 341, 406–9, 420, 429, 493–503, 519, 621, 625, 630, 648–50, 655–6, 663, 685–6, 688–9

X-ray diffraction 43
X-ray monochromator 42–4, 193, 195, 255–7
X-ray notation 85–6, 258
X-ray sources 39–43, 248, 265
  dual-anode 40–41
  line energies 39, 248, 266, 280
    widths 39, 248
  satellites 40, 243, 258, 261
  window material 40
X-ray (excited) Auger electron spectroscopy (XAES) 2, **238–313**
  energy resolution 55, 280
  fine structure 250, 263–5, 280–3
  operation 247–9
  physical principles 11–12
  surface specificity 22
X-ray photoelectron spectroscopy (XPS) 2, 202, **238–313**
  angle resolved 202, 290
  applications
    adhesion 301–4
    catalysis 294–300, 459–60
    corrosion and passivation 285–94
    surface reactions 303–11, 679
  asymmetry parameter 274–5, 277
  CHA 64, 238–9
  chemical shifts 259–61
    theory 269–73
  data processing 240–7, 296
  double-pass CMA 61
  energy resolution 55, 239
  intensities 272–4
  line shapes 265–8
  multiplet splitting 261–2, 264
  nomenclature 267–9
  operation 238–47
  physical principles 11–12
  quantification 274–8, 285
  relative sensitivity factor 276–7, 285
  scanning 251–3
  spatial resolution 249–57
  spectral interpretation 258–65
  surface specificity 22
  theory 265–74
  valence band spectra 261, 263

ytterbium, oxidation, appearance of giant resonance 151

zeolites, surface treatments, FABMS 588–92
zero-order desorption 676–7
zinc sulphide, CLS spectra 212–13